D0085134

ELECTRICAL FUNDAMENTALS FOR TECHNICIANS

ROBERT L. SHRADER
Former Chairman of Electronics
Laney College

McGraw-Hill Book Company
Gregg Division

New York St. Louis Dallas San Francisco Auckland Bogotá Düsseldorf
Johannesburg London Madrid Mexico Montreal New Delhi Panama
Paris São Paulo Singapore Sydney Tokyo Toronto

Library of Congress Cataloging in Publication Data
Shrader, Robert L
Electrical fundamentals for technicians.

Includes index.
1. Electric engineering. I. Title.
TK146.S557 1976 537 76-13838
ISBN 0-07-057141-4

ELECTRICAL FUNDAMENTALS FOR TECHNICIANS, second edition

4567890 DODO 8321

The editors for this book were George J. Horesta
and Mark Haas,
the designer was Charles A. Carson,
the cover was designed by Sullivan-Keithley Inc.,
the art supervisor was George T. Resch,
and the production supervisor was Iris A. Levy.
It was set in Memphis Light by Progressive Typographers.
Printed and bound by R. R. Donnelley & Sons Company.

CONTENTS

CHAPTER 19

CHAPTER 20

CHAPTER 21

CHAPTER 22

CHAPTER 23

CHAPTER 24

PREFACE

This text is for school or home-study students who expect to become technicians in any of the wide variety of fields having electricity as a basis; particularly electronics, electricity, and radio. It uses a "spiral" approach to learning. Basic information is given to provide a working vocabulary in the first 16 simple, one-a-day overview chapters of electrical fundamentals. More advanced chapters follow this and cover the same basic material in more detail, plus a wide variety of other fundamental electrical subjects. All chapters are in semi-programmed format, with immediate comprehension check-up quizzes every few pages.

The spiral method of presentation aids learning by providing reinforcement of the subject matter. For a heterogeneous group of students with dissimilar backgrounds, it provides all with a similar vocabulary in a minimum of time, allowing students equal opportunity to benefit from the more advanced subjects.

Each of the first 16 chapters is expected to require two to three hours of study and lecture. Depending on the final objectives, the advanced chapters will require more time. Subject matter has been developed in many smaller chapters to allow selection of topics deemed necessary for desired goals. (House wiring information may have little application in electronics; the *j* operator may be of minimal importance to many electrical fields.)

Students should skim the whole of each assigned chapter briefly, then go back and study, with notes, up to a quiz; take the quiz, check it, and refer back to the text for items missed. Then proceed to the next quiz. The chapter tests are particularly important to home study as a review of the whole chapter. Answers to each check-up quiz will be found on the next overleaf. Answers to all chapter tests preceed the appendixes.

Answers to problems should agree within one or two digits of the third significant figure. In some cases, all possible answers are not included in quiz answers since these questions are merely a check on basic understanding. The author would greatly appreciate hearing from instructors or readers regarding errors found so that corrections can be made in the next printing. Please address your correspondence to Robert L. Shrader, c/o Gregg Division, McGraw-Hill Book Company, 1221 Avenue of the Americas, New York, N.Y. 10020.

The author would like to acknowledge the valuable help he received during the development of the first edition from the late Emery Simpson. Others whose help the author would like to acknowledge are Edward J. Null, John Brown, Norman C. Harris, Russell Bentson, Gil Roach, and, of course, wife Dorothy, son Douglas, and daughter Patricia for their cooperation during the writing and production periods. Ralph Ameroli, Sonoma County, California, Building Inspector, aided with suggestions for the second edition of the chapter on wiring practices.

ROBERT L. SHRADER

ELECTRICAL
FUNDAMENTALS
FOR TECHNICIANS

CHAPTER OBJECTIVE. To develop a usable understanding of these electrical vocabulary terms: component, circuit, pictorial diagram, schematic diagram, switch, dry-cell, positive, negative, neutral, source, load, closed circuit, current, electron, proton, neutron, polarity, atom, molecule, nucleus, electron orbit, gravity field, electrostatic field, charge density.

1-1 GETTING STARTED

The fundamental theory underlying all electric, electronic, and radio circuits is the same. First, it is necessary to find out what the many different electrical parts, commonly called *components*, look like, then how they function by themselves, and finally, how they interact when connected together in *electric circuits*.

In this chapter it will first be shown how three common components—a battery, a lamp, and a switch—plus connecting wires, can be connected to form a simple electric circuit. Then the basic makeup of substances and the tiny particles involved in any working electric circuit will be discussed briefly.

But first, a word about studying electricity. It has been found that we learn by feeling, smelling, seeing, hearing, thinking, imitating, and repeating. From a book alone, and even in a classroom, neither feeling nor smelling may be particularly applicable. However, learning by seeing is accomplished by reading, by examining all illustrations closely, and by taking notes on new words and ideas. Every time students take a note, they see the important words again. Learning by hearing is also aided by saying each new word aloud several times. (Look up new words in a dictionary to make sure of the proper pronunciation.) In the classroom the student may be aided by hearing the instructor, but even this is not going to ensure retention of facts unless students take notes on what is said. (It is recognized generally that lecture alone is not an efficient teaching device for certain types of obscure facts.) After reading a page in a book, students should examine their notes and think over what has been read. As soon as they have reproduced the information contained on the page in their own words, it is time to move on to the next page. Diagrams should be copied, always with the stress on *why* they are drawn as they are. Might they be drawn some other way?

No one will learn all the thousands of details that make up the fundamentals of electricity without much rereading, reexamination of notes, and redrawing of diagrams. If you want to learn, always have a notebook and pencil with you when you study, and use them! Do you have them with you now?

1-2 A SIMPLE CIRCUIT

One of the simplest of electric circuits is shown in *pictorial diagram* form in Fig. 1-1*a*. The same circuit drawn in *schematic diagram* form is shown in Fig. 1-1*b*. The

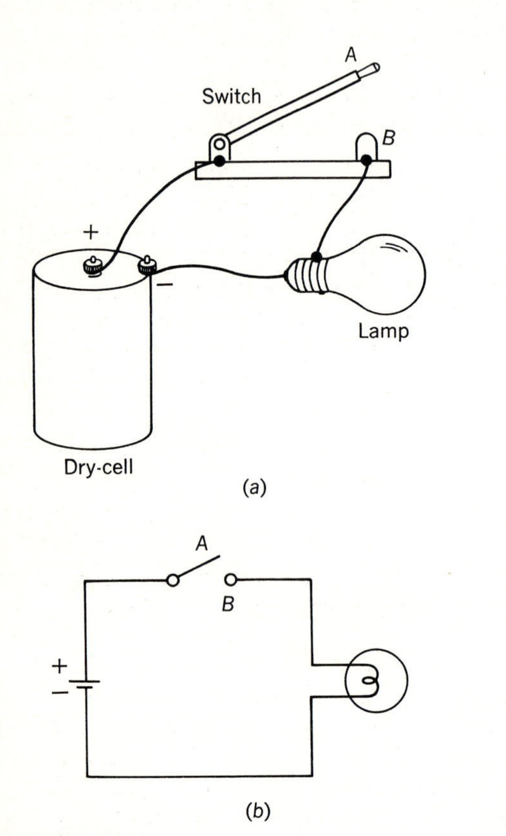

Fig. 1-1 A simple lamp circuit: (*a*) pictorial diagram; (*b*) schematic diagram.

schematic is far simpler to draw. For this reason, practically all electric circuits are represented schematically. The only way to learn to draw schematic diagrams is to practice each diagram until it can be reproduced without reference to the original.

Disregarding connecting wires, there are only three components used in this particular circuit: the dry-cell, the lamp, and the switch. Note the positive (+) and negative (−) marking on both pictorial and schematic symbols of the dry-cell. Note also how the symbols resemble the components in a simplified manner. The dry-cell supplies the energy and pressure that permits the circuit to work and may be termed the *source* of energy. Since the lamp accepts energy from the source, the general term *load* may be assigned to the lamp. A third component is the switch, acting as a *controlling device*. All electric circuits will have at least four things: a source, a load, a controlling device, and connecting wires.

When the switch arm is lowered and the switch contacts are closed, points *A* and *B* are connected. Electric current can now flow through the circuit, and the lamp glows. But what is an electric current? How can anything flow in solid copper wires? What makes such a current flow? Why does the lamp glow, and what determines how brightly it will shine? To explain what is occurring in this so-called simple circuit, it will be necessary to investigate more closely the makeup of substances used in electric circuits.

Following is the first of many check-up quizzes spaced every few pages throughout the book. They not only test understanding of the subject matter, but also develop important basic theory. Answers may be jotted in the spaces provided or written on a separate sheet of paper. When you have answered all the questions, compare your answers with those at the bottom of the next turned page.

Quiz 1-1. Test your understanding. Answer these check-up questions.

1. Which pole (+ or −) has its terminal at the center of the top of a dry-cell? __________ The outer can is which pole? __________
2. What relative length line of a dry-cell symbol in-

dicates the negative pole? _________ Positive pole? _________

3. When the circuit in Fig. 1-1*b* is not operating, is the switch open or closed? _________
4. What three types of components will always be found in operating electric circuits? _________ _________ _________
5. Which is easier to draw, a schematic or a pictorial diagram? _________
6. In the simple circuit shown in Fig. 1-1*b*, what is the source device? _________ Controlling device? _________ Load? _________
7. Without looking at the text, reproduce the schematic diagram of a simple circuit.

1-3 BASIC PARTICLES

There are three particles that must be considered in a basic study of electricity: the electron, the proton, and the neutron.

Generally speaking, the moving particles in electricity are *electrons*. When one or more electrons move in any one direction, they form an electric *current* in that direction. Each electron may be considered as having an invisible electric field of force emanating from it in all directions, as in Fig. 1-2*a*. The particular effect that the field of an electron has on anything near it has been arbitrarily designated as *negative*. Thus the electron is shown with a negative (—) sign on it.

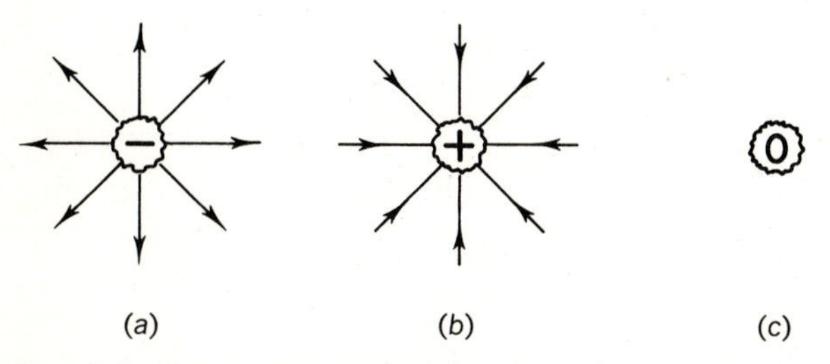

Fig. 1-2 Representation of (*a*) the field around an electron, (*b*) the field around a proton, and (*c*) the lack of field around a zero-charged neutron.

The *proton*, a much heavier particle than the electron, also has an invisible electric field surrounding it (Fig. 1-2*b*). The effect of its field of force on its surroundings differs from that of the electron, is termed *positive*, and is represented with a plus (+) sign.

Note that the terms *positive* and *negative* do not have a "more" or "less" meaning. These terms merely identify two opposite-effect electric fields or pressure directions.

While a proton has a mass, or weight, equal to about 1840 electrons, the negative field of force around an electron is exactly the same strength electrically as is the positive field surrounding a proton. Thus weight has no bearing on the electrical strength of the particles.

The *neutron* has a neutral, or zero, electric charge and is considered to have no electric field of force surrounding it. The neutron weighs slightly more than the proton, and it may be considered as a locked-together proton and electron.

To try to comprehend how incredibly small and light these basic atomic particles are, consider their approximate weights in grams (20 drops of water weigh about 1 g):

1 electron =
0.000 000 000 000 000 000 000 000 000 908 g

1 proton =
0.000 000 000 000 000 000 000 001 670 g

1 neutron =
0.000 000 000 000 000 000 000 001 670 9 g

Quiz 1-2. Test your understanding. Answer these check-up questions.

1. What would be the charge of an object if it had more electrons than protons? _________
2. What would be the charge of an object if it had the same number of electrons as protons? _________
3. What would be the charge of an object if it had equal numbers of protons and electrons but twice as many neutrons as electrons? _________

4. What would be the charge of a neutral object if it lost electrons? ________
5. Would 100 protons represent more, less, or the same magnitude of charge as 100 electrons? ________
6. Which is normally considered as the moving particle in electricity? ________
7. Is it known how a positive charge differs from a negative charge? ________

1-4 ATOMS AND MOLECULES

In nature, electrons, protons, and neutrons group together to form particles called *atoms*. All substances on earth are composed of one or more of about 93 different types of atoms. (Actually there are over 100 types of atoms identified so far, but the heaviest do not occur naturally and are produced only in laboratories.) Every substance on earth is some combination of these relatively few types of atoms.

Some substances or objects are composed of atoms of only one type. Examples are copper wires, iron wires, aluminum wires, carbon rods, and sulfur. Some substances are made up of two or more different types of atoms. In such cases the tiny atoms combine or link together chemically into more complex groupings called *molecules*. As examples, a molecule of table salt, chemically known as sodium chloride, or NaCl, consists of one sodium (Na) atom and one chlorine (Cl) atom. A molecule of water (H_2O) consists of two atoms of hydrogen (H_2) and one of oxygen (O). A molecule of common sugar ($C_{12}H_{22}O_{11}$) has 12 carbon (C), 22 hydrogen, and 11 oxygen atoms. Even the more complicated molecules are so small that they are not visible singly in normal microscopes, and only the largest molecules can be seen in powerful microscopes. No single atom is visible in any present-day microscope.

It is generally accepted that all atoms are made up of a spherical nucleus, or center, consisting of protons and neutrons, with electrons whirling around the electrically positive nucleus in orbits of various altitudes (Fig. 1-3). If there are six protons in the nucleus of an atom, there will be six electrons in orbit around it. Each positively charged proton is able to attract and hold one negatively charged electron in orbit around the nucleus. As a result of an equal positive-negative attracting effect, every atom normally has a zero electric charge, regardless of whether it consists of one proton and one electron or of twenty protons and twenty electrons. (Under certain circumstances the number of neutrons in the nucleus may differ, but since neutrons have no charge, there is no *electrical* effect on the atom.)

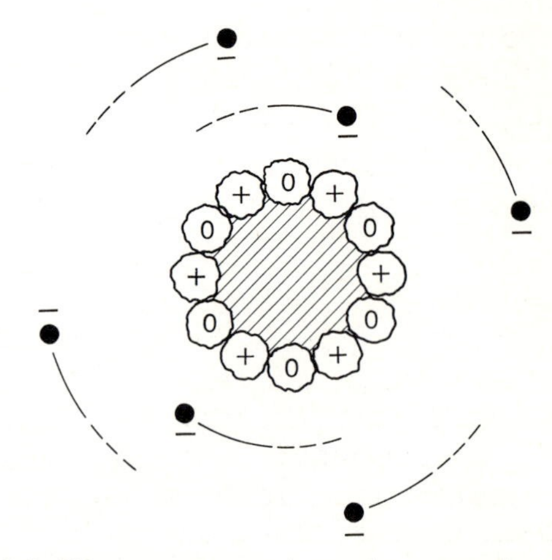

Fig. 1-3 Representation of a carbon atom having a nucleus of six protons and six neutrons, with six orbiting electrons in two orbital levels.

ANSWERS TO CHECK-UP QUIZ 1-1

1. (Positive) (Negative) 2. (Short) (Long) 3. (Open) 4. (Source, load, control device) 5. (Schematic) 6. (Dry-cell) (Switch) (Lamp) 7. (See Fig. 1-1*b*)

Notice that it is stated that protons can attract electrons. From this fact comes one of the fundamental laws of electricity:

Unlike charges attract; like charges repel.

Any two electrons will repel each other because they have similar polarity. Any two protons will repel each other for the same reason. However, a negative electron will attract a positive proton, or a proton will attract an electron, because they have opposite polarity.

No orbiting electron in an atom will ever touch another, since they all have the same negative polarity. This raises the question: What allows the positive protons to remain grouped together in the nucleus? A simplified theory is that protons are separated by zero-charged neutrons in the nucleus and also have their positive charges neutralized by the electrons they hold in orbit around themselves.

What prevents electrons from pairing off with protons and breaking up the atom? One explanation is that the rapidly moving electrons have acquired energy of motion and are moving so fast that they cannot fall into the field of the heavier, more or less stationary, fields of force of the protons in the nucleus. This is roughly similar to gravity not being able to pull down a satellite orbiting the earth unless the satellite loses some of its energy of motion by friction between itself and air molecules. Since there is no friction in the absolutely empty space between the particles of an atom, orbiting electrons do not spontaneously lose their energy.

Quiz 1-3. Test your understanding. Answer these check-up questions.

1. What physical effect would a positively charged object have on a negatively charged object? ________ On a positively charged object? ________ On a zero-charged object? ________
2. Does the chemical formula HCl represent an atom, a proton, a molecule, a neutron, or an electron? ________
3. The nucleus of any atom (except hydrogen atoms) is composed of what two atomic particles? ________ ________
4. What electric charge does a normal oxygen atom have? ________ A normal copper atom? ________
5. How many different kinds of atoms occur in nature? ________
6. Do electrons ever touch other electrons? ________
7. When two or more atoms form a molecule, what parts of them must be interweaving to hold them together? ________

1-5 REPRESENTING FIELDS OF FORCE

If a rock is held up in the air, it is in the gravity field of the earth; that is, it is being attracted to any and all mass near it. This can be indicated in an illustration by drawing *lines of force* from the lighter rock toward the heavier earth, as in Fig. 1-4. Arrowheads can be placed on the

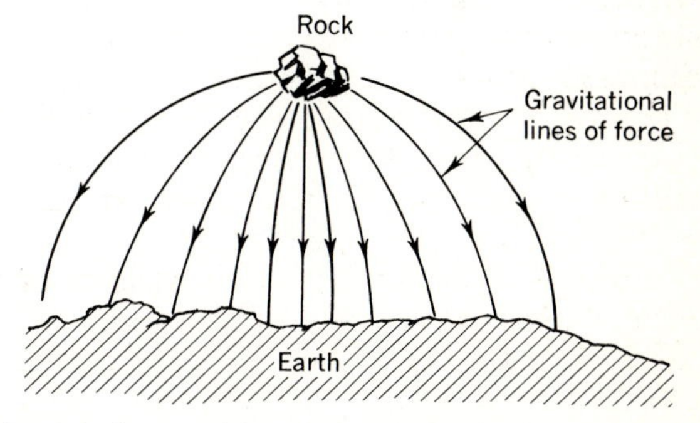

Fig. 1-4 Lines of force of gravity acting between a lighter and a heavier object. Arrowheads indicate direction of pull on the lighter object.

lines to indicate the direction of the attracting force that the heavier mass of the earth is exerting on the lighter rock. When the rock is released, it will follow the pull of the theoretical gravitational lines of force until it strikes the heavier earth.

Actually, the rock is attracting the earth too, but because of the great difference in the two masses, it is the rock that apparently does all the moving.

If a metal ball has deposited on itself in some way a million extra electrons, it attains a negative electric charge. If another metal ball has a million electrons taken from it, this ball becomes positively charged. As long as these two charged objects are held near each other, an attracting negative-positive *electrostatic* field of force will exist between them (Fig. 1-5). This electric field can be indicated by drawing lines of force between the two objects. If a negative electron is released at point A in this field, it will be repelled by the negative charge and will be attracted to the positive one. Thus both charges will tend to move the electron in the direction of the lines of force between the two objects. The arrowheads in Fig. 1-5 indicate the direction of motion that would be taken by the electron if it were in different areas of the electrostatic field.

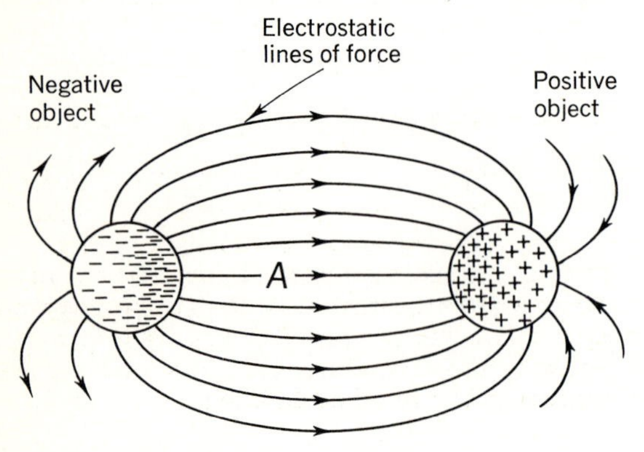

Fig. 1-5 Representation of electrostatic lines of force between unlike electrically charged objects.

This text uses the electron current theory, in which the movement of electrons from a negative charge toward a positive charge forms the current. When electricity was first investigated, the terminals on a battery were arbitrarily labeled "positive" and "negative." Not knowing anything about electrons, researchers decided that current would flow from what was called the positive terminal through the circuit to the negative terminal. Unfortunately they picked the wrong current direction. Later, when vacuum tubes were used, it was proved that electrons actually flow from the negative terminal to the positive. By this time all texts were written with the "conventional" current flow. Since electrical circuits behave the same regardless of which way the current is considered to be traveling, engineering texts still use conventional current. Their electrostatic field arrows will be reversed from those shown in this text.

Quiz 1-4. Test your understanding. Answer these check-up questions.

1. If a rock were thrown upward, would the arrowheads illustrating its gravitational lines of force be shown pointing up or down? ______
2. If two heavy bodies in outer space were moved farther apart, would a drawing of the lines of gravitational force between them show more, fewer, or the same number of lines? ______
3. If a proton were thrown in the field in Fig. 1-5, would it try to follow a path shown by the lines of force? ______ Would it move in the direction shown by the arrowheads? ______
4. Why would the negative charge density be greater on the surface of the negative object in Fig. 1-5 than inside the object if electrons are free to move anywhere in the object? ______

ANSWERS TO CHECK-UP QUIZ 1-2

1. (Negative) **2.** (Zero) **3.** (Zero) **4.** (Positive) **5.** (Same) **6.** (Electron) **7.** (No)

ANSWERS TO CHECK-UP QUIZ 1-3

1. (Attract) (Repel) (None) **2.** (Molecule) **3.** (Protons, neutrons) **4.** (Zero) (Zero) **5.** (93) **6.** (No) **7.** (Outer orbiting electrons)

5. Would the right or the left side of the negative object in Fig. 1-5 have the greater density of electrons? ________ Why? ________
6. What is the only possible practical method of charging an object positively? ________

Following is the first of the tests that appear at the end of every chapter. Answer each question in one or two words if possible. The answer section appears at the end of the book.

CHAPTER 1 TEST • AN ELECTRIC CIRCUIT

1. In the simple electric circuit, where is the energy if the circuit is not operating? Where is the energy expended when the circuit is operating?
2. If an electric circuit is opened, does this mean that it begins to work or stops working?
3. What is the general term that might be applied to the lamp in the simple circuit?
4. What is the name of the controlling device used in the simple circuit?
5. Is the short or the long line in the symbol for a dry-cell the positive terminal?
6. The center pole of a dry-cell always has what electric polarity?
7. What is the name of the atomic particle that has a positive charge? Negative? Zero?
8. Which is heavier, the positive or the negative atomic particle? How much heavier?
9. Which has the greater electric charge, the positive or the negative particle?
10. What electric charge does a piece of copper normally have?
11. If 10 million electrons were taken from a piece of copper, what electric charge would the copper have?
12. Does "H_2O" represent an atom, a proton, a molecule, a neutron, or an electron?
13. What two particles are found in the nucleus of most atoms?
14. Is there air between the electrons and protons in an atom?
15. What is the direction given to the lines of force between two unlike electric charges? What is the proper designation of such lines?
16. A positive and a negative object are near each other. Where would the greatest concentration of positive charge be on the positive object?
17. Draw a schematic diagram of a simple electric circuit and label all components.

ANSWERS TO CHECK-UP QUIZ 1-4

1. (Down) **2.** (Less) **3.** (Yes) (No, opposite) **4.** (Electrons repel each other) **5.** (Right) (Positive object attracts electrons) **6.** (Removing electrons from it)

2

CURRENT AND VOLTAGE

CHAPTER OBJECTIVE. To develop a usable understanding of these vocabulary terms: current, free electrons, induction, ampere, ammeter, series, resistance, electrolyte, coulomb, electron drift, metric units, kilo-, centi-, milli-, impulse velocity, volt, voltage, electromotive-force (emf), potential difference, voltmeter, parallel, shunt.

2-1 ELECTRIC CURRENT

In Chap. 1 a simple basic circuit was discussed. As a preliminary to understanding circuit operation, the basic atomic particles—electrons, protons, and neutrons—were examined, and the electrostatic field of force that exists between charged atomic particles was considered.

A most interesting phenomenon exists when a short length of copper wire is in the electrostatic field between negative and positive objects, even though the wire does not touch either object (Fig. 2-1). It will be noted that the path of some of the lines of force goes through the copper wire rather than through the air. This indicates that the copper wire must be a better medium for the electrostatic lines of force to exist in than the air is. The closer the wire is brought to the charged objects, the more lines of force the wire will hold and the fewer there will be in the air.

If it were possible to see into the wire, it would be found that some of the outer-orbit or outer-layer electrons of the copper atoms (called *free electrons*) that are under the influence of the electrostatic field have moved along the field lines to the far end of the wire nearest the positive object. At the same time, the copper atoms in the wire near the negative object would be found to lack the same number of electrons. Thus, while one end of the wire is made negative by what may be termed *electrostatic induction*, the other end becomes positively charged by an *induced* lack of electrons. Another way of explaining this electron shift is that some of the free electrons at the left end of the wire are repelled by the nearby negative object and are attracted to the positive object at the other end of the wire.

Now, if both of the charged objects are touched to the wire simultaneously, electrons from the negative object will rush into the empty gaps in the outer electron rings of the nearby copper atoms, and electrons will rush out of the right-hand end of the wire to the positive areas in the positive object. (Remember, an object is positive only because atoms in it lack electrons.) In this way electrons can be conducted or transferred from left to right along the wire *conductor* until all the atoms in the wire and in the two charged objects have an equal number of electrons. Such a movement of electrons rep-

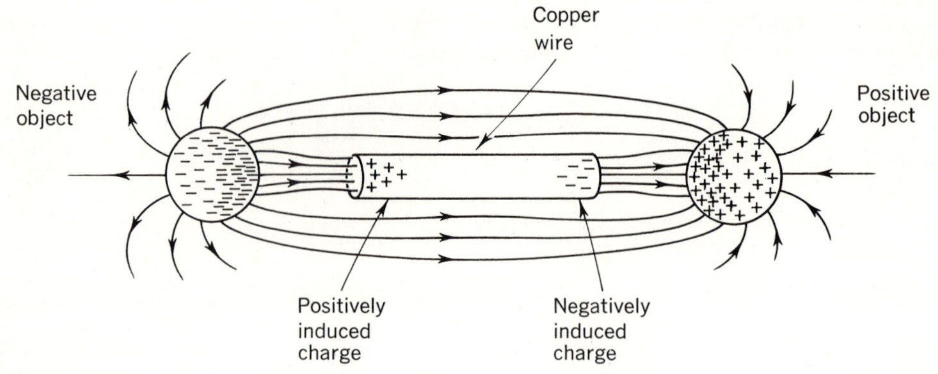

Fig. 2-1 Electrostatic induction will induce a charge into the two ends of a conductor held between two charged objects.

resents an *electric current*. The current in this case will cease as soon as the copper wire and the charged objects all have their normal number of electrons, or as soon as there is no difference of charge throughout the whole circuit. With no difference in electric charge anywhere, there is no longer an electric field acting as a source of pressure to move any free electrons from copper atom to copper atom. The circuit is completely discharged, and current no longer flows.

In Fig. 2-2, the conductor is not in the space *between* the positive and negative objects, and therefore it might be assumed to be out of the electrostatic field. However, if the wire were close to the objects, it would be found that some of the field lines passed into the wire, because it is a much better medium for electrostatic fields than air is. Regardless of how the wire may be bent, it will act as a guiding path for the electrostatic lines. When the wire is touched to the charged objects, current (electrons) flows through the wire, just as it did through the wire that was in a direct line between the two charged objects.

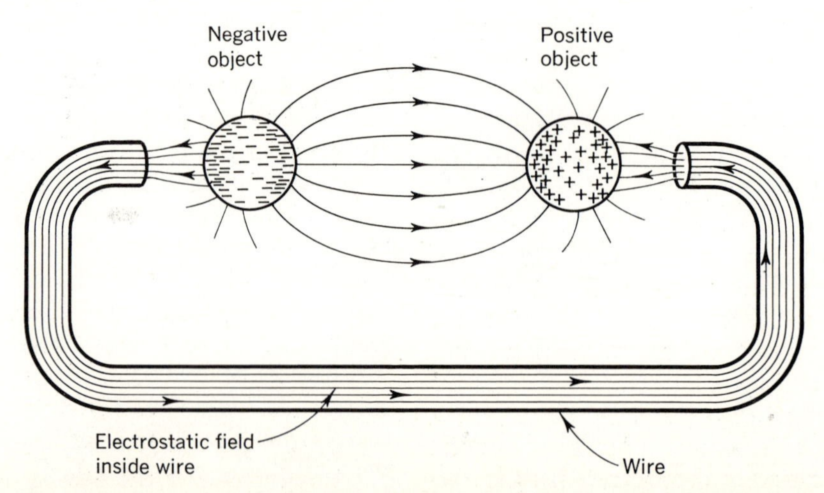

Fig. 2-2 A conductor or wire will act as a guide for electrostatic lines.

The unit of measurement of current flow is the *ampere*. A steady stream of electrons flowing along a wire and moving past any single point in the wire at a rate of approximately 6 million million million electrons per second represents one *ampere* of current.

The symbols used to indicate intensity of flow, or current, are either I or i. The symbol used to indicate amperes is A. If a current of 10 A flows in a circuit, this can be expressed as "the I is 10 A," or perhaps "the i is 10 A." Whether to use the uppercase or lowercase symbols will be discussed later.

A meter used to measure current is called an *ammeter* (note the spelling: no letter "p"). Since an ammeter must indicate the number of electrons flowing per second, it must be inserted in *series* with one of the connecting wires (Fig. 2-3). That

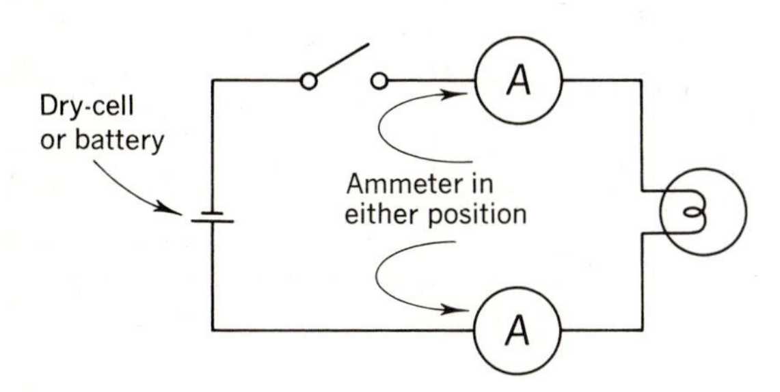

Fig. 2-3 An ammeter in any position in a simple circuit will read the same.

is, the wire is cut, and the ammeter is inserted at this point. In any circuit having several components in series (connected end to end), there must always be exactly the same value of current flowing in all components at any given instant. Therefore it makes no difference at what point an ammeter is placed in such a circuit. Because an ammeter should not change the current value in a circuit into which it is connected, the meter should be constructed of materials that have negligible opposition, called *resistance*, to current flow.

The dry-cell in Fig. 2-3 is a device in which chemical energy can be converted to electrical energy. The chemicals inside the cell pull electrons from the positive pole (shown as the longer symbol line), forcing them to flow through the internal *electrolyte* (a chemical liquid capable of conducting electrons) and onto the negative pole (the shorter line). When the back pressure of electrons repelling each other on the negative pole becomes as great as the internal chemical pumping pressure, all electron motion ceases. In the normal dry-cell this is said to be at a 1.5 V pressure. ("Volt" is abbreviated "V.") If a load is connected across the cell by closing the switch, a path for the electrons that are under pressure is provided, and electrons begin moving from the negative pole through the load into the positive pole. This relieves the back pressure in the cell, and the electrolyte chemicals start pumping, or working to maintain a constant current flow as long as the load is connected and the chemicals hold out. Actually, electrons from the circuit start moving into the positive terminal of the dry-cell at exactly the same time and rate as electrons start moving out of the negative terminal.

Quiz 2-1. Test your understanding. Answer these check-up questions.

1. If a proton were released at point A in Fig. 1-5, in what direction would it move? ____________ In what direction would a neutron be pulled? ____________
2. What other field of force have you heard of besides the gravitational and the electrostatic? ____________
3. Which is a better medium for electrostatic lines of force, air or copper? ____________
4. What does "$I = 9.4$ A" mean? ____________
5. Is an ampere a rate of flow, a quantity of elec-

trons, or a force that moves electrons? __________

6. Will any electrons flow into your body if you touch the negative terminal of a dry-cell or battery? __________
7. In Fig. 2-3, when the switch is closed, will the current first flow through the top ammeter, the lamp, the bottom ammeter, the switch, or the cell? __________
8. Mark the proper polarity signs (+ and −) on the dry-cell in Fig. 2-3.
9. In Fig. 2-3, which terminal of the upper ammeter would have a + marking on it if the meter is reading correctly? __________
10. What is the direction of current flow (clockwise or counterclockwise) through the dry-cell? __________ Through the circuit as a whole? __________

2-2 THE COULOMB

One of the basic units in electricity is the *coulomb* (kōō-lŏm), abbreviated C. The coulomb is the unit of electrical quantity or charge, and it is usually assigned the letter symbol Q. While it is common to think of a coulomb as being about 6 million million million negative electrons (actually 6.25×10^{18}), the same number of positive proton charges would also be a coulomb. However, since only electrons can be removed from atoms, it would be highly unlikely to find a coulomb of protons. It is quite possible to have an object lose a coulomb of electrons and thereby attain a positive charge equivalent to a coulomb of protons.

Normally, everything has a neutral charge. If 1 C of electrons is moved from a neutral object to another similar object, the difference in charge between the two objects is then 2 C. (One object is +1 C and the other is −1 C.)

Note that 6 million million million electrons is a quantity of charge, whereas 1 C of electrons moving past a point in a circuit in 1 s (second) is a rate of flow called an ampere. (In addition to being used as a means of determining the ampere, the coulomb is also used to determine the unit of measurement of capacitance, called the *farad.*)

Quiz 2-2. Test your understanding. Answer these check-up questions.

1. If there is a difference in charge of 2 C between two objects, approximately how many electrons will move if the objects are connected by a wire conductor? __________ Exactly how many electrons? __________ If it took 1 s for the charges to neutralize, what average current would have been flowing? __________ If it took only 0.5 s? __________
2. If a current of 3 A flows through a meter for 1 min, how many coulombs actually pass through the meter? __________ How many electrons is this? __________
3. Is a coulomb a rate of flow, a quantity of electrons, or a force that moves electrons? __________
4. How many electrons are actually indicated by "Q = 4 C"? __________

2-3 VELOCITY OF CURRENT

One of the properties of electricity is its speed. The velocity of an electric impulse is approximately 300,000,000 m/s (meters per second), which is approximately 186,000 mi/s (miles per second), or about 7 times around the world in a second.

The international standard of measurement, called the *metric system,* is based on the meter, a length supposedly equal to one ten-millionth of the distance between the North Pole and the equator. A listing of the metric system units is included in Appendices I and J. A few frequently used equivalencies are shown in Table 2-1.

Table 2-1 ***EQUIVALENT UNITS OF MEASUREMENT***

1 meter (m)	= 39.37 in.	= 3.28 ft
1 kilometer (km)	= 1000 m	
	= 3280 ft	= 0.621 mi
1.61 kilometers	= 5280 ft	= 1 mi
1 centimeter (cm)	= 0.01 m	= 0.3937 in.
30.48 cm	= 1 ft	= 12 in.
1 millimeter (mm)	= 0.001 m	= 0.03937 in.

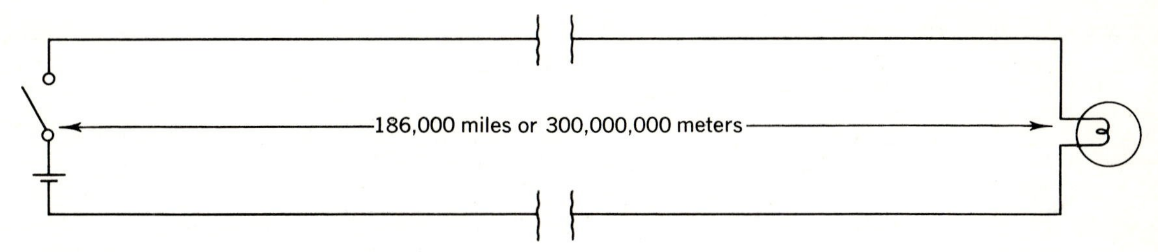

Fig. 2-4 It takes 1 s after the switch is closed before electrons start to move through the lamp.

Because the electrical impulse travels so fast, electricity may be considered as operating instantaneously in circuits involving only a few feet of wire. In longer circuits, however, time may become a factor. Consider the theoretical circuit shown in Fig. 2-4. It will take one full second after the switch is closed before current begins to flow through the lamp.

Actually, individual free electrons in the molecules of the wires may travel (drift) only a fraction of an inch in 1 s, but the forward impulse that they produce, the effect of their movement on other electrons ahead of and behind them, travels at the speed of light (300,000,000 m/s).

Quiz 2-3. Test your understanding. Answer these check-up questions.

1. In Fig. 2-4, after the switch is closed, how long would it be before electrons began entering the positive terminal of the source? ________ Before electrons began leaving the negative terminal? ________ Before electrons began flowing through the load? ________
2. About how long, in seconds, does it take an electrical impulse to travel 1 mi? ________ 1 m? ________
3. After the switch is opened, how long will electrons move in the Fig. 2-4 circuit? ________
4. Do individual electrons travel along a wire at a speed of 300 million m/s? ________
5. In what direction would current flow in the lamp of Fig. 2-4 (clockwise or counterclockwise)? ________
6. If the moon is 395 million m from the earth, how long would it take a radar wave to be transmitted, bounce off the moon, and return to earth? ________ How long would it take a laser beam of light transmitted toward the moon from earth to become visible to a viewer on earth? ________
7. What fraction of a meter equals 1 ft? ________
8. What is a kA? ________ What is a mA? ________

2-4 ELECTRIC PRESSURE OR VOLTAGE

Two important concepts have been presented: (1) the progressive movement of free electrons from one atom to another along a wire forms an electric current; (2) an electric difference of polarity, negative and positive, produces an electrostatic field between the poles which represents an electron-moving force. This electron-moving force, properly termed *electromotive-force,* abbreviated "emf," is measured in units called *volts.*

When electrical pressure (voltage) is applied across a suitable electron-conducting substance, a movement of electrons (current) will result. Symbols used to indicate emf or electrical pressure are E or e. However, the unit of measurement of emf, the volt, is indicated by V. For example, in Fig. 2-5, $E = 1.5$ V.

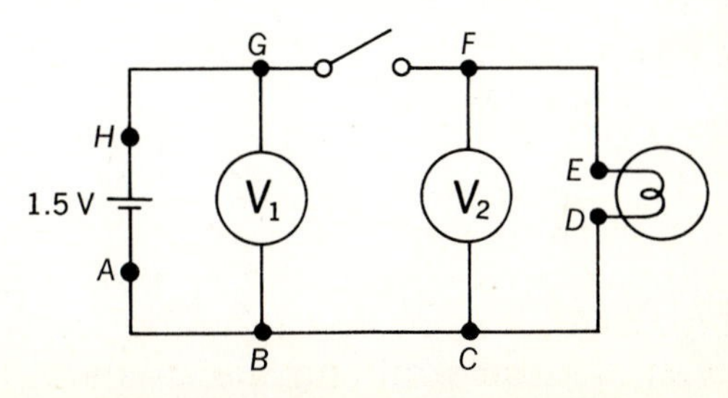

Fig. 2-5 Voltmeters and a load in parallel.

A *voltmeter* is used to measure the emf that is present across the two wires of a circuit—across the *potential difference*. Figure 2-5 illustrates the placement of two voltmeters, V_1 and V_2. V_1 is connected to measure the emf across the lines from the source at all times, and it should read 1.5 V. With the switch open, there is no current flowing in the load and there is no difference of potential across the lamp. Therefore V_2 reads 0 V.

As soon as the switch is closed, V_2 and the lamp are also connected across the dry-cell and the line. V_1, V_2, and the lamp all have 1.5 V across them now. V_1 and V_2 will read 1.5 V, and the lamp will glow. Under this condition the two meters and the lamp are said to be connected in *parallel* across the source. It may also be said that they are *shunted* across the source. Note that the same emf is across any and all components if they are connected in parallel.

Usually voltmeters require relatively insignificant amounts of current flowing through them to get a reading. (In most practical cases voltmeter currents are so small that they are disregarded.) In this circuit the current through the lamp will be many times greater than the currents flowing in the meters. In circuits with two or more parallel-connected branches or loads, the voltage across each component may be the same, but the currents in the various branches will differ considerably, as they do in the voltmeters and load in this case.

In all the simple series circuits (one source, one load) shown thus far, the dry-cell produces 1.5 V of emf by its internal chemical action. The wires of the circuit allow a current of electrons to flow through them if the circuit is complete. Only with the switch closed does a completed path for electrons exist. Free electrons in the wire may then move into the positive attracting terminal, move out of the negative repelling terminal, and move through the lamp, producing an equal value of current throughout all parts of the series circuit. With the switch in the closed position, the voltage keeps pumping electrons through the circuit. When the chemicals in the dry-cell become depleted, the cell can no longer pump electrons rapidly enough. The emf of 1.5 V decreases, and the current decreases. Eventually all current ceases as the chemicals wear out.

ANSWERS TO CHECK-UP QUIZ 2-1

1. (Toward the negative) (None) **2.** (Magnetic) **3.** (Copper) **4.** (Current is 9.4 A) **5.** (Rate of flow) **6.** (Relatively few) **7.** (Current starts at same time in all assuming wires are short) **8.** (Negative sign on short line) **9.** (Right-hand terminal) **10.** (CW) (CW)

ANSWERS TO CHECK-UP QUIZ 2-2

1. (6 million million million) (6.25×10^{18}) (1 A) (2 A) **2.** (180 C) (1130 million million million, or 1.125×10^{21}) **3.** (Quantity) **4.** (25 million million million, or 2.5×10^{19})

ANSWERS TO CHECK-UP QUIZ 2-3

1. (At once) (At once) (1 s) **2.** (0.00000556 s) (0.000 000 00333 s) **3.** (1 s) **4.** (No) **5.** (CCW) **6.** (2.63 s) (2.63 s) **7.** (0.3048 m) **8.** (1000 A) (1/1000 A)

Quiz 2-4. Test your understanding. Answer these check-up questions.

1. In Fig. 2-5, with the switch open, what is the direction of lines of force across the switch? ________ Are there lines of force in the air between points *B* and *G*? ________ *C* and *F*? ________ *D* and *E*? ________
2. With the switch open, is there any current flowing between points *B* and *G*? ________ *C* and *F*? ________ *G* and *F*? ________
3. With the switch closed, are there lines of force in the air between points *C* and *F*? ________ *D* and *E*? ________ *G* and *F*? ________

4. With the switch closed, is any current flowing directly between *B* and *G*? __________ *C* and *F*? __________ *G* and *F*? __________

5. Is voltage a rate of flow, a quantity of electrons, or a force that moves electrons? __________

6. Should we normally consider current to be flowing through an operating voltmeter? __________ Through an operating ammeter? __________

7. In circuits having parallel loads, is the voltage across all loads and the source voltage the same value? __________ Will the current through all the branches be the same? __________

8. If it is said that a wire is shunted across an ammeter, in what circuit configuration are the ammeter and wire connected? __________

9. What does "$E = 3$ V" mean? __________

10. On scratch paper, reproduce the diagram of the circuit used in the first four questions above. Name the parts without referring to the diagram in the text.

CHAPTER 2 TEST • CURRENT AND VOLTAGE

1. What are the names of the three basic atomic particles discussed so far?
2. Which would be permeated more easily by electrostatic lines of force, air or copper?
3. A wire is between a + and a − object, but not touching them. Which end of the wire would have a + charge induced in it?
4. When a wire is acting as a conductor, is it conducting lines of force, atoms, electrons, or protons?
5. Arrowheads on electrostatic lines of force are shown pointing toward what polarity?
6. Two ammeters are in a simple series circuit. Each reads 2 A. What is the current value of the circuit?
7. In question 6, if one meter reads 2 A and the other reads 2.2 A, what would you consider the circuit current value to be?
8. What is the term used for the quantity 6 million million million electrons?
9. If 12 million million million electrons pass a point in a circuit in $1/5$ s, what value of I is flowing?
10. Which should have the lesser resistance to electric current flow, a voltmeter or an ammeter?
11. What are the velocities in meters and in miles, of radio waves, light waves, and electric-current impulses?
12. In what direction do electrons flow through a dry-cell in an operating circuit? In an "open" (open-switch) circuit?
13. What is the unit of measurement of electrical pressure?
14. What does "emf" stand for?
15. If a lamp across a dry-cell burned out (if it were open-circuited), would current flow through it? Would it have a high or low resistance to current flow? Would electrostatic lines of force be developed across the open part of the filament?
16. Would the current values in all components in a series circuit always be the same?
17. Would the voltage values across all components in parallel in a circuit always be the same?
18. Would the potential difference across all components in a series circuit always be the same?
19. Would the current value through all components in a parallel circuit always be the same?
20. What are three common electrical terms that mean the same as potential difference?

ANSWERS TO CHECK-UP QUIZ 2-4

1. (*F* to *G*) (Yes) (No) (No) **2.** (Meter current only) (No) (No) **3.** (Yes) (Yes) (No) **4.** (Meter only) (Meter only) (Yes) **5.** (Force) **6.** (No) (Yes) **7.** (Yes) (Not necessarily) **8.** (Parallel) **9.** (The emf is 3 V) **10.** (See Fig. 2-5)

3 RESISTANCE AND CONDUCTANCE

CHAPTER OBJECTIVE. To develop a usable understanding of these electrical vocabulary terms: battery, series-aiding, series-bucking, vector arrow, polarity markings, short-circuit current, resistance, conductance, ohm, omega, wire gages, mil versus milli, Nichrome, Constantan, resistance-temperature coefficient, metric conversions, resistor, ohmmeter, ionization, insulator, semiconductor, doping, ion, ionization potential, electrolyte, rheostat, potentiometer.

3-1 ELECTRON SOURCES IN SERIES

In Chap. 2 current and electromotive-force were discussed. So far the only source of emf for most explanations has been a single 1.5-V dry-cell (Fig. 3-1*a*). When more voltage is desired, two or more cells can be connected in *series* (Fig. 3-1*b*). Two or more cells connected in this manner form a battery of cells, or simply a *battery*.

A single dry-cell produces an emf of 1.5 V, but a two-cell battery in a *series-aiding* configuration, as in Fig. 3-1*b*, operates as a 3-V source. If the two cells are connected, *series-bucking*, as in Fig. 3-1*c*, the emf of each cell will be operating in opposite directions, which provides a zero-voltage source. No current can flow through any load connected across them. In any series-bucking circuit, the effective emf value will be the voltage *difference* between the two opposing emf values, and it will have the polarity of the greater emf.

The symbol for a battery is usually two cells (in series), regardless of the actual number of cells in the battery. The voltage value can be indicated on the schematic diagram next to the battery symbol, as in Fig. 3-1*d*. The "vector" arrows on the diagrams indicate the direction either of emf in the circuit or of current flow.

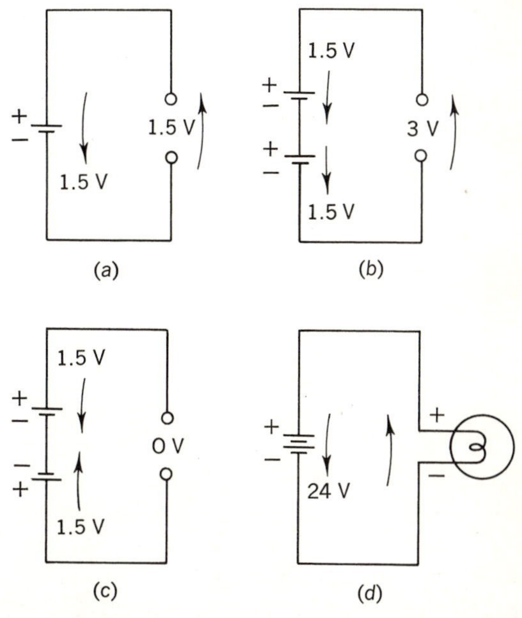

Fig. 3-1 (*a*) A single 1.5-V dry-cell. (*b*) Two cells connected series-aiding. (*c*) Two cells connected series-bucking. (*d*) Symbol for a 24-V battery across a lamp load.

It is usual to think of a field of force capable of moving electrons as existing only between positive and negative poles, but an electron-moving force is also developed between either two negative charges or two positive charges of different values. In Fig. 3-2*a*, one battery has 6 V and the other 9 V, with positive terminals connected together. The difference in potential across the lamp is the difference between −6 and −9 V, or 3 V. The value of current that flows is theoretically the same as if the lamp were across a 3-V source. The current will flow away from the higher and toward the lower negative potential, as indicated by the arrow, because free electrons in the circuit will be repelled with more force by the greater negative charge.

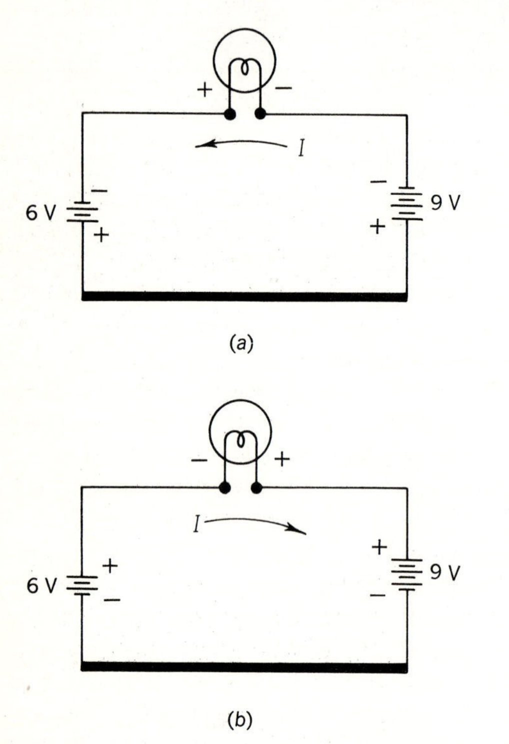

Fig. 3-2 Current flows (*a*) from a higher negative potential and (*b*) toward a higher positive potential.

In Fig. 3-2*b*, the two negative terminals are connected together. Current will flow from the lower toward the higher positive potential, as indicated by the arrow, because free electrons in the circuit will be attracted to the most positively charged point. (These are series-bucking sources with opposing emf values.)

The + and − polarity markings drawn on sources and loads, as in Figs. 3-1 and 3-2, do not indicate that the *load*, if removed from the circuit, would act as a source of emf with these polarities. Polarity markings on loads merely indicate the direction in which the current is moving through the loads (negative to positive). Remember, current flows from a negative source to a positive source in any electric circuit. Applying this theory to any load in a circuit will indicate what polarities may be marked on the load.

Quiz 3-1. Test your understanding. Answer these check-up questions.

1. If two 1.5-V cells are connected in series, + to +, what is the total emf developed across the two? ________
2. In what direction (+ to −, or − to +) does current flow through any lamp? ________ Through a battery? ________ From a battery? ________
3. What letter symbol is used to indicate electrical pressure? ________ What is used to indicate the unit of measurement of electrical pressure? ________
4. Object *A* has a deficiency of electrons, and object *B* has a greater deficiency. If the two objects were connected by a conductor, would current flow? ________ If so, in what direction? ________
5. Object *A* has an excess of electrons, and object *B* has a neutral charge. If the two objects were connected by a conductor, would current flow? ________ If so, in what direction? ________
6. Object *A* has a deficiency of electrons, and object *B* has a neutral charge. If the two objects were connected by a conductor, would current flow? ________ If so, in what direction? ________

7. In Fig. 3-1c, if the top battery had 45 V and the lower 9 V, what would the terminal voltage be? __________ What polarity marking should the top output-terminal carry? __________
8. In Fig. 3-2a, if the 6-V battery were reversed in the circuit, what would be the voltage across the load? __________ Which end of the load would be the + end? __________

3-2 RESISTANCE DECREASES CURRENT

Nothing has been mentioned to indicate that there might be any limit to the number of electrons that could flow in a simple circuit. Actually, there are many practical limiting factors. One is that the chemicals of a dry-cell cannot produce limitless current. A small, type AA, 1.5-V pen-light cell can produce only 1 A or so with a wire "shorted" (connected directly) between its + and − terminals. A type D, 1.5-V flashlight cell is able to produce a short-circuit current of only about 6 A. A large, #6, 1.5-V doorbell-ringing dry-cell can produce about 30 A. After a few minutes of operation at their maximum values, the current output of these cells decreases materially. To produce efficient operation of such cells, the continuous-current-drain values should not exceed 1/100 to 1/1000 of their short-circuit current capabilities.

A second limiting factor is that no electron movement along wires can be accomplished without overcoming some of the resistance that the wires always have to a movement of electrons through them. The result of this resistance is a loss of some of the energy supplied to the moving electrons by the emf of the source. Some of this energy imparted to the moving electrons is converted to heat in the resistance of the wires. A small wire of little mass with 1 A flowing through it will heat appreciably. The conversion of the energy of the electrons into heat in the wire causes a resistance effect to any progressive electron flow and results in a movement of fewer electrons. Fewer electrons moving per second is less current. If the wire is long enough, the resistance can limit the current to a very small value. With larger, more massive wires, more atoms are available, with more free electrons. More free electrons result in less opposition to a progressive movement of current along the wire with a given emf or pressure. The same value of current that produced considerable heat and resistance in the smaller wire may produce almost no appreciable heat or resistance in the larger wire with its better conductance, or conducting ability.

A third current-limiting factor is the load. The wire filament that glows when a lamp is in operation is usually made of tungsten, a metal which has a high resistance value to electron movement through it. As a result, much of the energy of the moving electrons is converted to heat in the high-resistance filament wire. The filament becomes so hot that it glows. The connecting wires will not warm appreciably, because they are made of metals that have much less resistance than the lamp filament. The connecting wires act as efficient conductors of energy from source to load. The lamp, being an inefficient conductor, heats and glows, losing energy in the form of heat and light.

3-3 RESISTANCE AND WIRES

The unit of measurement of electrical resistance is the *ohm*, represented by the Greek letter omega (ō-mē′ga), **Ω**. The letter symbol for resistance is *R*. If the resistance of a lamp is 35 ohms, this can be expressed as $R_L = 35\ \Omega$, where the *R* with the subscript *L* indicates load resistance.

The resistance required to limit current flow to 1 A when a source with an elec-

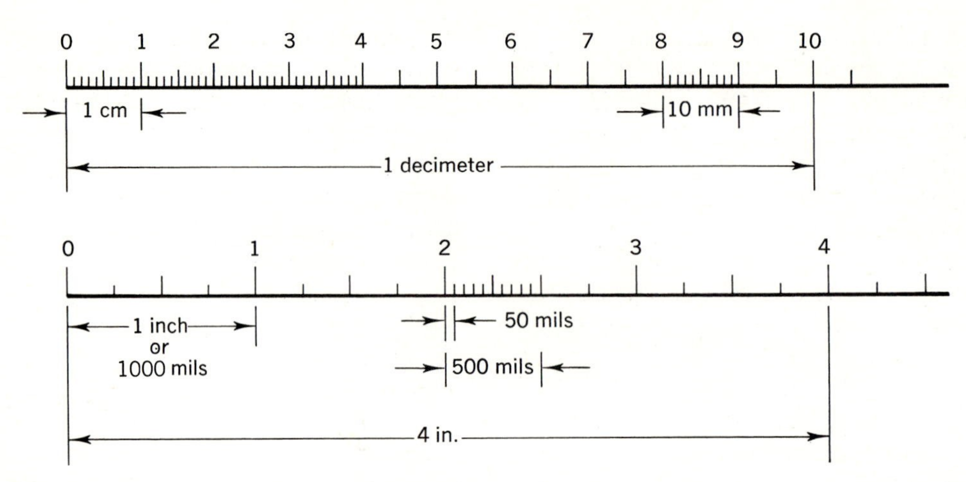

Fig. 3-3 Comparison of the metric decimeter (dm), centimeter (cm), and millimeter (mm), with the U.S. and British inch and mil.

tromotive-force of 1 V is applied to the circuit is 1 **Ω**. One ohm of resistance can also be considered to be the value of resistance a round copper wire with a diameter of 1 mm (millimeter) (or 1/1000 m) would have if it were 45.6 m long. According to the American Wire Gage, a round wire of 1 *mil* (1/1000 in.) diameter and 1.15 in. long would have 1 **Ω**.

Wires made of different metals will have different resistance values. Table 3-1 shows the lengths of various wires that would have 1 **Ω** of resistance.

Table 3-1 ***WIRE LENGTHS FOR 1 OHM***

Millimeter wire gage, required for 1 Ω with 1 mm diameter		American Wire Gage, required for 1 Ω with 1 mil diameter	
Silver	= 47.9 m	Silver	= 1.21 in.
Copper	= 45.6 m	Copper	= 1.15 in.
Aluminum	= 24.6 m	Aluminum	= 0.62 in.
Tungsten	= 14.3 m	Tungsten	= 0.362 in.
Zinc	= 12.9 m	Zinc	= 0.326 in.
Iron	= 6.05 m	Iron	= 0.153 in.
Lead	= 5.46 m	Lead	= 0.138 in.

ANSWERS TO CHECK-UP QUIZ 3-1

1. (0) **2.** (− to +) (+ to −) (− to +) **3.** (*E*) (V) **4.** (Yes) (*A* to *B*) **5.** (Yes) (*A* to *B*) **6.** (Yes) (*B* to *A*) **7.** (36 V) (+) **8.** (15 V) (Left)

Comparisons of the metric millimeter (1/1000 m), the centimeter (1/100 m), and the decimeter (1/10 m) with the U.S. inch and mil (1/1000 in.) are illustrated in Fig. 3-3.

All metals increase in resistance value when heated. For example, if a given length of copper wire has 1 **Ω** at 0°C (freezing temperature of water), it would have about 1.43 **Ω** at 100°C (boiling temperature of water) and 1.85 **Ω** (nearly twice the resistance as it had at freezing) at 200°C.

A special wire, Nichrome, made of nickel, chromium, iron, and manganese, has about 60 times the resistance of copper. Another, Constantan (copper and nickel) has about 30 times the resistance of copper. Besides high resistance, these wires also have the advantage of not changing resistance to any great extent when heated. They are said to have a low *resistance-temperature-coefficient.*

One common size of copper hookup wire used to connect components in electronic equipment is #20 gage AWG (American Wire Gage). It has a diameter of 0.813 mm (32 mils) and a resistance of about 33.2 Ω/km (10.4 Ω/1000 ft). It will carry a current of 1 A without heating noticeably. A heavier gage of wire used in house wiring is #10. It has a diameter of 2.59 mm (102 mils) and will handle 30 A without noticeable heating.

A component manufactured to have resistance is called a *resistor*. A resistor may be made of a high-resistance wire such as Nichrome, Constantan, or Manganin, wound on a nonconducting form, or it may be fabricated from special materials that have high resistance, such as carbon or graphite (Fig. 3-4).

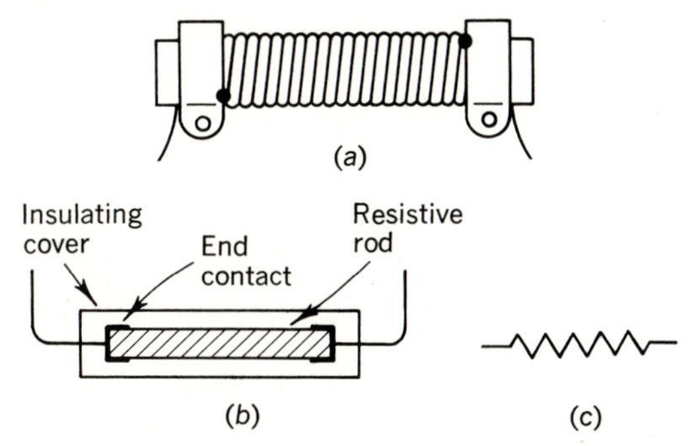

Fig. 3-4 (*a*) Wire-wound resistor, exterior view. (*b*) Carbon-type resistor, cross section. (*c*) Symbol for resistance or a resistor.

A resistance meter, called an *ohmmeter*, is used to measure the internal resistance of components or devices (Fig. 3-5). The scale of an ohmmeter is calibrated in ohms. There is an important requirement to remember when using an ohmmeter, however. Voltage must never be across any device while its resistance value is being measured. Erroneous readings will result, or the delicate mechanism of the ohmmeter may be damaged.

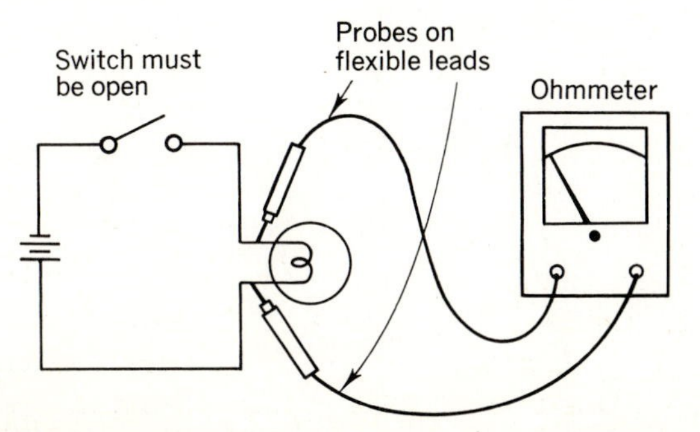

Fig. 3-5 Measuring the resistance of a lamp with an ohmmeter.

Quiz 3-2. Test your understanding. Answer these check-up questions.

1. What is the unit of measurement of emf? ________ Current? ________ Resistance? ________
2. What is always developed in a resistor when current flows through it? ________
3. Which has the greater resistance, a large- or a small-diameter 1-ft-long copper wire? ________
4. Does a metal conductor have more or less resistance when it becomes hot? ________
5. In a simple lamp circuit, will a greater current flow at the instant the switch is closed or 1 s later? ________ Why? ________
6. What is the name of the meter used to measure resistance? ________ What is the requirement to remember when using this type of meter? ________
7. Which has the greatest resistance: aluminum, copper, iron, lead, or silver? ________ Which has the least? ________
8. Would a #12 gage wire be larger or smaller in diameter than a #10 wire? ________ Than a #20 wire? ________
9. Name two types of materials that have high resistance but do not change resistance appreciably when heated. ________ ________

3-4 CONDUCTORS AND IONIZATION

Metals such as silver, copper, aluminum, iron, and lead have electrons in the outer orbits of their atoms or molecules that are relatively free to move. Such metals are good conductors of electrons at normal room temperatures. Increasing the temperature of these metals produces greater electron activity and resistance.

A second group of materials, such as sulfur, glass, rubber, ceramics, plastics, mica, and cotton, has practically no free

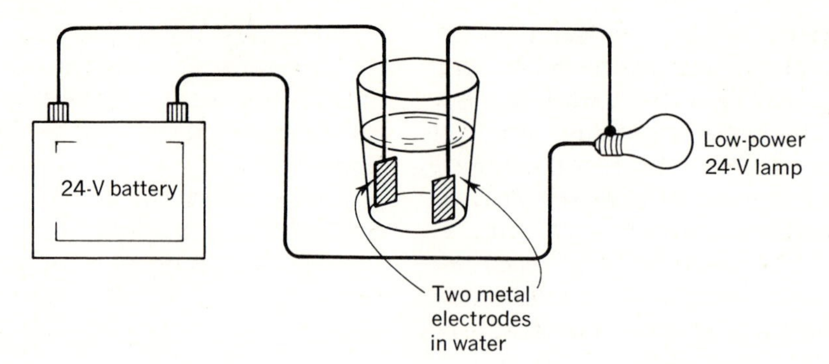

Fig. 3-6 Pure water is a nonconductor. As salt is added, the solution ionizes, current flows, and the lamp will light.

electrons in the outer orbits of the molecules and will not conduct any current at room temperatures. Such materials are called *insulators*. At extreme temperatures, however, electrons may be driven free of the outer orbits, and even these materials may then act as conductors.

A third group of substances—pure germanium, silicon, and carbon—conduct current poorly at room temperature and are called *semiconductors*. Unlike conductors, semiconductors will conduct electric current *better* when they are heated. Semiconductors that are *doped*, or mixed with small amounts of other substances, may become much better conductors. On hot days their conductance value may increase considerably.

Gases, such as neon, argon, and hydrogen, have no free electrons and will not support current until the electrostatic lines of force across the gas become intense or dense enough to tear outer-orbit electrons from the gas atoms. Any atom in this minus-an-electron condition has a positive charge and is termed a positive *ion* (ī-ŏn). Up to the "ionization potential," the gas acts as an insulator. When an emf equal to or greater than the ionization voltage is applied across the gas, the gas ionizes, develops free electrons, and becomes a good conductor.

Contrary to popular belief, pure water is an insulator, but when contaminated with acids, bases, or salts, the materials introduced into it ionize, or separate chemically and the solution may then become a very good conductor. (There is no ionization potential in *electrolytes*, the proper term for conductive liquids.) The circuit pictured in Fig. 3-6 consists of a 24-V battery, a low-power 24-V lamp, and two metal plates immersed in a glass of pure water. Essentially no current will flow in the circuit. However, if table salt is mixed in the water, the salt ionizes and current begins to flow. The more salt there is, the more ionization, the less the resistance of the solution, and the brighter the lamp lights.

Now, if one of the electrodes immersed in the ionized liquid is lifted free of the surface, this will act as a switch and cut off the current flow. It would also be found that as the plate was being removed from the liquid, the lamp would dim, indicating that the resistance must be increasing as

---

ANSWERS TO CHECK-UP QUIZ 3-2

1. (Volt) (Ampere) (Ohm) **2.** (Heat) **3.** (Small) **4.** (More) **5.** (At closing) (Filament resistance increases when heated) **6.** (Ohmmeter) (Circuit must be dead) **7.** (Lead) (Silver) **8.** (Smaller) (Larger) **9.** (Nichrome, Constantan, Manganin)

---

the plate area decreases. A variable resistor that acts like this is called a *rheostat* (rē-ō-stat), and it might be useful in dimming lights, for example. Rheostats are made in many forms, one common type consisting of a circular wire-wound resistor with an adjustable contact arm, as in Fig. 3-7*a*. Note that the rheostat has only two external connections. A somewhat similar device is the *potentiometer* (pō-tĕn"shē-ŏm'itr), except that it has a third connection on the far end of the resistance. Potentiometers are usually used in "voltage-divider" circuits, discussed in later chapters.

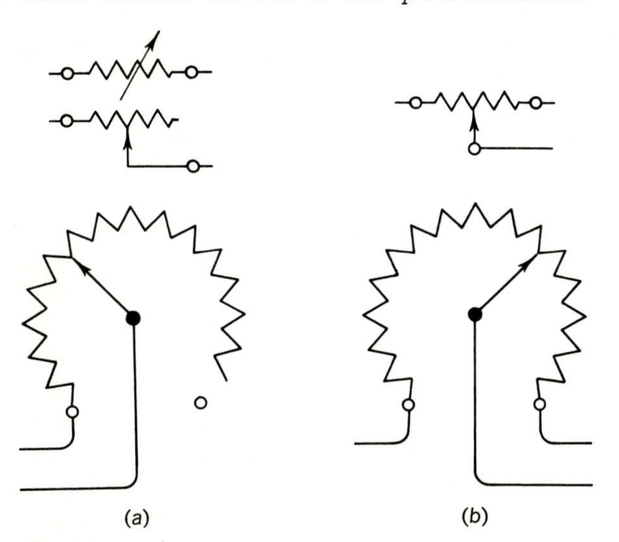

Fig. 3-7 (*a*) Two symbols for a rheostat or variable resistor and a pictorial representation. (*b*) Symbol for a potentiometer and pictorial representation.

Dry earth is normally considered a poor conductor, but it becomes a very good conductor when dampened. If solid enough connections are made to the earth, there may be less than 1 Ω of resistance between two cities 100 mi apart.

Quiz 3-3. Test your understanding. Answer these check-up questions.

1. What is a substance called (insulator, metal conductor, or semiconductor) that has less resistance when heated? __________ More resistance when heated? __________ Extremely high resistance? __________
2. In what condition is a solution when it is capable of carrying electric current? __________
3. Would you expect seawater to be an insulator, a semiconductor, or a conductor? __________ Mountain lake water? __________ A carbon rod? __________
4. Would a carbon-filament lamp circuit draw more current at the instant the switch is closed or 1 s later? __________
5. Could a rheostat be used in place of a potentiometer? __________ Could a potentiometer be used in place of a rheostat? __________

CHAPTER 3 TEST • RESISTANCE AND CONDUCTANCE

1. What is the total effective voltage if a 24-V and a 36-V battery are connected series-aiding? Series-bucking?
2. The symbol for a battery has two long lines and two short lines. Which length of line represents the + pole? Would you know the battery voltage value from such a symbol?
3. What two things might vector arrows drawn on schematic diagrams indicate?
4. Object *A* has a deficiency of electrons, and object *B* has a neutral charge. If a conductor is connected between them, would current flow? If so, in what direction?
5. Current is flowing through a resistor from left to right. Which end of the resistor would be the + end?
6. What may a 0-Ω resistance load across a battery be termed?
7. If it were possible to have a perfect conducting material, would it warm up when current flowed through it?
8. Does a larger gage number indicate a larger or a smaller diameter of wire? A greater or lesser *R* value?
9. Which is the more efficient current-carrying device, connecting wires or electric lamp filaments?
10. What metal is usually used in lamp filaments?

11. What change in resistance occurs when any metal is heated? When a semiconductor is heated?
12. What does the Greek letter Ω mean in electrical circuitry? What is the name of this letter?
13. What are two advantages in using Nichrome or Constantan to make wire-wound resistors?
14. A 1-mil 10-Ω copper wire would be approximately how long?
15. What is the name of the meter used to measure the resistance of a component or circuit? What is it important to remember when using such a meter?
16. Name two substances which might have less resistance on a hot summer day than they would on a winter day.
17. What is a device called that is manufactured to have an extremely high electrical resistance?
18. If a molecule of table salt loses an electron when in water, what is the electronless particle called?
19. What is a proper term to designate liquids that can conduct electric current?
20. A wire-wound resistor has a contact that can be moved along the resistor. What are three names that might describe such an electrical component?

ANSWERS TO CHECK-UP QUIZ 3-3

1. (Semiconductor) (Metal) (Insulator) **2.** (Ionized) **3.** (Conductor) (Insulator) (Semiconductor) **4.** (1 s later) **5.** (No) (In most cases, yes)

4

OHM'S LAW IN SERIES CIRCUITS

CHAPTER OBJECTIVE. To develop a usable understanding of these electrical vocabulary terms: Ohm's law for current, Ohm's law for voltage, Ohm's law for resistance, voltage-drop, Kirchhoff's voltage law, Ohm's law triangle. Also, to develop a facility to insert given values into simple formulas to solve for an unknown, and to rearrange basic formulas.

4-1 OHM'S LAW

In the preceding chapters the three basic or fundamental concepts underlying electricity, electronics, and radio have been discussed:

1. A movement of free electrons from atom to atom forms an electric current.
2. Electrostatic lines of force between two unlike charges produce a pressure or electromotive-force that can move electrons.
3. All substances oppose the movement of electrons to some extent and are said to have resistance.

These three factors are always present in any operating electric circuit. It is possible to incorporate them all in one inclusive statement:

> **The value of current that will flow in any circuit depends on the value of the electromotive-force and the value of the resistance.**

A closer look, however, will show that, whereas an increase in emf or pressure in a circuit will increase the current flow, an increase in resistance in the circuit will *decrease* the current flow. Therefore a more accurate statement of an operating electric circuit might be:

> **The value of current that will flow in a circuit will be directly affected by the value of the electromotive-force but will be inversely affected by the value of the resistance.**

This is a word statement of *Ohm's law*, probably the most important law in electricity.

With Ohm's law it becomes possible, for example, to determine how much current will flow in a circuit if the voltage value and the resistance value of the circuit are known. As a practical example, what current would the ammeter in Fig. 4-1 read if

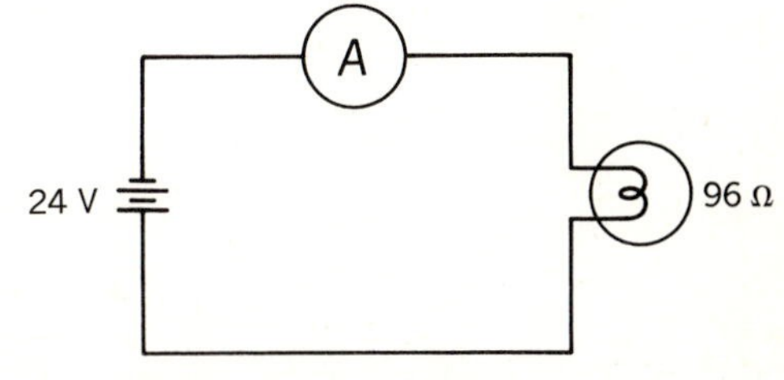

Fig. 4-1 A simple circuit in which $I = E/R = 24/96$, or 0.25 A.

the source voltage were 24 V and the lamp had a filament resistance of 96 Ω?

Ohm's law, stated simply, says:

The current is directly proportional to the voltage and inversely proportional to the resistance.

Written out in a more mathematical form, this might read:

$$\text{Current} = \frac{\text{voltage}}{\text{resistance}}$$

or in formula form,

$$I = \frac{E}{R}$$

where I = current, in amperes
E = emf, in volts
R = resistance, in ohms

In Fig. 4-1, with a source voltage of 24 V and a load resistance of 96 Ω, the circuit current would be

$$I = \frac{E}{R} = \frac{24}{96} = \frac{1}{4}\ \text{A}$$

The Ohm's law formula must be committed to memory, as it will be used time after time in problems relating to working circuits as well as in explaining electrical theories.

As a suggestion, when working electrical problems, if no schematic diagram is given, always sketch one to help visualize the circuit under discussion.

Quiz 4-1. Test your understanding. Answer these check-up questions.

1. What is the name of the law that states that the I is directly proportional to the E and inversely proportional to the R? __________
2. State the law of question 1 in electrical letter symbols. __________
3. An electric lamp has a hot-filament R of 50 Ω when it is connected across a 120-V power line. What value of current does it draw? __________ What current would flow if the lamp were across a 100-V line? __________
4. An automobile starter motor has 0.1 Ω of resistance. What current flows when you step on the starter if this connects the motor across a 12-V battery? __________
5. If your television receiver has an internal R of 80 Ω, what current does it draw from the 120-V line? __________

4-2 SOLVING FOR VOLTAGE

The basic Ohm's law formula is useful when it is required to determine an unknown current if a known resistance is across a source of known voltage. If the current and resistance are the given values, the same formula can be rearranged to solve for the voltage. This can be done by multiplying both sides of the Ohm's law formula or equation by R; thus

$$I = \frac{E}{R} \qquad \text{or} \qquad I(R) = \frac{E(R)}{R}$$

In the right-hand side of this second equation, the lower R divides into the upper R once, leaving the formula

$$IR = E \qquad \text{or} \qquad \mathbf{E = IR}$$

where E is in volts, I is in amperes, R is in ohms.

This voltage form of Ohm's law is useful for determining the value of emf that is being applied in a simple circuit, such as that of Fig. 4-2, if the current value flowing

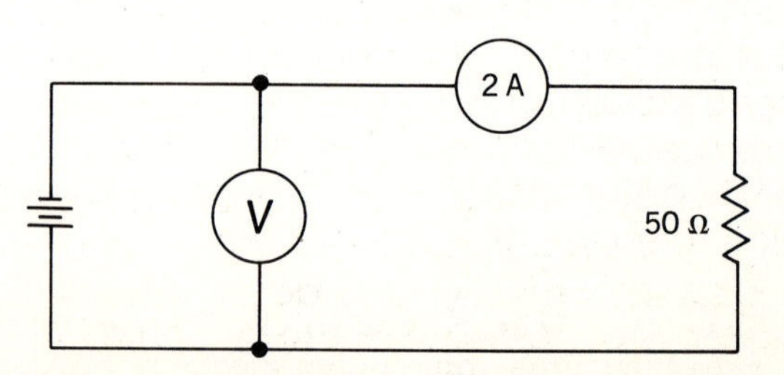

Fig. 4-2 A simple circuit in which the source-voltage value is determined by $E = IR$.

through source and load and the load resistance value are both known. In this particular case, $E = IR$, or 2(50), or 100 V, would be indicated by the voltmeter V across the source and load.

While the ability to solve for the source voltage in a simple circuit is valuable, the voltage form of Ohm's law tells us a fact that may be even more significant in electrical theory: the difference of potential, or *voltage-drop*, across any resistance that has current flowing through it can be determined by multiplying the resistance value by the current value. In Fig. 4-3, two

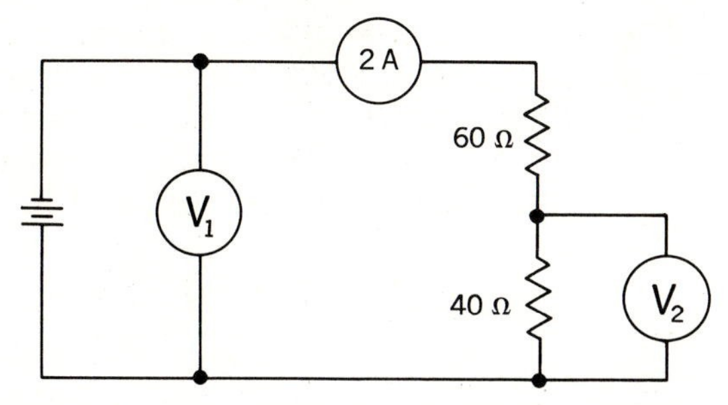

Fig. 4-3 A series circuit in which all E, I, and R values can be determined by formulas of Ohm's law.

voltages are being measured by voltmeters. In this circuit there are two resistors, of 40 Ω and 60 Ω, in series, resulting in a total resistance of 100 Ω. The current flowing through the whole series circuit, according to the ammeter, is 2 A. Therefore the total voltage (the source voltage) must be $E = IR$, or 2(100), or 200 V, shown by V_1.

What about the second voltmeter, V_2? It must be reading the voltage-drop (which may be assigned the symbol V instead of E) across the 40-Ω resistor, or $V = IR$, or 2(40), or 80 V. This means that of the total 200 V of pressure produced by the source, 80 V of pressure is lost across the 40-Ω resistor. The remainder of the 200 V, that is, 200 − 80 = 120 V, must be across the 60-Ω resistor. This can be proved by applying the voltage form of Ohm's law to the 60-Ω resistor; $V = IR$, or 2(60), or 120 V.

Note that a voltage-drop is developed across a resistor when current flows through the resistor. Technically, voltage is emf from a source. However, the word *voltage* is often used in place of *voltage-drop*.

A most important concept for electric-circuit theory has just been presented. It is simple enough, but it is so important that it has been given a special name, *Kirchhoff's voltage law*. It can be stated as:

The sum of the voltage-drops around a circuit will equal the source voltage.

In this case, 80 + 120 = 200 V. Kirchhoff's voltage law is often stated as:

The algebraic sum of the voltage-drops around a circuit plus the source voltage will equal zero.

This is true when the voltage-drops are assigned negative (−) mathematical values and the source voltage is given a positive (+) value.

Quiz 4-2. Test your understanding. Answer these check-up questions.

1. Express the current form of Ohm's law in letter symbols. ________
2. Express the voltage form of Ohm's law in letter symbols. ________
3. What law states that the sum of the voltage-drops around a circuit equals the source emf? ________ State this law in its algebraic form. ________
4. An 80-Ω resistor is connected across a power supply. A 1.5-A current is found to be flowing through it. What must the source-voltage value be? ________
5. In Fig. 4-4, what is the total resistance of the circuit? ________ What is the "*IR*-drop" across the 10-Ω *R*? ________
6. In Fig. 4-4, what is the total voltage-drop across the three resistors in series? ________

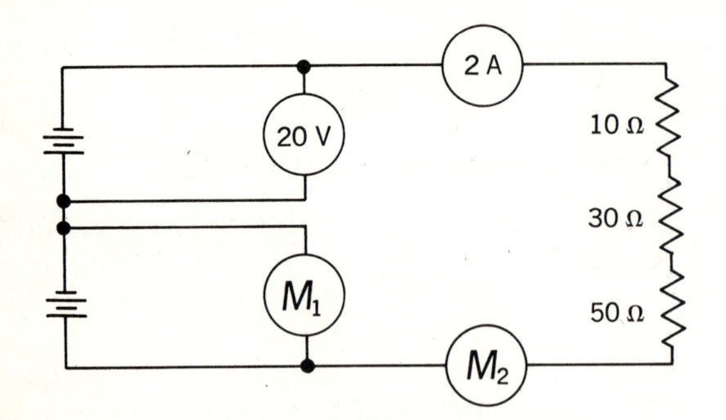

Fig. 4-4 A series circuit that can be completely solved by Ohm's and Kirchhoff's laws.

7. In Fig. 4-4, what value would M_1 read? ______ M_2? ______

8. In Fig. 4-4, what value would a voltmeter read if it were connected across the 10-Ω and 30-Ω resistors in series? ______

9. In Fig. 4-4, if the 30-Ω resistor were shorted out, what would the total R value be? ______ What would M_2 read? ______ What would M_1 read? ______

10. In Fig. 4-4, if the 30-Ω R were open-circuited, what would the total R value be? ______ What would M_2 read? ______ What would M_1 read? ______

4-3 SOLVING FOR RESISTANCE

It was possible to rearrange the current form of Ohm's law to solve for the voltage if current and resistance were known by multiplying both sides of the formula by R. However, if both sides of the voltage form, $E = IR$, are divided by I, a third Ohm's law formula results. Thus

$$E = IR \quad \text{or} \quad \frac{E}{I} = \frac{IR}{I}$$

The I values on the right-hand side of this equation cancel, leaving the formula

$$\frac{E}{I} = R \quad \text{or} \quad R = \frac{E}{I}$$

where R is in ohms, E is in volts, and I is in amperes.

This formula enables us to determine the resistance value of a load if the source voltage and the load current are known. As an example, what is the resistance value of the load in Fig. 4-5? According to the meters, the source voltage is 100 V, and the circuit current is 4 A. The total resistance must be $R = E/I$, or 100/4, or 25 Ω. If the total load value is 25 Ω, and if one of the two resistors has a value of 10 Ω, then the other resistor, R, must have a value of 25 − 10 = 15 Ω. Thus by reasonably simple steps it is possible to determine many things about electric circuits by applying Ohm's law in one form or another—along with a little common sense.

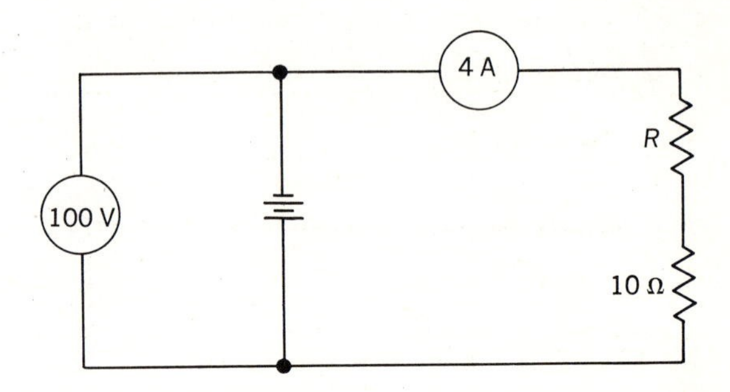

Fig. 4-5 The voltage-drop across R will be the difference between the source voltage (100 V) and the IR-drop across the 10-Ω resistor (40 V).

In all three forms of Ohm's law it is assumed that two things are known before the third can be found. When faced with a problem in electrical circuitry, first jot down all given values of E, I, and R. Then pick out two things about one part of the circuit that can be used to solve for a third value according to one of the formulas known. Sometimes several steps may be needed. Sketch a schematic diagram of a problem circuit and place all known values on it before starting computations.

ANSWERS TO CHECK-UP QUIZ 4-1

1. (Ohm's law) **2.** ($I = E/R$) **3.** (2.4 A) (2 A) **4.** (120 A) **5.** (1.5 A)

Quiz 4-3. Test your understanding. Answer these check-up questions.

1. State the current form of Ohm's law in symbols. __________
2. State the voltage form of Ohm's law in symbols. __________
3. State the R form of Ohm's law in symbols. __________
4. A radio receiver draws 0.8 A when it is connected across a 120-V power line. What must its effective "internal" resistance be? __________

Questions 5 to 8 refer to Fig. 4-6.

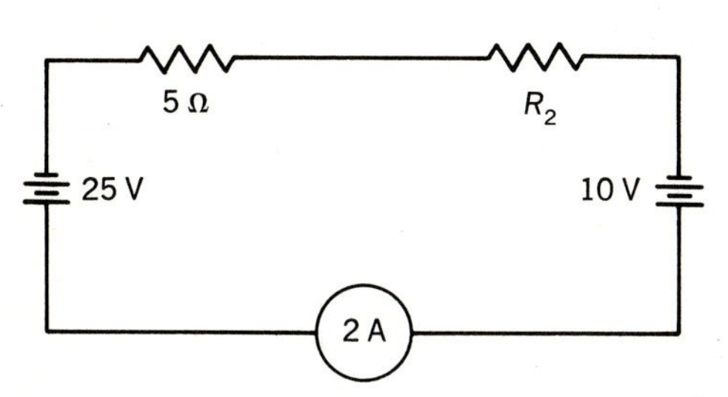

Fig. 4-6 Series-bucking sources in a practice problem.

5. What is the total effective E? __________ In which direction (clockwise or counterclockwise) does current flow? __________
6. What is the total circuit R? __________
7. What is the voltage-drop across the 5-Ω resistor? __________ Across R_2? __________
8. Find the resistance value of R_2 from the formula $R = E/I$. __________

4-4 THE OHM'S LAW TRIANGLE

To remember the Ohm's law formula, use the triangle shown in Fig. 4-7. Hold a finger over the letter to be solved for. The proper formula stands out. For example, what is the formula to solve for I? Place a finger over I. The correct formula is seen to be E/R. A finger covering E shows that the formula to solve for E is IR, or I times R. It is necessary to remember only that E stands for excellence and therefore should be at the top. The order of the other two letters makes no difference.

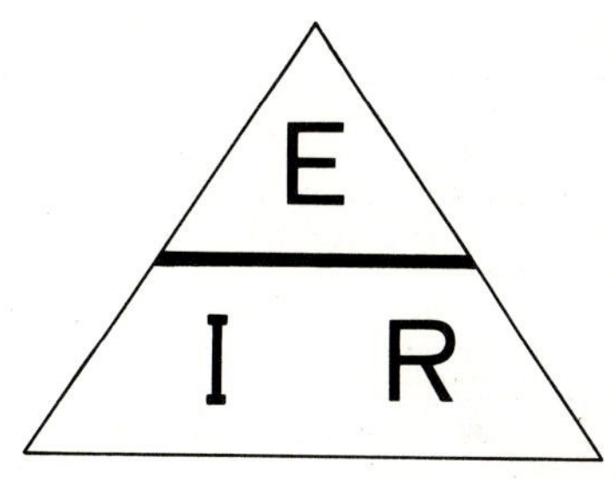

Fig. 4-7 The Ohm's law triangle. Cover the letter to be solved.

CHAPTER 4 TEST • OHM'S LAW IN SERIES CIRCUITS

1. Give the E form of Ohm's law in letter symbols.
2. Give the R form of Ohm's law in letter symbols.
3. Give the I form of Ohm's law in letter symbols.
4. What is the name of the law that states that the algebraic sum of the voltage-drops and source emf around a circuit is zero?
5. Draw a diagram of a 100-Ω rheostat as a load on a 6-V battery. What current flows when the rheostat is set to maximum resistance? When it is set to three-fourths of the resistance?
6. Draw a diagram of a 100-Ω rheostat, a 50-Ω lamp, and a 120-V source, all in series. What is the maximum I that can be made to flow through the lamp? The minimum I?
7. In question 6, what are the minimum and maximum voltage-drops that can be produced across the lamp if the rheostat contact arm never leaves the resistance wire?
8. What must the hot resistance of a vacuum-tube filament be if a 0.3-A current flows when 6.3 V is across the filament?
9. A 10-Ω, a 20-Ω, and a 30-Ω resistor are all in series across a 15-V source. What is the value of the current in the 20-Ω resistor? What is the voltage-drop across it?
10. What is the resistance value of a load resistor that is burned out? Of one that is shorted (short-circuited)?

ANSWERS TO CHECK-UP QUIZ 4-2

1. ($I = E/R$) **2.** ($E = IR$ or $V = IR$) **3.** (Kirchhoff's law) ($E_S = -V_1 + -V_2$) **4.** (120 V) **5.** (90 Ω) (20 V) **6.** (180 V) **7.** (160 V) (2 A) **8.** (80 V) **9.** (60 Ω) (3 A) (160 V) **10.** (Infinite Ω) (Zero) (160 V)

ANSWERS TO CHECK-UP QUIZ 4-3

1. ($I = E/R$) **2.** ($E = IR$ or $V = IR$) **3.** ($R = E/I$) **4.** (150 Ω) **5.** (15 V) (CW) **6.** (7.5 Ω) **7.** (10 V) (5 V) **8.** (2.5 Ω)

CHAPTER OBJECTIVE. To develop a usable understanding of these electrical vocabulary terms: energy, photon, joule, watt, wattmeter, wattsecond, kilowatthour meter. Also, to rearrange the basic energy and power formulas and combine power and Ohm's law formulas to compute E_n, E, I, R, and P values in electric circuits.

5-1 ELECTRICAL ENERGY

The more common electrical devices carry ratings in volts, and perhaps in amperes, on a nameplate. A few may carry resistance ratings. However, most equipment will have a voltage rating, and also a power rating in watts. The power value indicates the *rate* at which the device uses electrical *energy*.

The source of most of the energy in the world can be traced to light and heat energy waves from the sun, called *photons*. When photons strike the atoms or molecules of any matter, energy is added to that already present in the orbiting outer-ring electrons of the substance. This can cause several reactions, one being photoemission, in which electrons are given off by the material. Absorption of energy may also result in a speedup of chemical reactions, as in human nerve endings.

The pressure that can move electrons has already been described as electromotive-force, or voltage, which can be developed by chemical action in batteries. When electrons are forced into motion, they possess energy. Electrical energy involves both voltage and a quantity of electrons (measured in coulombs). The formula for electrical energy is therefore

$$E_n = EQ$$

where E_n is in *joules*, E is in volts, and Q is in coulombs. If a 100-V pressure is able to move 3 C of electrons through a resistor, 300 J (joules) of energy have been expended. In this case the electrical energy is completely transformed to heat energy, resulting in heating of the resistor.

From Sec. 2-2, 1 A is 1 C/s. Thus the number of joules of electrical energy a 6-V battery would release in 1 s if it were producing a 2-A current flow would be $E_n = EQ$, or 6(2), or 12 J. In this computation of energy, using 1 s as the period of measurement canceled out time since 2A is 2 C/s. Energy is always timeless.

5-2 ELECTRICAL POWER

The *rate* at which energy is being expended is not the same thing as energy alone, just as "kilometers" and "kilometers per hour" are not the same. If 1 J of energy is expended in 1 s, the rate of expending energy is one *watt*, abbreviated W. If a joule equals *EQ* (volt-coulombs) then 1 J/s equals 1 VC/s, also known as a watt (rate of energy expenditure). In formula form, watts of power is energy divided by time, or

$$P = \frac{EQ}{T}$$

where P is in watts, E is in volts, Q is in coulombs, and T is in seconds. Unfortunately, this is not a very handy formula to use, because there is no "coulombmeter" with which coulombs can be measured.

As previously explained, 1 W equals 1 V times 1 C/s, or $P = EQ/T$. Also, 1 A is 1 C/s, or $I = Q/T$. If I is substituted for Q/T in the power formula above, then

$$P = E\frac{Q}{T} \quad \text{or} \quad P = EI$$

where P is in watts, E is in volts, and I is in amperes. This is a useful formula. Both voltmeters and ammeters are common measuring instruments. This basic power formula ranks in importance with Ohm's law in computing operating electric circuits. For example, if a television receiver draws 1.5 A across a 120-V power line, it must be expending energy at the rate of $P = EI$, or 120(1.5), or 180 W.

From $P = EI$, the power in the circuit of Fig. 5-1 would be 100(2), or 200 W. The wattmeter W is a two-in-one meter. One part of it responds to the voltage across the circuit (terminals A and B). The other part responds to the current flow through the load (terminals A and C). Both values help determine how far the meter pointer swings. Greater source voltage or greater load current, or both, will result in a greater wattage reading.

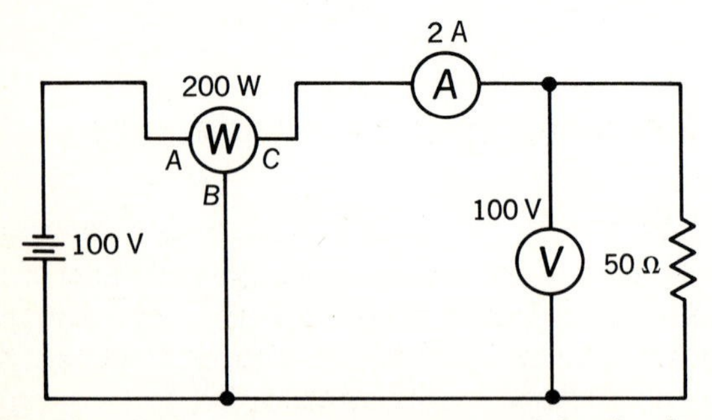

Fig. 5-1 The power being dissipated by the load can be determined either by multiplying the voltmeter and ammeter readings or by the wattmeter directly.

Just as the Ohm's law formula can be rearranged to solve for I, E, or R, the basic power formula can be algebraically rearranged by dividing both sides of the equation by E or I, resulting in three useful formulas:

$$P = EI \qquad E = \frac{P}{I} \qquad I = \frac{P}{E}$$

As examples, if 2 A is flowing through a load resistor and converts 50 W of power into heat, the source voltage must be $E = P/I$, or 50/2, or 25 V. Also, if a source of 120 V produces a 60-W power dissipation in an electric light globe, the lamp current must be $I = P/E$, or 60/120, or 0.5 A.

Quiz 5-1. Test your understanding. Answer these check-up questions.

1. What is the basic power formula, in letter symbols? ________
2. What is the formula for determining voltage when current and power are known? ________
3. What is the formula for determining current when voltage and power are known? ________
4. What power is used by a load in which 5 A is made to flow by 80 V? ________
5. What power is used by a load when 100 V moves 10 C through it in 0.5 s? ________
6. How much energy is developed in a load by 20 V forcing 50 C through the load? ________
7. A 10-W 24-V lamp will require how much current to make it glow at full brilliance? ________
8. What emf is required to make a resistor dissipate 3 W when 50 mA (0.05 A) flows through it? ________

5-3 COMBINING POWER AND OHM'S LAW

The Ohm's law formula $I = E/R$, the power formula $P = EI$, and two combinations of these formulas round out the mathematical tools required to solve most of the simpler E, I, R, and P problems.

Consider the power formula $P = EI$. It is possible to substitute known expressions for the E and for the I in this formula. For example, it is known that $E = IR$. Substitute the expression IR in place of E in the power formula, as

$$P = (E)I \qquad P = (IR)I \qquad P = IIR \qquad \mathbf{P = I^2R}$$

With this formula the power in a resistor can be determined if the current through it is known. For example, how much heat does a 200-Ω resistor dissipate if a 2-A current is flowing through it? *Answer:* $P = I^2R$, or $2^2(200)$, or 800 W.

From this formula, power is proportional to the *square* of the current. If the current is increased by a factor of 2, the power will increase by a factor of 4; if the current increases 3 times, the power will increase by a factor of 9; and so on. In the example above, if the current through the 200-Ω resistor is increased to 4 A, then $P = 4^2(200)$, or 3200 W.

As with Ohm's law and the basic power formula, the formula $P = I^2R$ can be algebraically rearranged to solve for I or R by dividing both sides of the equation by I^2 or by R, respectively.

$$P = I^2R \qquad \text{or} \qquad \mathbf{R = \frac{P}{I^2}} \qquad \text{or} \qquad I^2 = \frac{P}{R}$$

The last of these is a formula to find the square of the current. To solve for I (not I^2), it is necessary to take the square root of both sides of the formula:

$$\sqrt{I^2} = \sqrt{\frac{P}{R}} \qquad \text{or} \qquad \mathbf{I = \sqrt{\frac{P}{R}}}$$

As an example, what value of current must be flowing through a 2-Ω resistor when it is dissipating 228 W? *Answer:* $I = \sqrt{P/R}$, or $\sqrt{288/2}$, or $\sqrt{144}$, or 12 A.

There are four methods of determining square roots: (1) Using a slide rule, (2) using a pocket calculator, (3) using logarithms, and (4) using the long-division-like method found in mathematics texts.

Quiz 5-2. Test your understanding. Answer these check-up questions.

1. What is the square root of 1? ________ 10? ________ 100? ________ 1000? ________ 10,000? ________ (Do you see a pattern here?)
2. What is the square root of 2? ________ 20? ________ 200? ________ 2000? ________ 20,000? ________ (Is there a pattern here?)
3. What is the square root of 5? ________ 50? ________ 500? ________ 5000? ________ 50,000? ________
4. A wire-wound resistor is marked "500 Ω, 0.25 A." What is the maximum power that it can safely dissipate? ________ How much voltage would be required across it to make it dissipate this value of power? ________
5. It is known that a 100-W lamp requires 0.833 A flowing through it. What is its hot-filament resistance? ________
6. What value of current is required to make a 1000-Ω resistor dissipate 1000 W? ________ 2000 W? ________ 5000 W? ________

5-4 SUBSTITUTING E/R IN THE POWER FORMULA

As in Sec. 5-3, it is possible to substitute the expression E/R from the Ohm's law formula $I = E/R$ in place of the I in the power formula $P = EI$. This results in another very useful formula:

$$P = EI \qquad P = (E)\frac{E}{R} \qquad P = \frac{EE}{R} \qquad \mathbf{P = \frac{E^2}{R}}$$

With this last formula it is possible to solve for power when the voltage and resistance of a circuit are known. For example, what is the power in a 500-Ω resistor when it is connected across a 50-V power line? *Answer:* $P = E^2/R$, or $50^2/500$, or 2500/500, or 5 W.

From the formula $P = E^2/R$, by algebraic rearrangement it is possible to derive the formula to solve for E if P and R are known. This is accomplished by multiplying both sides of the formula by R.

$$P = \frac{E^2}{R} \qquad P(R) = \frac{E^2(R)}{R} \qquad E^2 = PR$$

To solve for E alone, the square root of both sides of the formula must be taken:

$$\sqrt{E^2} = \sqrt{PR} \qquad \mathbf{E = \sqrt{PR}}$$

With this formula it is possible to determine the required voltage to produce a given power dissipation in a known resistance value. For example, how many volts are required to develop 200 W of dissipation in a 50-Ω resistor? *Answer:* $E = \sqrt{PR}$, or $\sqrt{200(50)}$, or $\sqrt{1\,00\,00}$, or 100 V.

Another power formula, also a derivation of $P = E^2/R$, can be developed by first multiplying both sides by R:

$$P = \frac{E^2}{R} \qquad PR = \frac{E^2R}{R} \qquad PR = E^2$$

Then, dividing both sides of the last formula by P,

$$\frac{PR}{P} = \frac{E^2}{P} \qquad \mathbf{R = \frac{E^2}{P}}$$

This last formula can be used to determine, for example, the resistance value of a 150-W lamp that is rated for operation on a 120-V power line: $R = E^2/P$, or $120^2/150$, or 14,400/150, or 96 Ω.

Quiz 5-3. Test your understanding. Answer these check-up questions.

1. When a 6-V battery is connected across a 10-Ω resistor, what is the power? ______
2. What is the highest value of voltage that can be applied across a 20-W 8000-Ω wire-wound resistor without heating it excessively? ______
3. What is the hot-filament resistance value of a 60-W home light globe? ______

5-5 PRACTICAL ENERGY FORMULAS

The formula $E_n = EQ$ for energy in joules, given in Sec. 5-2, is an accurate explanation of energy, but it is simpler to consider the joule as a *wattsecond* (Ws). There seems to be an inconsistency here; energy is explained as being timeless, but the method of measuring it is to multiply power by time. Actually, multiplying by time is the means by which time is removed from the power formula. Consider the formula $P = EI$. Time is hidden in the current, inasmuch as an ampere is 1 C/s. Thus, if $P = EI = EQ/T$, then when power is multiplied by time, $P = EQT/T$, the time values T cancel, leaving only energy EQ, or volt-coulombs. Thus a wattsecond actually has no time in it; it is a measurement of energy.

To determine energy, measure the voltage across a load, then multiply this by the current measured flowing through the load, and then multiply by the number of

ANSWERS TO CHECK-UP QUIZ 5-1

1. ($P = EI$) 2. ($E = P/I$) 3. ($I = P/E$) 4. (400 W) 5. (2000 W) 6. (1000 J) 7. (0.417 A) 8. (60 V)

ANSWERS TO CHECK-UP QUIZ 5-2

1. (1) (3.16) (10) (31.6) (100) 2. (1.41) (4.47) (14.1) (44.7) (141) 3. (2.24) (7.07) (22.4) (70.7) (224) 4. (31.25 W) (125 V) 5. (144 Ω) 6. (1 A) (1.41 A) (2.24 A)

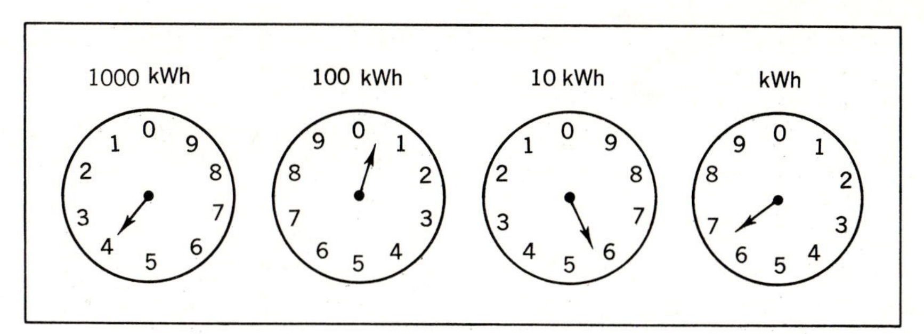

Fig. 5-2 Kilowatthour meter dials. It is read from left to right.

seconds the load is connected. The answer is energy in wattseconds or joules. For example, if a 100-V source develops 2 A in a light for 1 min, the energy is *EIT*, or 100(2)(60), or 12,000 Ws.

Other units of measurement for energy are *watthours* (Wh) and *kilowatthours* (kWh). The so-called "power companies" do not sell power: they charge for the *energy* delivered. Every home and industrial plant has a kilowatthour meter in its power lines. A kilowatthour meter is actually a tiny electric motor, often with four continuously rotatable indicator dials (Fig. 5-2), each geared down 10 times from the one driving it. With no load connected across the lines, the motor has no current flowing through one of its two windings, and it does not turn. As soon as a load is connected, the load's current flows through the meter winding, and its motor starts to turn the indicator dials. The heavier the current or the longer it is on, the farther the indicators turn, and the more the consumer pays. Energy costs between 0.25 and 11 cents (U.S.) per kilowatthour.

Quiz 5-4. Test your understanding. Answer these check-up questions.

1. If a car radio draws 3 A from a 12-V battery for 15 min, what is the power? ________ Energy? ________
2. An electric clock has 5000 Ω of internal resistance and is used across a 120-V line. What is its power rating? ________ How much energy in kilowatthours does it consume per day? ________ If electricity costs 4¢/kWh, what is the cost to run the clock for 30 days? ________
3. An electric motor across 240 V draws 6 A. What power is it drawing from the line? ________ What energy does it require in 30 min? ________
4. How many kilowatthours does the meter in Fig. 5-2 indicate? ________
5. Read the kilowatthour meter at your residence. How many kilowatthours does it indicate? ________ Tomorrow read the meter again and determine the number of kilowatthours used. ________ At 4c/kWh, what was the cost of electricity for that day? ________

CHAPTER 5 TEST • ENERGY AND POWER

1. What are the waves of light energy from the sun called?
2. Under what condition are electrons considered to contain energy?
3. One unit of measurement of electrical energy is the wattsecond. What is another name for the same unit?
4. Which is timeless, energy, current, or power?
5. How much energy would be developed in a 50-Ω resistor if it were held across 100 V for half a minute? What is the power?
6. $P = EI$ is the basic power formula. What are the two other formulas which can be derived from it?

7. If a 35-W soldering iron draws 300 mA from a power line, what is the line voltage?
8. A 12-V light globe across a 12-V battery dissipates 25 W. What current is flowing? What is its hot-filament resistance?
9. What formula determines power if E and R are known? What are the two other formulas that can be derived from it?
10. What formula determines power if I and R are known? What are the two other formulas that can be derived from it?
11. What value of current must be flowing in a 200-Ω resistor when it is dissipating 500 W? What is the voltage-drop across it?
12. If the internal resistance of a television receiver is 242 Ω when it draws 200 W from the power line, what is the voltage?
13. In what units does the consumer pay for electrical power?
14. What are the three basic Ohm's law formulas?

ANSWERS TO CHECK-UP QUIZ 5-3

1. (3.6 W) **2.** (400 V) **3.** (With 120 V, 240 Ω)

ANSWERS TO CHECK-UP QUIZ 5-4

1. (36 W) (32,400 J, or 9 Wh) **2.** (2.88 W) (0.0691 kWh) (8.29¢) **3.** (1440 W) (2,592,000 J, or 720 Wh) **4.** (4056 kWh)

---

6 PARALLEL CIRCUITS

CHAPTER OBJECTIVES. To develop a usable understanding of these electrical vocabulary terms: conductance, parallel circuits, reciprocal-of-the-sum-of-the-reciprocals formulas, prodivisum formulas, leaning-ladder graphing, Kirchhoff's current law. Also, to compute simpler series and parallel resistor circuits.

6-1 REVIEW OF SERIES CIRCUITS

The loads and cells considered in previous chapters have been connected in series. The total resistance of two or more resistors in series is the simple sum of the resistances,

$$R_t = R_1 + R_2 + R_3 \cdots$$

where R_t is the total resistance and R_1, R_2, and R_3 are the resistors connected in series.

If two 1.5-V cells are connected in a series-aiding circuit, they produce a total of 3 V. Fifteen cells in series form a 22.5-V battery.

Kirchhoff's voltage law states that the sum of all the voltage-drops across the resistors in a series circuit will always equal the source voltage. This is true even if the source is made up of several batteries in series, connected either series-aiding or series-bucking.

Figure 6-1 shows a more complex series circuit. The source batteries of 30 V and 10 V are series-bucking. The total working voltage of the circuit, shown by voltmeter V_1, would be 30 − 10, or 20 V. The total resistance of the circuit is 6 + 16 + 18, or 40 Ω. By Ohm's law, the current read by the ammeter would be $I = E/R$, or 20/40, or 0.5 A.

With 0.5 A flowing through the 6-Ω resistor, the voltage-drop across it, shown by V_2, would be found by Ohm's law to be $V = IR$, or 0.5(6), or 3 V.

Voltmeter V_3 is connected across the 30-V battery and the 18-Ω resistor in series, and therefore it reads the sum of the voltage of this battery plus the voltage-drop across this resistor. To determine whether to *add* the 30-V source voltage to the

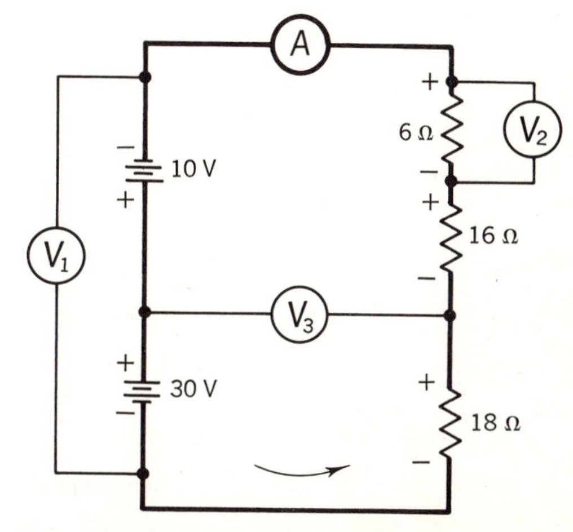

Fig. 6-1 Series circuit with series-bucking sources and three load resistors in series.

voltage-drop across the 18-Ω resistor or to *subtract* one from the other, it is first necessary to label the 18-Ω load with the proper polarities. Since the 30-V battery is greater than the 10-V battery, the current in the circuit must be flowing counterclockwise, as shown by the arrow. Current in any operating circuit flows from negative to positive through any load. Therefore the polarity markings on the 18-Ω resistor are − at the bottom and + at the top. Comparison of the polarities of the 30-V source and the 18-Ω resistor shows that they are series-bucking. Therefore the third voltmeter will read the *difference* of these two voltages. The voltage across the 18-Ω resistor is $V = IR$, or 0.5(18), or 9 V. V_3 will read 30 − 9, or 21 V.

This last voltage answer can be proved by adding up the voltage-drops across the 16-Ω and the 6-Ω resistors (11 V) and then adding this sum to the 10-V source. Since the polarities of this battery and the voltage-drops are all series-aiding, the total is 11 + 10, or again 21 V.

Quiz 6-1. Test your understanding. Answer these check-up questions.

All questions refer to Fig. 6-2.

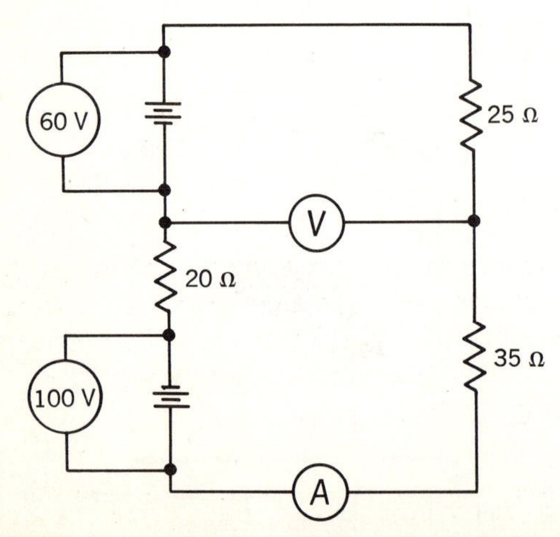

Fig. 6-2 Series circuit used in Check-up Quiz.

1. What is the total operating emf in this circuit according to the voltmeter readings? ________
2. What is the direction of I (clockwise or counterclockwise)? ________
3. What polarity marking should be made at the bottom of the 25-Ω resistor? ________
4. What is the R_t value? ________
5. What value should the ammeter, A, read? ________
6. What is the voltage-drop across the 25-Ω resistor? ________ Across the 35-Ω resistor? ________ Across the 20-Ω resistor? ________ Does the total of these three voltage-drops agree with the source E? ________
7. Using the upper source and the 25-Ω resistor, what is the voltage-drop across the voltmeter V? ________
8. Add the voltage-drops across the 20-Ω and 35-Ω resistors to that of the lower source. What is this voltage value? ________
9. What would the ammeter A read if it were connected between the 25-Ω resistor and the upper source? ________

6-2 CONDUCTANCE AND RESISTANCE

A copper wire has many free electrons, which makes it a good conductor. When an electrostatic field is applied between the ends of a wire, a flow of electrons results. However, copper is not a perfect conductor; it resists the flow of current to some extent. An iron wire has fewer free electrons and hence has more *resistance* R, or less *conductance* G, than copper. The ability to conduct and to resist is present in any material. The unit by which opposition is measured is the ohm, as discussed previously.

Since conductance is the opposite of resistance, it is given an opposite unit name and an opposite mathematical value.

Resistance:

$$R = \frac{1}{G} \qquad \text{(Unit: ohm, } \Omega\text{)}$$

Conductance:

$$G = \frac{1}{R} \qquad \text{(Unit: mho or siemens, S)}$$

If a resistor has 20 Ω of resistance, it has 1/20 mho, or 0.05 mho of conductance. A 0.4-Ω resistor has a conductance of 1/0.4 mho, or 2.5 mhos, or 2.5 S.

When a device has a conductance value of 25 mhos, it has a resistance of 1/25 Ω, or 0.04 Ω. A conductance of 0.0008 mho represents a resistance of 1/0.0008 Ω, or 1250 Ω. Resistance is the *reciprocal* of conductance, and conductance is the reciprocal of resistance.

Although circuits are usually considered from the viewpoint of their resistance, it would also be possible to think of them from a conductance viewpoint.

6-3 PARALLEL CIRCUITS

In dealing with loads connected in parallel it is advantageous to consider them from the conductance viewpoint. In Fig. 6-3, R_1 has twice the resistance of R_2. This

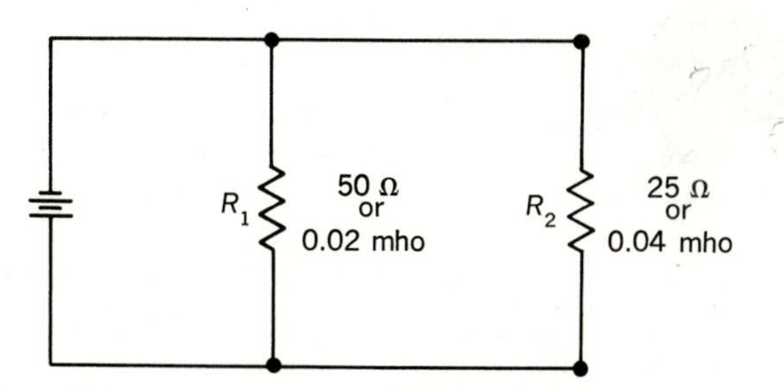

Fig. 6-3 Circuit with a 0.02-mho conductance in parallel with a 0.04-mho conductance and a G_t of 0.06 mho.

means that R_1 must have *half* the conductance of R_2. What is the total resistance, or R_t, of this circuit? Will R_t be equal to 50 Ω plus 25 Ω? It would be if this were a series circuit, but this is a parallel circuit. If the total *conducting* ability of this parallel circuit is determined, the reciprocal of the conductance value will be the resistance value. Note that both the resistors are conducting as far as the source is concerned. Therefore the total conductance of the circuit is 0.02 mho for the first branch plus 0.04 mho for the second branch, or 0.06 mho of total conductance. If the total conductance value is 0.06, the total resistance must be 1/0.06, or 16.7 Ω. (This is far different from the R_t value if the 50-Ω and 25-Ω resistors had been in series.)

The total resistance of parallel resistors is determined by finding the conductances of each parallel branch separately, adding these conductances, and then dividing 1 by this answer to obtain the resistance value. The parallel-resistance formula that does this is

$$R_t = \frac{1}{G_1 + G_2 \cdots} = \frac{1}{1/R_1 + 1/R_2 \cdots}$$

where R_t is the total resistance and R_1 and R_2 are the resistors in parallel. This is called the *reciprocal-of-the-sum-of-the-reciprocals formula*. The value $1/R_1$ converts the first branch resistance to conductance; this is added to the conductance of the second branch, $1/R_2$, and then 1 is divided by the sum of these two conductances to give the total *resistance* value of the parallel circuit. The same formula can be used for two, three, or as many parallel resistors as are desired.

There are two methods of working this formula. One is using decimal fractions:

$$\begin{aligned} R_t &= \frac{1}{1/R_1 + 1/R_2} \\ &= \frac{1}{0.02 + 0.04} = \frac{1}{0.06} \\ &= 16.7\ \Omega \end{aligned}$$

The second method employs the addition and division of common fractions and involves finding the common denominator and inverting the lower fraction and multiplying by 1:

$$\begin{aligned} R_t &= \frac{1}{1/50 + 1/25} = \frac{1}{1/50 + 2/50} = \frac{1}{3/50} = \frac{50}{3} \\ &= 16.7\ \Omega \end{aligned}$$

Use the method that seems simplest with the values given. Usually, a simple common denominator is not apparent, and it is better to use the decimal-fraction method.

The basic formula for parallel resistances can be algebraically manipulated into a simpler formula for use when there are only two resistances in parallel. This two-resistor parallel-circuit formula is

$$R_t = \frac{R_1 R_2}{R_1 + R_2}$$

Working the same problem as before but using this product-divided-by-sum, or *prodivisum*, type of formula

$$R_t = \frac{R_1 R_2}{R_1 + R_2} = \frac{50(25)}{50 + 25} = \frac{1250}{75} = 16.7\ \Omega$$

A prodivisum-type formula can also be used to solve for the total resistance of three parallel resistors. First the parallel resistance of two of the resistors is determined. This parallel-group resistance is then considered as a single value in parallel with the third resistor, and the total of these two resistances is in turn computed by the prodivisum formula.

The prodivisum formula can also be expressed as a single formula to solve for three parallel resistors:

$$R_t = \frac{R_1 R_2 R_3}{R_1 R_2 + R_2 R_3 + R_3 R_1}$$

Note that in all parallel circuits the *same voltage* will always appear across all the loads that are in parallel. This is the type of circuit used for lights and equipment in home and industry. All loads are in parallel with each other and have the same voltage across them. The idea of the same *voltage* across parallel circuits is a most important point to remember. In series circuits, the same *current* flows in all parts of the circuit.

ANSWERS TO CHECK-UP QUIZ 6-1

1. (40 V) **2.** (CW) **3.** (+) **4.** (80 Ω) **5.** (0.5 A) **6.** (12.5 V) (17.5 V) (10 V) (Yes) **7.** (72.5 V) **8.** (72.5 V) **9.** (0.5 A)

Quiz 6-2. Test your understanding. Answer these check-up questions.

1. What is the conductance of a 1000-Ω resistor? ________ What is the G value of a 4.7-Ω resistor? ________
2. What is the resistance of a device having a conductance value of 80 mhos? ________ A G value of 0.004 mho? ________

Questions 3 to 9 refer to Fig. 6-4.

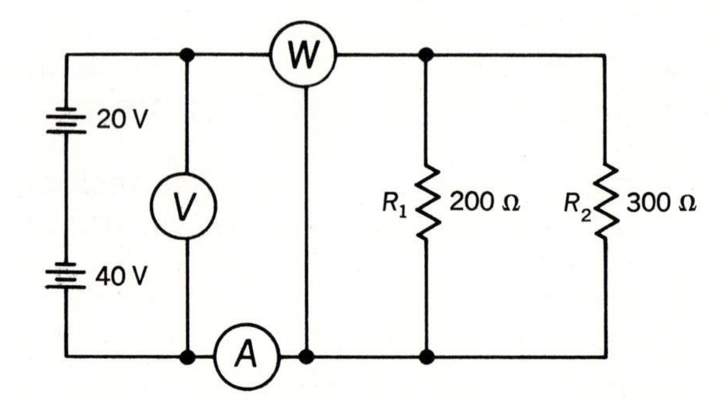

Fig. 6-4 Parallel-resistor circuit used in the Check-up Quiz.

3. What value would the voltmeter, V, read? ________
4. What is the conductance of the 200-Ω resistor? ________ The 300-Ω resistor? ________ What is the total G value? ________
5. What is the total R value? ________
6. What is the R_t value by the fraction method? ________
7. What value would the ammeter, A, read? ________
8. What value would the wattmeter, W, read? ________
9. If the voltage of the source were doubled, what effect would this have on the total conductance? ________

10. What would be the total resistance of two 400-Ω resistors in parallel? __________ Two 10-Ω resistors in parallel? __________
11. From the answers to question 10, what would be a simple formula to determine the total resistance of two *similar-valued* resistors in parallel? __________
12. What would be a simple formula to determine the total R of three similar-valued parallel resistors? __________
13. A 40-Ω, a 24-Ω, and a 60-Ω resistor are connected in parallel. Which has the greatest conductance? __________ What is the R_t value? __________ Which resistor would dissipate the greatest power according to the formula $P = E^2/R$? __________

6-4 LEANING-LADDER GRAPHS

Computing by formulas is one way to determine the total of two or more resistances either in series or in parallel. It is also possible to *graph* two or more resistance values to determine their total value. The accuracy is as good as the precision with which the graphs are measured.

Graphing two (or more) *series* resistors consists of adding one vector arrow, having a length proportional to the first resistor value in ohms, to a second vector having a length proportional to the second resistor (Fig. 6-5*a*). Thus, a 20-Ω resistor in series with a 30-Ω resistor has a total of 50 Ω.

When two resistances are in parallel, the two vector arrows with lengths proportional to their resistance values are drawn parallel to each other (Fig. 6-5*b*). A dashed line is drawn from the tip of each arrow to the base of the other. The distance between the base line and the intersection of the dashed lines is equal to the parallel resistance value of the two resistors. In the figure, a 20-Ω resistor is in parallel with a 30-Ω resistor, resulting in a total resistance of 12 Ω.

Figure 6-5*c* illustrates how three parallel resistors, R_1, R_2, and R_3, can be determined. First the parallel resistance of R_1 and R_2 is found. The resultant, R_r, is then considered in parallel with R_3, and the total, R_t, is the final answer.

Check the answers to some problems involving parallel resistances and determine how close you can come to the correct total resistance value by this *leaning-ladder* graphing method.

6-5 KIRCHHOFF'S CURRENT LAW

In circuits having two (or more) loads connected in *series*, it was found that the sum of all the voltage-drops across the

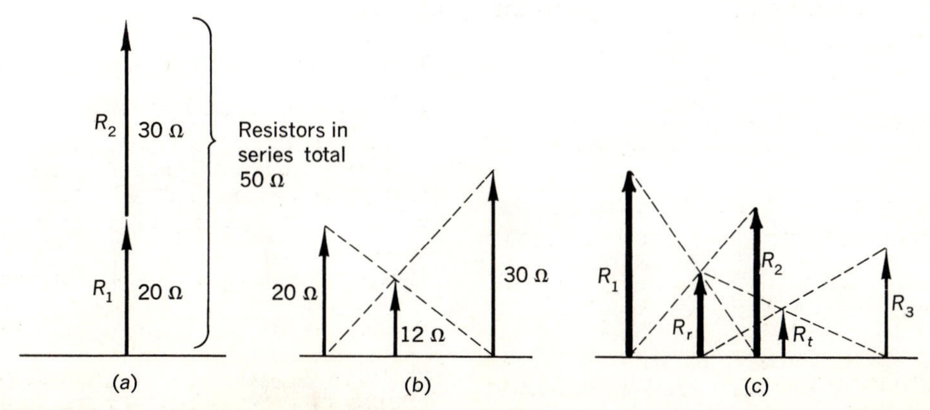

Fig. 6-5 Graphing (*a*) series resistor, (*b*) two parallel resistors, and (*c*) three parallel resistors.

loads equals the source voltage. This was identified as Kirchhoff's voltage law.

In circuits having two (or more) loads in parallel, as in Fig. 6-6, the source current

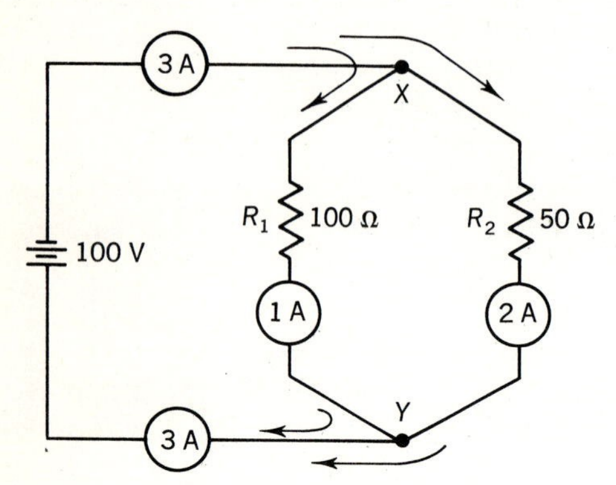

Fig. 6-6 Illustration of Kirchhoff's current law.

depends on the conductance (or resistance) of the two loads. In the circuit shown, the current through R_1 would be 1 A, and the current through R_2 would be 2 A. Therefore the source current would have to be 3 A. If a 3-A current flows out of the negative terminal of the 100-V source, when it reaches point X in the circuit, it has to split into two components, one through the 100-Ω load (1 A) and one through the 50-Ω load (2 A). At point Y the two currents recombine into a current having a value of 3 A. This rather simple idea illustrates *Kirchhoff's current law:*

> **The sum of the currents leaving a point in a circuit equals the sum of the currents entering the point.**

This might also be stated as:

> **The algebraic sum of the currents entering and leaving any point in a circuit is zero.**

As simple as Kirchhoff's laws seem, they will be found in later chapters to be powerful tools in working with complex circuits.

CHAPTER 6 TEST • PARALLEL CIRCUITS

1. What is the output voltage if three 6-V batteries are connected series-aiding? If they are connected in parallel, with positives to positives?
2. What law states that the source emf equals the sum of the voltage-drops around a circuit?
3. If the 10-V battery in Fig. 6-1 were reversed in polarity, what value would V_1 read? What value would V_2 read? What value would V_3 read?
4. A 24-Ω resistor has what conductance value?
5. A 0.000125-mho conductor has what resistance value?
6. When resistors are in series, are their resistance values or their conductance values added to determine the total? When resistors are in parallel, what is always added?
7. Write the prodivisum formula for parallel resistors.
8. Write the reciprocal-of-the-sum-of-the-reciprocals formula for three parallel resistors.

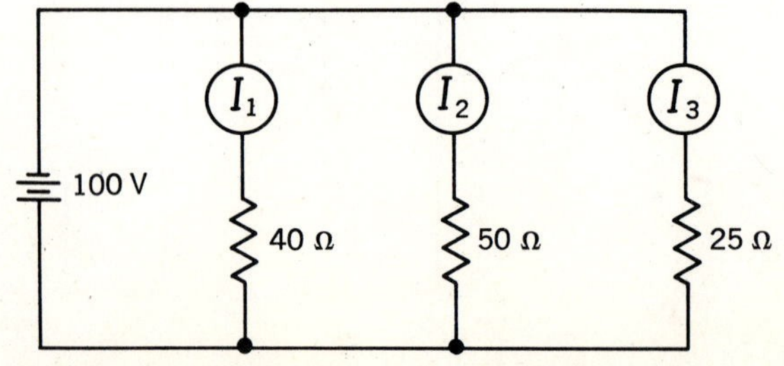

Fig. 6-7 Circuit for questions in the test.

ANSWERS TO CHECK-UP QUIZ 6-2

1. (0.001 mho) (0.213 mho) **2.** (0.0125 Ω) (250 Ω) **3.** (60 V) **4.** (0.005 mho) (0.00333 mho) (0.00833 mho) **5.** (120 Ω) **6.** (120 Ω) **7.** (0.5 A) **8.** (30 W) **9.** (None) **10.** (200 Ω) (5 Ω) **11.** ($R_t = R/2$) **12.** ($R/3$) **13.** (24-Ω resistor) (12 Ω) (24-Ω resistor)

---

9. A 300-Ω resistor and a 600-Ω resistor are in parallel. What is their total conductance? Their total resistance?
10. If the parallel-resistor group of question 9 were across 100 V, which resistor would carry the greater current value? What would this *I* value be? What would be the power dissipation for the resistor that dissipates the greater amount of power?
11. In Fig. 6-7, what is the value of I_1? I_2? I_3? I_s (source current)?
12. In Fig. 6-7, determine the total resistance of the circuit by Ohm's law. By the parallel-resistor formula.

7 SERIES-PARALLEL CIRCUITS

CHAPTER OBJECTIVES. To develop a usable understanding of these electrical vocabulary terms: series-parallel circuit, branch, leg. Also, to present practice in working Ohm's law and power formulas in series, parallel, and series-parallel circuits, and to mentally approximate answers in some of these circuits.

7-1 BASIC CIRCUIT FORMS

In the previous chapters three basic circuit forms have been discussed. The first, and simplest, consists of a source, a load, and connecting wires, as in Fig. 7-1*a*. (A controlling device, such as a switch, might have been added.) This may be termed a *simple circuit*. The second circuit form consists of a source with two (or more) loads connected in series across it, as in Fig. 7-1*b*. This is known as a *series circuit*. The third form of circuit consists of two (or more) loads in parallel, or shunt, as in Fig. 7-1*c*; this is a *parallel circuit*.

Another possible circuit form has two (or more) parallel groups in series with

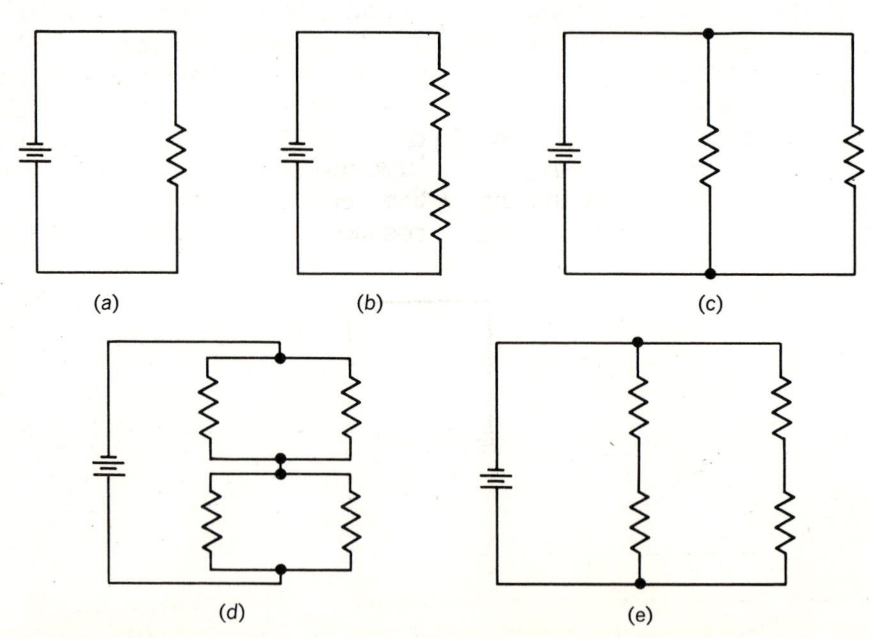

Fig. 7-1 (*a*) A simple circuit. (*b*) A series circuit. (*c*) A parallel circuit. (*d*) A seriesed-parallel circuit. (*e*) Paralleled-series circuits.

one or more other resistors, as in Fig. 7-1*d*. Still another possible form of load connection is two (or more) series branches in parallel with each other, as in Fig. 7-1e. It might be said that these form parallel-seriesed and series-paralleled groups. However, these are only a few of the many complex combinations of loads that are possible. To simplify terminology, any combination of loads that is not simple, series, or parallel will be termed a *series-parallel circuit*. This chapter concerns the fundamentals of solving the simpler types of series-parallel circuits.

7-2 SERIES-PARALLEL CIRCUITS: FIRST FORM

Most of the essentials of the first form of series-parallel circuits can be explained by reference to Fig. 7-2*a*. Sufficient information is given to determine the total resistance, the currents in each resistor, the voltage-drops across each resistor, and the power dissipation in each resistor.

The most likely point of trouble for the beginner is the voltage-drop across the parallel 20- and 30-Ω resistors. These two resistors in parallel have an equivalent resistance value of $R_t = R_1R_2/(R_1 + R_2)$, or 20(30)/(20 + 30), or 12 Ω. Therefore the circuit might be redrawn as shown in Fig. 7-2*b*. Such a circuit will be recognized as a series circuit of 8 and 12 Ω. This in turn is the equivalent of a simple circuit of 20 Ω across the 40-V source (Fig. 7-2c).

In the simple circuit, the current value must be $I = E/R$, or 40/20, or 2 A. Thus there must be 2 A flowing through the 8-Ω resistor and also through the equivalent resistance of 12 Ω of the parallel circuit.

The voltage-drops across the two series resistances are determined by Ohm's law. The voltage-drop across the 8-Ω resistor is $V = IR$, or 2(8), or 16 V. The voltage-drop across the 12-Ω equivalent resistance must be 2(12), or 24 V. And here is a most important point. There is a 24-V drop across the 12-Ω resistance. Therefore there must be a 24-V drop across both the 20-Ω resistor and the 30-Ω resistor in parallel. This is the same 24-V drop across these two resistors—not similar 24-V drops. There is 24 V across the 20-Ω resistor and the *same* 24 V across the 30-Ω resistor.

If there is a 24-V drop across the 20-Ω resistor, the current flowing through it must be $I = E/R$, or 24/20, or 1.2 A. Applying the same theory, the current through the 30-Ω resistor must be $I = E/R$, or 24/30, or 0.8 A. According to Kirchhoff's current law, the sum of these two currents equals the current flowing through the 8-Ω resistor, which is 2 A. Adding the two

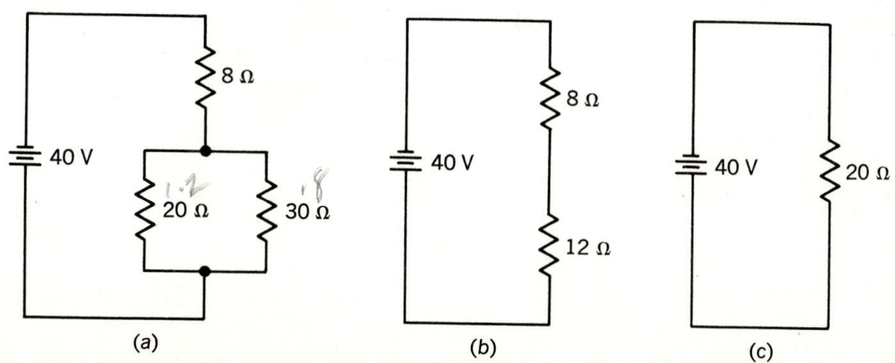

Fig. 7-2 (*a*) A series-parallel circuit and (*b*) its series equivalent. (*c*) The total equivalent circuit.

branch currents, 1.2 A and 0.8 A, results in the required 2 A.

It should be noted that the sum of the voltage-drops across the 8-Ω resistor plus the voltage-drop across the 20-Ω resistor will equal the source voltage. Similarly, the sum of the voltage-drop across the 8-Ω resistor plus the voltage-drop across the 30-Ω resistor equals the source-voltage value.

Since E, I, and R of all parts of this circuit have been solved, the power dissipated in any component can be determined by any of the power formulas.

Notice that in this one circuit can be found applications for Ohm's law, both Kirchhoff's laws, and the power formulas.

Quiz 7-1. Test your understanding. Answer these check-up questions.

Questions 1 to 4 refer to Fig. 7-3*a*.

1. What is the equivalent resistance of the parallel group? ________ What is the total resistance? ________
2. What value would A read? ________ What value would V_1 read? ________ What value would V_2 read? ________
3. What is the power dissipation in the 20-Ω resistor? ________ The 60-Ω resistor? ________ The 120-Ω resistor? ________
4. If the 120-Ω resistor were open-circuited, what value would the ammeter read? ________ What value would V_1 read? ________ What value would V_2 read? ________

Questions 5 to 8 refer to Fig. 7-3*b*.

5. What is the voltage-drop across the 10-Ω resistor? ________ The 20-Ω resistor? ________ What is the current through the 20-Ω resistor? ________ What is the resistance value of R_1? ________
6. What value would the voltmeter read? ________
7. What value would W_1 read? ________ What value would W_2 read? ________
8. If the 20-Ω resistor were short-circuited, what value would the ammeter read? ________ What value would the voltmeter read? ________ What value would W_1 read? ________ W_2? ________

7-3 SERIES-PARALLEL CIRCUITS: SECOND FORM

The first form of series-parallel circuits is a parallel group in series with one resistor. In the second form two resistors in series are in parallel with a third resistor, as in Fig. 7-4*a*. Examination of the circuit in Fig. 7-4*a* shows that the first branch consists of a 30-Ω resistor and a 90-Ω resistor in series, offering 120 Ω as a total resistance for this branch. The circuit

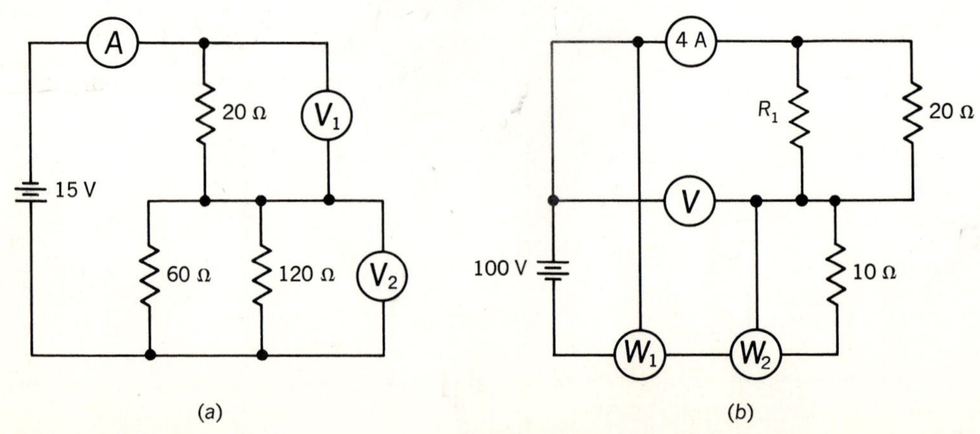

Fig. 7-3 Circuits used ... heck-up Quiz 7-1.

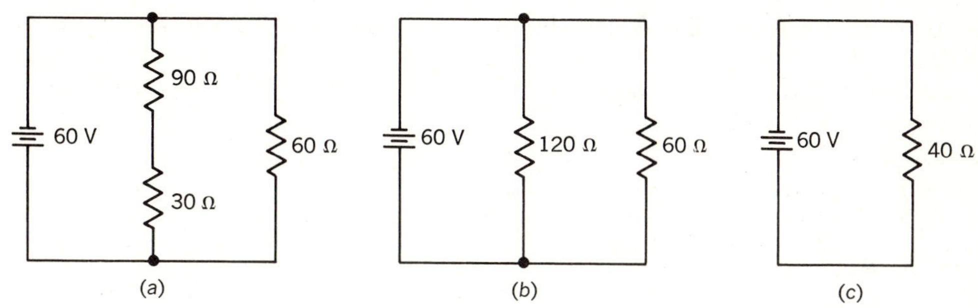

Fig. 7-4 (*a*) A second form of series-parallel circuit and (*b*) the parallel circuit equivalent. (*c*) The total equivalent circuit.

could therefore be drawn as in Fig. 7-4*b*, as a 120-Ω resistor and a 60-Ω resistor in parallel across the 60-V source. As computed previously, the equivalent resistance of 120-Ω and 60-Ω resistors in parallel is 40 Ω. Therefore the series-parallel circuit appears to the source as a single 40-Ω resistor (Fig. 7-4*c*).

The total source current can be determined as $I = E/R$, or 60/40, or 1.5 A.

The currents flowing in the separate legs can be determined from the parallel drawn circuit. For the first branch the current is $I = E/R$, or 60/120, or 0.5 A. The current in the second branch is $I = E/R$, or 60/60, or 1 A. Note that the sum of the first branch current, 0.5 A, and the second branch current, 1 A, equals the total source current, 1.5 A, as it should.

The voltage-drops across the two series resistors are still unknown. Since the current through this branch is known to be 0.5 A, the voltage-drop across the 90-Ω resistor is $V = IR$, or 0.5(90), or 45 V. The drop across the 30-Ω resistor is 0.5(30), or 15 V. The sum of these two voltages, 45 + 15, equals the source voltage of 60 V, as it must.

Since all the E, I, and R values are known, the power dissipation of each of the components can be determined by any one of the three basic power formulas. For example, the power dissipation in the 30-Ω resistor is

$$P = EI, \text{ or } 15(0.5), \text{ or } 7.5 \text{ W}$$
$$P = I^2R, \text{ or } 0.5^2(30), \text{ or } 0.25(30), \text{ or } 7.5 \text{ W}$$
$$P = E^2/R, \text{ or } 15^2/30, \text{ or } 225/30, \text{ or } 7.5 \text{ W}$$

Quiz 7-2. Test your understanding. Answer these check-up questions.

Questions 1 to 5 refer to Fig. 7-5*a*.

1. What is the source-voltage value? ________ In which direction does current flow in the load resistors (up or down)? ________
2. What R_t is seen by the source? ________ What value would A_1 read? ________ What value would A_2 read? ________
3. What value would V_1 read? ________ What value would V_2 read? ________
4. What is the power dissipation of the 60-Ω resistor? ________ The 10-Ω resistor? ________ The 30-Ω resistor? ________
5. If the 10-Ω resistor were short-circuited, what R_t would be seen by the load? ________ What

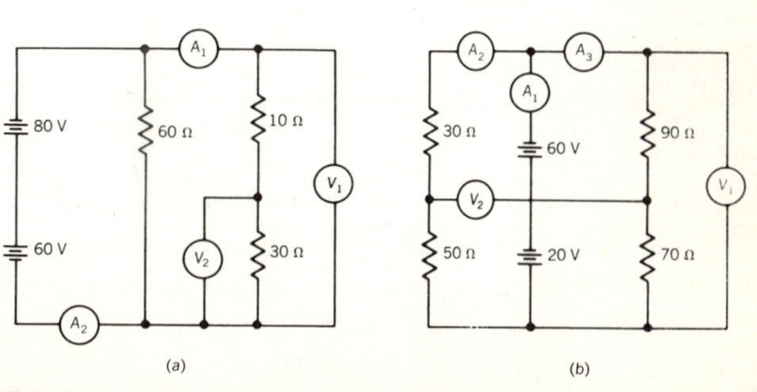

Fig. 7-5 Circuits used in the Check-up Quiz.

value would A_1 read? ________ What value would A_2 read? ________

Questions 6 to 10 refer to Fig. 7-5*b*. (Redraw to simplify?)

6. What value would V_1 read? ________ What R_t is seen by the source? ________ What value would A_1 read? ________
7. What value would A_2 read? ________ What value would A_3 read? ________
8. What polarity marking should be placed on the top of the 50-Ω resistor? ________ On the bottom of the 70-Ω resistor? ________
9. What is the voltage-drop across the 50-Ω resistor? ________ Across the 70-Ω resistor? ________
10. What value would V_2 read? ________

7-4 MENTAL APPROXIMATIONS

With series resistors the total resistance is easily determined. It is only necessary to add the resistance values.

In parallel circuits, it is usually a good idea to work out mentally an approximation of the total resistance of any parallel group before doing the mathematical computations. If the computed value varies more than about 10% from the approximated value, the mathematics of the problem may be in error. This is an important concept for beginning workers in electricity.

As pointed out in Chap. 6, two 100-Ω resistors in parallel form a 50-Ω network. Similarly, the total value of two nearly equal resistances, such as a 90- and 110-Ω parallel group, would also be about 50 Ω (49.5 Ω).

Three parallel 100-Ω resistors will have a total value of 33.3 Ω. Therefore the value of a 90-, 100-, and 110-Ω parallel group would be approximately 33 Ω (actually 33.1 Ω).

What is the approximate total value if a 52-Ω resistor is connected in series with a 70- and 80-Ω parallel group? The value of the parallel group would be about 37.5 Ω (half of 75), which, when added to 52 Ω, would be approximately 89.5 Ω (89.3 Ω).

When one parallel resistance is 2 times the other, the total will be two-thirds the value of the lower-valued resistor. Thus a 30-Ω resistance and a 60-Ω resistance have a parallel value of $\frac{2}{3}(30)$, or 20 Ω. The value of a 48-Ω resistance and a 95-Ω resistance would be about $\frac{2}{3}(48)$, or 32 Ω (31.8 Ω).

When one parallel resistance is 5 times the other, the total value will be five-sixths the lower value. When one resistance is 10 times the value of another parallel resistance, the total value will be $\frac{10}{11}$, or approximately 90% of the lower-valued resistance. (Do the $\frac{2}{3}$, $\frac{5}{6}$, and $\frac{10}{11}$ ratios indicate a pattern?) When one resistance is 100 times the value of another parallel resistance, the higher value has almost no effect on the total parallel value.

Quiz 7-3. Test your understanding. Answer these check-up questions.

1. If one parallel R is 4 times the other, the resultant will be what fraction of the lower resistor? ________
2. A 550-Ω resistor and a 475-Ω resistor are in parallel. What is their approximate total resistance value? ________
3. A 255-Ω resistor and a 520-Ω resistor are in parallel. What is their approximate total resistance value? ________
4. An 82-Ω resistor and an 820-Ω resistor are in parallel. What is their approximate total resistance value? ________
5. A 700-Ω R is in series with a 1000- and 900-Ω par-

ANSWERS TO CHECK-UP QUIZ 7-1

1. (40 Ω) (60 Ω) **2.** (0.25 A) (5 V) (10 V) **3.** (1.25 W) (1.67 W) (0.833 W) **4.** (0.1875 A) (3.75 V) (11.25 V) **5.** (40 V) (60 V) (3 A) (60 Ω) **6.** (60 V) **7.** (400 W) (160 W) **8.** (10 A) (0 V) (1000 W) (1000 W)

---

allel group. What is the approximate total resistance value? ________

6. Two 300-Ω resistors are in series, and in parallel with this group is a 700-Ω resistor. What is the approximate total resistance of the circuit? ________
7. Two 20-Ω resistors and a 10-Ω resistor are all in parallel. What is the exact resistance value by the approximation method? ________
8. A 100-, 120-, and 50-Ω parallel group has approximately what total resistance value? ________
9. A 700-, 60-, and 10,000-Ω parallel group has what approximate resistance value? ________

CHAPTER 7 TEST • SERIES-PARALLEL CIRCUITS

1. What are the names of the four basic types of circuits mentioned in this chapter?
2. A 450-Ω resistor is in series with two parallel resistors of 600 Ω and 800 Ω each. What would you approximate as the R_t value? What is the computed R_t value?
3. If the circuit in question 2 is across a source of 100 V, what is the value of the current in the 450-Ω *R*? The 600-Ω *R*? The 800-Ω *R*?
4. In the circuit of question 2, what is the voltage-drop across the 450-Ω *R*? The 600-Ω *R*? The 800-Ω *R*?
5. A 34-Ω resistor and a 24-Ω resistor are in series across a 100-V source, and a third resistor of 75 Ω is connected across the source. What would you approximate as the resistance seen by the source? What is the computed value?
6. In the circuit of question 5, what is the voltage-drop across the 34-Ω *R*? The 24-Ω *R*? The 75-Ω *R*?
7. In the circuit of question 5, what is the power dissipation of the 34-Ω *R*? The 24-Ω *R*? The 75-Ω *R*?
8. For an 80-, 40-, and 20-Ω parallel group, what would you approximate as the total *R* value? What is it by computation?
9. For a 1200-, 600-, and 40-Ω parallel group, what would you approximate as the total *R* value? What is it by computation?

ANSWERS TO CHECK-UP QUIZ 7-2

1. (20 V) (Down) **2.** (24 Ω) (0.5 A) (0.833 A) **3.** (20 V) (15 V) **4.** (6.67 W) (2.5 W) (7.5 W) **5.** (20 Ω) (0.667 A) (1 A) **6.** (80 V) (53.3 Ω) (1.5 A) **7.** (1 A) (0.5 A) **8.** (−) (+) **9.** (50 V) (35 V) **10.** (15 V)

ANSWERS TO CHECK-UP QUIZ 7-3

1. (4/5) **2.** (255 Ω) **3.** (171 Ω) **4.** (75 Ω) **5.** (1173 Ω) **6.** (323 Ω) **7.** (5 Ω) **8.** (26.1 Ω) **9.** (55 Ω)

8 MAGNETISM

CHAPTER OBJECTIVES. To develop a usable understanding of these electrical vocabulary terms: mass, magnetic field, line of force, left-hand coil rule, magnetic flux, ϕ, weber, microweber, ampere-turn, magnetomotive-force, mmf, field intensity, H, flux density, tesla, B, permeability, leakage lines of force, temporary magnet, permanent magnet, retentivity, domain, saturation, the knee of a curve, ferromagnetic, magnetic compass, magnetic versus geographic poles, relay, armature. Also, to practice solving some simple magnetic-circuit problems.

8-1 THE MAGNETIC FIELD

In Chap. 1 two types of invisible fields of force were discussed. One field known to everyone is gravity, an invisible attractive force developed between two bodies having weight, or *mass* (the quantity of matter in a body), which tends to pull them together.

The second field of force that was discussed in Chap. 1 is the electrostatic field that exists between positively and negatively charged bodies or poles. A body becomes positively charged when electrons are taken away from it and negatively charged when an excess of electrons is placed on it. Figure 8-1 illustrates a negative and a positive body and the electrostatic lines of force that are assumed to exist between them. If an electron, a negative particle, were released in the air between the negative and positive poles, at point *A*, it would be repelled by the negative body and attracted to the positive body. It would travel along the lines of force, on a track indicated by the arrowheads. As it starts to move, the electron absorbs energy from the electrostatic field. Once it starts moving, the new energy of motion is stored in the area around the electron in another form—a magnetic field. Thus a *moving* electron not only has its own electrostatic charge and field, but it also develops a *magnetic* field around it. Note that the magnetic field develops at right angles to the direction of motion of the electron and in a circular form around the electron.

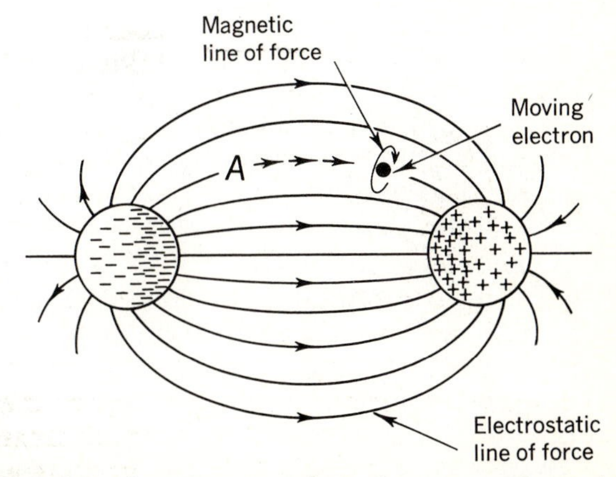

Fig. 8-1 An electron released at point *A* would travel along the electrostatic line and develop a magnetic field around itself as it traveled.

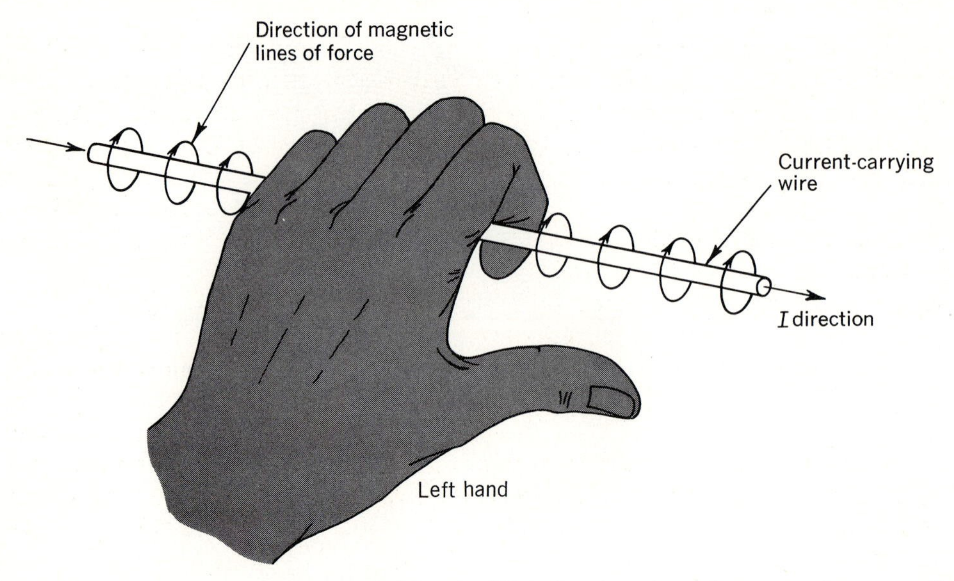

Fig. 8-2 The left-hand current, or magnetic field, rule. With the thumb in direction of current, the magnetic field direction is shown by the fingers of the left hand. (With conventional current theory, the right hand is used.)

If a single moving electron develops a magnetic field around its path, a wire carrying a current of electrons will develop a magnetic field around the wire along its length. From this it can be seen that any time there is a working circuit and an electric current is flowing, a magnetic effect will be present. It is impossible to separate the theories of magnetism and electricity, since one produces the other.

A convenient method of determining the assumed directivity given to the lines of force of a magnetic field is the *left-hand magnetic-field rule*. This is illustrated by Fig. 8-2. When the *left* hand grasps a wire with the thumb in the direction of current flow (in the direction of the electrostatic field), the fingers will indicate the direction assigned to the lines of magnetic force that develop around the wire. It must be pointed out that these lines indicating the direction of magnetic force are theoretical only. No such lines can be seen, nor do they exist as lines.

The idea that any moving electron is surrounded by an energy field of magnetic force is more far-reaching than might be suspected at first. Since any atom consists of a nucleus with moving electrons in orbit around it, all matter will have some magnetic effect. Some metals, such as iron, nickel, and cobalt, can be highly magnetic because the orbital paths of their electrons happen to be just right to produce "north" and "south" magnetic poles at the opposite sides of the atoms. Most other substances (copper, aluminum, glass, etc.) may be considered to have orbiting electrons with paths that cancel each other's magnetic effects, resulting in almost no external magnetic effect.

8-2 MAGNETISM OF A COIL

When a wire has a current flowing through it, a magnetic field forms around the wire. If the wire is wound into a coil, as in Fig. 8-3, the magnetic fields of adjacent turns add to form a concentrated field down the center, or core, of the coil. The core lines of force complete themselves outside the coil, where they are not confined and the field is far less concentrated.

Note that all the lines of force have an outward direction at one end of the coil and an inward direction at the other end. The end of the coil having the lines of force coming *out* of it is called the *north pole* of the coil; the *south pole* has the lines of force going into it.

The lines of force coming out of the north pole of the core (or going into the south) are said to be the *magnetic flux* of the core. The metric unit of measurement of flux is the *weber* (Wb). One weber is equal to 100 million (or 10^8) lines of force. Since this is a rather large number of lines of force, a more convenient unit is the *microweber* (μWb), one-millionth of a weber, or 100 lines of force. The symbol used to represent flux in magnetic circuitry is the Greek letter phi (ϕ). It might be said, for example, that a certain coil with a current flowing in it has $\phi = 35\ \mu$Wb surrounding it.

If the fingers of the left hand are placed around a coil in the direction in which electrons are moving (Fig. 8-3*b*), the thumb indicates the direction of the flux lines in the core and also the north end of the coil. This is called the *left-hand coil rule*. Check the flux polarity in Fig. 8-3*a* by this rule.

Basically, flux is produced by current flowing in a wire. The more turns or the tighter the wire is coiled, the more concentrated the lines of force in the core area become. The product of the current times the number of turns of the coil gives a unit called *ampere-turns* (amp-turns or *NI*), which is known as the *magnetomotive-force* (mmf) being applied to the core area. In formula form,

$$F = NI$$

where F = magnetomotive-force, in ampere-turns
N = number of turns
I = current, in amperes

If a coil with a certain number of ampere-turns of magnetomotive-force is

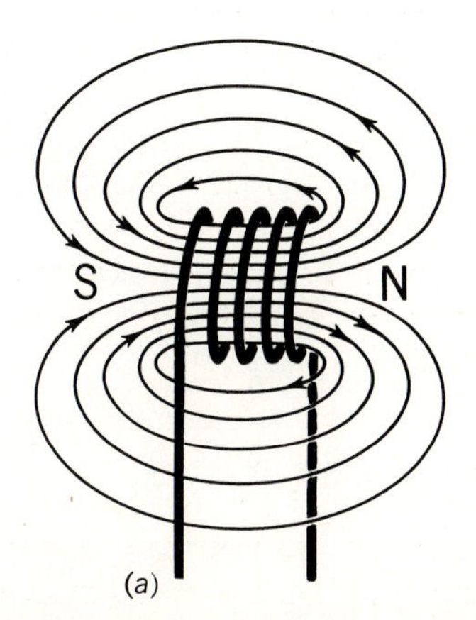

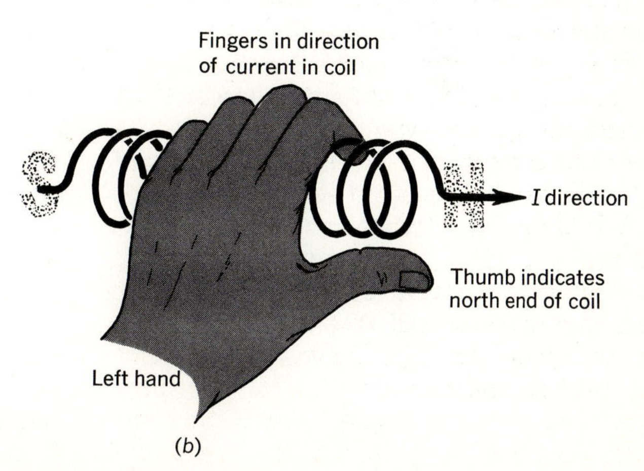

Fig. 8-3 (*a*) Coil-carrying current develops a magnetic field around it. The north pole is where the arrowheads point outward. (*b*) The left-hand coil rule (right-hand with conventional current). With the fingers in the direction of current in a coil, the thumb indicates the north end of the coil.

stretched out to twice its original length, the intensity of the field, that is, the concentration of lines of force in the core, will be half as great. From this reasoning, the field intensity of a coil will be directly proportional to the magnetomotive-force but inversely proportional to the length of the coil. Expressed as a formula,

$$H = \frac{NI}{\text{length}}$$

where H is field intensity in ampere-turns per meter and length is in meters.

The field intensity of a 40-turn, 8-cm-long coil, with 2 A flowing through it, is $H = NI/\text{length}$, or 40(2)/0.08, or 1000 ampere-turns/m. If the same coil is stretched out to 16 cm, the wire length and the current remain the same, but the field intensity in the core of the coil will be $H = 40(2)/0.16$, or 500 amp-turns/m. (What would happen to the field intensity if the coil turns were jammed together into a coil 4 cm long?)

If both the total flux at the end of a coil and the cross-sectional area (in square meters) of the core of the coil are known, the number of lines divided by the area produces a measurement known as the *flux density* of the core. The metric unit of measurement of flux density is webers per square meter, or *tesla* (Wb/m^2, or T). The letter symbol for flux density in any system of measurement is B. For an air-core coil with a cross-sectional area of one-thousandth of a square meter (0.001 m^2) and a flux count of 35 μWb, the flux density B is 0.000035/0.001, or 0.035 T.

The formula that gives the basic meaning of flux density is

$$B = \frac{\phi}{A}$$

where B = flux density, in webers per square meter
ϕ = flux, in webers
A = cross-sectional area of the core, in square meters

Quiz 8-1. Test your understanding. Answer these check-up questions.

1. What are the names of the three different fields of force discussed so far? __________ __________ __________
2. A moving electron has what two fields of force? __________ __________
3. What is the angular relationship between the direction of current and the magnetic field produced? __________
4. What is the shape of the magnetic field around a moving electron? __________
5. In the left-hand magnetic-field rule, in what direction does the thumb point? __________ What does the direction of the fingers indicate? __________
6. Would 1 A flowing through a 2-m length of wire made into a single loop produce more, the same, or less magnetism if it were wound into a coil 2 cm in diameter and 4 cm long? __________
7. In question 6, would the core field of the coil have a greater magnetomotive-force than the single loop? __________ Greater field intensity? __________ Greater flux density? __________
8. If the lines of force are drawn with arrowheads pointing into the left-hand end of a coil, what magnetic polarity will the right-hand end of the coil have? __________
9. An air-core coil 2 cm in diameter has 50 turns, is 5 cm long, and has 0.5 A flowing through it. What is the mmf of this coil? __________ What is its field intensity? __________
10. If a 30-turn air-core coil 6 cm long has 2500 amp-turns/m, would the flux density increase, decrease, or remain the same if the coil were pulled out until it was 8 cm long? __________
11. Do magnetic lines of force rotate in the direction shown by the arrowheads? __________

8-3 PERMEABILITY OF MATERIALS

Magnetic lines of force can be developed much more easily in some materials than in others. Actually, air can be thought of as resisting the setting up of lines of force in it. As a result, when air surrounds a current-carrying wire, the magnetic lines of force tend to be pushed back toward the wire. The area of greatest concentration will be at the surface, with many of the lines of force inside the wire.

Some materials, such as the elements iron, cobalt, and nickel, have electron orbits that are able to conform to any magnetic field in which they happen to be placed. As a result, when such materials are used as the core inside a coil of wire, many of the lines of force that would have remained inside or near the wire surface will expand out into the more permeable core material. A permeable core material is one which can be permeated, or pervaded, more easily than air or space. Iron, for example, has hundreds to thousands of times the *permeability*, or μ, of air.

Figure 8-4 is a coil, 5 cm long, having 10 turns. Suppose it has only an air core. With the switch closed a current of 2 A flows in it. From past information, the magnetomotive-force would be 2(10), or 20 amp-turns. The field intensity would be the number of ampere-turns per meter, 20/0.05, or 400 amp-turns/m. A certain number of lines of force would be produced in the core, perhaps 50. Now, if an iron core were slipped into the coil, the mmf and the field intensity would remain the same, but the flux density would be found to be several thousand times greater, perhaps 25,000 lines in the core area. In this case the iron would have a permeability of 5000 times that of air ($\mu = 5000$). Thus the use of an iron core instead of an air core in an electromagnet will increase the effectiveness of the magnet several thousand times. Most electromagnets, relays, etc., are made with iron cores.

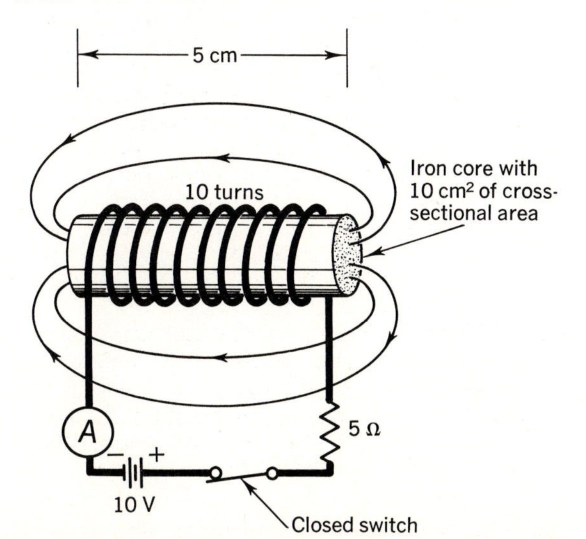

Fig. 8-4 With 10 turns and 2 A flowing, the magnetizing force producing flux in the core is 20 amp-turns.

Although the lines of force in an iron core may be very concentrated, a short distance out past the end of the core the lines begin to spread apart. The farther from the end of the core, the less the flux density and the weaker the magnetic field is. To produce an area of high flux density, the core may be bent into a horseshoe, or C, form, as in Fig. 8-5. A north

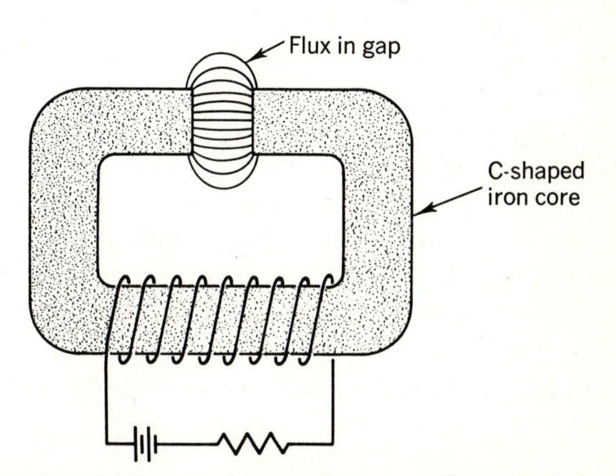

Fig. 8-5 Reducing the gap between the poles of an electromagnet increases the field strength in the gap and reduces leakage lines of force.

pole is formed at one end of the C and a south pole at the other. With this core shape, practically all the lines of force form between the two core ends. Because only a few of the lines of force leak out into the air around the gap, the field in the gap area is extremely concentrated. In electronics it is quite common to increase flux density in magnetic circuits by reducing the air gap.

Quiz 8-2. Test your understanding. Answer these check-up questions.

1. According to the left-hand rules, what polarity does the left end of the core in Fig. 8-4 have? __________ What polarity does the end of the magnet at the right of the gap have in Fig. 8-5? __________
2. Which would have more leakage lines, a bar magnet or a horseshoe magnet? __________
3. When the iron core is pulled out of an electromagnet, does the coil have a greater, smaller, or the same value of magnetizing force? __________ Of flux density? __________ Of field intensity? __________
4. If the core in Fig. 8-5 were made of copper instead of iron, would there be a concentration of lines of force in the gap area? __________ If the core were cobalt? __________
5. In Fig. 8-4, if the resistance value were doubled, what effect would this have on mmf? __________ On H? __________
6. What is the symbol letter for permeability? __________

8-4 TEMPORARY AND PERMANENT MAGNETS

A core of air can be magnetized by forming a coil around the core area. As soon as the current stops in the coil, the lines of force that were in the air core collapse back into the wire from which they originally expanded, much as if they were elastic bands that were suddenly released. Air is said to have no *retentivity*, or ability to retain a magnetized state. Pure uncrystallized iron also has practically no retentivity. Pure or "soft" iron, as it is called, can be used only when temporary magnets are desired.

When iron is crystallized, as in steel, its magnetic capabilities change. It may decrease in permeability, but when it is subjected to a magnetizing force, small groups of iron molecules, in the core of a current-carrying coil, align themselves with the lines of force and lock into little magnetic islands called *domains*. When the magnetizing force is removed, the domains tend to retain their magnetic alignment, and the steel retains some of the lines of force that were originally produced by the magnetizing force. The steel core can now be removed from the coil, and the core continues to exhibit a magnetic field. It has become a *permanent magnet*.

When all the molecules and domains of a piece of iron or steel are aligned in the direction of the magnetizing force, the

ANSWERS TO CHECK-UP QUIZ 8-1

1. (Gravity) (Electrostatic) (Electromagnetic) **2.** (Static) (Magnetic) **3.** (90°) **4.** (Circular) **5.** (Current) (Magnetic lines) **6.** (Same) **7.** (Yes) (Yes) (Yes) **8.** (N) **9.** ($NI = 25$) (500NI per meter) **10.** (Decrease) **11.** (No)

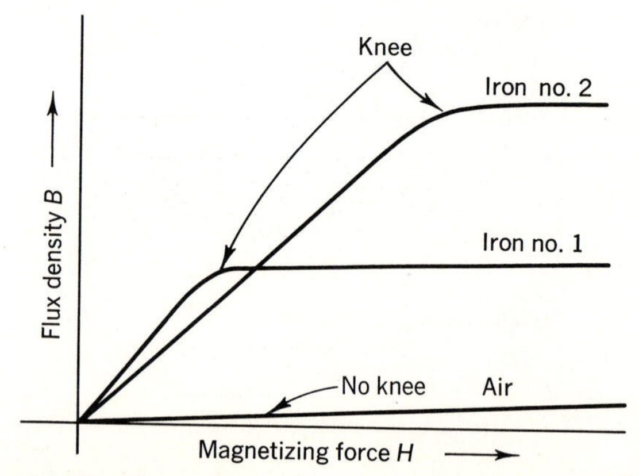

Fig. 8-6 Graph of magnetizing force versus flux density for air and two types of iron.

material is said to be magnetically *saturated*. Above the saturation point, iron and steel will have a permeability no greater than air. A graph of this condition is shown in Fig. 8-6. Here the magnetizing force (or field intensity, H) is graphed against the flux density, B, in lines per square meter. The curves show that the flux density of air increases little with an increase in H. Iron number 1 increases rapidly in flux density with an increase in H, reaching saturation with relatively little H, whereas iron number 2 requires considerably more H before it develops a "knee" and becomes saturated.

8-5 MAGNETIC ATTRACTION AND REPULSION

It is well known that permanent magnets will attract pieces of iron and steel but have no attractive effect on most other materials. Figure 8-7*a* shows a copper bar in the field of a permanent magnet. The copper has practically the same permeability as air, and therefore the magnetic lines pass through it as though it were air. In copper there is no domain development.

Figure 8-7*b* shows an iron bar placed in the field of the permanent magnet. Here the lines of force find a material which they can pervade far more easily than they can the surrounding air or the copper. As a result, the lines of force crowd into the iron and develop domains in it. Most of the lines of force extend from the N pole to the S pole through the iron bar. The elasticity of the lines of force will cause them to collapse back into the wire when the magnetizing force is removed. This tendency of the lines of force to shorten themselves produces an attractive force on the iron bar which pulls the bar toward the magnetic poles of the horseshoe magnet, holding it there snugly.

Figure 8-8*a* illustrates the shape of the magnetic field around a bar magnet. If a piece of iron, or any other ironlike (*ferromagnetic*) material, were placed in the field, the lines of force would permeate the iron. As the lines tried to contract, they would pull the iron toward the N pole of the magnet (Fig. 8-8*b*).

When two similar magnetic poles are placed near each other, as in Fig. 8-9*a*, the similar-direction lines of force repel each other, and the two magnets try to

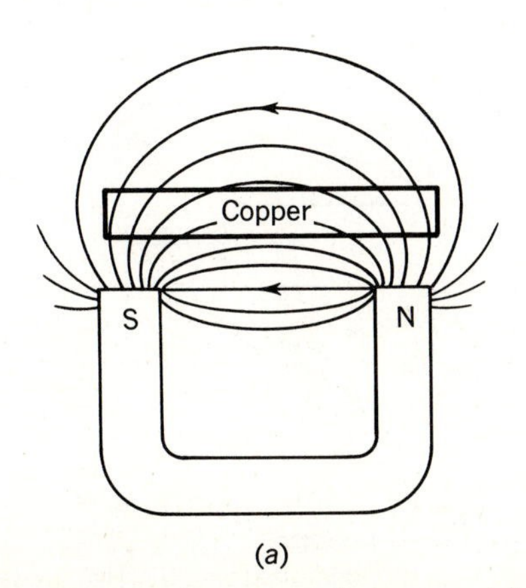

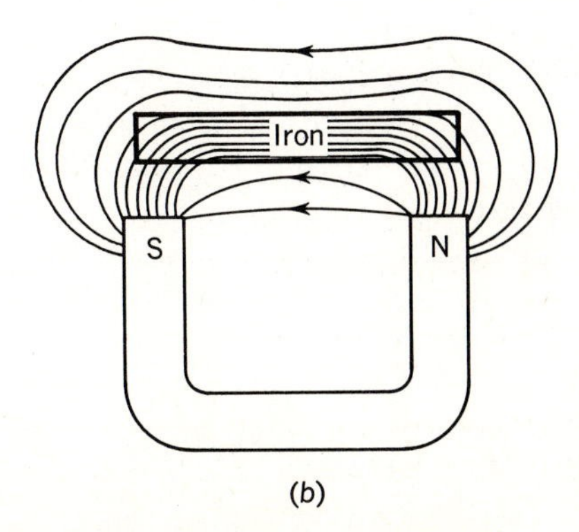

Fig. 8-7 The copper bar does not affect any of the lines of force and is not attracted to the magnet. The iron bar traps many lines of force which contract and pull the iron to the magnet.

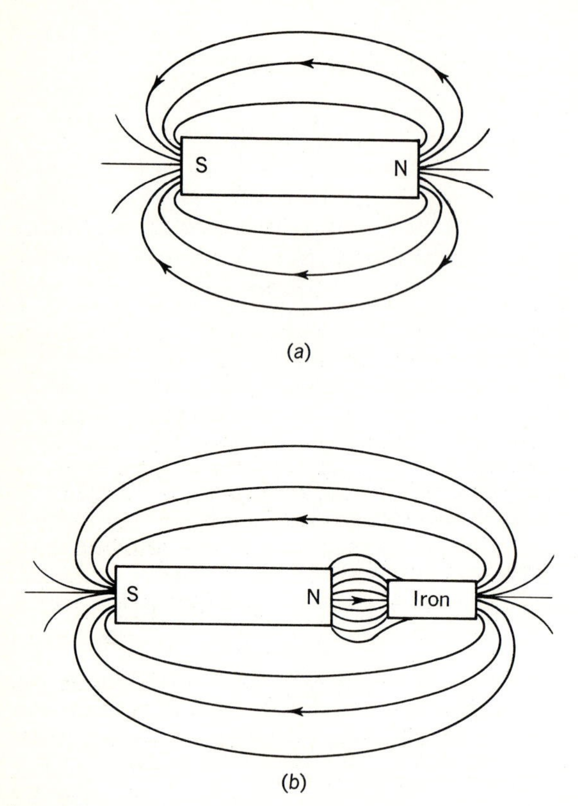

Fig. 8-8 (*a*) The field around a permanent magnet. (*b*) A piece of iron in the field is pulled toward the nearest magnetic pole.

move apart. If the N and the S poles of two magnets are placed near each other, as in Fig. 8-9*b*, the adjacent external field lines will be dissimilar, since they are opposite in direction, and will link together to form long loops. These long, continuous lines try to contract and pull the two magnets together.

In electrostatics, unlike poles (+ and −) attract, whereas like (+ and +, or − and −) poles repel. The same is true in magnetism: unlike poles (N and S) attract, and like poles (N and N, or S and S) repel each other. An example of magnetic attraction is the navigator's magnetic compass. The earth has a magnetic core, with its *magnetic south* pole under the surface of the ground near the *geographic north* pole, and the magnetic north pole below the surface near the geographic south pole, as illustrated in Fig. 8-10 (a somewhat confusing set of circumstances). The compass needle is a long, thin permanent magnet. Its so-called "north-seeking," or pointing, end has a *north magnetic* polarity. When it is free to move on its central bearing point, the needle aligns its magnetic field with the magnetic field of the earth, with

---

ANSWERS TO CHECK-UP QUIZ 8-2

1. (S) (S) **2.** (A bar magnet) **3.** (Same) (Smaller) (Same) **4.** (No) (Yes) **5.** (Halved)(Halved) **6.** (μ)

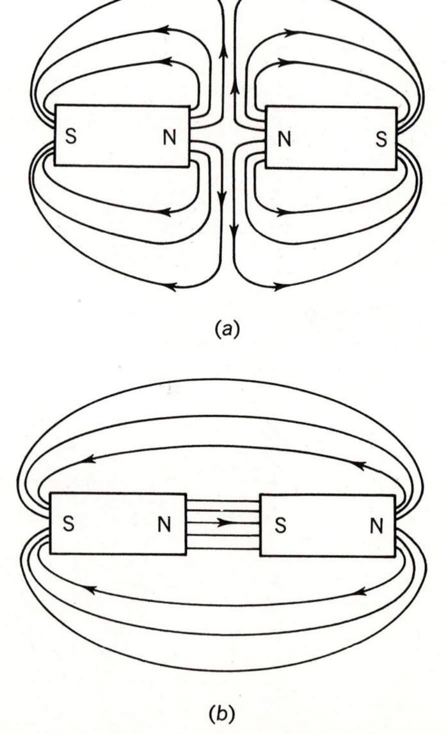

Fig. 8-9 (*a*) Fields from like poles repel, tending to push the magnets apart. (*b*) Unlike poles interconnect their lines and pull together as the lines try to contract.

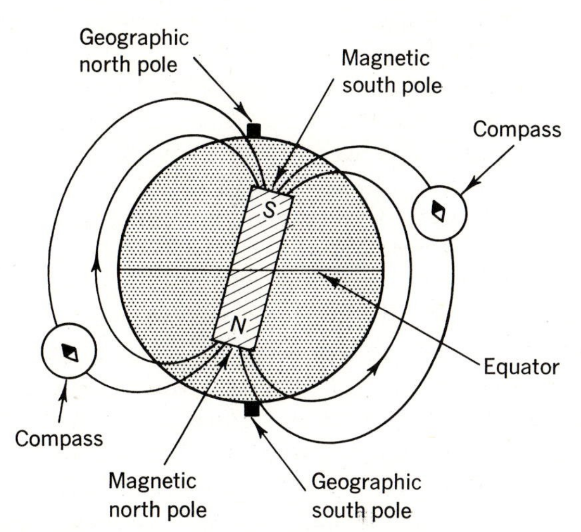

Fig. 8-10 The magnetic poles of the earth are just the opposite of the geographic poles. Therefore the north end of a compass points to the geographic north pole.

its north end pointing toward the earth's magnetic south pole, commonly termed the *geographic north pole* (because opposite magnetic poles attract).

Quiz 8-3. Test your understanding. Answer these check-up questions.

1. Is the retentivity of air low, high, or zero? __________ What is the relative retentivity of iron? __________ Steel? __________

Questions 2 to 9 refer to Fig. 8-11.

2. When the coil-circuit switch of the "relay" is closed, what value of current flows in the coil? __________ How much magnetomotive-force is developed in the iron core? __________
3. To make the relay armature move down against the spring tension, of what materials might it be made? __________
4. If the spring were tightened, would more, less, or the same battery voltage be needed to operate the armature? __________
5. If the stop were pulled upward a little, would more, less, or the same battery voltage be needed to operate the armature? __________
6. With the battery circuit switch open, is there an electrical connection between *A* and *B*? __________ With the switch closed? __________
7. Why might the relay not operate satisfactorily if the armature and core were steel? __________
8. What two adjustments could be made to make this type of relay more "sensitive"? __________ __________
9. Would the top of the core and the armature have the same or opposite polarities when the relay is energized with dc? __________ If the battery leads were reversed? __________
10. Does the magnetic north end of a compass needle point to the geographic north pole or south pole? __________
11. Name the three ferromagnetic elements. __________ __________ __________
12. In Fig. 8-7*b*, by induction which end of the iron bar is north? __________

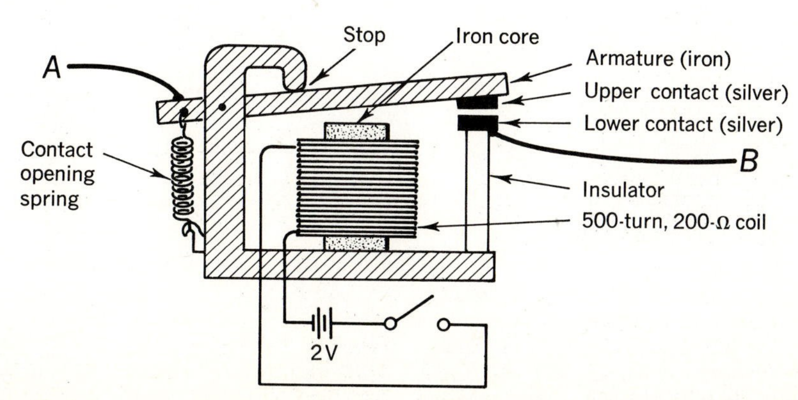

Fig. 8-11 Electromagnetic relay used to control high currents and/or high voltages operating from a low-current or low-voltage source.

CHAPTER 8 TEST • MAGNETISM

1. Under what condition is an electron considered to have a magnetic field around it?
2. When a wire is grasped with the left hand, with the fingers in the direction of the magnetic lines of force, what does the thumb indicate?
3. If a piece of wire 5 m long has 1 A flowing in it, is more magnetism developed if the wire is coiled? More magnetomotive-force? More field intensity? More flux density?
4. Which magnetic pole has lines of force emanating from it?
5. How many lines of force are there in a weber? In 570 μWb? What is the symbol for flux?
6. If the fingers of the left hand are placed around a coil in the direction in which current is flowing, what will the thumb indicate?
7. What is the unit of measurement of mmf?
8. What is the field intensity of a 12-cm-long 100-turn coil with 0.5 A flowing in it? What is the mmf?
9. What are the two units of measurement of flux density? What is the symbol for flux density?
10. Does the word *permeability* refer to the ability of a material to become a permanent magnet?
11. What are the three natural elements that are known to be ferromagnetic?
12. Which of the materials mentioned in the text has the greatest permeability?
13. What core shape will produce the greatest concentration of lines of force between its poles?
14. What type of electromagnet core has absolutely no retentivity? Which of the materials mentioned in the text has the greatest retentivity?
15. When molecules of a ferromagnetic material align themselves into little fully magnetized islands, what are these islands called?
16. When all the molecules of a ferromagnetic material are aligned by some magnetizing force, in what state is the substance said to be?
17. What effect do two parallel magnetic lines of force have on each other if they have the same directivity? If they have the opposite directivity?
18. What effect does the S pole of one magnet have on the S pole of another nearby magnet?
19. What is the magnetic polarity of the "north-pointing" end of a magnetic compass needle?
20. Of what kind of metal must a relay arm be made? Should the metal have a high or low retentivity?
21. Does the direction of the current flowing in the relay coil of Fig. 8-11 have any effect on the operation of the relay?
22. What opens the contacts of an electromagnetic relay?

ANSWERS TO CHECK-UP QUIZ 8-3

1. (Zero) (Low) (High) **2.** (0.01 A) (5 amp-turns) **3.** (Iron) (nickel) (cobalt) **4.** (More) **5.** (More) **6.** (No) (Yes) **7.** (Both would become permanently magnetized and might not pull apart) **8.** (Less spring tension) (Push stop down a bit) **9.** (Opposite) (Still opposite) **10.** (North) **11.** (Iron, nickel, cobalt) **12.** (Left)

9 ALTERNATING CURRENT

CHAPTER OBJECTIVES. To develop a usable understanding of these electrical vocabulary terms: direct current, dc, amplitude, pure dc, pulsating dc, varying dc, square wave, sawtooth wave, alternating current, ac, direction vectors, left-hand generator rule, sine wave, sinusoidal ac, alternator, slip-rings, brushes, frequency, hertz, Hz, cps, kHz, MHz, peak value, effective value, average value, peak-to-peak value. Also, to compute 30° and 60° sine values, and to convert from peak to effective to average to peak-to-peak values.

9-1 DIRECT CURRENT

Up to this point all electrical voltages were developed by batteries or other sources producing electrical pressure in the same direction at all times. The circuit of Fig. 9-1*a* is one in which a unidirectional current would flow in the load when the switch is closed. From examination of the diagram it can be seen that a potential difference of 10 V is developed across the load resistor, and a current of 0.5 A will flow through it when the switch is closed.

The graph of Fig. 9-1*b* indicates that at some instant in time the switch is closed. At this instant current flow is established, and it continues to flow at the same amplitude, magnitude, or strength as long as the switch remains closed. This unidirectional unvarying current is commonly termed *direct current*, or *dc*. The source is said to be producing a *pure dc voltage* (a voltage that does not vary).

The graph of Fig. 9-1*c* indicates that the switch is being opened and closed periodically, causing pulses of current to flow through the resistor. This form of current is

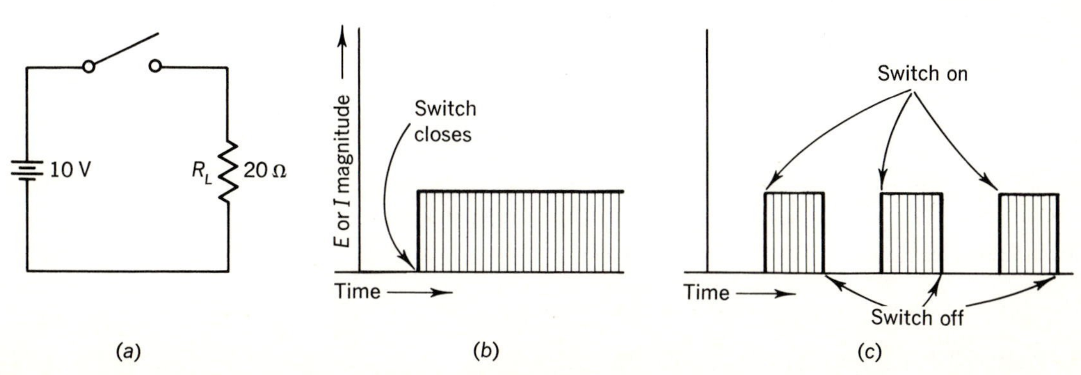

Fig. 9-1 (*a*) A simple dc circuit with switch open. (*b*) When switch closes, a constant-amplitude current starts to flow instantly. (*c*) Pulsating dc can be produced by alternately opening and closing the switch.

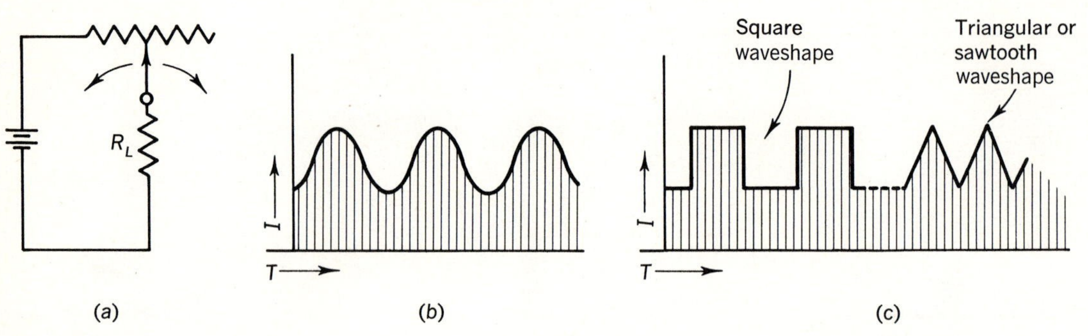

Fig. 9-2 (*a*) A circuit in which current can be varied. (*b*) Graph of varying dc. (*c*) Graph of square-wave and sawtooth-wave varying dc.

known as *pulsating dc*. The source produces a constant-amplitude voltage, but the making and breaking of the circuit by the switch feeds a pulsating, constant-amplitude voltage to the resistor, resulting in the pulsating dc flow in the circuit. There are many electric and electronic circuits in which pulsating dc may be the only type of current flowing.

In the circuit of Fig. 9-2*a*, a rheostat, or variable resistance, is the controlling device. As long as the rheostat arm does not run off the end of the resistance wire, which would open the circuit, there will be current flowing in the load resistor, R_L. If the rheostat arm is constantly moved back and forth, the resistance of the circuit will continually change, causing a changing-strength current flow through the load, as in Fig. 9-2*b*. This current can be termed *varying dc*. One difference between pulsating dc and varying dc is that varying dc never drops to zero. Varying dc is the main type of current flowing in many electronic circuits.

It can be reasoned that by proper manipulation and turning of the rheostat control it would be possible to produce a *square-wave* shape for the graphed current, or perhaps a *sawtooth-wave dc*, as shown in Fig. 9-2*c*. In electronics both square-wave and sawtooth-wave dc, as well as many odd-shaped dc voltages and currents, are quite common.

9-2 ALTERNATING CURRENT

When a current flows through a load in first one direction and then the other, it is said to be an *alternating current*, or *ac*. An alternating current can be developed in a load by using two batteries and a potentiometer connected as shown in Fig. 9-3*a*. When the variable arm is at the midpoint of the potentiometer, a condition of balance occurs as far as the load R_L is concerned. The load is connected between the center of a 20-V source and the center of a 20-V voltage-drop. Therefore no difference in potential exists across R_L, and as a result the current through the load is zero. However, if the variable arm of the potentiometer is moved up to the + end of the potentiometer, R_L is then directly across the upper 10-V battery, and current flows through it from *left to right*. As the arm is moved down toward the − end of the potentiometer, it moves through the point of 0 V at the midpoint and then on to the bottom. At this last point R_L is directly across the lower battery, and current flows through it from *right to left*.

Figure 9-3*b* indicates the current as starting at zero, increasing in a positive

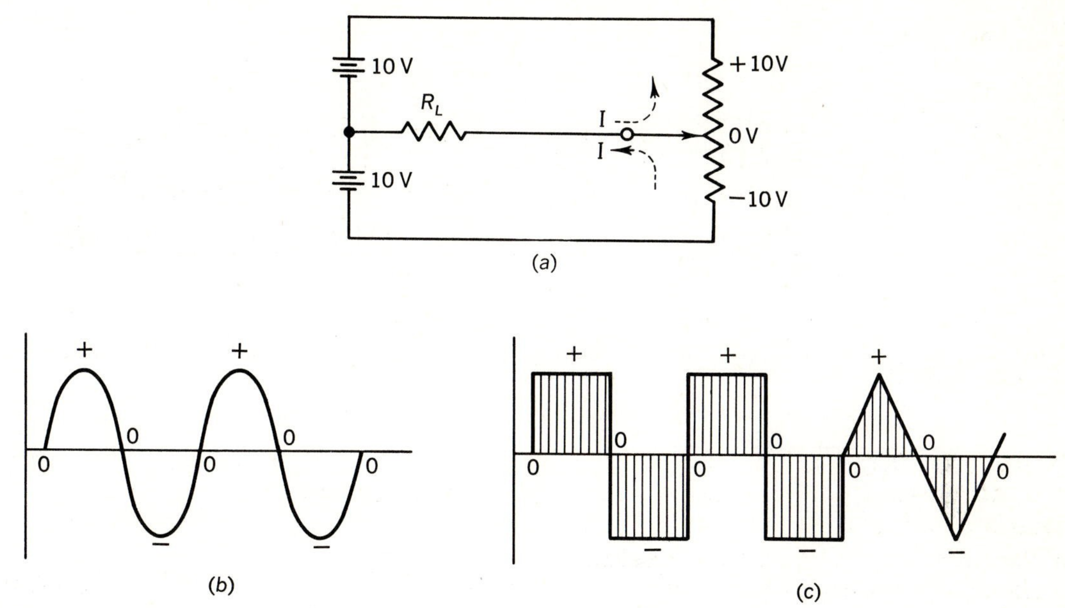

Fig. 9-3 (*a*) A circuit capable of producing an alternating-current flow in the load. (*b*) Graph of ac. (*c*) Graph of square-wave and sawtooth-wave ac.

direction to a maximum, decreasing to zero, and then increasing to the maximum negative value, returning to zero, and so on. The current through the load is alternating in direction and is called alternating current, or ac.

If the arm of the potentiometer remained for a time at first the + and then the − ends, the current might graph as a square-wave ac, as in Fig. 9-3*c*. If the arm moved at a different rate, it might also be possible to produce a sawtooth-wave ac, as shown.

9-3 INDUCING EMF

One of the important concepts in electricity is the generation of an emf when magnetic lines of force are cut by a conductor. In Fig. 9-4*a* a wire (small circle with a dot in it) is indicated as moving upward through lines of force. An emf will be induced in it with such a polarity as to move electrons along the wire in a direction outward from the page.

The direction of the emf induced in a conductor can be determined by the *left-hand generator rule* (Fig. 9-4*b*). The thumb and the first and second fingers of the left hand are held at right angles to each other, as shown. The first finger is made to point in the direction of the lines of force (magnetic lines have a direction from north to south). The thumb is pointed in the direction of the wire's motion. The second finger will then indicate the direction of the induced emf. Check this rule with the indicated emf direction in the illustration by orienting your left fingers and thumb according to the rule.

The dot in the wire in the illustration is an indicator representing the point of a vector arrow moving toward the reader. A wire with an emf or current-vector arrow moving away from the reader would have a cross (×) in it, representing the

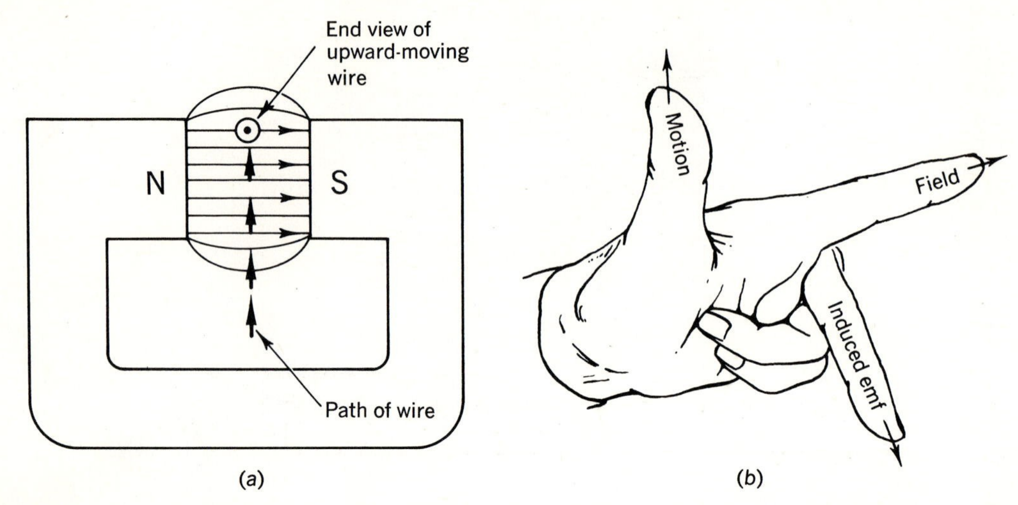

Fig. 9-4 (*a*) The direction of induced emf or current in a conductor moving through a magnetic field can be determined by (*b*) the left-hand generator rule (right-hand with conventional current).

end view of the "tailfeathers" of a vector arrow.

Quiz 9-1. Test your understanding. Answer these check-up questions.

1. A wire lying in an east-west direction is lifted upward from the ground. What magnetic lines of force is it cutting? ________ By the left-hand generator rule, toward which end of the wire would electrons flow? ________
2. If the wire in Fig. 9-4 were moved downward, in what direction would an emf be induced in it? ________
3. If the wire in Fig. 9-4 were moved up and down at a rate of 10 times per second, what kind of emf would be induced in it? ________
4. If the wire in Fig. 9-4 were moved from the N pole to the S pole along a line of force, what would be induced in it? ________
5. If you move an upright finger from left to right across the N pole of a permanent magnet that is facing you, would the polarity induced in the end of your finger be positive or negative? ________ Why would you feel no electric shock? ________
6. A current stops and starts 10 times per second. What kind of a current is this? ________
7. A current changes in value from 3 to 4 A at a rate of 10 times per second. What kind of current is this? ________
8. What is the variable resistor called that has a movable arm and a connection at only one end of the resistance? ________ A movable arm and a connection at both ends of the resistance? ________
9. What is another name that might be given to a dc that varied in a triangular-wave fashion? ________
10. In the left-hand generator rule, what does the first finger indicate? ________ The thumb? ________ The second finger? ________

9-4 GENERATING AN AC

The underlying principle of an ac generator, properly called an *alternator*, is represented in Fig. 9-5*a*. In this circuit a current is flowing in the field coil in such a direction as to produce the magnetic poles shown. A wire is inserted in the magnetic field between the two poles and is made to rotate counterclockwise, from positions 1 to 2, to 3, to 4, etc., around to position 1 again. According to the generator rule, the polarity of the voltage induced in the wire as it is rotated into these 12 different positions is indicated by the crosses and dots. In both the first and the seventh positions the wire is moving par-

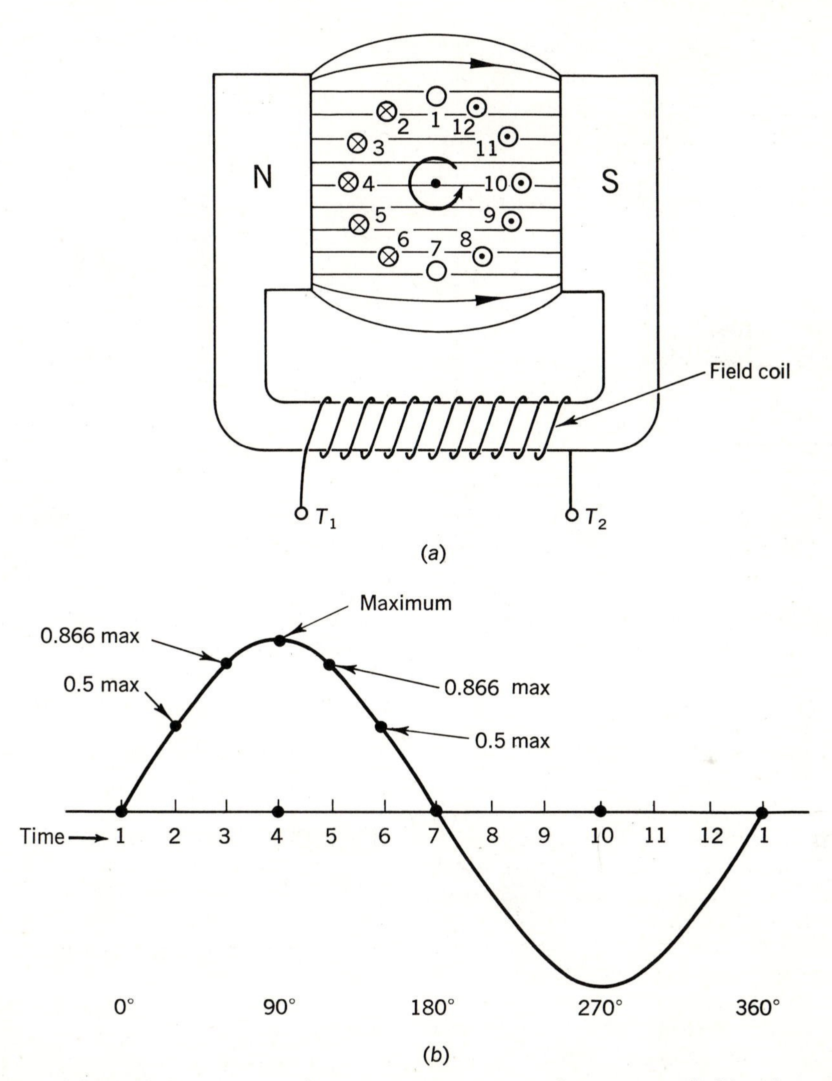

Fig. 9-5 (*a*) Rotation of a conductor in a circle in a constant-strength magnetic field induces (*b*) a sinusoidal changing voltage in the conductor.

allel to the lines of force, and no emf is induced, since no lines of force are being *cut*.

It will be assumed that the wire is rotating at a constant velocity. Thus equal time intervals may be marked along the time line in the graph (Fig. 9-5*b*), indicating positions of the wire as it rotates. At positions 1 and 7 the induced voltages are zero. If the flux density is constant between the pole faces, it will be found that at positions 4 and 10 (90° and 270° of rotation) the induced emf's are at maximum value, because the wire is cutting the greatest number of lines of force in a given time interval. However, these two emf's are of opposite polarity.

As the wire passes position 2 (30° from the zero-voltage or starting point), the emf induced in the wire will be exactly half, or

0.5, of the maximum voltage, as indicated in Fig. 9-5*b*. As the wire passes point 3 (60° from the zero-voltage point), the induced emf will be 0.866 of the maximum value. As the wire passes position 4, maximum voltage is induced. At position 5 (which is again 60° from a zero point) the emf value is again 0.866 of maximum. At position 6 the emf is once more 0.5 of maximum. At point 7 it is again zero. Now, as the wire moves *upward* across the same lines of force, the polarity of the induced voltage reverses, as shown by both the dot in the wire and the graphing of the emf on the lower side of the time line. At point 8 the emf is 0.5 of maximum, at point 9 it is 0.866 of maximum, and at point 10 it is maximum. The emf is 0.866 of maximum at point 11 and 0.5 of maximum at point 12.

In one complete cycle the wire has traveled 360°, and the induced emf has risen to a maximum in one direction, gone back to zero, risen to a maximum in the other direction, and returned to zero. This represents one cycle of alternating current or voltage.

Those familiar with trigonometry will recognize the values 0.5 for 30°, 0.866 for 60°, 1.0 for 90°, etc., as the *sine* values for these angles. Since this is an explanation of a theoretically perfect induced ac emf cycle, it is only natural to call the perfect ac waveshape a *sine wave*. This is the reason we speak of sine-wave ac, or sinusoidal ac waveforms. Whenever ac is mentioned, it is assumed to be sine-wave ac unless stated otherwise (such as square-wave or sawtooth-wave ac).

In actual alternators, of course, a single wire is not rotated. A many-turn coil is wound and rotated. Each turn of the coil acts as two wires in series. Thus the voltage induced in a 100-turn coil would be equivalent to that produced in 200 single wires. Then, to shorten the air gap and reduce flux leakage between the two pole faces, a soft-iron core, called a *rotor*, is used inside the coil. The rotor and its coil are driven around by some outside motor force. To allow continuous connection to the turning coil, it is necessary to mount two brass *slip-rings* on, but insulated from, the shaft of the rotor. The ends of the coil are terminated at the slip-rings. Contact is made to the slip-rings by two brass or carbon *brushes* held against the slip-rings by springs. In this way, as the coil rotates, the generated ac is led out of the machine through the two stationary brushes. This is discussed in Chap. 21.

ANSWERS TO CHECK-UP QUIZ 9-1

1. (Earth's) (East) **2.** (Away from you) **3.** (Ac) **4.** (Nothing) **5.** (Negative) (Insufficient emf generated) **6.** (Pulsating dc) **7.** (Varying dc) **8.** (Rheostat) (Potentiometer) **9.** (Sawtooth-wave dc) **10.** (Flux direction) (Conductor motion) (Induced emf direction)

9-5 FREQUENCY OF AC

The number of complete cycles of ac produced in 1 s is known as the *frequency* of the ac. For example, most public utilities in the United States have alternators that turn fast enough to produce 60 complete ac cycles each second. The ac is said to have a frequency of 60 cycles per second. In the past it was often said (incorrectly) that a "60-cycle frequency" was used. To conform to international terminology, the term *hertz* (Hz) is now recommended in place of cycles per second. We now say the power frequency is 60 Hz, although the longer term "60 cycles per second" (cps) is still correct. Many countries use 50 Hz as their power-line frequency.

Since the term *kilo* means "1000 times," a *kilohertz*, or kHz, is a frequency of 1000

Hz (or cps). Radio broadcast stations generate ac of about 1000 kHz (previously stated as 1000 kilocycles per second, or kc), and with an antenna can radiate energy at this frequency. When radio receivers are tuned to this frequency, these broadcast programs become audible.

The term *mega* means "1 million times." It is possible to express 1 million cps as 1 million Hz, as 1000 kHz, or as 1 MHz (previously 1 megacycle per second, or 1 Mc).

A question that often arises is why low-frequency ac is always used to transport electrical power from one place to another but high frequencies are employed in radio broadcasting. The answer is simple. The magnetic fields that build up and collapse back into the conductor during each half-cycle have time to get back to the wires provided low enough frequencies are used. When high-frequency ac flows in a wire, some of the expanding magnetic energy is radiated outward away from the wire and is lost into space when it cannot return to the wire as the cycle reverses. High-frequency ac would be uneconomical for power transmission, but it produces the desired radiation effect necessary for radio and television broadcasting.

Quiz 9-2. Test your understanding. Answer these check-up questions.

1. Express the following frequencies in kilohertz: 2340 Hz ________ 14,500 Hz ________ 276 Hz ________ 8,490,000 Hz ________ 60 MHz ________
2. Express the following frequencies in megahertz: 6,350,000 Hz ________ 28,400,000 Hz ________ 458,000 Hz ________ 63,000 kHz ________ 450 kHz ________ 60 Hz ________
3. Express the following frequencies in hertz: 2350 kHz ________ 68 MHz ________ 85 kHz ________ 9.30 kHz ________ 54 kHz ________ 2.7 MHz ________
4. How many alternations are there in one cycle? ________ How many in 2.1 kHz? ________
5. If the maximum induced emf in an alternator is 100 V, what is the emf value when the rotating coil is at an angle from the 0 emf point of: 30°? ________ 60°? ________ 90°? ________ 150°? ________ 180°? ________ 240°? ________ 270°? ________ 330°? ________ 360°? ________
6. If the maximum induced emf in an alternator is 30 V, what is the emf value at 30°? ________ 120°? ________ 180°? ________ 210°? ________
7. If the maximum induced emf in an alternator is 220 V, what is the emf value at 30°? ________ 120°? ________ 180°? ________ 270°? ________
8. In Fig. 9-5*a*, according to the magnetic polarity of the poles shown, is current flowing into or out of terminal T_2? ________
9. In Fig. 9-5*a*, if the polarity of the magnetic field were reversed and the direction of wire rotation were also reversed, in what way, if any, would this affect the ac wave produced? ________
10. In Fig. 9-5*a*, if the field coil current value were reversed, would the shape of the output-emf wave change? ________ Would the maximum amplitude of the output emf change? ________
11. What are the names of the two sets of parts used to lead the generated emf out of an alternator? ________
12. What is the name of the part in an alternator on which the rotating coil is wound? ________
13. What is the name of the perfect ac wave? ________
14. How might leakage lines of force be reduced in a practical alternator? ________

9-6 PEAK, RMS, AND AVERAGE VALUES

An ac cycle starts at 0 V, rises to a peak, and falls to zero again. If the maximum value is 10 V, for example, the ac may be called a 10-V *peak ac*. It is more likely to be called a 7-V *effective* or *rms* ac because a 10-V peak ac will do the same job of heating a resistor that a 7-V dc will. So, effectively, 10-V peak ac equals 7-V dc workwise.

The effective value can be determined mathematically. One half-cycle is exactly

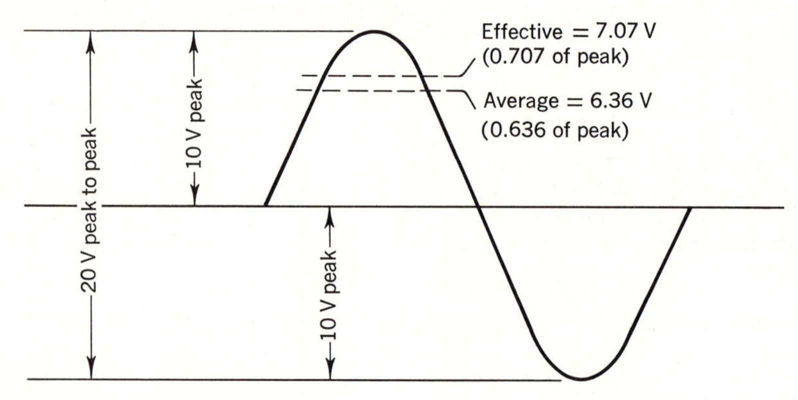

Fig. 9-6 Relative values of a sine-wave ac, showing peak, peak-to-peak effective, and average.

the same as the next half-cycle. If voltages are measured at each degree along the time line for one half-cycle, the result is 180 voltage values. If these are all added and then divided by 180, the resultant voltage is the *average value* of the cycle. With sine-wave ac, this always computes to be 0.636 of the peak or maximum value. The average value is not used very often, but it will be referred to in the discussion of meters.

If each of the 180 voltage values is squared and the square root is taken of the average of the squared values, the resultant, the square root of the mean of the squared values, is called the *root-mean-square* (rms) or *effective value*. With sine-wave ac, this is always 0.707 of the peak or maximum. When ac is discussed, it is assumed to be the rms or effective value (the heating value) unless stated otherwise.

In some cases the *peak-to-peak* (or p-p) *value* is used. For example, a 10-V peak ac has a 20-V peak-to-peak value (see Fig. 9-6).

Conversion factors from one ac value to another are as follows:

Peak × 0.707 = rms	Rms × 1.414 = peak
Peak × 0.636 = avg	Avg × 1.57 = peak
Rms × 0.9 = avg	Avg × 1.11 = rms

ANSWERS TO CHECK-UP QUIZ 9-2

1. (2.34 kHz) (14.5 kHz) (0.276 kHz) (8490 kHz) (60,000 kHz) **2.** (6.35 MHz) (28.4 MHz) (0.458 MHz) (63 MHz) (0.45 MHz) (0.00006 MHz) **3.** (2,350,000 Hz) (68 million Hz) (85,000 Hz) (9300 Hz) (54,000 Hz) (2,700,000 Hz) **4.** (Two) (4200) **5.** (50 V) (86.6 V) (100 V) (50 V) (0 V) (86.6 V) (100 V) (50 V) (0 V) **6.** (15 V) (26 V) (0 V) (15 V) **7.** (110 V) (191 V) (0 V) (220 V) **8.** (Into) **9.** (None) **10.** (No) (No) **11.** (Slip-rings, brushes) **12.** (Rotor) **13.** (Sine wave) **14.** (By reducing magnetic gaps)

Quiz 9-3. Test your understanding. Answer these check-up questions.

1. When a power company sells "120-V ac" energy, is this the peak, average, or rms value? RMS
2. What is the peak value of 120-V effective? ________
3. What is the peak value of 3.5-A effective? ________
4. A 60-V peak ac has what rms value? ________ What average? ________
5. An average ac of 16 A has what effective value? ________ What maximum value? ________
6. A 440-V rms ac has what average value? ________ What is the peak-to-peak value? ________

CHAPTER 9 TEST • ALTERNATING CURRENT

1. Identify the following currents: Changes in amplitude periodically but does not reach zero at any time. Changes in amplitude but reaches zero periodically. Continues to flow in the same direction with constant amplitude. Rises to a sharp peak and then falls to a sharp minimum before rising again.
2. What is the general term for current that flows in one direction and then reverses its direction of flow?
3. What are the three basic waveshapes discussed in the text for dc and ac?
4. In the left-hand generator rule, which finger indicates the direction of the induced emf? Which should be pointed in the direction of the magnetic field? Which should be pointed in the conductor motion direction?
5. What does a small dot in the center of the cross section of a wire indicate? What does a cross indicate?
6. In Fig. 9-7, which pole, P_1 or P_2, would have a north polarity? If the wire is rotated according to the arrows 7 times in 0.5 s, what value of emf, in hertz, is generated? When the wire is moving downward, should it be shown with a dot or a cross in it?
7. In Fig. 9-7, at which lettered point(s) will the induced voltage be at a maximum? At zero?
8. If the angular velocity (rotational speed) of the wire in Fig. 9-7 were constant, what waveshape would be induced in the wire?
9. What is a variable resistor called that has a movable arm and a connection at only one end of the resistance?

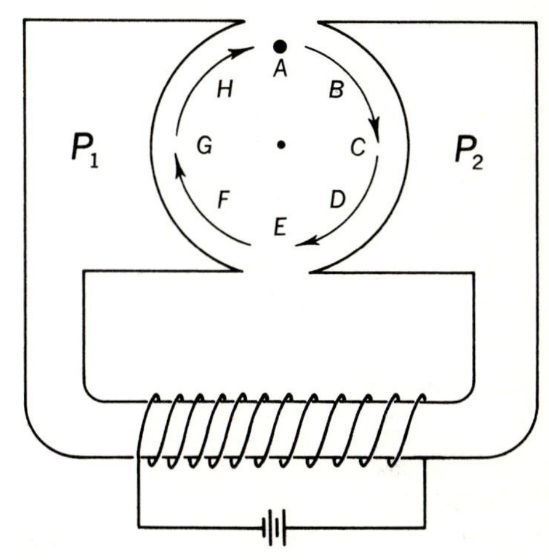

Fig. 9-7 Circuit for Chap. 9 test.

10. What is a term that is synonymous with cycles per second?
11. If an ac has a peak of 150 V, what is the effective value? Average value? Rms value? Peak-to-peak value?
12. If an ac has an rms value of 100 V, what is the peak value? Average value? Peak-to-peak value?
13. What does rms mean?
14. If the peak, effective, average, or peak-to-peak value of a 100-V ac is not stated, which should be assumed? Which would be considered equivalent to the dc value for heating applica-ions?

ANSWERS TO CHECK-UP QUIZ 9-3

1. (Rms) **2.** (170 V) **3.** (4.95 A) **4.** (42.4 V) (38.2 V) **5.** (17.8 A) (25.1 A) **6.** (396 V) (1244 V)

10 INDUCTANCE AND TRANSFORMERS

CHAPTER OBJECTIVE. To develop a usable understanding of these electrical vocabulary terms: inductance, transformer primary, transformer secondary, counter-emf, transient, reactance, henry, H, mH, μH, inductor, shorted-turn effect, transformer, tertiary, toroidal core, transformer voltage ratio, current ratio, power ratio, hysteretic loss, eddy current loss, lamination, fuse. Also, to solve simple transformer turns ratio, voltage, and current problems.

10-1 DEVELOPING COUNTER-EMF

Wires and coils have been discussed in previous chapters with respect to their ability to conduct current and develop magnetic fields. Another important property is the ability of a wire or a coil to oppose any *change* in current. This property is known as *inductance*.

The ability of a conductor to oppose current flow due to its resistance has been discussed. If a coil has 5 Ω of resistance, it limits the value of current that can flow when a 10-V dc or effective ac source is applied across it to a value of 2 A. Also, a moving field crossing a conductor induces an emf in the conductor. When a varying dc is flowing through a wire, the magnetic field expands and contracts with the variations of current amplitude. When the current is at maximum, the magnetic field extends out farther than when the current is at minimum. In the case of ac, the magnetic field lines not only expand and contract on each alternation but also reverse their direction on successive alternations. This results in an expansion, a contraction, an expansion in the opposite direction, and a second contraction for each ac cycle.

In Fig. 10-1*a*, a dc source, a resistor, a switch, and the wire marked "primary" form the basic circuit. Near the primary portion of the circuit lies a second conductor, marked "secondary." Since there is no complete circuit, there is nothing happening. But what happens when the switch is closed?

At the instant the switch makes contact (Fig. 10-1*b*), current starts to flow in the primary, and magnetic lines of force start to expand outward from the primary wire. As they move outward, some of the lines of force cross the secondary wire and induce an emf in the secondary. It is possible to determine the direction of this emf by the left-hand generator rule. According to the illustration, the directivity given to the lines of force as they cross the secondary wire is into the page. The *relative* direction of crossing is as if the secondary wire were moving to the *left* across stationary lines of force. The direction of the induced emf would be upward in the secondary wire. (Check this by Fig. 9-4.) This

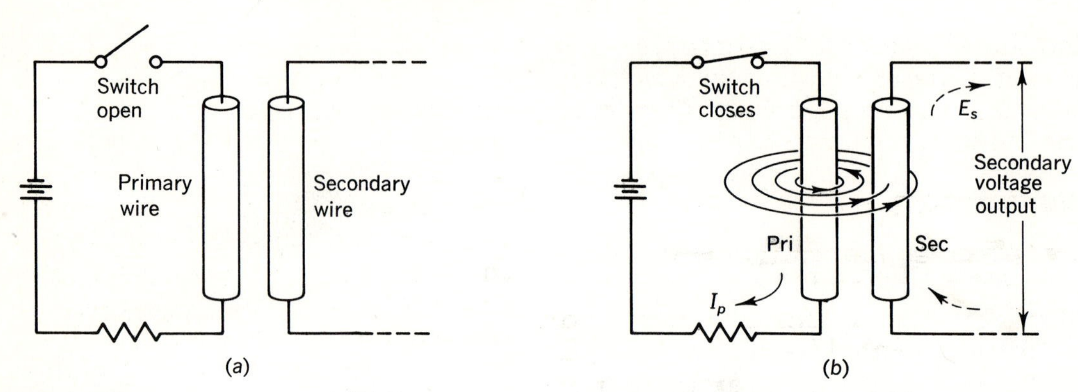

Fig. 10-1 (*a*) Primary circuit open. (*b*) Primary circuit closes, and magnetic lines of force start expanding from the primary.

produces a secondary-circuit emf with negative polarity at the top and positive polarity at the bottom.

One of the important ideas being developed is that the direction of the induced emf is opposite to that of the current in the primary that produced it. (Current increasing in a downward direction in the primary produces an upward emf in the secondary.)

To carry this idea one step further, if there were only the primary wire, as current started to flow, magnetic lines of force would expand outward across the *surface* of the primary wire and would induce in this primary wire a reverse-direction emf, called a *back-emf*, or a *counter-emf*. The counter-emf is counter to the source emf and results in a reduction of current flow in the primary wire as long as the lines are expanding. When the lines have expanded to a maximum extent, they will no longer be moving, and the counter-emf value will drop to zero. Since the counter-emf occurs only while the magnetic lines of force are in the process of expanding, it may be called a *transient* effect. Another important result of this counter-emf is the slowing of the buildup of current in the wire. When the switch is closed, the current cannot reach maximum instantaneously because of the counter-emf.

If the primary current is stopped, the magnetic field will immediately start to collapse. Now, the lines of force are moving in the opposite direction across both the primary and secondary conductors. As a result, an opposite-direction emf is induced in each of the conductors. This induced emf will be in the same direction in the primary as the source emf and will try to keep the primary current flowing. In the secondary wire, Fig. 10-1*b*, the induced emf will be downward.

The effects of counter-emf and the induced emf are always in a direction contrary to what the primary circuit would like to do. If the current tries to increase, the counter-emf opposes it. If the current tries to decrease, the induced emf opposes the decrease by developing an emf in the direction of the current. From this comes the definition:

The property of a circuit to oppose any change in current is called *inductance*.

When current in the wire is alternating, it is continually changing and therefore continually producing an induced emf opposing the source emf. This opposition to ac is called *reactance*. It acts like resistance in that it lowers the value of the current in the circuit.

10-2 INDUCTANCE

How effective the counter-emf developed in a conductor will be in a circuit is measured in units of *henrys*, abbreviated as H. The letter symbol for inductance is L. For example, the value of inductance in a certain circuit might be $L = 10$ H.

A circuit that will produce an average of 1 V of counter-emf when an average current change of 1 A/s is occurring in the circuit has an inductance of 1 H. In electronic circuits this is a relatively large value. Although there are many cases in which several henrys of inductance may occur, it is probably more usual to have values in *millihenrys* (thousandths of a henry), abbreviated as mH, or in *microhenrys* (millionths of a henry), abbreviated as μH.

Actually, a short piece of wire has a very small value of inductance. The same wire made into a coil will have much more inductance. Each turn of the three-turn coil in Fig. 10-2 is shown with only a single magnetic line of force. According to the left-hand rule, the current must be flowing downward in the turns where the magnetic fields are shown. The field of loop 1 will induce a counter-emf in itself as current increases, but as the field expands outward, it will also induce a counter-emf in turns 2 and 3. All the counter-emf's will be in the same direction and will be additive. Therefore the current in this one turn will induce about 3 times the counter-emf that it would be able to induce if the wire were straight. Turn 2 will also induce an emf into the wires adjacent to it, for a total of 3 times the emf that it would induce in itself alone. Turn 3 will induce 3 times as much emf as it would induce in itself alone. Thus for 3 turns, the total induced emf will be increased by a factor of 9. When a conductor is made into a coil, then, its inductance will be roughly proportional to the *square* of the number of turns ($3^2 = 9$).

Fig. 10-2 As lines of force expand from one turn, they cut across other adjacent turns, inducing a counter-emf in all turns.

Coiling a wire increases not only the inductance value but also the ability of the wire to react against an alternating emf that is trying to produce current flow in it. The faster the emf alternates—that is, the higher its frequency—the greater the reactance to current flow produced by the same coil. If the inductance value is increased, the counter-emf value increases, resulting in a slower buildup and fall-off of current flow through a coil or *inductor* across a source of ac emf. The greater the inductance value, the greater its reactance against alternating or varying currents.

Increasing the number of turns in a coil is one method of increasing the total inductance. If still more inductance is desired, an iron core can be added. Using iron with a permeability (μ) of several thousand times that of air will result in an inductor with several thousand times as much inductance. Thus inductance is directly proportional to the permeability of the iron core used.

The two usual types of inductors are iron-core and air-core. An inductor wound on an iron core is termed an iron-core inductor. If it is wound on any insulating material having a permeability approximately that of air ($\mu = 1$), the coil is called an air-core coil.

Quiz 10-1. Test your understanding. Answer these check-up questions.

1. A wire lies in a north-south direction. In what direction would an emf be induced in it if current began to increase in a N direction? ________ If the current decreased in value? ________ If current increased in a S direction? ________
2. What is it that inductance opposes? ________
3. Will the magnetic field produced by a current reach maximum at the instant the current reaches maximum or some time later? ________
4. If the current in a primary wire is varying dc, is the induced emf in the secondary ac or varying dc? ________
5. If the current in a primary wire is ac, is the emf induced in the secondary ac or varying dc? ________
6. If a switch in a circuit containing a dc source and an inductor is closed and then opened, an emf will be developed in each case. Which of these emf's will have the greater amplitude? ________ Why? ________
7. When magnetic lines of force collapse, do they reverse in direction? ________
8. What is the name for the resistancelike opposition that results when ac flows through a coil? ________
9. What important effect does inductance have on the buildup of magnetic fields when current begins to flow? ________
10. What is the basic unit of measurement of inductance? ________ What are two subunits of measurement? ________ ________
11. If a 5-cm-long coil of 10 turns were rewound to have 20 turns, what increase in inductance would result? ________ What increase would result if it were rewound with 100 turns? ________
12. What is the effect on the inductance of a coil if the ac used is increased in frequency? ________
13. What is the effect on the reactance of a coil if ac frequency increases? ________
14. Give at least two effects of using iron as the core of an inductor. ________ ________

10-3 THE SHORTED-TURN EFFECT

An important point relating to inductance is the *shorted-turn effect.* A shorted turn around an inductor is shown in Fig. 10-3*a*. Since the shorted turn encompasses the coil, when a current starts to flow in the coil, magnetic lines of force expand outward, cut across the shorted turn, and develop a counter-emf in it (counter to the current in the primary). The counter-emf in the shorted turn produces a heavy flow of current in the direction of the counter-

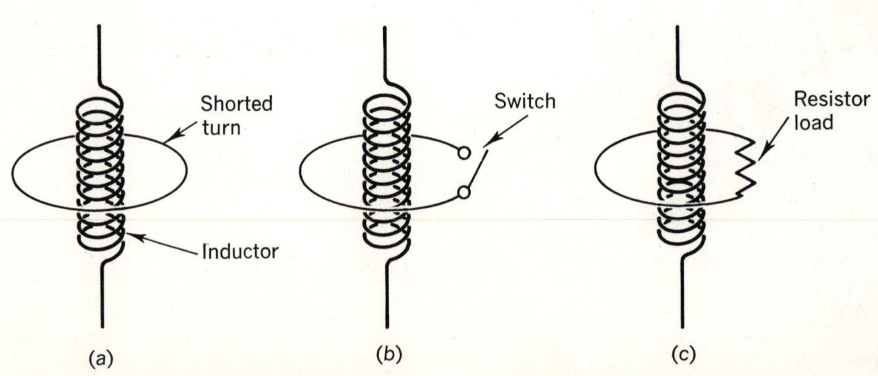

Fig. 10-3 A coil, or inductor, with (*a*) a shorted turn around it, (*b*) a shorted turn opened by a switch, and (*c*) a shorted turn with a resistor in series with it.

emf. This current also develops a magnetic field around itself which expands in all directions from the shorted turn. Part of this secondary magnetic field "expands" inward, crossing the inductor turns and inducing what may be termed a *counter-counter-emf* in this coil. The counter-counter-emf will have the same direction as the source emf and will partially cancel the counter-emf in the primary coil, speeding the buildup of the magnetic field around the coil. If the shorted turn increases the speed of the magnetic field buildup of the coil, then the shorted turn must decrease the inductance value of the coil.

In Fig. 10-3*b*, if the shorted-turn switch is open, the shorted turn will have no effect on the inductance of the coil. If the coil were across an ac source, its inductance (counter-emf) would be opposing the current, and only a relatively small current would flow. If a resistor were used instead of the switch (Fig. 10-3*c*), the shorted-turn current and effect would be present, but to a lesser degree, resulting in a less than normal inductance value of the coil, but not as low as if the shorted turn had no resistance. This is the basic theory of operation of a *transformer*. With nothing connected across the secondary (an open turn around the primary), the primary inductance is high, and little primary ac flows. When a resistance load is connected in series with the secondary, current flows in the secondary and its load resistor, which cancels some of the primary inductance. This allows greater primary current to flow because of the lessening of the opposition in the primary. The increased primary current is the source of energy for the secondary current.

10-4 TRANSFORMERS

The previous descriptions of the functioning of primary and secondary wires and the effect of a partially shorted turn around a primary coil represent the basic ideas of the theory of transformers. There are many different types of transformers. Some have iron cores and others have air cores, but the underlying principles are the same.

The transformer in Fig. 10-4*a* has an iron core with many primary turns but few secondary turns. The symbol for an iron-core transformer is shown in Fig. 10-4*b*. The symbol for the air-core transformer, Fig. 10-4*c*, differs only in that it has no straight lines to indicate an iron core.

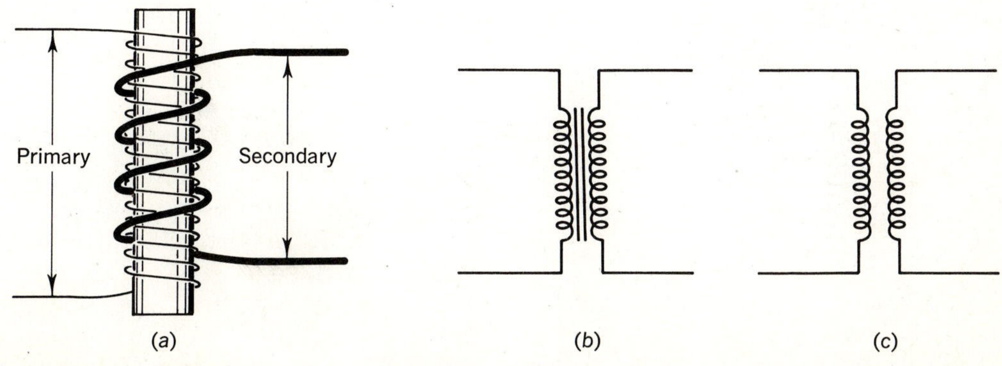

Fig. 10-4 (*a*) A basic transformer wound on an iron core. (*b*) Symbol for an iron-core transformer. (*c*) Symbol for an air-core transformer.

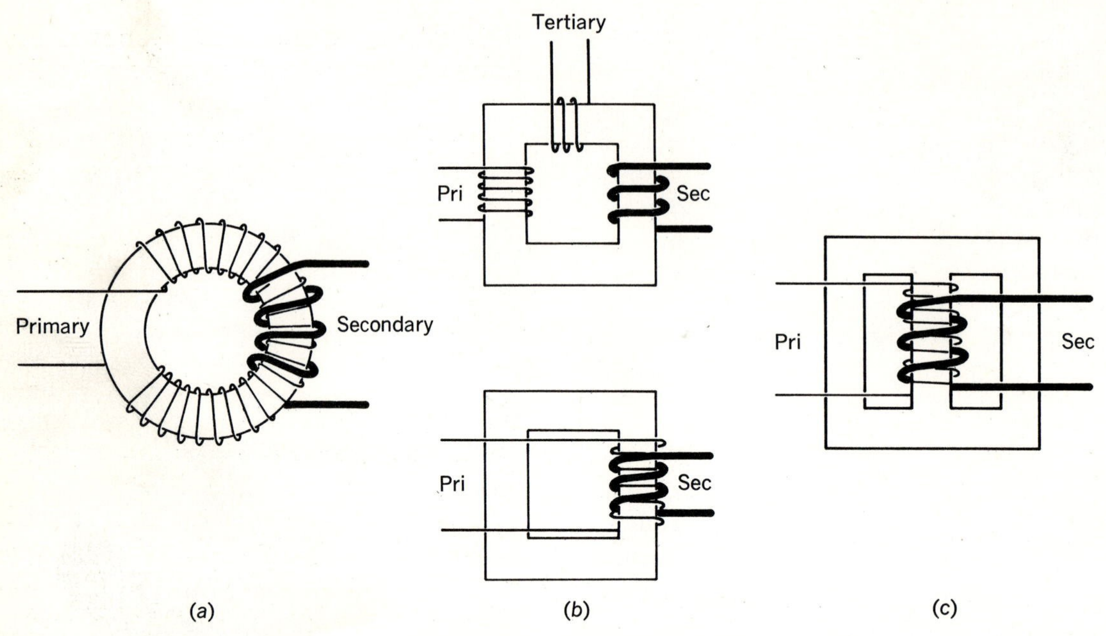

Fig. 10-5 (*a*) Toroidal transformer essentials. (*b*) Two methods of winding single-window or modified-toroid transformers. (*c*) One method of constructing a double-window core transformer.

In any transformer either winding could be the primary. The other winding would then be the secondary. If there are three windings on a transformer, the third is the *tertiary*.

While iron-core transformers could be manufactured in the shape shown in Fig. 10-4*a*, the leakage lines of force from pole to pole would be excessive. Instead, practical iron-core transformers may be wound on a toroidal (ring-shaped) form (Fig. 10-5), on a modified toroid with square corners, or on a form having double windows. Toroidal transformers are more efficient, since they have no sharp corners at which leakage lines can develop, but are more costly to construct. The double-window type is probably the most frequently seen. It is common practice to surround transformers with a high-permeability iron case to prevent lines of force from expanding outward from it.

Transformers are used to change (transform) one ac voltage value to another. For example, a transformer is constructed with 400 primary and 400 secondary turns. If the primary winding has an ac source emf of 100 V connected across it, each primary turn will have a 100/400, or a 0.25-V, voltage-drop across it. Each secondary turn will also have 0.25 V induced in it. Therefore the 400-turn secondary would produce a 100-V ac output. A 2000-turn secondary would produce 2000(0.25), or 500 V output. A 40-turn secondary would produce a 10-V output. Thus the voltage ratio of a closely wound transformer will be directly proportional to the

---

ANSWERS TO CHECK-UP QUIZ 10-1

1. (S) (N) (N) **2.** (*Change* in *I*) **3.** (Simultaneously) **4.** (Ac) **5.** (Ac) **6.** (Switch open) (Open because the field has no opposition and collapses quickly) **7.** (No) **8.** (Reactance) **9.** (Slows the buildup) **10.** (H) (mH, μH) **11.** (4*L*) (100*L*) **12.** (None) **13.** (Increases) **14.** (Increases *L*, increases buildup time of *I*, smaller coil size needed, fewer turns necessary)

turns ratio of primary to secondary windings. A turns ratio of 3:1 can be used either to step up the voltage by a factor of 3 or to step down the voltage to ⅓ the primary voltage. In formula form,

$$\frac{E_{pri}}{E_{sec}} = \frac{T_{pri}}{T_{sec}}$$

where E_{pri} = primary voltage
E_{sec} = secondary voltage
T_{pri} = primary turns
T_{sec} = secondary turns

As an example, a transformer has a 400-turn primary, a 1200-turn secondary, and an 80-turn tertiary. What are the output voltages if the input to the primary is 120 V ac? The ratio of primary to secondary turns is 400/1200, or 1:3. The secondary voltage will be 3(120), or 360 V. The ratio of primary to tertiary is 400/80, or 5:1. The tertiary voltage will be 120/5, or 24 V.

10-5 POWER AND CURRENT RATIOS IN TRANSFORMERS

Large iron-core transformers are almost 100% efficient. Under this condition the power delivered to the primary will be equal to the power delivered to the load connected across the secondary (Fig. 10-6).

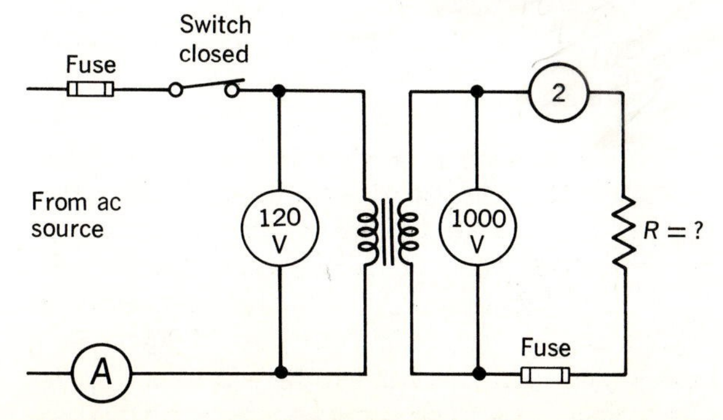

Fig. 10-6 Resistor load coupled to a power line through a step-up transformer.

According to the values shown in the diagram, the secondary voltage is 1000 V and the load is drawing 2 A. The load resistance must have a value of $R = E/I$, or 1000/2, or 500 Ω. The power delivered to and dissipated by the load must be $P = EI$, or 1000(2), or 2000 W. If the load is drawing 2 kW from the secondary, then the primary must be accepting at least 2 kW from the power lines. With the power-line voltage of 120 V as indicated, the primary current can be found by simply rearranging the basic power formula $P = EI$ to read $I = P/E$, or 2000/120, or 16.7 A.

Actually, 100% efficiency is not possible. There will always be some losses due to resistance in the primary and secondary wires, *hysteresis* losses produced in the core as the iron undergoes alternate magnetizations, and a core loss due to *eddy currents*. Because of these losses, the primary power will always be greater than the secondary-load power value.

Eddy currents are produced in iron cores of transformers when magnetic fields expand and contract through the iron, which is a fairly good conductor. The emf induced in the core sets up whirlpoollike currents in the iron that heat it. This loss can be reduced by *laminating* the core iron (slicing it into thin sheets) and painting an insulation on each lamination. The eddy currents developed in the thin insulated iron sheets will be very much smaller, and the power loss produced by these currents will be reduced.

Note the *fuse* shown in the primary and secondary circuits of Fig. 10-6. The current rating for fuses is usually 1½ to 2 times the maximum expected current in the circuit. If an excessive current should flow in the circuit (from a short circuit or overloading), the fuse wire will melt, opening the circuit and stopping all operation. This protects the transformer, the load, and the power lines.

The current ratio of a transformer is usually considered to be the inverse of the voltage ratio. That is, if the voltage step-up ratio is 1:3, the current ratio will be a step-down of 3:1. If the voltage is stepped up in the secondary from 120 to 360 V, then when 1 A is drawn by a load across the secondary, the primary will draw 3 A from the power line. In formula form,

$$\frac{I_{\text{pri}}}{I_{\text{sec}}} = \frac{T_{\text{sec}}}{T_{\text{pri}}}$$

When there is no load on the secondary (secondary current is zero) this formula does not apply.

Air-core transformers are not used in power-line applications but are used in high-frequency ac circuits, particularly in tuned circuits, which will be discussed later.

Quiz 10-2. Test your understanding. Answer these check-up questions.

1. What effect will a shorted turn around an inductor have on the inductance value? __________ Would the shorted turn be more effective around the top or the middle of the coil? __________
2. When a shorted turn is around an inductor, what does it induce in the inductor? __________
3. Would a shorted turn with low resistance heat? __________
4. Is a shorted turn essentially the same as a secondary winding with the load short-circuited? __________ What would a transformer do if its many-turn secondary winding were short-circuited? __________
5. How does the symbol for an iron-core transformer differ from an air-core? __________
6. What is the name given to a second secondary winding in a transformer? __________
7. A secondary winding delivers 600 W at 300 V to a load, and a second secondary delivers 50 W at 25 V to another load. What power does the primary draw from the power line? __________
8. Would a transformer operate if a dc source were connected across its primary? __________ Why? __________
9. Would a transformer operate if a varying dc were connected across the primary? __________ Why? __________
10. Would a transformer operate if a pulsating dc source were connected across the primary? __________ Why? __________
11. What are transformer cores said to be when made of thin, insulation-covered sheets of permeable iron? __________
12. What shape of core is most efficient for a transformer? __________ Does such a core have any external field? __________
13. A 4:1 step-up transformer has 520 turns on its primary. How many turns does it have on the secondary? __________ If a 24-V ac is fed to its primary, what is the output voltage? __________
14. Which would represent the heavier load on a transformer secondary, a 100-Ω resistor or a 300-Ω resistor? __________ Why? __________
15. Does a transformer with no load on the secondary draw any current from the power line? __________ Why? __________
16. A 7:1 step-up transformer has what primary-to-secondary-voltage ratio? __________ Current ratio? __________ Power ratio? __________
17. A 12:1 step-down transformer has 450 turns on the primary and 120 V across it. What is the number of secondary turns? __________ A 2.5-Ω load will draw how much secondary current? __________ How much primary current? __________
18. Would a transformer operate well with a high-retentivity iron core? __________ Why? __________
19. In Fig. 10-6, what current rating would the primary fuse have? __________ The secondary fuse? __________
20. What kind of transformer core has no eddy current or hysteresis losses? __________

CHAPTER 10 TEST • INDUCTANCE AND TRANSFORMERS

1. What is the property of a circuit called that opposes current flow through it? What is the property of a circuit called that opposes any change in the current in it?
2. A horizontal wire has a current flowing in it toward the right. In what direction is a voltage induced into it when the current is increasing? Decreasing? At maximum?
3. If an alternating current is flowing in an inductor, what is the current value when the magnetic field is at maximum? Zero? When the induced voltage is at maximum? Zero?
4. What kind of emf is developed in the secondary of a transformer if the primary current is varying dc? Pure dc? Ac?
5. What is the unit of measurement of inductance?
6. How is the inductance related to the number of turns in a coil?
7. How can the inductance of a coil of wire be increased other than by increasing the number of turns?
8. What is the resistancelike opposition effect called that is always produced in an inductor when a varying-amplitude current is flowing in it?
9. What effect does a shorted turn around a coil have on the inductance value of the coil?
10. What is the proper term for the third winding on a transformer?
11. A transformer has 510 turns in the primary and 1530 turns in the secondary. What is the voltage ratio? The expected current ratio? The power ratio?
12. What are the names of the two types of core losses in iron-core transformers? Which, if any, of these losses occur in air-core transformers?
13. What is the name of the electrical component that is made to burn out when an excessive current flows through it?
14. What is the proper name for a ring-shaped core of an inductor or transformer?
15. Under what condition is the current ratio of a transformer ($I_{pri}/I_{sec} = T_{sec}/T_{pri}$) greatly inaccurate?
16. A highly efficient 4:1 step-down transformer has a 100-V ac across the primary and a 50-Ω load across the secondary. What is the value of E_{sec}? I_{sec}? The load power dissipation? The primary power input? I_{pri}?

ANSWERS TO CHECK-UP QUIZ 10-2

1. (Decreases it) (Middle) **2.** (Counter-counter-emf) **3.** (Yes) **4.** (Yes) (Burn out the primary or secondary, heat, blow fuses) **5.** (Straight lines between windings) **6.** (Tertiary) **7.** (650 W plus transformer losses) **8.** (No) (No moving magnetic field) **9.** (Yes) (Expansion and contraction of magnetic field) **10.** (Yes) (Same as question 9) **11.** (Laminated) **12.** (Toroid) (No) **13.** (2080) (96 V) **14.** (100 Ω) (Draws more current) **15.** (Yes) (Some core-magnetizing *I*) **16.** (1:7) (7:1) (1:1) **17.** (37.5) (4 A) (0.333 A) **18.** (No) (Hysteretic loss heats core, overloads primary, and burns primary out) **19.** (20 to 25 A) (3 to 4 A) **20.** (Air)

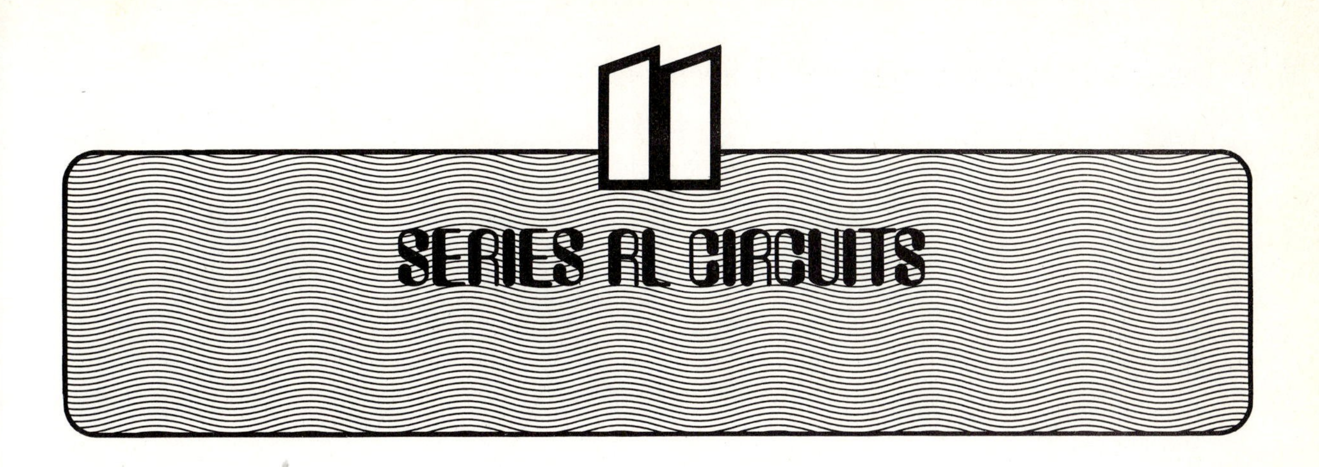

CHAPTER OBJECTIVE. To develop a usable understanding of these electrical vocabulary terms: linear, internal resistance, *RL* time constant, T_c curve, in phase, lagging current, DPDT switch, polarity-reversing circuit, inductive reactance, sinusoidal current, leading voltage, purely inductive circuit, purely resistive circuit, resistive component, reactor, graph of *Z*-*X*-*R*, phase angle, θ (theta), reactive energy, true power, apparent power, voltamperes, power factor, PF. Also, to graph *R*, *Z*, *X*, and phase angle of simple series *RL* circuit, and to compute the impedance of resistor-reactor circuits.

11-1 RESISTANCE-INDUCTANCE WITH TIME

Chapter 10 pointed out that one of the important electrical concepts is the ability of an inductor (coil) to slow the buildup of a current starting to flow through it. To find out how much it slows a current increase (or decrease), it is necessary to consider both the *resistance* of the coil and the circuit.

Figure 11-1*a* shows a circuit consisting of a source of emf, a switch, and a noninductive resistor, all in series. Current starts to flow at a maximum value just as soon as the switch is closed. The current value is dependent on only the *E* and *R* of the circuit, and it may be computed simply by Ohm's law. A graph of current flow from the time the switch is closed, T_0, is shown above the circuit.

Figure 11-1*b* represents another circuit, this one having a source, switch, and coil, all in series, with no resistance in the circuit. Theoretically, when the switch is closed, the current will start to flow and a magnetic field will start expanding from the coil. As the field expands, counter-emf is developed in the coil. With no resistance in the circuit, the counter-emf will almost equal the source emf, and a very small current starts to flow. The current will increase in equal proportions as time passes. If the *I* value after 1 s is 1 A, it will be 2 A after 2 s, 5 A after 5 s, and so on to infinity. The current increase in a pure inductance is "linear"—that is, it graphs as a straight line. Of course, a circuit or a source with zero resistance is not possible.

The circuit shown in Fig. 11-1*c* illustrates the actual resistances that would be found in a practical circuit. R_w represents the resistance of the connecting wires. The two resistances marked R_i represent the internal resistances of the source and of the coil. When the switch is closed and current starts to flow, counter-emf is developed in the coil, and current starts to increase linearly, as in a purely inductive circuit. However, because of the increasing voltage-drops across the circuit resistances, the current cannot continue to build up in a linear manner but increases

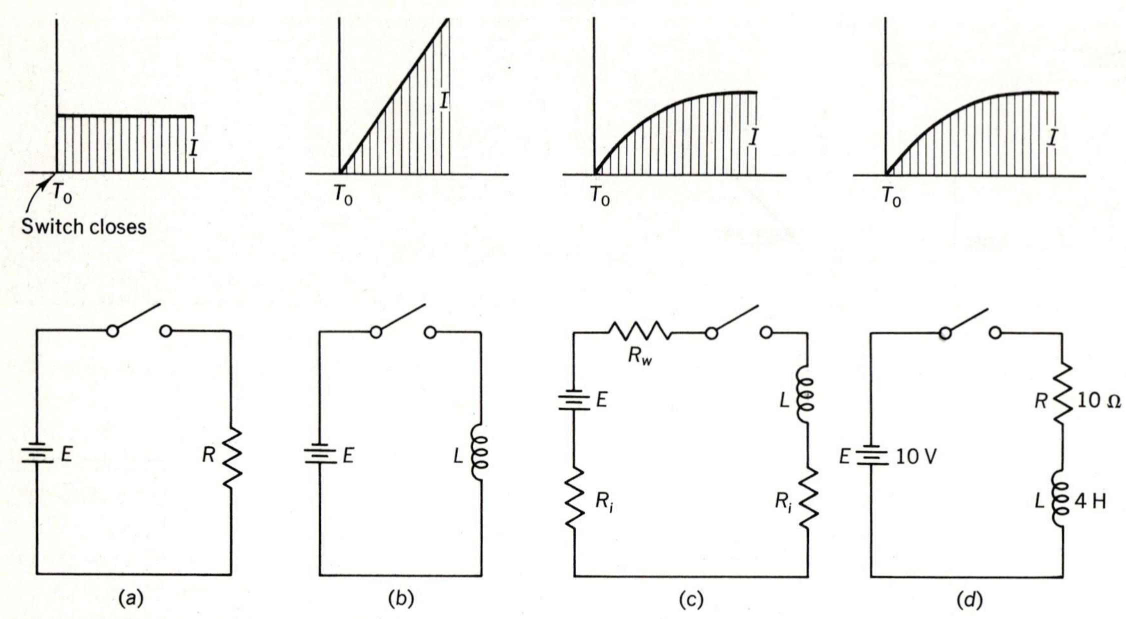

Fig. 11-1 (*a*) Current rise is immediate in a purely resistive circuit after T_0. (*b*) I rises linearly in a purely inductive circuit after T_0. (*c*) Practical internal and wire resistances in what might be thought of as a purely reactive circuit. (*d*) An *RL* circuit.

in a curved form as indicated. With the increasing current, the magnetic field builds outward, eventually reaching a maximum value as determined by the circuit resistances, at which time the counter-emf has decreased to zero. When the current in the circuit reaches maximum, its value can then be calculated by Ohm's law, $I = E/R$. According to the values shown in Fig. 11-1*d*, this will be 1 A.

When ohms and henrys are used as the units of measurement, the formula for determining the time for current to build up is

$$T_c = \frac{L}{R}$$

where T_c = time, in seconds
L = inductance, in henrys
R = resistance, in ohms

The T_c computed by this formula is the time required for the current to build up to approximately 63% of the maximum value. This is called the *RL time-constant* formula.

Consider the *RL* circuit (Fig. 11-1*d*) during the transient state, as the field is expanding. If the inductance is 4 H and the resistance is 10 Ω, the time required for the current to reach 63% of its eventual maximum value of 1 A is $T_c = L/R$, or 4/10, or 0.4 s. After the first time interval the circuit still contains only 4 H and 10 Ω, so the time-constant formula is still applicable. In the next 0.4 s the current will build up 63% more toward the 1-A maximum. In the third time interval the current will build up 63% more, and so on for an infinite number of time intervals. However, after five time intervals the current will reach more than 99% of the absolute maximum value. (This is a satisfactory maximum for general use, since many meters in electric circuits may have an accuracy of only 2%.) It is customary to consider the time required for an *RL* circuit to allow current

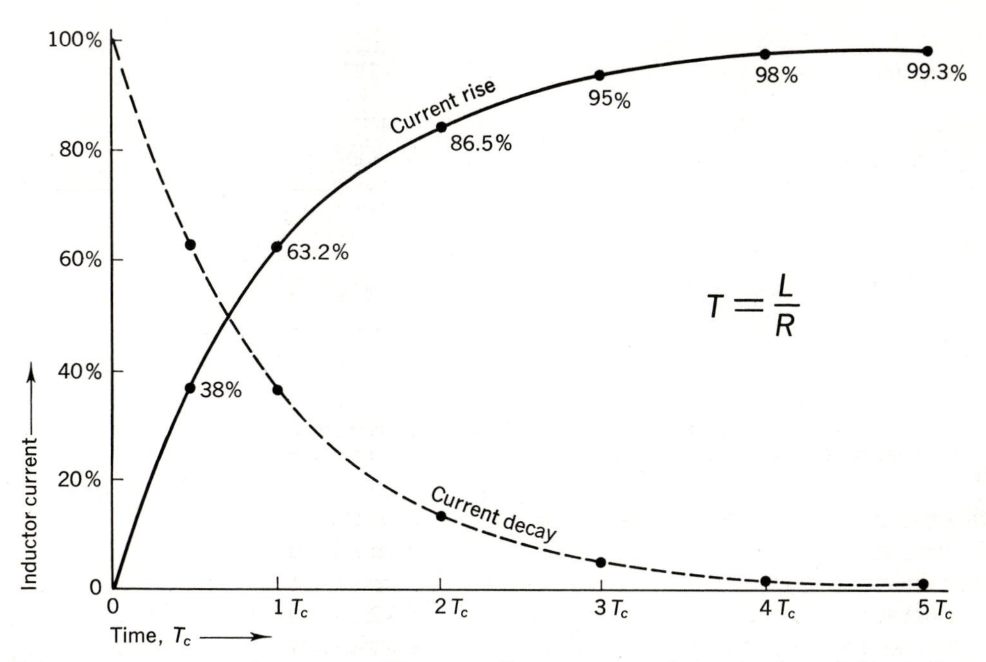

Fig. 11-2 Time-constant curves for a series *RL* circuit. Current rise is shown by the solid line. Current decay, or voltage-drop across coil, is shown by the dashed line.

to rise to essentially maximum to be

$$T_{\max} = \frac{5L}{R}$$

The graph in Fig. 11-2 plots the rise in current against time, and is called a time-constant curve.

In Fig. 11-3, with the switch in the *A* position, after a period of five time constants the field around the coil will be at maximum. If the switch is then thrown to position *B*, the magnetic field no longer has a source to maintain it, and the field begins to collapse. As the lines of force move back into the coil, they induce a voltage in the turns in the same direction as the original source voltage. This induced emf produces a current in the coil and resistor in the same direction as the source current, which attempts to maintain the magnetic field.

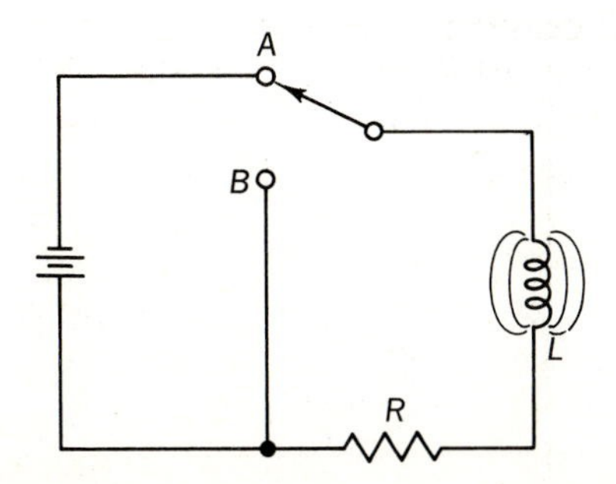

Fig. 11-3 When the switch is moved to *A*, the current and magnetic field in the coil builds up to maximum in five time-constant periods. When the switch is moved to *B*, the current and field decays to zero in five time-constant periods.

Because of the resistance in the circuit, the field decreases, and the current flow in the coil drops to zero at a rate determined by the time-constant formula, $T_c = L/R$. In a time interval equal to $T_c = 5L/R$ the current would be essentially at zero. The decay of current in a discharging *RL* circuit drops off in a mirror image (dashed curve, Fig. 11-2) of the rise-of-current curve.

Quiz 11-1. Test your understanding. Answer these check-up questions.

1. What is the property of a circuit to oppose the flow of current called? ________
2. What is the property of a circuit to oppose any change in current called? ________
3. How long would it take for the current in a purely inductive circuit to build up to maximum after being connected across a 10-V source? ________ How much power would be dissipated in the inductor? ________
4. How long will it take a current to build up to 63% of maximum after being connected across a 20-V source if the circuit L is 5 H and the R is 150 Ω? ________ If L is 5 mH? ________
5. How long will it take for a 3-A current to decay to essentially zero in a circuit having 2 H and 4000 Ω of R if the source is suddenly short-circuited? ________
6. From the curve of Fig. 11-2, approximately how many time-constant periods would it take for current to build up to a value of 50% of maximum? ________ 80% of maximum? ________ 90% of maximum? ________
7. When the circuit in Fig. 11-1*d* arrives at a steady-state condition, how much power is being dissipated in the R? ________ In the L? ________
8. Figure 11-2 indicates how the current increases in an RL circuit after the circuit is completed. What E-drop in the circuit would be represented by the dashed line? ________ The solid line? ________

11-2 CURRENT LAGS VOLTAGE

In a purely resistive circuit, an increasing voltage produces a current that builds up at exactly the same time. The current is in step, or, more correctly, *in phase*, with the voltage of the source.

In an inductive circuit in which the source emf is pure dc, the current flow in the circuit cannot attain maximum until after five time-constant periods. The current is said to *lag* the voltage. If the voltage in the circuit drops to some lower value, the induced emf in the coil prevents the current in the coil from decreasing to the new value immediately.

If the switch in Fig. 11-4*a* is alternated back and forth between points A and B, remaining at each point for five time-constant periods, the voltages that would be applied across the RL circuit are indicated by the dashed lines in Fig. 11-4*b*. The current would build up and decay according to the time-constant curves, shown by the solid curve lines. Besides causing the current to lag behind any voltage change in a circuit, an inductance in a circuit may modify the current waveshape. In this case a square-wave pulsating dc voltage produces sawtooth-wave varying dc.

A square-wave alternating emf can be developed by using a double-pole double-throw (DPDT) switch in a polarity-reversing circuit (Fig. 11-5*a*). If the switch is held for five time-constant periods at the A contacts and five at the B contacts, the source voltage and current flowing in the RL circuit will be as shown in Fig. 11-5*b*. A

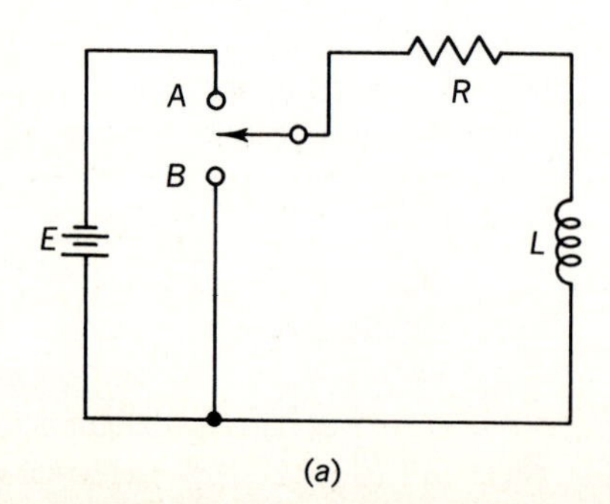

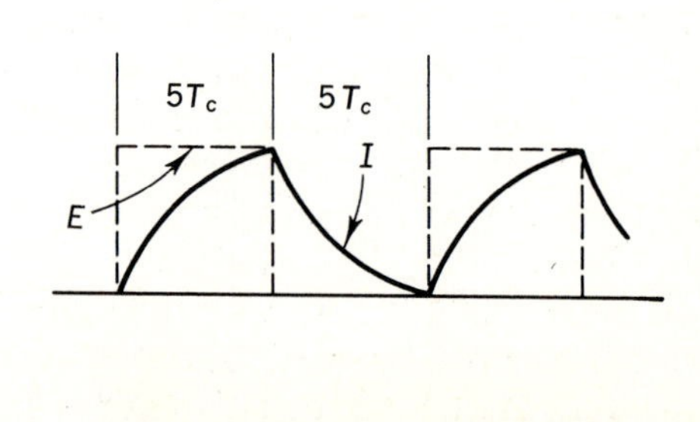

Fig. 11-4 (*a*) RL circuit to develop a square-wave pulsating-dc emf. (*b*) I and E in a 5-T_c circuit.

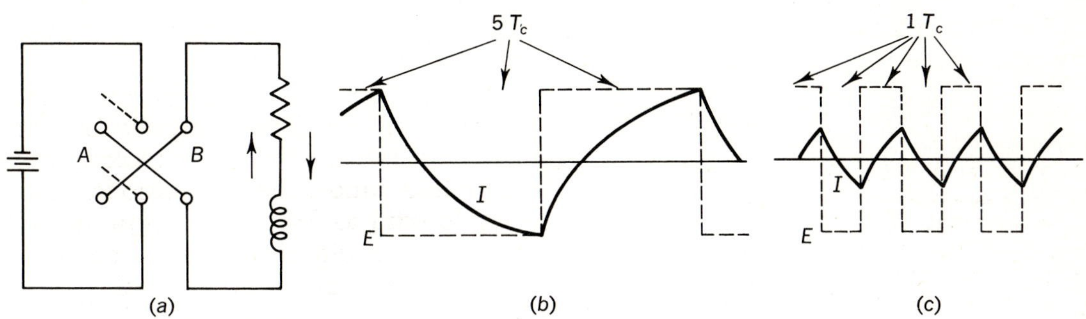

Fig. 11-5 (*a*) Circuit to produce square-wave ac. At *A*, *I* flows up in load. At *B*, current flows down. (*b*) Ac *E* and *I* if 5 T_c is used. (c) Ac *E* and *I* if 1 T_c is used.

square-wave ac voltage produces a saw-tooth-wave ac if there is inductance in the circuit.

If the speed of switching is increased, a higher frequency of square-wave ac is produced. If the contacts are held for only one time-constant period, the current can build up to only 63% of the amplitude that it would attain in five T_c periods. As the frequency of the ac increases still more, the current peaks will be reduced still further. This ability of an inductance to react against the buildup of current is called *inductive reactance*. The higher the frequency, the more inductive reactance a coil will have, and the more the current flow will be limited.

The ac in Fig. 11-5 is square wave. If the emf cycle is made more sine-wave-shaped, the shape of the current cycle becomes less saw-toothed and more sinusoidal. When a perfect-sine-wave emf is applied to an *RL* circuit, a perfectly sinusoidal current results. If there is no resistance in the circuit, but only inductance, the phase of the voltage and current will be exactly 90°, with the current lagging (Fig. 11-6). If the frequency of the ac is increased, the phase remains at 90°, although the current amplitude decreases because of the increased reactance.

With a purely resistive circuit there is zero phase difference between applied emf and resulting circuit current. With a purely inductive circuit the *E* and *I* are 90° out of phase, with *I* lagging (or *E* leading). When a circuit has both resistance and inductance in series, the current will lag by some angle between 0° and 90°.

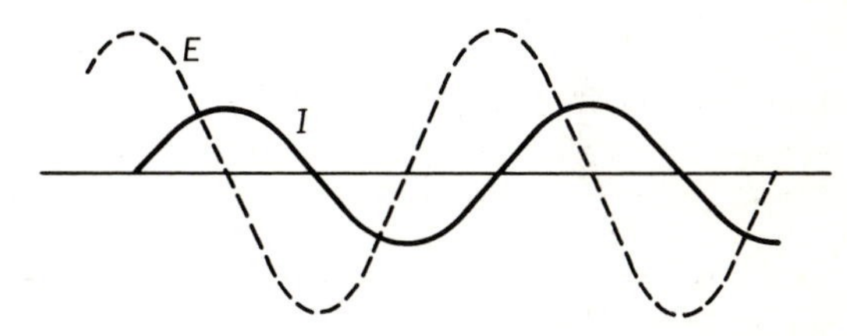

Fig. 11-6 When the source emf is sine-wave-shaped, the current is also sinusoidal. (*I* lags *E* by 90° here.)

Quiz 11-2. Test your understanding. Answer these check-up questions.

1. What name is given to the waveshape developed by an *RL* current-increasing curve followed by an *RL* current-decaying curve if the source is producing square-wave pulsating dc? __________ Sine-wave ac? __________
2. In an *RL* circuit, is the circuit current in phase with the source voltage, lagging the source voltage, or leading the source voltage? __________
3. An *RL* circuit is across a sinusoidal ac source. If the frequency of the ac is decreased, does the phase angle of *E* and *I* increase, decrease, or remain the same? __________ Does the current value increase, decrease, or remain the same? __________
4. What would a SPST switch be? __________ A 3PDT? __________
5. What is the resistancelike effect called that a coil has to ac? __________

11-3 INDUCTIVE REACTANCE

The ability of inductance to limit current flow in an ac circuit has been defined as the reactance of the inductor. More correctly, it is *inductive reactance*. In a completely inductive circuit (no resistance), as in Fig. 11-7, the opposition effect of inductive reactance can be converted to a usable value of ohms. This will allow the use of the Ohm's law formulas in ac circuits involving only inductance. The formula for inductive reactance is

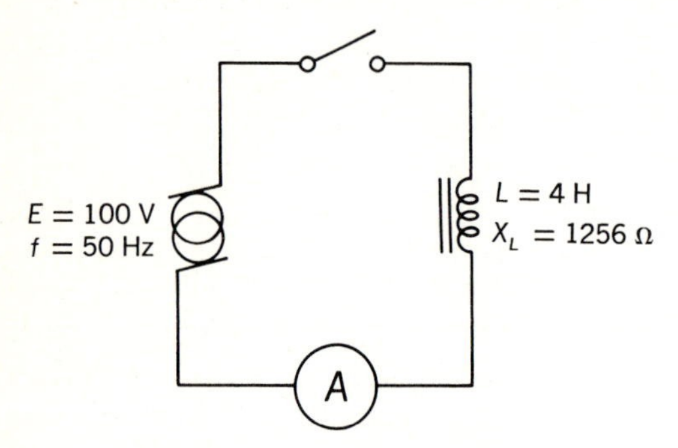

Fig. 11-7 A purely reactive circuit with an ac alternator as the source.

$$X_L = 2\pi f L$$

where X_L = inductive reactance, in ohms
$\pi = 3.14$
f = frequency, in hertz
L = inductance, in henrys

ANSWERS TO CHECK-UP QUIZ 11-1

1. (Resistance) **2.** (Inductance) **3.** (Infinite if no R) (No power if no R) **4.** (0.0333 s) (0.0000333 s) **5.** (0.0025 s) **6.** (0.7 T_c) (1.8 T_c) (2.6 T_c) **7.** (10 W) (Zero) **8.** (V across coil) (V across R)

ANSWERS TO CHECK-UP QUIZ 11-2

1. (Sawtooth) (Sine) **2.** (Lagging) **3.** (Remain the same) (Increase) **4.** (Single-pole-single-throw) (Triple-pole-double-throw) **5.** (Inductive reactance)

$2\pi f$ is often found in sinusoidal ac formulas.

In the circuit of Fig. 11-7, it is possible to determine the inductive reactance of the 4-H coil and then, by Ohm's law, to determine the current flowing in the circuit. The two interlocked rings represent an ac generator or alternator.

From the formula $X_L = 2\pi f L$, the inductive reactance is 2(3.14)(50)(4), or 1256 **Ω**. The current flowing in the circuit is $I = E/X_L$, or 100/1256, or 0.0796 A or 79.6 milliamperes (mA).

If the frequency is increased to 60 Hz, the value of the reactance of the same 4-H coil increases to $X_L = 2\pi f L$, or 2(3.14)(60)(4), or 1507 **Ω**. With greater opposition in the circuit, the current value drops to $I = E/X_L$, or 100/1507 or 0.0664 A, or 66.4 mA. Neither the source-emf nor the inductance values have changed to produce this reduction in current flow in the circuit; only the frequency has changed.

In a purely resistive circuit the current is dependent only on resistance and emf values.

In a purely inductive circuit with an ac source, the current is determined by the inductance (inductive reactance), the frequency, and the applied voltage. The current is directly proportional to the voltage, but the current is inversely proportional to both the frequency and the inductance value.

As long as the inductive reactance is 10 or more times greater than any series resistance, a circuit may be considered to be *purely* inductive and its current may be solved by $I = E/X_L$. Similarly, if a series circuit has a resistance value that is 10 or more times the reactance value, the circuit may be considered to be purely resistive, and its current can be solved by the formula $I = E/R$. These formulas will usually produce satisfactory answers.

Quiz 11-3. Test your understanding. Answer these check-up questions.

Questions 1 to 6 refer to Fig. 11-7.

1. If the circuit had a 1256-Ω resistor instead of the coil, what value would the ammeter read? __________
2. If the inductor were 8 H instead of 4 H, what value would the ammeter read? __________
3. If the frequency of the alternator were increased to 400 Hz, what would the reactance of the inductor be? __________ What value would the ammeter read then? __________
4. If the internal resistance of the components totaled 60 Ω, would the circuit be considered to be purely resistive or purely reactive? __________
5. What do the two parallel lines next to the coil indicate? __________
6. What do the two interlocked circles symbolize? __________
7. What is the formula for determining the inductance value if X_L and f are known? __________
8. A coil has 1000 Ω of X_L in a 50-Hz circuit. What is its inductance value? __________

11-4 IMPEDANCE

When series circuits have reactance and resistance values that differ by less than a factor of 10:1, they may no longer be considered to be either resistive or reactive but are said to be *reactive with a resistive component*. The total opposition effect of such reactive circuits can be determined several ways. The opposition of reactance plus resistance is known as *impedance*, symbolized by Z. The unit of measurement is the ohm. The symbol is Ω. One method of solving for the impedance value of an RL circuit is by graphing.

The diagram shown in Fig. 11-8 represents an RL circuit having equal values of resistance and reactance. Since it is a series circuit, the same value of current flows in the R, the X_L, the meter, and the source. The voltage-drop, $V = IR$, across any resistor is always exactly in phase with the circuit current, but the voltage-drop developed across any inductive reactance, $V = IX_L$, is always 90° leading the current. (A voltage leading the current by 90° is the same as a current lagging the voltage by 90°.) As a result, the sum of the two voltage-drops, $V_R + V_{X_L}$, will be the resultant (source) value provided they are added at 90° (at right angles) as in Fig. 11-8*b*. This graph indicates that the source emf E_s is equal to the sum of $V_R + V_X$ but only if they are added at 90°, as shown in Fig. 11-8c. This graph shows that the total impedance Z of the circuit is equal to the sum of 400-Ω resistance plus 400-Ω reactance added at 90°. From observation it can be seen that the value of Z is greater than either R or X_L, but is not as much as the arithmetical sum of the two.

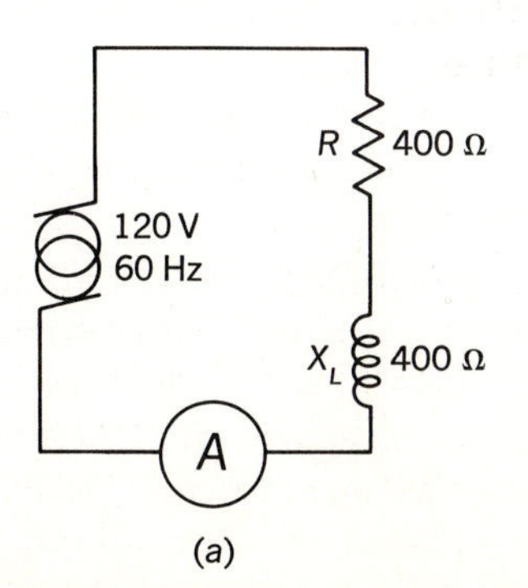

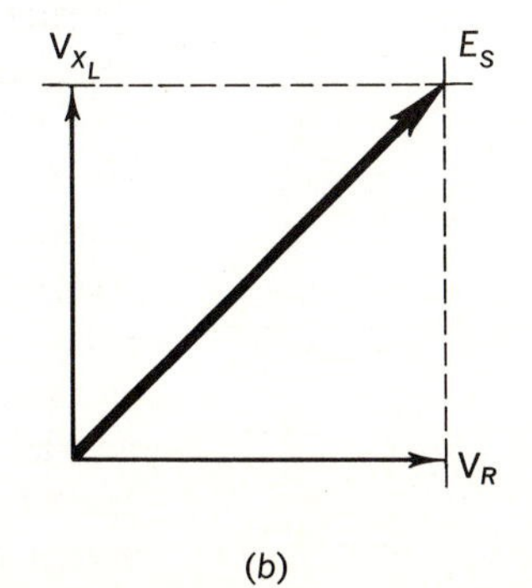

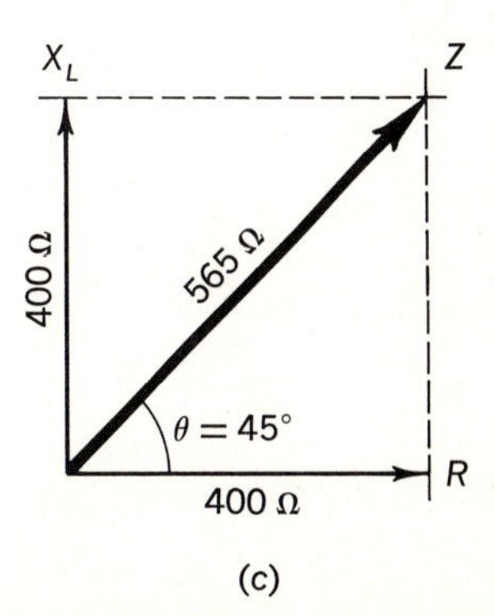

Fig. 11-8 (*a*) Series RL circuit across an ac alternator. (*b*) Plotting E_R and E_{X_L} to determine E_S. (c) Plotting R and X_L to determine Z.

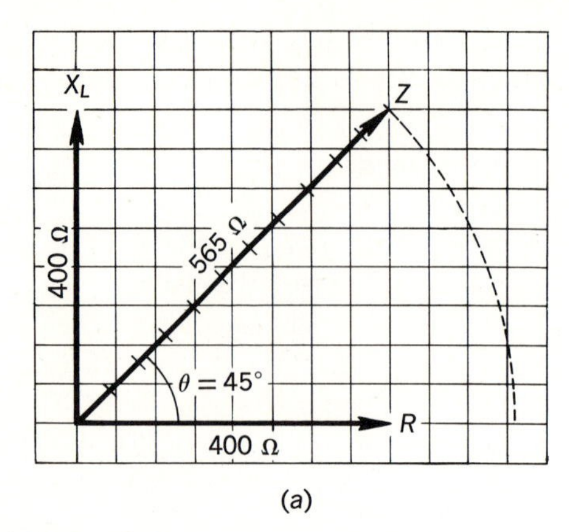

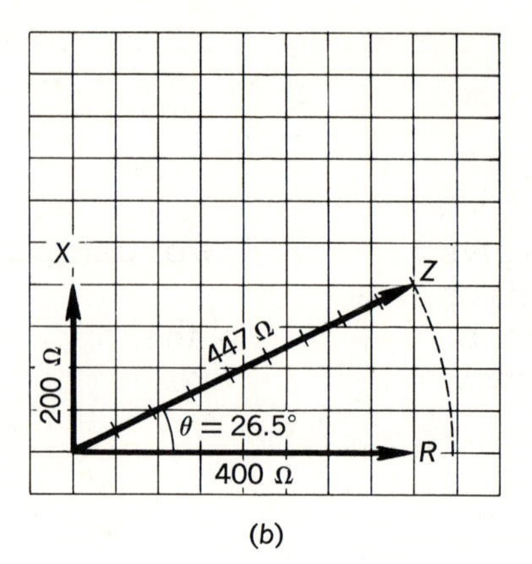

Fig. 11-9 (*a*) *RXZ* graph to determine Z and phase angle when *R* and X_L are known. (*b*) A similar graph but with less X_L.

The nonmathematical graphic method of determining the value of Z in a circuit having a 400-Ω X_L and a 400-Ω *R* is to draw two vector arrows at right angles on graph paper, as in Fig. 11-9*a*. In this particular case each square has been assigned a value of 50 Ω, making 400 Ω equivalent to eight squares. There are three methods of measuring the length of the resultant vector using: (1) a ruler, (2) a similar piece of graph paper laid along the resultant, (3) a compass, drawing an arc from the vector arrow point to the horizontal *R* line (dashed arc). In this case it is found that the Z vector has a value of a little more than 11 units, or a little more than 560 Ω (565 by the formula $Z = \sqrt{R^2 + X_L^2}$, discussed later).

The angle between the *R* and the Z vectors, 45°, is of interest because it will be the *phase angle* of the *E* and *I* in the circuit and source. The phase-angle symbol is the Greek letter theta, θ.

ANSWERS TO CHECK-UP QUIZ 11-3

1. (79.6 mA) **2.** (39.8 mA) **3.** (10,048 Ω) (9.95 mA) **4.** (Purely reactive) **5.** (Iron core) **6.** (Slip-rings of an alternator) **7.** ($L = X_L/2\pi f$) **8.** (3.18 H)

In Fig. 11-8*a*, if the impedance is 565 Ω, the ammeter must read $I = E/Z$, or 120/565, or 0.212 A.

How much power is being dissipated in this circuit? It may take energy to build up a magnetic field around a coil, but all this "reactive" energy is returned to the circuit when the magnetic field collapses. Thus a pure reactance never dissipates any power. The only loss of power in a circuit is that lost in heat by the resistance of the circuit. The power loss with 400 Ω of resistance and 0.212 A is, from the power formula $P = I^2R$, $0.212^2(400)$, or 18 W.

When the power formula $P = EI$ is used for the whole circuit, it gives a value of 120(0.212), or 25.4 W. This is not the same as the 18 W computed first. Which is correct? In one respect, both are. However, the formula $P = I^2R$ gives the *true power*, or P_t. The formula $P = EI$ gives the number of *voltamperes*, or VA, of the circuit, sometimes termed the *apparent power*, or P_a. The true power tells how much power is dissipated in heat. The apparent power is the product of the voltmeter and ammeter values. These quantities are related through the phase angle, as will be discussed in subsequent chapters. (If the voltage-drop across the resistor is computed, and this value is used in the formula $P = EI$, the power value is 18 W. Any power formula, $P = EI$, $P = E^2/R$, $P = I^2R$, applied to a resistor alone will give the true power dissipated by that resistor.)

The ratio of the true to the apparent power is known as the *power factor* (PF) of a circuit. A purely resistive circuit has a power factor of 1, because the apparent and true powers are the same. A purely reactive circuit has a power factor of 0, because while there may be a voltampere reading of the meters, there is no loss of energy in the expansion and contraction of the magnetic fields produced by the current alternating in the circuit.

In Fig. 11-9*b*, R is 400, but X_L is only 200. Since the Z vector is slightly less than 9 units long, the impedance must be approximately 450 Ω (actually 447 Ω). The *phase angle*, or θ, the angle by which the current will lag the voltage in the source in this inductive circuit, is the angle between the R and Z vectors. It can be determined by using a protractor. It will be found that the angle between the resistance and impedance vectors is 26.5° in this case.

Quiz 11-4. Test your understanding. Answer these check-up questions.

1. Express the three basic Ohm's law formulas used with purely resistive circuits. ________ ________ ________
2. Express the three basic Ohm's law formulas used with purely inductive circuits. ________ ________ ________
3. Express the three basic Ohm's law formulas used with *RL* circuits. ________ ________ ________
4. Which of these groups, (*a*) $R = 675$, $X_L = 8950$, (*b*) $R = 48$, $X_L = 180$, (*c*) $R = 400$, $X_L = 14$, could be considered purely resistive? ________ Purely reactive? ________ An *RL* circuit? ________
5. Use graph paper to find the Z value of a series *RL* circuit having 30-Ω R and 40-Ω X_L. ________ Using a protractor, find the phase angle of this circuit. ________
6. What is the approximate Z value in a series *RL* circuit having an 800-Ω R and a 1500-Ω X_L? ________ What is θ? ________
7. If R is 4.5 and X_L is 2.5, what is Z? ________ θ? ________
8. Given $\theta = 40°$ and $R = 60$ Ω, construct a graph to determine Z. What is the X_L value? ________ The Z value? ________
9. What is the power factor of Fig. 11-9*a*? ________
10. In a series *RL* circuit the E_s is 100 V, R is 200 Ω, and X_L is 150 Ω. Draw a schematic diagram of the circuit. What is the value of Z? ________ θ? ________ I? ________ P_t? ________ P_a? ________ PF? ________

CHAPTER 11 TEST • SERIES *RL* CIRCUITS

1. If a 10-Ω resistor is connected across a 10-V dc source, theoretically, how long would it take for the current to rise to 1 A?
2. If a 10-H coil with zero resistance is connected across a 10-V dc source with no resistance, theoretically, how long would it take for the I to reach maximum?
3. When a 6-H inductor with 10 Ω of internal resistance is connected across a dc source with 2 Ω of internal resistance, how much time is required for the current to reach 63% of the eventual current maximum? Essentially maximum value?
4. When an inductive load on an ac source is purely reactive, what is the phase angle between E_s and I_s? Which leads?
5. What current waveform is produced by a square-wave pulsating dc emf in an *RL* circuit with a long time constant? By a square-wave ac emf? By a sine-wave ac emf? Increasing the frequency of the square-wave pulses has what effect on the current?
6. What is the inductive reactance of a 0.3-H coil in a circuit in which the source frequency is 60 Hz? 1 kHz? 5 MHz?
7. A series circuit has 850 Ω of reactance and

10,000 Ω of resistance. For simple computations would this be considered a purely resistive or a purely reactive circuit? What would be the assumed phase angle?

8. When an inductive reactance of 580 Ω is in series with a 580-Ω resistor across a source, what is the phase angle of the E_s and I_s? Z as seen by the source?
9. A series circuit consists of an 8-Ω resistor and a 10-Ω inductive reactance. What is the impedance? The phase angle?
10. A 4500-Ω inductive reactance is in series with a 3000-Ω resistor. What is the value of Z? θ?
11. In question 10, if the source frequency were changed from 500 to 250 Hz, what would be the value of X_L? Z? θ?
12. A 30-Ω R is in series with a 40-Ω X_L across a 100-V source. What is the value of Z? I? The power dissipated by the resistor? The apparent power? The power factor?
13. If the X_L in a circuit is increased, how does this affect the power factor?
14. If the reactance value in an RL circuit is greater than the resistance value, is the phase angle always more or less than 45°?

ANSWERS TO CHECK-UP QUIZ 11-4

1. ($I = E/R$) ($R = E/I$) ($E = IR$) **2.** ($I = E/X$) ($X = E/I$) ($E = IX$) **3.** ($I = E/Z$) ($Z = E/I$) ($E = IZ$) **4.** (c) (a) (b) **5.** (50 Ω) (53°) **6.** (1700 Ω) (61.9°) **7.** (5.12 Ω) (29°) **8.** (50.3 Ω) (79 Ω) **9.** (0.707) **10.** (250 Ω) (37°) (0.4 A) (32 W) (40 VA) (0.8)

CHAPTER OBJECTIVE. To develop a usable understanding of these electrical vocabulary terms: capacitor, electrostatic field, dielectric, condenser, farad, microfarad, μF, $\mu\mu$F, picofarad, pF, axial leads, radial leads, mica capacitor, paper capacitor, tubular capacitor, ceramic capacitor, titanium dioxide, variable capacitor, rotor plates, stator plates, vacuum capacitor, formed electrolytic capacitor, tantalum electrolytics, dielectric strength, dielectric constant. Also, to compute parallel and series capacitors, and capacitance between two conductive plates.

12-1 THE CAPACITOR

The two basic circuit components discussed so far have been the resistor and the inductor. The third, and last, is the *capacitor*. The inductor, or coil, operates on an electromagnetic principle. Current flowing through it produces a magnetic field in which energy is stored. The capacitor however, is considered to be an *electrostatic* type of component. A voltage impressed across it develops electrostatic lines of force in it, and energy is stored in the electrostatic field. Inductors may be thought of as basically current-operated and capacitors as basically voltage-operated.

A simple capacitor consists of two metal plates separated from each other by air or some other insulating material, as in Fig. 12-1. The insulating material between the plates is given the special term *dielectric*. It is possible to have an air dielectric, a mica dielectric, a paper dielectric, a ceramic dielectric, etc., in a capacitor. The kind of dielectric material used names the type of capacitor.

Originally, capacitors were called *condensers*, a term still in use in the automotive trade. It may be used in some areas of electronics, such as in the *condenser microphone* of radio and television stations.

12-2 THE FARAD

One of the functions of a capacitor is to store energy in electrostatic form in its dielectric. In Fig. 12-1*b*, the 10-V source is connected to the top and bottom plates of the simple capacitor. Electrons from the negative terminal of the source flow to the top plate. At the same time, electrons are pulled from the bottom plate to the positive terminal and through the battery. This results in a negative top plate and a positive bottom plate. As between any two dissimilarly charged bodies, the charge is represented by electrostatic lines of force between − and +. As soon as enough electrons have moved onto the top plate and have left the bottom to produce an emf equal to the source, the capacitor is charged, and no more electrons flow. In Fig. 12-1*b*, the emf across the capacitor would be 10 V.

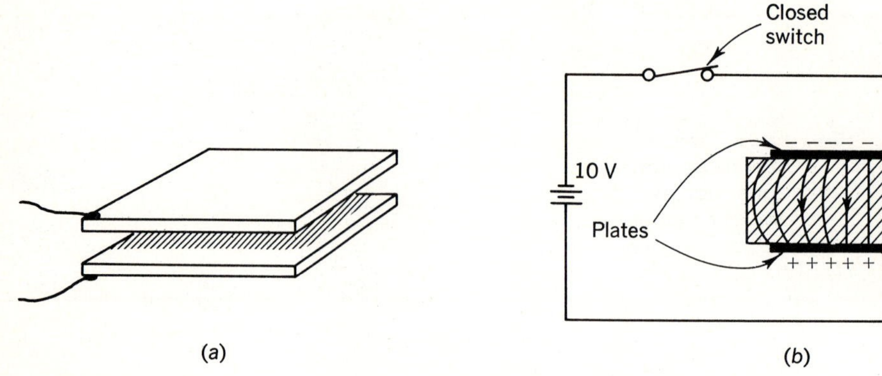

Fig. 12-1 (*a*) Basic capacitor, two metal plates separated by an insulating area. (*b*) The capacitor charges negative on one plate and positive on the other, and lines of electrostatic force form through the dielectric material.

If the capacitor had larger plates, more electrons would be needed to produce the density of electrostatic lines of force that represent 10 V. Thus a larger capacitor can store more energy in electrostatic form. If the source voltage is increased across any capacitor, the field density, or number of lines of force per unit area (square centimeter), will increase.

If the charged capacitor in Fig. 12-1*b* is disconnected from the charging source by opening the switch, the electrons are trapped on the top plate. The nearness of the electrons on the top plate holds the bottom plate as positive as when it was connected to the source. For this reason, even though it is no longer connected across the 10-V source, the capacitor will retain its 10-V charge until its electrons can find some leakage path from the negative around to the positive plate. Such paths might be through an imperfect dielectric or over the surface of any materials holding the plates apart.

If a capacitor can hold a charge of 1 coulomb (6.25×10^{18} electrons) when across a charging source of 1 V, it has one *farad* (F) of capacitance. Thus a 1-farad capacitor is able to store 1 coulomb per volt, or

$$C = \frac{Q}{E} \qquad \text{or} \qquad Q = CE$$

where C is in farads, Q is in coulombs, and E is in volts.

While the farad is the basic unit of measurement of capacitance, this is an extremely large value. More practical units of measurement are the *microfarad* (μF), one-millionth, or 10^{-6}, of a farad; or the *picofarad* (pF), a millionth of a millionth, or 10^{-12}, of a farad.

Capacitors used in power supplies of electronic equipment may have values ranging from a few to several thousand microfarads. Other capacitors in the internal circuits of such equipment may vary from a few to several thousand picofarads.

Quiz 12-1. Test your understanding. Answer these check-up questions.

1. Where is energy stored in coils? ________ In what form? ________
2. Where is energy stored in a capacitor? ________ In what form? ________
3. What is the proper term to use to identify the nonconductor between the plates of a capacitor? ________

4. What is the old name for a capacitor? ________
5. If the dimensions of the plates of a capacitor were changed from 10 to 5 cm^2, what effect would this have on the capacitance? ________
6. If the 10-cm^2 plates above were moved closer together, would the plates store more, fewer, or the same number of electrons across 10 V? ________ Would the capacitor have more, less, or the same capacitance? ________ Besides changing plate area and plate separation, what is another way that capacitance can be changed? ________
7. How many electrons would be stored by a 1-F capacitor across 10 V? ________ A 1-μF capacitor across 5 V? ________ A 10-pF capacitor across 400 V? ________
8. How many pF are there in 0.0015 μF? ________ How many pF are there in 0.4 μF? ________
9. How many μF are there in 350 pF? ________ How many μF in 4,500,000 pF? ________
10. Express 0.05 μF in picofarads. ________ Express 0.00045 μF in farads. ________

12-3 TYPES OF CAPACITORS

Capacitors will be found in many shapes and sizes (see Figs. 12-2 and 12-3). The more common fixed-value capacitors are manufactured in tubular (cylindrical) form with either axial leads (one lead out each end) or radial leads (both leads extending out at right angles from the body, or both leads out one end), in flat oblong form with axial or radial leads, flat disk radial-lead form, bathtub style (hermetically sealed in metal containers), plus specialized forms. Normally capacitors are named by the type of dielectric used in them.

Mica capacitors will always be flat because mica cannot be bent. They are constructed of alternate layers of aluminum foil and mica, with alternate plates connected (Fig. 12-4*a*). The whole capacitor is then encapsulated with a plastic protective covering. These capacitors usually have thick enough mica to withstand 500 V without breakdown of the dielectric. Some are made to operate with thousands of volts by increasing the thickness of the mica. They range in capacitance from a few picofarads to about 0.1 μF.

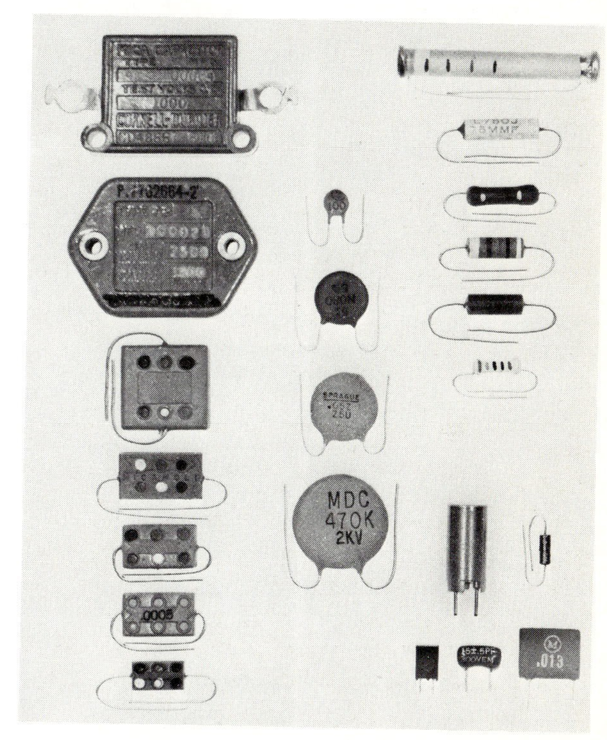

Fig. 12-2 Fixed capacitors. Left: mica-dielectric capacitors; middle: disk ceramic; upper right: axial- and radial-lead ceramics; lower right: plastic and ceramic capacitors for printed circuits.

Paper capacitors are normally constructed of two long aluminum-foil plates with oiled or waxed paper between them. To reduce their size, the capacitors are rolled into tubular form, as in Fig. 12-4*b*. Leads may be either axial or radial. Paper capacitors range in breakdown-voltage ratings from about 100 V to several thousand volts, depending on the thickness of the paper used. They are available in capacitance values from a few picofarads to several microfarads.

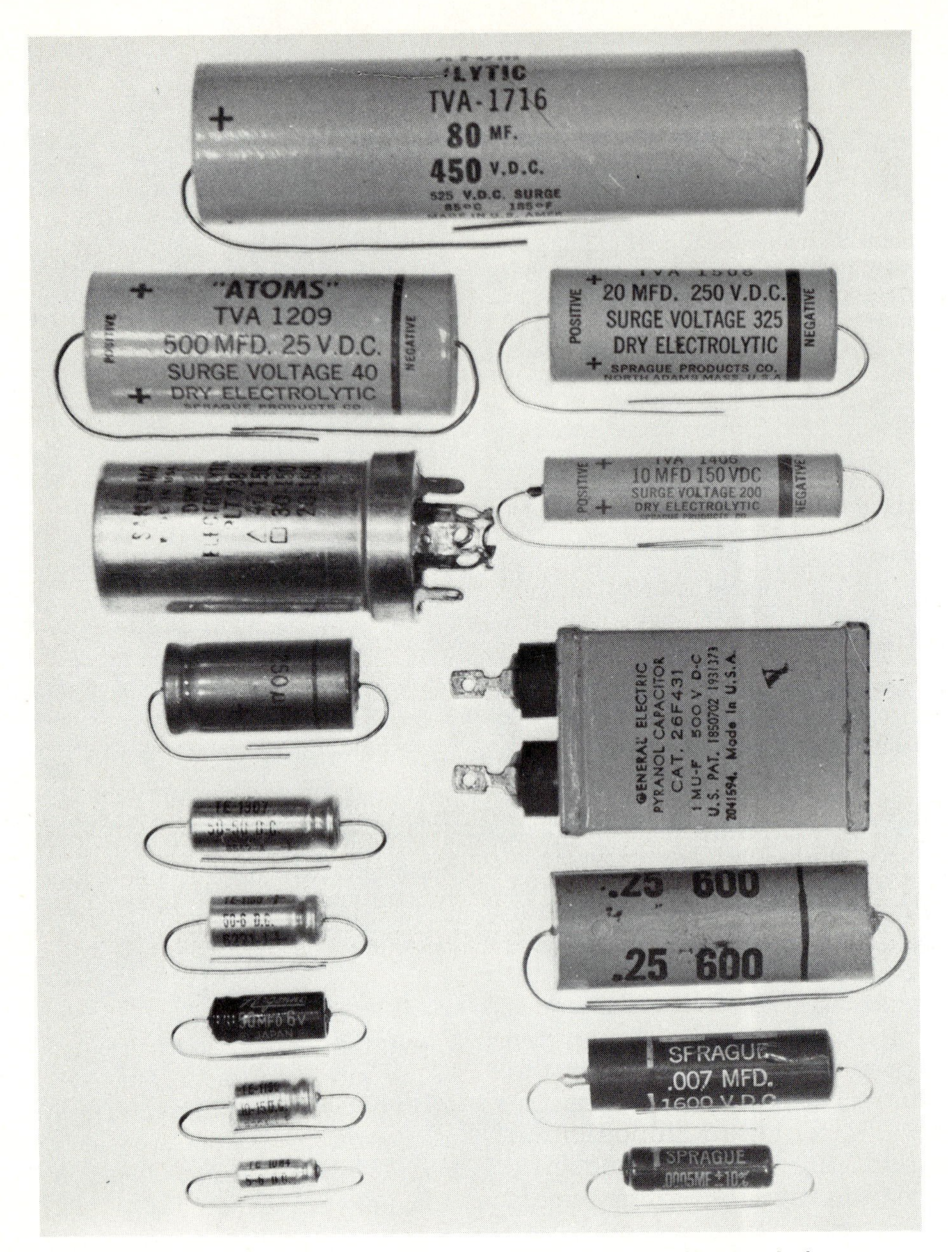

Fig. 12-3 Left column; fixed capacitors, electrolytic or tantalum; right column: tubular paper, metal-encased oil-filled, and electrolytics.

ANSWERS TO CHECK-UP QUIZ 12-1

1. (Around coil) (Magnetic) **2.** (Dielectric) (Electrostatic) **3.** (Dielectric) **4.** (Condenser) **5.** (Halve C) **6.** (More) (More) (Change dielectric material) **7.** (6.25×10^{19} electrons) (3.13×10^{13} electrons) (2.51×10^{10} electrons) **8.** (1500) (400,000) **9.** (0.00035) (4.5) **10.** (50,000) (0.000 000 000 45)

Plastic capacitors (Mylar, etc.) are similar to paper types but use any of several plastics as the dielectric material.

Ceramic capacitors may be in flat disk or hollow tubular form. The ability of their titanium dioxide dielectric to store electrostatic lines of force is much greater than that of paper or mica (a ratio of

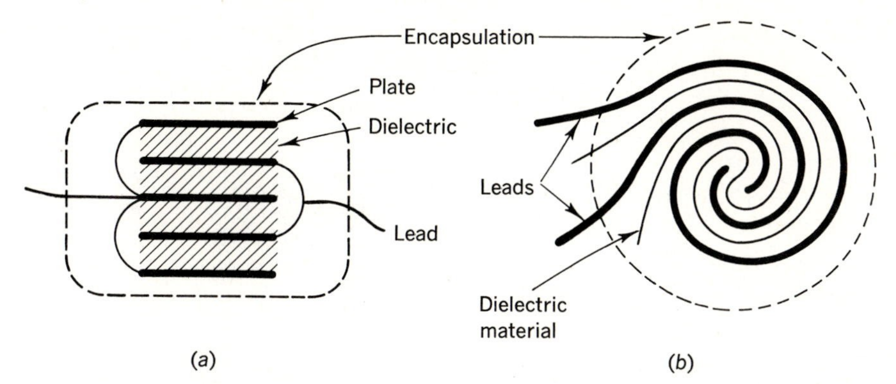

Fig. 12-4 (*a*) Construction of layer-type capacitors such as mica. (*b*) Essentials of roll-up construction of tubular capacitors such as paper and plastic.

about 500:1). These capacitors are relatively small. They are available in breakdown voltages from about 50 to 1000 V, with capacitance values from about 5 pF to 0.1 μF.

Air-dielectric capacitors are usually of the variable-capacitor type used to tune radio receivers or transmitters. They have one set of stationary aluminum *stator* plates, held apart by metal washers or spacers. A second set of *rotor* plates, insulated from the stators, can be rotated in such a way that the two sets of plates intermesh, but always with an air space between them. With the plates completely meshed, a maximum capacitance results. Air-variable capacitors may range from maximums of about 5 pF to about 400 pF. Their breakdown voltages range from about 100 V to many thousand volts, depending on the spacing between the plates.

Vacuum-dielectric capacitors look like glass-encased air capacitors. However, after they are constructed, all the air is removed from the glass envelope. The dielectric is then vacuum instead of air. With no air between the plates, a very small distance of separation between plates will allow very high voltage operation without sparking across. Whereas air will ionize and arcs over at about 10,000 V per 3 mm (about 0.1 in.) of separation, a vacuum capacitor will withstand more than 20,000 V/mm (about 0.002 in.). Vacuum capacitors are made in fixed and variable forms. They range in working voltages from 5 to 50 kV or more, and from about 20 pF to several hundred picofarads.

Electrolytic capacitors are tubular capacitors roughly similar in construction to paper capacitors. They consist of one true aluminum plate, a porous paper soaked with a borax or other chemical electrolyte solution, and a second aluminum contact plate, whose only function is to make electrical contact with the electrolyte. The capacitor is connected across a low-current source of dc to be *formed*. After a period of time, an aluminum oxide forms on the aluminum plate connected to the positive terminal of the source. The aluminum oxide (an insulator) becomes the dielectric material for the capacitor. Since the dielectric can be formed as a very thin covering on the + plate, the spacing between the plate on one side of the dielectric and the electrolyte on the other side is very small. With the two plates (the oxided aluminum and the electrolyte) very close together, the capacitor has relatively high capacitance. Electrolytics are much smaller in size per microfarad than any of the capacitors discussed above. They are sealed in aluminum cans to prevent evaporation of the electrolyte, with axial leads or radial leads at one end for printed circuit boards. They range in value from 1 to 50,000 μF. Working voltages range from about 6 to 600 V. They will work satisfactorily at any voltage below their rating. If electrolytics are operated at lower than the rated voltage, the oxide film may deform to the lower voltage value, the layer becomes thinner, and the capacitance value increases. Applying a reverse-polarity voltage to an electrolytic will deform it. It will pass heavy current, heat, develop steam, and may explode.

Tantalum capacitors are a newer form of the electrolytic capacitor, using a thin, solid tantalum layer as the formed plate. The dielectric of these capacitors has a slightly higher dielectric "constant" than the aluminum oxide of the original electrolytics, but is limited in working voltage to about 80 V. They range in capacitance value from 0.3 μF to a few hundred microfarads. They are smaller than standard electrolytics, but more expensive.

Both electrolytic and tantalum capacitors must be connected in circuits with their + terminals to the positive polarity of the circuit. None of the other capacitors mentioned above are polarized (have + and − markings), and they may be connected in circuits without regard to polarity. However, nonelectrolytic tubular capacitors may have a black line around one end. The line indicates that the lead at this end is connected to the outermost plate and should be connected as near to ground potential as possible.

12-4 DIELECTRIC STRENGTH AND CONSTANT

The ability of a dielectric material to withstand electrostatic pressure is its *dielectric strength*. This is measured in volts per millimeter, or volts per mil. A list of a few common materials and their approximate dielectric strengths is given in Table 12-1. If a simple two-plate air-dielectric capacitor had a sheet of mica slipped between its plates, it would withstand nearly 25 times as many volts across it without rupturing the dielectric.

Table 12-1 **DIELECTRIC STRENGTHS**

Substance	V/mm	V/mil
Air	3000	80
Mica	80,000	2000
Ceramics	40,000	1000
Oiled paper	60,000	1500

The *dielectric constant* of a dielectric material is a numerical expression of the ability of a dielectric material to be permeated with electrostatic lines of force compared with air or vacuum. The approximate dielectric constants of a few materials are listed in Table 12-2. If a simple two-plate air-dielectric capacitor had a sheet of mica slipped between its plates, it would not only withstand more voltage, but it would also have 6 times the capacitance. A formula to compute the capacitance of a two-plate capacitor, in picofarads, is

Table 12-2 **DIELECTRIC CONSTANTS**

Substance	Dielectric constant
Air or vacuum	1
Mica	6
Ceramics	500–5000
Oiled paper	3

$$C_{pF} = \frac{0.088kA}{s}$$

where k = dielectric constant
A = area of one of the plates, in square centimeters
s = the spacing between plates, in cm

The same formula, using inches instead of centimeters, is

$$C_{pF} = \frac{0.225kA}{s}$$

Quiz 12-2. Test your understanding. Answer these check-up questions.

1. Name five types of dielectrics used in capacitors that would work satisfactorily in ac circuits. ________ ________ ________ ________ ________
2. Name two types of dielectrics used in capacitors that would not be usable in ac circuits. ________ ________
3. A capacitor is tubular. What type of dielectric would it probably *not* have? ________
4. A capacitor is manufactured in a flat package. What are the three dielectrics most likely to be used in this form of capacitor? ________ ________ ________
5. What type of capacitor uses an aluminum oxide as the dielectric? ________ What type uses titanium dioxide? ________
6. What three types of dielectrics besides air or a vacuum might be usable in variable or adjustable capacitors? ________
7. Of the four substances listed in the text, which has the highest dielectric strength? ________ The highest dielectric constant? ________
8. A 2-μF capacitor is to be used in a 2000-V circuit. What type of dielectric might be most practical? ________
9. A 200-μF capacitor is to be used in a 450-V circuit. What type of dielectric might be most practical? ________
10. A 50-μF capacitor is to be used in a small space across a 10-V dc. What type dielectric might be most practical? ________

12-5 CAPACITORS IN PARALLEL

When two capacitors are connected in parallel, as in Fig. 12-5 (switch closed), the

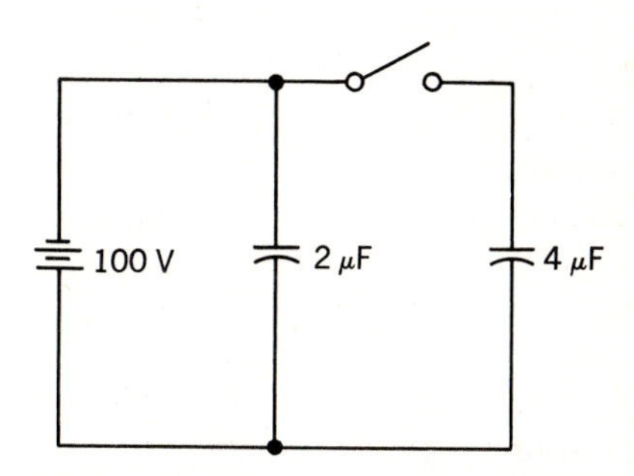

Fig. 12-5 Switch open, 2 μF across source. Switch closed, 6 μF across source.

total capacitance is the sum of the two separate capacitances. With the switch closed, this circuit has a total of 6 μF. With the switch open, there is only 2 μF across the source. For parallel capacitors

$$C_t = C_1 + C_2 + C_3 + \cdots$$

where C_t is the total capacitance, C_1 is the capacitance of the first capacitor, C_2 is the capacitance of the second, etc.

As with other parallel-circuit components, the same voltage would be across all parallel capacitors. In Fig. 12-5, each capacitor would charge to 100 V with the switch closed.

12-6 CAPACITORS IN SERIES

When two or more capacitors are connected in series, as in Fig. 12-6 (switch open), the total capacitance can be computed by the prodivisum or the reciprocal-of-the-sum-of-the-reciprocal formulas,

$$C_t = \frac{C_1 C_2}{C_1 + C_2} \quad \text{or} \quad \frac{1}{\frac{1}{C_1} + \frac{1}{C_2} + \frac{1}{C_3}}$$

In the illustration, with the switch open the total capacitance is $C_t = 10(10)/(10 + 10)$, or 100/20, or 5 μF. With the switch closed, the capacitance across the source is 10 μF.

When capacitors are in series, the source voltage divides across the capacitors. With two similar-value capacitors, the voltage division is equal. In Fig. 12-6, each capacitor would charge to 50 V. With two capacitors in series the dielectric separation across the circuit is greater, resulting in less total capacitance.

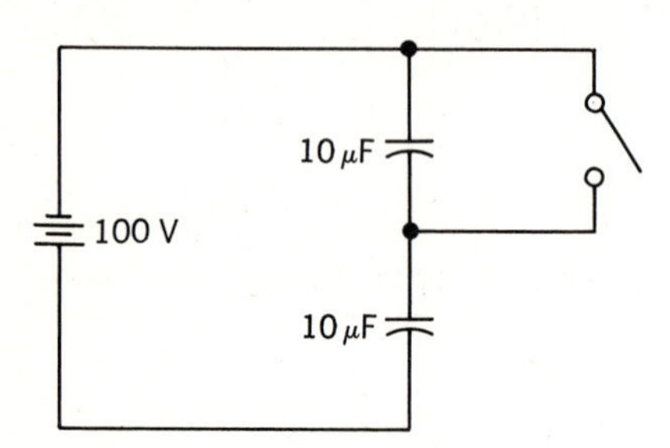

Fig. 12-6 Switch open, 5 μF across source. Switch closed, one capacitor shorted, 10 μF across source.

When two series capacitors do not have the same values of capacitance, the voltage divides in inverse proportion to their capacitances. That is, the greater voltage-drop occurs across the *smaller* capacitor. In Fig. 12-7, with a 1-μF and a 2-μF capacitor in series, $C_t = 1(2)/(1 + 2)$, or ⅔ μF. It can be assumed that the 2-μF capacitor has twice the plate area of the 1-μF capacitor. Since the two capacitors are in series, the same number of electrons must flow into each. Therefore the larger capacitor has the same number of electrostatic lines of force, but they are spread over twice the area. The density of the lines of force in the larger capacitor must be half as great as in the smaller. Half the density of lines of force could only be the result of half the voltage across this capacitor. The voltage distribution must be 33.3 V across the 2-μF capacitor, and 66.7 V across the 1-μF capacitor.

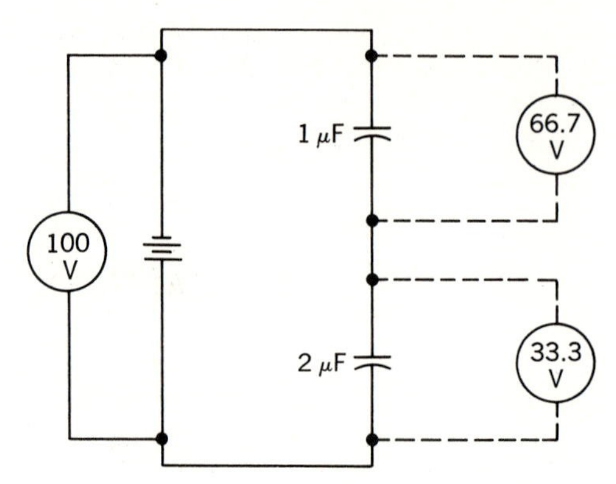

Fig. 12-7 Voltage-drop across series capacitors is inversely proportional to capacitance.

ANSWERS TO CHECK-UP QUIZ 12-2

1. (Mica, ceramic, paper, plastic, air, vacuum) **2.** (Aluminum oxide, tantalum oxide) **3.** (Mica) **4.** (Mica, ceramic, paper, sometimes radial-lead tantalum or Mylar plastic) **5.** (Electrolytic) (Ceramic) **6.** (Mica, ceramic, plastic) **7.** (Mica) (Ceramic) **8.** (Oiled paper) **9.** (Electrolytic) **10.** (Tantalum or electrolytic)

Quiz 12-3. Test your understanding. Answer these check-up questions.

1. A 2-μF capacitor is across a 100-V dc. How many electrons are involved in charging this capacitor? ________ How many coulombs are there in the capacitor? ________
2. A 2-μF capacitor and a 5-μF capacitor are in parallel across 100 V. What is the total capacitance across the circuit? ________ What would be the total capacitance if the two capacitors were in series? ________
3. Capacitors of 1.5 μF, 45,600 pF, and 250,000 pF are all in parallel. What is the total capacitance? ________
4. A 10-μF capacitor and a 5-μF capacitor are in series across 30 V. What is the total capacitance value? ________ The voltage-drop across the 10-μF capacitor? ________

5. A variable capacitor with plates meshed is charged to 100 V and is then disconnected from the charging source. What voltage is present across the capacitor? __________ Would the voltage across the capacitor increase, decrease, or remain the same if the plates were rotated to a position of less capacitance? __________
6. In Fig. 12-7, would the voltage distribution be the same if the source were 100 V rms? __________

CHAPTER 12 TEST • CAPACITANCE

1. What was the old name for the capacitor?
2. In what form does a capacitor store energy? In what part of a capacitor would you consider energy to be stored?
3. How many coulombs would be stored by a 2.5-μF capacitor if it is across 500 V? How many electrons does this represent?
4. What must be the value of a capacitor that can store 0.005 C when it is across 1000 V?
5. How many μF are there in 850 pF? How many pF in 0.0145 μF? How many μF in 0.35 F?
6. Name five types of dielectrics used for tubular capacitors.
7. What type dielectric listed has the greatest dielectric strength?
8. What type of dielectric was indicated as having the greatest dielectric-constant value?
9. What are the materials used for the positive plates in the two types of electrolytic capacitors mentioned?
10. What is the capacitance, in picofarads, of a two-plate air-dielectric capacitor if the plates each have a 15-cm^2 area and are separated by 1.1 mm? What voltage would this capacitor stand?
11. What capacitance would the capacitor in question 10 have if mica were used as the dielectric? What voltage would the capacitor then stand?
12. What will happen to an electrolytic capacitor if its positive terminal is connected to the negative terminal of a dc source and its negative lead to the positive terminal? What would happen to such a capacitor if it were connected across an ac source?
13. What is the total capacitance value if a 0.004-μF capacitor and a 0.008-μF capacitor are connected in parallel? In series?
14. What voltage will two 0.5-μF 500-V capacitors stand if in parallel? Series?
15. Capacitors of 47, 56, and 89 pF are all in series. If they are connected across 100-V dc, across which capacitor will the greatest voltage-drop appear? Would the same voltage-drop appear if the source were 100-V ac?

ANSWERS TO CHECK-UP QUIZ 12-3

1. (1.25×10^{15} electrons) (0.0002) **2.** (7 μF) (1.43 μF) **3.** (1.7956 μF) **4.** (3.33 μF) (10 V) **5.** (100 V) (Increases to 200 V at $\frac{1}{2}C$, from $Q = CE$ or $E = Q/C$) **6.** (Yes)

13 RC CIRCUITS

CHAPTER OBJECTIVE. To develop a usable understanding of these electrical vocabulary terms: *RC* time constant, *x*-axis, *y*-axis, *IR*-drop, current lead, capacitive reactance, X_C, capacitive impedance, Z, Kirchhoff's voltage law for ac circuits, parameters. Also, to learn to compute simple *RC* time-constant and capacitive-reactance problems, and to solve for Z, I_S, I_C, P, VA, θ, and E_R in simple series *RC* circuits by graphing.

13-1 RESISTANCE-CAPACITANCE TIME

Chapter 12 indicated that a capacitor across a source of emf charges, and electrostatic lines of force form in the dielectric material between the plates. It might be assumed that a capacitor charges instantaneously when connected across a source. However, all practical circuits have some resistance in them to limit the value of current that can flow into the capacitor. The time required to charge a capacitor through a resistor is therefore determined by how large the capacitor is and how much resistance is in series with the circuit. The value of voltage is not important in this case.

In Fig. 13-1*a*, the switch is in the *A*, or open, position. The capacitor has no charge. If the switch is moved to the *B* position, the capacitor will charge immediately, since there is no resistance shown. If it is moved to the *A* position again, the capacitor will retain its 10-V charge. If it is moved to the *C* position, a short circuit is across the capacitor, and it will discharge instantaneously.

In Fig. 13-1*b*, resistors R_1 and R_2 have

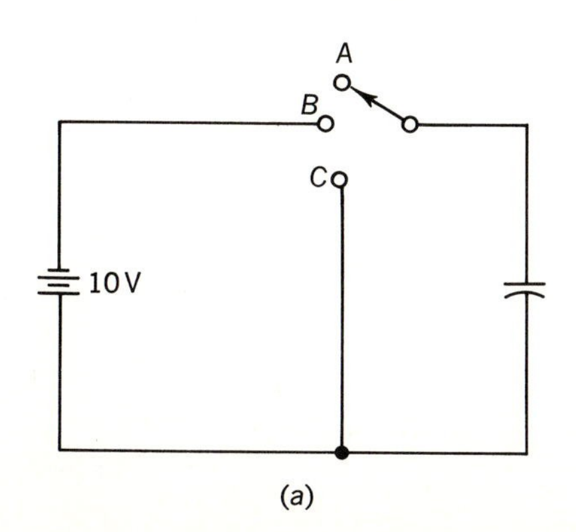

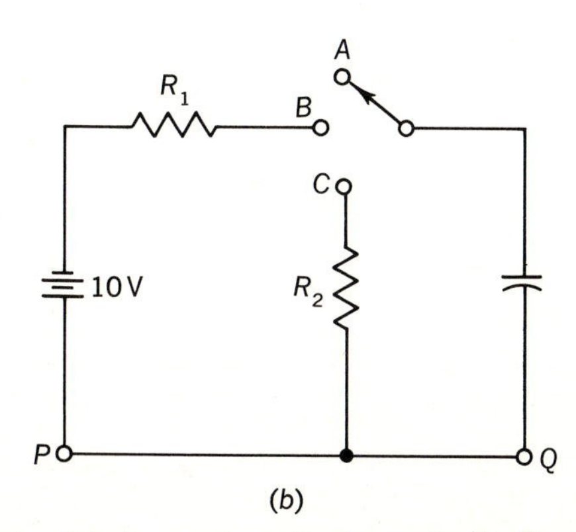

Fig. 13-1 (*a*) A circuit in which a capacitor can be instantly charged or discharged. (*b*) Time of charging and discharging depends on values of R_1 and R_2.

been added. With the switch in the A position the capacitor has no charge, as before. If it is moved to the B position, it charges to 10 V, but because of R_1 it will now require a certain amount of time before it attains its full charge. Partial charging time is

$$T_c = RC$$

where R = series resistance, in ohms
C = capacitance, in farads
T_c = time constant, which is the time required for the capacitor to charge to 63% of the source voltage

In each time-constant interval the capacitor increases its voltage 63% of what is left to charge. After five time-constant intervals the capacitor is essentially fully charged.

In Fig. 13-1*b*, if R_1 has 50,000 Ω and C is 1 μF, in a single time-constant interval, $T_c = RC$, or 50,000(0.000001), or 0.05 s, the capacitor would charge to 6.3 V. In five time-constant intervals, or 0.25 s, it would charge to 9.9 V, or to essentially full charge.

If the switch is now moved to the C position, the capacitor will discharge through R_2. The time required to discharge will be determined in exactly the same way as with the time of charge. With R_2 equal to 50,000 Ω, in 0.25 s the capacitor could be considered completely discharged.

Capacitor charge time, when graphed with time on the horizontal or *x*-axis and percentage of full charge on the vertical or *y*-axis, appears exactly the same as the *L/R* time-constant curve except that the *y*-axis is voltage instead of current (Fig. 13-2).

The charging voltage is graphed with the solid line. The value of the current flowing into the capacitor is shown as a

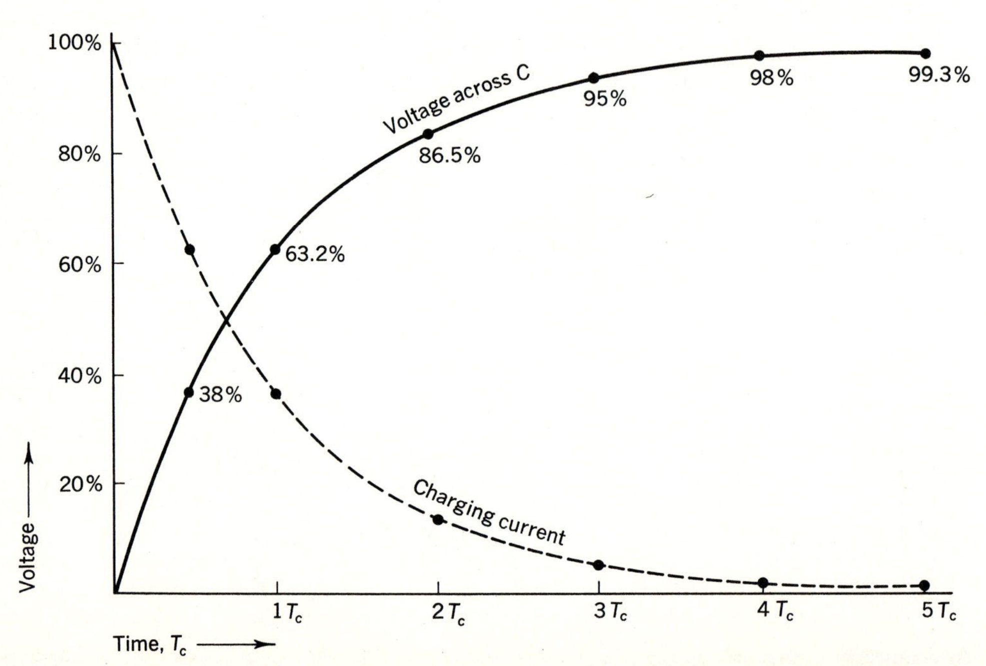

Fig. 13-2 Graph of voltage charge and current flow into a capacitor in an *RC* circuit.

dashed line. The shape of the current-decrease curve is just the reverse of the voltage-rise curve. The dashed curve shows that at the instant the switch is closed, the current value is at maximum and is the value that would be computed by Ohm's law, $I = E/R$. If E is 10 V and R is 50,000 Ω, the current at the instant of switch closing would be 10/50,000, or 0.0002 A, or 200 μA. After one time-constant period the current would have dropped 63%, to 37% of 200 μA, or to about 74 μA. After five time-constant intervals the current flowing into the capacitor would be essentially zero.

Quiz 13-1. Test your understanding. Answer these check-up questions.

1. Under what condition would a capacitor take zero time to charge to the source value? __________
2. In Fig. 13-1*b*, if the switch moves from *A* to *B* to *C*, what kind of current flows through point *P*? __________ Point *Q*? __________
3. In Fig. 13-1*b*, if R_1 is 500 Ω and the capacitor is 0.4 μF, how long would it take to charge the capacitor to 6.3 V? __________ Essentially 10 V? __________ Absolute full charge? __________
4. In Fig. 13-1*b*, if R_2 has 3 MΩ and the capacitor has 500 μF and is fully charged, how long will it take to discharge the capacitor? __________
5. The dashed line in Fig. 13-2 represents the current flowing into the capacitor as the switch in Fig. 13-1*b* is moved from contact *A* to contact *B*. What voltage or voltage-drop could it represent in the circuit? __________
6. As the switch is moved from contact *B* to *C*, what *E* or *I* in the circuit would follow the dashed curve? __________ The solid line? __________
7. What would be the designation for the type of switch shown in Fig. 13-1*a*? __________

13-2 PHASE ANGLE

With inductors the current rises to a maximum some time after the voltage is applied. The circuit current lags the voltage in an inductive circuit.

The circuit shown in Fig. 13-3*a* represents a series *RC* circuit. If the switch is alternated between contacts *A* and *B*, the capacitor will alternately be charged through *R* and then discharged through *R*. Thus the current flowing through *R* will be ac. Figure 13-3*b* illustrates the waveshape if the switch is held on *A* for five time-constant periods and is then immediately switched to *B* and held for five time-constant periods, etc.

Whereas the emf across this *RC* circuit is square-wave dc, because of the time-constant effect the actual current that flows in the *RC* circuit is sawtooth-shaped ac.

It will be remembered that with an inductor the current will increase relatively slowly after the emf is impressed across the inductor. With inductors, the current

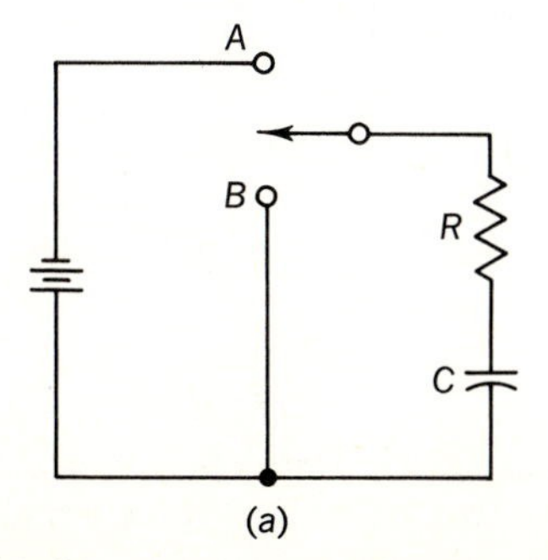

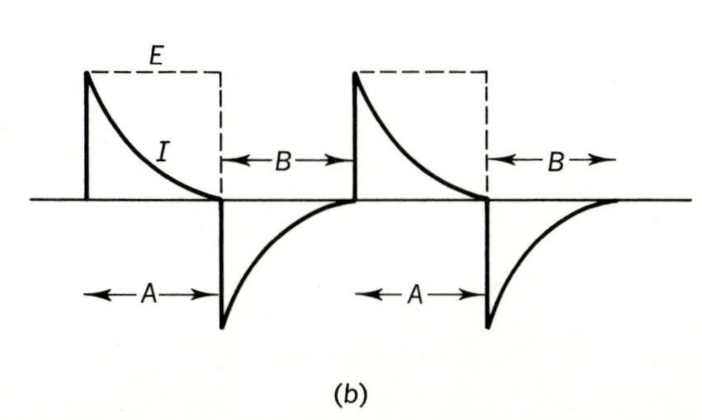

Fig. 13-3 (*a*) A circuit to produce square-wave ac emf across the *RC* part of the circuit. (*b*) Voltage and resulting charging and discharging currents in the *RC* circuit.

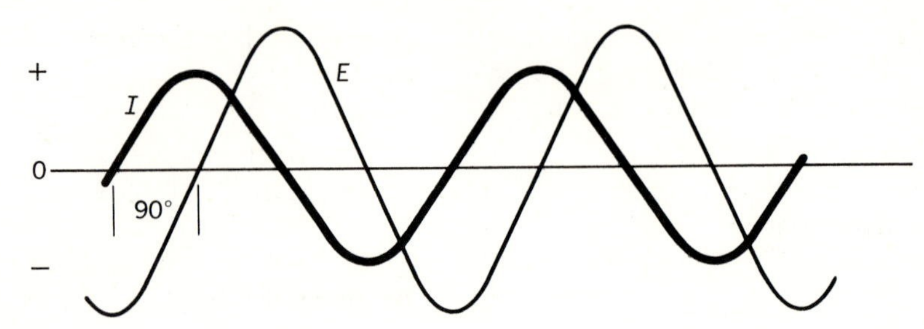

Fig. 13-4 In a capacitive circuit the current leads (crosses the time line) and reaches the peak value 90° ahead of the voltage.

lags the voltage. With capacitors, maximum current starts to flow as soon as an emf is impressed across it. As electrons flow into the capacitor, the voltage across the capacitor builds up toward maximum. Thus current leads the voltage, because there is no voltage developed across a capacitor until an electron current flows into it.

If a source is producing a perfect *sinusoidal* emf, the current flowing into and out of a capacitor across the source will also be sine-wave-shaped, but the current will *lead* the voltage by 90°, as illustrated in Fig. 13-4.

13-3 CAPACITIVE REACTANCE

It should be noted that with a square-wave emf, in which emf rises instantaneously to a maximum, the current into a capacitor is at a maximum value. If the emf rose more slowly, the pressure pushing electrons into the capacitor would be less, and fewer electrons per second would be flowing (I would be less). Thus with low-frequency sinusoidal ac, the current flow (the electrons per second) into and out of a capacitor is relatively small. If the frequency of the ac is increased, the rate of increase of the emf is considerably faster, and as a result more electrons *per second* flow into the capacitor. Therefore the higher the frequency of the ac, the greater the current value. This is just the opposite of the reactive effect produced by inductors, where the higher the frequency, the greater the counter-emf developed and the less the current flow.

The ability of a capacitor to react against current flow as frequency is lowered is called *capacitive reactance*. The capacitive-reactance effect can be converted into usable units of measurement, so that Ohm's law can be employed, by the formula

$$X_C = \frac{1}{2\pi fC}$$

where X_C = capacitive reactance, in ohms
f = frequency, in hertz
C = capacitance, in farads

As an example, the capacitive reactance of a 2-μF capacitor to 100 Hz is $X_C = 1/(2\pi fC)$, or $1/[6.28(100)(0.000002)]$, or approximately 800 Ω. The same capacitor would have 80-Ω reactance to a 100-Hz frequency.

ANSWERS TO CHECK-UP QUIZ 13-1

1. (No R in circuit) **2.** (Single pulsation of dc) (Single cycle of ac) **3.** (0.0002 s) (0.001 s) (Theoretically, forever) **4.** (7500 s or 2.08 h) **5.** (E_R) **6.** (I_C and E_C) (None) **7.** (Single-pole triple-throw, or SP3T)

If 2π is divided into 1, the resultant constant is 0.159. Therefore the capacitive-reactance formula may also be expressed as

$$X_C = \frac{0.159}{fC}$$

In a purely reactive circuit, such as Fig. 13-5, the current value is determined by

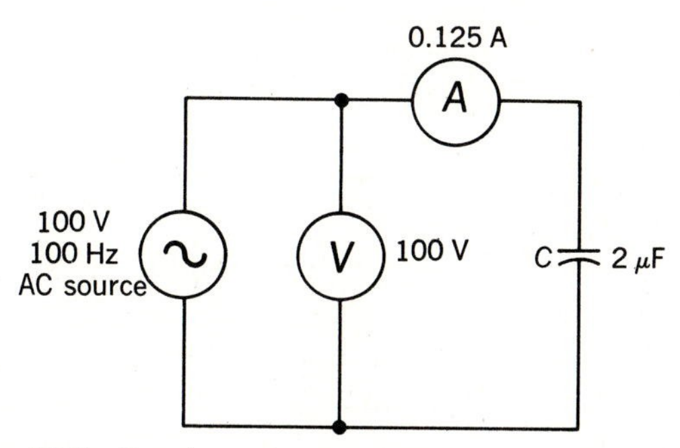

Fig. 13-5 Purely reactive circuit with a capacitor alone across the source. (Meters are considered lossless.)

the source emf value (rms) and the capacitive reactance of the capacitor. In this case the reactance is 800 Ω, and the source voltage is 100 V. The current is determined by Ohm's law, $I = E/X_C$, or 100/800, or 0.125 A. The voltmeter would read 100 V and the ammeter 0.125 A. The product of the two meter readings, the voltamperes of the circuit, or the apparent power, would be 100(0.125), or 12.5 VA. Since there is no resistance shown in the circuit, there must be no dissipation of power. So the true power is zero, apparent power is 12.5 VA, and the power factor, PF = P/VA, is zero. The current flowing in the circuit is 90° out of phase with the voltage of the source and is leading the voltage by this angle.

With the circuit having a power factor of zero, all the energy being fed into the circuit by the source is used to establish the electrostatic lines of force, or the field, between the plates of the capacitor. All this energy is returned to the circuit each time the source emf alternates and discharges the capacitor. The loss of energy to the source is zero.

Quiz 13-2. Test your understanding. Answer these check-up questions.

1. In a purely inductive circuit, what is the phase relationship in degrees between the circuit E and I? ________Which lags? ________
2. In a purely capacitive circuit, what is the phase relationship in degrees between E and I? ________ Which leads? ________
3. If the source of emf in an RC circuit is square wave, what will be the waveshape of the circuit current? ________ What would it be if the source were sinusoidal? ________
4. A battery, a coil, and a switch are in series. What is at maximum value in the circuit as the switch closes? ________ What is at minimum? ________
5. A battery, a capacitor, a resistor, and a switch are in series. What two things are at maximum value in the circuit at the instant of closing? ________ ________ What is at a minimum value? ________
6. If the frequency of the source ac is increased but the voltage remains constant, will the circuit current increase, decrease, or remain constant in an inductive circuit? ________ In a capacitive circuit? ________
7. Why does a capacitor have a higher reactance value to a lower frequency of ac? ________
8. What is the capacitive reactance of a 0.03-μF capacitor to a frequency of 400 Hz? ________ A frequency of 1 MHz? ________
9. From the capacitive-reactance formula, what value of capacitance would have 1000-Ω reactance to a frequency of 500,000 Hz? ________
10. What type of circuit has a power factor of 1? ________
11. Does the VA (or P_a) value of a circuit tell anything about the amount of power being dissipated? ________

13-4 IMPEDANCE IN CAPACITIVE CIRCUITS

When a capacitor and a resistor in series are across a source of ac, the total

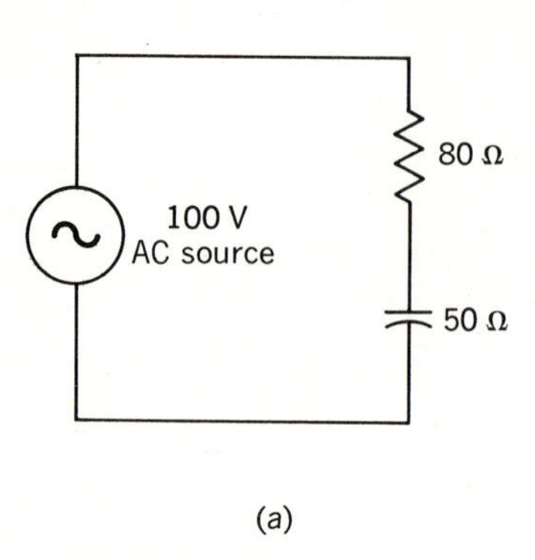

(a)

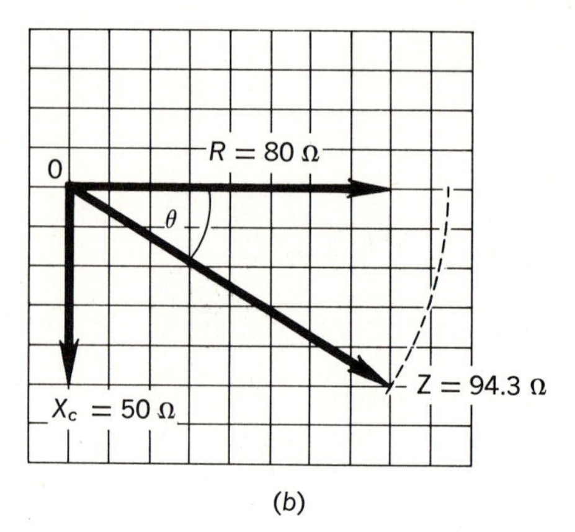

(b)

Fig. 13-6 (*a*) A series *RC* circuit across an ac source. (*b*) Graph of *R* and X_C vectors to produce the impedance and phase angle of the circuit. (Using E_R and E_X results in E_S.)

opposition of the *R* and X_C is called *impedance*, symbolized by Z. The unit of measurement is the *ohm*, symbolized by Ω. If the resistance value is 10 or more times the reactance value of the capacitor, the circuit is considered to be essentially resistive. If the reactance of the capacitor is 10 or more times the resistance value, the circuit is considered to be essentially capacitive. The circuit currents would be computed by $I = E/Z$, or in the case of an essentially resistive circuit, by $I = E/R$, and in the case of an essentially capacitive circuit, by $I = E/X_C$.

When the resistance and the capacitive-reactance values have less than a 10:1 ratio, it is necessary to either graph or compute the impedance value of the circuit before the current can be determined by $I = E/Z$. While a method of graphing an *RC* circuit is similar to the method discussed for *RL* circuits, the capacitive-reactance-value vector is drawn downward, as in Fig. 13-6 (the X_L vector is drawn upward). As indicated in the graph, the phase angle, θ, the number of degrees by which the current in the circuit leads the source voltage, is the angle developed between the 80-Ω *R* and the 50-Ω Z vectors, as with inductive circuits. Measurement of the Z vector arrow shows it to be 94.3 units, or 94.3 Ω of impedance.

With an impedance value of 94.3 Ω and a source emf of 100 V, the current through the resistor and the capacitor is $I = E/Z$, or 100/94.3, or 1.06 A.

If it is known that a current of 1.06 A is flowing through an 80-Ω resistor, the voltage-drop across the resistor can be determined by Ohm's law as $V = IR$, or 1.06(80), or 84.8 V. Similarly, the voltage-drop across the capacitive reactance must be $V = IX_C$, or 1.06(50), or 53 V. It is fairly obvious that the sum of 84.8 V and 53 V is not equal to 100 V. However, if these two values are laid out on graph paper and thereby added at right angles, the resultant will be 100 (try it). This resultant represents the source voltage.

ANSWERS TO CHECK-UP QUIZ 13-2

1. (90°) (*I* lags) **2.** (90°) (*I* leads) **3.** (Sawtooth) (Sinusoidal) **4.** (Counter-emf) (*I*) **5.** (*I*, and *E* across *R*) (*E* across *C*) **6.** (Decrease) (Increase) **7.** (Fewer electrons per second flow into it) **8.** (13,270 Ω) (5.3 Ω) **9.** (0.000318 μF, or 318 pF) **10.** (Purely resistive) **11.** (No)

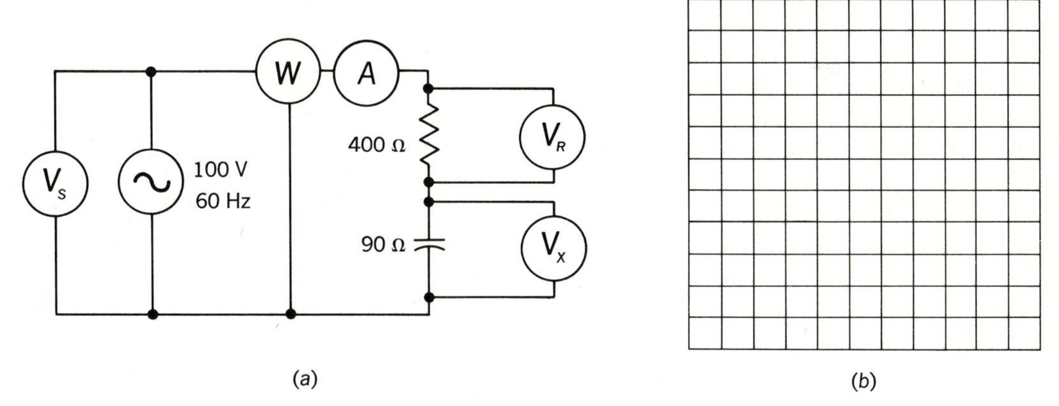

Fig. 13-7 (*a*) Circuit to be solved in Check-up Quiz. (*b*) Graph to be used.

Kirchhoff's voltage law states that the sum of the voltage-drops around a resistive circuit equals the source-voltage value. For ac circuits, Kirchhoff's voltage law states:

The sum of the voltage-drops around a simple *R* and *X* circuit will equal the source voltage, provided the resistive and reactive voltages are added vectorially (at right angles).

In purely resistive circuits the voltages and voltage-drops can be handled by simple additions. In purely reactive circuits the voltages and voltage-drops can also be handled by simple additions. That is, if two capacitive reactances, one 40 Ω and the other 60 Ω, are in series across 100 V, the total reactance is 100 Ω. With 100 V and 100 Ω, the circuit current is 1 A. With 1 A flowing through the capacitors, the voltage-drop across the 40-Ω reactor will be 40 V, and across the 60-Ω reactor 60 V, totaling 100 V.

In ac circuits in which *R* and *X* values differ by less than 10 : 1, it is necessary to add the voltage-drops vectorially, by graphing, or to compute them with methods to be described in Chaps. 31 and 32.

Quiz 13-3. Test your understanding. Answer these check-up questions.

The following questions refer to Fig. 13-7.

1. Draw vector *R* and X_C on the graph provided. What is the value of *Z*? ________
2. What value would the ammeter read? ________
3. What value would V_S read? ________
4. What value would V_R read? ________
5. What value would V_X read? ________
6. What value would the wattmeter read? ________
7. What is the apparent power value? ________
8. What is the power factor of this circuit? ________
9. What is the phase angle of this circuit? ________ Does the source *E* lead or lag the source *I*? ________
10. What is the phase angle of the *E* across and the *I* through the resistor? ________
11. What is the phase angle of the *E* across and the *I* into and out of the capacitor? ________
12. If the resistor were 90 Ω and the X_C were 400 Ω, would the value of *I* remain the same, increase, or decrease? ________ The value of *Z*? ________ The value of the E_R? ________ *θ*? ________ PF? ________

CHAPTER 13 TEST • *RC* CIRCUITS

1. When a capacitor is connected across a source of dc emf, what determines how fast it will charge?
2. What is the formula to compute the time required for a capacitor to completely discharge through a resistor?
3. What would be the formula for determining how the voltage would increase across a capacitor in series with a resistor?
4. A 500-Ω resistor is in series with a 0.4-μF capacitor. If these are connected across a 150-V source, how long would it take to charge the capacitor to 63% of 150 V? To charge to essentially full source voltage?
5. What is the phase angle of the E_S and I_S of a purely capacitive circuit? A circuit having equal R and X_C values? A circuit having a 500-Ω R and a 700-Ω X_C? Would the I lead, lag, or be in phase with the E_S in the last question?
6. What is the capacitive reactance of a 0.008-μF capacitor to a frequency of 450 Hz? 260 kHz?
7. Does an inductor have greater reactance to a rapidly or to a slowly rising current? Does a capacitor have greater reactance to a rapidly or to a slowly rising current?
8. A source of square-wave pulsating dc emf will produce what current waveshape in a long-time-constant *RC* circuit? Why?
9. What is the waveshape of the current in an *RC* circuit if the source emf is sinusoidal?
10. If a source is developing 10 A of current in a load having 20-Ω X_C, what power is being dissipated in the capacitor? What is the apparent power? The power factor?
11. An *RC* circuit having a 250-Ω R and a 500-Ω X_C in series is across a 100-V ac source. What is the value of Z? I_S? I_C? P? VA? θ? E_R?

ANSWERS TO CHECK-UP QUIZ 13-3

1. (410 Ω) **2.** (0.244 A) **3.** (100 V) **4.** (97.6 V) **5.** (21.96 V) **6.** (23.8 W) **7.** (24.4 VA) **8.** (0.975) **9.** (12.7°) (Lags) **10.** (0°) **11.** (90°) **12.** (Same) (Same) (Decrease) (Increase) (Decrease)

14

SERIES LCR CIRCUITS

CHAPTER OBJECTIVES. To develop a usable understanding of these electrical vocabulary terms: *LC* circuit, *LCR* circuit, resonant circuit, series resonant circuit, Q of a series circuit, Q of a resonant circuit, bandwidth. Also, to compute *Z*, *I*, *E*, *P*, *VA*, and θ in simple series *LCR* and resonant circuits.

14-1 SERIES *LC* CIRCUITS

The last few chapters have explained briefly how an inductor and resistor and then how a capacitor and resistor in a series circuit can affect the current, voltages, power, power factor, and phase angle. The next logical step is to add all three fundamental circuit components—inductors, capacitors, and resistors—and see how they affect circuit values when working together.

Capacitors and inductors both have reactance values which allow them to oppose the flow of ac current in a circuit. In almost all other ways capacitors and inductors behave in exactly opposite manners. Solving a circuit composed of a series of inductive reactances is similar to solving a circuit composed of a series of capacitive reactances, or of a group of resistances in series. However, in a circuit having only inductive reactance, the current lags the circuit voltage. In a capacitively reactive circuit, the current leads the circuit voltage. In a purely resistive circuit, *E* and *I* are in phase. When plotting X_L, the vector is drawn upward. For X_C the vector is drawn downward. For resistance the vector is drawn horizontally and to the right. How is the circuit solved if X_L and X_C are in series across a source of ac with no resistance present?

The circuit shown in Fig. 14-1*a* represents a 200-Ω X_L in series with a 150-Ω X_C across a 100-V ac source. Since X_L is plotted upward and X_C downward, Fig. 14-1*b* shows the two reactances plotted in vector form, 200 Ω up and 150 Ω down. In this case, then, the source sees the resultant of these two vectors as a 50-Ω X_L, as in Fig. 14-1c. If the source sees a 50-Ω X_L, the current value in the source and in the whole series circuit must be $I = E/X_t$, or 100/50, or 2 A.

If it is known that a current of 2 A is flowing through a 200-Ω X_L, the voltage-drop across the coil must be $V = IX_L$, or 2(200), or 400 V, shown by voltmeter V_L. The voltage-drop across the capacitor must be $V = IX_C$, or 2(150), or 300 V, shown by voltmeter V_C. It appears that the total voltage-drop across the circuit would be $400 + 300 = 700$ V. However, it is known that the source voltage is only 100 V. Remember that this is a series circuit, and in all series circuits the same current value flows through all components at any

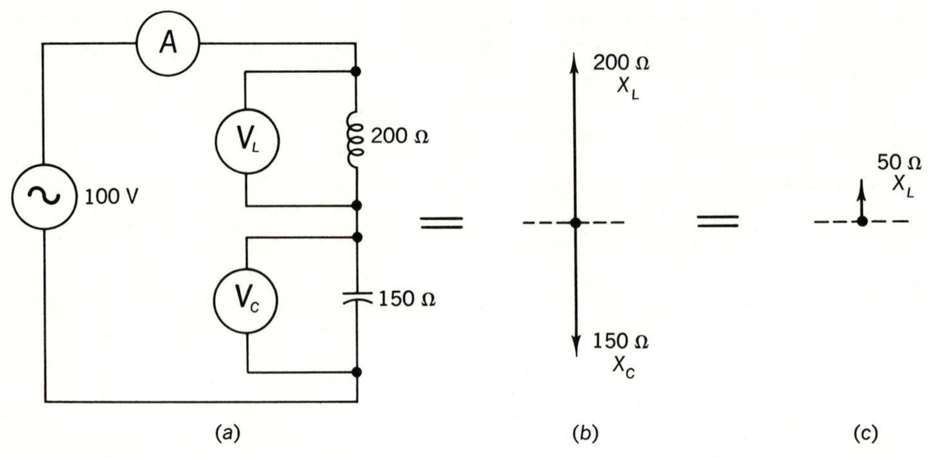

Fig. 14-1 (*a*) An inductor and a capacitor in series across a source of ac. (*b*) Vector representation of the reactances. (*c*) Resultant sum of reactive vectors.

given instant. Using the current as the common value in the circuit, any voltage-drop across an inductor is known to be leading the current by 90°, while any voltage-drop across a capacitor is known to be lagging the current by 90°. The sum is 90 + 90 = 180°. If the two voltages are 180° out of phase, they tend to cancel each other. For this reason the 400 V across the inductor and the 300 V across the capacitor cancel each other, resulting in 400 − 300 = 100 V, which, of course, is the actual voltage of the source.

14-2 SERIES RESONANCE

In the special case where the inductive reactance exactly equals the capacitive reactance, an interesting set of circumstances exists. Such a circuit is said to be *resonant*, although this is usually designated as *series resonant*. In this case, $X_L = X_C$, or

$$2\pi fL = \frac{1}{2\pi fC}$$

The circuit in Fig. 14-2*a* is a resonant circuit with a 100-Ω resistor added in series. If the reactances are graphed in vector form, as in Fig. 14-2*b*, it is seen that they are exactly equal. Therefore the sum of these two reactances in series is 0 Ω. If there is no impedance across the two reactances in series, then there must be no voltage-drop between points *A* and *B* in the diagram. If there is no loss of voltage across the reactances, the full 100 V of the source must be across the 100-Ω resistor. 100 V and 100 Ω in a circuit result in a 1-A current flow. It can be seen that in a resonant circuit the total source-voltage value appears across whatever resistance is in the circuit, and that the resistance is the total impedance.

Quiz 14-1. Test your understanding. Answer these check-up questions.

1. A 35-Ω X_C and a 400-Ω X_L are in series across a 120-V 60-Hz ac line. What type and value of reactance does the line see? __________ __________
2. In question 1, would it be correct to ask what "impedance" the line sees if there is no resistance present? __________
3. In question 1, if the frequency of the line increased, would the impedance of the series circuit increase, decrease, or remain the same? __________ Why? __________

4. In question 1, would it be possible to change the line frequency to make this circuit resonant? ________ If so, would the frequency have to be raised or lowered? ________
5. In Fig. 14-2, what is the voltage-drop across the coil? ________ The capacitor? ________
6. In Fig. 14-2, what is the phase angle of this resonant circuit as seen by the source? ________ Would all resonant circuits have the same θ? ________
7. Is the circuit in Fig. 14-2 inductively reactive, capacitively reactive, or resistive? ________ Is this true of all resonant circuits? ________
8. In Fig. 14-2, what is the true power of the circuit? ________ What is the apparent power? ________ What is the power factor? ________ Would all resonant circuits have the same PF? ________
9. In Fig. 14-2, if the source frequency increased 1 Hz, would the circuit be resonant? ________ Would it be inductively reactive, capacitively reactive, or resistive? ________
10. In Fig. 14-2, if the frequency dropped to 30 Hz, what would be the X_L value? ________ The X_C value? ________ Try graphing X_L, X_C, and R to determine the Z value. What is the Z value? ________

14-3 SERIES *L*, *C*, AND *R*

The series-resonant circuit discussed above was a special form or type, in which $X_L = X_C$. As soon as the frequency shifts one cycle, the circuit is no longer resonant but becomes what is commonly known as an *LCR* circuit.

Only one extra step is required to solve a series *LCR* circuit over that of solving a series *RL* or *RC* circuit. Figure 14-3*a* represents a basic series *LCR* circuit. When vectored, the *R* vector is always drawn horizontally, on the *x*-axis, and to the right. X_L vectors are drawn upward on the *y*-axis. X_C vectors are drawn downward on the *y*-axis. In Fig. 14-3*b*, the two reactances partially cancel, leaving a total reactance of 30-Ω X_C (Fig. 14-3*c*). The resultant of the 30-Ω X_C and the 50-Ω R is the total Z value. Before continuing any further, try vectoring this circuit on a piece of graph paper to determine the Z value.

After the Z value has been determined, Ohm's law can be applied to the circuit as a whole. With a source voltage of 100 V and a total impedance value of 58.3 Ω, the circuit current is $I = E/Z$, or 100/58.3, or 1.72 A.

A current of 1.72 A flowing in the circuit produces a voltage-drop across the re-

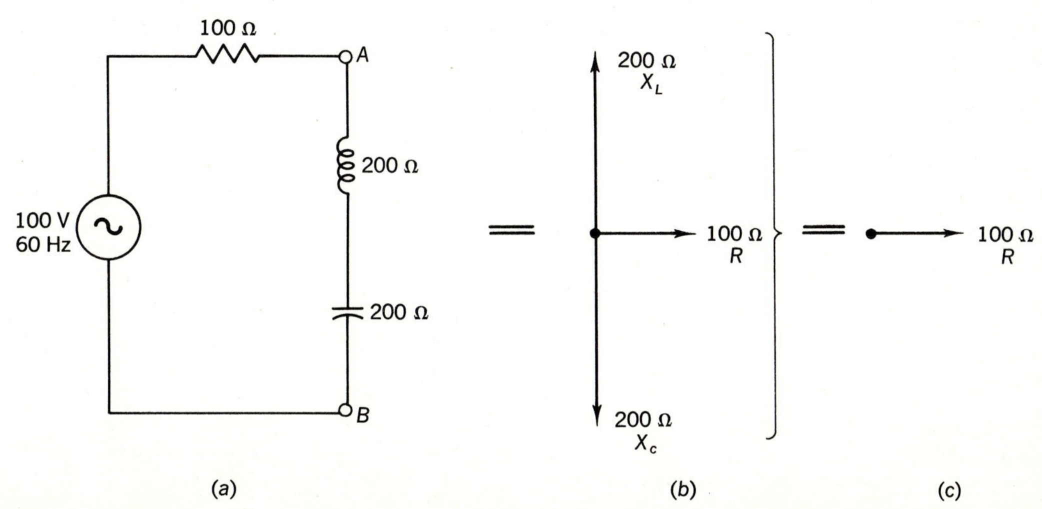

Fig. 14-2 (*a*) A resonant circuit. (*b*) Vector representation of reactances and resistance. (*c*) The resultant is pure resistance seen by the load.

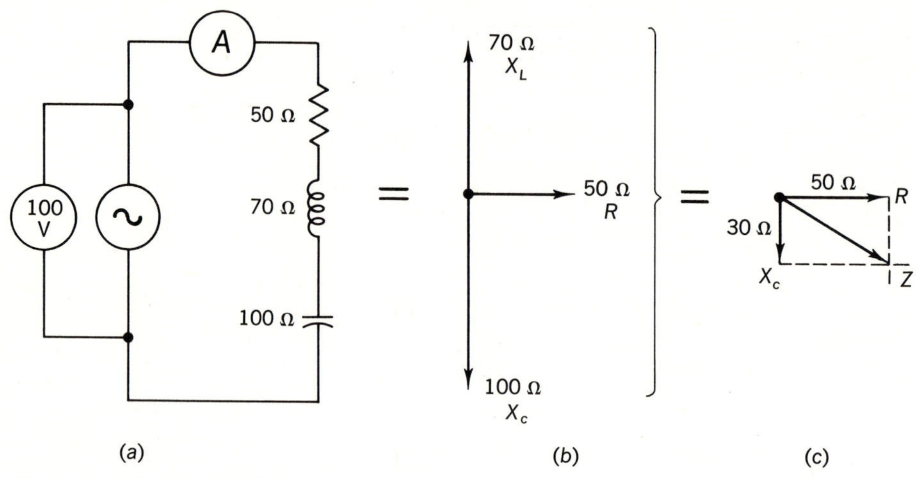

Fig. 14-3 (*a*) A series *LCR* circuit. (*b*) Vector representation of reactances and resistance. (*c*) The resultant reactance and resistance and graph of the impedance.

sistor of $V = IR$, or 1.72(50), or 86 V. The voltage-drop across the coil is solved similarly, by $V = IX_L$, or 1.72(70), or 120.4 V. Using Ohm's law again, the voltage-drop across the capacitor must be $V = IX_C$, or 1.72(100), or 172 V.

The power being dissipated in the circuit is the power value computed for the resistor. It has been determined that the current flowing through the 50-Ω resistor is 1.72 A, and the voltage-drop across it is 86 V. Therefore the power dissipation for the resistor (and the whole circuit) can be found by using R, E_R, and I_R values with any of the power formulas, $P = I^2R$, or $P = E^2/R$, or $P = EI$. Try computing the power all three ways. The answer is approximately 148 W.

The apparent power for the circuit is 100(1.72), or 172 VA. From this and the true power above, the power factor is $PF = P/VA$, or 148/172, or 0.86. This tells us that 86% of the energy fed into the circuit is being dissipated in heat by the resistor. Also, 14% of the energy fed to the circuit goes into developing electrostatic and electromagnetic fields in the reactances but is being returned to the circuit on the alternate half-cycles of ac.

Graphing R and the total X to determine the impedance value and then measuring the phase angle with a protractor should result in a current leading the voltage (a capacitive circuit, because X_C predominates) by an angle of 31°.

ANSWERS TO CHECK-UP QUIZ 14-1

1. (X_L) (365 Ω) **2.** (Yes) **3.** (Increase) (X_L increases, X_C decreases) **4.** (Yes) (Lowered) **5.** (200 V) (200 V) **6.** (0°) (Yes) **7.** (Resistive) (Yes) **8.** (100 W) (100 VA) (1) (Yes) **9.** (No) (Inductively reactive) **10.** (100 Ω) (400 Ω) (316 Ω)

Quiz 14-2. Test your understanding. Answer these check-up questions.

1. Does Fig. 14-4*a* show an X_L, X_C, or R circuit? ________ Would the circuit I lead or lag the source E? ________ Draw a schematic diagram of the circuit shown by the vectors.
2. Does Fig. 14-4*b* show the resultant of the circuit indicated in Fig. 14-4*a*? ________
3. In Fig. 14-4*b*, use graph paper and a protractor to find the Z value. ________ What is the value of θ? ________
4. In Fig. 14-4*c*, knowing this to be an *LCR* circuit, from the information given on the vector dia-

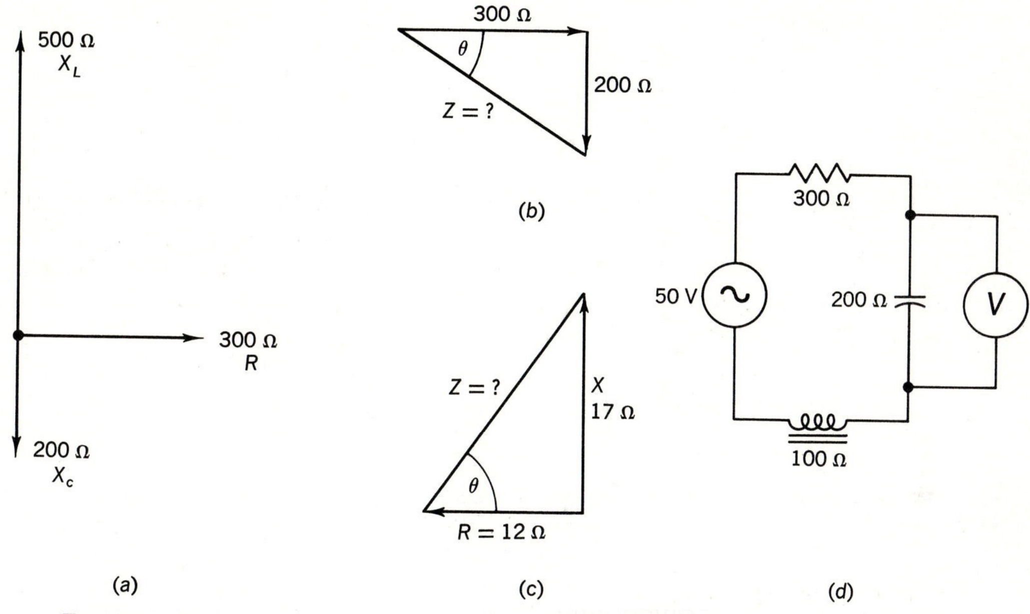

Fig. 14-4 Vector and circuit diagrams to be used in the Check-up Quiz questions.

gram, do you know the value of X_C in the circuit? ________ X_L? ________ R? ________

5. In Fig. 14-4*c*, what is the Z value? ________ What is the θ value? ________ Does E or I lead in this circuit? ________
6. In Fig. 14-4*d*, what do the two lines under the coil indicate? ________ What is the total reactance value? ________
7. In Fig. 14-4*d*, what is the value of Z? ________ I? ________ VA? ________ P? ________ PF? ________ What is the voltmeter reading? ________

14-4 THE Q OF A SERIES CIRCUIT

A term commonly used is the "Q of a circuit." Q can be thought of as meaning "quality" in the sense that an inductive circuit having no resistance or losses in it is all inductance and is therefore "high Q." Conversely, if an inductive circuit has an inductive reactance of 500 Ω and an internal or added resistance of 50 Ω, the ratio of the inductive reactance to the resistance of the circuit is 500/50, or 10. The circuit would be said to have a Q of 10; this is a fairly low Q value. It is possible to produce LC circuits with Q values up to several hundred.

The formula for the Q of a circuit is

$$Q = \frac{X_L}{R} \qquad \text{or} \qquad Q = \frac{X_C}{R}$$

Since the losses in most capacitors are so small, when an inductor and a capacitor are in series, the Q of the circuit will normally be considered as the Q value of the coil. The only significant resistance may be the internal resistance of the coil. If some other resistance is added to the circuit, this resistance will have to be added to the internal resistance value of the coil to determine the circuit Q.

The Q of series-resonant circuits is often encountered. Figure 14-5*a* shows a series LC circuit with a representation of the internal resistance R_i. The frequency of the ac source can be varied across the resonant frequency of the L and C used. An ammeter registers the current flow in the

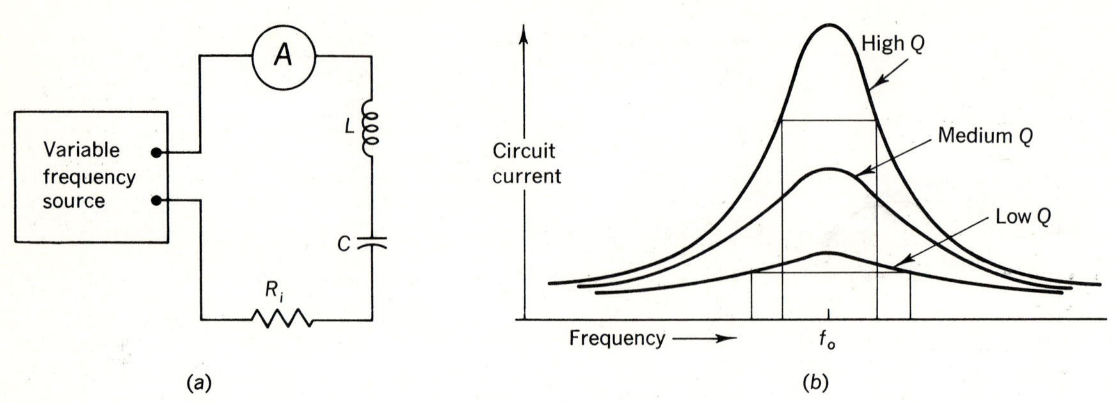

Fig. 14-5 (*a*) A variable-frequency source across an *LCR* circuit. (*b*) Response of the circuit current as the frequency is swept across the frequency of resonance f_0. Lightweight lines indicate the relative bandwidths.

circuit. Figure 14-5*b* shows a series of graphs that might have been made with three similar *LC* circuits having different values of internal resistance. The symbol f_0 represents the resonant frequency of the *L* and *C* used.

If the internal resistance were relatively high in value, it would limit the current flow in the circuit to some relatively small value. This is represented by the resonant curve marked "Low Q." If the internal resistance were insignificant in value, the current at resonance would rise sharply, as indicated by the curve marked "High Q." An intermediate value of internal resistance would result in a medium value of Q for the circuit.

The resonance curves show something else of considerable importance in electronics. This is the relative *bandwidth* of high- and low-Q circuits. The source is decreased in frequency from f_0 until the response of the current curve drops to 70% of what it was at the peak, and this frequency is noted. Then a similar point on the high-frequency side of f_0 is determined, and its frequency is noted. The difference in *frequency* between these two points is called the bandwidth of the circuit. The lighter-weight lines on the resonance curves of the high-Q and the low-Q circuits show that the high-Q circuit has a narrower bandwidth than the low-Q circuit. When narrow bandwidth is required, high-Q circuits will have to be used. It is important to know what a circuit must do before it is possible to say whether it should have high- or low-Q circuits.

ANSWERS TO CHECK-UP QUIZ 14-2

1. (Inductive circuit) (*I* lags) (See Fig. 14-4*d*, except $X_L = 500\ \Omega$ and no meter required) 2. (No; *X* value wrong; reactive vector is in wrong direction) 3. (360 Ω) (33.7°) 4. (No) (No) (Yes) 5. (20.8 Ω) (54.8°) (*E* leads) 6. (Iron core) (100-Ω X_C) 7. (316 Ω) (0.158 A) (7.9 VA) (7.49 W) (0.95) (31.6 V)

Quiz 14-3. Test your understanding. Answer these check-up questions.

1. A capacitor with a 60-Ω X_C has a 4-Ω resistor in series with it. What is the Q of this circuit? ________
2. A series-resonant circuit with a capacitive reactance of 1400 Ω has a coil with 60-Ω internal resistance. What is the Q of this circuit? ________
3. What must be done to the internal resistance of a resonant circuit if a narrower bandwidth response is desired? ________
4. What does the symbol "f_o" stand for in connection with resonant circuits? ________

5. How high up a resonant curve are the bandwidth points? ________
6. A certain coil is in series with a variable capacitor. Would the frequency of resonance be greater with the plates completely meshed or completely unmeshed? ________ If the Q of the circuit were 20 with the plates meshed, would the Q be greater, less, or the same with the plates unmeshed? ________

CHAPTER 14 TEST • SERIES *LCR* CIRCUITS

1. What is the total impedance of a 20-Ω *R*, a 30-Ω *R*, and a 40-Ω *R*, all in series?
2. What is the total impedance of a 14-Ω X_C, a 38-Ω X_C, and a 57-Ω X_C, all in series?
3. What is the total impedance of a 10-Ω X_C, a 34-Ω X_L, and a 43-Ω X_C connected in series?
4. In what direction and at what angle from the *R* vector line is an X_C vector drawn? An X_L vector?
5. It is known that 3 A is flowing through a series *LCR* circuit having 10-Ω *R*, a 20-Ω X_C, and a 30-Ω X_L. What is the voltage-drop across the capacitor? The coil? The resistor?
6. What is the source voltage in question 5? What is the total impedance of the circuit?
7. What is the impedance of a series *LCR* circuit having a 45-Ω X_C, a 5-Ω *R*, and a 45-Ω X_L? What is the special name for this type of circuit?
8. What is the Q of the circuit in question 7?
9. If the circuit in question 7 is connected across a 120-V ac line, what is the voltage-drop across the *R*? What is E_C? E_L? *P*?
10. What inductance value is needed to produce a resonant circuit at 100 Hz if the capacitance value is 0.159 μF?
11. What is the power factor of a resonant circuit? What is the phase angle?
12. A series *LCR* circuit has an 80-Ω X_C, a 120-Ω X_L, and a 30-Ω *R*. What is *Z*? θ?
13. How far down from the peak current (or peak voltage) point of a resonant-circuit curve is the bandwidth measured?
14. How can the bandwidth of a resonant circuit be decreased?
15. What must be done to the capacitance of a resonant circuit to increase the frequency of resonance? What effect should this have on the Q of the circuit? Why?
16. What must be done to the inductance of a resonant *LC* circuit to reduce the frequency of resonance? What are two practical methods of producing this effect?

ANSWERS TO CHECK-UP QUIZ 14-3

1. (15) **2.** (23.3) **3.** (Decrease R_i) **4.** (Frequency of resonance) **5.** (70% of the way to maximum current) **6.** (Unmeshed) (Greater, because at higher frequency X_L is greater but R_i is essentially the same)

15 PARALLEL LCR CIRCUITS

CHAPTER OBJECTIVE. To develop a usable understanding of these electrical vocabulary terms: capacitive susceptance, B_C, inductive susceptance, B_L, mho, antiresonance, parallel resonance, oscillating LC tank, damped oscillations, parallel LCR circuit, admittance, Y, current vectoring, flywheel effect. Also, to compute f, B_C, B_L, G, Y, Z, P, and θ of simple LCR parallel circuits.

15-1 PARALLEL-REACTANCE CIRCUITS

The total resistance value of two resistors in series is simply $R_1 + R_2$. Two *similar* reactances in series also have a total value equal to the simple sum of the reactances. However, the total reactance of a capacitive and an inductive reactance in *series* is the vector sum of the two. The reactors can be considered as being 180° out of phase with each other. It is necessary only to subtract the smaller value from the larger, giving the name of the larger to the resultant. For example, a 100-Ω X_C plus a 60-Ω X_L is $100 - 60$, or a 40-Ω X_C.

The first step in determining the total resistance value of two *parallel* resistors is to convert the resistance values to their equivalent conductance values ($G = 1/R$). The parallel conductances are added to find the total conductance. The reciprocal of the total conductance is the total resistance, or

$$R_t = \frac{1}{1/R_1 + 1/R_2}$$

A similar method is used to determine the total reactance value of two similar reactances (two X_C's or two X_L's). Whereas the reciprocal of resistance ($1/R$) is conductance, the reciprocal of reactance ($1/X$) is *susceptance*, symbolized by B. Thus $B = 1/X$. The unit of measurement is the *mho* (ohm spelled backward). (If the unit *siemens* is used instead of mho, then the letter symbol S may be used.) To determine the total reactance,

$$X_t = \frac{1}{1/X_1 + 1/X_2}$$

which actually states that

$$X_t = \frac{1}{B_1 + B_2} \qquad \text{or} \qquad B_t = B_1 + B_2$$

As with parallel resistors, it is possible to simplify the formula for two similar parallel reactances to

$$X_t = \frac{X_1 X_2}{X_1 + X_2}$$

As an example, if a 40-Ω and a 50-Ω capacitive reactance are in parallel, the total reactance is

$$X_t = \frac{40(50)}{40 + 50} = \frac{2000}{90} = 22.2\text{-}\Omega\ X_C$$

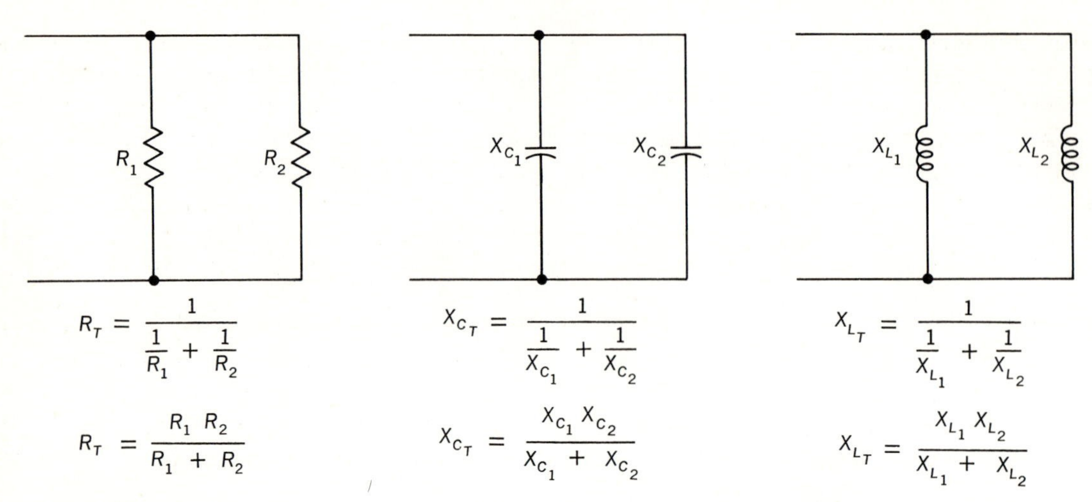

$$R_T = \frac{1}{\frac{1}{R_1} + \frac{1}{R_2}} \qquad X_{C_T} = \frac{1}{\frac{1}{X_{C_1}} + \frac{1}{X_{C_2}}} \qquad X_{L_T} = \frac{1}{\frac{1}{X_{L_1}} + \frac{1}{X_{L_2}}}$$

$$R_T = \frac{R_1 R_2}{R_1 + R_2} \qquad X_{C_T} = \frac{X_{C_1} X_{C_2}}{X_{C_1} + X_{C_2}} \qquad X_{L_T} = \frac{X_{L_1} X_{L_2}}{X_{L_1} + X_{L_2}}$$

Fig. 15-1 Parallel resistance and parallel similar-reactance circuits are computed similarly.

Figure 15-1 illustrates parallel resistors and parallel similar-type reactances with formulas to solve the circuits.

15-2 PARALLEL *LC* CIRCUITS

When two dissimilar reactances, X_C and X_L, are in parallel, as in Fig. 15-2, the total

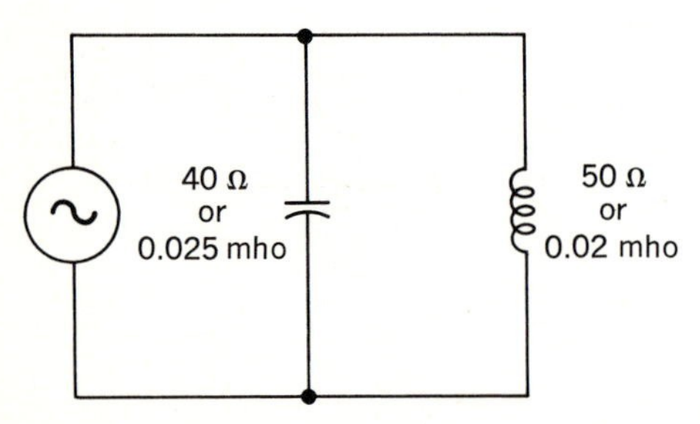

Fig. 15-2 To determine total reactance, it is necessary to convert to susceptances, subtract these, and then convert back to reactance.

reactance may be determined according to the basic parallel formula, provided it is remembered that when dissimilar reactances are added, they must be given opposite polarity signs and are then actually subtracted (the smaller from the larger). Thus with a 40-Ω X_C and a 50-Ω X_L in parallel, the reactances would first be converted to susceptance values, as

$$40\ \Omega = 1/40 = 0.025 \text{ mho}$$
$$50\ \Omega = 1/50 = 0.020 \text{ mho}$$

Adding these susceptances algebraically gives $B_t = B_1 + B_2$, or $0.025 - 0.02$, or 0.005-mho B_C. Since the capacitive susceptance is greater, the total susceptance is labeled *capacitive susceptance* B_C.

The total reactance is the reciprocal of the total susceptance, $X_t = 1/B_t$, or $1/B_C$, or 1/0.005, or 200-Ω X_C. Unlike the case of parallel resistors or parallel similar reactors, in which the total is always less than the lowest value, with two parallel dissimilar reactors the answer will always be *greater* than any of the separate reactor values alone.

Care must be exercised to determine the proper label on the total susceptance or reactance value. One method of determining the proper label is to ascertain which reactance would draw the greater current from the source. With the 100-V source shown, the 40-Ω X_C draws 2.5 A, whereas the 50-Ω X_L only draws 2 A. The source sees this parallel group as a capacitor since it must feed a capacitive (leading) current to it.

A simple formula to determine the total of two dissimilar parallel reactances, somewhat similar to the prodivisum formula, is

$$X_t = \frac{X_L X_C}{X_L - X_C}$$

This formula divides the product of the two reactances by the *difference* of the two reactances. The answer will carry the label of the *lower*-value reactance, X_C in the case above. Try solving the total reactance of a 40-Ω X_C and a parallel 50-Ω X_L.

Quiz 15-1. Test your understanding. Answer these check-up questions.

1. A 20-Ω X_C and a 40-Ω X_C are in series. What is their X_t value? ________ What would it be if they were in parallel? ________
2. A 200-Ω X_L, a 400-Ω X_L, and a 120-Ω X_L are in parallel. What is X_t? ________
3. A 100-Ω X_C and an 80-Ω X_L are in parallel. What is the susceptance of the X_C? ________ Of the X_L? ________ What are the B_t value and label? ________ ________ What are the X_t value and label? ________ ________
4. In question 3, what is the phase angle of the circuit? ________ Why is the PF zero? ________
5. In question 3, does the source current lead or lag its voltage? ________
6. In question 3, if the source frequency were increased, would it be possible to reach a point at which X_C and X_L were equal? ________ If so, what would the B_t value be? ________ What would the X_t value be? ________
7. What is the reciprocal of conductance called? ________ What is the reciprocal of inductive reactance called and what is its letter symbol? ________ ________
8. What is the X_t of a 40-Ω X_C in parallel with a 50-Ω X_L? ________

15-3 ANTIRESONANCE

A parallel *LC* circuit can produce the condition of $X_L = X_C$ at any desired frequency by choosing the proper inductance and capacitance values. It is also possible to use any value inductor and any value capacitor, and by varying the frequency of the source the condition of $X_L = X_C$ can be produced. This is possible because as frequency is increased, X_L increases, but X_C decreases. At some frequency X_L must equal X_C.

If a *parallel LC* circuit has X_L and X_C equal, this special case is called *antiresonance*. (Just as resonance is often termed series resonance, the condition of antiresonance is often called *parallel resonance*.)

The circuit shown in Fig. 15-3 has two reactors with equal values of 1000 Ω each across a 100-V source. Since the capacitor has 1000 Ω and is across 100 V, the cur-

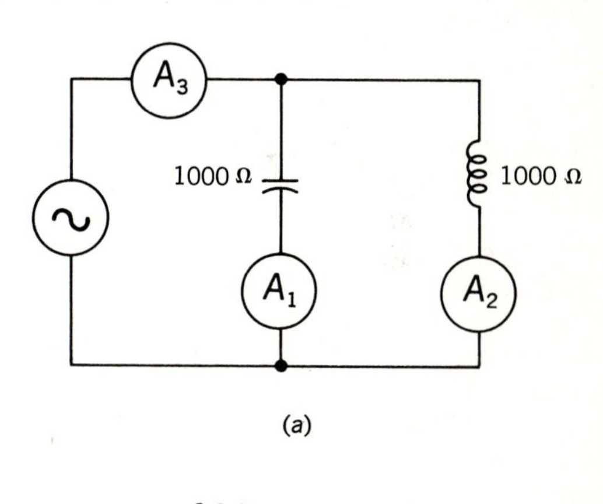

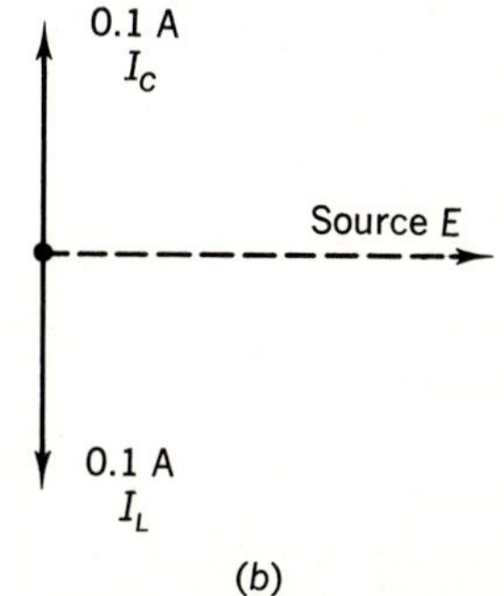

Fig. 15-3 An antiresonant circuit has all its current oscillating back and forth between the capacitor and the inductor, and none flows in the source.

rent in the capacitive branch, as read by ammeter A_1, would be 0.1 A and would be leading the source voltage by 90°. Similarly, the current flowing through the inductive branch, as read by ammeter A_2, would also be 0.1 A, but in this case the current would be *lagging* the source voltage by 90°.

When graphed 90° from the source voltage, as in Fig. 15-3*b*, the two currents are in opposite directions. This means that if there is 0.1 A flowing down in the capacitor, at that same instant there must be 0.1 A flowing up through the coil. At this instant, then, how much current is being fed to the *LC* circuit by the source? The answer is zero current. One half-cycle later there will be 0.1 A flowing up in the capacitor and 0.1 A flowing down in the inductor—still requiring no current from the source.

If a load requires no current from a source, what is the impedance of the load? From Ohm's law, $Z = E/I$, or 100/0, meaning the impedance must be infinitely high. Thus an antiresonant circuit with no resistance in it represents no load at all on the source after the source emf is once able to establish a current flow in the circuit. (Compare this with a resonant circuit, which has 0-Ω impedance, being essentially a short circuit across the source if there is no resistance in it.)

15-4 THE OSCILLATING *LC* TANK CIRCUIT

Once an antiresonant circuit is connected across a source of ac, current flows at the value demanded by the reactances for this particular source emf, and the source is no longer required to feed energy to the *LC* circuit. The *LC* circuit acts as a tank in which electrons are constantly being transferred from the capacitor to the coil and back again. Since there is no resistance, there is no loss of energy, and it is theoretically possible to disconnect the *LC* tank circuit from the source and electrons will continue oscillating back and forth between the reactances at the frequency to which the circuit is antiresonant. By formula this frequency is

$$2\pi fL = \frac{1}{2\pi fC}$$

$$f = \frac{1}{2\pi \sqrt{LC}} = \frac{0.159}{\sqrt{LC}}$$

where f is in hertz, L is in henrys, and C is in farads.

Figure 15-4 illustrates five steps covering one half-cycle of electron oscillation in an *LC* tank circuit. In Fig. 15-4*a*, the capacitor has been charged to maximum voltage by touching it for an instant across the source. One plate is negative and the other is positive, and energy is stored in electrostatic form in the dielectric of the capacitor.

In Fig. 15-4*b*, electrons from the negative plate of the capacitor start flowing through the coil to the positive plate. As they flow through the coil, they develop a magnetic field, storing some of their energy in this field.

In Fig. 15-4*c*, a time is reached when there are as many electrons present on the top plate of the capacitor as on the bottom, and the emf across the capacitor, and the circuit, is zero. At this instant the current reaches a maximum (maximum number of electrons moving per second). All the energy has now left the capacitor

---

ANSWERS TO CHECK-UP QUIZ 15-1

1. (60-Ω X_C) (13.3-Ω X_C) **2.** (63 Ω) **3.** (0.01 S) (0.0125 S) (0.0025 S) (B_L) (400-Ω X_L) **4.** (90°) (No resistance) **5.** (I lags) **6.** (Yes) (Zero) (Infinite) **7.** (Resistance) (Inductive susceptance) (B_L) **8.** (200-Ω X_C)

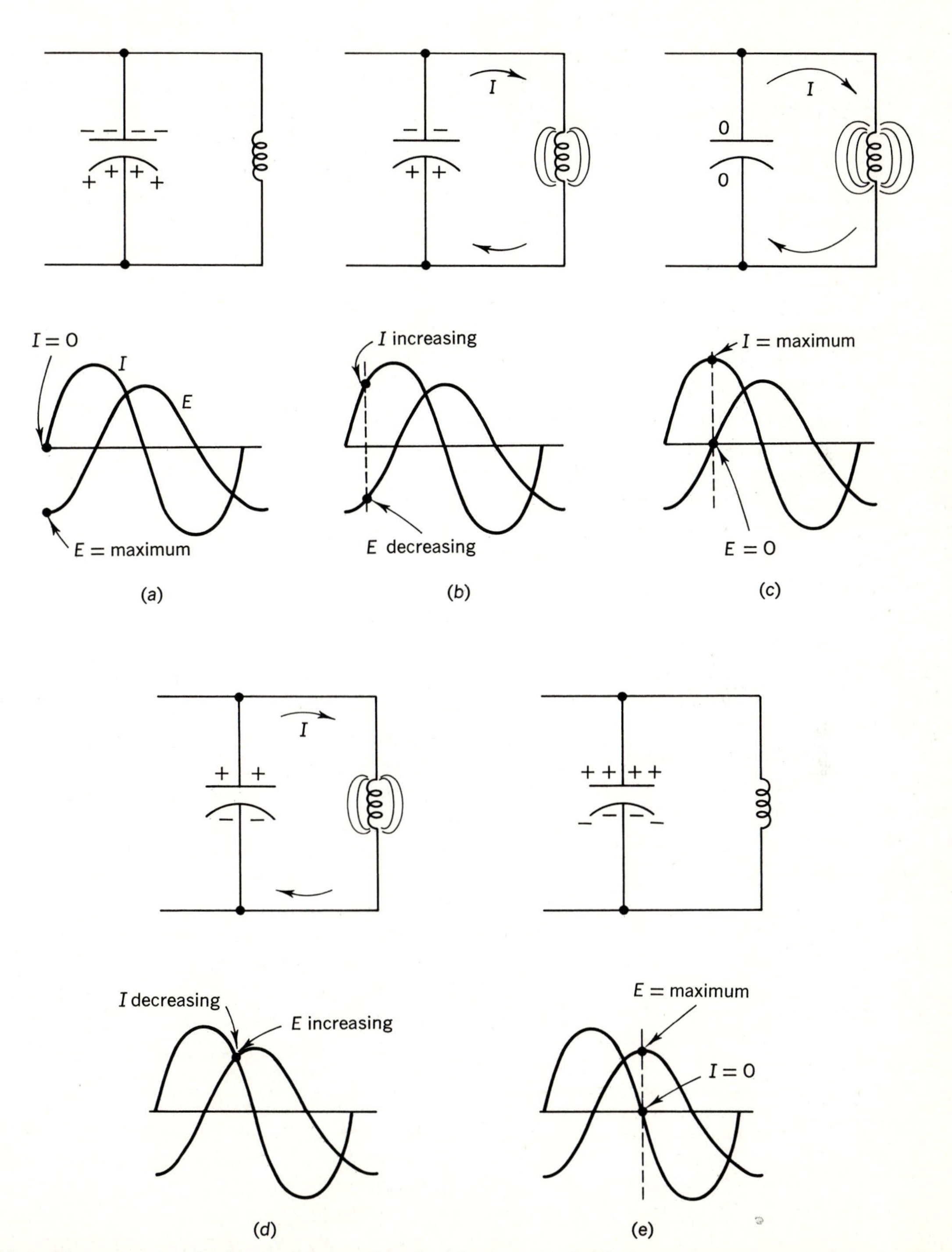

Fig. 15-4 Five steps in which energy stored in the electrostatic field of a capacitor is transferred to the magnetic field of a coil and then back to electrostatic fields in the capacitor. This forms one half-cycle of sinusoidal ac.

and is stored in the magnetic field of the coil.

In Fig. 15-4*d*, with no emf to support current, the magnetic field begins to collapse. This induces an emf in the coil, which forces electrons onto the bottom plate of the capacitor, taking them from the top plate, producing an increasing emf across the capacitor.

In Fig. 15-4e, the magnetic field collapses to zero. The full charge of electrons is now placed on the capacitor. The capacitor again has all the circuit energy stored in its electrostatic field. The capacitor is again charged to the source potential, but now the bottom plate is the negative one. One half-cycle of ac has occurred. Now electrons start back up the coil to complete the cycle.

Theoretically, electrons should continue to oscillate back and forth through the *LC* tank circuit indefinitely. However, if there is any resistance in the circuit, energy will be lost in heating the resistance as current flows through it. The tank-circuit oscillations will become weaker each half-cycle and will "damp out."

Quiz 15-2. Test your understanding. Answer these check-up questions.

1. What is the impedance seen by a source when a resonant circuit is connected across it? ________ When an antiresonant circuit is connected across it? ________
2. A resonant circuit with 50-Ω resistance in series with it appears as what impedance to a source? ________
3. If 50-Ω resistors were connected in series with both legs of an antiresonant circuit, would the circuit appear to a source as a higher or lower impedance? ________ Why? ________
4. A zero-resistance 500-Hz antiresonant circuit is connected across a source. If the source frequency were raised to 501 Hz, would the circuit appear to have lower or higher impedance? ________ Which would be greater, the inductive current or the capacitive current? ________ Would the circuit appear inductive, capacitive, or resistive? ________
5. A zero-resistance 500-Hz resonant circuit is connected across a source. If the source frequency were raised to 501 Hz, would the circuit appear to have lower or higher impedance? ________ Would the circuit appear inductive, capacitive, or resistive? ________
6. Would an antiresonant circuit be considered a reactive or a resistive circuit? ________ Would a resonant circuit be considered to be reactive or resistive? ________
7. Would it be correct to say that an antiresonant circuit allows energy to oscillate from electrostatic to electromagnetic form? ________
8. If a zero-resistance antiresonant circuit has a secondary coil coupled to its inductor, will a resistor connected across the secondary increase or decrease the impedance seen by the source? ________
9. What is the antiresonant frequency of a 2-H coil and 0.4-μF condenser? ________

15-5 PARALLEL *R*, *L*, AND *C*

To solve a circuit consisting of a resistor, an inductor, and a capacitor all in parallel across an ac source, as in Fig. 15-5*a*, it is first necessary to determine the total reactance value of the circuit and label it properly. It will be remembered from Fig. 15-2 that a 40-Ω X_C and a 50-Ω X_L in parallel appear as a 200-Ω X_C to the source. After the total reactance is determined, the circuit should be simplified by redrawing, as in Fig. 15-5*b*.

The *LCR* circuit appears to the source as a parallel *RC* circuit, and it can be computed by graphing the parallel conductance and susceptance at 90°. The 400-Ω *R* has a conductance value of 1/400, or 0.0025 mho. The 200-Ω X_C has a susceptance of 1/200, or 0.005 mho. These two values, 25 ten-thousandths and 50 ten-thousandths, can be graphed as 25 and 50, as illustrated in Fig. 15-6. Note that whereas X_C is graphed downward, its reciprocal value, B_C, is graphed in the opposite direction, or upward. Following this

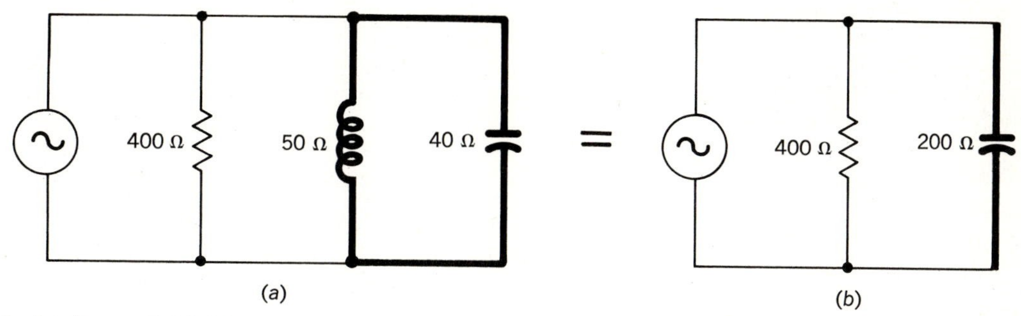

Fig. 15-5 (*a*) A parallel *LCR* circuit and (*b*) the equivalent circuit as seen by the source and as it would be computed.

same reasoning, a B_L vector would be graphed downward.

A new symbol appears in Fig. 15-6. The

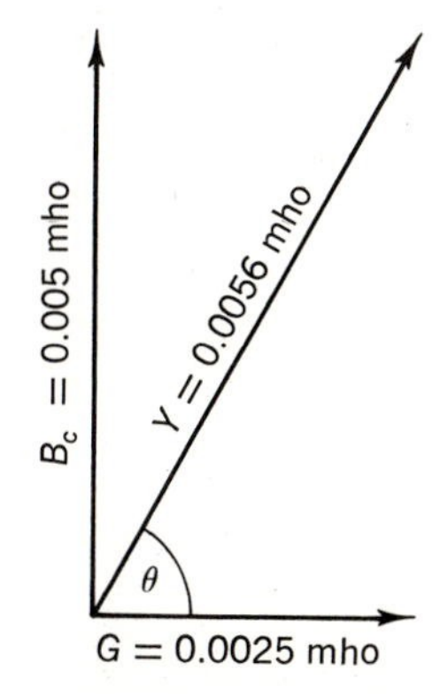

Fig. 15-6 Vector graph of Fig. 15-5, with the conductance and susceptance used to find the admittance and phase angle.

Y stands for the reciprocal of impedance. That is, $Y = 1/Z$, or $Z = 1/Y$. Y is the symbol of *admittance*, the reciprocal of impedance. Since impedance uses the ohm as the unit of measurement, admittance, or Y, uses *mho* (or siemens, S) as its unit of measurement.

The phase angle of the circuit as seen by the source is the angle between the G and Y vectors. A protractor shows the phase angle to be 63.5°. The length of the Y vector on the graph is measured as 56 units, or 0.0056 mho. The reciprocal of 0.0056 mho is 179 Ω of impedance, with the current leading by 63.5° (a capacitively reactive circuit).

15-6 CURRENT VECTORING A PARALLEL *LCR* CIRCUIT

There are other methods used to solve parallel *LCR* circuits. Any source-emf value can be selected that will result in reasonably simple current values. For example, in the circuit of Fig. 15-5*a*, if the source voltage is 100 V, the currents through the branches will be

$$I_R = 0.25 \text{ A}$$
$$I_L = 2.0 \text{ A}$$
$$I_C = 2.5 \text{ A}$$

In a parallel circuit the same voltage is across all the branches. The currents through the different branches may not be the same, however. It is known that the current in an inductive branch *lags* the source voltage by 90°, and that the current in a capacitive branch *leads* the source voltage by 90°. Thus the inductive and the capacitive currents are in opposite directions at all times and tend to cancel each other as far as the source is concerned. The source sees only the resistive current plus the *difference* between the two reactive currents. In this problem the source sees an I_R of 0.25 A and a reactive current of $2.5 - 2 = 0.5$ A capacitive.

To solve the circuit, it is only necessary to graph the two known currents, a resistive and a capacitively reactive current, which will be 90° out of phase, or at right angles (Fig. 15-7). This is essentially the

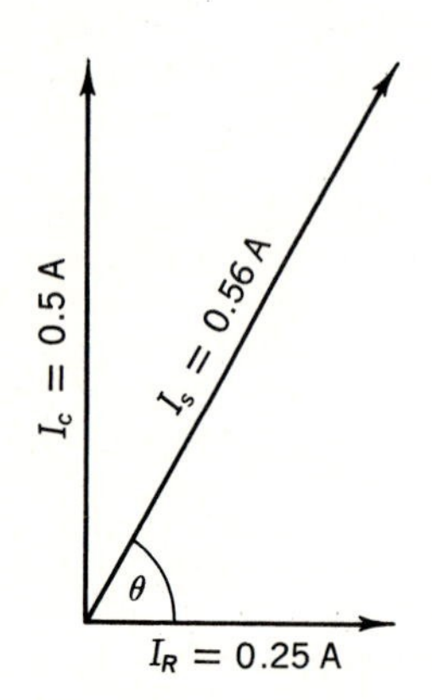

Fig. 15-7 Vector graph of Fig. 15-5, with resistive and capacitive currents used to find the source current and phase angle.

same figure as the *GBY* graph, or triangle, even to the same numbers obtained by choosing the particular source emf of 100 V. Another source voltage would have given a similar ratio to the vector-arrow lengths, but not the same numerical values.

Once the source current has been determined, the impedance can be found by Ohm's law, $Z = E/I$, or 100/0.56, or 179 Ω.

Note that the capacitive-current vector I_C is drawn upward. This is because the capacitive current leads the resistive current by 90°. I_R is always in phase with the source voltage and is always drawn on the *x*-axis from the point of origin to the right.

ANSWERS TO CHECK-UP QUIZ 15-2

1. (Low) (High) **2.** (50 Ω) **3.** (Lower) (I^2R loss in R_s) **4.** (Lower) (I_C) (Capacitive) **5.** (Higher) (Inductive) **6.** (Resistive) (Resistive) **7.** (Yes) **8.** (Decrease) **9.** (178 Hz)

Which is the better method to use, the *GBY* or the current-vector method? Most beginners seem to prefer the current-vector method, but it is important that the terms *conductance*, *susceptance*, and *admittance* be understood, as they occur in electrical literature.

Quiz 15-3. Test your understanding. Answer these check-up questions.

The following questions refer to Fig. 15-8.

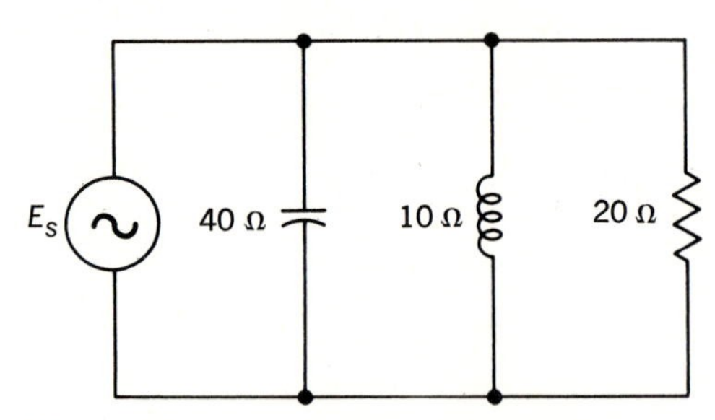

Fig. 15-8 *LCR* circuit for Check-up Quiz.

1. What is the value of B_C? ________ B_L? ________
2. What is the value of B_t? ________ G_t? ________ Y? ________ Z? ________ The phase angle? ________
3. Which would be the best voltage to assume in computing the parallel circuit by current vectors, 10, 20, 40, or 80 V? ________
4. With 20 V as E_S, what is the value of I_C? ________ I_L? ________ I_R? ________
5. What is the total I_X value? ________
6. Graph the current values. What is the I_t value? ________ With the assumed 20 V, what is the impedance value of the circuit? ________
7. If the source frequency were doubled, what would be the value of X_C? ________ X_L? ________ What would be the Z value of the whole circuit? ________ What is the special name given to the circuit in this case? ________

CHAPTER 15 TEST • PARALLEL *LCR* CIRCUITS

1. To determine the total reactance of two similar reactors in series, should the reactance or the susceptance values be added?
2. To determine the total reactance of two similar reactances in parallel, should the reactance or the susceptance values be added?
3. What is the letter symbol that indicates conductance? Susceptance? Admittance?
4. What are the two possible units of measurement of susceptance?
5. A 20-Ω X_L is in parallel with a 25-Ω X_L. What is the total susceptance value? The total X_L value?
6. A 50-Ω X_C is in parallel with an 80-Ω X_L. What is the total B value? The total X value?
7. A 10-Ω R, a 20-Ω X_L, and a 40-Ω X_C are all in parallel. What is the total G value? The total B value? The total Y value? The total Z value?
8. In question 7, does the source see the circuit as demanding more inductive or more capacitive current?
9. A 50-Ω X_L, a 0.04-mho G, and a 0.02-mho B_C are all in parallel. What is the total B value? What special type of circuit is this? What is the impedance of this circuit? The phase angle?
10. If the frequency of the ac in question 9 is halved, does the circuit appear inductive, resistive, or capacitive to the source? Why?
11. Why is a high-Q antiresonant circuit said to have a "flywheel effect"?
12. What is the impedance value of an antiresonant circuit with no resistance in it? If 100 Ω of resistance is added to one branch, will the circuit impedance increase, decrease, or remain the same?
13. What would be the waveshape of the ac flowing in a high-Q antiresonant circuit?
14. A 50-Ω R and an 80-Ω X_L are in parallel across a 200-V ac source. What is the value of I_R? I_X? I_S? θ? Z by Ohm's law?

ANSWERS TO CHECK-UP QUIZ 15-3

1. (0.025 mho) (0.10 mho) **2.** (0.075-mho B_L) (0.05 mho) (0.090 mho) (11.1 Ω) (56.3°) **3.** (10 V) **4.** (0.5 A) (2 A) (1 A) **5.** (1.5-A X_L) **6.** (1.8 A inductive) (11.1 Ω) **7.** (20 Ω) (20 Ω) (20 Ω) (Antiresonant with shunt resistance)

16 TUNING CIRCUITS AND FILTERS

CHAPTER OBJECTIVE. To develop a usable understanding of these electrical vocabulary terms: tuning circuit, antenna, electromagnetic wave, electrostatic wave, frequency selection, frequency rejection, attenuate, pass band, bandpass, loose coupling, critical coupling, tight coupling, overcoupling, coefficient of coupling, bandwidth, frequency response, inductive coupling, capacitive coupling, double-tuned transformer, bandpass filter, low-pass filter, *m*-derived filter, constant-*k* filter, *LC* filter, *RC* filter, π-type filter, high-pass filter, bandstop filter, band elimination filter, wavetrap.

16-1 TUNING CIRCUITS

There are countless applications where it is required to pass only one particular frequency, or a small band of adjacent frequencies, to some electronic circuit. For example, the air is constantly filled with radio waves from thousands of different stations on thousands of different frequencies. If a radio or television receiver accepted all these frequencies, the output of the receiver would be a hopeless jumble of music, speech, and noise. It is necessary to devise some circuit which will accept a desired frequency and at the same time reject all others. The circuit shown in Fig. 16-1 represents an antenna wire connected to an air-core transformer which can select a desired frequency.

It is desired to tune in a radio station transmitting a signal on 1000 kHz. Electromagnetic and electrostatic "waves" developed and transmitted by this station, as well as thousands of other stations on other frequencies, are constantly passing across the antenna wire and inducing voltages into it. The antenna circuit consists of an antenna wire having inductance, a variable capacitor, a primary coil with inductance, and capacitance between the antenna wire and the ground, making up a complex series *LC* circuit resonant to some frequency. The only adjustable part of this circuit is the variable capacitor. With the fixed value of inductance in the circuit, there must be some capacitance value that will bring the antenna circuit into resonance at 1000 kHz. By tuning the antenna capacitor to the proper value, the circuit will resonate at 1 MHz and produce a maximum current in the antenna circuit at this frequency. At the same time, the 1-MHz resonant circuit will tend to reject all other frequencies. Even a strong local signal 100 kHz away may not be able to produce any appreciable current flow in the antenna circuit and the tuning transformer primary.

Coupled to the primary, or antenna coil,

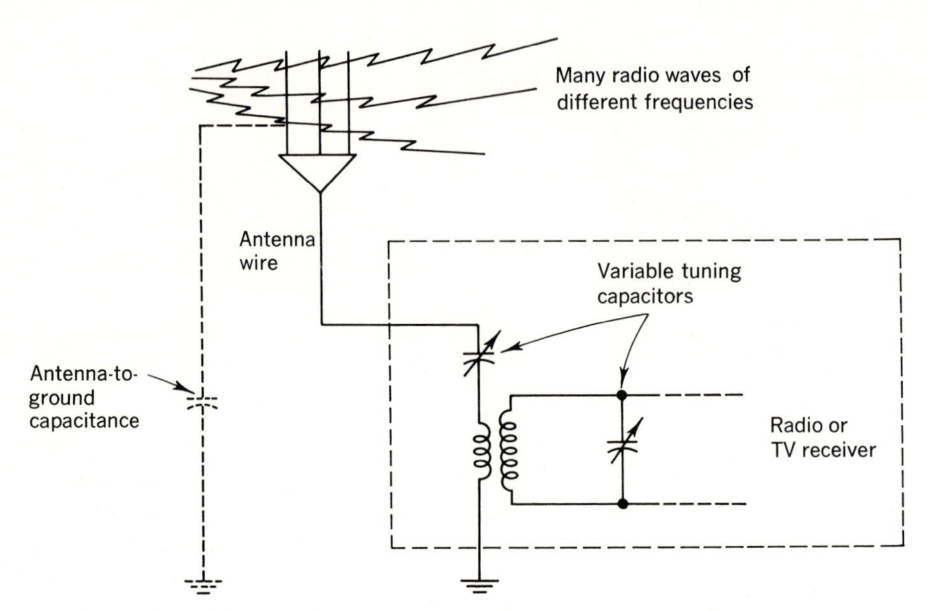

Fig. 16-1 A possible antenna circuit feeding into a radio receiver.

is a second tuned circuit. This is actually a series resonant circuit, because the ac induced into it is induced into individual turns, and not across the whole coil at once.

If the secondary circuit is tuned to the same frequency as the primary, two circuits are selecting 1000 kHz and attempting to reject all other frequencies. The 1000-kHz ac developed in the secondary tuned circuit is passed on to the receiver circuits, where it is detected, amplified, and made audible by a loudspeaker.

The higher the Q of the tuned circuits, the narrower the band of frequencies that is passed without attenuation.

To tune in another station on another frequency, perhaps 1200 kHz, it is necessary to tune both the antenna circuit and the secondary circuit to the new frequency. Both tuned circuits will then reject the 1000-kHz signal that they previously accepted.

Quiz 16-1. Test your understanding. Answer these check-up questions.

1. Would a tuning circuit be practical if it consisted of a variable inductor and a fixed capacitor? ________
2. In what two forms are waves transmitted from radio stations? ________ ________
3. When a radio wave crosses an antenna wire, what happens? ________
4. What could be done to tuned circuits to make them reject off-frequency signals better and still pass the desired frequency? ________
5. Is the secondary tuned circuit of Fig. 16-1 a resonant or antiresonant circuit? ________ What is the primary? ________
6. Express 1200 kHz in MHz. ________

16-2 DOUBLE-TUNED TRANSFORMERS

There are many cases in which a fixed band of frequencies must be passed and all other frequencies must be attenuated. For example, it is often desired to pass a band of frequencies from 450 to 453 kHz.

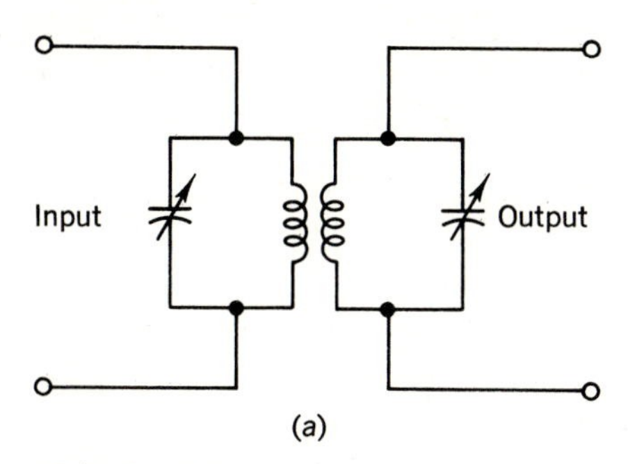

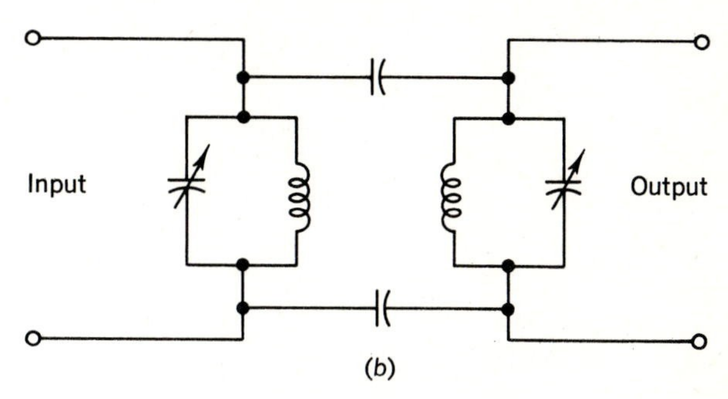

Fig. 16-2 Double-tuned transformers, forming bandpass-type filter circuits: (*a*) inductively coupled form; (*b*) capacitively coupled form.

The circuit must have a "pass band," or a *bandpass*, of 3 kHz. This can be accomplished by using a series of double-tuned transformers. In Fig. 16-2*a*, signals are fed to the input antiresonant circuit. Signals at the antiresonant frequency produce maximum current flow in this circuit. The current produces magnetic fields that "couple," or cross over, to the resonant secondary circuit. If the coupling is *loose*, only relatively few primary lines of force induce emf into the secondary, and the value of the output voltage is low, as shown in Fig. 16-3.

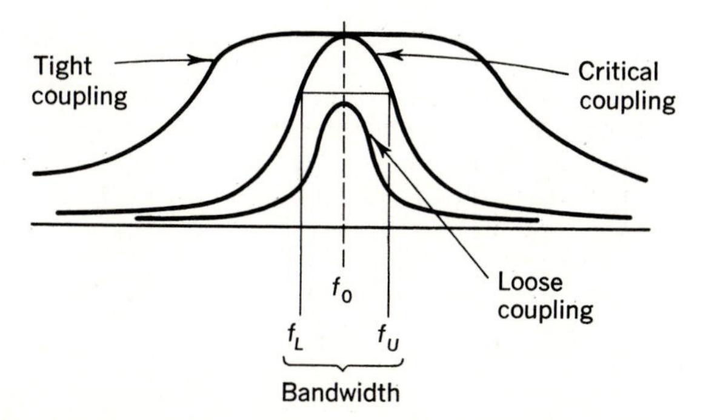

Fig. 16-3 Frequency responses of a double-tuned transformer with loose, critical, and tight coupling. The bandwidth, f_L to f_U, is shown on the critically coupled curve.

If the degree of coupling between the primary and secondary circuits is increased, more lines of force cut the secondary, and a greater output voltage is produced, but also, more adjacent frequencies can force their way through the system. *Critical coupling* is the lowest coupling value that will produce maximum output signals at f_o.

With *tight coupling* the voltage across the output does not increase over that of critical coupling, but the circuit becomes very *broad*, accepting frequencies relatively far removed from the resonant frequency of the circuits. Tight coupling (*overcoupling*) produces a broad bandwidth response, whereas loose coupling produces a narrow bandwidth. Although the terms *loose* and *tight* are often used, the proper terms are *low coefficient of coupling* and *high coefficient of coupling*. From Fig. 16-3, even with a low coefficient of coupling some frequencies far removed from f_o will appear in the output, although they may be attenuated considerably.

The *bandwidth* of a circuit is considered as the number of hertz or kilohertz between points on the resonant curve that are above 70% of the peak response. The bandwidth is indicated on the critical-coupling curve in Fig. 16-3, where f_L is the lower-frequency bandwidth point and f_U is the upper-frequency bandwidth point.

It is evident that bandwidth is a function of coupling between the primary and secondary circuits of a double-tuned transformer. The Q of the coils used in a double-tuned transformer also helps to determine the bandwidth. The lower the resistance of the wires in the circuit, the higher the Q of the circuit and the narrower the bandwidth. If the output circuit operates into a relatively low resistance, this reflects back as a heavy load on the primary and lowers its Q, broadening the bandwidth of the whole circuit.

The circuit of Fig. 16-2*b* shows capacitive coupling between two tuned circuits. The coils are not close enough to have any magnetic induction. Small coupling capacitors result in loose coupling, and larger capacitors produce tighter coupling and broadening of the pass band of the circuit.

A double-tuned transformer may be termed a *bandpass filter*, since it filters out the undesired frequencies and passes only the desired frequencies in its pass band.

Quiz 16-2. Test your understanding. Answer these check-up questions.

1. How far up the resonant curve is the bandwidth of the circuit measured? ________
2. What are two other terms that mean the same as "low percentage of coupling"? ________ ________
3. What are two other terms that mean the same as "high percentage of coupling"? ________ ________
4. What is meant by "critical coupling" between two circuits? ________
5. What type of a filter system is produced by using double-tuned transformers? ________
6. What are three methods by which the bandwidth of a double-tuned transformer circuit can be narrowed? ________ ________ ________
7. What is another term that means the same thing as "electrostatic coupling"? ________
8. What is meant by "f_o" in speaking of filters? ________
9. What is meant by "f_L" in speaking of filters? ________
10. In Fig. 16-2*a*, is the secondary a resonant or an antiresonant circuit? ________
11. In Fig. 16-2*b*, is the secondary a resonant or an antiresonant circuit? ________
12. Could the bandwidth of a resonant circuit be expressed the same way as it is with double-tuned transformers? ________ Could this terminology be used with an antiresonant circuit? ________

ANSWERS TO CHECK-UP QUIZ 16-1

1. (Yes) **2.** (Electromagnetic, electrostatic) **3.** (It induces an emf in the wire) **4.** (Raise the circuit Q) **5.** (Resonant) (Resonant) **6.** (1.2 MHz)

16-3 LOW-PASS FILTERS

The frequencies involved in transmission of intelligible human speech are usually from 200 Hz to about 3 kHz. While frequencies higher and lower than these are audible, they are not required for intelligibility. A telephone company uses many 3-kHz low-pass *filters*—devices that will pass frequencies below 3 kHz but not frequencies above this.

There are many circuits that will discriminate against higher frequencies but pass lower frequencies. Figure 16-4 illustrates some basic filtering circuits and the response curves for them. In Fig. 16-4*a*, the capacitor has lower reactance to the higher frequencies and acts as a short circuit to any higher frequencies fed into the filter. As a result, only low frequencies will pass relatively unattenuated in this *RC* low-pass filter. The curve below the diagram indicates, in a relative fashion, what frequencies will actually be allowed to pass through the filter.

In Fig. 16-4*b*, the inductance in series with the line increases its reactance at higher frequencies but passes the lower frequencies with little attenuation.

In Fig. 16-4*c*, both shunt capacitors and

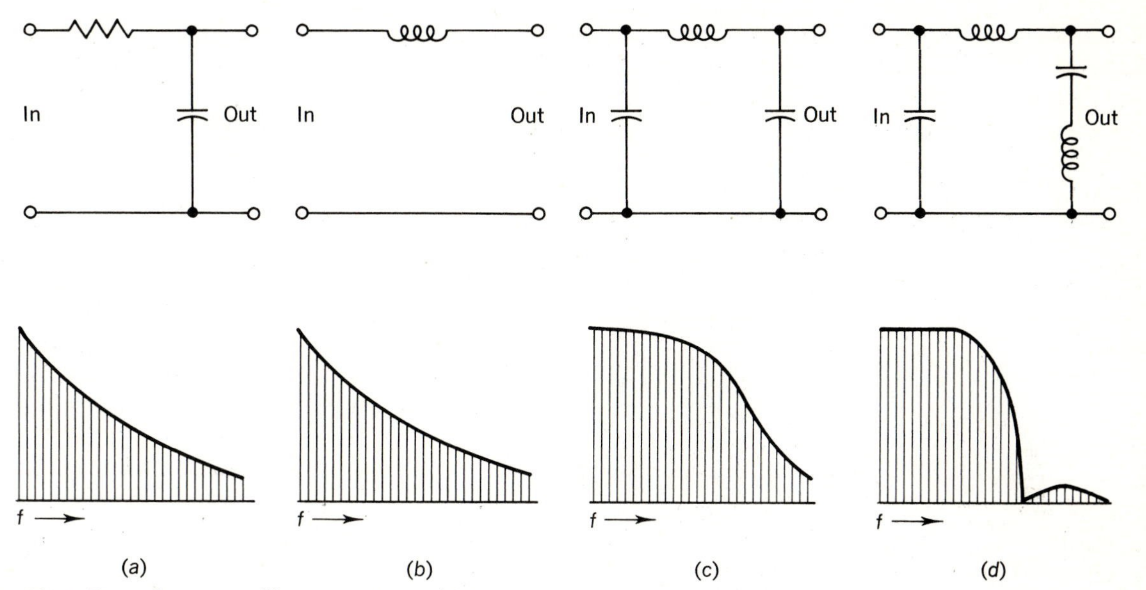

Fig. 16-4 Basic low-pass filter circuits and frequency responses. (*a*) *RC* L-type. (*b*) Series inductor. (*c*) π-type constant-*k*. (*d*) π-type *m*-derived.

a series coil are employed to encourage a sharper cutoff of frequencies above a certain frequency. The filter is a π-type (from its shape) low-pass unbalanced constant-*k* filter. If it had another similar coil in the lower line between the capacitors, it would be a balanced filter. In general *LC* filters without resonant circuits are called *constant-k* filters.

In Fig. 16-4*d*, the series resonant circuit in the output makes this a π-type low-pass unbalanced *m-derived* filter. The resonant circuit across the output of the filter acts as a short circuit to the frequency to which it is resonant. This one frequency is prevented from passing through the filter. In a 3-kHz filter this resonant circuit might have a resonant frequency of about 3.3 kHz.

All power supplies that change ac to pulsating dc and then reduce the pulses to smooth dc use some form of a low-pass filter, since the pulse frequency is higher than dc (zero frequency).

16-4 HIGH-PASS FILTERS

When all frequencies are to be passed above a given cutoff frequency, a high-pass filter is used. High-pass filters and low-pass filters are opposite configurations. For example, Fig. 16-5*a* shows a shunt coil across the filter. An inductor has a high impedance to high frequencies and a low impedance to low frequencies. The coil acts as a short circuit to low frequencies but allows high frequencies to pass along from input to output.

In Fig. 16-5*b*, a capacitor has high impedance to low frequencies and low impedance to high frequencies. Therefore it passes the high frequencies but not the lows.

In Fig. 16-5*c*, the π-configuration of a balanced high-pass constant-*k* filter is shown. In most balanced filters of this type, the coils would be center-tapped and grounded (connected to the metal chassis of the equipment, which in turn is

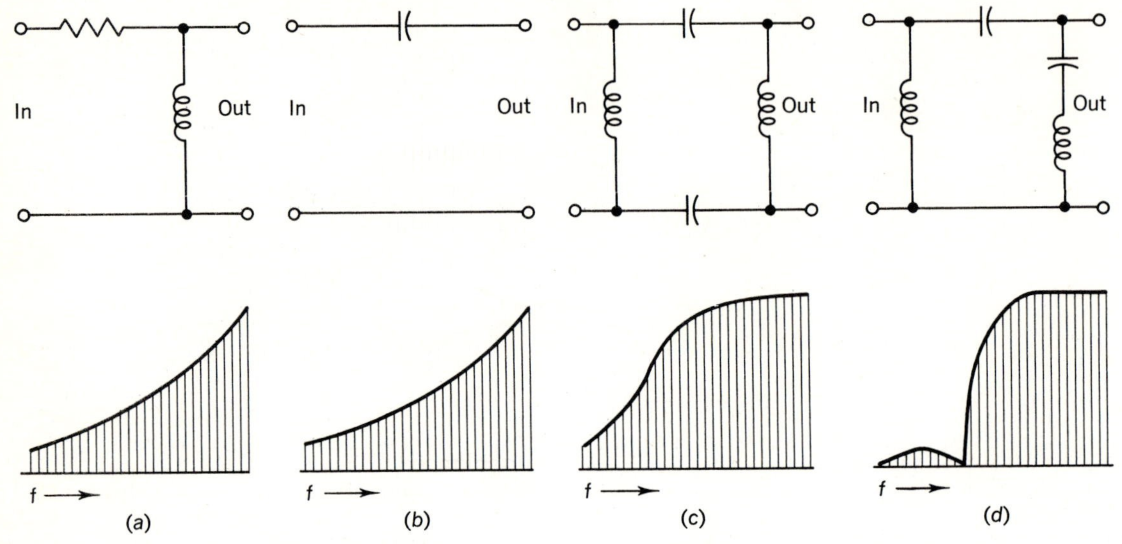

Fig. 16-5 Basic high-pass filter circuits and frequency responses.

often connected to a water pipe or to a pipe driven into the earth).

In Fig. 16-4*d*, the resonant circuit will be tuned to a frequency just lower than the lowest frequency desired to be passed by the filter to sharpen the cutoff characteristics of this *m*-derived filter circuit.

Most television receivers have some form of high-pass filter between the antenna and the first stage of the receiver. This filter is tuned to cut off all frequencies lower than about 54 MHz, which is the lowest television transmitting frequency. The filter prevents energy of nearby transmitters on lower frequencies from interfering with received television signals.

ANSWERS TO CHECK-UP QUIZ 16-2

1. (0.7 of max) **2.** (Loose coupling) (Low coefficient of coupling) **3.** (Tight coupling) (High coefficient of coupling) **4.** (Max response at f_0 with least coupling) **5.** (Bandpass) **6.** (Raise Q) (Decrease coupling) (Light load on secondary) **7.** (Capacitive coupling) **8.** (Resonant frequency) **9.** (Lower bandwidth frequency) **10.** (Resonant) **11.** (Antiresonant) **12.** (Yes) (Yes)

16-5 BANDSTOP FILTERS

A fundamental type of filter, although it is used less frequently than others, is the *bandstop*, or band-elimination, filter. It is designed to pass a wide band of frequencies but to stop a small segment of these frequencies. A basic single-section type of bandstop filter is shown in Fig. 16-6. The antiresonant circuit in the upper line should have a very high impedance to its antiresonant frequency, stopping energy of this frequency from being passed. If the Q of this circuit is not high enough to stop the frequency entirely, the resonant circuit that follows will act as a short circuit to this frequency and should eliminate it completely.

If the antiresonant and resonant circuits are tuned to two slightly separated frequencies, all frequencies between them, as well as the antiresonant and resonant frequencies, will be effectively attenuated in the output. If there is too wide a separation in frequency, the in-between frequencies may be able to pass.

If only one antiresonant or resonant cir-

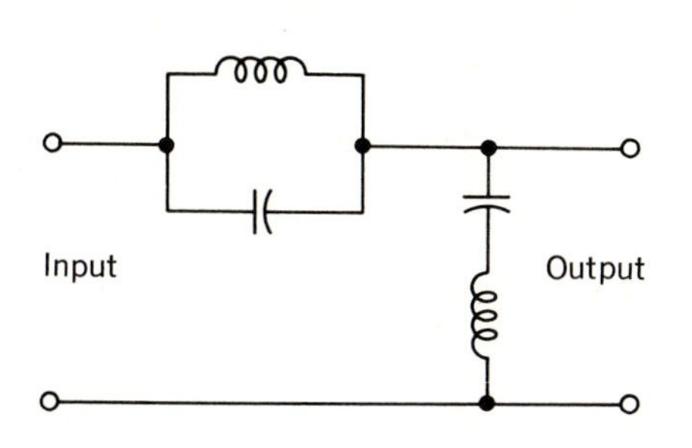

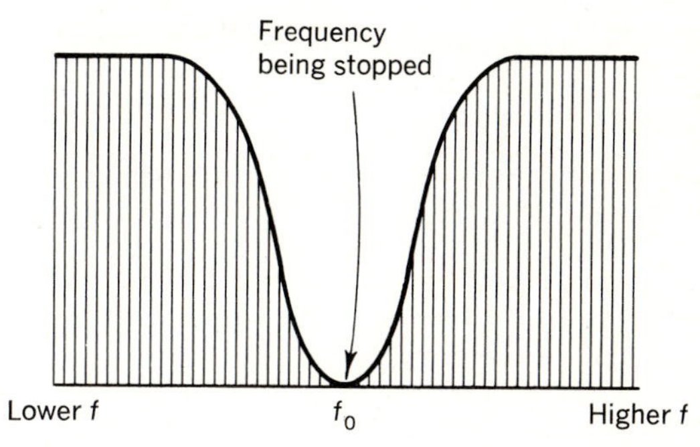

Fig. 16-6 An L-type bandstop filter. If either tuned circuit is used alone, they would be called wavetraps.

cuit is used, it is known as a *wavetrap*, and is fairly effective in stopping a single frequency. Every TV receiver has at least three wavetraps in it to stop undesired frequencies from appearing where they are not wanted.

Quiz 16-3. Test your understanding. Answer these check-up questions.

For the filter circuits shown in Fig. 16-7 identify (1) the configuration (L, T, or π types); (2) the pass type (high-pass, etc.); (3) whether the circuit is constant-k or m-derived; (4) whether the circuit is balanced or unbalanced.

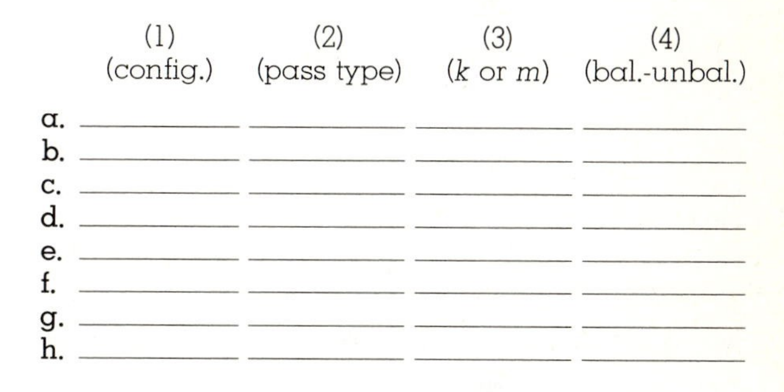

	(1) (config.)	(2) (pass type)	(3) (k or m)	(4) (bal.-unbal.)
a.				
b.				
c.				
d.				
e.				
f.				
g.				
h.				

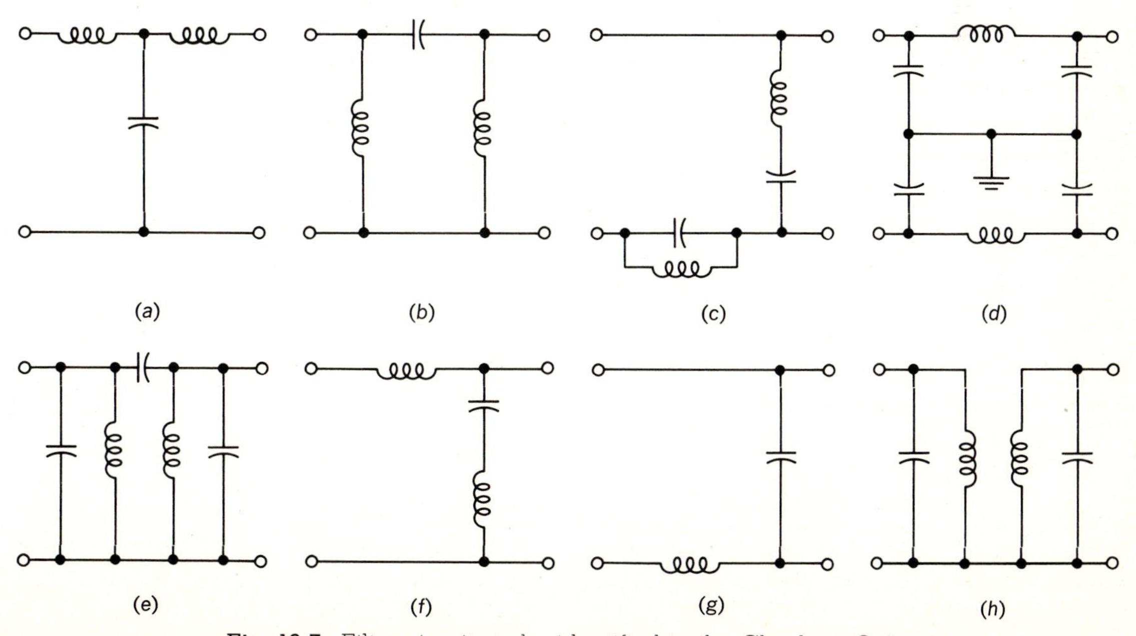

Fig. 16-7 Filter circuits to be identified in the Check-up Quiz.

CHAPTER 16 TEST • TUNING CIRCUITS AND FILTERS

1. If a parallel *LC* circuit is in series with an antenna leading to a receiver, would this circuit stop, pass, or have no effect on the frequency to which it is resonant?
2. Would any wire, regardless of its length, be resonant to some frequency?
3. Can resonant circuits be used to pass certain frequencies, stop certain frequencies, or both?
4. A tuned primary circuit is loosely coupled to a tuned secondary. What would be the expected relative pass band of this circuit? If the coupling coefficient were increased, what effect would this have on the pass band?
5. What degree of coupling would result in maximum output voltage of a double-tuned transformer with minimal bandpass?
6. Name two basic methods of coupling energy from one tuned circuit to another.
7. What can be done to a tuned circuit to reduce its bandwidth characteristics?
8. What does the "over" in "overcoupling" indicate?
9. When speaking of filters, what is signified by f_o? By f_L?
10. What is the effect on the Q of an antiresonant circuit if resistance is added in series with a branch? If resistance is connected across the circuit?
11. What basic form of filter would be used by a telephone company to limit frequencies to 3 kHz? In a television receiver to prevent local lower frequencies from entering the receiver? To allow signals from 450 to 460 kHz to get through?
12. What type of filter produces a faster cutoff of frequencies than is obtainable with a constant-*k* type?
13. What is the basic form of filter that has a coil across and a capacitor in series with the line?
14. What are the type and configuration names of a filter having an input *L* across a series *C*, and on output *L* across the line? A series capacitor, a shunt coil, and a series capacitor?
15. What is the special name for the simplest type of bandstop filter?

ANSWERS TO CHECK-UP QUIZ 16-3

a. (T, low-pass, const.-*k*, unbal.) **b.** (π, high-pass, const.-*k*, unbal.) **c.** (L, bandstop, *m*-derived, unbal.) **d.** (π, low-pass, const.-*k*, bal.) **e.** (π, bandpass, *m*-derived, unbal.) **f.** (L, low-pass, *m*-derived, unbal.) **g.** (L, low-pass, const.-*k*, unbal.) **h.** (Not applicable, bandpass, not applicable, unbal.)

CHAPTER OBJECTIVE. To describe a few of the control devices and basic circuits and to develop a usable understanding of these electrical vocabulary terms: diode, passive, active, plate, anode, cathode, heater, filament, space charge, plate current, rectify, rectification, triode, grid, bias, photoemissive cell, photoelectric, amplifier, transducer, audio frequency, I_p cutoff, tetrode, pentode, phanotron, mercury-vapor diode, thyratron, ignitron, ignitor, solid-state diode, semiconductor, doped, impurity atom, *N*-germanium, *P*-germanium, hole, barrier, forward bias, LED, transistor, emitter, base, collector, collector current, *NPN*, *PNP*, FET, IC, MSI, LSI, SCR, gate junction, halfwave, fullwave, triac, thyristor, LASCR, thermistor.

17-1 THE DIODE VACUUM TUBE

The chapters up to this point have dealt with components of a *passive* nature, such as resistors, inductors, and capacitors, transformers, and relays. An *active* device is one which can cause change. In this chapter some active devices will be considered from the standpoint of their controlling abilities. From a historical viewpoint, the first active device was the *vacuum-tube diode* (Fig. 17-1). It consists of a glass envelope in which two *elements* are placed, a *plate* or *anode* (an electron collector), and a *cathode* (an electron emitter). All the air is pumped out of the envelope, making a vacuum inside.

One common diode has a *heater*, which is a thin filament of wire under a metal cathode. The cathode surface is coated with barium or strontium oxide, which liberates and emits electrons when heated. The other element in the vacuum tube is the metal plate. (Note that a coated metal cathode and its heater wire are considered as only one element.) The cathode, when hot, liberates electrons, so that they are free to travel across the vacuum space to the plate. This will occur only if the cathode is hot *and* the plate is positive in respect to the cathode.

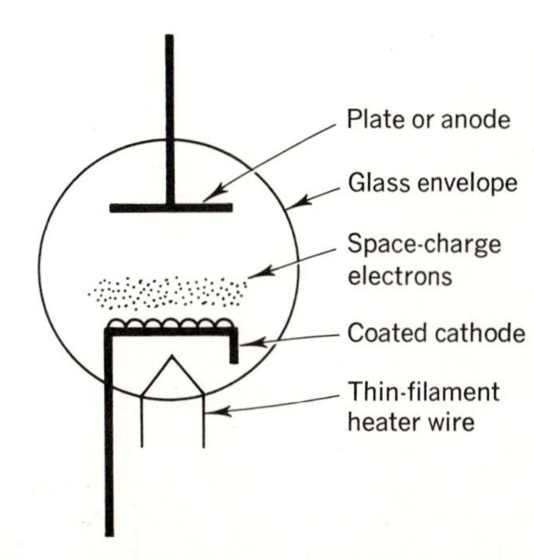

Fig. 17-1 Essentials of a vacuum-tube diode. Space-charge electrons are available only if the cathode is heated by the heater wire inside it.

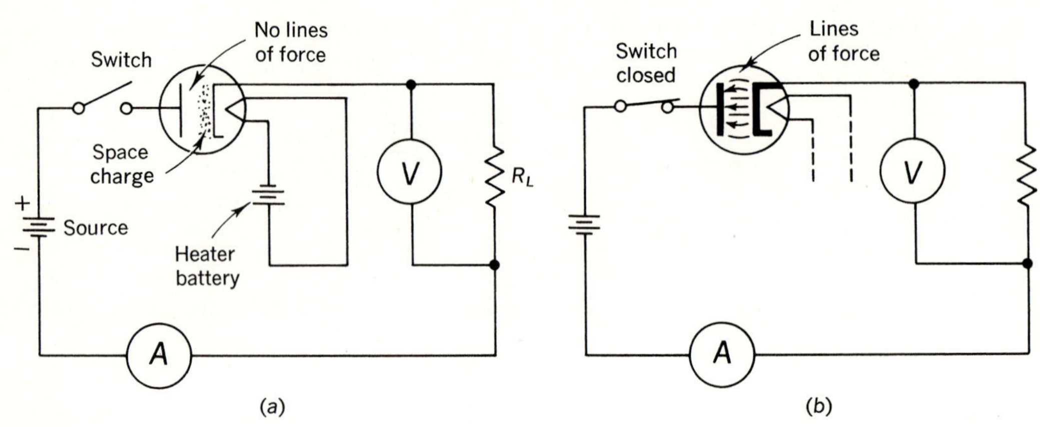

Fig. 17-2 Diode circuits. (*a*) Switch open, no electrostatic lines of force to pull electrons to plate. (*b*) Switch closed, space-charge electrons follow lines to plate.

To heat the cathode, a source of emf must be connected across the heater wire (Fig. 17-2*a*). Current flowing through this resistance wire heats it, in turn heating the nearby cathode metal. The hot cathode excites outer electrons of the coating material on its surface. The orbital motion of these electrons is accelerated to such an extent that the electrons actually leave their atoms and fly out into the space of the vacuum. Normally such liberated electrons would circle around and return to their atoms (which had been made positive ions). However, if the cathode remains hot, there will always be a cloud of these free electrons above the surface of the cathode coating, called a *space charge*. Since it is composed of electrons, a space charge has a negative polarity.

The circuit of Fig. 17-2*a* shows a vacuum diode in series with a battery source, an ammeter, a load resistor, and a switch. The switch is open. The heater wire is connected to another battery, enabling the heater to warm the cathode and develop a space charge. With the switch open, no current can flow in the circuit containing the load resistor.

In Fig. 17-2*b*, the switch has been closed. Completing the circuit places the diode and resistor in series across the battery. Electrostatic lines of force appear between cathode and plate. The diode becomes a vacuum-dielectric capacitor. Unlike the usual capacitor, this one has free electrons in the dielectric area. These electrons are free to follow along the electrostatic lines of force to the plate and around the complete circuit. The flow of current can be measured by the ammeter and is termed *plate current*. Plate current flowing through the load resistor develops a voltage-drop across the resistor. If the resistance to current flow presented by the diode is a low value, the voltage-drop across the load will be nearly the same in value as the plate-circuit battery voltage. A diode is a good electrical conductor when its plate is positive in respect to its cathode. But what if the plate is negative?

The circuit shown in Fig. 17-3*a* is similar to the previous circuits, except that the plate-circuit battery is reversed in polarity. Now, the electrostatic lines of force between plate and cathode are reversed. The diode is still a vacuum-dielectric capacitor, but its space-charge electrons

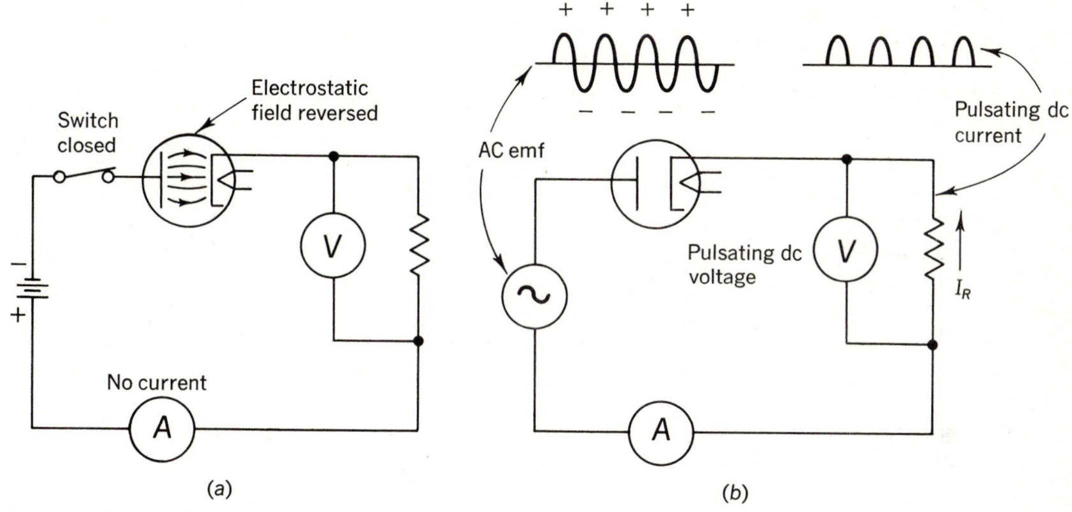

Fig. 17-3 (*a*) With the plate-supply polarity reversed, no plate current flows. (*b*) With an ac plate supply, current flows during half of a cycle. The circuit acts as a rectifier.

cannot move against the lines of force and are driven back to the cathode. Inasmuch as the plate is not hot, it does not liberate any electrons, so no current flows in the circuit. The ammeter reads zero. The voltmeter across the load also reads zero.

In Fig. 17-3*b*, an ac source has been substituted for the plate-circuit battery. With an ac emf, half the time the plate is positive and half the time it is negative. Half the time electrostatic lines of force will support plate current, but half the time they will not. The result is a pulsating dc flow. The ac is *rectified* to pulsating dc.

Quiz 17-1. Test your understanding. Answer these check-up questions.

1. How many elements would a *diode* have according to the term itself? ________ How many would a *triode* have? ________
2. What is the name given to the electron emitter in a vacuum tube? ________ The electron collector in a VT? ________
3. What prevents air molecules from interfering with the free movement of electrons in a diode VT? ________
4. What polarity must the plate have to allow plate current to flow in a VT? ________ What polarity must the electron emitter have in respect to the plate? ________
5. What is meant by the term *rectify* in speaking of diode circuits? ________
6. If the heater circuit were open in a VT diode, could plate current flow? ________ Could the tube rectify? ________
7. If ac were used in the heater circuit, would the emission of electrons have a constant value? ________
8. In Fig. 17-3*b*, what is the polarity of the top of the load with respect to the bottom? ________
9. In Fig. 17-3*b*, if the diode were reversed, what type of current, if any, would flow in the load? ________ What would be the polarity of the top of the load resistor? ________

17-2 THE TRIODE

When a third element is added between the cathode and the plate of a vacuum tube, a *triode* is formed. This in-between element is a gridwork of closely spaced fine wires. Note the electrostatic lines of force in Fig. 17-4*a*. With the grid connected to the cathode, there is no difference of potential between cathode and grid. Space-charge electrons from the

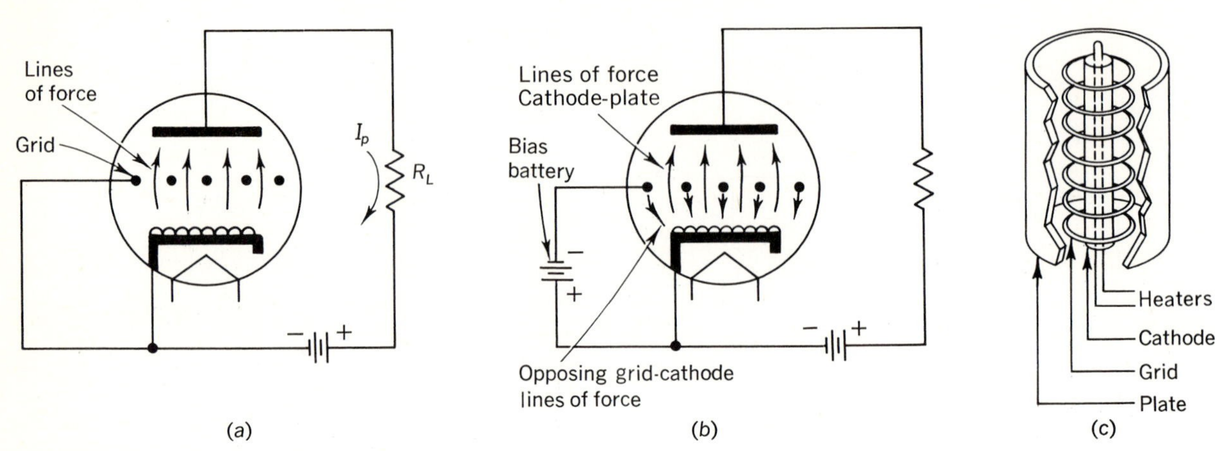

Fig. 17-4 (*a*) With the grid at cathode potential, space-charge electrons follow lines of force to the plate. (*b*) The reverse-direction field between grid and cathode tends to drive electrons back toward the cathode, decreasing plate current.

cathode move through the grid wires to the plate to establish a plate current.

In Fig. 17-4*b*, a *bias battery* has been added between cathode and grid, with the negative terminal to the grid. The bias voltage produces electrostatic lines of force between the grid and the cathode in opposition to those between cathode and plate. As a result, space-charge electrons between the grid and the cathode can be driven back toward the cathode if the negative potential on the grid is sufficient. If the negative potential does not completely neutralize the cathode-plate lines of force, some plate-current electrons can follow the lines of force to the plate. In a triode the amount of plate current depends on how negative the grid is made with respect to the cathode—highly negative, no plate current (I_p); slightly negative, some I_p; zero charge, heavy I_p.

ANSWERS TO CHECK-UP QUIZ 17-1

1. (2) (3) **2.** (Cathode) (Anode or plate) **3.** (Vacuum area has no molecules) **4.** (+) (−) **5.** (Change ac to pulsating dc) **6.** (No) (No) **7.** (Yes; cathode surface heats and cools slowly) **8.** (+ half the time) **9.** (Pulsating dc) (−)

It is significant that *voltage changes* in the grid circuit can produce *current changes* in the plate circuit. Current flowing through the load resistance represents power ($P = I^2R$). Thus any voltage changes in the grid circuit control the power in the plate circuit. Since there is no current flowing in the grid circuit (unless the grid is positive), the power in this controlling circuit is $P = EI$, or $E(0)$, or 0 W.

A triode tube can be used as a controlling device, as shown in Fig. 17-5. At the left, an electric lamp serves as a source of light. The light is beamed to a *photoemissive cell* in a box. This type of cell emits electrons from one portion of it to another portion, but only if excited by light energy. Any photoemitted electrons form a weak dc, downward through the 1,000,000-Ω resistor in the grid circuit. From the direction of the current flow, the top of the resistor becomes negative, which biases the triode to the point of nonconduction, and no plate current flows through the relay coil.

If some opaque object passes between the light source and the light-sensitive

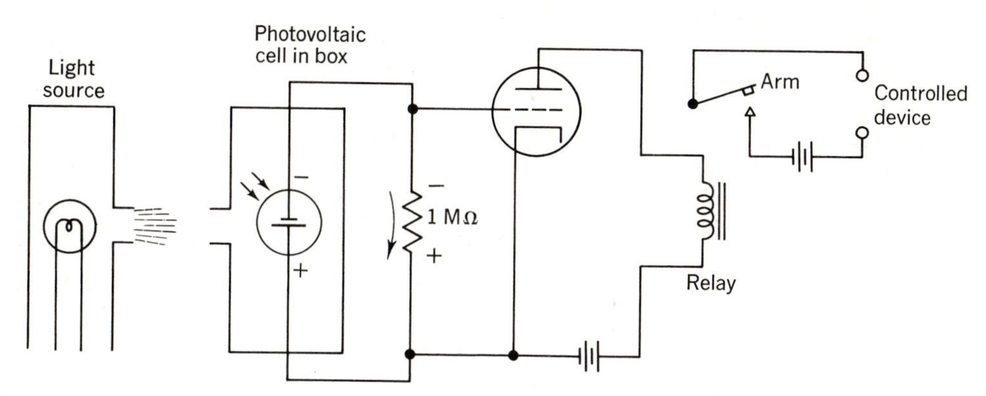

Fig. 17-5 Photoelectric control circuit. The relay contacts close when the light beam is blocked.

cell, the current through the cell stops. With no current there is no bias, and plate current flows through the relay coil. Current in the coil produces a magnetic field which pulls the iron relay arm toward the core, closing the relay contacts. The contacts can be connected to any form of controlled device. A person walking between the light source and the cell will actuate the relay to start a motor that opens a door. A row of cans on a production line passing through the light beam can actuate an electronic counter every time the relay closes, etc.

Using a triode as an "amplifier" is shown in Fig. 17-6. Sound waves striking the diaphragm of a microphone-type *transducer* (changer of one form of energy to another form) produce weak *audio* (audible) *frequency* ac emfs. These are fed to a step-up transformer to develop a higher-voltage ac in the secondary at the same frequency as the sound waves hitting the diaphragm. The triode is biased just enough to give half-maximum plate current (halfway to I_p *cutoff*). The audio-frequency (AF) ac voltage added to the dc bias produces a varying dc voltage between grid and cathode, which results in large I_p variations. The I_p variations are coupled through a transformer to a loudspeaker. The ac in the output transformer actuates the diaphragm of the loudspeaker, producing a sound of the same

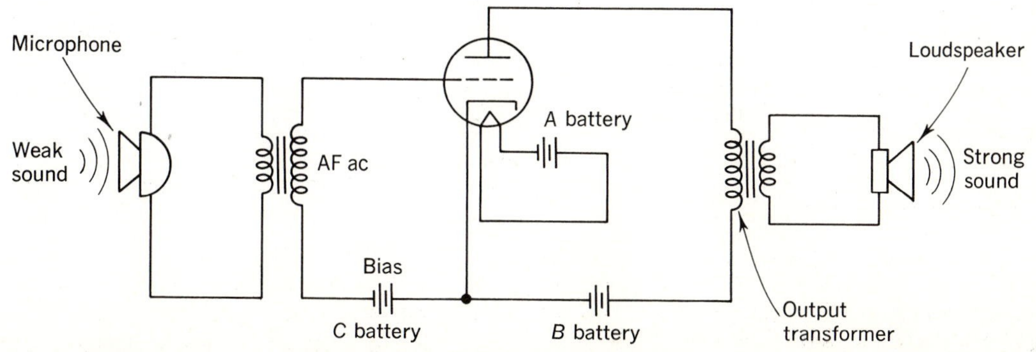

Fig. 17-6 A triode in a circuit to amplify weak sounds into stronger sounds.

frequency as that picked up by the microphone, but at a greater power level. Thus the triode tube works as an AF amplifier.

If one tube amplifies 40 times, but this is not enough, another similar stage can be added, and the amplification is then 40(40), or 1600 times. With a series of a few triodes tremendous amplification is possible.

There are tubes with two grids, called *tetrodes*, which amplify more than triodes. Those with three grids are called *pentodes* and amplify still more. Besides these basic types (Fig. 17-7), there are hundreds

Fig. 17-7 Vacuum tubes. Across top: three 8-pin octal-base tubes (high-voltage rectifier, glass envelope voltage regulator, metal envelope pentode), subminiature tetrode, pencil triode, and lighthouse triode. Across bottom: two 7-pin miniatures, a 9-pin noval-base, and three 9-pin noval-base tubes. Metal caps are plate connections in high-voltage tubes.

of variations and special types of tubes available.

Quiz 17-2. Test your understanding. Answer these check-up questions.

1. What potential charge should a plate-circuit battery place on the anode of a triode? _________ The bias battery on the grid? _________
2. If its heater is cold, can a triode still amplify? _________ When the heater is first turned on, the triode cannot immediately start to amplify. Why? _________
3. If the grid wires are closely spaced, would it require more or less grid bias to produce plate-current cutoff? _________ Would it require more or less signal voltage on the grid to control the plate-current variations? _________ Would such a triode amplify more or amplify less than one with widely spaced grid wires? _________
4. What is the name for a light-sensitive cell that emits or gives off electrons when struck by light energy? _________
5. What type of device converts electrical variations to mechanical sound waves? _________
6. What is the name of devices that can convert sound waves to electrical variations or alternations? _________
7. A single-triode amplifier stage has a gain of 50. What would be the gain of three stages in "cascade" (one following another)? _________ If a 20-μV signal were fed to the first grid, what would be the output voltage of the third plate circuit? _________
8. What plate-current waveform is produced when a sinusoidal ac is added to the bias in the grid circuit of a triode? _________ When a sawtooth ac is applied? _________
9. Would a triode amplify if the cathode-plate current did not vary? _________ If the grid-cathode voltage did not vary? _________
10. Could a hot filament wire alone, without a coated metal cathode, be used as an electron emitter in a VT? _________

17-3 GASEOUS TUBES

A part of electronics is the theory of current flow through gases. One rectifier tube used for higher-voltage and higher-current applications is constructed like a vacuum diode, but after evacuation, mercury-vapor gas is introduced into the envelope. A positive plate voltage of more than 15 V produces the required ionizing potential. When the mercury vapor ionizes, it turns bright blue, and current flows through the tube. The mercury will not ionize with the plate negative, resulting in a rectification of ac. Within normal operating conditions for the tube, the voltage-drop across mercury-vapor diodes (sometimes called *phanotrons*) is 15 V regardless of the current value. In comparison, high-vacuum tubes may have only a few volts drop across them if the load draws low current, but the voltage-drop may increase to several hundred volts with a heavy load.

Mercury-vapor diodes operate much cooler than do high-vacuum types, but they must be held in an upright position to allow mercury to pool at the bottom of the envelope, and the mercury must be warmed by the filament before plate voltage is applied.

A gas-filled triode is known as a *thyratron*. A thyratron is a mercury-vapor (or other gas) diode, but with a cylindrical grid element between cathode and anode, as in Fig. 17-8*a*. Although shown

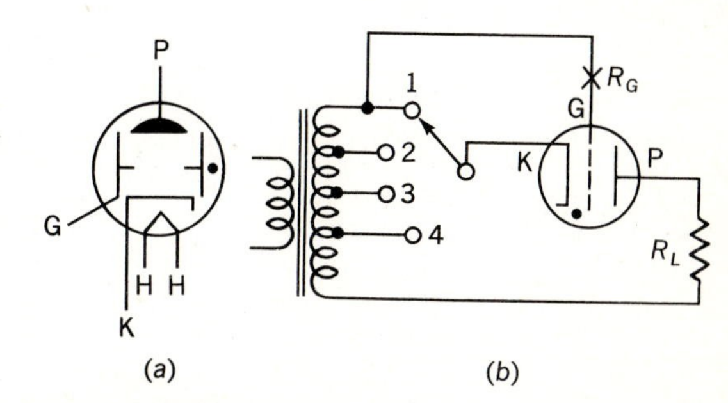

Fig. 17-8 (*a*) Basic thyratron construction. (*b*) Negative-control thyratron circuit.

with a heater cathode, thyratrons may have wire-filament cathodes.

A mercury-vapor thyratron is shown in series with an ac source and load in Fig.

17-8*b*. With the switch in position 1, the grid and cathode are at the same potential. If the grid opening is relatively large, as soon as the plate potential exceeds the normal 15-V ionizing point of mercury vapor, the gas ionizes and current flows. This results in almost 180° of current flow per cycle of source ac. Most thyratron tubes use a negative bias to determine the time during the cycle when the tube fires. It requires only a few volts of negative bias to hold off the firing of these thyratrons completely.

In the circuit of Fig. 17-8*b*, with the switch in the position shown, current flows through R_L for half the cycle. By moving the switch to position 2, a negative bias is developed between grid and cathode at any time a positive voltage is being developed between plate and cathode. If position 2 adds just enough negative bias, the thyratron may not fire for perhaps 60° after the start of the cycle, and current will flow through the load for only 120°. Switch position 3 may produce only 90° of current flow. In this way, the average current in the load is controlled by the position of the switch. To prevent heavy grid-current flow during the "off" half-cycle, a resistor (R_G) of about 1 MΩ is added in series with the grid, at *X*. The black dot in the symbol indicates gas in the envelope.

Mercury-vapor thyratron turn-on times range from 10 to 20 μs, with turn-off times from 100 to 1000 μs. The lighter the gas in a thyratron, the faster the turn-on and turn-off. When fast turn-on and turn-off is required, hydrogen thyratrons are used, but current-handling ability is reduced.

A lower-voltage, higher-current mercury triode is the *ignitron*. The ignitron consists of a pool of mercury at the bottom of the tube, as in Fig. 17-9*a*, a highly resistive *ignitor* probe with its point in the mercury, and a carbon or graphite anode. A positive plate voltage alone can produce no ionization of the liquid mercury. If a pulse of voltage is fed to the ignitor, tiny arcs develop between ignitor and mercury, starting ionization and vaporization of the mercury. Electrostatic lines of force develop across the high-resistance ignitor rod, producing an ionized atmosphere that moves up the ignitor. If the plate is positive, the ionization continues on up to the plate, and the tube conducts. It con-

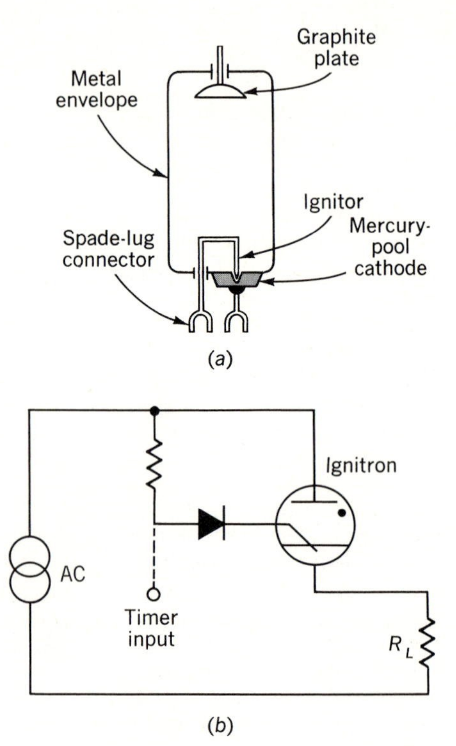

Fig. 17-9 (*a*) Basic construction of an ignitron. (*b*) Simple ignitron firing circuit.

ANSWERS TO CHECK-UP QUIZ 17-2

1. (+) (−) **2.** (No) (Time required to heat the cathode metal) **3.** (Less) (Less) (More) **4.** (Photoemissive transducer) **5.** (Loudspeaker, earphones, or transducer) **6.** (Microphone or transducer) **7.** (125,000) (2.5 V) **8.** (Sinusoidal varying dc) (Sawtooth varying dc) **9.** (No) (No) **10.** (Yes)

tinues to conduct until the plate voltage decreases to the deionization potential of the mercury vapor. Figure 17-9*b* illustrates one simple method of starting the ignitron. This circuit produces halfwave rectification, with current flowing for almost 180° of the source cycle. The diode prevents ignition of the mercury during the nonconducting half-cycle. The ignitron can also be fired by applying a positive timing pulse to the "timer input."

Quiz 17-3. Test your understanding. Answer these check-up questions.

1. What is a phanotron? __________
2. What is the visible indication when a mercury-vapor diode is conducting? __________
3. What is the ionization potential of mercury vapor? __________
4. What are two advantages of mercury-vapor over high-vacuum diodes? __________ __________
5. What does the dot in the symbol of a tube indicate? __________
6. In Fig. 17-8*b*, what occurs with the switch in No. 4 position? __________
7. What is the advantage gained by using low-atomic-number gases in thyratrons? __________
8. What elements do thyratrons have that an ignitron does not have? __________ __________
9. What does an ignitron have that a thyratron does not have? __________
10. What is the advantage of an ignitron over a thyratron? __________
11. For what might thyratons and ignitrons be used? __________ __________

17-4 SOLID-STATE DIODES

Most modern electronics uses solid-state devices in place of vacuum or gaseous active devices. Some materials, such as metals, have many free electrons in their atoms at room temperature, and as a result make good conductors. Other substances, glass, rubber, sulfur, etc., have almost no free electrons in their outer orbits at room temperature and are good insulators. There are several materials which have only a few free electrons at room temperature and are called *semiconductors*. Some of the more common semiconductors are silicon, germanium, and carbon, as well as compounds such as gallium arsenide.

Germanium, manufactured in pure crystalline form in laboratories, is a fairly good insulator. However, if about one in a million of the atoms making up the germanium crystal is replaced during manufacture by an arsenic *impurity atom*, the resultant *doped* crystal will have one in a million atoms with an electron that is only loosely held. As a result, such a crystal becomes a reasonably good electrical conductor. It is known as *N-germanium*, although it does not actually have a negative charge.

If germanium crystals are manufactured with gallium as the doping impurity, there is one area in a million that appears to have a *hole* in its outer electron ring and to lack an electron—or at least it seems attractive to electrons. This substance is known as *P-germanium*, although it does not actually have a positive charge. Because of the electronless holes that can loosely hold electrons, *P*-germanium also acts as a fairly good conductor.

In Fig. 17-10*a* there is no emf applied across the *N*- and *P*-germanium (or silicon) unit. At the junction some of the positive holes and some of the freely moving electrons of the *N* material combine and discharge this area. The zero-charged area acts as a barrier to any further combination of holes and electrons.

If a germanium *NP* junction is *biased* with an external emf, with the negative end of the bias battery to the *N* material and the positive to the *P* (Fig. 17-10*b*), any external emf greater than approximately

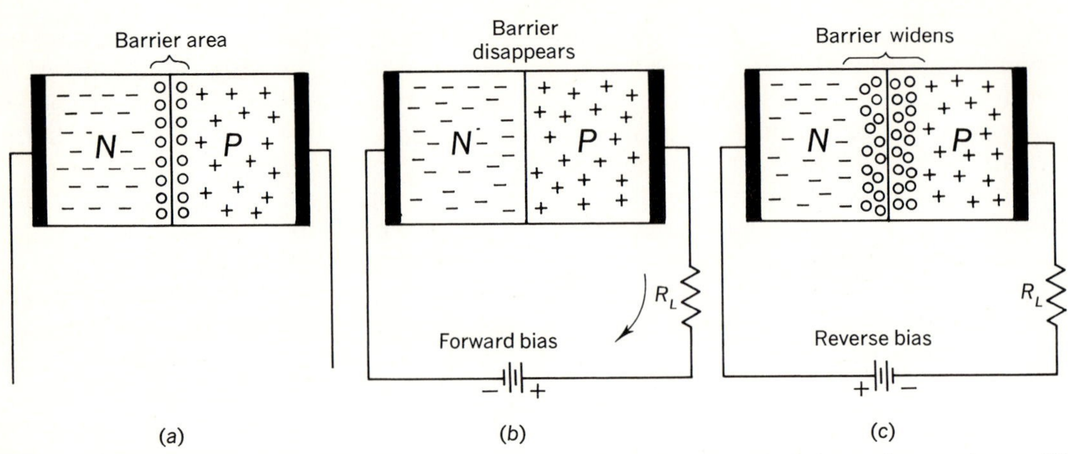

Fig. 17-10 Solid-state junction actions. (*a*) A barrier area is developed with no bias voltage. (*b*) With forward bias (− to *N*, + to *P*), the barrier disappears and the junction conducts. (c) With reverse bias, the barrier widens and no current flows.

0.3 V will overcome the barrier effect at the junction, and the whole unit becomes a reasonably good conductor. (The barrier voltage of a silicon junction is about 0.6 V.) The junction is said to be *forward-biased* into conduction, and current flows through the load resistor R_L.

In Fig. 17-10*c*, the junction is reverse-biased, negative to *P* and positive to *N*. This results in electrons from the negative terminal of the battery filling the holes in the positive semiconductor material, widening the barrier area at the junction. No current can flow through the junction or through the load. If the external source is an ac voltage, current can flow through the load for only half of each cycle, during the time that the junction is forward-biased. This is *rectification*. Thus when rectification is required, a vacuum-tube, a gaseous-tube, or a solid-state diode might be used.

Solid-state diodes can be made very small. A diode the size of a match head, for example, might be considered large (Fig. 17-11).

A germanium or silicon diode junction with current flowing through it radiates energy at an infrared frequency. A gallium-arsenide junction radiates at a red frequency, and a gallium-phosphide junction gives off green light. A diode made to give off light is called a *light-emitting diode*, or an LED.

ANSWERS TO CHECK-UP QUIZ 17-3

1. (Mercury-vapor diode) **2.** (Gas glows blue) **3.** (15 V) **4.** (Higher I_p) (Run cooler) **5.** (Gas) **6.** (Probably stop I_p flow) **7.** (Faster on-off times) **8.** (Grid) **9.** (Ignitor) **10.** (Higher I_p) **11.** (Motor controlling, high-current rectifiers)

17-5 TRANSISTORS

A transistor is similar to a vacuum-tube triode in some respects and very different in others. A transistor is composed of three areas of semiconductor material in contact (Fig. 17-12). Notice that in a transistor there are two junctions. From the bottom, first there is an *NP* junction, and above this a *PN* junction. The bottom area is termed the *emitter*. The very thin middle section is called the *base*, and the upper part is the *collector*. The emitter must emit

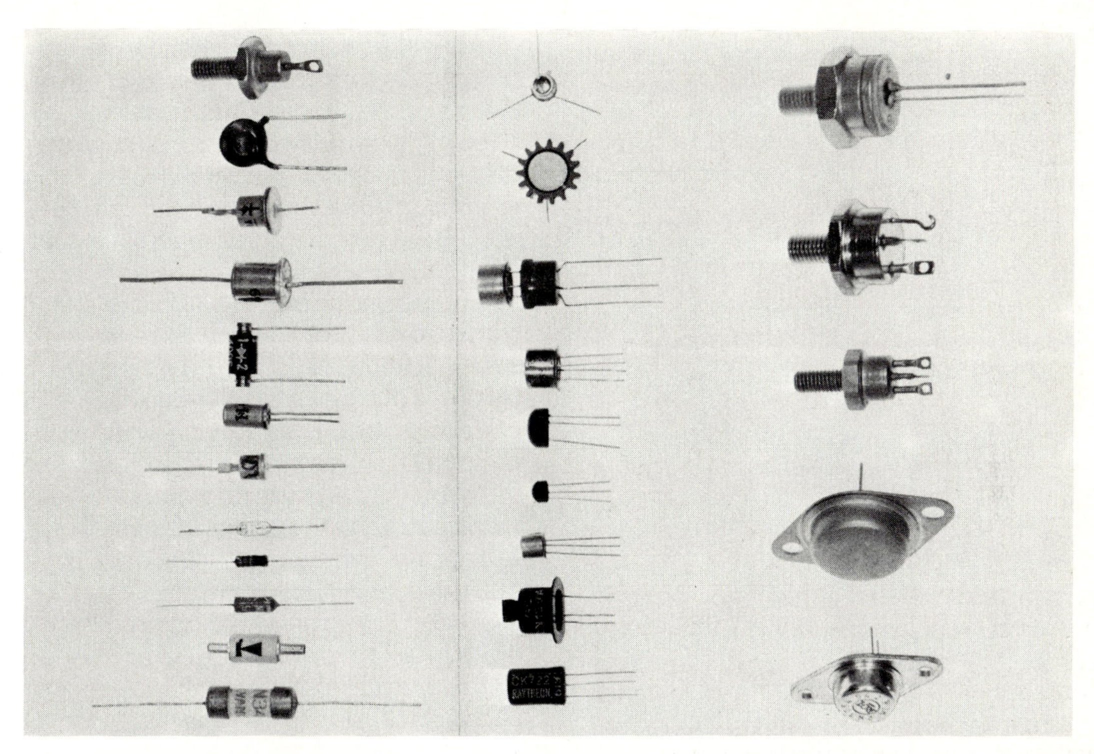

Fig. 17-11 Semiconductor devices. Left column, from top: five power-supply diode types, two voltage-regulator zener diodes, and five high-frequency and general-purpose diodes. Middle column, from top: phototransistor, transistor with a heat-sink around it, transistor in a socket, and six varieties of transistors. Right column: silicon controlled rectifiers and power transistors.

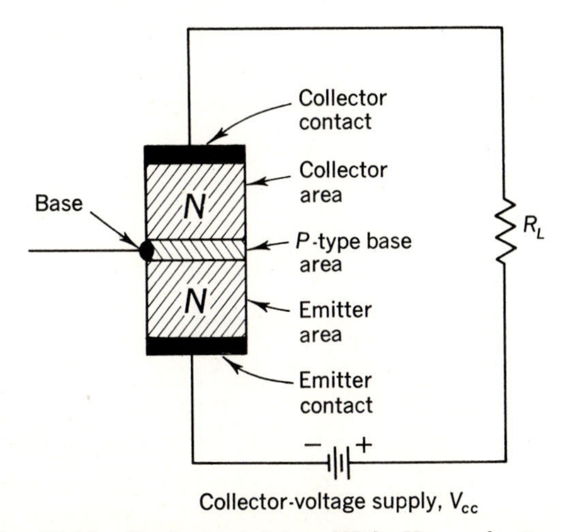

Fig. 17-12 Basic transistor. With V_{CC} polarity as shown, the *NP* lower junction is forward-biased, but since the *PN* upper junction is reverse-biased, no current can flow in R_L.

electrons into the base and is therefore somewhat analogous in function to the cathode in a vacuum tube. The collector is used to collect any electrons that can make their way through the base area. It is somewhat analogous to the plate of a vacuum tube. When a little current flows from the emitter to the base, through the emitter-base junction, an action takes place that allows many electrons from the emitter to pass through the base to the collector. Thus a small emitter-base current may produce a relatively large emitter-collector current flow. If the base current decreases a little, the emitter-collector current decreases considerably. For this reason a transistor is considered to be a *current-operated* amplifying device, whereas a vacuum triode is voltage-operated.

The symbol for the *NPN* type of transistor described above is shown in the circuit diagram of Fig. 17-13. In this circuit,

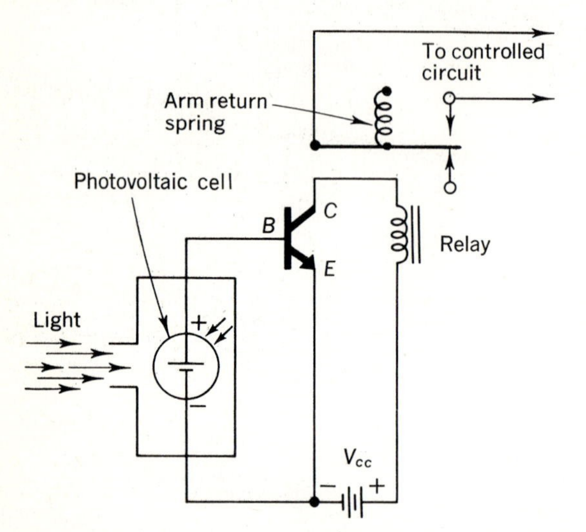

Fig. 17-13 Simple light-operated relay circuit using a transistor and light-sensitive cell.

with light shining into the light-sensitive photoemissive cell, the current developed by the cell is of such polarity as to forward-bias the emitter-base *NP* junction, allowing collector current to flow through the relay coil, pulling the relay arm (armature) downward. The normally open, or arm-down, contact on this relay is not being used. If the light beam is interrupted, the forward bias on the base ceases, collector current ceases, the relay arm is pulled back by its return spring, and the arm-up contact closes and actuates any controlled device desired.

An application of a transistor in an audio amplifier circuit is shown in Fig. 17-14. The base is slightly forward-biased by the voltage divider resistors across the collector-circuit battery. As a result, the collector is drawing a medium value of current. When sounds strike the microphone, an ac current flows in the microphone transformer secondary. This changes the forward bias of the base, thereby changing the collector current. Collector-current changes in the primary of the output transformer induce ac in the secondary and the speaker diaphragm vibrates, producing sound frequencies identical to those striking the microphone, but of much greater amplitude or loudness.

These are only two simple examples of the use of transistors. There are literally thousands of different types of transistors. The basic transistor discussed here is the

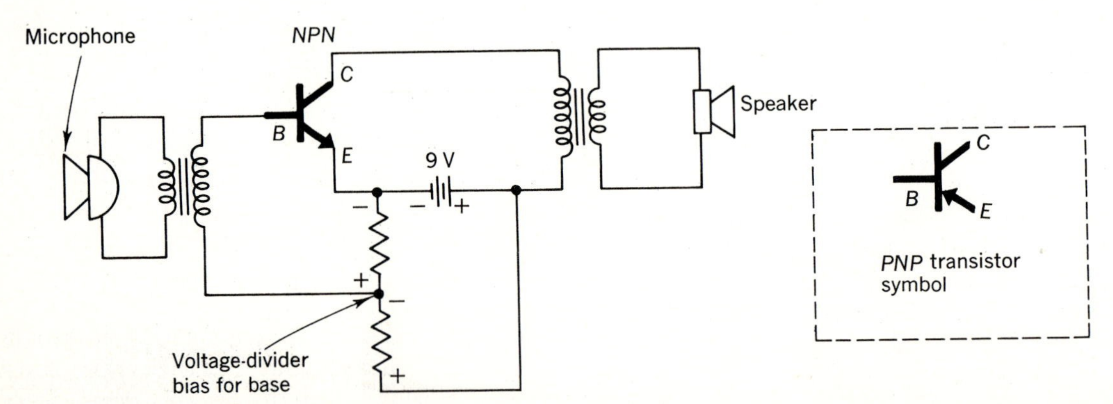

Fig. 17-14 Single stage of amplification using a transistor to amplify weak microphone signals to louder signals from loudspeaker.

NPN. The other major type is the *PNP*. The circuitry for a *PNP* transistor is the same as that for an *NPN* transistor, except that the polarity of the collector supply is reversed. A negative-biased base turns on a *PNP* transistor. Collector current flows through a *PNP* device from collector to emitter. The symbol for a *PNP* transistor is shown next to the circuit in Fig. 17-14.

One difficulty with transistors is that they become better conductors if they heat. The more they heat, the more current they pass. The greater the collector current (I_c), the more they heat. As a result, they can burn themselves out if they are not protected with *heat-sinks* (air-cooling fins thermally attached to the collector) or special circuits involving thermistors.

Besides bipolar *NPN* and *PNP* transistors there are *unipolar field effect transistors* (FETs) of either *junction* (JFET) or *insulated gate* (IGFET, MOSFET, COS/MOS, etc.) types. Figure 17-15*a* shows essentials of an *N*-channel JFET. With current flowing from source (*S*) to drain (*D*), adding a negative potential to the gate (*G*) expands the depletion areas further into the *N*-channel, tending to pinch off *S-D* channel current (I_{DD}). Forward bias must not be applied to the gate of a JFET or gate current will flow, decreasing gate-source circuit impedance drastically. A possible JFET amplifier is shown in Fig. 17-15*b*. JFETs are also made in *P*-channel form.

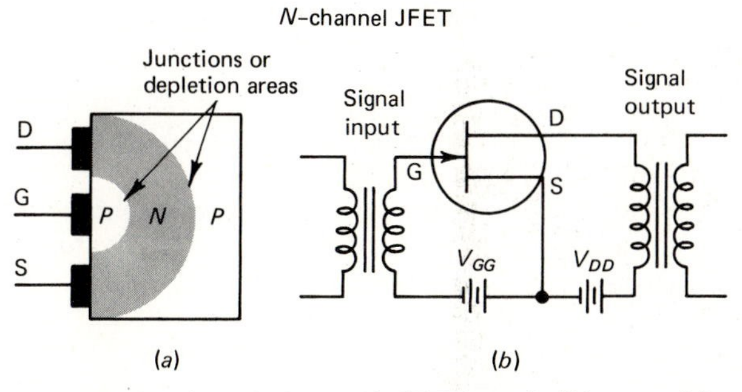

Fig. 17-15 (*a*) *N*-channel JFET and (*b*) possible amplifier.

Figure 17-16*a* illustrates the essentials of an *N*-channel *depletion*-type MOSFET (metal-oxide semiconductor FET) and a possible amplifier circuit. A medium value of I_{DD} flows from *S* to *D* through the *N*-channel with no bias on the gate. The gate is insulated from the channel by a very thin and delicate oxide. A contact is made to the *P*-type bulk or base (*B*). If *G* is made more negative than *S* (reverse bias), the electrostatic field that results pinches off, or depletes, the carriers in the channel, reducing I_{DD}. A positive (forward) bias increases I_{DD}. The gate insulation is so thin that more than 15 to 30 V may puncture it, ruining the device. MOSFETs are also made in *P*-channel form.

Figure 17-16*b* represents an *enhancement* MOSFET. With no bias voltage on the gate, which is insulated from the *N* and *P* areas, no I_{DD} can flow. If the gate is made positive (forward-biased), it attracts electrons from the *P* material. These electrons act as carriers between *S* and *D* areas, a channel is formed, and I_{DD} can flow. Varying the gate voltage varies the I_{DD}. Note how FET symbols relate to the elements in the devices. With *P*-channel devices, reverse all battery polarities and arrowhead directions from those shown.

The size of a single low-power transistor or diode may be far smaller than the head of a pin. When several diodes, transistors, and resistors are constructed into a more or less complicated circuit on a single semiconductor chip and are encapsulated, the unit is called an *integrated circuit* (IC). ICs are developed into general amplifying (called linear) circuits, or into on-off (called binary or digital) circuits. ICs are also made into

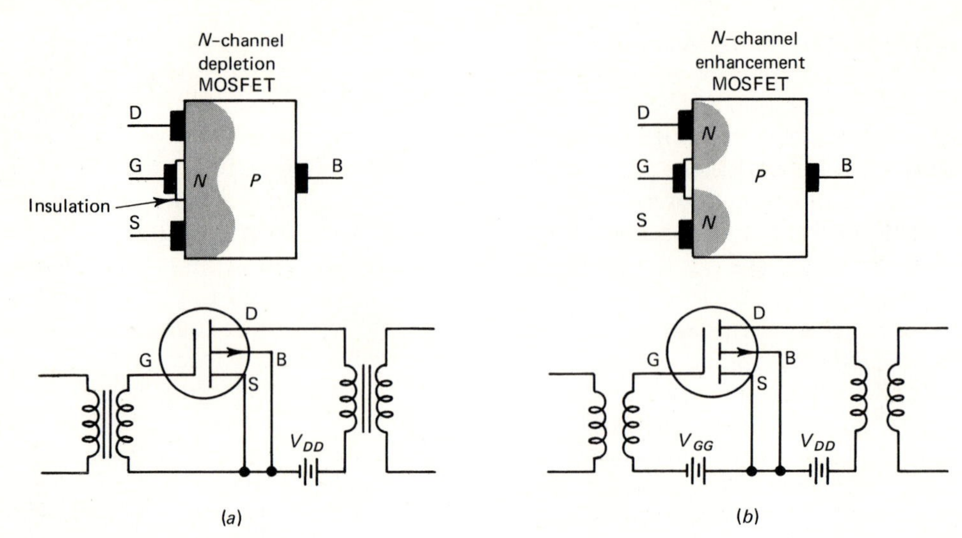

Fig. 17-16 (*a*) *N*-channel depletion MOSFET and amplifier circuit. (*b*) *N*-channel enhancement MOSFET and amplifier circuit.

medium-scale integration (MSI) and large-scale integration (LSI) units.

Quiz 17-4. Test your understanding. Answer these check-up questions.

1. What is germanium doped with arsenic called? ________ If doped with gallium? ________
2. Name a semiconductor material having electron holes. ________ Having some relatively free electrons. ________
3. What is the electrical charge of *P*-germanium? ________ Of *N*-germanium? ________
4. To make the barrier vanish in a *PN* junction, what polarity of emf would be connected to the *P* material? ________ How is the junction said to be biased in this case? ________
5. About what voltage value does the barrier area develop in a germanium junction? ________ In a silicon junction? ________
6. List two advantages of a solid-state over a VT diode. ________ ________
7. What are the two basic types of bipolar transistors? ________ ________ Which requires a collector potential similar in polarity to that of a vacuum triode plate? ________
8. What is the name of the area in a transistor that is analogous to a vacuum-tube grid? ________ Plate? ________ Cathode? ________
9. A vacuum tube is said to be "voltage-operated." What is a transistor said to be? ________
10. How many junctions are there in a transistor? ________
11. If the grid of a triode is made less negative, how is the plate current affected? ________ If the base of an *NPN* transistor is made less negative, how is the collector current affected? ________ With a *PNP* transistor? ________
12. With zero volts of bias on the base-emitter junction, would a transistor develop high, low, or zero output-circuit current? ________
13. With zero volts of bias on the grid, would a VT develop high, low, or zero output-circuit current? ________
14. When it is operating as an amplifier, should a transistor base be forward- or reverse-biased? ________
15. When it is operating as an amplifier, is the bias on a triode similar to forward or to reverse transistor biasing? ________
16. What device might be subject to "thermal runaway"? ________ How is this prevented? ________
17. When a group of transistors and diodes are all formed on one chip, what is such a unit called? ________
18. Name the three FETs. ________ ________ ________

17-6 SILICON CONTROLLED RECTIFIERS

A device that finds use in such fields as motor-speed control, high-voltage generation, high-power switching with low currents, electric-light dimming, etc., is the *silicon controlled rectifier*, or SCR. It is a solid-state thyratron, one of several semiconductor devices that are made in a *PNPN* form.

Consider the *PNPN* device in Fig. 17-17*a* (not an SCR). It is in series with an ac

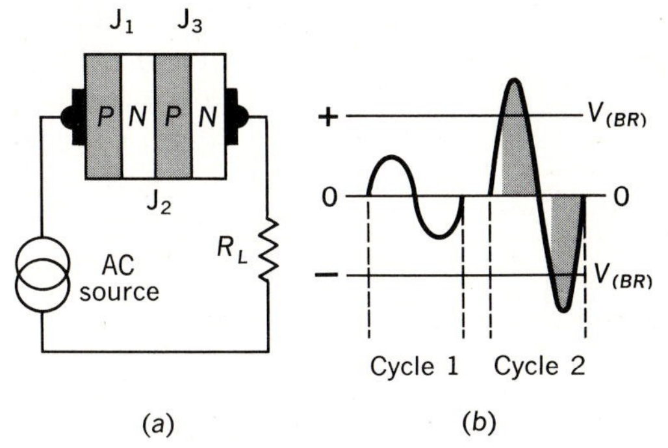

Fig. 17-17 (*a*) Current flows in both directions in the *PNPN* device when $V_{(BR)}$ is exceeded. (*b*) Current flow times are shown shaded.

source and a load resistor. Regardless of polarity, at least one of the three junctions will be reverse-biased, preventing current flow through the load. With a relatively low ac emf, in cycle 1, Fig. 17-17*b*, no current should flow through R_L. However, if the ac voltage is raised enough to reach the breakdown voltage ($V_{(BR)}$) of the junctions, cycle 2, current starts to flow as $V_{(BR)}$ is reached. Current continues to flow even after the voltage decreases below this value, until the junctions are cleared of carriers. This type of *avalanche* current does not burn out the device unless the applied voltage greatly exceeds the $V_{(BR)}$ value. Current flows on both half-cycles of the ac.

A *PNPN* SCR is shown in Fig. 17-18, using a dc supply, V_{AA}, and a resistive load. The *N*-type end area, with a metal contact on it, is called the cathode (K). The thin, adjacent *P*-type segment also has a

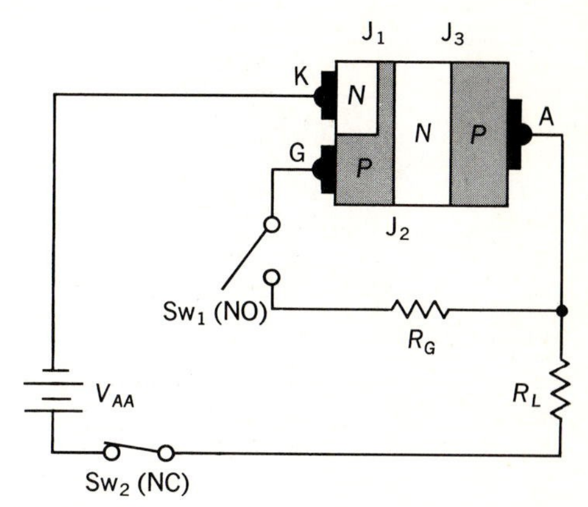

Fig. 17-18 SCR in a dc circuit.

metal contact and is called the gate (G). The far *P* segment with the metal contact is the anode (A).

Assuming V_{AA} is some value below $V_{(BR)}$, no current flows in the circuit, because junction J_2 is reverse-biased. If the normally open (NO) switch (Sw_1) is closed for an instant, the first *P*-type segment, acting like the base of an *NPN* transistor, is forward-biased through R_G and R_L, allowing current to flow through J_2. Since J_3 is forward-biased already, current flows through the whole device and circuit.

If the gate switch is opened, current continues to flow, because J_2 loses control of the current carriers in it when a heavy current flows through the thin *P* area. The gate can be reverse-biased, and it still cannot stop current flow through J_1 or J_2. It is necessary to reduce V_{AA} to almost 0 V, or open Sw_2 before current will stop in the circuit.

A lamp-dimming circuit, using ac and an SCR, is shown in Fig. 17-19. With the switch at "1," the gate is connected directly to the cathode (K). Since the ac voltage is less than both $V_{(BR)F}$ (forward breakdown voltage) and $V_{(BR)R}$ (reverse breakdown voltage), no current can flow during cycle-1, switch-1 conditions.

If the switch is at "2," during the half-

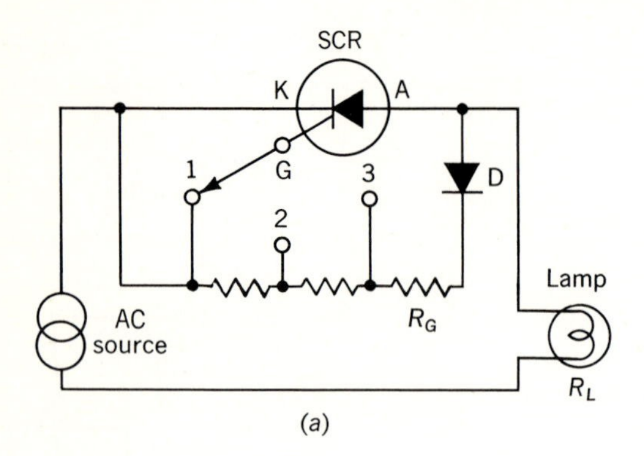

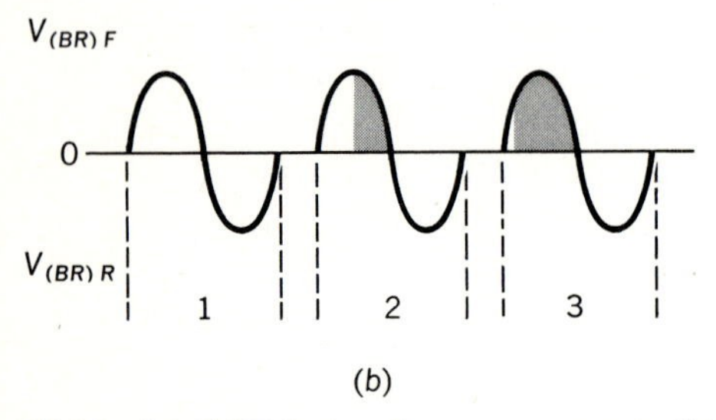

Fig. 17-19 (*a*) SCR lamp-dimming circuit. (*b*) "On" times for three different bias conditions.

cycle that the anode is positive and the cathode negative, a positive gate voltage develops across the resistance between G and K, firing the SCR as the ac approaches its peak value. Anode current, I_A, shown shaded in cycle 2, flows for only about 90°. During the second half-cycle, conduction ceases and the gate bias voltage is reversed, preventing SCR refiring.

With the gate switch at "3," the gate bias voltage is greater, causing the SCR to fire earlier in the cycle and allowing current to flow for nearly 180°. During the second half-cycle, no current flows, hence the name "silicon controlled *rectifier*." The current in the lamp is pulsating dc.

If a potentiometer is used instead of the switch, the lamp can be controlled from no glow to maximum. In a similar fashion, an electric motor can be made to vary its speed from off, to slow, to high speed.

A circuit allowing almost 180° of *phase control* of the SCR current is shown in Fig. 17-20*a*. During the half-cycle that the SCR

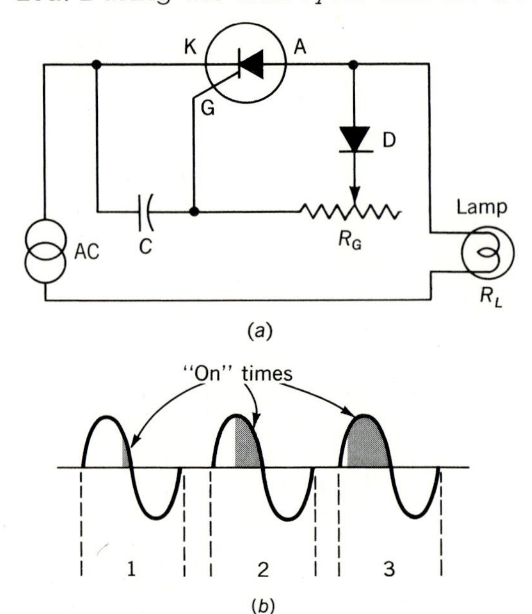

Fig. 17-20 (*a*) *RC* phase control circuit produces (*b*) almost 180° of controlled "on" time.

can fire, capacitor *C* charges through R_L and R_G, until it builds up to enough voltage to fire the SCR. If R_G is a high resistance, *C* charges slowly, and the SCR is fired late, resulting in only a few degrees of current (cycle 1, Fig. 17-20*b*). With a low value of R_G, the capacitor charges rapidly, the SCR is fired immediately, and almost 180° of current flow is produced. This is a *halfwave* circuit because only half of the ac cycle is used.

ANSWERS TO CHECK-UP QUIZ 17-4

1. (*N*-germanium) (*P*-germanium) **2.** (*P*-germanium) (*N*-germanium) **3.** (Neutral) (Neutral) **4.** (+) (Forward) **5.** (0.3 V) (0.6 V) **6.** (Smaller, no heater required) **7.** (*NPN*) (*PNP*) (*NPN*) **8.** (Base) (Collector) (Emitter) **9.** (Current-operated) **10.** (2) **11.** (Increased) (Increased) (Decreased) **12.** (Zero) **13.** (High) **14.** (Forward slightly) **15.** (Reverse) **16.** (Transistors) (Heat sinks, special circuits, thermistors) **17.** (IC) **18.** (JFET) (Depletion MOSFET) (Enhancement MOSFET)

Once the gate of an SCR is triggered "on," anode current will continue to flow until the anode or cathode circuits are interrupted by switching or until the anode voltage drops to zero or is reversed. If the power supplied is ac, the SCR automatically turns off as the ac cycle reverses.

A *fullwave* control of load current is shown in Fig. 17-21. With the switch open,

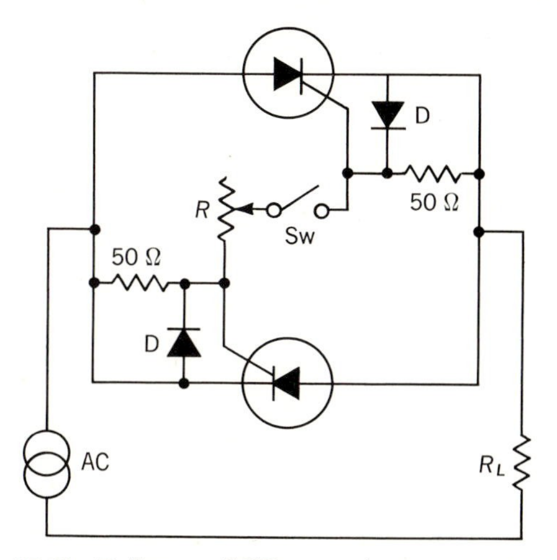

Fig. 17-21 Fullwave SCR control circuit.

neither SCR can fire. With Sw closed, the value of *R* determines the conduction portion of both halves of the ac cycle.

The diodes between gate and cathode of the two SCRs are polarized so that they conduct only when the voltage-drop across the gate resistors applies a reverse voltage to the gates, thereby protecting the SCRs against excessive reverse gate voltage.

Quiz 17-5. Test your understanding. Answer these check-up questions.

1. What does the R stand for in SCR? ________ Why would you assume that SCRs are not made with germanium? ________
2. What is meant by $V_{(BR)}$? ________ By $V_{(BR)F}$? ________ By $V_{(BR)R}$? ________
3. What is the name of the terminal of an SCR labeled K? ________ Labeled G? ________
4. What is an advantage of using ac as the power source when using SCRs? ________
5. What is the function of the diode shown in Fig. 17-11? ________
6. Why are SCRs better for current control than rheostats? ________
7. What is the range of current control for the circuit in Fig. 17-19? ________ With the same source and load, how could twice the current be fed to the load? ________
8. In Fig. 17-20, how many degrees of control are obtainable? ________ In Fig. 17-21? ________
9. What do the diodes prevent in Figs. 17-19, 17-20, and 17-21? ________

17-7 THE TRIAC

Another breakdown semiconductor, or *thyristor* device, is the *triac*. It is a two-way SCR of the *PNPN* type, shown in Fig. 17-22.

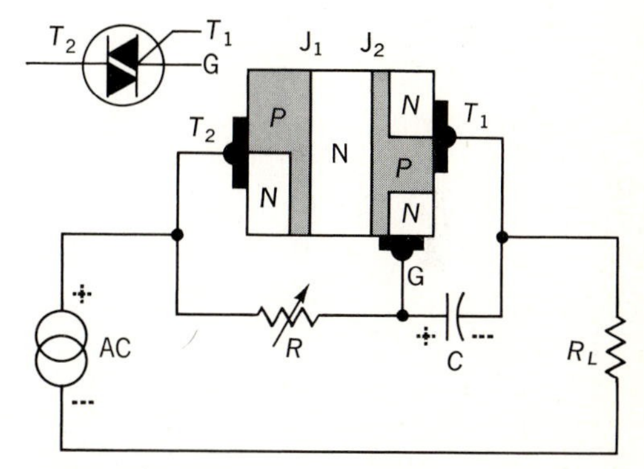

Fig. 17-22 Triac and fullwave phase control circuit. Symbol shown above.

When the source emf attempts to produce current flow downward through R_L, it is opposed by junction J_1, which is reverse-biased. J_2 prevents current flow when the emf reverses. As T_2 becomes positive (dotted polarity), *C* begins to charge through R_L and *R*. The voltage across *C* produces a forward bias for the *NP* junction next to T_1. The gate acts as a *P*-type base of an *NPN*-type transistor. A positive gate injects carriers across J_2, and current can flow across the top of the triac from T_1 to T_2. Once conducting, there is essentially no voltage-drop across the triac, and C discharges.

As the source ac reverses, current through the triac ceases, and C starts charging, with negative toward the gate. When the gate NP junction is biased with a negative voltage, carriers are injected into the adjacent junctions, and current flows across the bottom of the triac. The same R and C determine how soon the current starts flowing during this half-cycle also. The triac acts as two SCRs back to back, in reverse direction, producing fullwave current flow under the control of the one gate.

17-8 LIGHT-ACTIVATED SCRS

A light-activated SCR, or LASCR, has a glass lens in the metal case surrounding the semiconductor layers (Fig. 17-23*a*). This allows light to fall on the gate-cathode junction. The diagram of Fig. 17-23*b* is a possible circuit that will turn on an LASCR when its gate region is light-activated. The two arrows pointing to the symbol indicate light, and are part of the LASCR symbol.

On one half-cycle, the ac source is polarized properly to produce current flow from cathode (K) to the LASCR anode (A), but R_G holds the gate (G) at cathode potential and the LASCR in nonconduction.

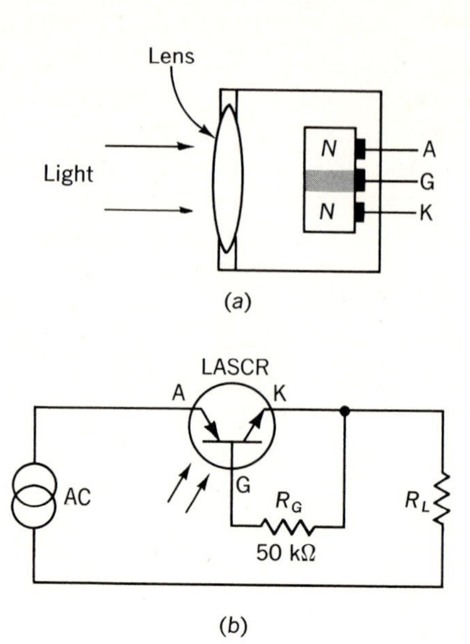

Fig. 17-23 (*a*) LASCR. (*b*) Light-operated LASCR circuit.

If the gate area is illuminated with light, photons of energy free carriers in the GK junction, allowing current flow from cathode to anode. With an ac source the circuit turns off each time the cycle reverses. When an LASCR is used in a dc circuit, some means must be provided to turn off, or "unlatch," it.

The power-handling capabilities of an LASCR are low. When a heavy load current must be controlled by a light source, the LASCR can be coupled to an SCR as in Fig. 17-24.

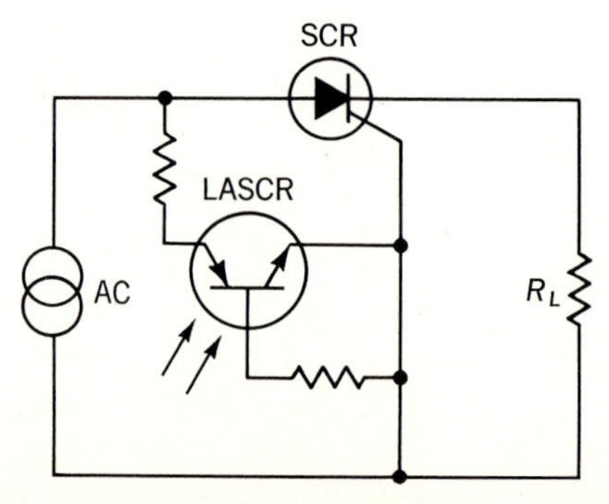

Fig. 17-24 LASCR-controlled SCR.

ANSWERS TO CHECK-UP QUIZ 17-5

1. (Rectifier) (S stands for silicon) **2.** (Breakdown voltage of a junction) (Forward breakdown voltage) (Reverse breakdown voltage) **3.** (Cathode) (Gate) **4.** (Self-extinguishing) **5.** (Prevents reverse voltage from appearing across the gate) **6.** (Rheostats produce great heat loss and are bulky for heavy currents) **7.** (180°) (Parallel another SCR circuit reversed across the first) **8.** (180°) (360°) **9.** (Buildup of reverse gate voltage)

The LASCR differs from a *light-activated transistor* which has no base connection. Light-activated transistors lower their emitter-collector resistance value in direct proportion to the intensity of illumination of the base-emitter junction.

17-9 THERMISTORS

A thermistor is a temperature-sensitive device. If it increases resistance with an increase in temperature, it has a +TC (positive temperature coefficient of resistance). If it decreases resistance with an increase in temperature, it has a −TC. Since semiconductors have a −TC, a solid-state diode may be used as a thermistor. −TC thermistors may be tiny beads, or wafers, made of semiconductor materials with contacts on two sides. + TC thermistors are constructed of oxides of special metals or alloys of two or more metals. Thermistors are useful, for example, as electronic thermometers, etc.

Quiz 17-6. Test your understanding. Answer these check-up questions.

1. Is a triac a halfwave or fullwave control device? ________ An LASCR? ________
2. Why is a diode not used between gate and T_1 of a triac? ________
3. What is a thyristor? ________
4. To what is a triac similar in operation? ________
5. How does a light-activated transistor differ in operation from an LASCR? ________
6. Which TC would be produced by using: Copper wire? ________ Nichrome wire? ________ A bead of germanium? ________ A bead of silicon? ________ A pure carbon rod? ________
7. What is a thermistor? ________
8. Try to identify all the symbols shown in Fig. 17-25.

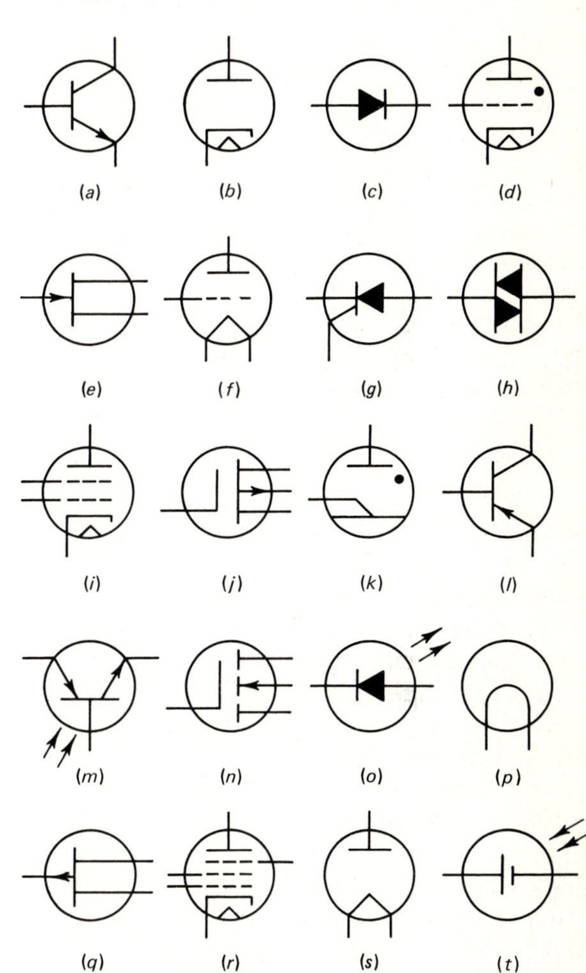

Fig. 17-25 Identify these symbols. (See Check-up Quiz 17-6 answers.)

CHAPTER 17 TEST • ACTIVE DEVICES

1. Name four passive electric circuit devices.
2. Name five active electric circuit devices.
3. Name the electron emitter in a VT. In an *NPN* transistor. In an SCR. In a triac.
4. Name the electron collector in a VT. In an *NPN* transistor. In an LASCR. In a triac.
5. What is the direction of lines of force across a VT to produce conduction?

6. What is meant by *rectification?*
7. List three means of increasing I_p in a VT diode.
8. Name the three-element VT. The four-element VT. The five-element VT.
9. What potential must a VT grid have to reduce I_p?
10. Normally a VT grid is not driven +. If it is, how would I_p be affected? What would occur in the grid circuit?
11. Name the transducer that converts sound waves to electrical variations. Light waves to electrical current. Electrical variations to sound waves.
12. In the VT applications discussed, did the VT do anything other than control current?
13. What is a phanotron?
14. What is a thyratron?
15. What is the ionizing potential of mercury vapor?
16. What is an ignitron?
17. What element does an ignitron have that a thyratron does not?
18. To turn off I_p in gaseous triodes, what must be done?
19. Why is hydrogen used in some thyratrons?
20. A silicon diode is constructed with what two types of materials?
21. If a silicon diode is forward-biased, to which material is the + terminal of the bias battery connected?
22. What is the approximate barrier-voltage value of a silicon junction? A germanium?
23. What are the three elements in all transistors?
24. In normal operation as an amplifier, should the bias be forward or reverse for a transistor? For a VT?
25. With zero bias voltage, would output current be high or low in a VT? In a transistor?
26. What is an IC?
27. What is meant by LSI?
28. What does FET mean? To what is this device similar in operation?
29. To what device is an SCR similar in operation?
30. Name the control element in a triode. In a transistor. In an SCR. In a thyratron. In an FET.
31. For what might SCRs be used?
32. To what is a triac similar in operation?
33. What is an LASCR? What are its element names?
34. What is the TC of a solid-state diode?

ANSWERS TO CHECK-UP QUIZ 17-6

1. (Fullwave) (Halfwave) **2.** (Device fires on both half-cycles) **3.** (Device using junction breakdown) **4.** (Back-to-back SCRs) **5.** (Does not latch on and handles more current) **6.** (Slight +TC) (Zero TC) (−TC) (−TC) (−TC) **7.** (+ or −TC resistor) **8.** (*a*) NPN, (*b*) VT diode, (*c*) Solid-state diode, (*d*) Thyratron, (*e*) *N*-channel JFET, (*f*) Filamentary VT triode, (*g*) SCR, (*h*) Triac, (*i*) VT tetrode, (*j*) *P*-channel depletion MOSFET, (*k*) Ignitron, (*l*) *PNP*, (*m*) LASCR, (*n*) *P*-channel enhancement MOSFET, (*o*) LED or solid-state lamp, (*p*) Filamentary lamp, (*q*) *P*-channel JFET, (*r*) VT pentode, (*s*) Filamentary VT diode, (*t*) Photocell or solar cell.

18 DC AMMETERS

CHAPTER OBJECTIVE. To discuss the construction and operation of the moving-coil-type ammeter, simple shunting, an Ayrton shunt, meter damping, linear and nonlinear meter scales.

18-1 MOVING-COIL METERS

The basic meter movement is the *moving-coil*, or *D'Arsonval* meter. This one meter can be used as a dc ammeter, a milliammeter, a microammeter, a voltmeter, or an ohmmeter, and with rectifiers it can indicate alternating currents and voltages.

The moving-coil meter (Fig. 18-1) is an electromagnetic device consisting of:

1. A horseshoe-shaped permanent magnet
2. A round iron core piece between the magnet poles
3. A rotating mechanism that is free to move about 90°, which includes:
   (*a*) A lightweight coil
   (*b*) A pointer attached mechanically to the coil
   (*c*) Two delicate spiral springs to return the pointer to zero
   (*d*) Two precisely ground jewel bearings
4. A calibrated paper or metal scale
5. A metal, Bakelite, or plastic case

The coil and pointer assembly are shown removed from their normal position surrounding the iron core piece between the poles of the magnet. The iron core does not move, but the coil assembly

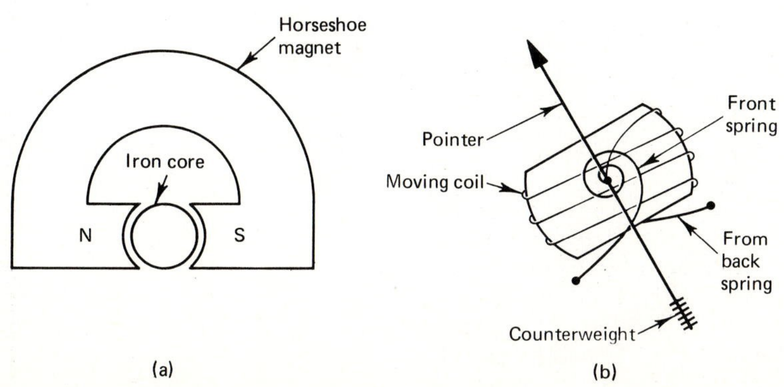

Fig. 18-1 Essentials of a moving-coil meter. (*a*) Magnet and core piece. (*b*) Removed from the space between the magnet poles and the iron core are the coil, springs, and pointer.

rotates in the narrow spaces between the core and the magnet poles.

The spacing between the magnet poles and the soft-iron core is important. The closer these parts are, the fewer the leakage lines of force, the stronger the field in which the moving coil operates, and the more sensitive the meter is to weak currents flowing in the coil. Fewer leakage lines of force also mean less interaction between the meter magnet and external magnetic fields, and thus less chance of meter error.

Each of the two spiral bronze springs is connected electrically to an end of the moving coil and mechanically to an insulated point on the shaft of the coil form. The other ends of the springs are attached mechanically and electrically to two insulated points on the magnet and pole-piece assembly. In addition to being used to zero the indicating needle, the springs provide the only path by which current is fed into and out of the moving coil. The coil assembly is insulated electrically by front and back jewel-type bearings.

When a current of electrons flows through the coil, it becomes an electromagnet with a north pole and a south pole. These poles are repelled by the permanent-magnet field, and the indicator moves clockwise against the spring tension. If the current is small, the indicator-needle deflection will be small. Greater current flow results in greater needle deflection. If the current flows through the coil in the opposite direction, the meter needle will try to move counterclockwise, to the left of the zero point. The meter terminal that should be connected to the positive side of the source of emf is usually marked +.

Since the coil is made of many turns of very fine wire, care should be taken never to feed an excessive current through it. The current necessary for full-scale deflection will not damage the coil, but a 100% or greater overload may burn out the coil, take the temper out of the springs, burn them out, or bend the delicate aluminum needle as it is forcibly driven past the end of the scale. When current ceases in the coil, the springs return the needle to the zero setting.

Another type of meter has a central, round, permanent magnet mounted so

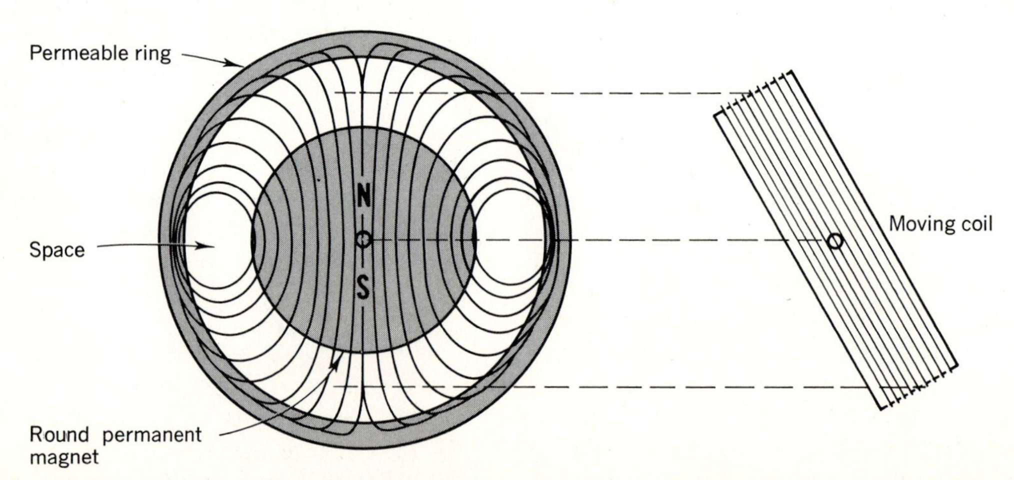

Fig. 18-2 Meter movement with central permanent magnet. Moving coil has been removed from its position in the space between magnet and permeable ring.

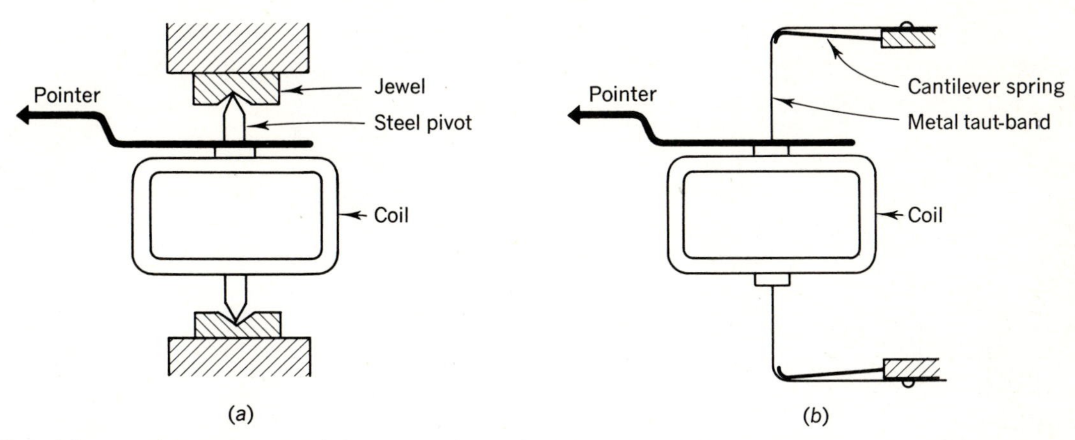

Fig. 18-3 Meter coil suspensions: (*a*) pivot and jewel; (*b*) taut-band. Return springs are required on pivots in pivot-and-jewel suspensions (not shown). No springs are needed in taut-band.

that the moving coil moves around it, as in Fig. 18-2. To provide a high-permeability return path for the permanent-magnet field, a round iron ring is installed concentrically around the space in which the coil moves. This construction concentrates the magnetic field in a narrow gap, providing a strong and constant field strength in which the moving coil rotates. For clarity the coil is shown outside the magnetic field.

Besides the pivot and jewel-type suspensions of most meters (Fig. 18-3), some moving-coil meters suspend the moving coil in position by a thin torsion wire, or *taut-band*.

On many meters an adjustment screw is brought through the front glass or case. By rotating this screw, more or less torque, or twisting force, can be placed on the front spring, and the pointer can be accurately set to zero. This is the *zero-adjust control*.

Well-balanced meters will read the same whether installed vertically or horizontally. Others may require a zero-adjustment correction if their position is changed. Better meters overcome this by having not only a counterweight at the bottom of the indicator needle, but two additional *quadrantal* weights at right angles to the needle (Fig. 18-8). When these weights are properly adjusted, the meter will give the same reading regardless of its orientation.

Meters are delicate. They should be handled gently. If subjected to strong magnetic fields, the permanent magnets may weaken, and low readings will result. With proper treatment, meters have been known to function satisfactorily for 60 years or more.

Meters are made to be mounted on steel (iron) panels or on nonmagnetic panels (aluminum, Bakelite, etc.). If mounted on the wrong type of panel, the internal magnetic fields may be affected, and inaccurate readings will result. Meters to be used on iron panels are usually encased in iron. A few common meters are shown in Fig. 18-4.

18-2 METER ACCURACY

The accuracy of a meter is expressed in percent-of-full-scale reading. A 2% 100-mA full-scale meter is guaranteed to be within 2% of the full-scale value at any point on the scale when sold. This means ±2 mA

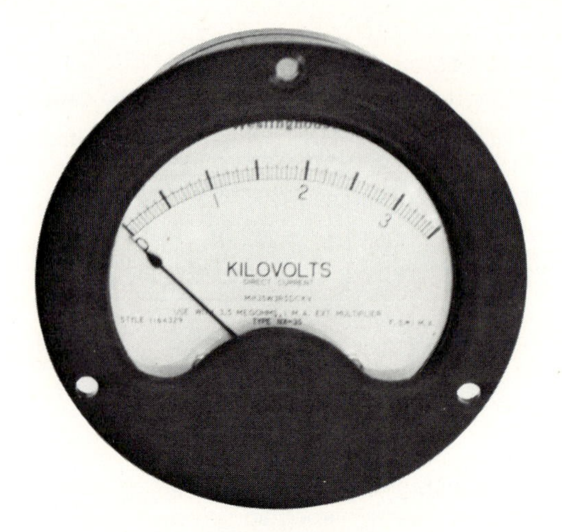

Fig. 18-4 Various panel-mounting meters.

in a 0- to 100-mA meter. An indication of 98 mA means an actual current value between 96 and 100 mA. A 10-mA reading may also be ±2 mA, or between 8 and 12 mA. Less expensive meters may have 3% accuracy; more expensive ones may be accurate to 1%. Some laboratory meters with special antiparallax mirrors under the indicators can provide consistent readability within 0.1%. Many electronic meters are accurate to 0.01% or better.

A meter should have a full-scale value almost twice that of the measurement expected to be made. Meters are usually calibrated to give most accurate readings at about two-thirds scale.

Criteria for meter selection are:

1. Function (ammeter, voltmeter, wattmeter, etc.)
2. Movement (moving-coil, taut-band, dynamometer, etc.)
3. Full-scale range (0–1, 0–3, 0–5, 0–10, etc.)
4. Accuracy (in percent or parts per million)
5. Type of scale (linear, logarithmic, expanded)
6. Scale length in centimeters or inches
7. Dimensions and material of case

For more sophisticated meters, the choices are:

1. Vacuum-tube-aided meters (VTMs, etc.)
2. Transistor-aided meters (TVMs, etc.)
3. Digital meters (with numerical readouts)

Quiz 18-1. Test your understanding. Answer these check-up questions.

1. What approximate full-scale-reading meter should be chosen to measure a current known to be between 5 and 7 mA? __________ Between 1 and 1.2 A? __________ Between 20 and 30 μA? __________
2. A 0- to 100-mA meter is accurate to 2% and reads 45.5 mA in a circuit. What are the highest and lowest possible current values flowing through it? __________ __________
3. What is another name for a moving-coil meter? __________
4. What does the iron core accomplish in the usual moving-coil type of meter? __________ Does it rotate when current flows through the meter? __________
5. What parts of a moving-coil meter are likely to be damaged if excessive current flows through the meter? __________
6. In what direction does the indicator normally move in a meter when current of the correct direction flows through it? __________ Will 80 mA in the wrong direction damage a 100-mA meter? __________
7. What are the two balance weights called that are on the right and left sides of the pointer? __________
8. Will a meter made to be mounted on a nonferrous-metal panel read correctly if it is mounted on a wooden panel? __________ A steel panel? __________
9. What is the screw called that protrudes through the meter front of some meters? __________
10. What parts are missing in a taut-band meter that would be found in most moving-coil meters? __________ __________

18-3 CHANGING AMMETER RANGES

Moving-coil meters are fairly sensitive. Many can be driven to a full-scale reading with a current of only 1 mA or less flowing through the coil. When such a meter is used in a circuit in which 1 A flows, a resistor must be shunted across the meter so that most of the current will go through the shunting resistor. The problem is to determine what value the *shunt resistor* should have to protect the meter and to give the desired range of indications.

The circuit in Fig. 18-5 illustrates a 0- to 5-mA meter having 25 Ω of internal resis-

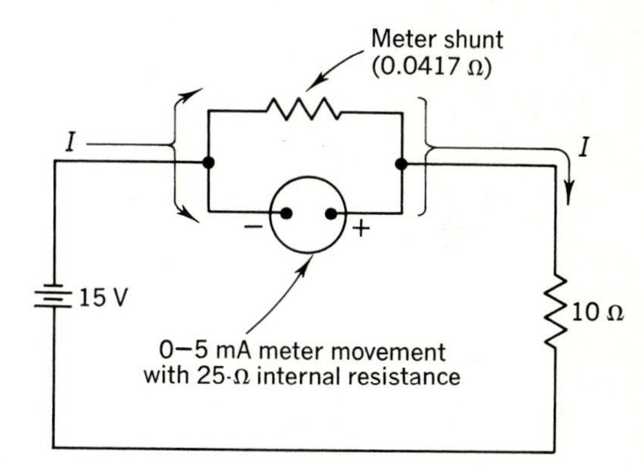

Fig. 18-5 Shunted meter. Circuit current divides, part flowing through the shunt and a small part through the meter.

tance (coil and springs) in a circuit with a 10-Ω load and a 15-V source. From Ohm's law, the circuit current should be about 1.5 A. Without the shunt on the meter, the current through the meter would be about 286 mA, which would burn out the coil and springs of the meter. If the shunt resistor had zero resistance, it would take all the current, and the meter could not indicate. What is the required resistance of the meter shunt?

Since the circuit current is to be 1.5 A, any full-scale value from about 2 to 3 A would give good, readable results. If 3 A is chosen, the shunt value can be determined by Ohm's law. The voltage-drop across the meter movement whenever it reads full scale must be $V = IR$, or 0.005(25), or 0.125 V. Thus, whenever the voltage-drop across the meter is 125 mV, the meter will be passing 5 mA and indicating full scale.

If the full-scale current through the meter movement plus the shunt is 3 A (3000 mA), the meter current will be 5 mA, and the shunt current will be 2995 mA. The source current divides 2995 mA through the shunt and 5 mA through the meter. This develops 125 mV across both the shunt and the meter. The resistance value required to pass 2.995 A and develop 0.125 V across it is $R = E/I$, or 0.125/2.995, or 0.0417 Ω.

The 0- to 3-A meter of Fig. 18-5 actually consists of 0- to 5-mA meter movement with a 0.0417-Ω shunt resistor across it. The shunt is usually inside the meter case and cannot be seen unless the meter is taken apart. Since the meter movement came with 0- to 5-mA scale markings, the scale will have to be replaced by a new scale calibrated from 0 to 3 A.

When a technician changes the range of a meter, no dust, particularly metal particles, must be allowed to get into the narrow spaces between moving coil and permanent magnet, as this will hinder movement of the coil and make the meter unusable. Meters should be worked on only in dry, dust-free, draft-free areas.

Determination of the shunt value has been explained as being a relatively simple two-step Ohm's law problem. It is also possible to compute the shunt by considering that the lower the shunt resistance value, the greater the proportion of the total current that will flow through it. That is, the ratios of the shunt and meter resistances are inversely proportional to the current ratios, or

$$\frac{R_{sh}}{R_m} = \frac{I_m}{I_{sh}}$$

where R_{sh} = shunt resistance
R_m = meter resistance
I_m = meter current
I_{sh} = shunt current

To solve for the shunt resistance, both sides of the equation can be multiplied by R_m:

$$R_{sh} = \frac{I_m R_m}{I_{sh}}$$

For the 0- to 3-A meter discussed above,

$$R_{sh} = \frac{0.005(25)}{2.995} = 0.0417\ \Omega$$

The type of wire used in meter shunts is important. If copper wire is used, the indications will not be reliable. For example, with 2.995 A flowing through the shunt, the power dissipation in the shunt is about 0.374 W. This would heat the copper wire. Because copper has a positive temperature coefficient of resistance, the shunt resistor would increase as the temperature increased. It is necessary to use one of the special zero-TC wires such as Manganin or Nichrome for shunt resistors.

ANSWERS TO CHECK-UP QUIZ 18-1

1. (10 mA) (2 A) (40 μA) **2.** (43.5 to 47.5 mA) **3.** (D'Arsonval, but Weston and galvanometer are also used) **4.** (Shortens the flux gap) (No) **5.** (Coil, springs, pointer) **6.** (CW) (No) **7.** (Quadrantal) **8.** (Yes) (No) **9.** (Zero-adjust) **10.** (Spiral springs, jeweled bearings)

Quiz 18-2. Test your understanding. Answer these check-up questions.

1. In Fig. 18-5, if the shunt short-circuited, what current value would flow in the meter? ________ If the shunt open-circuited, what current would flow in the meter? ________ What would the meter then read? ________
2. A 1-mA meter with 50-Ω internal resistance is to be converted to a 0- to 100-mA meter. What is the full-scale voltage-drop across the meter as a 0- to

1-mAmeter? __________ As a 0- to 100-mA meter? __________ What is the voltage-drop across it when used as a 0- to 100-mA meter if 50 mA is flowing in the circuit? __________ What shunt resistance value would it have? __________

3. Which would be the more sensitive meter, a 0- to 1-mA or a 0- to 10-mA meter? __________ Why? __________
4. What is the shunt resistance value required to convert a 40-μA meter with 2000-Ω internal resistance to a 0- to 5-mA meter? __________
5. What value shunt is required to convert a 0- to 100-mA meter with 0.5-Ω internal resistance to read 0 to 50 A? __________ What power dissipation would the shunt have at full-scale reading? __________ At half-scale reading? __________
6. What percentage of accuracy do you think most general-purpose meters might have? __________

18-4 MULTIRANGE AMMETERS

It is possible to use a single ammeter and, by switching shunts across it, measure a wide range of current values. The circuit shown in Fig. 18-6 uses a 0- to 1-mA

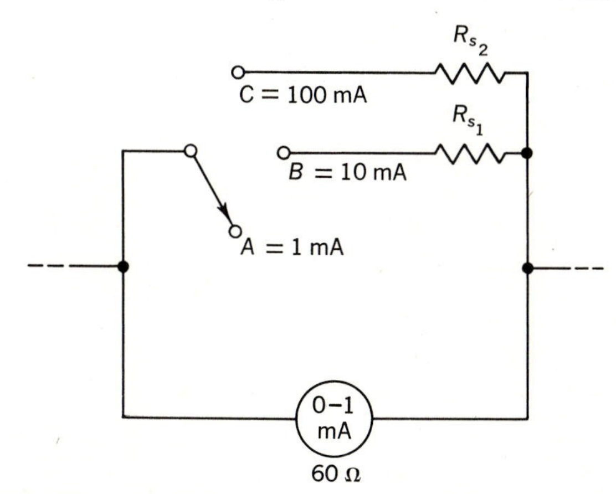

Fig. 18-6 Simple multirange ammeter. The meter alone may be used, or two shunt values may be selected.

meter movement. With the switch in the *A* position, there is no shunt across the meter, and it can be used to measure currents up to 1 mA. By moving the switch to the *B* position, a relatively high resistance shunt is connected across the meter. This shunt might be of a value to make the meter read 10 mA full-scale. The switch in the C position connects another relatively lower resistance shunt across the meter and might be the value to make the meter read 100 mA full-scale. In this way, one meter can be used to read three ranges of current. Since the full-scale ranges are all multiples of 10, the single scale can be used for all three ranges. That is, the 0.6-mA mark on the first range would indicate 6 mA on the second and 60 mA on the third.

There is one difficulty in this simple switching circuit. Note that if the switch were moved from *B* to *C* while 10 or more mA was flowing in the circuit, for an instant, between switch connections, or if there were loose or dirty contacts, the meter movement would be alone in the circuit, and a surge of current might flow through it that would burn it out or bend its pointer.

A safe switching multirange shunt circuit is the *universal*, *ring*, or *Ayrton* shunt, shown in Fig. 18-7. The three full-scale ranges selected for this discussion are 1, 10, and 100 mA. The basic movement is a 0- to 0.5-mA full-scale 100-Ω internal resistance meter. In the *A* position (1 mA), half the current must flow through the shunt (0.5 mA through the meter and 0.5 mA through the shunt equals 1 mA total). Therefore the total shunt value ($R_1 + R_2 + R_3$) must equal the meter resistance, 100 Ω.

When the switch is moved to the *B* position, the full-scale deflection is to be 10 mA. Under this condition, 0.5 mA must be flowing through both R_1 and the meter to make the pointer deflect to full scale, while 9.5 mA must be flowing through the shunt resistors R_2 and R_3. Figure 18-7*b* illustrates the circuit with the switch at point *B*. Note that the two series resistors,

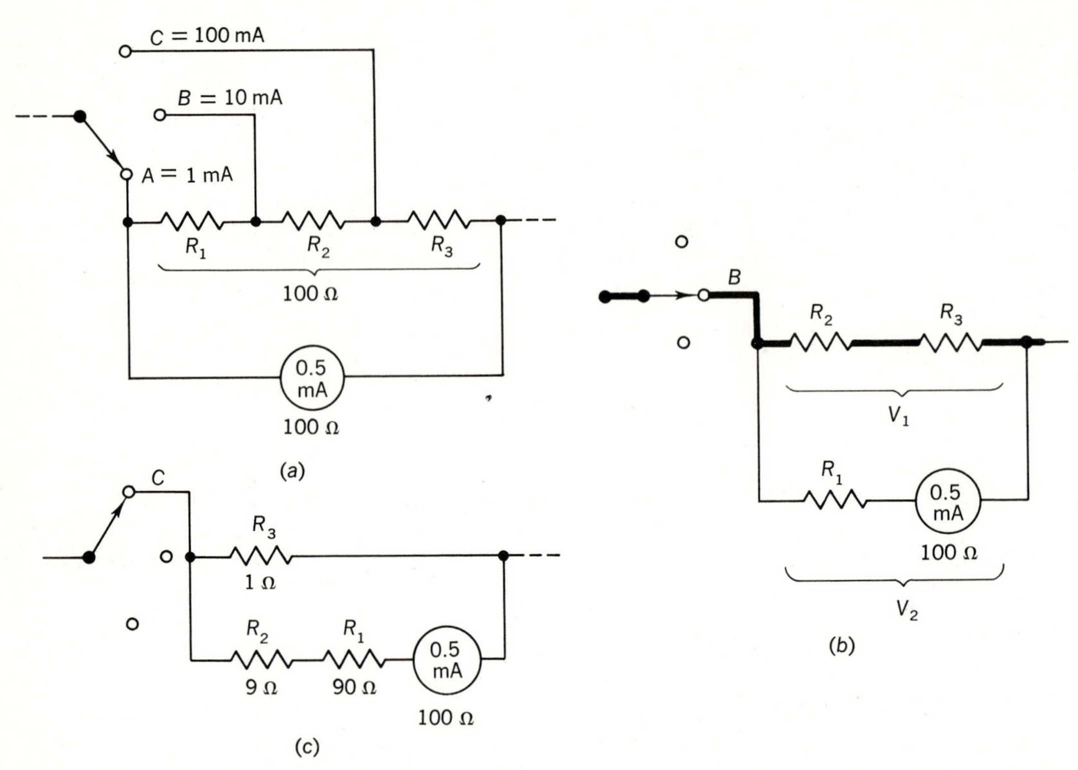

Fig. 18-7 Universal or Ayrton shunt circuit: (*a*) Circuit for meter set to measure 0 to 1 mA; (*b*) to measure 0 to 10 mA; and (*c*) to measure 0 to 100 mA.

R_2 and R_3, are in parallel with R_1 and the meter. As a result, the voltage-drop across $R_2 + R_3$ must be the same as that across R_1 plus the meter, or $V_1 = V_2$. It is then possible to set these two voltages equal and use two voltage ($V = IR$) equations

$$\overbrace{I_m(R_m + R_1)}^{V_1 =} = \overbrace{I_{sh}(R_2 + R_3)}^{V_2 =}$$
$$0.0005(100 + B_1) = 0.0095(R_2 + R_3)$$

It is known that $R_1 + R_2 + R_3 = 100\ \Omega$. From this it follows that $R_2 + R_3 = 100 - R_1$. If $(100 - R_1)$ is substituted for $(R_2 + R_3)$ in the equation above,

$$0.0005(100 + R_1) = 0.0095(100 - R_1)$$
$$0.05 + 0.0005R_1 = 0.95 - 0.0095R_1$$
$$0.0005R_1 + 0.0095R_1 = 0.95 - 0.05$$
$$0.01R_1 = 0.9$$
$$R_1 = \frac{0.9}{0.01} = 90\ \Omega$$

If R_1 is known to be 90 Ω, the total resistance of the meter movement plus R_1 must be 190 Ω. The current for full-scale deflection of the meter is 0.0005 A. The voltage across R_1 and the meter must be $V = IR$, or 0.0005(190), or 0.095 V. This must also be the voltage-drop across $R_2 + R_3$. The current through these resistors is known to be 0.0095 A. Therefore the total resistance of

---

ANSWERS TO CHECK-UP QUIZ 18-2

1. (0 A) (0.286 A) (Zero. Burned out) **2.** (0.05 V) (0.05 V) (0.025 V) (0.505 Ω) **3.** (0- to 1-mA) (Less I required for full-scale reading) **4.** (16.1 Ω) **5.** (0.001 Ω) (2.495 W) (0.6238 W) **6.** (1 to 3%)

$R_2 + R_3$ must be $R = E/I$, or 0.095/0.0095, or 10 Ω.

When the switch is thrown to the C position (Fig. 18-7c), the full-scale deflection is to be 100 mA. R_3 alone is the main shunt resistance, and R_2 and R_1 are in series with the meter movement. The voltage-drop across R_3 will equal the voltage-drop across the meter and the two resistors. The current through the meter circuit must still be 0.0005 A full scale, but the current through the R_3 shunt must be 99.5 mA, or 0.0995 A. As before, an equal-voltage equation can be set up:

$$I_m(R_m + R_1 + R_2) = I_{sh}(R_3)$$

Substituting the known factors, we have

$$0.0005(100 + 90 + R_2) = 0.0995(R_3)$$

Since it is known that $R_2 + R_3 = 10\ \Omega$, then $R_3 = 10 - R_2$. Now $10 - R_2$ can be substituted in the equation for R_3.

$$\begin{aligned} 0.0005(190 + R_2) &= 0.0995(10 - R_2) \\ 0.095 + 0.0005R_2 &= 0.995 - 0.0995R_2 \\ 0.0005R_2 + 0.0995R_2 &= 0.995 - 0.095 \\ 0.1R_2 &= 0.9 \\ R_2 &= \frac{0.9}{0.1} = 9\ \Omega \end{aligned}$$

If it is known that R_2 is equal to 9 Ω and that $R_2 + R_3$ is 10 Ω, then R_3 must be 1 Ω, and all resistor values are known.

If the values computed are correct, then V_1 should equal V_2, by Ohm's law. Solving for V_1, we have $V = IR$, or 0.0995(1), or 0.0995 V. For V_2, we have $V = IR$, or 0.0005(199), or 0.0995 V.

Quiz 18-3. Test your understanding. Answer these check-up questions.

1. In Fig. 18-6, what are the correct values for the shunt resistors R_{sh_1} and R_{sh_2}? __________ __________
2. In Fig. 18-7, if the meter movement is a 0.1-mA meter with 1000-Ω internal resistance, what would be the shunt resistor values to produce the same full-scale indications of 1, 10, and 100 mA? __________ __________ __________

18-5 METER DAMPING

The most desirable pointer action for a meter is to rise rapidly to the proper value and stop at that value when a current is applied to the meter.

If the moving coil of a meter were wound on a nonconductive coil form, when a current suddenly flows through the meter, the indicator would overshoot, fall back below the proper value, overshoot again, continuing to oscillate over and under for several seconds. The meter is said to have no *damping*.

There are several methods by which needle oscillation can be prevented. One, the most common, is to wind the moving coil on an aluminum bobbin or coil form. The coil form acts as a shorted turn when moving across magnetic lines of force. If a current starts to flow through the meter, the coil and indicator start moving through the lines of force of the permanent-magnet field. According to Lenz's law, this motion induces an emf and a current in the coil form that produces a magnetic field around the form with a polarity opposite to the permanent-magnet field that produced the current. As a result of the opposite magnetic fields, the pointer is prevented from moving rapidly. It approaches the correct value more slowly and should not overshoot. However, even with this form of damping many meters still overshoot somewhat and are said to be *underdamped*.

When a shunt resistor is connected across the moving coil, as in an ammeter, another effect is also present. The moving coil, as it moves through the magnetic

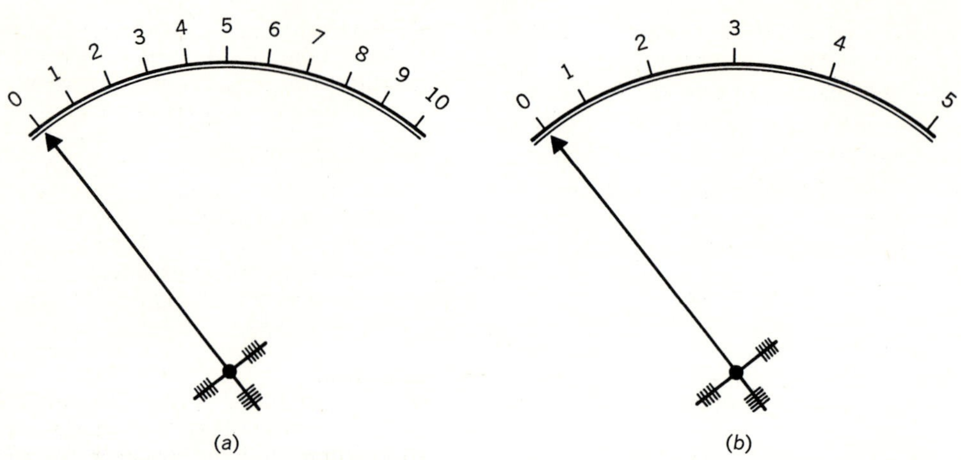

Fig. 18-8 (*a*) Linear meter scale. (*b*) Nonlinear scale. Quadrantal counterweights are shown on the pointer assemblies.

field, has a counter-emf induced in it. The counter-emf sees the shunt as a load and develops current in it. This circulating current through the shunt and moving coil produces a magnetic field around the moving coil. Again, according to Lenz's law, this magnetic field opposes the field that produced it. Any shunt damps the movement of the indicator needle.

Because of the shunt-current effect, an underdamped 1-mA meter with a shunt on it to make it a 10-mA meter may turn out to be perfectly damped. This is termed *critically damped*. That is, there is fast motion of the needle, but no overshoot. The same 1-mA movement, however, with a lower value of shunt, to make the meter a 100-mA meter, for example, may now be overly damped. The result of this will not be an overshoot, but an excessively slow needle motion. If the same meter is made into a 1-A meter, the indicator motion may be excessively slow. A simple multirange ammeter will not have the same damping characteristics on all scales. The Ayrton shunt, however, has relatively constant damping characteristics, making it a more desirable multirange circuit than the simpler shunt.

A third type of damping incorporates light aluminum paddles attached to the pointer assembly. The motion of these paddles through an enclosed air chamber prevents rapid rotation of the coil and pointer. Such air damping is used only on large long-distance-viewing panel meters.

ANSWERS TO CHECK-UP QUIZ 18-3

1. (6.67 Ω) (0.606 Ω) **2.** (99.89 Ω) (9.99 Ω) (1.1 Ω)

18-6 LINEAR AND NONLINEAR METER SCALES

Most of the moving-coil-type meters discussed so far have good scale linearity. That is, if the first milliampere moves the needle 3 mm, then the second milliampere moves it another 3 mm, the third another 3 mm, and so on, as in Fig. 18-8*a*.

If the magnetic field between the horseshoe-shaped magnets does not provide a constant field strength across the whole pole face, the scale of the meter will not be linear. The values at either the high-current or the low-current ends of the scale may be spread apart.

In most of the basic ac meters, the low-current values are close together and the high-current values are spread apart (Fig. 18-8*b*). Some meter scales are said to be *current-squared*. In these meters if 1 mA gives 1 mm of needle movement, 2 mA gives 4 mm, 3 mA produces 9 mm, 4 mA gives 16 mm, etc.

Quiz 18-4. Test your understanding. Answer these check-up questions.

1. What are two reasons why the Ayrton shunt is to be preferred over simpler multirange switching methods? ________ ________
2. What are the three methods by which damping is accomplished on meters? ________ ________ ________
3. What is the name of the law which states that the current induced into a moving wire will always produce a current whose magnetic field opposes the field that helped produce the current? ________
4. What are two other names used for the Ayrton shunt circuit? ________ ________
5. If a meter reading changes when the position of the meter is changed, what are two things that might be wrong? ________ ________
6. A meter pointer rises very rapidly but does not overshoot. What degree of damping is this? ________ What degree is it if it overshoots and oscillates? ________
7. Without looking at the illustration, draw a diagram on a separate sheet of paper of a three-range ammeter using an Ayrton shunt, and then check against the text illustration.

CHAPTER 18 TEST • DC AMMETERS

1. What are the maximum and minimum possible current values when a 500-mA meter reads 350 if guaranteed accurate within 1%?
2. What is the common name for a D'Arsonval movement meter?
3. What would be the advantage of having an iron case around a meter?
4. Of what material is the coil form made in most meters? Why?
5. By what means is current led into and out of the moving coil of a meter?
6. What must a meter have to ensure proper readings regardless of the physical position of the meter?
7. What does the central iron core accomplish in moving-coil-type meters?
8. What is physically moved when the zero-adjust control screw is turned?
9. In what two ways is the operation of a meter affected when a shunt is connected across it?
10. A 500-μA meter movement with 80 Ω of internal resistance is to be changed to a 1-A ammeter. What value of shunt resistor must be connected across it? Why might such a meter not be considered satisfactory?
11. If an ammeter shunt were made of copper, after the meter had been used for a time, would the readings be higher or lower than they should be? What type of wire should be used for meter shunts?
12. Why do meter movements with higher sensitivity have higher internal resistance?
13. A meter needle rises as fast as possible without overshooting. What degree of damping is this meter said to have?
14. What type of meter scale will give greatest accuracy over the whole indicated range? What type of meter scale will give greatest accuracy above midscale?
15. A 50-μA meter with 4000 Ω of internal resistance is to be connected in an Ayrton shunt circuit and is to have full-scale values of 5, 50, and 500 mA. What are the resistance values of the required shunting resistors?

ANSWERS TO CHECK-UP QUIZ 18-4

1. (Safe switching) (No damping change) **2.** (Metal bobbin) (Shunted moving coil) (Air paddles) **3.** (Lenz's) **4.** (Universal) (Ring) **5.** Misadjusted quadrantal weights) (Strong external magnetic field) **6.** (Critical) (Underdamped)

19 OTHER DC METERS

CHAPTER OBJECTIVE. To describe moving-coil voltmeters, computation of multiplier resistors, voltmeter loading of a circuit, high-voltage measuring problems, composition of ohmmeters, decade boxes, bridges, VTVMs, TVMs, electrostatic voltmeters, DVMs, dc wattmeters, and ampere-hour meters.

19-1 DC VOLTMETERS

There are four basic voltmeters. The most common is a sensitive moving-coil microammeter or milliammeter in series with a multiplier resistor. The vacuum-tube or transistor voltmeters use a common meter movement but operate on a balanced-bridge principle. The digital voltmeter is an electronic voltmeter with a numerical readout. Electrostatic voltmeters are used in some laboratories to measure high voltages.

19-2 MOVING-COIL VOLTMETERS

At one time all dc voltmeters were the moving-coil-with-multiplier type. Such meters consist of sensitive ammeter movements with current-limiting *multiplier* resistors in series with them, and a scale calibrated in the number of volts required to make the needle move, rather than in the value of current flowing through the meter.

The voltmeter in Fig. 19-1 consists of a 0- to 1-mA meter movement with a 20,000-Ω multiplier resistor in series with it. If the resistance of the movement is neglected, the resistance of the meter, from A to B, is 20,000 Ω. If the meter is connected across the 20-V battery, the value of current flowing in the meter will be $I = E/R$, or 20/20,000, or 0.001 A, or 1 mA. Under this condition the meter indicator will be driven to full scale. Full scale for this voltmeter would be marked 20 V.

If the meter is connected across the 10-V source, the current flow through the meter will be only half as much, or 0.5 mA. This represents a half-scale deflection of the meter. Where the pointer rests should be marked 10 V.

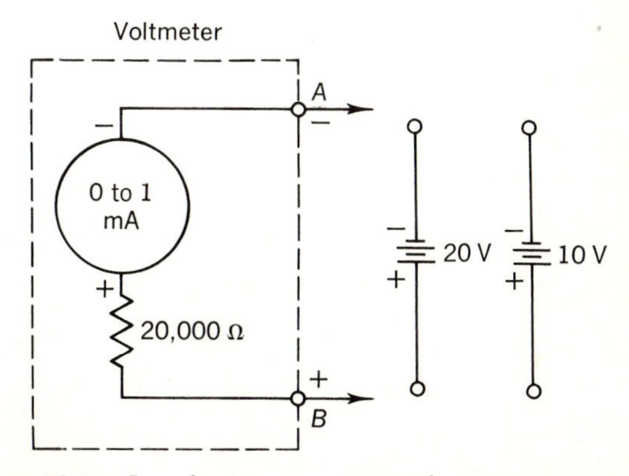

Fig. 19-1 A voltmeter consists of a sensitive ammeter and a multiplier (usually inside case). This meter reads full scale across 20 V, and half-scale if across 10 V.

If the 20,000-Ω multiplier resistor is replaced by a 50,000-Ω multiplier, more voltage will be needed across the meter to make it read full scale. In this case, $E = IR$, or 0.001(50,000), or 50 V full scale. If the multiplier has a value of 100,000 Ω, the result will be a 0- to 100-V full-scale meter. If 20 kΩ makes the meter a 0- to 20-V meter, 50 kΩ produces a 50-V meter, and 100 kΩ results in a 100-V meter, any 0- to 1-mA meter movement can be said to have a *sensitivity* of 1000 ohms per volt (1000 Ω/V). A 0- to 1-V meter would have a 1000-Ω multiplier.

When a more sensitive meter movement is used, such as a 50-μA meter, the multiplier resistance will have a greater value. For example, to produce a 0- to 20-V meter, the multiplier resistor would have to be $R = E/I$, or 20/0.00005, or 400,000 Ω. A 0- to 100-V meter would require a multiplier with a value of 2 MΩ (2,000,000 Ω). The sensitivity for a 0- to 50-μA meter movement is 20,000 Ω/V, or 20 times the sensitivity of the 0- to 1-mA meter movement.

The sensitivity of meters can be expressed in either the value of current required to drive the indicator to a full-scale deflection, or the ohms-per-volt rating of the meter. With ammeters, the full-scale current value volt rating may be more desirable.

Quiz 19-1. Test your understanding. Answer these check-up questions.

1. What is the name given to the resistor connected across an ammeter? ________ In series with a voltmeter movement? ________
2. A multirange voltmeter is shown in Fig. 19-2. What would be the value of R_1? ________ R_2? ________ R_3? ________ R_4? ________
3. In question 2, why could the resistance of the meter be neglected in the 0- to 100-V and 0- to 1000-V ranges? ________
4. If the meter in question 2 had a 100-μA sensitivity and a 500-Ω R_i, what value would R_1 be? ________ R_2? ________ R_3? ________ R_4? ________
5. What are the Ω/V sensitivities of meters with a full-scale 0 to 1 mA? ________ 0 to 0.0005 A? ________ 0 to 20 μA? ________ 0 to 0.01 A? ________
6. Would there be danger of meter burnout in the circuit of Fig. 19-2 while the switch is being moved from one contact to the next? ________

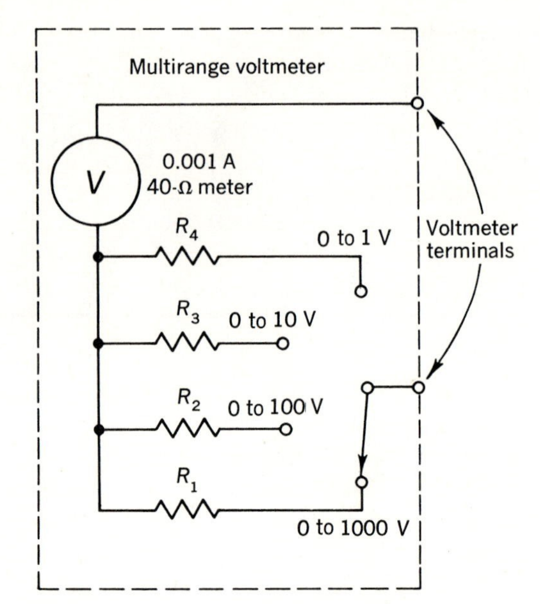

Fig. 19-2 A four-range voltmeter circuit.

19-3 VOLTMETER LOADING OF A CIRCUIT

Either a 1000-Ω/V or a 20,000-Ω/V meter could be manufactured to measure voltages from a fraction of a volt to 10,000 V or more. Which would be the better to use? One consideration in selecting a voltmeter is the cost. Usually a 50-μA meter movement will cost about 50% more than a 1-mA movement. A second consideration is accuracy. The more expensive meters usually have movements of 1% accuracy and resistors of better than 1% tolerance. Another consideration is the current required for full scale. A 1-mA movement requires 1 mA from the circuit being measured at full-scale readings. The 50-μA meter would draw only one-twentieth as much current. If current drain in the circuit is not important, the lower-sensitivity meter might be quite satisfactory.

When voltage-drops are being tested in a circuit, it is important that during the time the meter is connected across a component, the meter give the most accurate reading possible of the voltage that existed across the part before the meter was connected.

The loading effect of a voltmeter can be demonstrated by considering two meters of different sensitivity used to measure voltage in the circuit shown in Fig. 19-3. This is a series circuit across a 60-V source. From the ratio of the two resistors across the source, 50 and 100 kΩ, the voltage-drop across the 100-kΩ resistor should be two-thirds of 60, or 40 V (check by Ohm's law).

First, the 1000-Ω/V meter will be used to measure the voltage across the 100-kΩ resistor. As soon as the leads of the meter are touched to A and B, the resistance between these two points is no longer 100 kΩ. When the 100-kΩ R_i of the meter is in parallel with the 100-kΩ series resistor, there is only 50 kΩ between A and B. Now, the source is across two 50-kΩ resistances in series. There must be only half of 60, or 30 V, across A and B. This is the voltage that the 1000-Ω/V meter would indicate. It is loading the circuit. A 30-V reading represents an error of 10/40 (the difference of voltages to the true voltage), or 25%.

With the 20-kΩ/V meter across the 100-kΩ series resistor, the resistance between A and B is 100 kΩ and 2 MΩ in parallel, or approximately 95,000 Ω. The series circuit is now 50 kΩ + 95 kΩ, or 145 kΩ. By Ohm's law, the current will be $I = E/R$, or 60/145,000, or 414 μA. The voltage-drop across points A and B must be $V = IR$, or 0.000414(95,000), or 39.3 V. The meter would indicate 39.3 V. In comparison with the actual voltage-drop of 40 V from A to B with no meter, this meter indicates a voltage only 1.75% low. Since the usual meter is only accurate to

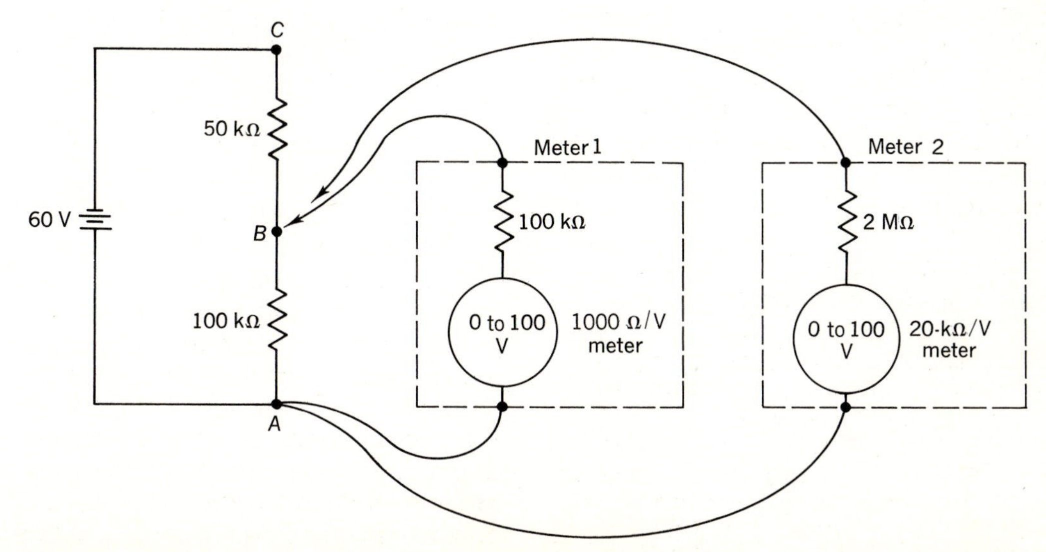

Fig. 19-3 Meters with different sensitivities give different voltage indications in a high-resistance circuit. Meter 1 reads 30 V from A to B; meter 2 reads 39.3 V.

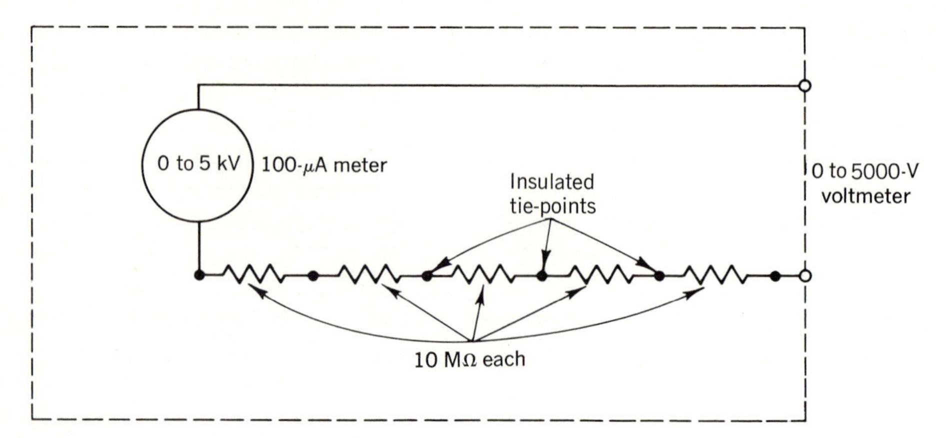

Fig. 19-4 A 0- to 5000-V meter. Five 10-MΩ resistors in series serve as the multiplier.

perhaps 2%, the 20-kΩ/V meter reads the voltage in this circuit quite accurately.

If the load resistances had been 50 and 100 Ω, the 0- to 1-mA meter with its 100,000 Ω in parallel with the 100-Ω load represents so little variation from 100 Ω as to be negligible. As a result, the 1000-Ω/V meter would indicate the voltage-drop to be 40 V, as would the 20,000-Ω/V meter. In low-impedance (low-resistance) circuits either meter would be satisfactory. In high-impedance circuits high-sensitivity meters are required to maintain the accuracy of voltage readings.

If the two load resistors were 5 and 10 MΩ, the 20-kΩ/V meter would read 24.5 V instead of 40 V, an error of about 40%. (The 1000-Ω/V meter would only indicate 1 V.) For very high-impedance circuits, some type of voltmeter other than the standard meter-and-multiplier may be necessary. Vacuum-tube, transistor, electrostatic, and digital voltmeters are possibilities.

ANSWERS TO CHECK-UP QUIZ 19-1

1. (Shunt) (Multiplier) **2.** (1 MΩ) (100 kΩ) (10 kΩ, or 9960 Ω) (960 Ω) **3.** (Neither meters nor resistors would normally be marked that accurately) **4.** (10 MΩ) (1 MΩ) (100 kΩ) (9500 Ω) **5.** (1000 Ω/V) (20 kΩ/V) (50 kΩ/V) (100 Ω/V) **6.** (No)

19-4 HIGH-VOLTAGE VOLTMETERS

When meters must read voltages above about 1000 V, it is necessary to consider the voltage arc-over possibilities of the multiplier resistors used. Suppose a 100-μA (10-kΩ/V) meter movement is to be made into a 5000-V meter. This means 10 MΩ of multiplier resistance per 1000 V, or 50 MΩ for a 5-kV meter. It is found that exceeding perhaps 1000 V across a resistor tends to break down or produce a spark that will arc-over across the resistor. Therefore the multiplier should be made up of five 10-MΩ resistors in series, each mounted on high-voltage insulated "tie points" (Fig. 19-4).

The usual test meters have flexible leads with rubber or plastic insulation capable of withstanding perhaps 2 kV. As these cords age, they wear, harden, and crack, and the insulating value lessens. When meters are used in high-voltage circuits, the meter leads must be insulated for at least a 100% safety factor. Needless to say, when measuring high voltages, technicians should be most careful, should not hurry, must not be tired or sleepy, and must keep their wits about them.

Quiz 19-2. Test your understanding. Answer these check-up questions.

1. In Fig. 19-3, in measuring the voltage from *B* to *C*, would the two meters give more, less, or the same accuracy of indications as in measuring *A* to *B*? __________
2. In Fig. 19-3, what value would the 1000-Ω/V meter indicate between *B* and *C*? __________ What value would the 20-kΩ/V meter indicate? __________
3. If the meter in Fig. 19-4 were a 0- to 1-mA meter, what would be the values of the five multiplier resistors? __________
4. What relative sensitivity should voltmeters have in high-impedance circuits? __________ Low-impedance circuits? __________
5. What relative sensitivity should ammeters have in high-current circuits? __________ Low-current circuits? __________

19-5 OHMMETERS

Resistance values of resistors can be determined by the use of an *ohmmeter*. Basically, an ohmmeter consists of a sensitive dc milliammeter or microammeter, a 1.5-V or 3-V battery, a current-limiting resistance, and a pair of flexible leads with pointed probes on the ends, as in Fig. 19-5*a*. This 50-μA meter is a 20,000-Ω/V meter. If the battery in the ohmmeter is 3 V, the current-limiting resistance should be 3 × 20,000, or 60 kΩ. To allow accurate zero adjustments, the limiting resistance might consist of a 50-kΩ resistor and a 15-kΩ rheostat.

Before resistance is measured, the two probes are held together (dotted probe) and the meter is checked to see if the probes-together condition registers 0 Ω (full scale). If it does not, the zero-adjust control is varied to a zero reading. If the probes are held across a 60,000-Ω resistor, the ohmmeter will read half-scale, because with twice the resistance in the meter circuit half the current flows and the pointer moves to only half-scale. When the probes are held across a resistance of higher value, less current flows through the meter, and the deflection falls farther to the left. With the probes open (not connected to anything), there is no current flowing through the meter, and it reads *infinity* (∞) ohms. This meter reads fairly accurately from about 1 kΩ to about

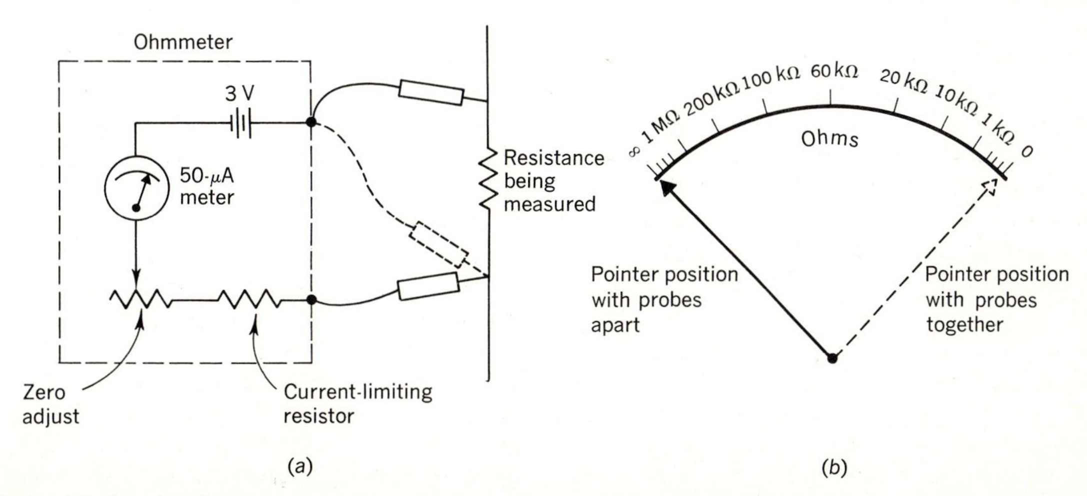

Fig. 19-5 (*a*) Simple ohmmeter circuit, with probes across the resistor being measured. Probe position for zero-adjust is shown dashed. (*b*) Ohmmeter scale, 0 Ω at right, infinite Ω at left.

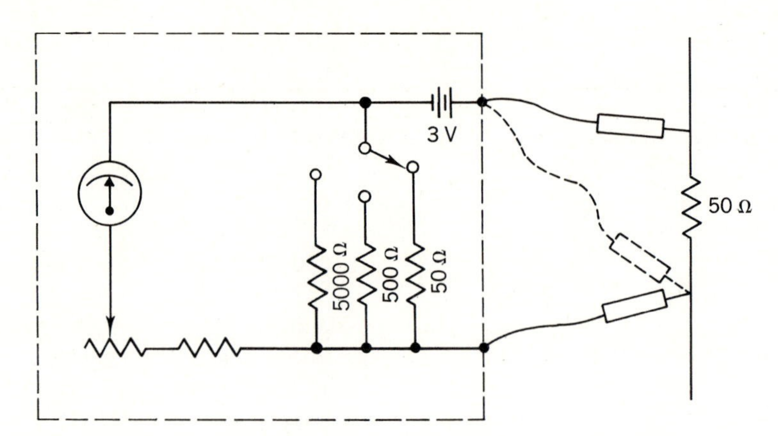

Fig. 19-6 Three-range ohmmeter circuit. With the internal resistor used as shown, the 50-Ω resistor being measured would produce a half-scale reading of the pointer.

200 kΩ. At the high-resistance end the graduations are crowded, and accuracy lessens.

A multiscale ohmmeter is shown in Fig. 19-6. With the switch in the position shown and with the probes shorted together, the meter, the battery, and the current-limiting resistances are in series. The internal 50-Ω resistor is in parallel with the 3-V battery. The meter is zeroed with the zero-adjust control. If the probes are now connected across a 50-Ω resistor, there are two similar resistors in series with the battery, and there is 1.5 V across each resistor. The meter is across only one of these resistors and therefore reads only half-scale. In this switch position the ohmmeter reads 50 Ω, or half-scale. It will give reasonably accurate readings from 1 Ω to several hundred ohms.

When the switch is moved to the 500-Ω position and the probes are across a 500-Ω resistor, the meter will read half-scale at 500 Ω. Values from about 50 to 10,000 Ω can be read easily.

ANSWERS TO CHECK-UP QUIZ 19-2

1. (Same) **2.** (15 V) (19.7 V) **3.** (1 MΩ each) **4.** (High) (Low or high) **5.** (Low) (High)

If the switch is moved to the 5000-Ω position and the probes are across a 5000-Ω resistor, the meter will read half-scale at 5000 Ω. Values from about 200 to 80,000 Ω are read fairly accurately.

Notice that when the probes are shorted on the 50-Ω switch position, the current flowing through the probes and battery is $I = E/R$, or 3/50, or 60 mA. This value of current flowing through some delicate electronic equipment might damage it.

Never try to read the resistance of a circuit in which there is current flowing. The current or voltage in the circuit may burn out the meter movement. (A small voltage might not burn out the meter but will cause incorrect resistance-value readings.)

19-6 DECADE BOXES

To produce an exact value of resistance, a *resistance decade* can be used. The circuit of a simple decade box is shown in Fig. 19-7. With the switches in the positions shown, the resistance across the whole decade is 0 Ω. (There is direct *continuity* between the bottom terminal and the top terminal.) If the "ones switch" is moved four positions to the right, there is 4 Ω

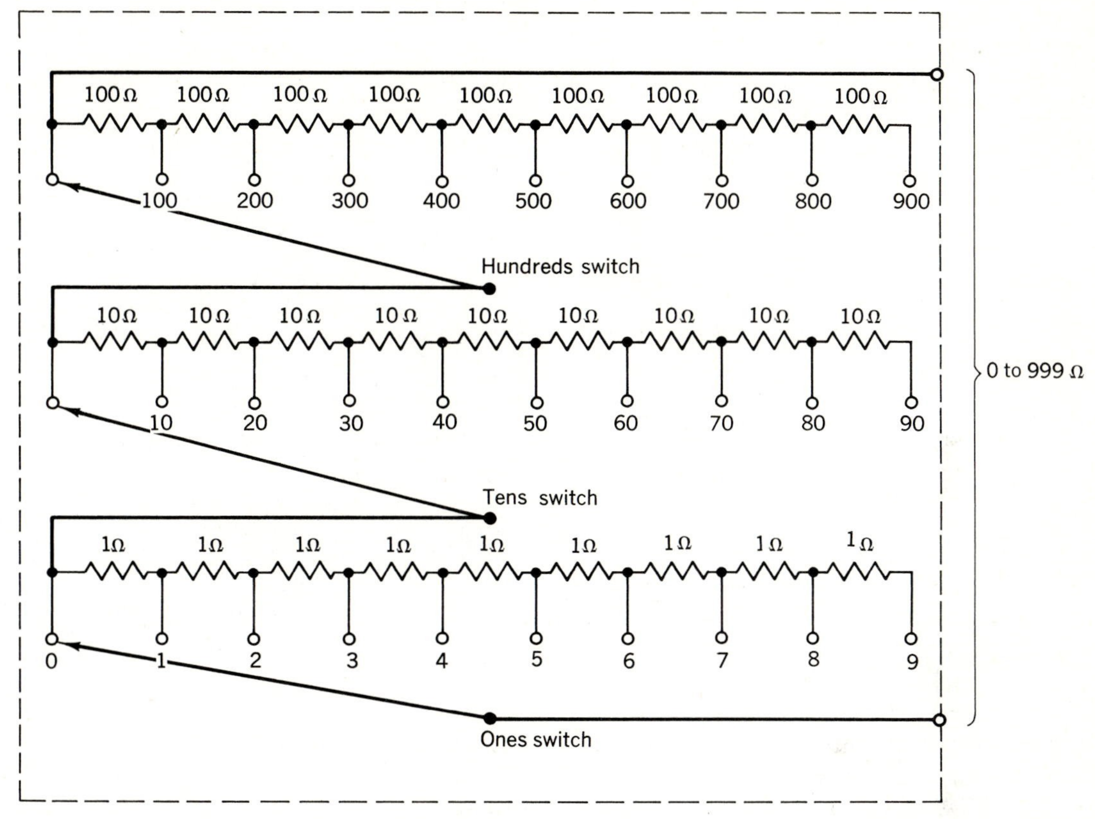

Fig. 19-7 Three-range resistance decade. Any value 0 to 999 can be selected in 1-Ω steps.

between decade terminals. If the "tens switch" is then moved three positions to the right, the total resistance of the decade is 30 + 4, or 34 **Ω**. If the "hundreds switch" is moved seven positions to the right, the total resistance is 734 **Ω**. It is possible to select any value of resistance from 0 to 999 **Ω** in 1-**Ω** increments with this decade.

Most decades will have a one-tenth-ohm switch, a thousands, and possibly ten-thousands and hundred-thousands switches.

There are also capacitor and inductor decades. The capacitor decades are not too accurate at lower values due to stray capacitances. They are subject to *hand-capacitance* effect, the small change of capacitance when a person's hand or other conducting mass is brought near the decade.

Inductor decades are not too accurate because of resistance in the coils.

19-7 BALANCED-BRIDGE MEASURING DEVICES

Many measuring devices utilize the principle of balancing a bridge circuit. One bridge-type device is the *Wheatstone resistance bridge* (Fig. 19-8). This bridge circuit is used to determine the resistance value of an unknown resistor R_x in the diagram. A decade resistance box forms one arm of the bridge. For this explanation the two resistors, R_1 and R_2, have the same value, 1000 **Ω**. The meter is a sensitive microammeter with a current-limiting

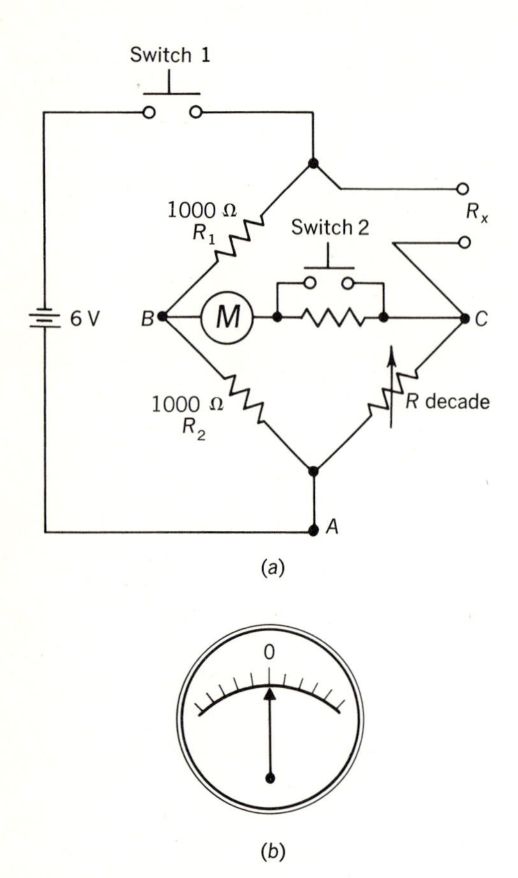

Fig. 19-8 (*a*) Wheatstone resistance bridge circuit. (*b*) Zero-center meter with no current flowing.

resistor in series with it. The meter has its springs adjusted to read zero current with the pointer at *midscale* (Fig. 19-8*b*).

The unknown resistor is connected across the terminals marked R_x. When switch 1 is closed (a push-to-close symbol is shown), current flows through R_1 and R_2, while another current flows through R_x and the decade box. One terminal of the meter is connected to the midpoint of R_1 and R_2, making this end of the meter 3 V more negative than point *A*. The other end of the meter is connected to the point between the decade and R_x. On the assumption that R_x and the decade do not have the same value, point *C* will not be 3 V more negative than point *A*, and the meter will be across a difference of potential. If point *C* has a greater potential than point *B*, the meter will move off its zero-center point, either to the right or to the left. If point *C* has less potential than point *B*, the meter will move in the opposite direction.

The decade is adjusted until the meter is exactly at zero center. Now there is no potential across the meter. R_x and the decade must have the same resistance value. The value shown by the decade switches is the resistance of the unknown resistor.

The zero-center indication may be sluggish if the current-limiting resistor is in the circuit. After an approximate zero reading is obtained, switch 2 can be pressed, which increases the current flow through the meter, and a more accurate zero-current reading can be obtained.

It is not necessary that R_1 and R_2 be equal. If R_1 were 10 times R_2, when the circuit was balanced (when the meter zero-centers), R_x would have a value 10 times that of the decade reading. (Multiply the decade by 10 for a correct reading.) Thus a simple decade with a means of switching R_1 to 10 times or to one-tenth of R_2 can produce a wide variation of accurate resistance measurements.

This same type of bridge can be used to determine the capacitance of an unknown capacitor by using an ac source, an ac meter (or earphones), a capacitor decade, and by placing the unknown capacitor across the terminals marked R_x. A zero current (or zero tone in the earphones) indicates the unknown capacitor has a capacitance equal to that of the decade. An inductance bridge can be developed along the same lines. For extreme accuracy the bridge circuits may have several other refinements built into them.

Quiz 19-3. Test your understanding. Answer these check-up questions.

1. In a balanced-bridge circuit, where R_1 is 1000 Ω, R_2 is 2000 Ω, and R_d is 4500 Ω, what is R_x? ________
2. In a balanced-bridge circuit, where $R_1 = 500$, $R_2 = 450$, and $R_d = 3400$ Ω, what is R_x? ________
3. In a balanced-bridge circuit, where $R_1 = 1000$, $R_2 = 10{,}000$, and $R_d = 520$ Ω, what is R_x? ________
4. In question 1, what would be the current value in the meter if the source were 6-V dc? ________
5. Why does a capacitor bridge require an ac source? ________
6. Would any usable result be obtained by using the bridge in Fig. 19-8 to measure an inductor? ________ If so, what? ________
7. What does *decade* mean? ________
8. In Fig. 19-8, which switch might be labeled "high-sensitivity"? ________ Which component protects the meter from burnout if the bridge is too far out of balance? ________
9. Without looking at the illustration, on a separate piece of paper, draw a diagram of a decade with units, tens, and hundreds switches. Check your drawing against Fig. 19-7.
10. Try drawing a diagram of a capacitor decade with units and tens switches. (It will not be the same as a resistance decade circuit.)

19-8 DC VACUUM-TUBE VOLTMETERS

A very sensitive voltmeter is produced by combining the amplifying ability of vacuum-tube triodes with a balanced-bridge circuit. To understand the vacuum-tube voltmeter (VTVM) circuit of Fig. 19-9, it must be remembered that a small voltage variation between the cathode and grid of a triode will produce a relatively large plate-current change.

With no voltage across the meter probes, both vacuum tubes VT_1 and VT_2 pass equal values of current. The total plate current for both tubes flows through R_1, developing a voltage-drop E_C of perhaps 3 V. Since both grids are connected to the negative end of E_C, both tubes have equal bias-voltage values of −3 V. The plate-circuit potentiometer, R_4, is set so that R_2 and R_3 have equal values.

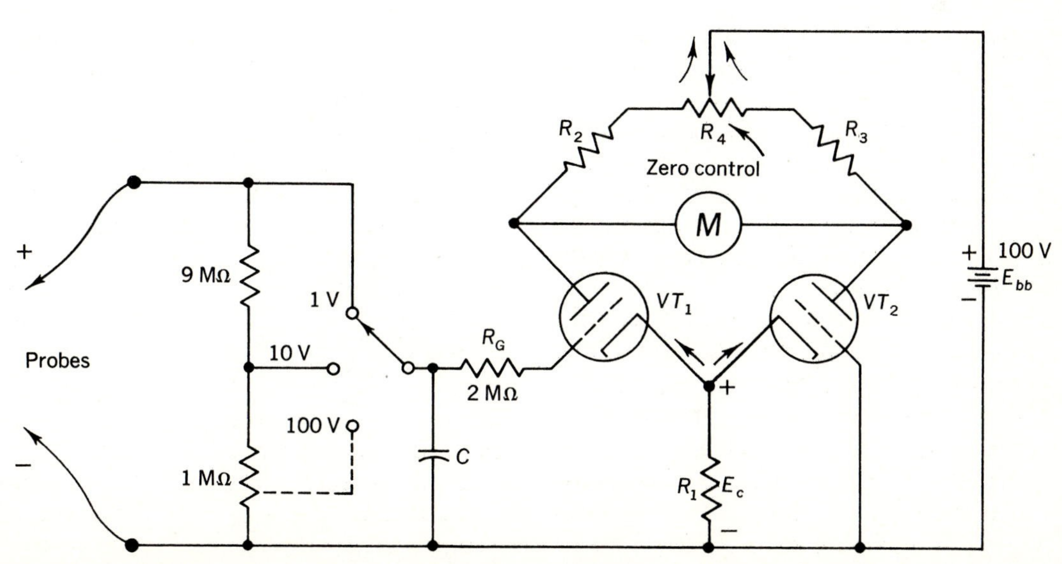

Fig. 19-9 Simplified schematic diagram of a three-range VTVM: 0 to 1, 0 to 10, and 0 to 100 V. The tubes and plate-circuit resistors form a bridge with the meter from corner to corner.

With similar resistors and plate currents, similar voltage-drops occur across R_2 and R_3. Being similar tubes with similar plate and bias voltages, the dc resistance of the two tubes will be the same. The meter is between the two branches of a balanced-bridge circuit, where R_2 and VT_1 is one leg and R_3 and VT_2 is the other leg. The meter should read zero. It it does not, a slight adjustment of the plate-circuit-potentiometer zero control will balance the currents and zero the meter. VTVM meters *usually* have the zero reading at the left end of the scale instead of having zero at the center as with other bridges.

What happens when the two probes are connected across a 1-V source? When the probes are connected across the voltage to be measured, there is +1 V in series with the −3-V bias on the VT_1 grid (but not in series with the VT_2 grid). With 1 V of grid bias canceled for VT_1, its plate current increases. More plate current through VT_1 and R_2 increases the voltage-drop across R_2. Now the meter is between two points of unequal voltage (an *unbalanced* bridge), current flows through the meter, and it deflects. If enough current can flow through the meter to give it full-scale deflection with 1 V applied to the VT_1 grid circuit, the VTVM circuit is a 0- to 1-V meter. The input impedance or resistance is 1 MΩ + 9 MΩ, or 10 MΩ. The sensitivity of the meter at this voltage range is 10,000,000 Ω/V.

The unbalance is greater than appears at first. As the VT_1 plate current increases through R_1, the voltage E_C increases and biases the VT_2 grid more negatively, resulting in a lessening of plate current for VT_2. With less voltage-drop across R_3 plus the greater voltage-drop across R_2, the meter is between almost doubly dissimilar voltage points.

If the input switch is adjusted to the 10-V scale position, the 1- and 9-MΩ resistors form a voltage divider. Only one-tenth of the voltage across the probes is now added to the grid circuit of VT_1. Thus a 10-V source will just produce full-scale deflection of the meter. It is now a 0- to 10-V meter with a sensitivity of 1,000,000 Ω/V. If the 1-MΩ resistor is divided into a 900,000- and 100,000-Ω voltage divider and the switch is set to this point, only one-hundredth of the probe voltage will affect the meter grid. The meter will now be a 0- to 100-V meter with a 100,000-Ω/V sensitivity. If it is made to read 0 to 1000 V full scale by further voltage division, it will have a sensitivity of only 10,000 Ω/V. Above 500 V this VTVM is not as sensitive as a good 50-μA, 20-kΩ/V voltmeter.

The resistor in series with the grid, R_G, is a protective device. If the probe is inadvertently placed across too high a positive voltage, grid current will flow. The voltage-drop developed across the resistor under this condition will prevent the grid from ever exceeding a fraction of a volt positive, preventing dangerously high I_p.

The capacitor, C, in the grid circuit has no effect on the operation of the meter with dc. However, when an ac probe circuit with rectifiers is used, the meter can then be used to measure ac voltages. The grid circuit capacitor charges to the peak voltage value of the ac being measured, but the voltage dividers for ac measurements are arranged so that the meter scale shows rms ac values.

A transistorized bridge-type voltmeter (TVM) using bipolar transistors makes a

---

ANSWERS TO CHECK-UP QUIZ 19-3

1. (2250 Ω) **2.** (3778 Ω) **3.** (52 Ω) **4.** (Zero) **5.** (No I flows through a C) **6.** (Yes) (Measures internal R) **7.** (Multiple of 10) **8.** (Sw_2) (Resistor in series with meter) **9.** (See Fig. 19-8) **10.** (*Hint:* nine capacitors have a common terminal for each switch)

good low-voltage meter for low-impedance circuits, but is not satisfactory for high-voltage measurements. A *field-effect transistor* (FET) is an electrostatically controlled solid-state amplifying device and is used in a transistor-voltmeter circuit almost identical to the VTVM. An FET may be used as a high-input-impedance amplifier ahead of a bipolar transistor bridge TVM, however.

19-9 ELECTROSTATIC VOLTMETERS

The moving-coil meter is magnetically operated and is therefore a current-indicating device. When voltage is applied across a capacitor, electrostatic lines of force develop between the positive and negative plates. These electrostatic lines of force are similar to magnetic lines of force in that they continually try to contract or shorten. Two plates of a charged capacitor are continually trying to pull together. A meter utilizing this + to − pulling effect is the *electrostatic voltmeter* (Fig. 19-10). It consists of a pair of stationary plates near a pair of rotatable plates, as shown. If a difference of potential is connected across the terminals, electrostatic lines of force develop between the plates. They are drawn together by the contraction of the lines of force, against the pull of the spring that tries to return the pointer to zero. Since the rotor plates are pivoted at the center, they cannot move up against the stator plates, but rotate parallel to the stators. The greater the voltage across the meter, the farther the rotor plates move. There is no flow of current through this meter. Also, the pointer moves in the same direction regardless of the polarity of the charge on the plates. For this reason the electrostatic voltmeter operates on dc but also indicates rms ac voltage. These meters are used in laboratories. They are rather bulky and delicate.

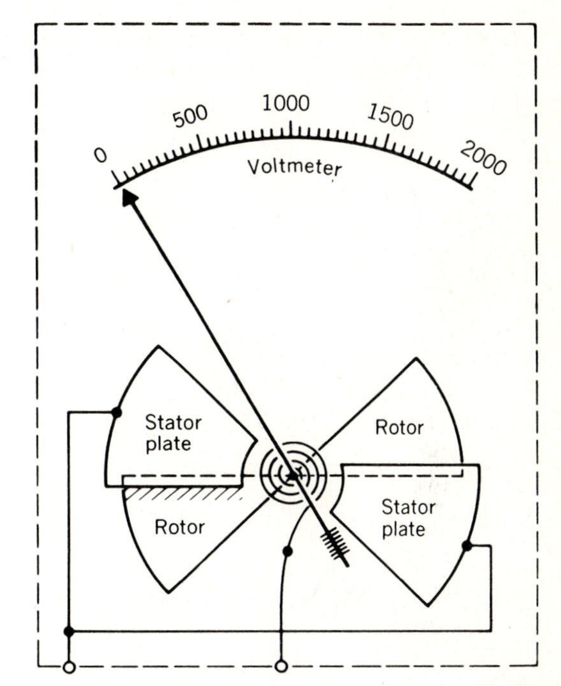

Fig. 19-10 Electrostatic voltmeter. When a voltage is applied between terminals, the rotor plates are pulled toward the stators, moving the pointer upscale.

Quiz 19-4. Test your understanding. Answer these check-up questions.

1. If a VTVM is turned on but the meter reads some value with no voltage across the probes, what should be adjusted? ________
2. If a VTVM meter were connected in the circuit in reverse, would the probe marked + work as the − probe, or would the VTVM not work properly? ________
3. In Fig. 19-9, if R_2 were a few ohms less than R_3, would the arm of the potentiometer in the diagram have to be moved to the right or to the left? ________
4. In Fig. 19-9, if the bias voltage on VT_2 is −3 V and R_1 has a value of 400 Ω, what is the I value flowing through the B battery when the meter reads 0 V? ________ What is the I_p value of VT_2? ________

5. In Fig. 19-9, what is the input (across probes) impedance of the VTVM on the 0- to 1-V range? ________ The 0- to 100-V range? ________
6. Which would load a circuit more, the VTVM shown or a 5000-Ω/V meter movement voltmeter when measuring 5 V? ________ 1000 V? ________ 2000 V? ________
7. Would the VTVM shown read if it were across an ac voltage? ________ Why? ________
8. Would the electrostatic voltmeter read across an ac voltage? ________ Why? ________
9. Why might a bias voltage be symbolized "E_c"? ________

19-10 DIGITAL VOLTMETERS

Digital voltmeters do not have a pointer, but register the voltage on little neon *Nixie* tubes or semiconductor *readout devices*, giving answers in glowing numbers. There are a variety of digital voltmeters (DVMs). One type has a series of decimal-counting unit (DCU) circuits coupled to a series of three, four, five, or six Nixies, all in a line. The right-hand DCU and Nixie indicates units; the Nixie and DCU to the left of this indicates tens; the next DCU and Nixie indicates hundreds; etc. The three-DCU register shown in Fig. 19-11*a* indicates a voltage of +748 V. The voltage registered on Fig. 19-11*b* is −6.05 V. The register has a decimal point that is placed electronically. The polarity, + or −, is also automatically registered.

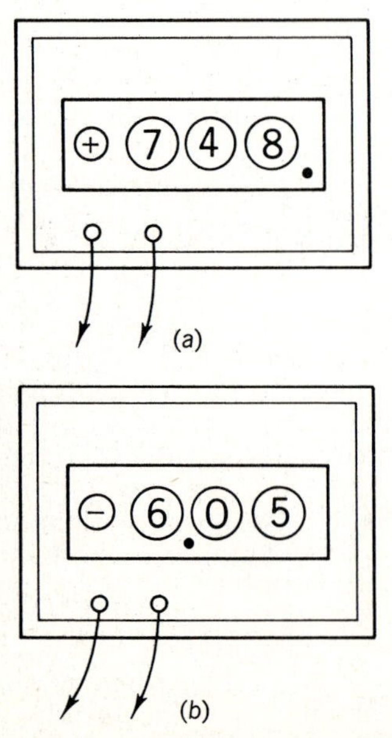

Fig. 19-11 Two digital voltmeter readouts: (*a*) +748 V, (*b*) −6.05 V.

One form of DVM has a self-contained transistor *oscillator*, which is an ac generator. It oscillates at a precise frequency of perhaps 100,000 Hz. The job of the DVM is to develop a linearly decreasing *ramp voltage* (Fig. 19-12) and continually compare the ramp voltage with the voltage being measured. Each time these two voltages are equal, a *null voltage*, similar to the balancing of a bridge circuit, occurs. The null starts the high-frequency oscillator feeding into the DCUs. The DCUs count the number of cycles being fed to them. As the ramp voltage decreases through zero into the negative region, a cutoff pulse develops that stops the counting of the oscillations. The number of ac cycles that occur between the null and zero times produces the readout on the

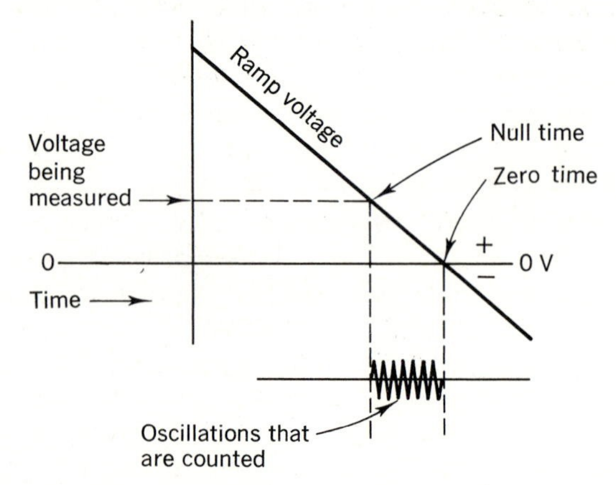

Fig. 19-12 When the decreasing ramp voltage reaches the level of the voltage being measured, oscillations and counting start. When the ramp voltage reaches zero, oscillations stop.

Nixies. If 748 cycles occurred between null and zero, the Nixies would register 748.

If the ramp voltage is made to decrease faster (to have a greater slope) by switching to a different RC constant in the equipment, the decimal will be moved over. Now, if there are 605 cycles between null and zero, the meter registers 6.05 V. The circuitry of these devices is quite sophisticated.

Some DVMs will recycle 10 times per second (develop 10 separate ramp voltages and displays per second); others may only make a single display per second. The displayed values can be made to register on printers, punch cards, or tapes, or record on magnetic tapes for a permanent record.

Digital voltmeters have the advantage over the normal *analog*, or pointer-type, meters of reducing human error (reading wrong values on multiscale meters), eliminating *parallax* error (reading slightly incorrect values when sighting the pointer and scale from either the right or left instead of perpendicular to the meter face), increasing reading speed, and automatically sensing + or − polarity without having to reverse the probes.

19-11 DC WATTMETERS

Power is determined by the product of voltage and current ($P = EI$). Therefore, a wattmeter must have two interacting magnetic fields, one proportional to the voltage across the line and the other proportional to the current flowing through the line. It uses a moving coil as one field but substitutes an electromagnet for the permanent magnet of D'Arsonval moving-coil meters (Fig. 19-13).

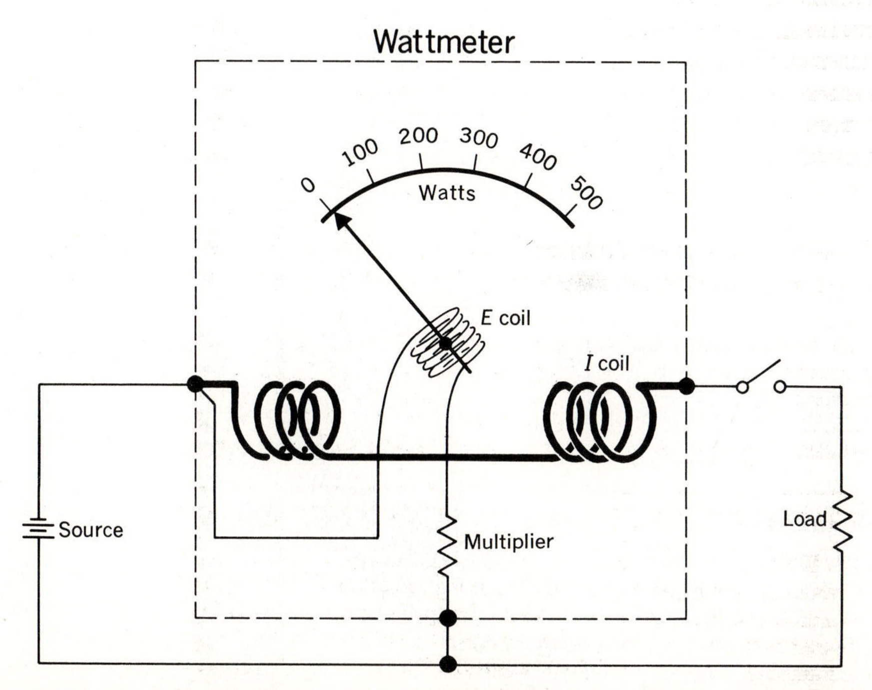

Fig. 19-13 A wattmeter in a circuit. Current coils are shown in heavy lines, voltage coil and multiplier in light lines.

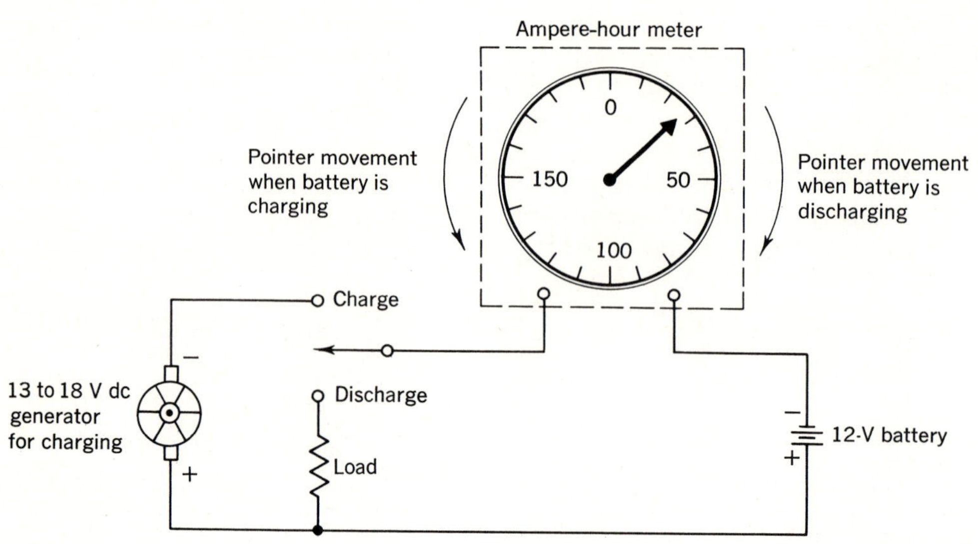

Fig. 19-14 An ampere-hour meter in a battery charging circuit. With the load connected, the battery discharges. In charge position the generator reverses circuit current, and meter pointer reverses.

If there is no current flowing through the load (switch open), there can be no movement of the moving coil, even though the moving coil and its multiplier resistor are across the source and are energized. When the line switch is closed, the field coils have load current flowing through them. Now, there are two fields that can interact: the load-current field and the line-voltage field (the moving-coil field). The pointer is driven up the scale to some value, against return-spring action. If the load resistance decreases (load current increases), the current field increases and the pointer is driven up further. If the source voltage increases, the moving coil field increases and the pointer moves upscale further.

The current field has an iron core, not shown, to make the meter more sensitive.

A wattmeter usually has the three contacts shown, although sometimes there are two current contacts and two voltage contacts brought out as four meter terminals.

19-12 AMPERE-HOUR METERS

An infrequently seen meter is the ampere-hour meter. It may be found in battery-charging circuits. It is a mercury-pool dc motor geared to a rotating pointer (Fig. 19-14). As a battery discharges, current flows through the meter, battery, and load in one direction, driving the pointer in a clockwise motion. When the battery is recharged, current is forced backward through the battery by some source of emf greater than the battery emf. The reversed-direction current flowing through the ampere-hour meter reverses the meter's pointer travel. When the pointer

---

ANSWERS TO CHECK-UP QUIZ 19-4

1. (Zero control) **2.** (Be the − probe) **3.** (Right) **4.** (7.5 mA) (3.75 mA) **5.** (10 MΩ) (10 MΩ) **6.** (5 kΩ/V) (5 kΩ/V) (Same) **7.** (No) (Pointer would try to move above and below zero, averaging zero) **8.** (Yes) (Pointer moves clockwise regardless of polarity applied) **9.** (Bias batteries are called "C batteries")

---

arrives back at the starting point, the battery has been charged as much as it had been discharged, and the charger can be turned off.

Quiz 19-5. Test your understanding. Answer these check-up questions.

1. What is the name given to any type of meter that displays information by means of a pointer? ______ By means of neon or other glow readouts? ______
2. What is meant by DCU? ______ DVM? ______ TVM? ______
3. A voltage is made to decrease linearly from some value down to zero. What term may be applied to this voltage? ______
4. If a meter is read from an angle rather than directly from the front, what type of error is produced? ______
5. The current that flows through the moving coil in a wattmeter is proportional to what parameter of the circuit? ______ To what is the current proportional that flows through the fixed field coils? ______
6. How many connecting terminals would be expected on a voltmeter? ______ An ammeter? ______ A wattmeter? ______ An ohmmeter? ______
7. The theory of operation of a wattmeter follows which power formula? ______
8. What does the symbol "∞" indicate on an ohmmeter? ______
9. On which side of the scale is the zero indication of a voltmeter? ______ An ohmmeter? ______
10. When an ohmmeter is to be used, what should be done with it first? ______
11. Would it be possible to make a linear-scale voltmeter? ______ Wattmeter? ______ Ohmmeter? ______
12. What precaution should be taken when measuring the resistance of a component in an electric circuit? ______
13. Would reversing the source polarity in a circuit make any difference in the indications given on a moving-coil voltmeter? ______ A wattmeter? ______
14. What is the name of the meter that can give an indication of the state of charge of a battery? ______ Reversing the current in this meter has what effect on the pointer? ______
15. What is a transistorized ac generator called? ______

CHAPTER 19 TEST • OTHER DC METERS

1. Is the meter movement of a moving-coil meter a current- or a voltage-operated device?
2. What is the sensitivity in ohms per volt of a 0- to 1-mA meter? A 0- to 50-μA meter? A 0- to 100-μA meter?
3. Under what condition is a higher sensitivity required in voltmeters?
4. Low-sensitivity voltmeters can be used to accurately indicate voltages in what types of circuits?
5. A 0- to 500-μA meter is to be made into a 0- to 5000-V meter. How many multiplier resistors should be used in series?
6. In the ohmmeters described, at which end of the scale (right or left) would 0 Ω be printed? ∞ Ω?
7. How many resistors would be used in a resistance decade box if it had "tenths," "units," "tens," "hundreds," and "thousands" switches?
8. What is measured by a Wheatstone bridge? What other less accurate device measures the same thing?
9. When a VTVM reads full scale, is its bridge balanced or unbalanced? What is the sensitivity of the VTVM when it is set to the 0- to 1-V range? The 0- to 1000-V range?
10. What is the name of the voltmeter that is actually a form of a variable air capacitor? If it were used with ac voltage, would it read at all, read effective values, read peak values, or read average values?
11. What does DVM mean?
12. While the ramp voltage is decreasing, what starts the oscillations in a DVM? What stops the oscillations?
13. What type of meters are known as analog?
14. What is the name of the error produced by looking at the indicator needle of a meter from an angle?

15. What is the basic difference in construction between a D'Arsonval meter and a wattmeter?
16. The current flowing through the moving coil in a wattmeter is proportional to what circuit parameter? To what is the field coil current proportional?
17. What meter is useful as an indicator of how much a battery has discharged as well as how much it has charged?
18. How many terminals are usually found on a wattmeter?
19. What is a transistorized ac generator called?
20. On a piece of paper, draw schematic diagrams of a single-scale ohmmeter; a two-range ohmmeter; a Wheatstone bridge circuit; a VTVM circuit; and a wattmeter in a circuit, including the source and load.

ANSWERS TO CHECK-UP QUIZ 19-5

1. (Analog) (Digital) **2.** (Decimal-counting unit) (Digital voltmeter) (Transistor voltmeter) **3.** (Ramp) **4.** (Parallax) **5.** (Voltage) (Load current) **6.** (Two) (Two) (Three or four) (Two) **7.** ($P = EI$) **8.** (Infinite ohms) **9.** (Left) (Right) **10.** (Short probes, adjust to accurate zero) **11.** (Yes) (Yes) (Not with an infinite reading at one side of scale) **12.** (No I flow in circuit) **13.** (Yes) (No) **14.** (Ampere-hour) (Reverses rotation) **15.** (Oscillator)

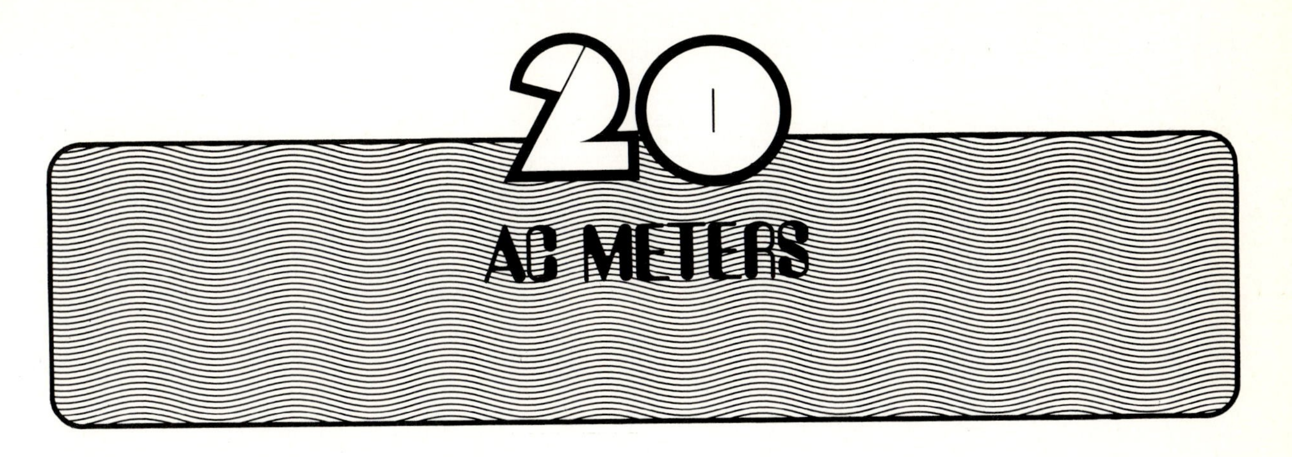

20 AC METERS

CHAPTER OBJECTIVE. To explain the operation and use of: ac rectifier meters, bridge rectifiers, peak-reading, dB and VU meters, electrodynamometers, voltage and current transformers, repulsion meters, ac wattmeters, watthour meters, thermocouple ammeters, frequency meters, grid-dip meters.

20-1 AC COMPONENTS WITH DC METERS

The dc moving-coil permanent-magnet field meter will be driven to a particular pointer indication by some constant value of current. What the meter indicates when the current through it is not constant is of interest because many circuits involve varying amplitude currents.

Figure 20-1*a* illustrates a 10-V source of emf that can be changed from pure dc to varying dc, pulsating dc, and ac. Both the voltmeter and the ammeter are dc moving-coil meters. What will they indicate?

In Fig. 20-1*b* the waveform is pure dc. The current through the load is $I = E/R$, or 10/10, or 1 A. The ammeter reads 1 A.

In Fig. 20-1c a low-voltage ac component has been added to the 10-V dc, resulting in a varying dc. During the peaks of voltage above 10, the meter pointer tries to rise higher than a 10-V reading, but during the valleys of voltage below 10 the pointer tends to fall below 10 V. The result, if peaks and valleys are equal, is a 10-V reading of the voltmeter. For the same reason, the ammeter would read 1 A with or without the addition of the *ac component*.

In Fig. 20-1*d*, the dc is chopped into square-wave pulsating dc, with "mark" (on) and "space" (off) periods of equal length. Half the time the pointer tends to be driven to 10 V, and half the time it tends to drop to zero. The result is an average reading of half of maximum, or 5 V. If the duration of the mark pulses were greater than the space times, the meter would read more than 5 V. If the space were greater, the meter would read less than 5 V. Meters always respond to the average values regardless of the waveform of the dc.

In Fig. 20-1e, a sinusoidal ac wave has been *fullwave-rectified*. That is, by use of a special four-diode circuit (Sec. 20-2), the negative half of each cycle has been reversed in polarity, forming a series of pulses of sinusoidal waveform. With this waveform the voltmeter indicates the average (0.636 maximum) of the pulse amplitudes, which is 6.36 V for the voltmeter and 0.636 A for the ammeter.

In Fig. 20-1*f*, a sinusoidal ac wave has been halfwave-rectified by use of a single diode, which allows only half of the ac wave to pass. Since halfwave pulses are present just half the time, the voltmeter

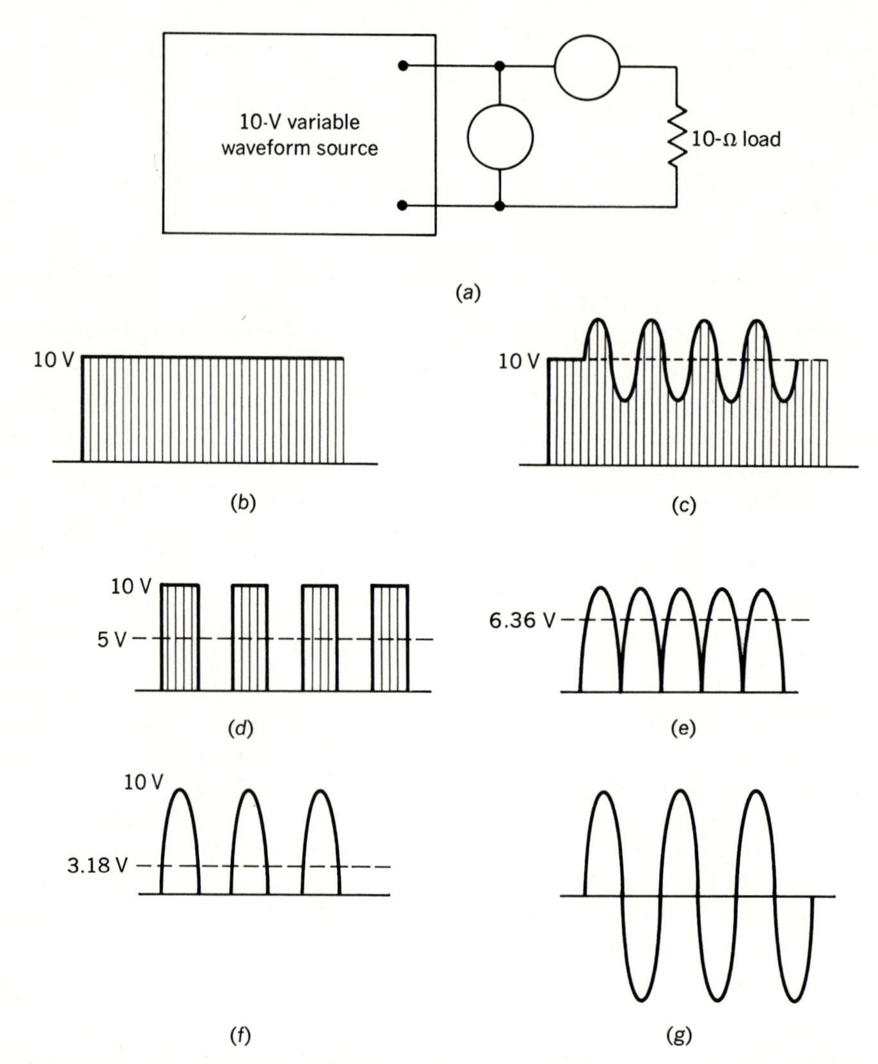

Fig. 20-1 Dc voltmeter readings across different waveforms. (*a*) Circuit. (*b*) Reads 10 V across 10-V dc. (*c*) With a small ac component the reading is 10 V. (*d*) With square-wave dc the reading is the average, or 5 V. (*e*) With fullwave-rectified ac, it reads 6.36 V. (*f*) With halfwave-rectified ac, it reads 3.18 V. (*g*) With ac, no reading.

will indicate only half of 6.36, or 3.18 V, and the ammeter 0.318 A.

In Fig. 20-1*g*, the source is producing a sinusoidal ac wave. If the ac had a frequency of 1 Hz, on one half-cycle the pointer would swing up, hitting a maximum of almost 10 V in the voltmeter and 1 A in the ammeter. On the other half-cycle, the pointer would be driven off scale past the zero point. If the frequency of the ac were 10 Hz, the pointer would swing up and down but would not reach the 10-V point before the next half-cycle started. As a result, it would only have time to oscillate between perhaps a 3-V peak reading and back past zero, depending on the damping factor of the meter. If the frequency were 20 Hz, the needle would probably vibrate slightly back and forth across the zero point. With 30 Hz and higher, the pointer might not show any vibration because of its inability to follow the rapid alternations.

It is significant that a dc meter will indicate nothing when connected in an ac circuit. It is possible to place a dc meter in an ac circuit and increase the voltage until the current through the meter burns out the moving coil or the springs, and never have the meter pointer move from zero.

To determine waveforms, or peak values, a device called a *cathode-ray oscilloscope* is required.

20-2 RECTIFIER-TYPE AC METERS

When ac is fed through a diode rectifier, the current that flows is unidirectional pulsating dc. Figure 20-2 illustrates a solid-state four-diode circuit that utilizes both halves of the ac cycle. Both halves produce current flow in the same direction through the load resistor.

In Fig. 20-2*a*, the top of the transformer secondary has a +10-V potential and attracts electrons. The path of the electron flow is out of the negative end of the transformer, through the diodes (*against* the arrowheads of the symbols), and through the load, as indicated by the heavy lines. With current flowing upward through the load, the upper end is positive (electrons travel − toward +).

In Fig. 20-2*b*, the ac cycle has reversed. Electrons from the negative end follow a path indicated by the dark lines. Again, the current is upward in the load. This produces the current waveform in Fig. 20-1e. The ac is said to be fullwave-rectified. This particular circuit is a *bridge rectifier*.

If a 10-V dc voltmeter were connected across the load, it would indicate a 6.36-V dc value. (This assumes no voltage-drop across the diodes.) If it were desired to calibrate the meter to read in *peak* ac volts, the point that the indicator of the meter now shows on its dc scale would be marked 10 V, because with a 10-V peak ac this is as far as the pointer will move.

Ac meters rarely carry peak-voltage calibrations. It is the effective value of the ac that is generally of interest. The effective, or rms, value is 0.707 times the peak value. For this reason, where the pointer indicates that a 10-V peak rectified ac is applied, the meter would normally be

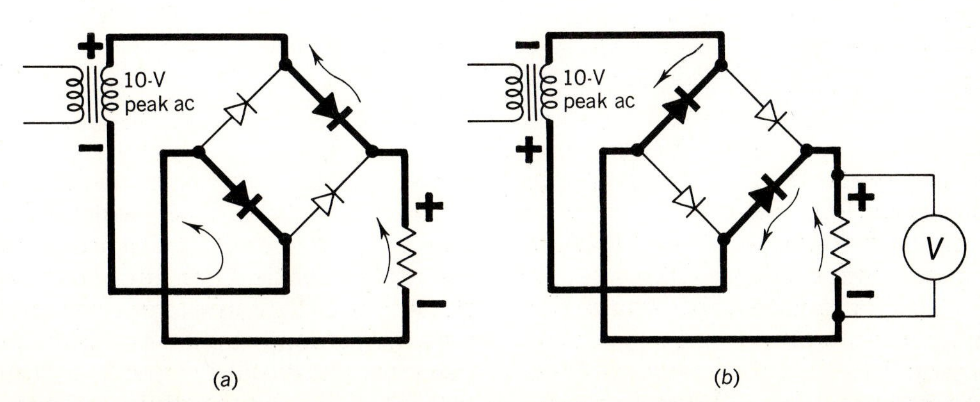

Fig. 20-2 Fullwave bridge rectifier. Path of current when (*a*) the top of the transformer secondary is positive, and (*b*) the top of the secondary is negative.

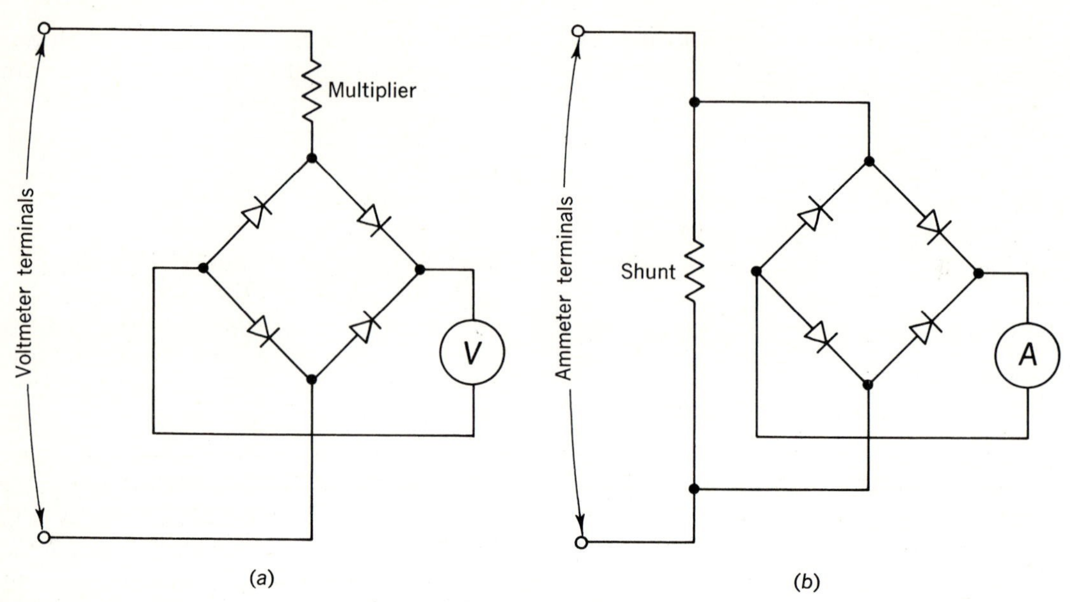

Fig. 20-3 (*a*) Rectifier-type ac voltmeter. Meter is calibrated in volts. (*b*) Rectifier-type ac ammeter. Calibrated in amperes, milliamperes, or microamperes.

marked 7.07 V. To convert the rms ac value to the peak value, multiply the rms by 1.414. The average value is 0.9 times the rms (or rms is 1.11 times average).

To utilize the existing dc voltage scale on a meter when a fullwave rectifier is used, the dc multiplier resistor may be replaced with a multiplier having 0.636 of its value (or it may be shunted by a resistor having 1.75 times its resistance).

A practical rectifier-type ac voltmeter is shown in Fig. 20-3*a*. The meter is the only load on the rectifier circuit. Because of the nonlinearity of the rectification process at low voltages, the scale of ac meters may be somewhat compressed near the zero readings but may be quite linear for the major portion of the scale. Figure 20-3*b* represents a rectifier-type ac *ammeter*.

To obtain an indication of ac when only a dc voltmeter is available, a single diode in series with the dc voltmeter is often used, as in Fig. 20-4. To convert the average reading shown by the dc meter to an ac rms value, the average reading shown on the dc meter scale is multiplied by *twice* 1.11, or 2.22. For example, if the dc meter reads 23 V, the rms ac value must be 23(2.22), or approximately 51 V. The diode used must be able to stand the reverse voltage of the peak ac value. This is the peak voltage that appears across the diode during the half-cycle when it is

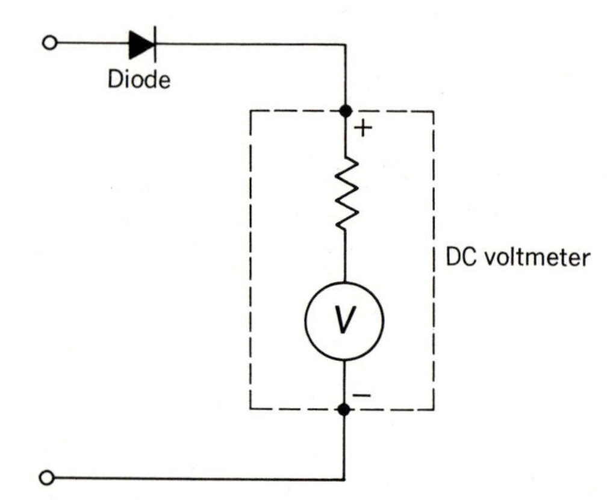

Fig. 20-4 A circuit for emergency ac readings when only a dc voltmeter is available.

not conducting, and it is 1.414 times the rms value.

Quiz 20-1. Test your understanding. Answer these check-up questions.

1. Would the pointer in Fig. 20-1c show any movement with a 10-Hz ac component? ________ 100 Hz? ________
2. In Fig. 20-1*d*, what value would the pointer indicate if the mark time is 0.01 s and space time is 0.02 s? ________ If the mark were 0.1 and the space were 0.01 s? ________
3. Is the waveform shown in Fig. 20-1e sinusoidal? ________
4. By what factor would an average dc voltage reading be multiplied to give the peak ac value if the meter had a fullwave rectifier? ________ A halfwave rectifier? ________
5. Why are vacuum-tube diodes not used in rectifier-type ac meters? ________
6. If a dc meter gives a steady upscale reading when across an ac circuit, what might this mean? ________
7. If square-wave ac were fullwave-rectified, what waveform would result? ________
8. If the circuit in Fig. 20-4 were connected across a 100-V rms ac source, what would be the maximum reverse voltage (inverse peak) that would appear across the diode? ________
9. What is the inverse peak voltage across each diode in Fig. 20-2*a*? ________
10. A 0- to 100-V dc milliammeter has a fullwave rectifier connected in conjunction with it, but no shunt resistor. It reads 85 mA. What is the rms current flow in the circuit? ________
11. What is the name given to the four-diode rectifier circuit? ________ Is it a halfwave or fullwave rectifier? ________

20-3 PEAK-READING AND dB METERS

When it is desired to find the highest peak voltage value, as when waveforms are nonsinusoidal, a *peak-reading voltmeter* is used (Fig. 20-5). The only differences between this voltmeter and the fullwave-rectifier voltmeter are the capacitor C and the resistor R_2. If the waveform in the circuit is as shown in Fig. 20-5*b*, the capacitor charges through R_1. The meter with R_2 acting as its multiplier indicates the voltage across the capacitor. The capacitor and resistor R_2 represent a long-time-constant circuit in comparison to the charging time constant of the capacitor and resistor R_1. As a result, the meter tends to linger at or near the peak values to which the capacitor charges, shown by the dashed lines in Fig. 20-5*b*. Meters of this type might be used in audio-

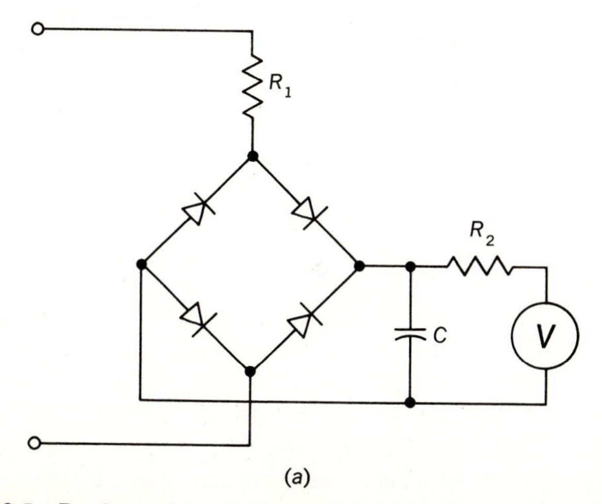

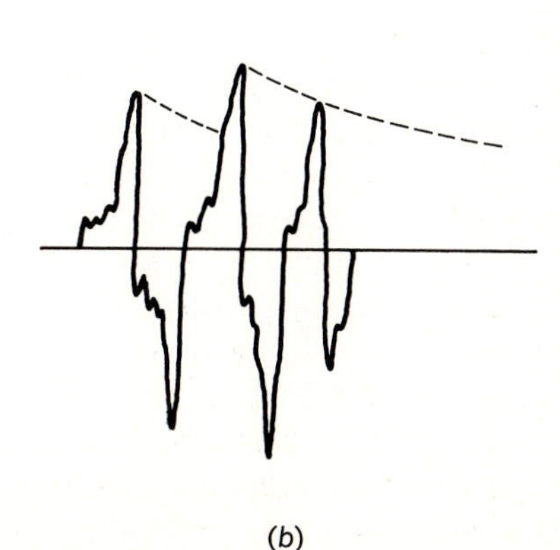

Fig. 20-5 Peak-reading voltmeter. (*a*) The meter and R_2 form a voltmeter to read the voltage to which capacitor C charges. (*b*) The meter indicates approximately the dashed values.

frequency amplifier systems to indicate the peaks of voice or music voltages.

The *volume unit* (*VU*) and *decibel* (*dB*) meters are peak-reading rectifier voltmeters. The *bel* (abbreviated B) is a "logarithmic" unit used in indicating relative sound levels. A 1-B change in audio-frequency ac strength represents a change of 10 times the *power*. A change of +1 B means an increase in power by a factor of 10; −1 B means a decrease in power to one-tenth of the original power.

A *decibel*, or 1/10 B, is one *logarithmic* tenth of a bel. Thus an increase of power by 1 dB is not an increase of 1/10 times the power, but an increase of 1.26 times the power. An increase of 2 dB is an increase of 1.26 × 1.26, or 1.59 times the power. An increase of 3 dB is an increase of 1.26 × 1.26 × 1.26, or twice the power. A listing of approximate power increases by decibels is given in Table 20-1. The formulas for determining decibel values accurately are given in Sec. 35-6.

Table 20-1 ***DECIBEL VS. POWER INCREASE***

1 dB = 1.26 increase	7 dB = 5 times increase
2 dB = 1.58 increase	8 dB = 6.3 times increase
3 dB = 2 times increase	9 dB = 7.9 times increase
4 dB = 2.52 increase	10 dB = 10 times increase
5 dB = 3.17 increase	20 dB = 100 times increase
6 dB = 4 times increase	30 dB = 1000 times increase

Logarithmic units are used for sound signals because the response of the human ear is very nearly logarithmic. The human ear can just barely detect a change in sound intensity of 1 dB, whether the level changes from 1 to 2 dB, from 50 to 51 dB, or from −8 to −9 dB.

Because dB units are not linear, graduations on VU and dB meters will not be linear, as shown in Fig. 20-6.

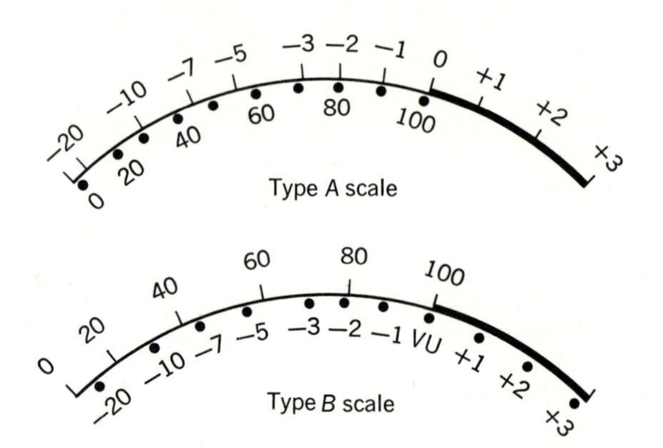

Fig. 20-6 Two VU meter scales. The 0 to 100 range indicates percentage of modulation in broadcast transmitters.

Actually, a VU meter is a dB meter, and both are forms of ac voltmeters. Not all dB meters, however, have the same scale graduations, meter sensitivity, or damping, nor are they necessarily calibrated for operation across lines of the same impedance. The VU meter is standardized. There will be no variation from one VU meter to another. VU meters are peak-reading meters, whereas dB meters may not be. The standardized "zero VU" voltage value is the amount of voltage developed across a 600-Ω line when 1 mW of power is being carried by the line, or $E\ \sqrt{PR}$, or $\sqrt{0.001(600)}$, or 0.775 V. (Voltages less than this are −VU values.) One milliwatt carried in a line of any other impedance would not produce the necessary voltage across the line for a VU zero-level indication.

In the past, several dB zero levels were used, 12 and 6 mW, as examples. In some dB meters the zero level is the voltage

ANSWERS TO CHECK-UP QUIZ 20-1

1. (Yes) (No) **2.** (3.33 V) (9.09 V) **3.** (Yes, fullwave-rectified sinusoidal) **4.** (1.57) (3.14) **5.** (They require heater power; higher R_i) **6.** (A dc component present, rectification occurring in poor connection, one half-cycle greater than the other) **7.** (Pure dc) **8.** (141 V) **9.** (14.1 V) **10.** (94.4 mA) **11.** (Bridge) (Fullwave)

developed with 1 mW in a 500-Ω line, and in others it is developed with 1 mW in 600-Ω lines. When dB values are referenced to 1 mW in 600-Ω lines, they may be called *dBm* values. Thus dBm and VU units of measurement are essentially the same.

Quiz 20-2. Test your understanding. Answer these check-up questions.

1. Will peak-reading rectifier-type meters indicate the peak value of nonsinusoidal ac? ________ Sinusoidal ac? ________
2. If a rectifier-type ac voltmeter is used across a sinusoidal ac, will the peak value always be the reading times 1.414? ________ If the ac is nonsinusoidal? ________
3. Would the sensitivity, damping, range, and zero level be standard for any dB meter? ________ Any VU meter? ________
4. What does VU stand for? ________
5. What type of ac is the VU meter expected to measure? ________ The dB meter? ________
6. The volume control of a stereo system is turned up from 0.7-W output to 7-W output. How many dB gain does this represent? ________ How many VU? ________
7. A VU meter will read correctly only if it is connected across a line of what impedance? ________

20-4 ELECTRODYNAMOMETERS

The *electrodynamometer*, or dynamometer, is used in low-frequency ac power circuits. It has a moving coil, but instead of permanent-magnet fields, dynamometers have electromagnet field poles, similar to the wattmeter. In fact, a wattmeter (Sec. 19-10) with the moving coil in series or in parallel with the field coils could operate as a dynamometer-type meter.

When current flows through the coils of a dynamometer (Fig. 20-7*a*), the fields oppose each other and the pointer is driven upscale. If the current reverses, it reverses through both sets of coils and the pointer is still driven upscale. Therefore dynamometer meters will indicate with either polarity of dc or with ac. The sensitivity of these meters is low. Some larger meters may require as much as 0.5 A to produce full-scale deflection.

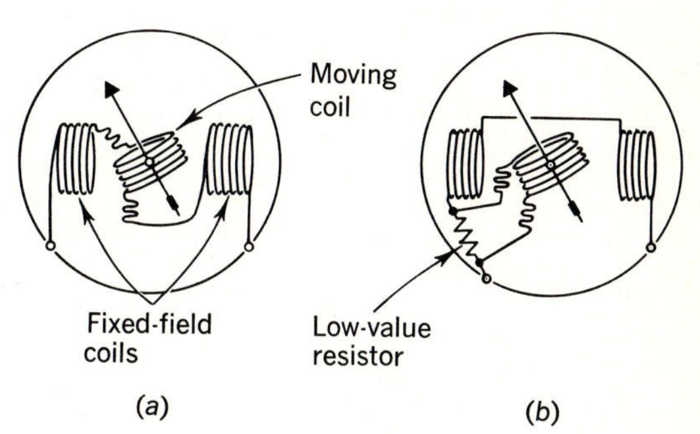

Fig. 20-7 Electrodynamometer connections: (*a*) low-current or voltmeter connection; (*b*) high-current ammeter connection.

Dynamometer voltmeters have many-turn field coils in series with the moving coil and multiplier. As an ammeter, the field coils will have fewer turns of heavy wire. For low-current ammeters, the two field coils may be connected in parallel with, or in series with, the moving coil. The usual high-current ammeter circuit connects the moving coil across a low-value resistor in series with the field coils (Fig. 20-7*b*) or across the field coils.

Since the inductive reactance of these meters is proportional to the frequency, these meters are calibrated at a specific frequency, such as 60 Hz, and are expected to be used at this frequency only.

With low current flow through the coils, neither the fixed field nor the moving-coil field is very strong, and the pointer hardly moves. As the current through the meter doubles, the strength of both fields doubles, and the pointer moves 4 times as far, resulting in current-squared meter scales.

For safety, when used to measure high-voltage circuits, ac meters may be connected across the secondary of a step-

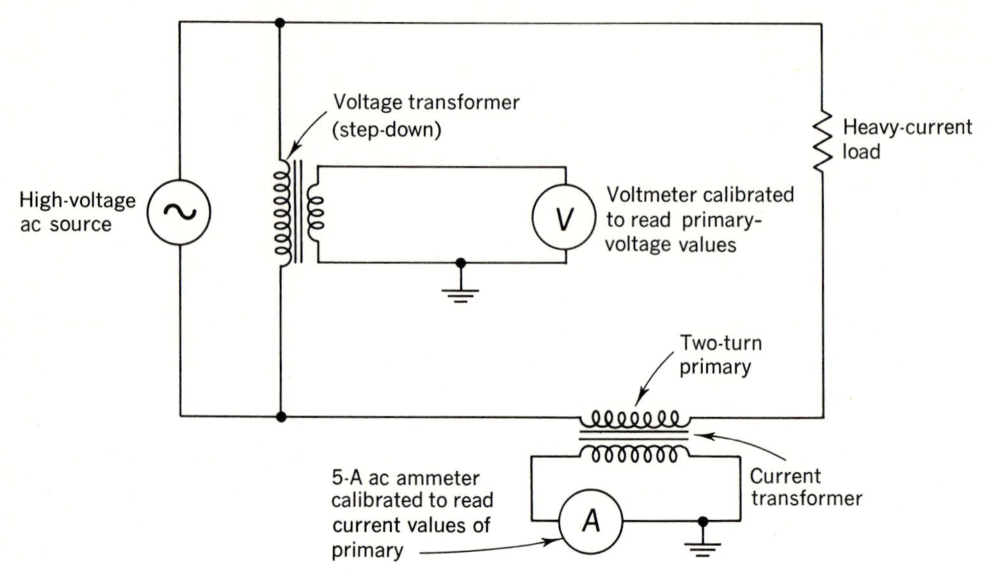

Fig. 20-8 A voltage transformer for high-voltage ac. A current transformer for heavy ac current values.

down-ratio *voltage transformer*, as shown in Fig. 20-8. When measuring heavy currents, a step-down *current transformer* may be connected in series with the line.

20-5 REPULSION-TYPE METERS

A *moving-vane* repulsion meter consists of a coil and, inside the coil, two highly permeable iron vanes. One of the vanes is fixed, and the other is rotary (Fig. 20-9). The pointer is attached to the moving vane and is returned to the zero-scale point by a spiral spring.

If either dc or ac current flows in the coil, both vanes are magnetized with *like polarity* at any instant. Since like magnetic poles repel, the moving vane is repelled from the fixed vane, against spring tension, carrying the pointer upscale. The greater the current, the greater the repulsion and the farther the indicator needle moves.

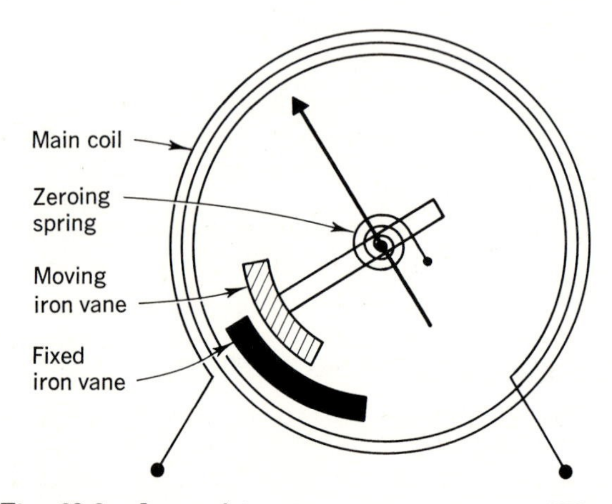

Fig. 20-9 A repulsion iron-vane ammeter. When current flows in the main coil, both vanes are magnetized alike and repel each other.

Iron-vane voltmeters have a coil consisting of many turns of relatively small wire, usually with a multiplier resistance in series with it. When made into ammeters, the coil consists of a few turns of very heavy wire. A shunt may be used across the coil.

ANSWERS TO CHECK-UP QUIZ 20-2

1. (Yes) (Yes) **2.** (Yes) (No) **3.** (No) (Yes) **4.** (Volume unit) **5.** (Audio frequency, nonsinusoidal) (AF, usually sinusoidal) **6.** (10 dB) (10 VU) **7.** (600 Ω)

While the iron-vane meter has a somewhat current-squared type of scale, by shaping the vanes it is possible to make it relatively linear over the major portion of its scale.

If the movable vane is hinged at one end (Fig. 20-10), maximum repulsion occurs at low current values with progressively less repulsion as current increases. This tends to produce a more linear scale.

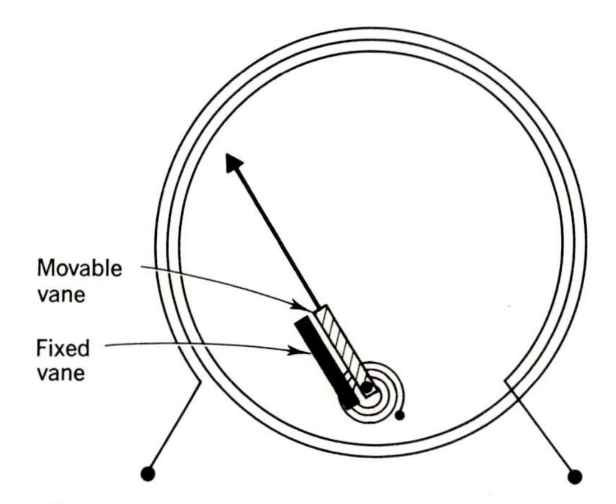

Fig. 20-10 A repulsion iron-vane meter.

An *inclined-vane* meter has its iron vane set on an inclined angle on an axle centered in the core of the coil. A pointer is attached to the axle. When current flows in the coil, the iron vane tries to align itself in the lines of force, pulling itself, the axle, and the pointer around against spring tension.

An *inclined-coil* meter is similar to an inclined-vane meter, except that a few shorted turns take the place of the inclined vane. When ac flows in the outer coil, an emf is induced in the inclined coil, producing current in it. According to Lenz's law, the magnetic field induced in the inclined coil opposes the field of the main coil. This results in a repulsion of the inclined coil and the pointer against spring tension. Unlike all other meters described, it will operate only on ac, because it depends on a continually induced current.

Repulsion meters are rather insensitive but are used in power-frequency applications.

20-6 AC WATTMETERS

An ac wattmeter is similar to a dc wattmeter, but the field and moving coils have reactance values that will vary with frequency. As a result, ac wattmeters are used only on the frequency for which they are calibrated. If a dc wattmeter were used on 60-Hz ac, the reactance of the meter coils would lower the current flow in the meter, and it would read low.

A point to remember about ac wattmeters is that they read only *true power*. For example, the circuit in Fig. 20-11 shows a 141-V source feeding a reactive load. Graphing X_L and R results in a total impedance of 141 Ω. With 141 V and 141 Ω the current is 1 A, shown by the ammeter. The voltmeter across the source would read 141 V. The apparent power, volts times amperes, is 141(1), or 141 VA. However, the wattmeter would read only the true power being dissipated by the resistor, $P = I^2R$, or $1^2(100)$, or 100 W.

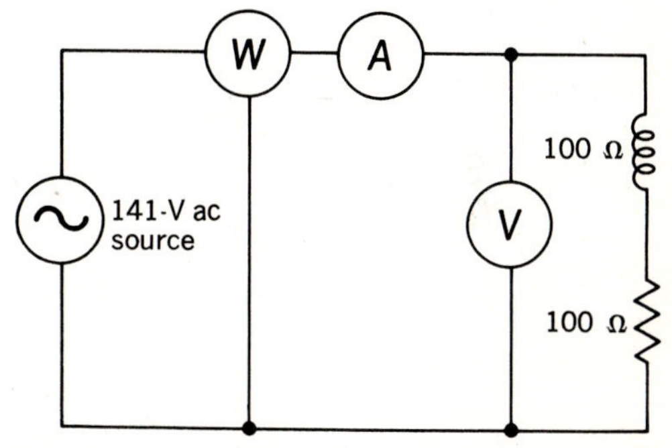

Fig. 20-11 An ac circuit in which the power, volt-amperes, and power factor can be found by meter readings.

The *power factor* of a circuit, the ratio of true power to voltamperes, can be found by dividing the wattmeter value by the VA. For Fig. 20-11, PF = P/VA, or 100/141, or 0.707. Since the reactance and resistance are equal in this circuit, the phase angle graphs as 45°. From a table of trigonometric functions (slide rule or pocket calculator), the cosine value of 45° is also equal to 0.707. The PF of an ac circuit may be found by P/VA, or by the cosine of the phase angle (cos ϕ).

20-7 WATTHOUR METERS

Electrical energy is measured in joules (J), wattseconds (Ws), or watthours (Wh). Energy-measuring meters are known as *watthour meters* or *kilowatthour meters*. These meters are discussed in Sec. 5-5.

The symbol for a kilowatthour meter in a diagram is "kWh" shown in a circle.

A kilowatthour meter will be found at the point where power company lines enter any home or commercial establishment. All the energy used by the consumer is indicated on this meter.

Quiz 20-3. Test your understanding. Answer these check-up questions.

1. A dynamometer voltmeter is calibrated at 60 Hz. Would it read the same, higher, or lower if it were operated at 600 Hz? ________ On dc? ________
2. Will reversing the current direction in a dynamometer reverse the pointer? ________
3. Would a moving-vane meter indicate at all if it were energized by dc? ________ Would the pointer reverse its motion if the dc polarity were reversed? ________
4. Would a dynamometer give more accurate readings at low-current, midscale, or high-current values? ________
5. A moving-vane meter calibrated from 0 to 10 has about 20 turns of #12 wire in its coil. Would it be an ammeter or voltmeter? ________
6. A 0- to 100-V ac meter is used to measure an 880-V ac line. What ratio transformer would be used for optimum readability? ________
7. A 0- to 5-A ac meter is used to measure the current in a line which should not exceed 100 A. What transformer ratio should be used? ________
8. What is a main point to remember about ac wattmeters? ________
9. Would an ac wattmeter read the correct value, a high value, or a low value if it were used with dc? ________
10. Is the charge for electrical power for the number of watts, amperes, volts, watthours, or coulombs? ________
11. Name two meters that have internal motors. ________ ________
12. To what is the cosine of the phase angle equal? ________
13. On a separate sheet of paper, draw a diagram of a low-current dynamometer, a high-current dynamometer, a voltage transformer with a moving-coil ac meter, and two types of moving-vane ammeters. Draw a diagram of a kilowatthour meter in a two-wire line between a source and load. Check your diagrams against the illustrations in the text.

20-8 THERMOCOUPLE AMMETERS

If wires of two different metals, such as iron and copper, are welded together and the joint is heated, the difference in electron activity in the two metals produces an emf across the joint. If a sensitive microammeter is connected across the joint, as in Fig. 20-12*a*, the meter will indicate a current.

A thermocouple junction can be heated in an oven, in sunlight, in a gas flame, or by passing current through it. Anything that heats the junction produces an emf. Both dc and rms ac current heat the junction the same amount. However, when using some thermocouple meters to measure dc current, it is necessary to take an average of the current flowing in each direction. Within limits, frequency has no effect on the heating ability of an ac, so the same thermocouple ammeter can be

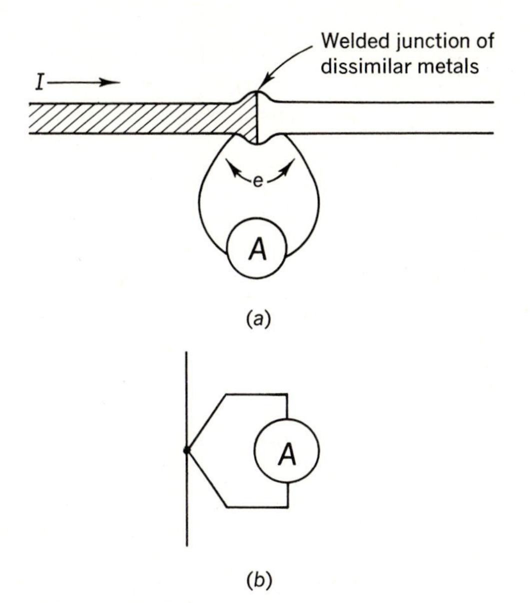

Fig. 20-12 (a) Thermocouple ammeter, showing junction and meter. (b) Symbol.

used to measure ac current from low frequencies up to 20 MHz or more without need of recalibration. Regardless of the waveform of the current, the meter always indicates rms values. The symbol for a thermocouple ammeter is shown in Fig. 20-12b. Disadvantages of thermocouple ammeters are crowded lower-scale readings and slow action of the pointer when current varies suddenly.

Thermocouples convert heat energy directly into electrical energy, and they might be used as a source of energy (charge batteries, etc.). The conversion efficiency (heat to electricity) is low.

20-9 FREQUENCY METERS

Five methods of measuring frequency will be discussed: vibrating reeds, electrodynamic frequency meters, digital counters, beat frequency, and RC circuit.

The *vibrating-reed* frequency meter may consist of seven or nine thin iron strips attached solidly at their bases but free to vibrate at the other ends. Each reed is cut to a length that will vibrate at a particular frequency. For example, one reed might vibrate at 58 Hz, the next at 58½ Hz, the next at 59 Hz, and so on. All reeds are surrounded by a single coil of wire in series with a current-limiting resistor and connected across the power line. If 60-Hz ac is fed to the coil, the 60-Hz resonant reed will vibrate more than the others. Whichever reed vibrates the most indicates the frequency of the ac in the coil (Fig. 20-13). If two reeds vibrate

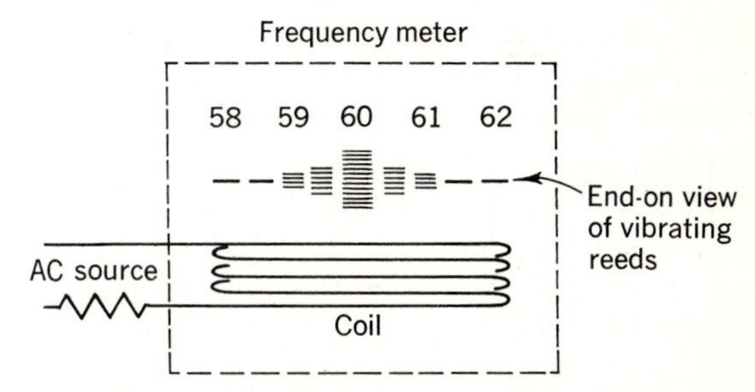

Fig. 20-13 A vibrating-reed frequency meter. The iron reed having a mechanical resonance closest to the frequency of the alternating magnetic field vibrates most, 60 Hz in this case.

equally, the frequency is halfway between the two reed frequencies. Only a narrow band of frequencies can be measured with this meter.

The *electrodynamic* type of frequency meter utilizes the principle of a vane and a pointer pulled in one direction by one coil field and in the opposite direction by another coil field (Fig. 20-14). One field is developed through a capacitor, and the other is developed through an inductor. At some frequency both fields will have equal pull on the pointer, and the pointer will indicate midscale. If the frequency of the ac increases, the current in the capacitive field increases and the inductive current decreases. The pointer is pulled in

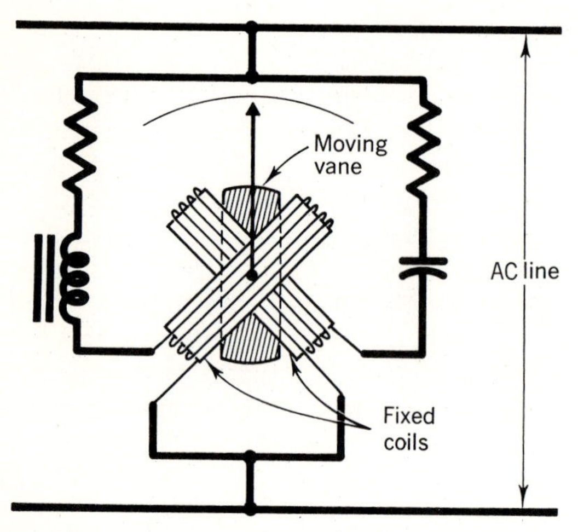

Fig. 20-14 An electrodynamic-type frequency meter connected across an ac line. When capacitive and inductive currents are equal, the pointer indicates midscale. If the frequency increases, the capacitor current increases, pulling the vane and pointer to the right.

one direction. If the frequency decreases, the inductive field increases and pulls the pointer in the opposite direction. A wider band of frequencies is measurable with this type of meter.

If the ac to be measured is passed through a 1-s gating circuit (Fig. 20-15), of a *digital counter*, the number of cycles that are registered on the decimal-counting units (DCUs) is the ac frequency. There is always the possibility of a one-cycle error in this type of frequency measuring, because the first cycle through the gate may be chopped short. It is possible to measure a frequency such as 9,843,762 Hz to within one cycle, which is an accuracy of approximately one ten-millionth, or 10^{-7}.

The *beat-frequency method* is not a meter, but operates on the principle that two different-frequency ac voltages fed into a nonlinear device, such as a diode, produce a difference frequency. If a known frequency of 4,000,000-Hz ac and an unknown frequency of approximately 4,000,500-Hz ac are mixed, there will be a 500-Hz resultant. The 500-Hz difference can be read out on other types of frequency meters, such as a digital frequency meter, or compared on an oscilloscope with known frequencies from a signal generator, or tuned in on a calibrated radio receiver if the beat is in the *RF* range.

An *RC* type of frequency meter feeds the ac to be measured into a *Schmitt trigger* circuit which develops a constant-amplitude (strength) and constant-width dc pulse each time the input ac starts to rise in positive polarity (Fig. 20-16). A 50-Hz ac produces 50 narrow pulses per second; a 1000-Hz ac produces 1000 narrow pulses per second. The dc pulses charge a capacitor, *C*, which is constantly being discharged by the meter across it. If the frequency of the ac is low, the capacitor never has a chance to charge to any appreciable voltage, and the meter has little current flowing through it. As the frequency of the ac increases, the number of charging pulses increases and the capacitor charges to a higher voltage. The meter indicates the relative voltage charge of the capacitor—low for low frequency and high for high frequency. The scale of the meter may be calibrated in hertz or kilohertz, depending on the *RC* time constants of the circuit.

20-10 GRID-DIP METERS

To determine the resonant frequency of a coil-and-capacitor tuned circuit, a *grid-*

ANSWERS TO CHECK-UP QUIZ 20-3

1. (Lower) (Higher) **2.** (No) **3.** (Yes) (No) **4.** (High) **5.** (Ammeter) **6.** (10:1) **7.** (25:1) **8.** (Read *true* power) **9.** (High) **10.** (Watthours) **11.** (Watthour) (Ampere-hour) **12.** (PF) **13.** (Symbol for kilowatt-hour meter is a circle with three leads and "kWh" in it)

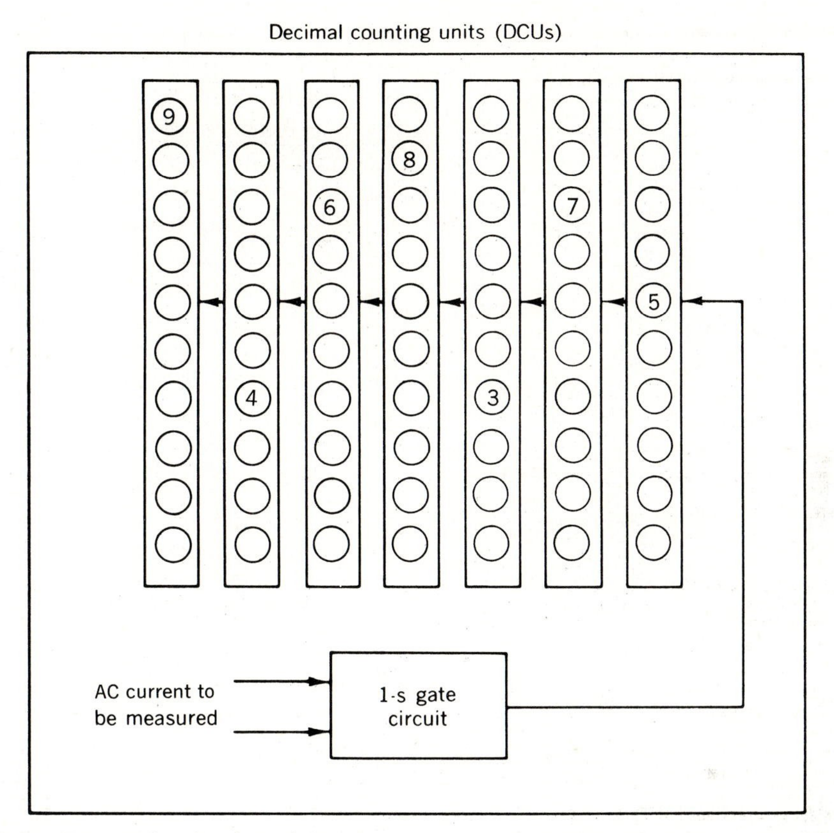

Fig. 20-15 Essentials of a digital frequency meter using DCUs and a 1-s gating circuit.

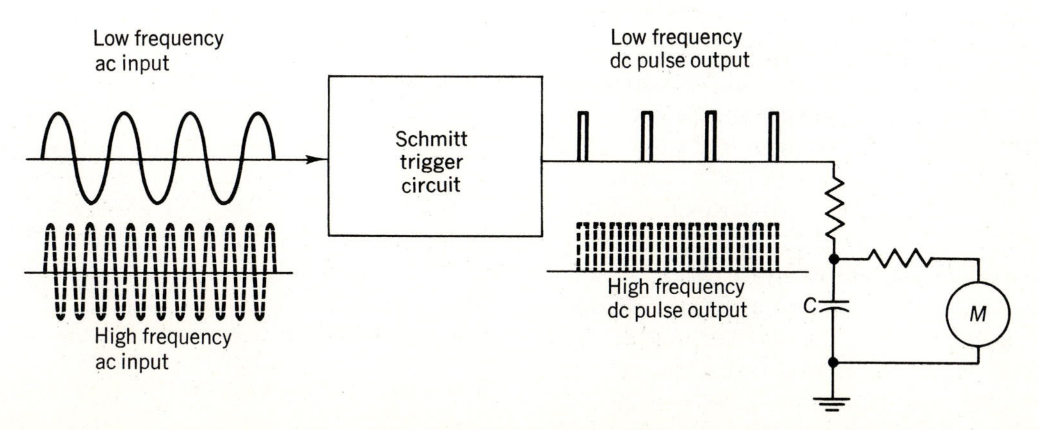

Fig. 20-16 One form of *RC* frequency meter. The Schmitt trigger produces narrow pulses for any input cycle. Pulses charge *C*, and the meter responds to the charge in the capacitor.

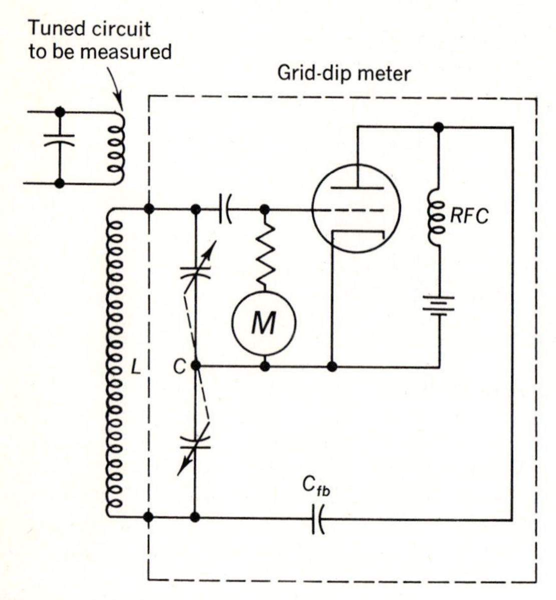

Fig. 20-17 Grid-dip meter. If the *LC* circuit is coupled to the tuned circuit and tuned across the frequency of the latter, the grid current will decrease.

dip meter (Fig. 20-17) is used. A grid-dip meter has a resonant circuit of its own, consisting of a coil *L* and two ganged series capacitors *C*. If electrons start oscillating in this *LC* circuit, an ac signal at the resonant frequency of the circuit is fed to the grid of the triode. The tube amplifies the signal and feeds it back to the tuned circuit through C_{fb}. This amplified energy keeps the *LC* circuit oscillating, or generating ac. As long as the *LC* circuit keeps oscillating, it feeds signals to the grid. The positive half-cycles of the signal attract electrons to the grid from the cathode, and current flows through the grid circuit milliammeter.

If the coil *L* is coupled to a tuned circuit that is not on the same frequency as the meter circuit, the grid current will not change. However, if the variable capacitors of the meter circuit are varied, as the meter circuit frequency approaches the resonant frequency of the *LC* circuit being measured, some of the ac will be induced into the *LC* circuit. This loss of energy reduces the strength of oscillation of the oscillator, and less signal will be fed to the grid. The result will be a decrease, or dip, in grid current as the oscillator circuit is tuned to the resonant frequency of the unknown *LC* circuit. The dial of the grid-dip meter is calibrated in frequency.

Quiz 20-4. Test your understanding. Answer these check-up questions.

1. What are two reasons why a thermocouple ammeter pointer would be very slow-moving? ________ ________
2. Why would two pieces of iron welded together not form a thermocouple junction? ________
3. Would a thermocouple ammeter be most likely to be found measuring power-line currents, radio-broadcast antenna currents, or transistor collector currents? ________
4. What are the five basic forms of frequency measuring? ________ ________ ________ ________ ________
5. A vibrating-reed frequency meter shows the 60-Hz reed vibrating as much as the 59.5-Hz reed. What is the frequency of the ac? ________
6. Would the vibrating-reed meter work if the reeds were made of copper? ________ Nickel? ________ Aluminum? ________
7. Would the electrodynamic frequency meter operate if one coil were fed through a resistor and the other through an inductor? ________ A capacitor? ________
8. What is a circuit called that turns on and then off to let ac signals through for a specific time? ________
9. How many cycles are being registered by the DCUs in Fig. 20-14? ________
10. What is the approximate accuracy of measurement if a frequency of 98,312 Hz is measured ±1 Hz? ________ Of 5015 Hz? ________
11. What will be the unknown frequency if a known 1.5-MHz ac is mixed with an unknown frequency and the resultant (or difference frequency) is 875 Hz? ________ (or) ________
12. Would thermocouple meters ever be used as voltmeters? ________
13. What is the unknown frequency when a variable-frequency oscillator of known fre-

quency is made to produce a frequency of 1000 Hz when beating against the unknown frequency? __________

14. In question 13, what would be known about the unknown frequency if it were brought to a zero-frequency beat (zero beat) when mixed with the variable-frequency ac generator? __________

15. If the resistor in series with the meter in an *RC* frequency meter is too low in value, what effect would this have on the frequency indications of the meter? __________

16. What does a dip in grid current indicate in a grid-dip meter? __________

CHAPTER 20 TEST • AC METERS

1. A 20-V rms 60-Hz ac is in series with 50-V dc. What value will a dc voltmeter across this circuit indicate?
2. A 50-V peak square-wave pulsating dc with 0.001-s mark pulses and 0.002-s space times will indicate as what value on a dc voltmeter?
3. A fullwave-rectified 80-V peak ac will register as how many volts on a dc voltmeter? What would it register if halfwave rectification had been used?
4. What value would be indicated on a 100-V dc voltmeter if it were across a 70-V sinusoidal ac?
5. How many diodes are required in a bridge rectifier? A halfwave rectifier?
6. It is desired to convert a 100-V dc meter to a 0- to 100-V ac meter by using a bridge rectifier and changing the multiplier. What value of multiplier should be used?
7. By what factor should the scale values be multiplied when a 0- to 10-V dc meter with a halfwave rectifier is used to give rms values? Peak values?
8. If the sound power output from a loudspeaker is 0.5 W, what is the power output if the sound is raised by 1 dB? 3 dB? 10 dB?
9. What is the most significant difference in the construction of electrodynamometer and D'Arsonval meters?
10. What is the name of the standardized sound-volume-indicating meter? What is its reference power-level value? Across what impedance line should it be used?
11. If a 0- to 150-V 60-Hz dynamometer meter were used across a 100-V dc, would it read accurately, high, low, or not at all?
12. Of what material must the vanes be made in repulsion-type meters? Will these meters operate on dc?
13. A 0- to 100-V rectifier-type ac voltmeter is connected across 60-V dc. Will polarity of the meter terminals be important? Will the meter read 60 V, more than 60 V, or less than 60 V?
14. The construction of a wattmeter is very similar to what other type meter? What is the minimum number of leads that a wattmeter will have?
15. An ac circuit has 120 V across it and 3 A flowing in the load, and the wattmeter connected to it reads 250 W. What is the power factor of this circuit?
16. On a watthour meter what is used to indicate the amount of energy used?
17. What are the names of the two types of electric meters that have motors in them?
18. What is the name of the only type of ammeter mentioned that could be used at frequencies from 1 to 20 MHz? What is the main disadvantage of this type of meter?
19. What type of frequency meter would probably measure the narrowest group of frequencies? What two types could be used over very wide bands of frequencies?
20. What type of frequency meter operates on a resonance principle?
21. What type of frequency meter may employ a Schmitt trigger circuit? A 1-s gate circuit?
22. What is the name of the meter that will indicate the frequency of resonance of an *LC* circuit coupled to it?
23. On a piece of paper, draw a diagram of a bridge rectifier circuit. A diagram of a high-current dynamometer meter. A block diagram of a digital-readout frequency meter. A diagram of an *RC* frequency meter. A VU meter face.

ANSWERS TO CHECK-UP QUIZ 20-4

1. (Time required to heat junction) (Low resistance shunt on microammeter) **2.** (No difference in electron activity when heated) **3.** (Antenna) **4.** (Vibrating reed, electrodynamic, digital, beat frequency, *RC* charge) **5.** (59.75 Hz) **6.** (No) (Yes) (No) **7.** (Yes) (Yes) **8.** (Gate or gating) **9.** (9,468,375 Hz) **10.** (1/100,000, or 10^{-5}) (1/5000, or 2×10^{-4}) **11.** (1,500,875 Hz or 1,499,125 Hz) **12.** (No) **13.** (Unknown is either 1 kHz higher or lower than the known) **14.** (Unknown is exactly same frequency as known) **15.** (Too low a frequency indication) **16.** (Meter is resonant to some coupled circuit)

CHAPTER OBJECTIVE. Explanations of rotating-coil, rotating-field, and inductor alternators, efficiency of machines, procedure to add an alternator to a line, 2-ϕ and 3-ϕ alternators, delta and star-connected circuits, transporting electrical power, and basic power distribution circuits.

21-1 ROTATING-COIL ALTERNATORS

Any machine that converts energy of any form into mechanical *torque* (twisting effort) is called a motor. Whether it is gasoline, steam, diesel, hydraulic, or electric powered, it may be called a *prime mover*.

Generators require some type of prime mover to rotate them so that they can produce an output emf, either ac or dc. The simplest of the mechanical-to-electrical-energy converters is the ac generator, or *alternator*.

The basic alternator consists of (1) a strong, constant magnetic field, (2) one or more conductors that can rotate across the field lines, and (3) some means of making a continuous connection to the conductors as they rotate (Fig. 21-1*a*). In this basic rotating-coil alternator the magnetic field is produced by current flowing through the stationary, or stator, *field coil*. Field-coil excitation can be controlled by varying the field rheostat or the dc excitation voltage. In some small machines the magnetic field is produced by a permanent magnet. There is no easy method of controlling the voltage output of such a machine.

The rotating part, or *rotor*, is supported at both ends of the shaft by bearings (not shown). The central part is a laminated iron rotor piece. A single turn of wire is around the rotor in the illustration. One end of this turn is connected to a brass *slip-ring*, which is insulated from the shaft. The other end of the turn is connected to a second slip-ring. Each time the rotor wire makes one complete revolution, one complete cycle of ac is developed in it. Instead of a single turn, a practical machine has the rotor slotted on opposite sides with several hundred turns wound into the slots. The result is several hundred times as much output voltage.

To allow the induced ac in the rotor or *armature* coil to be used, two carbon or brass brushes are spring-held against the slip-rings. The brushes against the slip-rings produce the necessary continuous connection between the ac induced in the coil and outside circuits.

The iron rotor shortens the magnetic field path between north and south poles, concentrating the lines into the narrow gaps in which the rotor wires travel, thereby increasing the output voltage. The rotor must be made of high-permeability iron having low retentivity because

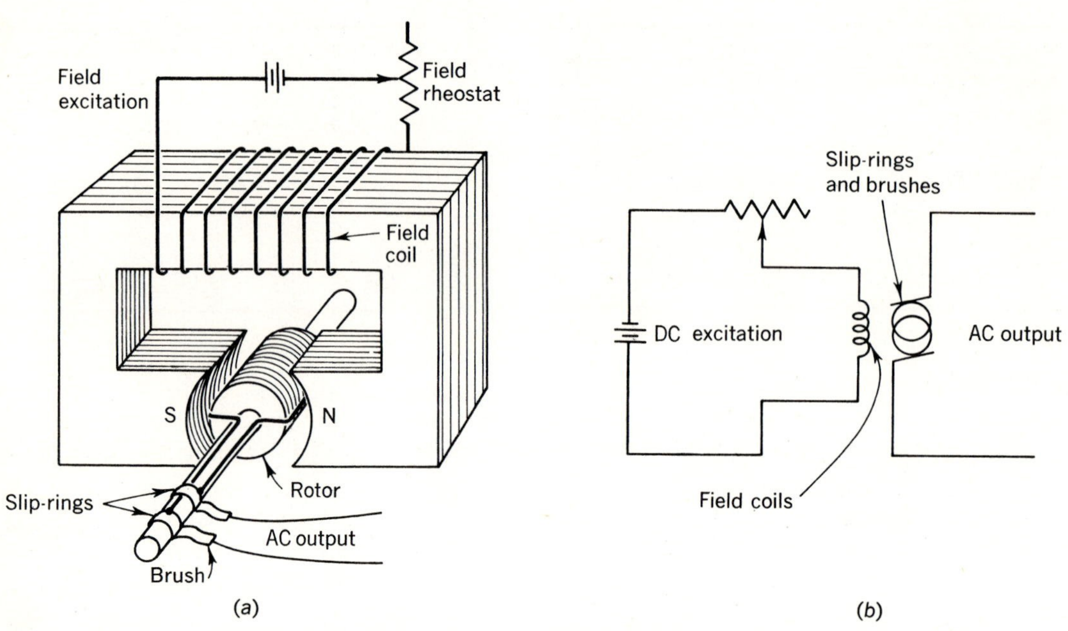

Fig. 21-1 Rotating-coil alternator: (*a*) simplified form; (*b*) schematic diagram.

the magnetism induced in it is rapidly changing as it rotates at perhaps 1800 revolutions per minute (rpm).

As the rotor is metal and is moving in a magnetic field, an internal emf is set up in the iron. This emf produces circulating *eddy currents* in the iron. Such currents heat the iron and result in a loss of power ($P = I^2R$) in the rotor. To reduce eddy currents, the core is laminated, as in a transformer, with the laminations punched out of thin sheets of iron. Each piece is insulated electrically from adjacent laminations. In this way the eddy currents are limited to very small loops, the thickness of a lamination.

If the electrical excitation of the field coil is increased by increasing the dc field supply voltage or by reducing rheostat resistance, the magnetic field increases in the gaps surrounding the rotor, and the rotor turns have a greater emf induced in them.

The symbol for the alternator in an electrical diagram is shown in Fig. 21-1*b*. (There is no indication of a prime mover in this illustration.)

If the rotational speed of the rotor is increased by increasing the prime-mover speed, the frequency of the ac will increase. Since more lines of force are being cut per second, the output voltage will also increase. In most applications a constant ac frequency is desired, requiring a constant speed of rotation. As a result, the voltage output must be controlled entirely by the field excitation.

With no load on the alternator, the rotor turns easily. When a load is connected across the brushes, the output ac voltage produces current flow in the load and in the rotor coil. This produces a magnetic field around the coil. By Lenz's law, this field is in a direction opposed to the magnetic lines of force that produced it (the field-pole lines). Under loaded conditions the alternator is hard to turn. The prime mover must exert enough mechanical energy to support the electrical-energy demand of the load.

Quiz 21-1. Test your understanding. Answer these check-up questions.

1. What is the word that means twisting effort? ________
2. Would a waterfall with a paddlewheel be considered a generator, an alternator, a prime mover, or a motor? ________
3. What is a common name for an electrical-to-mechanical-energy converter? ________ A mechanical-to-electrical-energy converter? ________
4. What are the names of the devices that bring the induced ac of an alternator out of the machine? ________ ________
5. What are some disadvantages when alternators have permanent-magnet field poles? ________
6. What are some advantages in using permanent-magnet field poles in an alternator? ________
7. Why are all iron rotors in electric machines laminated? ________ What is always in between laminations? ________ Is it necessary to laminate the field poles? ________
8. Would you expect the copper wire in the windings of an alternator to be rubber-covered, cotton-covered, plastic-covered, or enameled? ________ Why? ________
9. What are four methods of increasing the output voltage of an alternator? ________ ________ ________ ________
10. Which of the methods of question 9 would not be applicable if the frequency of the output ac had to be constant? ________
11. When an alternator is loaded suddenly, the frequency of the output may decrease. What is responsible for this? ________

21-2 ROTATING-FIELD ALTERNATORS

A disadvantage of the rotating-coil alternator is the slip-ring and brush contacts in series with the load. If these parts become worn or dirty, the flow of current to the load may be interrupted at times. If the field excitation is connected to the *rotor coil* and the machine is turned, the formerly stationary field coils will have an ac induced into them (Fig. 21-2). A load can be connected across these coils without

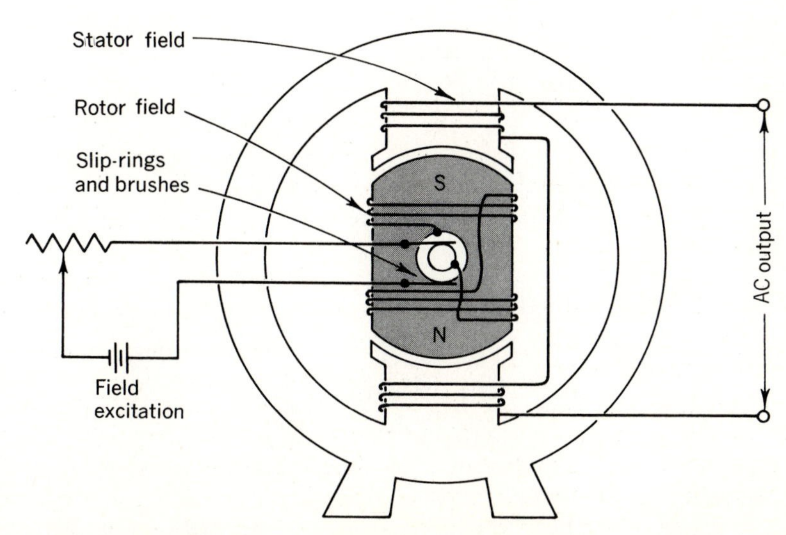

Fig. 21-2 Essentials of a single-phase rotating-field alternator.

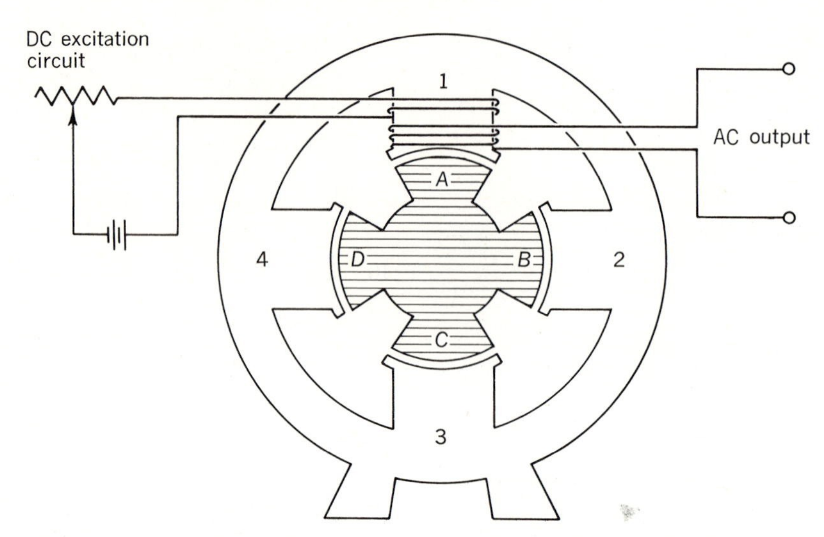

Fig. 21-3 Inductor alternator, with only one set of field pole windings shown.

having moving contacts in the output circuit.

Field excitation is fed to the rotating field through the slip-rings and brushes. Small interruptions of this dc will have much less effect on the magnetism of the rotating field pole, since this field is operating in a saturated condition. The current handled by armature brushes and slip-rings will be much less.

An advantage of the rotating-field is the ease of insulating rotating field coils, a decided advantage when high voltages are generated.

The illustration of the rotating-field alternator in Fig. 21-2 is more representative of the physical makeup of practical alternators than is Fig. 21-1. If pole pieces and rotor are properly dimensioned, the output from this type of alternator is sinusoidal.

To produce 60-Hz ac from an alternator of this type, it would be necessary to drive the rotor at a speed of 60 times per second, which is 3600 rpm. It is more practical to use alternators with four equally spaced field poles and a rotor with four poles, rather than two. The coils are wound so that adjacent rotor poles have opposite magnetic polarity. As the rotor makes one complete turn, the output coils have two complete cycles induced in them. This requires a prime mover with only 1800 rpm for 60-Hz output. For high-voltage output the stationary field coils may be connected in series. For high-current output the coils may be connected in parallel.

ANSWERS TO CHECK-UP QUIZ 21-1

1. (Torque) **2.** (Prime mover and motor) **3.** (Motor) (Alternator or generator) **4.** (Brushes) (Slip-rings) **5.** (Low power, no E control) **6.** (Small machines, no external excitation) **7.** (Reduce eddy-current losses) (Insulation) (Yes) **8.** (Enameled) (More compact, less deterioration with heat) **9.** (Increase excitation E) (Reduce rheostat R) (More turns on field coil) (Increase rotational speed) **10.** (Last one) **11.** (Prime mover slows when loaded)

21-3 INDUCTOR ALTERNATORS

A third type of ac generator, called an *inductor alternator*, is somewhat similar to a transformer, having both a primary and a secondary winding on each stator pole, as illustrated in Fig. 21-3.

The soft-iron toothed rotor (cross-hatched) is rotated by a prime mover. With the *A* tooth in position 1, as shown, the magnetic field produced by the dc excitation current completes itself through the toothed rotor, into tooth *A*, out *B* and *D*, into poles 2 and 4, and back to pole 1 through the iron case of the machine. All the field is in an iron path with few leakage lines. If the rotor is moved 45° clockwise, there is no completion of the magnetic circuit. As the *A* tooth rotates to pole 2, the magnetic field is once more contained in an iron path. The expanding and contracting lines of force as the magnetic circuit is completed and interrupted induce an alternating emf in the output coil.

Each complete rotation of the rotor produces four cycles of ac. If the machine had 20 poles and 20 teeth, there would be 20 cycles of ac produced for each rotation of the toothed rotor. Inductor alternators are used when ac up to several thousand hertz is desired.

Although the simplified illustration shows only one field pole with windings on it, each pole would have a similar primary and secondary winding. All primaries could be connected in series or in parallel, depending on the available source of excitation voltage. The secondary coils could be connected in series for high-voltage output, in parallel for high current, or in series-parallel for intermediate values.

21-4 THE EFFICIENCY OF MACHINES

The efficiency of any device is the ratio of its power output P_o to the power fed into it P_i, or

$$E_{eff} = \frac{P_o}{P_i}$$

This is the *coefficient of efficiency*. To express efficiency in percent, the coefficient is multiplied by 100. For example, a 2-hp motor (1 hp = 746 W) running at rated output acts as the prime mover for an alternator that has a load demand of 1.2 kW. What is the efficiency of the alternator?

$$\%E_{eff} = \frac{P_o}{P_i} \times 100 = \frac{1200}{1492} \times 100 = 80.4\%$$

Since the prime mover is supplying 1492 W but the alternator is delivering only 1200 W to the load, there must be a 292-W loss in the alternator. This may come from eddy currents in the iron, hysteresis losses in the iron, I^2R loss due to current flowing through resistance in the wires, frictional resistance of the bearing on the rotor shaft and the brush–slip-ring contacts, plus windage and cooling-fan losses. A total of 19.6% of the mechanical power supplied by the prime mover is lost in heat in the alternator.

21-5 ADDING AN ALTERNATOR TO A LINE

When two or more alternators are required to feed a system, there may be times when the load demand is light and only one alternator is needed on the line. If the load demand increases, the second alternator must be added at an instant when the voltage of the second alternator is at exactly the same amplitude and exactly in phase with the line voltage.

A circuit such as that in Fig. 21-4 can be used to add a second alternator. Alternator 1 is feeding the line. It is desired to add alternator 2. The prime mover for alternator 2 is started, and excitation is adjusted until the voltmeter across 2 reads the same as that of alternator 1 on the line (120 V).

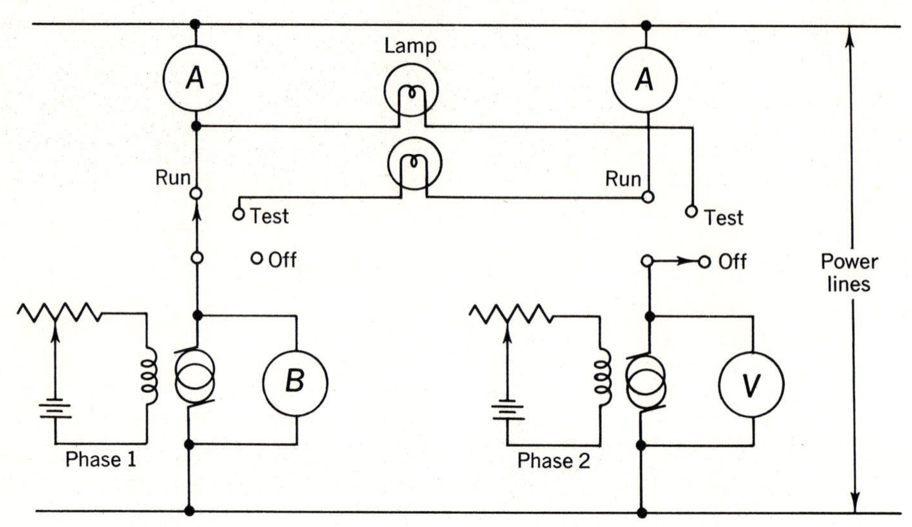

Fig. 21-4 Circuit to connect an alternator across power lines at the proper part of the ac cycle.

The alternator 2 switch is thrown to the "test" position. This connects a 240-V lamp (or a 300-V voltmeter) between the similar line terminals of the two alternators. If the alternators are in phase, the voltage *difference* between them will be zero. If 180° out of phase, the voltage difference will be 240 V, and the test lamp will light.

If the two machines are not generating ac of the same frequency, the lamp will alternately glow and darken as the two machines go into and out of phase. Adjustment of the alternator 2 prime-mover speed will regulate the frequency of this alternator. When the two are at nearly the same frequency, the lamp will light and fade very slowly. At the instant that the lamp goes out, the test switch is thrown to "run," and alternator 2 is on the line. Either alternator can be disconnected or added to the line.

Normally when an alternator is loaded, its output voltage tends to sag somewhat. As a result, it may be necessary to readjust the excitation control on alternator 2 until it is delivering half of the total line current (indicated by the two ammeters).

Alternators operate as inductors because of the many turns of their iron-core coils. With a purely resistive load the output voltage sags somewhat. If the load has an inductive component, this adds to the inductance of the alternator, and the voltage sags even more. However, if the load is slightly capacitive, this cancels some of the inductance of the alternator, and the output voltage tends to hold closer to the no-load value.

Once an alternator is connected across a line, it falls into synchronism with the line frequency. If its prime mover slows, the line ac begins to feed energy to the alternator and it begins to operate as a synchronous motor, which tends to increase the speed of its prime mover. However, it now acts as a load on the line instead of as an alternator supplying the line with energy.

Quiz 21-2. Test your understanding. Answer these check-up questions.

1. How could a rotating-field alternator be used as a rotating-coil alternator? ________

2. What does "operating in a saturated condition" mean with reference to magnetic poles? ______
3. Which would probably be more practical to use as the external excitation source, a battery or a dc generator running off the same prime-mover shaft? ______
4. When a rotating field is at a point in its movement farthest from a pole piece, is maximum, zero, or medium voltage induced in the field coils? ______
5. To produce a 50-Hz ac with a two-pole rotating-coil alternator, what would the prime-mover rpm have to be? ______ What would it be with a four-pole alternator? ______ With a 10-toothed inductor alternator? ______
6. Which winding in an inductor alternator would be analogous to a transformer primary? ______ Which winding in a rotating-field machine would operate as a secondary? ______
7. A 440-V alternator operating from a 30-hp prime mover turning at full capacity produces 20 kW into a load. What is the efficiency in percent of the alternator? ______ What is the resistance of the load? ______
8. A 5-kW alternator is 92% efficient at full load. What is the power requirement in hp for the prime mover? ______ What formula is used to solve for this? ______
9. Why should the toothed rotor of an inductor alternator be laminated? ______
10. If the prime mover of an alternator running on a line with other alternators suddenly slowed, what would the ammeter for the alternator do? ______

21-6 POLYPHASE ALTERNATORS

Observation of a "single-phase" sine-wave ac waveform shows that twice during each cycle, for an instant, there is no power in the circuit. In some applications this can be a disadvantage. When a continuous flow of power is required, it is possible to use a four-pole rotor in a two-pole two-phase (2-ϕ) rotating-coil alternator (Fig. 21-5). One pair of slip-rings is connected to the coils of poles 1 and 3, and the other slip-rings are connected to the coils of poles 2 and 4. When the rotor moves, first one coil will have a maximum voltage induced in it, and 90° later the sec-

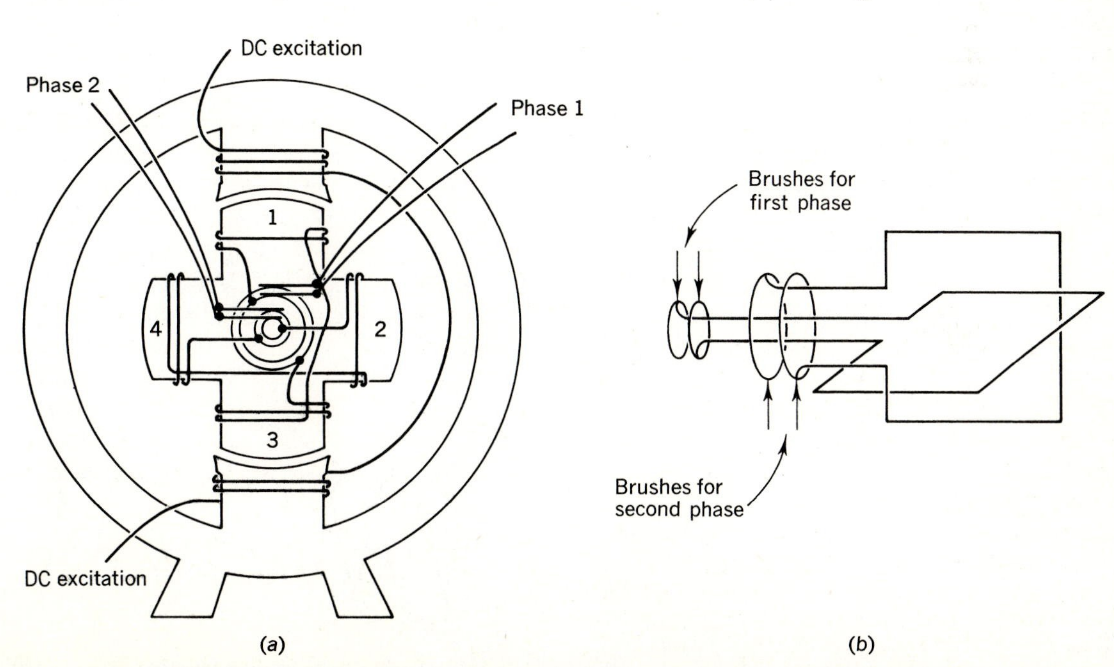

Fig. 21-5 (*a*) Basic 2-ϕ rotating-coil alternator. (*b*) Simplified schematic.

ond coil will have a maximum voltage. The result is two "phases" of sinusoidal ac, both equal in amplitude but displaced by 90° (Fig. 21-6). Voltage is available in the circuit at all times.

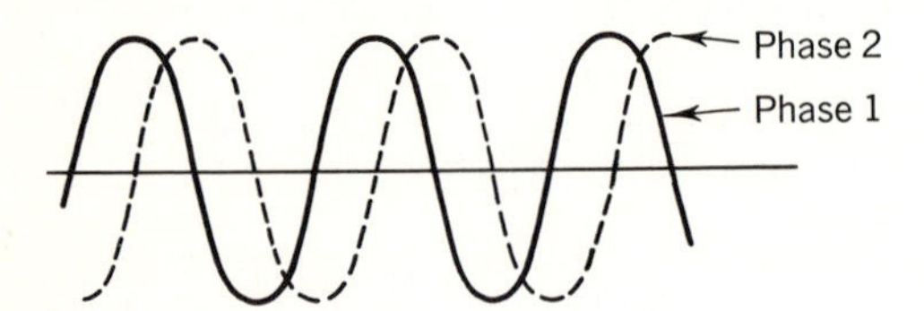

Fig. 21-6 In a 2-ϕ system there is always voltage available to perform work.

The two phases of 2-ϕ ac are normally separated by 90°, although any displacement, such as 45°, 60°, or 135°, is possible. A phase displacement of 180°, however, would result in two 0-V periods during each cycle and would be no improvement over 1-ϕ ac.

Multiple-phase or *polyphase* ac is useful for ac motors. Single-phase ac motors are not self-starting without special circuits.

Two-phase alternators have two separate internal circuits as illustrated in Fig. 21-7. To simplify circuit construction, it is easier to connect two of the four wires together and make a three-wire system for this 2-ϕ ac (Fig. 21-7*b*).

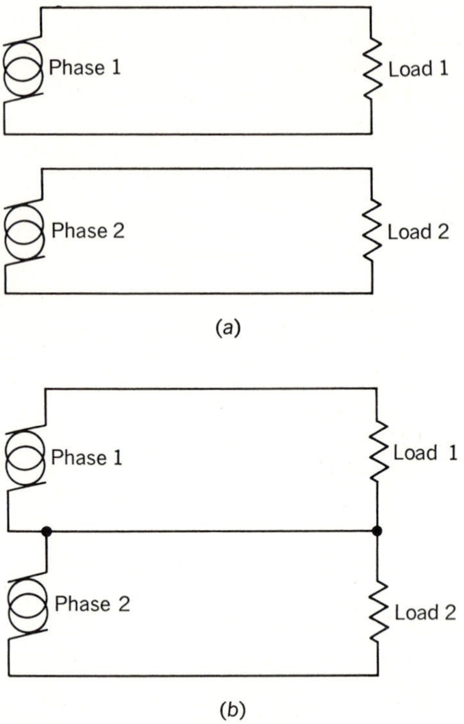

Fig. 21-7 In a 2-ϕ system each phase is connected to a separate load: (*a*) four wires; (*b*) three wires.

There are few 2-ϕ alternators and systems. Polyphase ac is usually three-phase. The displacement of 3-ϕ alternator fields is 120°. A basic machine might have three pairs of stator field poles, three pairs of rotor coils, and three pairs of slip-rings and brushes. With each complete turn of the rotor there will be three separate ac voltages produced. If plotted on the same time line, these voltages can be represented as in Fig. 21-8. At all times a high value of voltage is available in this system.

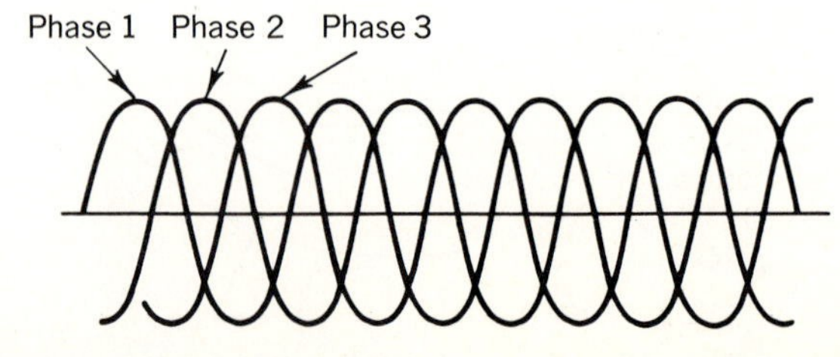

Fig. 21-8 In a 3-ϕ system each phase is 120° from the other two phases.

ANSWERS TO CHECK-UP QUIZ 21-2

1. (Excite stator coils, take ac from rotor coil) **2.** (Increasing field current will not increase number of lines of force) **3.** (Dc generator) **4.** (Max.) **5.** (3000) (1500) (300) **6.** (Exciter) (Output) **7.** (89.4%) (9.68 Ω) **8.** (7.28 hp) ($P_i = P_o/E_{eff}$, horsepower $= P_i/746$) **9.** (To prevent eddy-current loss) **10.** (Drop to zero and rise again as alternator becomes a motor)

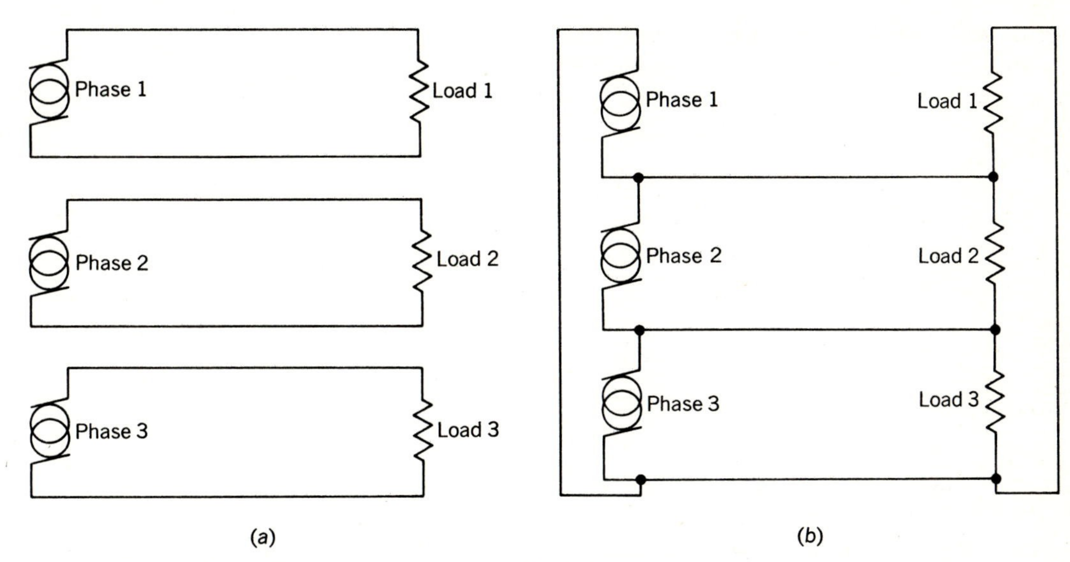

Fig. 21-9 (*a*) A 3-ϕ, six-wire system. (*b*) A 3-ϕ, three-wire Δ-connected system.

21-7 DELTA-CONNECTED ALTERNATORS

A basic circuit for a 3-ϕ system is shown in Fig. 21-9*a*. It is possible to use a single wire in place of two (Fig. 21-9*b*) to simplify the circuit to a three-wire transmission line to carry the energy from source to the loads. This particular method of connecting the source coils is called a *delta* (Greek letter Δ) circuit (Fig. 21-10).

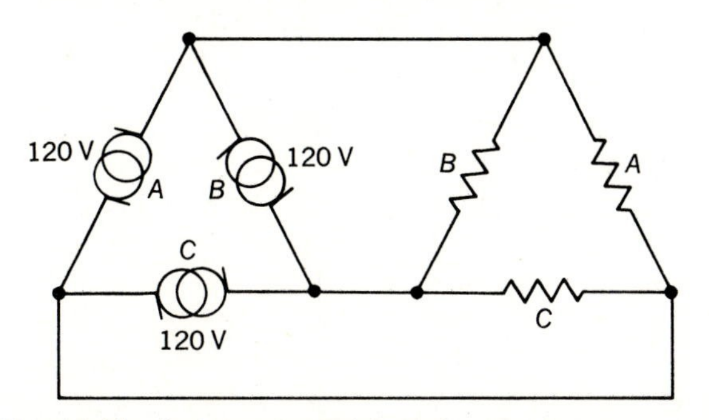

Fig. 21-10 A Δ-connected 3-ϕ alternator connected to Δ-connected loads. Source *A* is across load *A*, *B* is across *B*, and *C* is across *C*.

In the circuit shown, each source winding has 120 V (rms). Since each load is directly across one phase—that is, source *A* is connected across load *A*, source *B* across load *B*, and source *C* across load *C*—no more than 120 V can appear across any load. However, while source *A* is feeding energy to load *A*, what are sources *B* and *C* doing for load *A*? First, sources *B* and *C* are in series across load *A*. While they may not be in the same phase as *A*, these two voltages in series must have a vector sum exactly equal to the voltage across source *A* at all times. As a result, part of the energy being fed to load *A* is coming from sources *B* and C. In fact, *load A* can draw 1.73 times as much current as the wire size of *source A* would safely provide because of the additional energy fed to this load by sources *B* and *C* in series. The same is true for the other two phases. Since smaller wires can be used in the alternator, this will result in smaller machines and greater efficiency.

21-8 Y-CONNECTED ALTERNATORS

It is possible to connect the sources of a 3-ϕ system in another configuration (Fig. 21-11) called a Y, *wye*, or *star* system be-

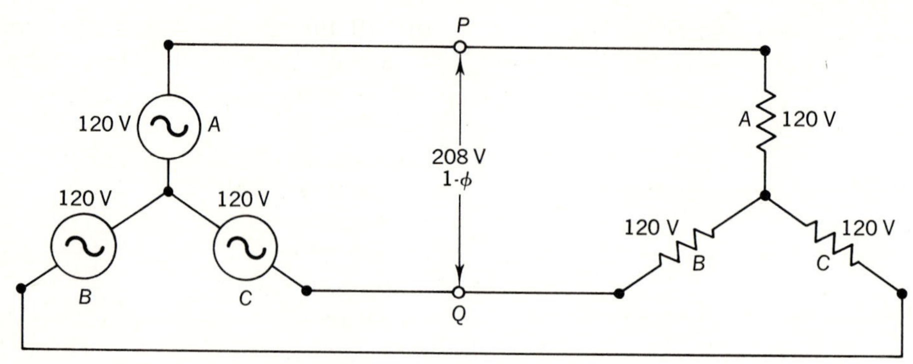

Fig. 21-11 Y-connected 3-ϕ sources connected to a star-type load arrangement. Each source produces 120 V, 1-ϕ, but between any two lines there is 208 V, 1-ϕ.

cause of its appearance in diagram form. In the wye configuration, if each source produces 120-V rms ac, each line would have a voltage equal to the vector sum of two sources (at 120°), which would be 1.73 times the voltage of either source alone. The rms voltage between points *P* and *Q* would be 120(1.73), or 208 V. When Y-connected, there are two loads in series across any one line, and the voltage across any one load will be 0.578 (the reciprocal of 1.73) of the line voltage, in this case 0.578(208), or 120 V. If the loads are balanced (equal current in all phases) with sources and loads Y-Y, the voltage across each load will be equal to the voltage of a source coil.

Three-phase systems for distribution of power and lighting in some industrial applications use the *grounded neutral Y* (Fig. 21-12). The system shown is a 208-V 3-ϕ system, with its neutral connected to a common ground point (water pipe, copper rod driven into the earth, etc.). Between any two legs of the three-wire 3-ϕ transmission line there is 208 V, which is ade-

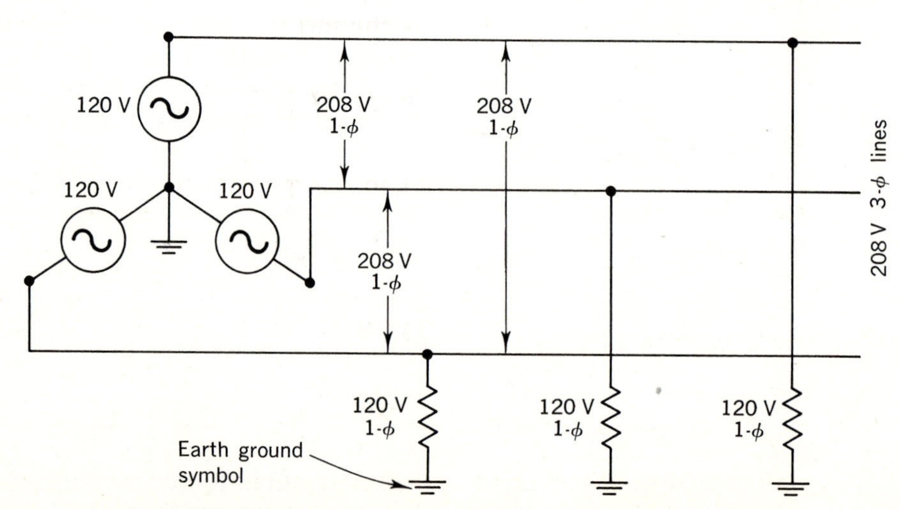

Fig. 21-12 A common power-distribution system, using a grounded neutral, 3-ϕ wye. This provides a 3-ϕ, 208-V ac, and three 1-ϕ, 120-V circuits.

quate to run relatively heavy-duty motors. Between any one of the three lines and ground there is 120 V 1-ϕ available for lighting and small machinery operation. An attempt must be made to balance the loads on the 120-V circuits to prevent overworking one of the phase sources.

Quiz 21-3. Test your understanding. Answer these check-up questions.

1. Is it possible to carry 2-ϕ energy in a single pair of wires? ________
2. What is the most desirable phase displacement for 2-ϕ ac? ________ 3-ϕ ac? ________
3. What does "polyphase ac" mean? ________
4. Is it possible to have 3-ϕ rotating-coil alternators? ________ To have 3-ϕ rotating-field alternators? ________
5. What is the minimum number of slip-rings that could be used on a 2-ϕ alternator? ________ A 3-ϕ delta-connected alternator? ________ A 3-ϕ star with a grounded neutral leg? ________
6. If each phase of a 3-ϕ Y-connected alternator with a grounded neutral had 1100 V, what would be the 3-ϕ transmission-line voltage? ________ The voltage between any wire and ground? ________
7. What are some advantages of 3-ϕ systems over 1-ϕ (single-phase) systems? ________ ________ ________
8. What are some advantages of delta-connected alternators? ________ ________ Y-connected alternators? ________ ________
9. On a separate sheet of paper, draw diagrams of: (*a*) Δ3-ϕ source, Δ3-ϕ load; (*b*) Δ3-ϕ source, Y3-ϕ load; (c) Y3-ϕ source, Δ3-ϕ load; (*d*) Y3-ϕ source with grounded neutral, star 3-ϕ load, and three 1-ϕ loads. Check your drawings against the illustrations in the text.

21-9 TRANSPORTING ELECTRICAL POWER

To reduce the I^2R losses that occur when high current flows through wires, power companies step up the voltages in their transmission lines so that the current is proportionally decreased. For a given amount of power, if the voltage is stepped up 10 times, the current is one-tenth as much. With one-tenth the current and the same wire-resistance value, the power loss is $P = I^2R$, or $0.1^2(R)$, or $0.01R$, which is one-hundredth the power loss.

Electrical power is usually transported from the mountains or over long distances as 3-ϕ ac at 120 kV, 240 kV, or more. When it reaches a city, it is stepped down to 3-ϕ 12 kV or other voltages for distribution. At the consumer's location, the 3-ϕ 12 kV is usually stepped down by power-pole transformers to three center-tapped 240-V three-wire 1-ϕ distribution lines. This provides a 120-V circuit on each side of the center tap. If a home is to be fed 120 V, the grounded neutral or center-tap wire and one of the "hot" 120-V wires are brought into the house. The next home will be fed a neutral and the other hot 120-V wire. Usually a home requires 240 V. The whole three-wire 1-ϕ circuit is fed to the home. In industrial plants the 12 kV may be connected to step-down transformers on the plant grounds. (See also Chap. 32.)

It has been found economical to transport large quantities of power by stepping 3-ϕ ac up to 500,000 V or more, fullwave-rectifying it to varying dc, and then transporting it as dc. The dc is then converted to ac by electronic means to produce ac for local distribution through transformers.

21-10 3-ϕ TRANSFORMER CONFIGURATIONS

It is possible to connect alternator sources and loads in Y-Y, in Δ-Δ, in Y-Δ, or in Δ-Y. Three-phase ac can also be stepped up or stepped down by transformers.

When it is required to reduce a 1200-V 3-ϕ line to a 120-V 3-ϕ line, either a 10:1 ratio 3-ϕ transformer or three 10:1 ratio 1-ϕ transformers may be used. Figure 21-13*a* shows the primaries in delta, with

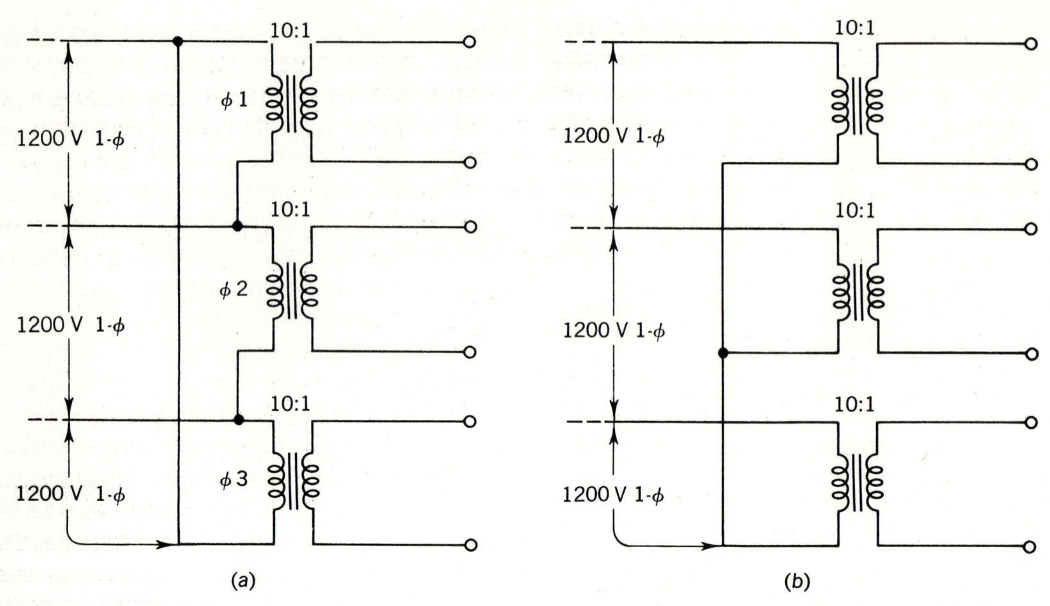

Fig. 21-13 (*a*) Three 1-ϕ transformers with primaries connected 3-ϕ Δ. (*b*) Three 1-ϕ transformers with primaries in 3-ϕ-star.

the secondaries unconnected. The secondaries can be delta-connected or wye-connected, or they may be used as three 1-ϕ sources. Figure 21-13*b* shows the primaries in star with secondaries unconnected.

Quiz 21-4. Test your understanding. Answer these check-up questions.

1. A transmission line is dissipating 0.7 kW in its wires. What saving in power would result if the line voltage were stepped up from 120 to 240 V? ________ To 1.2 kV? ________
2. What is the name of the meter that indicates the total energy used by a consumer over a period of time? ________
3. On a separate sheet of paper, redraw the circuit of Fig. 21-13*a*, with the secondaries connected in delta. What would be the output voltage E_o between any two output circuit wires in this case? ________
4. In your diagram for question 4, if the lower primary winding burned out (open-circuited), would the output circuit be 1-ϕ, 2-ϕ, or 3-ϕ? ________
5. Redraw the circuit of Fig. 21-13*a*, with the secondaries Y-connected. What is E_o in this case? ________
6. Redraw the circuit of Fig. 21-13*b* with the secondaries delta-connected. What is the E_o value? ________
7. Redraw the circuit of Fig. 21-13*b* with the secondaries Y-connected. What is E_o? ________

ANSWERS TO CHECK-UP QUIZ 21-3

1. (No) **2.** (90°) (120°) **3.** (Two or more phases) **4.** (Yes) (Yes) **5.** (3) (3) (4) **6.** (1900 V) (1100 V) **7.** (Efficiency, motors self-starting, smaller machinery, high and low voltages available with one system, less copper used) **8.** (High current, nothing grounded) (High voltage, three 1-ϕ power also available)

CHAPTER 21 TEST • ALTERNATORS

1. What is a motor device called when it is used to rotate a generator or alternator?
2. What part is excited by dc in a rotating-coil alternator? A rotating-field alternator?
3. What is the most practical method of decreasing the output emf of an alternator?
4. Under what condition would the rotor of an alternator turn most easily?
5. Why are alternators run at constant rpm?
6. Which would be most practical for generating 60-Hz ac, a rotating-coil, rotating-field, or an inductor alternator?
7. Why does loading an alternator often reduce the frequency of the ac?
8. A fully loaded 10-hp electric motor is driving a 120-V ac output alternator delivering 6.5 kW to a remote lighting system. If the transmission-line losses are 260 W, what is the approximate loss in watts in the alternator? What is the efficiency of the alternator?
9. What three parameters must be equal before an alternator can be added to a line?
10. What effect would making a load on an alternator slightly capacitive instead of inductive have on the output voltage?
11. Would it be practical to use a dc generator for field-excitation emf if it is operating from the same prime mover that is driving the alternator?
12. Would it be practical to rectify the output of an alternator and use this dc power to excite the field of the alternator?
13. A 50-kW alternator is 88% efficient at full load. What is the power requirement in horsepower for the prime mover?
14. When 2-ϕ alternators are used, what is the usual phase displacement?
15. What is the minimum number of wires that could be used in a transmission line between a 2-ϕ alternator and its load? A 3-ϕ alternator and its load?
16. Assuming 100-V 3-ϕ input ac and 1:1-ratio coupling transformers, what is the 3-ϕ output voltage if the transformers are connected delta-delta? Star-star? Delta-Y? Y-delta?
17. In question 16, what 3-ϕ transformer connections would result in three 100-V 1-ϕ circuits with a common ground connection, as well as a 3-ϕ circuit?
18. If a Y-delta-connected set of transformers burned out one of its secondary windings, would this result in 1-ϕ, 2-ϕ, or 3-ϕ output? What might be a term for such a circuit form?
19. If 12,000-V 3-ϕ ac is fed into a delta-Y step-down-transformer circuit having a 3:1 turns ratio, what is the output voltage? What would it be if connected Y-delta?
20. Why is it less expensive to transport electrical energy over long distances at very high voltages?
21. At what potential is ac energy transported from the mountains to a power station? From power stations to substations? From substations to pole transformers? From pole transformers to homes?
22. Draw diagrams: (*a*) A circuit to parallel a second alternator across a live line. (*b*) A 3-ϕ system with a grounded neutral that can provide three 1-ϕ circuits. (*c*) A 1-ϕ distribution system to two homes, one with 120 V, and the other with 240 V.

ANSWERS TO CHECK-UP QUIZ 21-4

1. (700 − 175 = 525 W) (693 W) **2.** (A kilowatthour meter) **3.** (120-V 3-ϕ) **4.** (3-ϕ; such a two-transformer circuit is called an "open delta") **5.** (208-V 3-ϕ) **6.** (69.4-V 3-ϕ) **7.** (120-V 3-ϕ)

22 GENERATORS AND MOTORS

CHAPTER OBJECTIVE. To describe the operation of separately excited, series, shunt, and compound electrical machines used as dc generators, and the same machines used as dc motors, with means of starting and controlling by resistors and thyratrons. The various ac motors and means of starting them are discussed, as are motor-generators and genemotors.

22-1 EXTERNALLY EXCITED DC GENERATORS

If a dc voltage is required, it is possible to use an alternator and rectify the output ac with diodes. It is also possible to *mechanically* rectify the ac. The result will also be pulsating dc.

Figure 22-1 represents an externally excited dc generator. This is an end-on view of the rotor of essentially the same machine as the basic alternator of Fig. 21-1*a*. The difference is the mechanical rectifying components, the *commutator* segments, in place of slip-rings. The commutator can be considered as a slip-ring cut in two, with the two parts insulated from each other. Each end of the rotor coil is connected to a commutator segment. The brushes press against the commutator segments from opposite sides of the rotating shaft.

The voltage developed in the dc generator rotor (*armature*) is the same ac developed in an alternator. With one rotation of the armature, the voltage developed is one cycle of ac. However, at the instant that the ac goes through its zero-voltage value (the crosses in Fig. 22-1*b*), the brushes move to the other commutator segment and reverse the polarity of the current coming out of the *brushes*. The dotted sine wave represents ac induced in the armature coil, and the heavy lines represent the output voltage at the brushes.

If the armature is rotating clockwise, when the wire on the rotor reaches point X, the commutator open area will be sliding under the right-hand brush. At the same time, the other insulated section between commutator segments will be sliding under the left-hand brush. As the ac cycle changes polarity (+ to −, or − to +), the output connections from the machine are reversed. The result will be current that always flows in the same direction out of one of the brushes and into the other brush. The current from the machine is pulsating dc.

If the prime mover of this generator is moving at 1800 rpm, the number of pulsations is 3600 per minute, or 60 per second. The pulsating dc "ripple frequency" is 60 Hz.

If the armature rotation speed is increased, the ripple frequency will increase, which will make it easier to filter (smooth), and the output voltage will also increase.

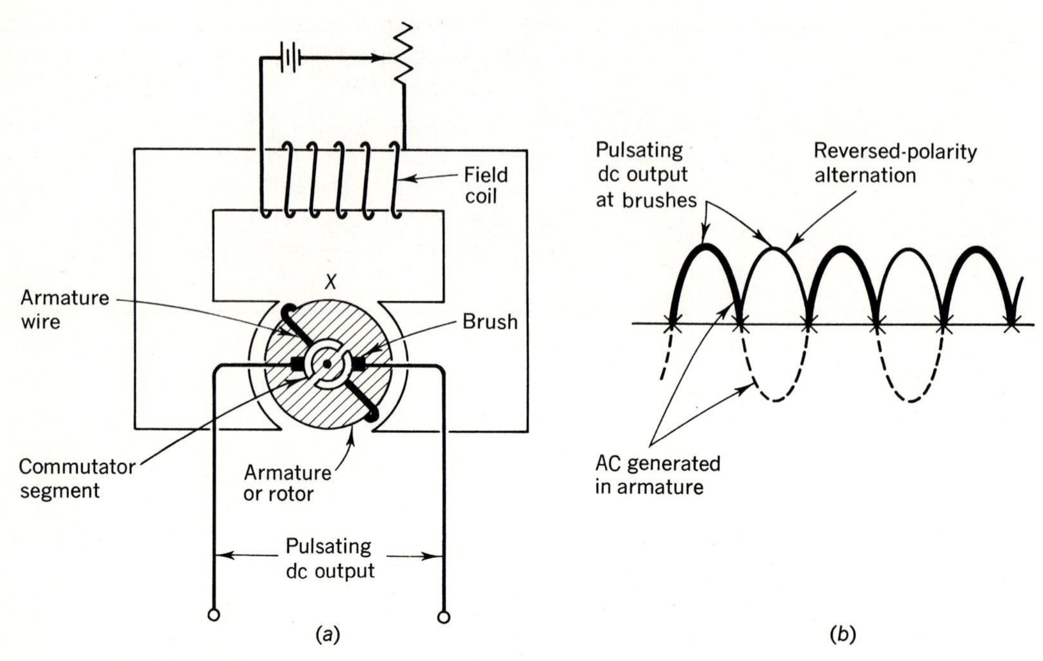

Fig. 22-1 (*a*) Simple dc generator, one armature turn, two commutator segments. (*b*) Ac developed in armature turn and pulsating dc at the brushes.

A more practical method of increasing the output voltage is to increase the field-excitation current by reducing the rheostat resistance value. A practical machine has many armature turns instead of the one shown.

22-2 VOLTAGE REGULATION

When an externally excited dc generator is loaded, the voltage will decrease, or sag, because of the voltage-drop produced when current flows through the resistance in the machine. How much sag represents the voltage regulation? The coefficient of voltage regulation is the ratio of voltage sag to full-load voltage, or

$$E_{\text{reg}} = \frac{E_{NL} - E_{FL}}{E_{FL}}$$

where E_{reg} = voltage-regulation coefficient
E_{NL} = no-load voltage
E_{FL} = full-load voltage

For voltage regulation in percentage, the coefficient is multiplied by 100. For example, the unloaded output of a generator is 100 V. Fully loaded, the output voltage is 92 V. The voltage regulation is 100 − 92 divided by 92, or 8/92, or 0.087, or 8.7%. If the voltage sagged more, the regulation value increases.

Quiz 22-1. Test your understanding. Answer these check-up questions.

1. What is said to have happened to the armature ac of a dc generator when it comes out of the machine brushes? ______
2. What two types of components could be used to

filter the output of the simple generator to smooth dc? __________ __________

3. In Fig. 22-1, which field pole would have N polarity? __________ What would happen to the output voltage if the battery were reversed? __________
4. What is the result of commutation in a generator? __________
5. In Fig. 22-1, is the waveform of the output voltage sinusoidal? __________
6. If the Fig. 22-1 prime mover rotates at 3000 rpm, what is the ripple frequency? __________
7. How is the output voltage normally changed in dc generators? __________ In alternators? __________
8. If a battery has 9 V with no load and 8.5 V under full load, what is its regulation? __________
9. A dc generator operates a load at 120 V. When the load is removed the output is 134 V. What is the regulation? __________
10. Why does the output voltage increase when the speed of the prime mover driving a generator increases? __________

22-3 THE EFFECT OF MORE COMMUTATOR SEGMENTS

Practical dc machines never have only two commutator segments. By using two coils and four commutator segments (Fig. 22-2), the peak voltage output remains the same. The brushes move from one segment to the other when the voltages are 0.707 of the peak value. As a result, the voltage becomes varying dc, with an average of nearly 90% of the peak value and with twice the ripple frequency.

Figure 22-2*b* represents the rectification present with a single coil and two commutator segments. Figure 22-2*c* illustrates the output waveform when two coils and four commutator segments are used. The resultant output is varying dc (solid line). The dotted lines represent the parts of the cycles that are developed in the coils. Practical generators have anywhere from eight to several hundred segments, to produce essentially pure dc.

In practical machines the armature coils are wound into slots in the rotor. Several methods of connecting the segments to the coils are used. One is the two-layer *drum winding*, shown in Fig. 22-3. The many-turn coils are usually prewound and preformed, are dropped into the armature slots, and are soldered to the commutator segments. The armature shown in Fig. 22-3 is an 8-segment type, with the brushes shown on the inside of the segments to simplify the illustration. In each slot one coil is on the bottom and the other is on top (two-layer winding). Each segment is connected to the adjacent segment through a coil. An advantage of

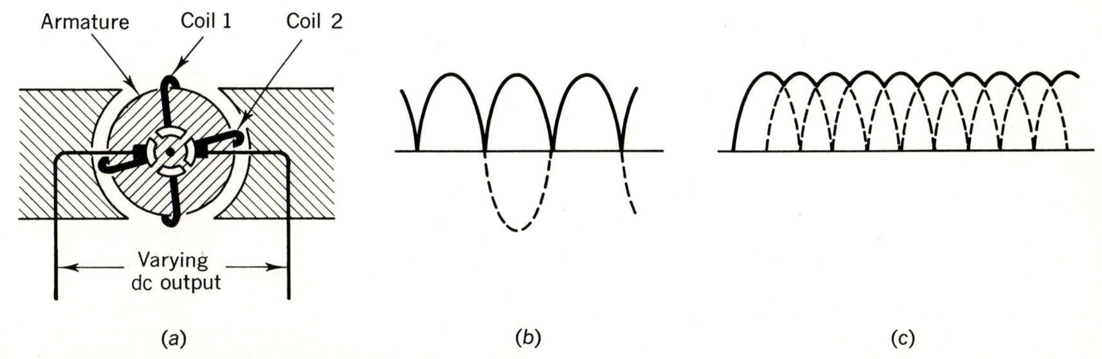

Fig. 22-2 A two-coil armature generator: (*a*) brushes, commutator segments, and coils; (*b*) waveform of one-coil generator output; (*c*) waveform of a two-coil generator voltage output.

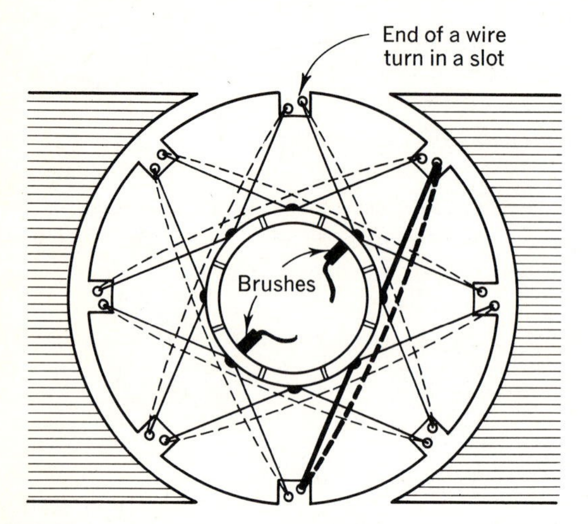

Fig. 22-3 A two-layer drum-wound armature for a two-pole generator. Each coil is represented by a single turn. One coil is shown in heavier lines.

drum winding is that all coils are carrying part of the load at all times, rather than overloading single coils.

22-4 COMMUTATORS AND BRUSHES

The slip-rings of the alternators are brass rings mounted on the rotor shaft over some form of insulation. The commutator of a generator is more complex (Fig. 22-4). The copper commutator segments are flared, one end locking into the commutator sleeve, the other end being clamped and held with a *V-ring*. A threaded collar screws over the threaded commutator sleeve and holds the V-ring in place. The coil connections are soldered into slots provided in the riser.

Each commutator segment is insulated with mica from adjacent segments, commutator sleeve, and V-ring. To allow the brushes to make good contact, the insulation between segments on the commutator should be *undercut* (cut down about 1 mm below the segment surface). Some automotive starter motors have copper ribbon commutator segments molded into a plastic sleeve and must not be undercut.

Voltages between segments should never exceed about 15 V. Thus high-voltage generators require more segments than low-voltage ones.

Brushes are made of carbon (graphite) or mixtures of powdered metal and carbon. Since they are softer than the hard-drawn copper segments on which they ride, they wear down and must be replaced periodically. The brush holder is a hollow tube (Fig. 22-5) in which the brush is free to slide. A spring presses the brush against the commutator. When brushes are replaced, they must be free to move but not so loose that they will rock in the holders. Rocking will result in uneven wear of the brushes and sparking at the commutator. New brushes should be shaped with sandpaper so that they fit the contour of the surface over which they must slide.

Quiz 22-2. Test your understanding. Answer these check-up questions.

1. What are some advantages of a two-coil armature over a single-coil type? ________
2. Does the output voltage ever reach zero with a two-coil armature? ________ A single-coil armature? ________
3. If the brushes were rotated around the armature and out of their proper operating position, what would be the results? ________
4. What is gained by having more commutator segments in a dc generator? ________
5. What is the advantage of drum-winding an armature? ________
6. Why is the drum-winding shown in Fig. 22-3 said to be a two-layer type? ________

ANSWERS TO CHECK-UP QUIZ 22-1

1. (Rectified) **2.** (Capacitors) (Inductors) **3.** (Right) (Reverse polarity only) **4.** (Ac changes to pdc) **5.** (Each pulse is half a sinusoid) **6.** (100 Hz) **7.** (By varying the field excitation) (Same) **8.** (5.88%) **9.** (11.7%) **10.** (More magnetic lines cut per second)

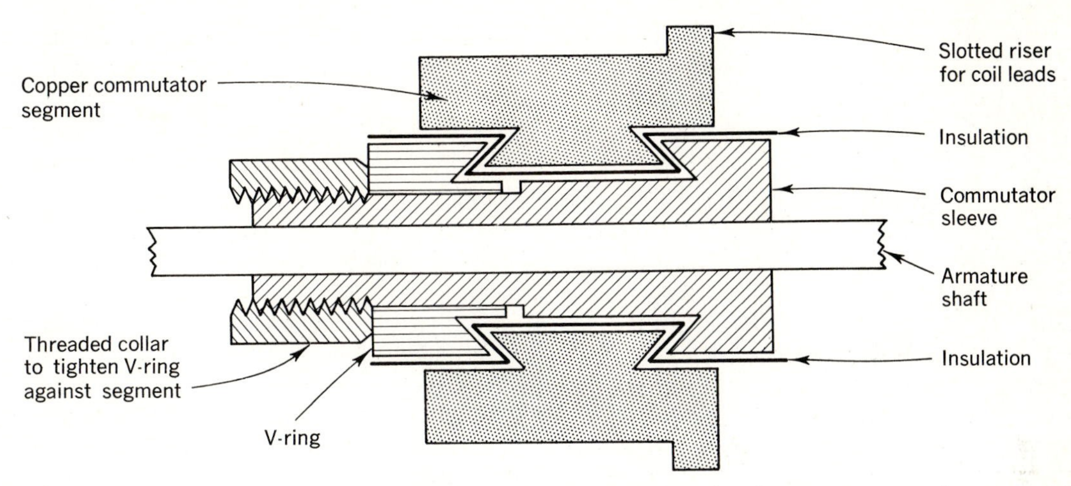

Fig. 22-4 How a commutator segment is held to the armature shaft by the commutator sleeve and V-ring. Each segment is insulated from both sleeve and V-ring.

7. Of what material are commutator segments usually made? ________ Of what are brushes made? ________
8. What does the V-ring do in the commutator assembly? ________
9. Voltages between commutator segments should not exceed what value? ________
10. What should be done to new generator brushes when they are put in the holders? ________
11. What is the process called when the mica insulation between commutator segments is lowered below the surface of the segments by about 1 mm? ________

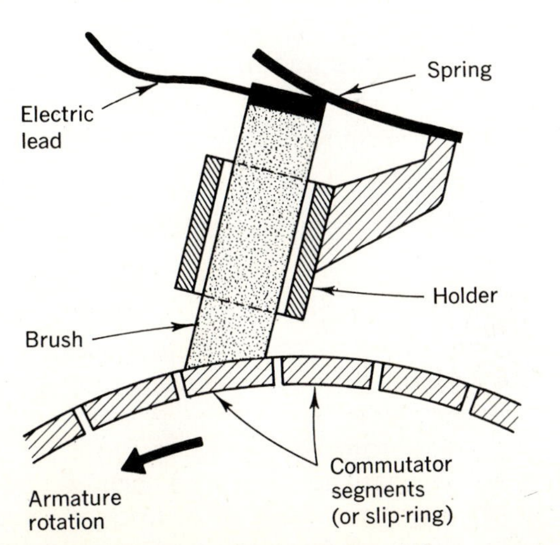

Fig. 22-5 Brush in holder, pressed by a spring against the commutator segment surface.

22-5 SHUNT GENERATORS

All generators so far have field windings excited by some external dc (battery). While this is one way in which generators can be operated, in most cases they will be self-excited. They use some of their own dc output to excite the field coils, as in the *shunt-wound generator* (Fig. 22-6).

When the armature first starts to rotate, its turns cut across the few lines of force

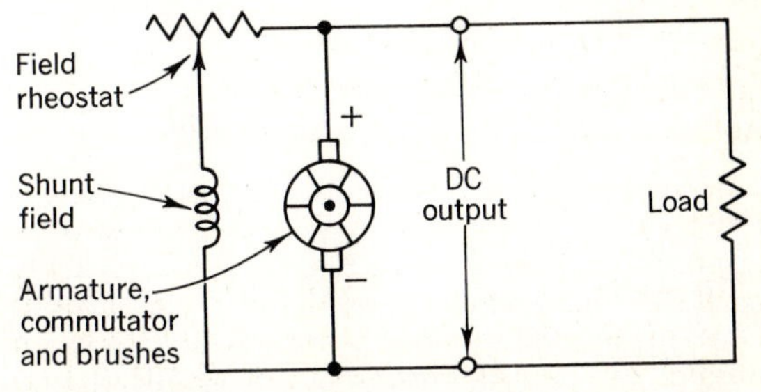

Fig. 22-6 Schematic diagram of a shunt-wound generator coupled to a load resistor. Dc output voltage is controlled by varying field rheostat.

remaining in the field poles from the last time the machine was operated. This develops a small voltage, which is fed to the shunt fields. The resulting weak current increases the magnetic field, which in turn increases the output voltage, exciting the fields more, and so on. In a short time the generator should attain full voltage output. If the field poles are unable to hold enough permanent magnetic lines of force, a battery may have to be "sparked" (touched) across the field coils to start the generating process.

The drop in voltage in the externally excited generator due to internal resistance is also present in the shunt-wound generator. Since this drop in output voltage lowers the excitation to the shunt fields, the voltage sags even more and voltage regulation percentage increases.

The field coils have many turns of rather fine wire with considerable resistance, which, in series with the field rheostat, limit the field current to a practical value. Variation of the field rheostat controls the output voltage of the shunt-wound generator. The two shunt field coils are usually connected in series, but may be connected in parallel.

22-6 SERIES GENERATORS

It is possible to wind the field coils with relatively few turns of heavy wire. If the armature and fields are connected in series, so that the full-load current flows through both (Fig. 22-7), the machine is called a *series-wound generator*.

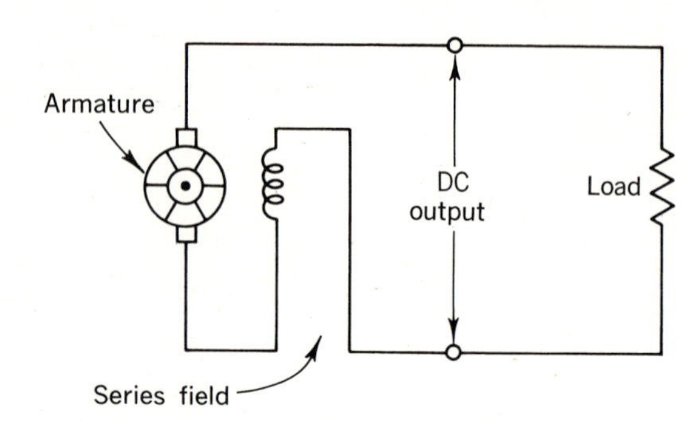

Fig. 22-7 Schematic diagram of a series-wound generator. Voltage output is directly proportional to load.

With a very light load on a series generator the current through the machine is low, the field excitation is low, and the output voltage is low. For example, a 120-V 50-W lamp is connected across the output of a series generator, and it lights properly. If a second, similar lamp is added in parallel, the load resistance is halved, excitation doubles, voltage output doubles, and both lamps burn out! The series connection is not practical for a dc generator, but is useful in dc motors.

However, the rising voltage characteristic under load is a feature which can be utilized.

22-7 COMPOUND GENERATORS

It is possible to balance the sagging-voltage effect of the shunt generator with the rising-voltage effect of the series machine by using both a shunt and a series winding on the field poles. A *compound-wound generator* is shown in Fig. 22-8.

The compound machine has a fine-wire, many-turn shunt winding on each field pole, and a heavy-wire, few-turn series winding. By selecting the proper

---

ANSWERS TO CHECK-UP QUIZ 22-2

1. (Higher average *E*, easier to filter) **2.** (No) (Yes) **3.** (Sparking, less efficiency) **4.** (Smoother dc, higher-voltage operation) **5.** (All coils carrying the load) **6.** (One coil lies over the other) **7.** (Copper) (Carbon, graphite, metal powder) **8.** (Holds commutator segment in place) **9.** (15 V) **10.** (Sanded to fit holder and commutator contour) **11.** (Undercutting)

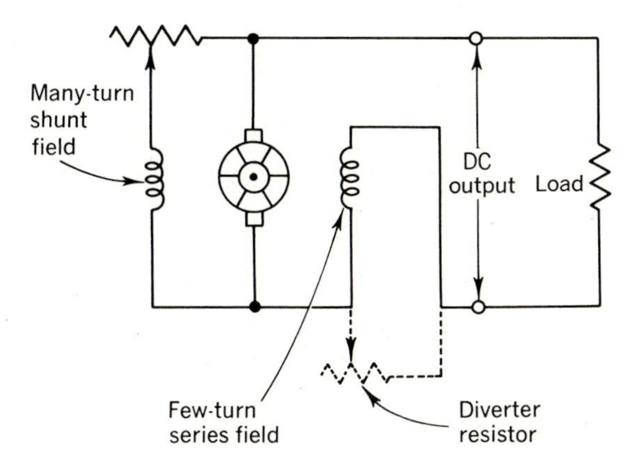

Fig. 22-8 A compound generator coupled to a load. A diverter resistor can be used to reduce compounding.

value of series winding, the machine will have a flat-compounded characteristic (constant voltage) over its entire operating load range.

Some machines are constructed with an excess number of series turns, resulting in an *overcompounding* (rising voltage under load). If the series coil has a *diverter* resistor connected across it, the degree of effectiveness of the coil is controlled, and the compounding can be adjusted to the desired value.

A long transmission line having a significant value of resistance results in a voltage-drop when current flows through the line, and an overcompounded generator may be desirable. At the generator the voltage rises as the load increases, but the line loss results in a relatively constant voltage at the distant load.

Quiz 22-3. Test your understanding. Answer these check-up questions.

1. What would be the main advantage of an externally excited generator over a shunt generator? ________
2. When is it sometimes necessary to spark the shunt field of a generator? ________
3. Which would have the highest percentage of voltage regulation, a shunt, a compound, or a separately excited generator? ________ Which would have the lowest percentage? ________
4. What *feature* does the compound generator borrow from the series generator? ________
5. How is the output voltage controlled in an externally excited generator? ________ A shunt generator? ________ A compound generator? ________
6. When might an overcompounded be preferable to a flat-compounded generator? ________
7. A 120-V-output two-pole shunt generator needs 1.5 A of field excitation. The field coils each have 25 Ω. What should be the approximate midscale resistance value of the field rheostat? ________ The total resistance range of the rheostat? ________
8. What device would be adjusted to change an overcompounded generator to a flat-compounded generator? ________

22-8 ARMATURE REACTION

Brushes must move from one commutator segment to the next during a time when there is no voltage difference between segments to prevent sparking at the brushes. When a generator is running with no load, there is almost no armature current, and lines of force form as straight lines between the north and south field poles (Fig. 22-9*a*). The four circles represent rotating conductors. Conductors 1 and 3 are in the maximum-emf generating positions. Conductors 2 and 4 are in the zero-emf positions. The dotted line represents the *neutral plane*, where the brushes must be placed to prevent sparking.

When a load is connected across the generator, armature current flows, and the armature conductors set up magnetic fields around themselves (Fig. 22-9*b*). These fields repel like-direction field-pole lines and attract opposite-direction lines, distorting the field as shown. As a result, conductors 2 and 4 will not develop zero emf in them until they are at the new neu-

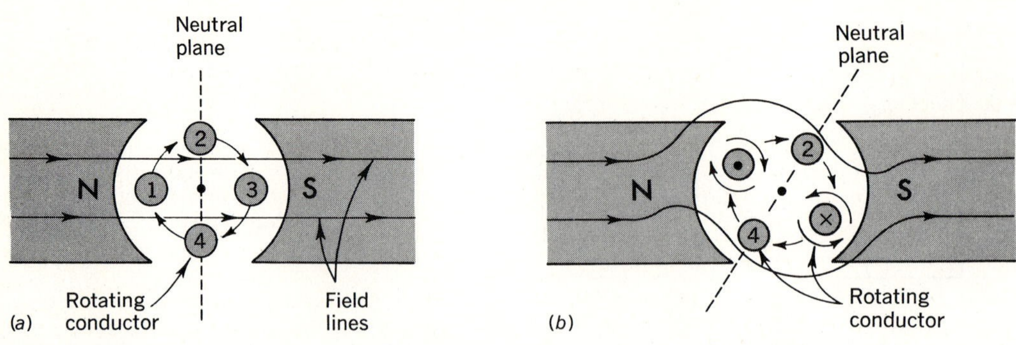

Fig. 22-9 (*a*) With no current flowing in the armature there is no field distortion. (*b*) When current flows, neutral plane shifts in rotation direction, requiring brush movement.

tral plane. The brushes must now be rotated to the new neutral plane to prevent sparking, which is not feasible under most operating conditions.

To counteract the shifting of the neutral plane, additional *interpoles* or *commutating poles* can be added between the field poles (Fig. 22-10). On each interpole a

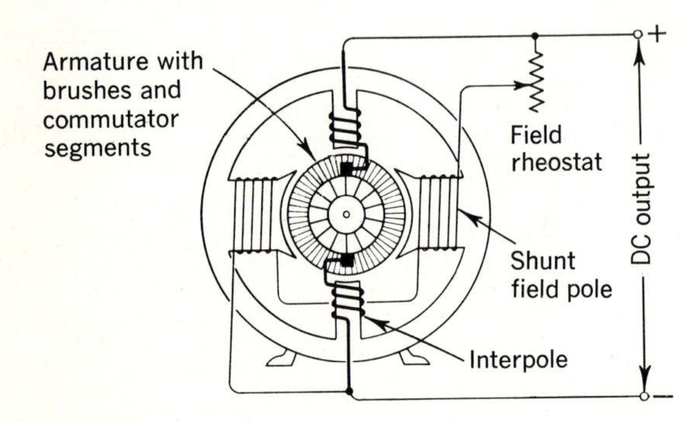

Fig. 22-10 Interpoles in a shunt-wound generator.

few turns of heavy wire are wound. Load current passes through the interpole windings. The greater the load current, the greater the interpole field strength. These fields have such position and polarity as to oppose the effect of the fields built up by the armature currents. With the interpoles working, the field-pole lines are repelled back into a position where the neutral plane is vertical again. Now, when the load varies, the armature fields vary, but the interpole fields have a counteracting effect, and the neutral plane does not shift. It is no longer necessary to shift the brushes as the load varies.

ANSWERS TO CHECK-UP QUIZ 22-3

1. (E_{reg}) **2.** (When it will not start generating) **3.** (Shunt) (Compound) **4.** (Rising *E* under load) **5.** (Vary field excitation or rheostat in all) **6.** (For feeding remote loads) **7.** (30 Ω) (0 to 60 Ω) **8.** (Diverter)

22-9 FOUR-POLE GENERATORS

In many generators only two field poles are used, but larger machines may use four poles. A four-pole generator will have four brushes, each displaced 90° from the adjacent brush, with the brushes on opposite sides of the commutator connected. Twice as many brushes means that each brush will only handle half as much current. Four-pole generators are made in shunt and compound types, and may also have interpoles.

22-10 MACHINE MAINTENANCE

When a machine produces sparking at the brushes, it may wear the brushes rapidly, pit the commutator or slip-rings, and cause interference to nearby radio receiving equipment. Sparking can be caused by worn or dirty commutators or slip-rings, worn or improperly fitting brushes, overloading, an open armature coil, a shorted interpole coil, improper neutral positioning, or segments worn down below the mica insulation.

Commutators and slip-rings are cleaned with a piece of heavy canvas soaked in a solvent. Bearings of rotary machines must be oiled periodically to prevent overheating. If a bearing overheats because of lack of oil, it should be flushed with light oil, while running, until cool. Do not allow oil to flow onto the commutator, brushes, field coils, or insulation.

Radio interference caused by generators can be reduced by connecting 0.1-μF capacitors from both output terminals to the frame of the machine and grounding the frame.

Quiz 22-4. Test your understanding. Answer these check-up questions.

1. Would the neutral plane shift in the direction of rotation or against rotation as a generator is loaded? ________
2. What are the two names for the poles that counteract armature reaction? ________ ________
3. In Fig. 22-10, which is the S field pole? ________
4. For what reason would four-pole generators be preferable to two-pole generators? ________
5. Would shunt fields be connected in series with each other or in parallel? ________
6. Why must the interpole wires be large gage? ________
7. How can interference to radio equipment from electric machines be reduced? ________
8. When a hot generator bearing is flushed to cool it, should the machine be stopped or left running? ________
9. On a separate sheet of paper, draw diagrams of: (*a*) The output waveform of a single-coil generator. (*b*) The waveform of a two-coil generator. (*c*) An 8-slot 8-segment two-layer drum-wound armature, showing coils and connections. (*d*) A shunt-generator circuit. (*e*) A series-generator circuit. (*f*) A compound-generator circuit. (*g*) A shunt generator with interpoles. Check your drawings against the illustrations in the text.

22-11 ELECTRIC MOTORS

The dc electric motor is the same machine as the dc generator. The types in use are the series-wound, shunt-wound, and compound-wound dc motors.

The electric motor principle can be illustrated by Fig. 22-11. A conductor lying in a magnetic field but having no current flowing in it has no reason to move (Fig. 22-11*a*). If current starts to flow outward in the conductor (Fig. 22-11*b*), the magnetic field produced by the current (field direction determined by left-hand current rule) reacts with the field in which the con-

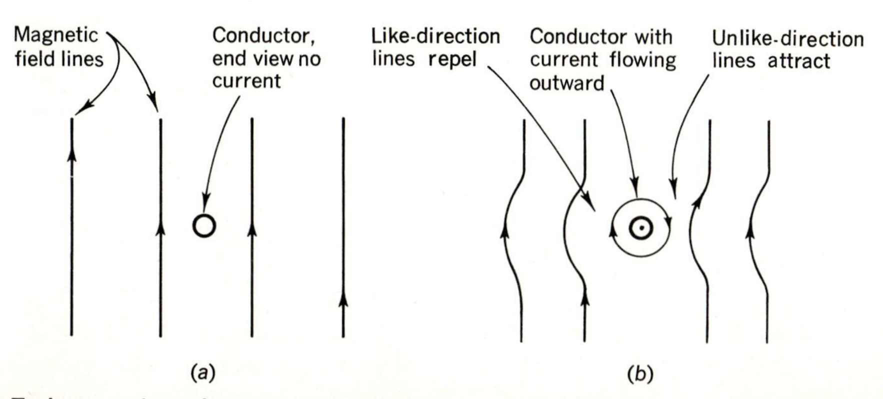

Fig. 22-11 (*a*) End view of conductor in stationary magnetic field. (*b*) When current flows outward in conductor, the field around the conductor repels and attracts the stationary field lines.

ductor is situated and the conductor will try to move. Where the lines of force around the conductor and from the stationary field are in the same direction, they repel. On the opposite side of the conductor the lines of force are in opposite directions and attract. Both effects try to push the conductor to the right.

The direction of conductor deflection can be determined by the *right-hand motor rule* (Fig. 22-12). Point the first finger

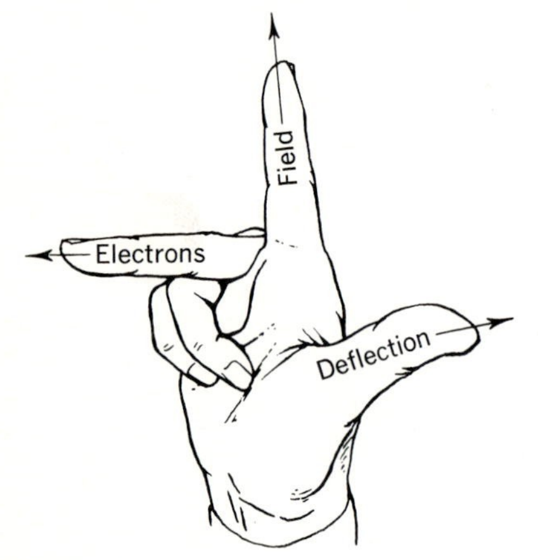

Fig. 22-12 Right-hand motor rule.

of the right hand in the direction of the lines of force of the stationary field. Point the second finger in the direction of the electron flow in the conductor. The thumb indicates the direction of deflection taken by the conductor. Check Fig. 22-11 by this rule.

22-12 SERIES MOTORS

A series-wound motor is the same machine as a series-wound generator, usually with some form of starting and speed-controlling device (rheostat in Fig. 22-13). Whereas the series generator was

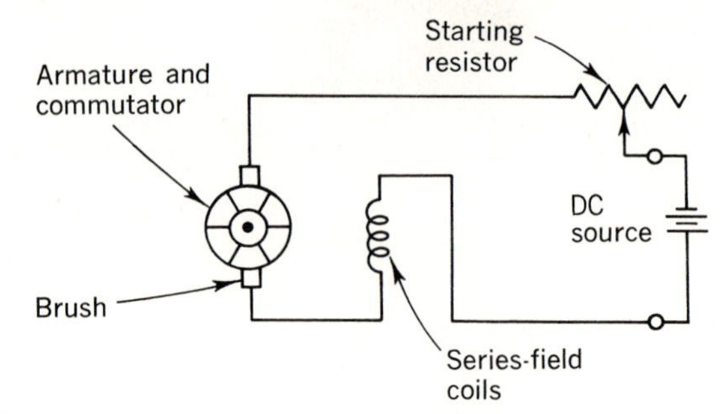

Fig. 22-13 Schematic diagram of a series-wound motor with a starting resistor.

rotated by a prime mover, and produced a dc output, the series motor with electrical (dc) excitation produces mechanical rotation of its armature and shaft.

Rotation of a series motor can be explained by Fig. 22-14. Three pairs of commutator segments are labeled. The only armature coil shown is for the *A* segments. Current flowing in the circuit produces an S magnetic field on the top field

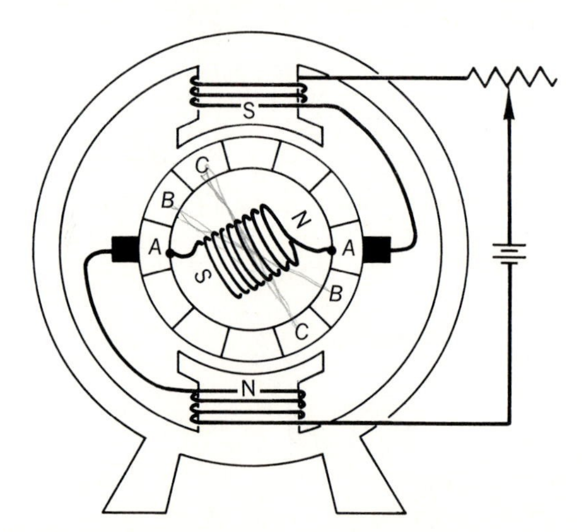

Fig. 22-14 A single coil of the armature in a series-wound motor develops fields to produce counterclockwise rotation.

ANSWERS TO CHECK-UP QUIZ 22-4

1. (In the direction of rotation) **2.** (Interpoles) (Commutator poles) **3.** (Right-hand) **4.** (Higher *I* output, smoother dc) **5.** (In series) **6.** (They carry the full output *I*) **7.** (Bypass output terminals with capacitors) **8.** (Running)

---

pole and an N on the bottom (left-hand rule). The same current produces an N polarity at the top of the armature coil and an S at the bottom. The armature coil fields are attracted toward the field poles, starting the armature rotating counterclockwise. As the armature rotates, the *B* commutator segments are brought under the brushes, and the *B* coil is then in the circuit. When the *B* coil (not shown) is energized, its field will be attracted to the field poles and continues the armature rotation until the *C* segments are under the brushes, and so on. A continuous rotating effect is produced, and the armature increases in speed.

As the armature coils rotate, they cut the lines of force of the field poles, and a counter-emf is induced in them which tends to limit the armature current. However, this counter-emf can never be as great as the source emf, so that the series motor may accelerate until it flies apart if not loaded in some way. If the motor is connected to fan blades or other mechanical load, the counter-emf plus the frictional loading will determine some top speed of rotation. If the load increases, the motor slows. Its *speed regulation* is poor. Speed regulation in a motor can be computed by

$$\% \text{ reg} = \frac{S_{NL} - S_{FL}}{S_{FL}} \times 100$$

where S_{NL} is the no-load speed in rpm and S_{FL} is the full-load speed.

An advantage of a series motor is its high *starting torque*, its low-speed twisting effort. Such motors are useful in electric buses, elevators, winches, etc., because of the fast takeoff they provide. After attaining a relatively low speed, the torque decreases rapidly.

Starting current is high, necessitating a series *starting resistor* or other device to reduce current in larger machines. In small machines, fans, etc., the number of field-coil turns usually provides all the resistance needed to prevent excessive starting current.

Quiz 22-5. Test your understanding. Answer these check-up questions.

1. What device might a series motor employ that might not be found in series generators? ________
2. What is the main disadvantage of series motors? ________
3. What is the main advantage of series motors? ________
4. If the source polarity is reversed, will this reverse the polarity of both the armature and the field? ________ The armature rotation direction? ________
5. What effect would be produced on the direction of rotation if the polarity of only the armature current were reversed? ________ If only the field-coil current were reversed? ________
6. Draw a diagram of a series motor with the necessary switch to reverse the direction of armature rotation (a DPDT switch is needed). What would this circuit do if it were used in a dc series generator? ________
7. What determines the speed of rotation of a series motor? ________
8. In a series motor would the field or the armature coils be wound with the larger wire, or would both windings be the same gage? ________

22-13 SHUNT MOTORS

Shunt motors are popular because of good speed control and regulation.

The shunt motor is a shunt generator with some form of starting-current limiting resistance. Figure 22-15 illustrates a hand starter (dashed oblong) connected to a shunt motor. As the starter switch is advanced to the second switch point, current flows in the armature circuit through the full starting resistance. Field-coil current is established through the arm-holding electromagnet and the field rheostat (set at minimum resistance for starting). The

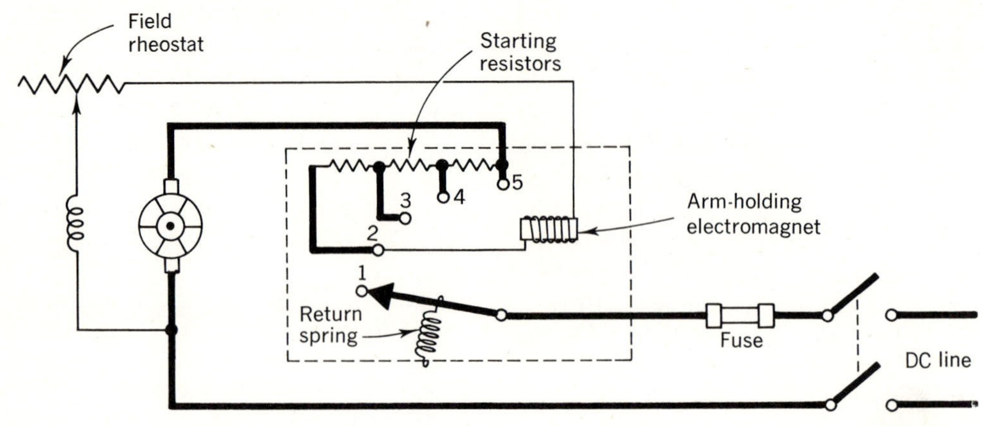

Fig. 22-15 A shunt motor with hand starter and speed-controlling rheostat.

motor starts to turn. After about 1 s the switch is moved to the third switch point, then to the fourth, and finally to the fifth. On the last contact the iron hand-switch arm is held by the electromagnet in the running position against the pull of the return spring.

Unlike the series motor, an unloaded shunt motor will reach an operating speed determined by the counter-emf developed in the armature coils as they turn through the magnetic field of the field poles. If the dc line is 120 V and the counter-emf reaches the same value, no further increase in rotation is possible, since a counter-emf higher than the source emf will reduce the speed.

If the field rheostat is *increased* in resistance value, the field current decreases. This results in a *speedup* of the motor. With less field current there is less field magnetism, and therefore less counter-emf developed in the armature coils. If the counter-emf is low, armature current increases, and rotation speed increases until the counter-emf and source-emf values balance again.

Shunt motors do not have a strong starting torque, but they maintain a nearly constant speed under changing load conditions. If the load increases, the armature tends to slow down. This decreases the counter-emf, the armature current increases, and the speed partially recovers.

The combination of series and shunt windings in a motor produces a *compound motor,* which may have better starting torque than the shunt motor and good speed regulation.

ANSWERS TO CHECK-UP QUIZ 22-5

1. (Starting resistor) **2.** (Speed regulation poor) **3.** (High starting torque) **4.** (Yes) (No) **5.** (Reverse rotation) (Reverse rotation) **6.** (See Fig. 22-18) (Reverse output E) **7.** (Mechanical load, internal R, excitation E) **8.** (Same gage would seem reasonable, but armature wire can be smaller because it is not carrying current all the time and can be overworked while it does)

22-14 EXCITING DC MOTORS WITH AC

Dc motors should operate when excited by ac. However, inductive reactance in the coils will severely limit the alternating current flow. The field and armature iron cores develop greater eddy-current and hysteretic losses with ac.

The *series motor,* wound with relatively few turns, with laminated iron poles and armature, operates satisfactorily as a low-power *universal* or ac/dc motor. If it is operated from a dc line, it will have more power than if operated from a 50- or 60-Hz ac line of the same voltage. If the frequency of the ac is increased, the efficiency and usefulness of the motor decrease.

The shunt motor operating across ac is quite inefficient. It is necessary to rectify the line ac to pulsating dc before these machines begin to function effectively.

The circuit shown in Fig. 22-16 illustrates a fairly simple ac-motor drive system.

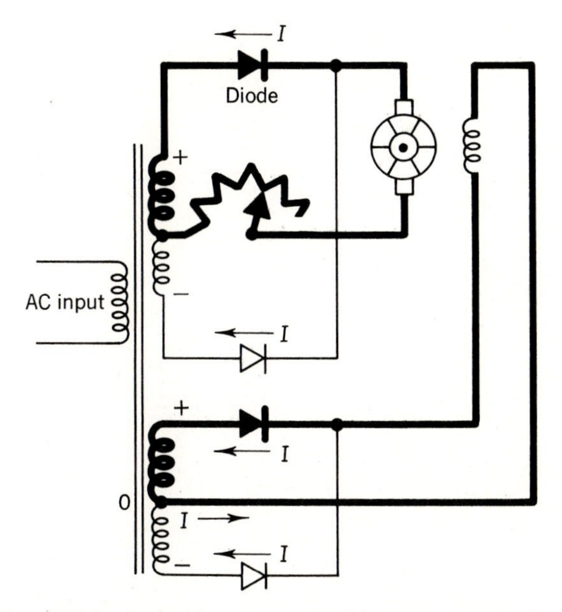

Fig. 22-16 A dc shunt motor with an ac excitation and control circuit. If only the heavy-line circuit is used, the ac is halfwave-rectified. If all circuitry is used, fullwave rectification results.

Consider the circuit in heavy lines. The diode in the armature circuit restricts the current flow in this circuit to counterclockwise only. The rheostat represents the starting resistance, but will also control the speed of the motor. Lower armature current results in less speed.

This circuit utilizes only half of each ac cycle. If the circuit in light lines is added, current flows through the armature in the same direction on the other half-cycles of the ac. The same is true of the field-coil-excitation circuit. These are called *center-tapped fullwave rectifier circuits.* Only half of each winding is used at any one time, but the whole cycle of ac is being used to produce power in the motor.

Quiz 22-6. Test your understanding. Answer these check-up questions.

1. What is the main advantage of a shunt motor over a series motor? ________ The main disadvantage? ________
2. What must a large shunt motor have that a shunt generator does not have? ________
3. What determines the running speed of an unloaded shunt motor operating at a given line voltage? ________
4. In Fig. 22-15, what would be the letter designation for the type of line switch shown? ________
5. In Fig. 22-15, would the starter switch arm fly back as soon as the line switch was opened? ________ Why? ________
6. If the field rheostat in Fig. 22-15 is decreased in resistance value, what effect will this have on the motor speed? ________
7. What type of motor can be used with both ac and dc? ________
8. What must be done to ac before it will operate a dc shunt motor satisfactorily? ________
9. Why is a separately excited dc motor not used? ________
10. What are three ways in which the speed of a dc shunt motor can be controlled satisfactorily? ________ ________ ________
11. What does "fullwave rectification" mean? ________
12. Where was fullwave rectification discussed before? ________ What name was used for that fullwave circuit? ________
13. In a shunt motor, would the field coils or the armature coils be wound with the larger wire, or would both windings be the same gage? ________
14. Why might a small series motor be termed "universal"? ________
15. When starting a shunt motor, should the field resistance be maximum or minimum? ________ Why? ________

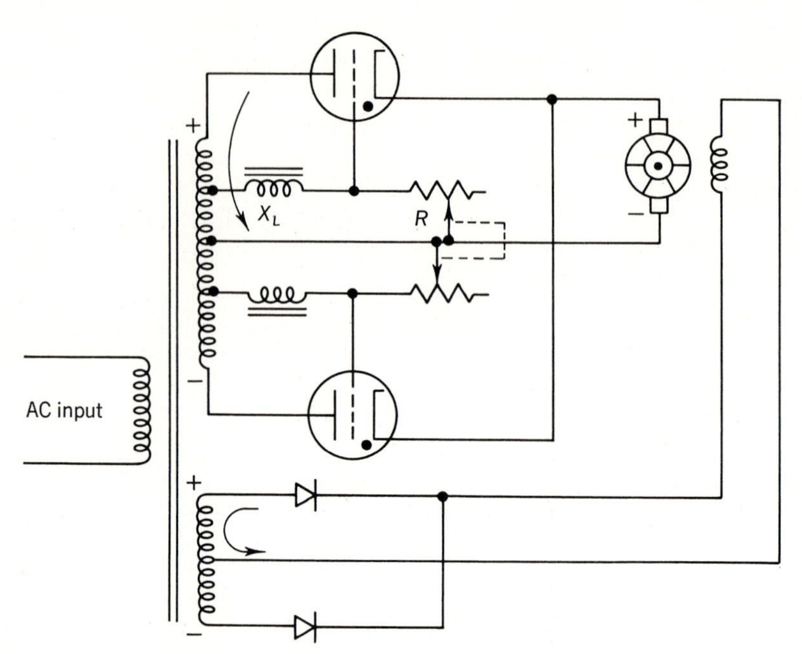

Fig. 22-17 Thyratron tubes act as rectifiers and control tubes operating a dc motor from ac. The phase of the grid voltage determines what portion of the plate-current pulse can flow.

22-15 THYRATRON MOTOR CONTROL

A circuit in which the *phase* of the voltage on the grid of a thyratron (Sec. 17-3) can control the armature current and thereby the speed of a motor is shown in Fig. 22-17. With the plate positive and no voltage on the grid, no current flows. If the grid is made a few volts positive, the gas between grid and cathode ionizes, and full plate current flows at once. The plate current will not stop until the plate voltage is either reduced below the ionizing potential of the gas or reversed in polarity. The thyratron acts as a voltage-operated off-to-on switch.

The grids of the thyratrons are connected to the center of a series RX_L (or RX_C) circuit. E and I are 90° out of phase in an X_L circuit and in phase in an R circuit. If X_L and R are equal, the grid voltage is 45° out of phase with the plate voltage. With zero R the grids are never positive with respect to the cathode, and the thyratrons cannot fire. With high resistance the thyratrons fire as soon as E_P is about +16 V. The phase of the grid voltage determines when the plate current may start to flow during the positive half-cycle. With a low percentage of plate current flow time, the average armature current is low. The higher the percentage of plate-current

---

ANSWERS TO CHECK-UP QUIZ 22-6

1. (Speed reg.) (Low starting torque) **2.** (Starting resistor) **3.** (Counter-emf or shunt field excitation) **4.** (DPST, for double-pole single-throw) **5.** (No) (Motor acts as generator as soon as excitation ceases) **6.** (Slows motor) **7.** (Series) **8.** (It must be rectified) **9.** (Would require two sources) **10.** (Source *E*) (Armature *I*) (Shunt field *I*) **11.** (Both half-cycles of ac used) **12.** (Ac meters) (Bridge) **13.** (Armature wire larger) **14.** (Operates on either dc or ac) **15.** (Minimum) (This starts motor at lowest speed)

---

flow time, the higher the average armature current and the faster the motor will run. The thyratrons act as the starting resistor and speed control for the motor.

The two rheostats are *ganged* (dotted lines). When one turns, the other rotates the same amount. This is a manually controlled system.

Although the field circuit is shown as a diode fullwave rectifier circuit, thyratron control can be applied to this circuit also, for a greater control of motor speed. By coupling voltages to the grids of the thyratrons from outside circuits, it is possible to produce automatic control. Somewhat similar circuits are used with solid-state SCRs.

22-16 REVERSING MOTORS

If the source polarity is reversed on a motor, both the field and the armature currents are reversed at the same time, and the motor continues to turn in the same direction. To reverse the direction of rotation, *either* the armature or the shunt field terminals must be reversed. Figure 22-18 shows a switching circuit that reverses the direction of motor rotation by reversing the armature current flow. With the switch in the *A* position, the current flows down through the armature. In the *B* position the current flows upward.

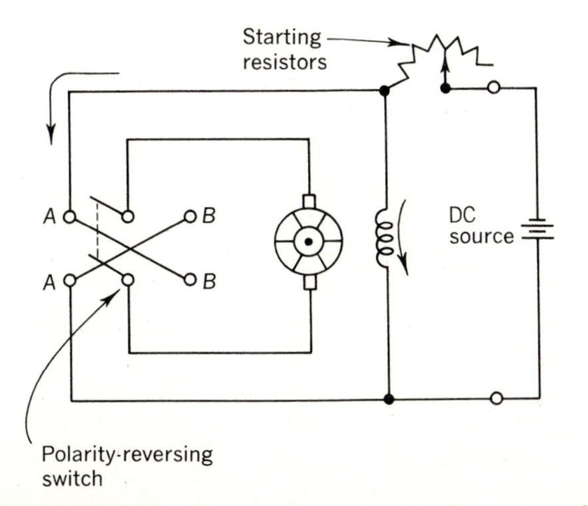

Fig. 22-18 Motor-reversing circuit in a shunt-wound motor.

22-17 MOTOR-GENERATOR SETS AND DYNAMOTORS

To change an available type of electrical power to some other form, a motor-generator set may be used. As examples: A 120-V ac motor drives a dc generator to convert the ac line voltage to 15-V dc. A 120-V dc line voltage operates a dc motor driving a 60-V ac alternator. A 12-V battery turns a 12-V dc motor driving a 500-V dc generator.

A *dynamotor*, or *genemotor*, converts one dc voltage to another. It consists of a single armature with a commutator at each end. The slots in the armature carry two sets of coils: one for the motor and one for the generator. The machine has one pair of field poles common to both motor and generator operations (Fig. 22-19). A dynamotor is usually used to convert low-voltage dc to high-voltage dc. To convert dc to ac, it has a dc motor commutator on one end of the armature and a pair of alternator slip-rings on the other end, and is called an *inverter*, or *rotary converter*.

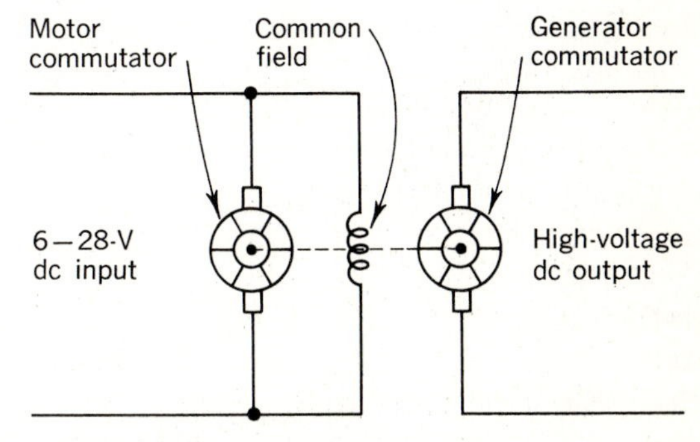

Fig. 22-19 Schematic of dynamotor.

Quiz 22-7. Test your understanding. Answer these check-up questions.

1. In the fullwave thyratron motor-control circuit of Fig. 22-17, would each thyratron pass a current pulse each alternation or each cycle? ________
2. In Fig. 22-17, if the diodes were reversed in the circuit, what effect, if any, would this have on the operation of the motor? ________
3. What term expresses the idea that two shafts are mechanically coupled so that they rotate together? ________
4. What would occur if a motor were switched into reverse without first being stopped? ________
5. What device could be used to convert low-voltage dc to high-voltage dc? ________ Ac to dc? ________ Dc to ac? ________
6. In an ac-motor–dc-generator set, would it be better to start the motor with the generator load on or off? ________ What adjustment should be used to change the voltage output? ________

22-18 SYNCHRONOUS AC MOTORS

The series-type motor can be operated on low-frequency ac. Another ac motor is the *synchronous* type. A rotating-coil alternator (Sec. 21-1) can be made to operate as a synchronous motor by applying dc to its field coil and ac to its slip-rings and rotor coil. However, it will not start. The rotor-coil field is alternating but tries to move in one direction on one half of the ac cycle, and in the other direction on the next half-cycle, resulting in no motion at all.

To produce 60-Hz ac with this two-pole alternator, it would be necessary to have the rotor turning at 3600 rpm. If some outside motor force is used to get the machine up to this speed and if ac and dc are then fed to it, the rotor coils will be passing the field poles at exactly the correct instants, and the machine will lock in step and will operate as a motor at precisely 3600 rpm. The speed regulation is perfect (0%). An ac electric clock is a form of synchronous motor. It maintains correct time as long as the *frequency* of the line ac is correct.

If heavily loaded, a synchronous motor may slow, fall out of synchronism, come to a halt, and possibly burn out.

A large synchronous motor is often started rotating with a dc motor mounted on a common shaft. When the motor is up to synchronous speed, ac is applied to it. The dc starting motor, now acting as a dc generator, is used to supply the dc field excitation for the synchronous motor. The load can then be coupled to the motor.

A synchronous motor operating with an overexcited field has a leading power factor and is said to be a *synchronous capacitor*. Industrial plants operating a large number of heavy-duty ac motors (normally inductive) will use some synchronous motors to raise the overall power factor of the lines. The higher the power factor, the less the charge for electricity.

22-19 SQUIRREL-CAGE AC MOTORS

Most motors that operate from ac lines have field poles, but the rotating part has no windings and is called a *squirrel-cage rotor*. The simplest form of such a rotor (Fig. 22-20) consists of copper bars welded

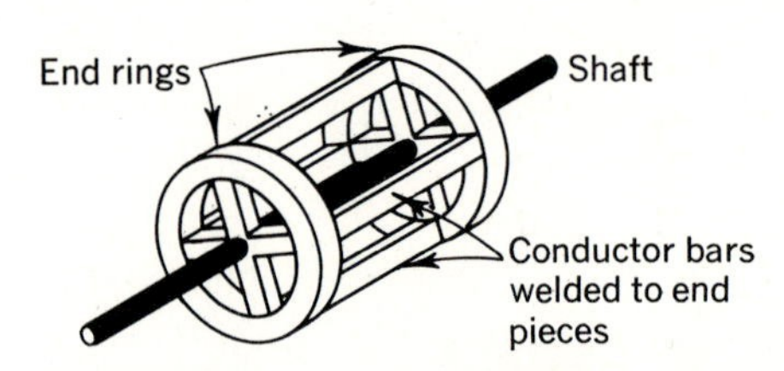

Fig. 22-20 Squirrel-cage rotor composed of copper bars welded to end-rings on a shaft.

to copper end-rings around a central shaft. Each pair of bars has voltage induced in it as the bars pass the field poles. Inasmuch as the bars form essentially shorted turns through the end-rings, induced current in them is quite high and the magnetic fields are very intense. According to Lenz's law, the magnetic fields produced by the induced voltages will be in opposition to the fields that produced the voltages. It is this repulsion effect that produces the rotation of a squirrel-cage rotor.

If ac is fed to the field poles of the machine while the rotor is stopped, voltages, currents, and magnetic fields will be produced, but they will merely repel the fields that produced them, and there will be no rotation of the rotor. Squirrel-cage motors require some method of making the polarity of the pole fields rotate to make the motors self-starting. How the pole fields are made to rotate names the motor.

A *two-phase squirrel-cage motor* uses a second pair of field poles, *X* and *Y*, in between the field poles *A* and *B* (Fig. 22-21). If the two sets of windings are fed 2-ϕ ac (voltages 90° out of phase), a rotating field will be produced. If maximum current is flowing in coils *A* and *B*, the *A* pole may have an N polarity. At this instant, zero current will be flowing in the phase windings, *X* and *Y*. A quarter-cycle later the *A* and *B* windings will have no current or magnetism, but now pole *X* of the phase windings will have maximum current and perhaps an N polarity, and so on. The magnetic N field is rotating around the machine. The rotor, with its shorted turns and induced fields, is being influenced by the *rotating field*. As a result, the induced fields try to follow the fields that produced them. Under these conditions the motor is self-starting.

This type of motor is not synchronous. There is a continual *slip* of the rotor behind the field rotation. If the field rotates at 3600 rpm, the rotor may be at only 3500 or 3400 rpm, depending on the load.

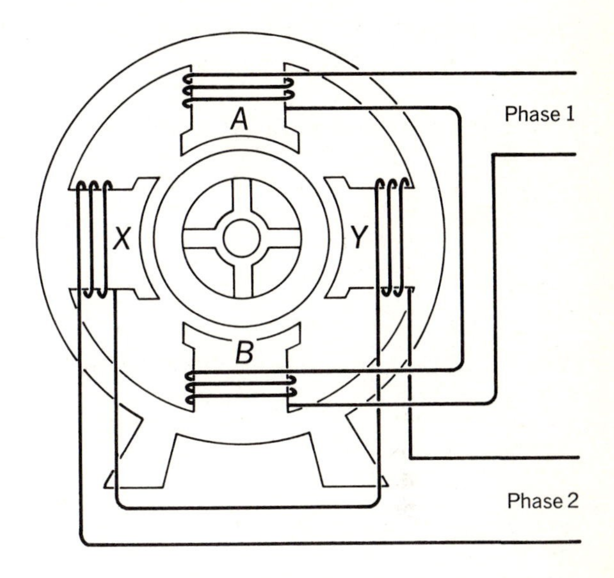

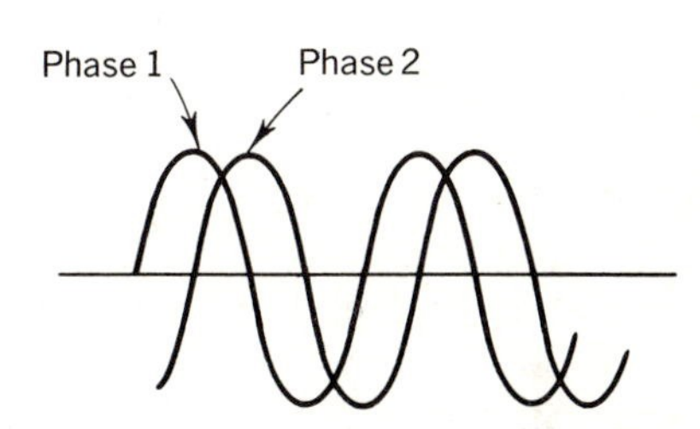

Fig. 22-21 A 2-ϕ motor with squirrel-cage rotor, capable of self-starting because its fields rotate in the machine.

A *split-phase squirrel-cage motor* has a somewhat similar construction but operates from 1-ϕ ac. If capacitance (or inductance) is used in series with the second, or *phase windings* (Fig. 22-22), current flowing through these windings will be out of phase with the main field, the field polarity rotates around the machine, and the motor is self-starting.

A centrifugal switch is in series with the phase, or *starter winding*. As the machine approaches running speed, the phase winding is disconnected by the switch, and the motor operates on the two running fields. The motor operates more efficiently with the starting winding discon-

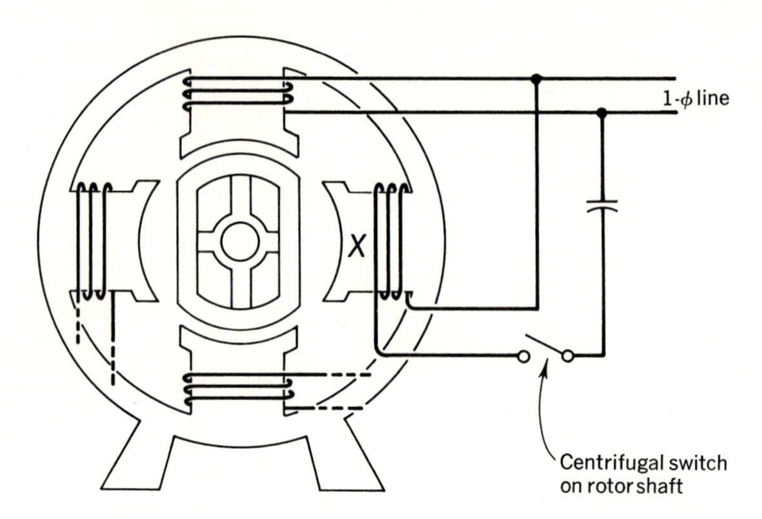

Fig. 22-22 Capacitor-start split-phase squirrel-cage motor. The rotor is shaped to force the machine into synchronous speed. The centrifugal switch opens as the machine approaches running speed. *B* and *Y* coils would be connected in series with *A* and *Y*.

nected. The capacitor has a value of about 50 μF. A series inductor could be used, the phase windings could be wound with more turns to give them high inductance, or a resistor could be substituted for the capacitor. Any method of producing an out-of-phase current in the starter winding can make the machine self-starting.

If two opposite sides of a squirrel-cage rotor are flattened (Fig. 22-22), the rotor operates more like a two-pole magnet and will follow the rotating field closely. There is no slip, and the motor becomes *synchronous*.

The area between the bars in a squirrel-cage rotor is not open as shown, but is usually filled with iron laminations to provide a single solid rotor piece. An exception to this is the centrifugal fan, in which the rotating blades themselves form a hollow squirrel cage that pumps air through the hollow rotor.

A *shaded-pole motor* (Fig. 22-23) produces a rotating field in a different manner. Each field pole is slotted, with a copper ring over one section of it. Ac is fed to the field winding. As an alternation is increasing in amplitude, the magnetic field expands and induces an emf and

ANSWERS TO CHECK-UP QUIZ 22-7

1. (Each cycle) **2.** (Reverse motor) **3.** (Ganged) **4.** (High armature *I*) **5.** (MG or dynamotor) (MG or rectifier) (MG or inverter) **6.** (Off) (Generator shunt field excitation)

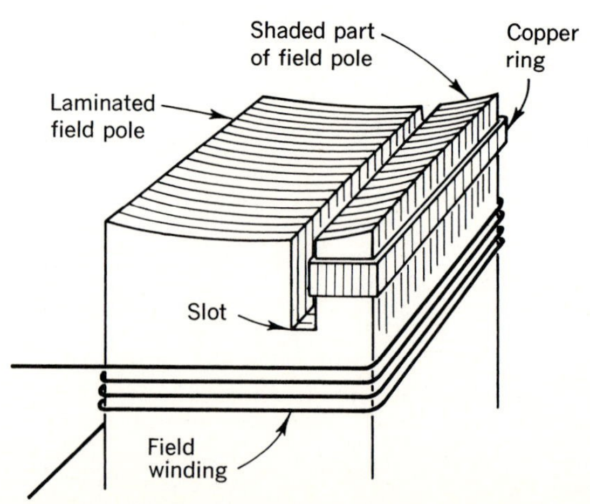

Fig. 22-23 Shaded pole, showing copper ring surrounding one part of the slotted field pole.

current in the copper ring. This produces a magnetic field around the ring that bucks the magnetism in the part of the pole surrounded by the ring (Lenz's law). A maximum magnetic field is developed in the unshaded part of the pole and a minimum in the shaded part. At maximum current the magnetic field is no longer increasing, the copper ring has no current induced in it, and maximum magnetic field is across the whole pole. As the alternation decreases in amplitude, the field collapses, inducing an emf and a current in the ring in the opposite direction, producing a maximum magnetic field in the shaded part. Thus maximum magnetic field moves from the unshaded to the shaded parts of the field pole as the cycle progresses. This is enough rotation of the field to make the motor self-starting. Because the copper rings cannot be disconnected, these motors operate with efficiencies of 20 to 50%. A squirrel-cage motor may be more than 75% efficient. Shaded-pole motors are manufactured only in smaller sizes. If the rotor is flattened on two sides, the machine becomes synchronous.

22-20 THREE-PHASE MOTORS

Most large motors operate from 3-ϕ ac. Since all field poles are excited 120° apart, a true rotating field is produced. They are always self-starting but will have slip unless the rotor is specially shaped.

A two-pole (two poles per phase) 3-ϕ motor operating on 50-Hz ac will have a speed of 3000 rpm. A four-pole 3-ϕ motor will rotate at 1500 rpm, neglecting slip.

In all motors, the starting current is high. Smaller motors have high resistance and inductance in their windings, allowing them to withstand instantaneous starting overloads. Heavy-duty motors include starting resistors in series with the line which are automatically shorted out by relays as the machine approaches running speed. Heavier fuses or breakers may be used for starting, with lighter ones for running. With motors up to 1 hp, "slow-blow" fuses may be used to withstand starting currents.

With no load, ac motors operate at almost synchronous speed. Voltages and currents induced in the rotor occur at such times that they do not cancel the inductance of the field windings, X_L is high, and the line current is low. When a load is applied, slip is produced. This develops fields in the rotor with a phase that tends to cancel the field-coil inductance and X_L, requiring the line current to increase. This is why a heavily loaded machine demands heavy current from the line.

Quiz 22-8. Test your understanding. Answer these check-up questions.

1. What is the name given to the ac motor made from a rotating-field or rotating-coil alternator? ________ What is another name by which it might be known? ________
2. If a two-pole ac motor is operated from 40-Hz ac lines, what would be the unloaded speed of rotation? ________ What would it be if the motor were a four-pole motor? ________
3. Large synchronous motors may have a dc machine on a common shaft. What are two functions of the dc machine? ________ ________
4. What determines the speed of an electric clock? ________
5. What is the name of the rotor used in most ac motors? ________
6. In a 2-ϕ motor, would the size of wire used for the second set of field windings be the same as that used for the first set? ________ In a split-phase motor? ________ In a 3-ϕ motor? ________
7. To produce a rotating field, what must be the relationship between the currents in the main and the secondary poles in a motor? ________
8. How would the direction of rotation be reversed in a 2-ϕ motor? ________ A split-phase motor? ________ A 3-ϕ motor? ________

9. What must be done to the rotor of an ac motor to make the machine run at synchronous speed? ________
10. Since all ac motors (except the universal) tend to be synchronous, how could a change of speed be produced? ________ Would this be very practical? ________
11. Why does a split-phase motor switch out the starter winding? ________ What is the name of the switch used to accomplish this? ________
12. If there is no slip in an ac motor, at what speed will it run? ________
13. What two things are done to the field poles of a shaded-pole motor? ________ ________
14. Why are large ac motors never shaded-pole types? ________
15. What is the mechanical advantage of the squirrel-cage type of motor over the universal and dc types? ________
16. A four-pole 3-ϕ motor running on 60-Hz ac will have what rotational speed? ________
17. When an ac motor is running unloaded, does it have maximum or minimum inductance in its field coils? ________

CHAPTER 22 TEST • GENERATORS AND MOTORS

1. In a single-coil single-field dc generator, what waveform and type of emf is developed in the rotor? Delivered at the brushes?
2. What is the waveform of the emf delivered at the brushes of any dc generator having two or more rotating coils? What are two advantages of having more coils?
3. What is used to control the output voltage on an externally excited dc generator? A shunt-type generator?
4. A dc generator has 132 V of output under no load and 114 V under full load. What is the regulation? Would it be higher or lower if the load were lighter?
5. Why should the brushes move from one commutator segment to the next when the two segments have the same potential?
6. An 8-segment dc generator is rotated at 2000 rpm. What is the ripple frequency?
7. What is the name of the type of armature winding in which all coils are interconnected?
8. What is the approximate maximum potential difference that should be allowed between commutator segments?
9. What is the main reason that series-type dc generators are not used?
10. When should the shunt field of self-excited dc generators be "sparked"?
11. What series-generator factor is useful when a series winding is added to a shunt machine? What is the resulting machine called?
12. A diverter resistor is connected across what in a dc generator?
13. What type of compounding would be desirable in a dc generator feeding a long transmission line?
14. What is the name of the type of field pole that counteracts armature reaction?
15. What happens in a dc generator if the brushes are not at the neutral plane?
16. How many brushes would be used in a four-pole dc generator? How would the brushes that are 180° apart be connected?
17. What is the main advantage of using a four-pole generator?
18. What is normally used with a shunt motor that is not used in a shunt generator?
19. Besides the counter-emf developed in the rotating armature of a series motor, what will limit the rotational speed of the motor?
20. What is the speed regulation of a motor that runs at 3550 rpm with no load and 3350 rpm loaded?
21. What is the main advantage of heavy-duty-type series motors?
22. What are the two main advantages of shunt-type dc motors?
23. What change must occur in the field-winding current in a shunt motor to increase the speed of rotation?
24. What two factors limit the speed of rotation of a shunt-type motor?
25. What is the name of the dc motor that has an armature, series field, and shunt field?
26. When a small series motor can be operated from either ac or dc, what is it called? If the frequency of the ac is increased, how does this affect the torque?

27. When a shunt motor is operated from an ac line through rectifiers, how would the operation be affected if the frequency were doubled? Which would operate more satisfactorily, halfwave or fullwave rectification?
28. What controls the strength of the plate-current pulses in a thyratron motor-speed-control system? How would motor rotation be reversed?
29. What is the name of the machine that has a common field for both the dc motor and dc generator armature windings?
30. If a machine has a common field with an armature carrying commutator segments on one end and slip-rings at the other end, what is it called?
31. When a rotating-field alternator is brought up to running speed and is then fed ac, what type of motor is it? Is its field excited with dc or ac? What would its speed regulation be?
32. What is the name given to the type of rotor used in most ac motors?
33. What must be added to the usual ac motor to make it self-starting when it is used on 1-ϕ ac?
34. Why are phase windings on ac motors disconnected when the motor reaches nearly operating speed?
35. What basic types of ac motors are normally self-starting without having special circuits built into them?
36. What is the minimum number of wires required to feed 2-ϕ ac to a motor? 3-ϕ ac?
37. How is a squirrel-cage motor made synchronous?
38. A record turntable motor requires constant speed of rotation, but efficiency is of little importance. What basic type of starting would it employ? How might it be made synchronous?
39. What would be the most practical heavy-duty ac motor for use in heavy industrial applications?
40. Do most ac motors appear inductive, capacitive, or resistive to the ac line? What are two ways that this could be at least partially counteracted to correct power factor?
41. On separate pieces of paper, draw diagrams of: (*a*) A dc shunt generator with means of controlling voltage output. (*b*) A dc series generator. (*c*) A dc compound generator with means of controlling voltage output. (*d*) A shunt motor operating from an ac line and using fullwave-rectifying circuits. (*e*) A thyratron motor-control circuit connected to a shunt motor.

ANSWERS TO CHECK-UP QUIZ 22-8

1. (Synchronous) (Synchronous capacitor) **2.** (2400 rpm) (1200) **3.** (Starter) (Field excitation) **4.** (Frequency) **5.** (Squirrel-cage) **6.** (Same) (Phase winding smaller wire) (Same) **7.** (Out of phase, usually 30° or more) **8.** (By reversing one phase) (By reversing the phase-winding connections) (By reversing any two of the three leads) **9.** (Flatten opposite sides) **10.** (Change frequency) (Not usually, but used in steamships, for example) **11.** (Prevents possible winding burnout and is more efficient) (Centrifugal) **12.** (Synchronous, or rpm $= 60f$) **13.** (Poles are slotted, section is ringed) **14.** (Efficiency) **15.** (No slip-rings, commutators, brushes, or radio interference) **16.** (1800 rpm) **17.** (Maximum)

23

WIRING PRACTICES

CHAPTER OBJECTIVE. To outline some of the methods used to interconnect components or devices in low-power electronic circuits and some of the wiring requirements when electrical equipment must be installed in buildings or in heavy-duty electronic devices.

23-1 PC BOARDS AND SOLDERING

Originally, vacuum tube and transistor circuits were assembled with separate (discrete) coils, capacitors, resistors, and transformers, together with tubes or transistors. Circuit design was often laid out with the components mounted on an insulating base or sheet, with wires first connected from terminal to terminal and then soldered. This was known as *breadboarding*. If the circuit was found to operate satisfactorily, it could then be rebuilt on a metal chassis, still using interconnecting wires soldered to the terminals of the components. Some laboratory experiments may use this type of wiring. Breadboarding is useful for potentials up to a few hundred volts. Precautions must be taken against shock to experimenting personnel when voltages exceed about 50 V.

Many electronic and small electrical circuits are now using *printed circuit* (PC) boards. A PC board is often a small oblong of 1.5-mm- (1/16-in.-) thick phenolic insulator sheeting with copper plating on one side. The discrete components are mounted with their wire leads pushed through holes drilled in the phenolic sheet. On the opposite (copper foil) side, connections are drawn from terminal hole to terminal hole using a *resist ink* or lacquer. The parts are removed, and the board is immersed in a ferric chloride solution that etches away all exposed copper. This leaves only the resist-ink lines with the copper under them. The resist ink is washed away with lacquer thinner, exposing the copper-line connections. The components are replaced on the board, and the leads are soldered to the copper connections. This is a modern form of breadboarding.

When PC boards are to be turned out in quantity, the resist-ink connections are first drawn out on paper and are then laid down on the copper foil by a printing or photographic process. After etching and washing, the parts are mounted. All leads can be soldered simultaneously by passing the copper side of the PC board over a *wave-soldering* machine. A wave of molten solder is developed by rotating a cylinder just under the surface of a small tub of molten solder. When the PC board (PCB) is moved over the top of the wave, all component leads are soldered to the copper lines, and the copper lines are given a protective coating of solder at the same time.

A two-sided PCB has interconnecting wiring developed on both sides with *plated-through* holes in which components may be fitted. The PCB is then wave-soldered. When two or more PCBs are constructed as a single plug-in unit, they are known as *multilayer* PCBs. A multilayer PCB is very difficult to service or alter. One form of PCB uses simple wiring designs on both sides, interconnected by plated-through holes, but with contact pads extending from the holes on one side. Components are wave-soldered into place, and interconnections are then made by an impulse-bonding (a 0.003-s 300-A welding process) of connecting wires between pads. Such PCBs are simpler to alter and act as a more sophisticated means of breadboarding.

Hand soldering is accomplished with 20- to 35-W soldering irons or guns on PC boards, and up to 100-W or more irons or guns on larger wire connections. Hollow wires of solder with rosin or other noncorrosive "flux" cores are used. The iron is held against the two parts to be soldered (copper foil and a lead, for example); at the same time the solder wire is pushed against the hot iron tip. This sputters flux on the heated joint, cleaning it, and allows the solder to flow over the heated connections. If the iron is not held long enough against the connection, a cold-soldered joint results. A cold-soldered joint may have high resistance, may break loose, or after a period of time may oxidize and become intermittent, or rectify, or even open-circuit. If the iron is left on the connection too long, it may overheat the copper foil and loosen it, or heat may travel up the lead to the component and damage the part. This is particularly true of transistors and diodes, which can be damaged easily by overheating. A pair of long-nose pliers or other *heat-sink* device may be clamped to the lead between component and soldered point to absorb heat while soldering.

When wires are to be soldered to a terminal strip, the wires should first make tight physical connection with the terminal and then be anchored with a coating of solder.

When a wire loop is to be held under a machine screwhead, it is important that the loop be turned clockwise under the head, so that when torque is applied by a screwdriver, screwhead rotation does not open the loop and loosen the connection.

Solder has higher resistance than copper and is not normally used to complete heavy-current electrical connections. It is used only to hold tightly twisted or clamped wires together and to provide a protective coating to the wires. In the case of a PCB, however, the solder does complete the circuit between component leads and copper foil. Although solder has some resistance, PCB currents are usually so small that I^2R heating is negligible and the solder is an adequate conductor for these connections.

23-2 CONDUCTORS AND HEAT

The selection of the proper wire size to use in any electrical circuit is determined by: (1) the current-carrying ability of the wire with its insulation, (2) how confined the space is in which it will be operating, (3) the voltage rating of the insulation, and (4) the length of the wire.

The current-carrying ability of a wire is basically determined by the mass (the diameter of a round wire). The larger the wire, the more current it will carry *without heating excessively*. Even a small copper wire will carry a surprising amount of current, but it may become hot. For example, a #32 American Wire Gage (AWG) copper wire has a diameter of 8 mils (0.008 in.). Its cross-sectional area in circular mils is 8^2, or 64 cir mil. If used in a transformer (a confined application) in which wire with enamel insulation may be rated at 1000 cir mil/A, it will carry 64 mA satisfactorily. At this current value the wire will heat somewhat, and after a time the transformer will become warm to the touch. If a more heat-resistant enamel insulation is used, the wire may be rated at 500 cir mil/A, or 128 mA, and the transformer may become hot to the touch. If still higher heat-resistant enamel is used and the wire is operated at 300 cir mil/A, the transformer may be too hot to touch unless cooled in some way.

If the same #32 wire is used out in free air and is carrying 200 mA, it is not warm to the touch. When carrying 2 A, it will be noticeably warm; when carrying 6 A, it will be hot enough to burn off low-temperature enamel insulation. If air is blown across the wire, however, the heat will be dissipated rapidly and the insulation will not burn. If 12 A is passed through the wire, it may melt unless it is air-cooled. (Small-diameter wires are sometimes used as fuses in electronic circuits.) It is possible that a #32 wire might carry 20 A if cooled adequately by outside means. Anything that prevents the dissipation of heat from a wire will derate the current-carrying capacity. Thus, the more confined the wire, the larger it will have to be to prevent I^2R heat from accumulating and possibly starting a fire. It is the danger of starting a fire that is responsible for much of the thinking pertaining to wiring practices.

23-3 ELECTRICAL WIRES IN BUILDINGS

Probably everyone has observed wiring of various types—open wiring, cabled wiring, wiring in conduits, perhaps in raceways, gutters, etc. Many older wiring practices once thought adequate are no longer considered safe. Modern wiring requires permits from a local building inspector's office as well as adherence to national and local requirements to safeguard persons and property from hazards arising from the use of electricity, and a final inspection of all wiring is necessary. Proper wiring information is spelled out in the National Electrical Code book, published by the National Fire Protection Association in Boston. However, the Code book is not intended as an instruction manual for untrained persons. Nor should anyone attempt any building wiring until taking a special comprehensive course in this field. The relatively few items of information in this chapter are meant to develop an appreciation of proper wiring requirements, to point out some of the complexities of wiring, and to make a study of the Code book somewhat easier. It will be found that there will be many exceptions to some of the necessarily generalized statements made here.

One of the electrical terms used in the construction field is *ampacity*. This means the ampere-carrying capacity of a wire, and it is dependent on ambient temperature, type of insulation, etc. Appendix B lists some information regarding wires from AWG sizes 34 to MCM 2000. As an example, the ampacity of a #8 AWG copper wire with a 60°C rated type TW (thermal- and water-resistant) insulation is 40 A. With a higher temperature-rated 75°C (type THW) insulation on the same wire, the ampacity is 45 A. With a special

200°C asbestos (type A) insulation, the ampacity of the wire is 70 A, although type A insulation can be used only where moisture is not a factor.

The ampacities in Appendix B are for not more than three wires in any specially constructed channel to hold wires, such as a raceway, conduit, etc., or for direct burial in the ground. For a single insulated copper wire in air, the ampacity rating is greater than that shown in the table by about 33% for a #14 wire, up to about 54% for a size 0000 wire. Because larger wires are very stiff, #8 and larger wires in raceways must be *stranded* (several smaller wires twisted together but having a circular mil area adequate for its gage size).

A next larger AWG wire size will always have a diameter 1.123 times greater (two gage sizes = 1.261 as in Appendix B). Conversely, the next smaller diameter wire will be 1/1.123, or 0.89 times the diameter of the first (two gages = 0.793 times). Wires larger than #0000 (or #4/0) are specified in MCM, meaning area in thousands of circular mils. (Note that M in electrical work may mean 1000 times, whereas in electronics it means mega, or 1 million times.)

Aluminum wire has been used instead of copper. The resistance of an aluminum wire is 1.64 times that of a similar-gage copper wire. Because of its higher resistance and therefore greater I^2R heating, an aluminum wire must be larger by two gage numbers than a copper wire of similar ampacity. Older types of aluminum wire have a high thermal expansion, and as a result connections with these wires tend to loosen, oxidize, increase resistance, and overheat when current flows. As a result, fires can be started at old, loose aluminum connections. New types of aluminum wires, when used with proper fittings, do not have this disadvantage.

Power lines that come to a building from the electrical utility company poles are known as the *service drop*. At the *weather head* the service-drop wires attach to the *service-entrance* wires, which are fed through a conduit to the kilowatthour meter and to the *service enclosure box*. Service-drop wires are supplied by the utility company and are of a size determined by them. From there on the contractor installs the wiring. As an example of service-entrance wires, a common 100-A, three-wire, single-phase residential service using copper wires may require only #4 wires (type THW), although this size wire in a conduit is rated at 85 A. This represents an 85% factor that assumes not more than 85% of the electrical equipment will ever be on at one time. For a 110-A service, a #3 wire would be used; for 125 A, a #2 wire; for 150 A, a #1 wire; and for 200 A, a #2/0 wire. In some cases a 100-A service-entrance line may be required to use a #3 wire which is rated at 100 A. Local inspectors may set their own minimum standards.

A *raceway* is a metal or other-material channel or duct constructed to hold wires that connect a service panel box to some other distribution or switch box. A *conduit* is a round, thin-walled or rigid (thick-walled) metal piping that encloses wire or cables that are connected between electrical boxes. The usual nonmetalic (NM) *cable* is two or more insulated wires enclosed in a plastic sheath, although two or more insulated wires bound or tied together by some means is also known as a cable. A commonly used NM cable has two insulated wires (hot and grounded) plus a third similar-size bare grounding wire. The three wires are formed into a flat-oval, polyvinylchloride (PVC) covered cable, with the bare grounding wire in between the two current-carrying wires. Other multiwire cables are made into round form, with the lightly twisted wires covered by an outer plastic sheath. Strands of fiber may be used as a filler to round out the cable. In some cases the cable is metal-sheathed with armor or braid for mechanical protection or shielding, and it may then be further protected against moisture with a PVC or other plastic or rubber outer coating. A lead-sheathed cable is impervious to moisture and corrosion, but it is seldom used.

Wires above ground potential and carrying current are known as *hot* or *ungrounded* wires. Those connected to ground but carrying current are known as grounded wires. Since metal boxes and metal equipment containers are usually required to be grounded, a "two-wire" cable may include a third uninsulated *grounding* wire.

The entrance-service enclosure or box is connected to a driven copper-clad steel rod, or some other earth ground. Any two-wire cables running from the entrance-service box may carry the third grounding wire, which is used to bond together metallic boxes or equipment of any type being fed from this line. A *neutral* wire is used in balanced systems in which the neutral is not expected to carry any significant current. It is usually grounded. (The grounding wire is sometimes referred to as a neutral.)

Color coding is used on wiring to enable easy identification wherever the wires must be connected to switches, outlets, or equipment. The color of the insulation on a wire (or added color coding such as colored taping, splashes of paint, printed color words, etc.) on neutral or grounded conductors must be either gray or white. Ungrounded or hot wires may be any color except white, gray, or green. Interconnecting grounding or bonding wires are green, green with a yellow tracer, or in some cases bare. Every 24 in. along the insulation there must be marked the wire gage number (or circular mil size), an Underwriters' Laboratories, Inc. (UL) approval, the insulation type (TW, THW, etc.), and the maximum voltage for which the insulation is rated.

When conductors are spliced together in buildings, the splice must be made in one of a number of types of metal or nonmetallic junction boxes, gutters, etc. Splices are never made inside a conduit or in a cable run without a box of some sort being used. Smaller wires, #22 to #10, can be joined with twist-on connectors (splicers), which are hollow plastic cones with either tapered internal threads or a tightly coiled spring on the inner surface. When two or more bared wires are pushed into the splicer and the plastic cap is turned, the wires twist together and lock themselves securely into the tapered threads or coiled spring. Another method of splicing is by twisting bared wires together, fitting a copper collar over them, and with a crimping tool making a four-indent crimp in the collar, which securely locks the wires together. Larger wires can be connected by using a *tap connector*, which is a large-diameter screw with a slit the length of the threads. The two wires to be connected are laid next to each other in the slit, and a nut is run down the threads until the wires are tightly squeezed together. Insulation over a splice must be as thick as that on the wires being connected. Usually about six layers of black plastic electrician's tape is adequate for 600 V of insulation, unless sharp points exist, which require additional layers of insulation. A tap connection to a wire can be made by removing insulation from the middle of a wire and squeezing the bared end of a second wire to the bared area of the first wire with a tap connector. Soldered connections are rarely used in building wiring.

When wire runs are long and current values are large, resistance becomes a factor. Larger wires must be used to prevent excessive voltage-drop ($V = IR$) in the lines. Conductors should be sized so that the maximum voltage-drop on any circuit does not exceed 3% on branch lines, or an overall 5%. Low voltage due to IR-drop on ac motors causes them to overheat and possibly burn out. This is particularly hard on capacitor-running motors because it may overheat and damage the capacitors. Reduced voltage also reduces the horsepower output, since hp (power) varies with the square of the voltage.

In general, electrical wiring inside a building is never smaller than #14 copper (#12 aluminum) except for some signaling and control circuit wiring.

All electrical wires, equipment, and parts should carry an Underwriters' Laboratories, Inc. (UL) approval marking.

Quiz 23-1. Test your understanding. Answer these check-up questions.

1. What is the term used for developing prototype electric or electronic circuits? ________
2. What is meant by PCB? ________
3. With what are circuit conductors laid out on PCBs? ________
4. What machine can solder hundreds of PCB connections simultaneously? ________
5. What is the duty of flux in soldering? ________
6. What carries current across a two-wire splice? ________ Between a PCB and a component? ________
7. Under what conditions will a small wire carry more current than a large one? ________
8. What does THW insulation mean? ________
9. What is meant by the ampacity of a conductor? ________
10. If a #10 wire has a diameter of 100 mils, what will be the diameter of a #8 wire? ________ A #12 wire? ________
11. Above what AWG size are wires in raceways usually stranded? ________
12. Why are aluminum wires always two AWG sizes larger for the same current value? ________
13. What is the color coding for ground neutral wires? ________ ________ For grounding wires? ________ ________ ________
14. What is the function of PVC around cables? ________ ________ ________
15. List four methods of joining wires. ________ ________ ________ ________
16. How much insulation should cover spliced connections? ________

23-4 RESIDENTIAL WIRING FUNDAMENTALS

A possible wiring circuit of a residence is shown in Fig. 23-1. This is a diagram of a simple 100-A, 120/240-V service.

The utility company power transmission line is shown as being 10 to 12 kV. This is reduced by a transformer on a nearby pole to 240 V center-tapped, providing one 240-V or two 120-V-to-ground circuits for the house.

The service-drop wires (#4 or #3 THW for 100 A) come in either overhead or underground and tie into a meter box with a breaker-disconnect switch (box, socket, and switch supplied by the contractor), although the switch box may be separate from the meter and may be mounted inside the building. The utility company supplies the kilowatthour meter, which is plugged into the meter box socket. The grounding wire, a #6 or larger wire (#8 if armored), is connected to a nearby cold-water pipe that runs underground, or to one or more 8-ft copper-clad steel pipes driven into the ground near the meter box (for a less-than-25-Ω connection to earth), or preferably to 20 ft or more of #4 copper

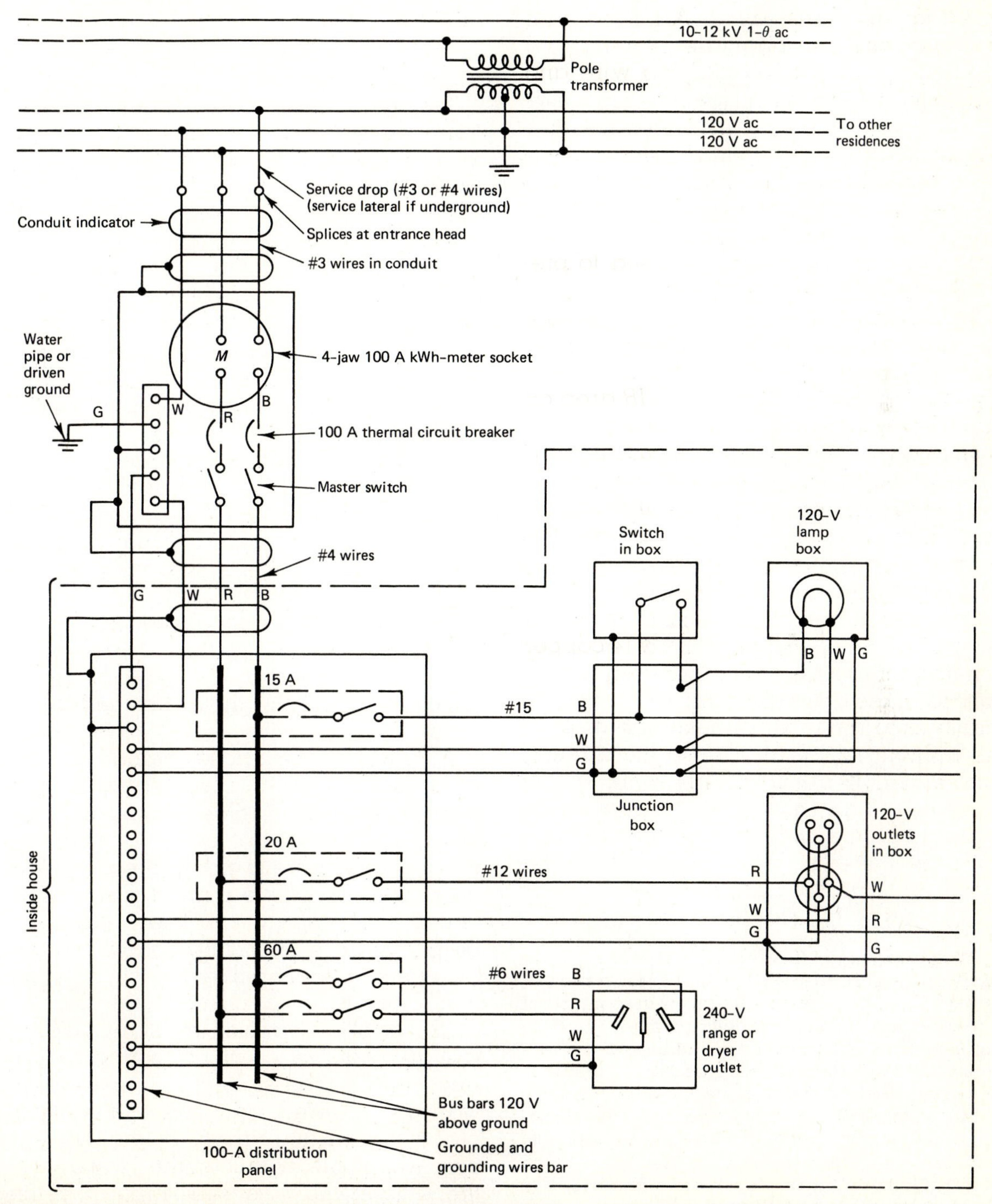

Fig. 23-1 Schematic wiring diagram of a 100-A, 120/240-V electrical service for a residence. Three clamp-on-type circuit breakers are shown dashed. Wires indicated as green (G) may be bare.

wire embedded in the concrete foundation. Special heavy-duty ground clamps are used for grounding connections.

The #4 service ground (white or gray), the two #4 hot leads (usually red and black), and a #4 grounding wire (green or bare) are fed into a 100-A distribution panel in the building. The red and black wires are connected to two rigid bus bars that are mounted vertically inside the panel. The incoming white or gray grounded wire and the green grounding wires are connected to a multiconnector ground bar attached to the metal panel box.

The lighting circuit, if using #14 wire, has a 15-A thermally (or magnetically) activated circuit-breaker-switch unit that clamps onto one of the hot 120-V bus bars. A two-wire NM cable is shown carrying the hot lead from one of the bus bars, one grounded wire, and one grounding wire to a junction box. If the junction box is metal, the grounding wire is connected to it. If it is nonmetallic, the grounding wire is spliced to the grounding wire of the cable that runs to the next connecting box, or to the lighting switch box, or to the lighting fixture box. If all boxes are nonmetallic and have no grounding screws, the grounding wire may be cut off and insulated if there is no equipment to be grounded. If the hot wire shorts to ground in the system, the 15-A circuit breaker will trip and open the circuit.

The receptacle outlet circuit is essentially the same as the light system except that it usually has #12 wires and a 20-A breaker-switch. Switches to control receptacle outlet circuits may be used for lamps, but most outlet circuits have no switches in them other than the service equipment circuit breaker. The 20-A breaker-switch unit is shown clamped to the other 120-V bus bar in an attempt to equalize the load on the two circuit legs.

The 240-V power circuit (for an electric range, dryer, etc.) has a double-pole circuit-breaker-switch unit that clamps to both bus bars. The two hot legs (#10 THW for 30 A, #6 THW for 60 A) and similar-size grounded and grounding wires are run in a cable to the power type outlet and to the outlet box.

A 200-A service may use either a larger 200-A distribution panel or two 100-A panels in different areas in the house fed in parallel from the meter box service circuits.

Although Edison-base (standard 120-V incandescent lamp type) screw-in *plug* fuses are found in older distribution services, the modern method is to use thermal or magnetic circuit breakers. With a breaker, excessive current snaps an internal connection open. The breaker can be put back in operating condition again by opening the breaker-switch fully and then closing it again.

NM cable (corrosion-resistant NMC in wet or corrosive locations) wiring may be run through wall studs if both sides of the wall are covered, and in the attic. At no place in living quarters of the house is any NM cable within sight. However, in the attic it may be run through ceiling joists, or over the joists at the juncture of

---

ANSWERS TO CHECK-UP QUIZ 23-1

1. (Breadboarding) **2.** (Printed circuit board) **3.** (Resist ink, photographic means) **4.** (Wave-soldering) **5.** (Clean joint) **6.** (Physical contact of wires) (Solder) **7.** (If cooled) (Higher temperature insulation) **8.** (Thermoplastic-, heat-, and water-resistant) **9.** (Current-carrying ability in amperes) **10.** (126.1 mils) (79.3 mils) **11.** (#8) **12.** (Heat more because higher R) **13.** (White, gray) (Green, green-yellow, bare) **14.** (Insulates) (Waterproofs) (Protects) **15.** (Splicing and soldering) (Plastic cones) (Crimped caps) (Tap connectors) **16.** (Six tape layers, or same thickness as on wire)

joists and rafters. When laid out along joists, it must be clamped to the joists every 6 ft. All wiring within 6 ft of the attic crawl hole must be protected. Exposed wiring running near a grounded area (basement) must be in grounded metal-covered flexible cable, in conduit, or protected by building construction.

At least one convenience outlet must be installed in bathrooms, basement, garage, laundry room, and outside. In all other rooms no point on any wall-floor junction may be more than 6 ft from a receptacle outlet.

Lighting for residential rooms should be a minimum of 3 W per square foot of floor area, with 0.5 W/ft² in garages.

23-5 GROUND FAULT PROTECTION DEVICES

A relatively recent aid to personnel safety is the ground fault circuit interrupter (GFCI or GFI) in residences and other buildings. Basically, a GFCI senses when there is a leak of current of 5 mA or more to or from the metal case of a piece of equipment being used, and it opens the circuit to the equipment.

The basic circuit of a GFCI is shown in Fig. 23-2. This GFCI is installed in place of a duplex outlet in an ordinary wall box, and it acts as a single outlet into which equipment may be plugged. The load (a hand drill, hair dryer, etc.) is plugged into the GFCI outlet. If the hot and the grounded wires are connected properly (and with the metal case of the load connected to the grounding wire circuit), both hot-line and grounded-line currents flowing to and from the load through the two heavy wire toroidal transformer windings are exactly equal. As a result, the GFCI pick-up coil winding has exactly equal and opposite voltages induced in it by expanding and contracting magnetic fields when current is flowing in the hot and grounded wire lines. This results in zero signal voltage to the semiconductor amplifier. Should a high- or low-resistance fault (R_F) occur between load and case, the sum of the grounded line current plus the fault current back to ground (dashed arrow) via the grounding circuit will equal the hot-line current. Now the grounded-line coil has less current induced in it than does the hot-line coil. The resulting difference of voltage in the pick-up coil is amplified, and if it is high enough, it will energize the relay coil, opening the normally closed (NC) contacts of the relay. (Note symbol for NC relay

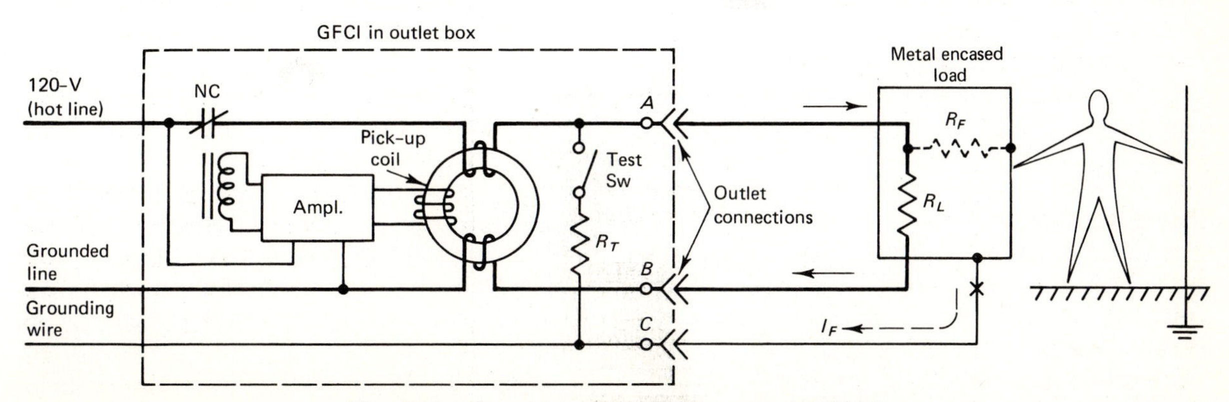

Fig. 23-2 Ground fault circuit interrupter (GFCI).

contacts.) If the GFCI had not been in the circuit, anyone touching both the case of the load and ground would be across any voltage-drop developed across any resistance in the grounding circuit. This would probably not be enough emf to cause a noticeable shock.

On the other hand, if the grounding connection to the load case broke loose (opened at *X*), anyone holding the load and touching ground would be between ground and the hot line through the fault resistance. If the fault resistance were as high as 10,000 Ω, the current through the person could be $I = E/R$, or 120/10 000, or 12 mA. This would be a sizable shock and might even be fatal to someone with a weak heart (25 to 50 mA is usually considered in the lethal range). With the circuit protected by a GFCI, at the instant the person touched the case and ground, the unbalance of the hot- and grounded-line currents would have been sensed and amplified, and the relay contacts would have broken immediately (about 0.025 s), protecting the person from a prolonged and therefore dangerous shock.

Once activated, the GFCI relay contacts latch open, and a switch (button) on the face of the device must be reset manually to put the GFCI outlet back in operating condition again. In some GFCI outlet devices, a red LED may be used to indicate that the relay has been tripped. Usually a GFCI has a test button that connects a test resistance (R_T) capable of producing a 5-mA current to the grounding circuit to check relay operation. Power to operate the amplifier and relay is taken from the ac input lines as shown. This GFCI protects only the load plugged into it. By paralleling wires from points *A*, *B*, and *C* to one or more other simple duplex outlets, all outlets will then be ground-fault-protected.

A GFCI may also be installed at a distribution panel and will protect all outlets on the circuit in which it is installed.

In residences, the Code requires that all outlets in bathrooms and outdoors have ground fault protection (GFP) devices. In addition, outlets on construction sites, near pools of all types, and circuits to underwater lighting fixtures must have GFP devices.

The GFCI explained is for a two-wire, 120-V circuit. For a three-wire, 240-V circuit (as to a range, etc.), another form of GFCI would be used, having NC double-pole single-throw relay contacts to open both of the hot lines simultaneously.

23-6 COMMERCIAL BUILDING WIRING

Commercial buildings may involve not only low-current 1-ϕ, 120/240-V lighting and outlet-type services as in a residence, but also high-current and high-voltage 1-ϕ as well as high- and low-voltage 3-ϕ circuits. The advantage of 3-ϕ is in motor starting and reduced motor and transformer sizes for a given power or horsepower. Usually 3-ϕ conductors run in metal conduits, raceways, etc., except on some large switching panels in which copper bus bars are used.

Some electrical terms regarding electrical installations are illustrated in Fig. 23-3. The utility service drop (or underground lateral) passes through the service-entrance wires to a kilowatthour meter and to a master breaker-switch to protect the meter and utility lines from short circuits in the building. These meters, in high-current applications, may operate from current transformers (Sec. 20-4) rather than be connected directly in the power lines. By *mains* the power is fed to the main distribution panel. From there some of the power goes to a lighting breaker panel via a lighting *feeder* circuit,

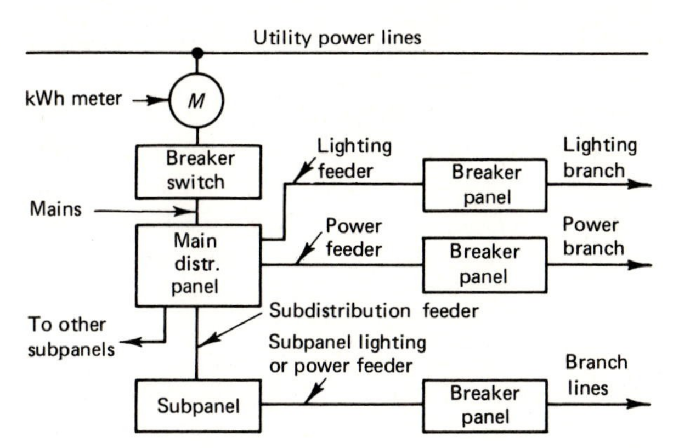

Fig. 23-3 Terminology of electrical wiring using block-diagram form.

and via *branch* line(s) to the lighting loads.

Power to operate motors and other heavy-duty equipment may be fed from the main panel via power feeders to a power-load breaker panel and via power branch lines to the loads.

If there are remote loads, the main distribution panel may have a subdistribution feeder to a subpanel, with subfeeders to the branch line breakers and loads.

The branch breakers must trip when currents are drawn that are higher than is safe for the branch wires. Main panel breakers must have a current rating that allows them to trip when currents occur that are greater than the wire sizes that the feeder lines will safely pass. Breakers protect the *wires* past them from carrying too much current, overheating, and starting a fire.

When 3-ϕ ac is used, it may be provided from star- (wye-), delta-, or open-delta-connected transformers. With the star circuit, a wire from the center of the Y would be grounded, but it might not carry significant current unless it is feeding lighting or other 1-ϕ circuits.

With the relatively simple delta or open-delta system shown in Fig. 23-4, one 240-V

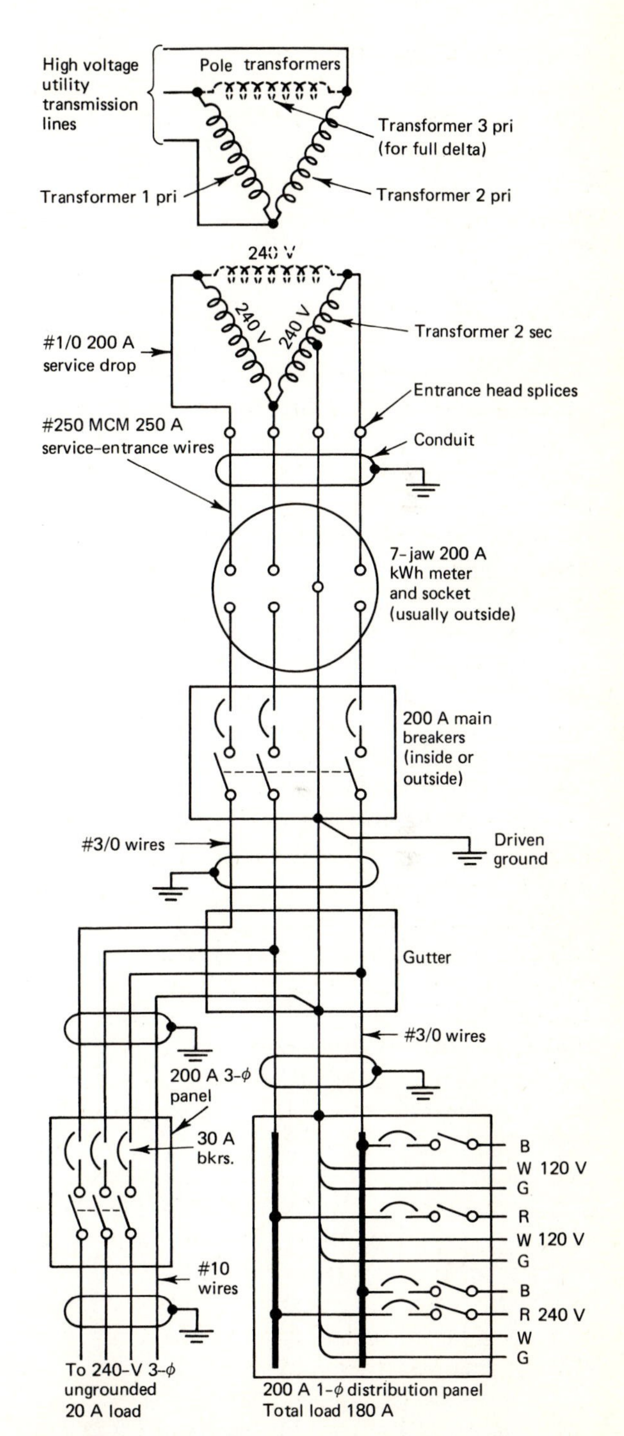

Fig. 23-4 A possible schematic wiring system to provide both single- and three-phase ac to a building.

leg may be center-tapped to provide two 120-V-to-ground circuits, one 240-V, 1-ϕ circuit, and the 240-V, 3-ϕ circuit. Two of the conductors are common to both 1-ϕ and 3-ϕ systems.

Figure 23-4 represents a system feeding an ungrounded 3-ϕ load plus grounded 1-ϕ lighting and outlet loads. Since the load is not heavy, only two transformers will be needed, connected open-delta, rather than three transformers (dashed) to provide the higher power capability of a full-delta system.

The service drop from the utility pole to the entrance head or cap at the building, being in free air and designed to carry 200 A, might be four #1/0 (THW) wires. These are spliced to four wires that will carry 200 A in an entrance rigid conduit (similar to iron pipe). With four conductors in the conduit, a derating of 80% is required (250-A capability), necessitating 250 MCM (THW) copper wires. The minimum inside-diameter conduit to provide the maximum allowable 40% fill of the conduit is found in Table 23-1 to be 2.5 in.

Table 23-1 ***NUMBER OF TW OR THW CONDUCTORS IN A CONDUIT (40% FILL)***

Gage AWG MCM	Conduit ID sizes in inches								
	1/2	3/4	1	1 1/4	1 1/2	2	2 1/2	3	4
14	6	10	16	29	40	65			
12	4	8	13	24	32	53	76		
10	4	6	11	19	26	43	61	95	
8	1	3	5	10	13	22	32	49	85
6	1	2	4	7	10	16	23	36	62
4	1	1	3	5	7	12	17	27	47
3	1	1	2	4	6	10	15	23	40
2	1	1	2	4	5	9	13	20	34
1		1	1	3	4	6	9	14	25
0		1	1	2	3	5	8	12	21
2/0		1	1	1	3	5	7	10	18
3/0		1	1	1	2	4	6	9	15
4/0			1	1	1	3	5	7	13
250			1	1	1	2	4	6	10
500				1	1	1	1	3	6

The 40% fill is required to allow room for conductors to be pulled through bends in the conduit and to allow some cooling.

Because the required load is only 200 A, the kilowatthour meter only needs to be rated at 200 A. To accommodate the four-wire meter, a seven-jaw meter socket is required. Part of (or separate from) the meter housing will be a three-pole, 200-A main breaker-switch. If the conduit nipples between the main breaker and the gutter and between the gutter and other panels are less than 24 in. in length, derating is not required, and size #1/0 (THW) wire may be used to the gutter and panels.

Except for knife switches, most switches must be enclosed in a protective enclosure, either metal or nonmetallic. A switch must never open a ground wire unless it also opens all other ungrounded legs of the circuit first or simultaneously. The blade of a knife switch must be connected in the circuit so that the blade is dead when the switch is open.

One enclosure used when splicing conductors or making a tap connection to a conductor is a metal *gutter* (perhaps 4 in. × 4 in. and 12 to 48 in. long) having a cover to completely enclose the wiring. The cross-sectional area of the conductors inside it must not exceed 20% of the internal cross-sectional area of the gutter (not to exceed 75% when considering splices, taps, and conductors).

To allow for higher current during starting, conductors to motors, according to Code, must have ampacities at least 1.25 times that of the running current value. For applications where the motor is started and stopped often, the factor may be as high as 2 times the running value. Overload devices in the motor feed lines normally should trip at 1.25 times the full-load current rating.

Quiz 23-2. Test your understanding. Answer these check-up questions.

1. What are three methods mentioned to obtain a good ground? ________ ________ ________
2. Of what does an NM cable consist? ________ ________ ________
3. In house wiring, to what is the grounding wire connected in branch circuits? ________ ________ ________
4. Why should lighting and outlet circuits be connected to different hot bus bars? ________
5. In house wiring, for what is 240 V usually used? ________ ________
6. In house wiring, where can NM cable sometimes be seen? ________ Where is it usually run? ________
7. What is the minimum recommended watts per square foot for lighting a home? ________ A garage? ________
8. What does GFCI mean? ________ GFP? ________
9. Where are GFCI outlets required in a home? ________ ________
10. Under what condition will a GFCI trip if a fault is present? ________
11. If a GFCI is tripped, how is it made operative again? ________
12. Why are GFCIs not used at all convenience outlets in a home? ________
13. What are the names of the three basic types of circuits in commercial building wiring? ________ ________ ________
14. What determines the maximum breaker current ratings on branch line distribution circuits? ________
15. When may an open-delta system be desired over a full delta? ________
16. What is the maximum fill value of a conduit? ________
17. For what is a gutter used? ________

CHAPTER 23 TEST • WIRING PRACTICES

1. On what is resist ink used? What does ferric chloride do?
2. When is solder used to complete a circuit connection?
3. What is the circular mil area of a wire 12 mils in diameter?
4. What wattage soldering iron is used on a PCB?
5. Basically, what derates the current-carrying ability of a wire?
6. What is meant by the "ampacity" of a wire?
7. How much more resistance does aluminum have than copper?
8. How many amperes can usually be carried by a #14 copper wire? By a #12 copper wire?
9. Why would flexible #16 gage cords have an ampacity of 10 A for three-wire but 13 A for two-wire cords?
10. What is the preferred ground in a residence?
11. An NM cable has a white, black, and bare wire. What are the three circuits indicated?
12. What is the name of the base of a standard 120-V incandescent lamp of the screw-in type?
13. What is the advantage of a circuit breaker over a fuse?
14. Basically, when must wiring be metal-covered?
15. How close to each other must convenience outlets be in a residence?
16. What does GFI mean? What are two places they are now required in residences?
17. Under what condition will a GFI device trip if there is an internal leakage and the grounding wire is open?
18. What might commercial buildings have as entrance circuitry that would not be used in residences?
19. If 240-V 3-ϕ plus two 120-V 1-ϕ circuits are required, what 3-ϕ transformer connection would be required?
20. For what is a gutter used in electrical wiring?
21. How short must a conduit or nipple be before derating is not required for the wires in it?
22. If 208-V 3-ϕ plus one or more 120-V 1-ϕ circuits are required, what 3-ϕ transformer connection would be used?
23. Draw a schematic diagram of a complete residential wiring circuit including a lighting and an outlet circuit.
24. Draw a schematic diagram of the open-delta 3-ϕ and 1-ϕ commercial building wiring circuit discussed in the text.

ANSWERS TO CHECK-UP QUIZ 23-2

1. (Water pipe) (Driven ground) (#4 wire in foundation) **2.** (Hot wire) (Grounded wire) (Bare grounding wire) **3.** (Switch box) (Junction box) (Fixture box) **4.** (Even loading) **5.** (Range) (Dryer) **6.** (Attic) (Through studs behind wall surfaces) **7.** (3) (0.5) **8.** (Ground fault circuit interrupter) (Ground fault protection) **9.** (Bathroom) (Exterior outlet) **10.** (If load or user touches ground) **11.** (Reset switch on GFCI surface) **12.** (Cost) **13.** (Mains) (Feeders) (Branch) **14.** (Branch wire size) **15.** (Low power 3-ϕ system) **16.** (40%) **17.** (Interconnect wiring)

24 BATTERIES

CHAPTER OBJECTIVE. To discuss the most used batteries of today, including methods of charging cells. Metal plating is also outlined.

24-1 ELECTROCHEMICAL ACTIONS

Some of the important principles regarding electricity lie in the electron action that takes place in *electrolytes* (solutions that current can flow through). The usual electrolytes are acids, alkalies, or salts in water.

If two carbon, two copper, or any other two conductor electrodes of a similar type are placed in pure water (Fig. 24-1) and a source of emf is applied across them, no current will flow between the electrodes. When an acid, a base (alkali), or a salt is dropped into the water, ionization occurs. The molecules of the acid, alkali, or salt break up into positively and negatively charged ions. Since one of the electrodes is negatively charged, it feeds electrons to the positive ions. The positive electrode attracts the negative ions, and a current flows through the electrolyte.

Bubbles form around both electrodes, pure hydrogen at the negative electrode and pure oxygen at the positive. Since water consists chemically of two atoms of hydrogen and one of oxygen (H_2O), twice as many bubbles form at the negative electrode. The action of current breaking down water into its constituent parts is called *electrolysis* (a method of obtaining pure oxygen and hydrogen gases).

24-2 ELECTROPLATING

The process of laying down a gold, silver, or copper plating onto a less expensive metal by electrical means, called *electroplating*, is not new. The Persians

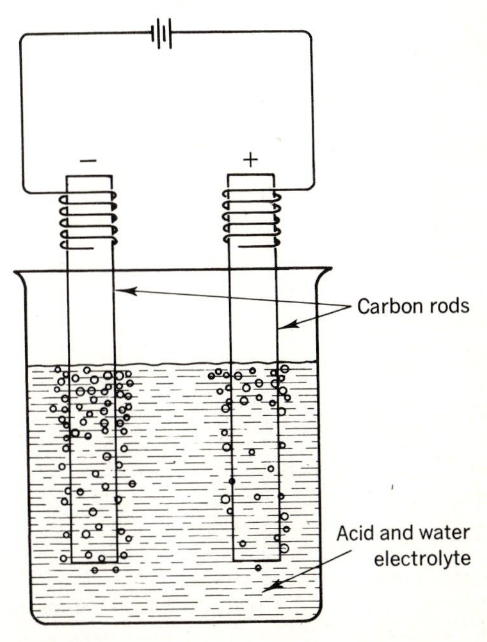

Fig. 24-1 No current flows if the liquid is pure water, but current flows and bubbles form if the liquid is acidic, basic, or a salt solution.

did it over 2100 years ago. Today electroplating equipment consists of an insulated open tank of electrolyte and a source of dc. The object to be plated, such as a carbon brush for a motor, is used as one of the electrodes. The second electrode is normally a solid piece of the metal that is to plate the other electrode. In the case of copper-plating a carbon brush, the second electrode would be copper. When copper sulfate ($CuSO_4$) is mixed into the water of the tank, it ionizes:

$$CuSO_4 \rightarrow Cu^{++} + SO_4^{--}$$

The copper becomes doubly ionized positively (loses two electrons), and the SO_4 ion is doubly ionized negatively, with two excess electrons. If the carbon brush is connected to the negative terminal of the source (Fig. 24-2), the positive copper ions will be attracted to it and will accept two electrons, making the copper ion a neutral copper atom. As copper atoms accumulate on the surface of the carbon brush, they form a solid film of copper on it.

The doubly ionized negative SO_4^{--} ion is attracted to the positive copper electrode. Each ion loses two electrons to the positive electrode and attaches itself to one of the copper atoms of the electrode. The result is the formation of a molecule of $CuSO_4$ to replenish the molecule that was originally ionized and used to plate the carbon brush. This process continues as long as there is a copper electrode and a source of emf.

If the second electrode is any conductor other than copper, the SO_4^{--} ion is attracted to it, but a chemical reaction takes place with the water to produce sulfuric acid, H_2SO_4, plus molecules of free oxygen. The acid remains as the electrolyte, and the oxygen escapes as a gas. The only copper available for plating is in the copper sulfate salts mixed into the electrolyte, which would soon be exhausted.

Objects to be plated must be free of any oil or other contaminating surface material, and they must be electric conductors. Excessive current during plating will speed the plating but may produce an inferior coating.

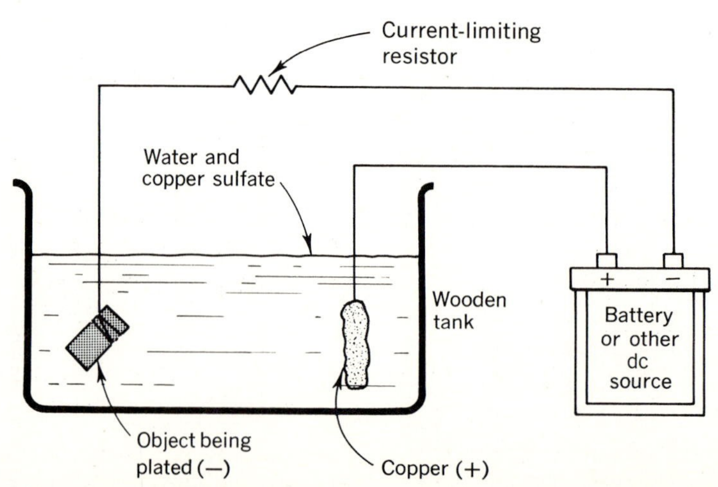

Fig. 24-2 An electroplating system for copper-plating an object.

Pure copper can be produced by using impure copper ingots to plate a negative electrode. Impurities (silver, etc.) drop to the tank bottom and may be recovered.

24-3 THE DEVELOPMENT OF THE CELL

If the ancient Persians were able to plate metal objects, they had an electrical source. What electrolyte they used is not known, but electric cells made by them consist of a copper electrode at the bottom of a jar and an iron electrode protruding down through an insulating stopper. By itself, copper in an electrolyte develops an oxidation-reduction potential of −0.337 V. Iron develops +0.44 V. The difference between these two metals in the same electrolyte represents about 0.777 V. However, the electrolyte helps determine the actual voltage developed between the two electrodes. This is the basic idea of a chemical electric cell—two *dissimilar* metal electrodes in some kind of an electrolyte. The cell produces an emf until the chemical action is complete.

In 1800 an Italian scientist, Volta, developed a cell using silver and zinc electrodes with a salt solution as the electrolyte. In 1866 a French scientist, Leclanché, developed the 1.5-V *dry-cell* that is still in use today. The electrodes are zinc for the negative pole and manganese dioxide for the positive pole, with an electrolyte of moist ammonium and zinc chloride salts.

Rechargeable "wet-type" automotive batteries were developed in the late 1800s.

In 1947 the mercury cell, another so-called dry-cell, was developed.

The usual method of categorizing the different types of cells is to list them as primary, secondary, and other types. Primary cells cannot be recharged. Secondary cells can be recharged. Other cells, such as light-sensitive, photoelectric, solar, nuclear, and fuel, do not fit in either of these categories.

24-4 THE LECLANCHÉ CELL

The most familiar dry-cell is called a *carbon-zinc* cell, although the carbon is only a means of making connection to the damp powdered manganese dioxide which serves as the positive electrode.

A cross section of a dry-cell is shown in Fig. 24-3*a*. The negative terminal and

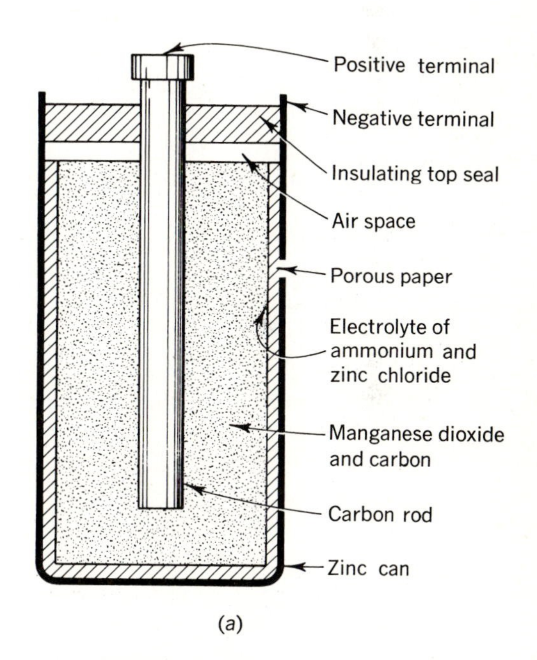

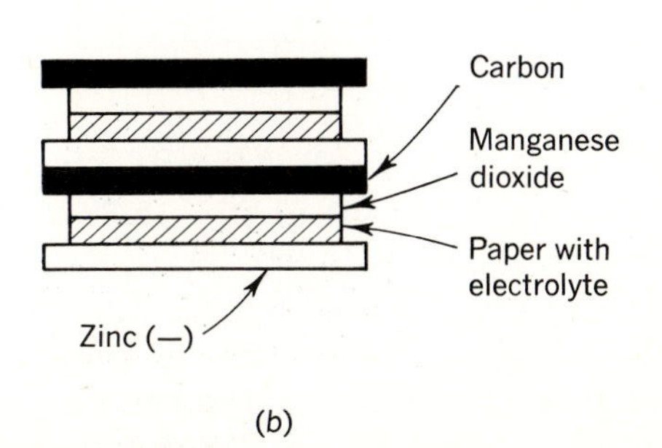

Fig. 24-3 (*a*) Leclanché dry-cell. (*b*) Layer-constructed cells in series form a battery.

electrode is a thin zinc container. Inside the zinc can is a porous paper saturated with electrolyte (solution of ammonium and zinc chlorides). The positive electrode is a damp paste of the electrolyte and manganese dioxide mixed with powdered carbon for better conductivity. The carbon rod contacts the manganese dioxide and serves as the positive terminal. Above the active materials is a small air space and an insulating air seal.

The oxidation-reduction potentials of the electrodes and electrolyte produce a no-load emf of 1.505 V. If a cell is discharged at a very slow rate, the voltage drops slowly until it reaches about 0.9 V. At this point, the internal resistance increases so much the cell is considered discharged.

When zinc-carbon cells are discharged at a rapid rate, the voltage drops below 0.9 V rapidly due to an accumulation of gas at the electrodes. The cell is said to be *polarized*. After a rest period the gas combines chemically with the manganese dioxide, and the cell becomes "depolarized" and will operate almost normally again.

Eventually the chemicals of any cell will combine even if the cell is not used. How long a cell can be stored before it begins to deteriorate is its *shelf life*. Shelf life can be increased by storing in a cool area, in most rooms, either on or near the floor. Although dry-cells may not be damaged by exposure to low temperatures, they will not produce their normal current when cold.

A large no. 6 dry-cell (6.7 × 16.5 cm) has a short-circuit current of about 32 A. The largest flashlight cell, type D, has a short-circuit current of about 6.5 A. The smaller penlite size has about 4.5 A (Fig. 24-4). Short-circuit tests are not recommended.

The state of charge of a dry-cell is normally determined by the output voltage with a light load. As soon as the voltage drops to about 0.9 V, the cell is usually considered dead.

Fig. 24-4 Cells and batteries. Across top: two button-type cells, a button battery, and a 9-V mercury battery. Across bottom: type-D, type-C, type-AA, and type-AAA carbon-zinc and alkaline cells.

If the zinc is replaced by magnesium and magnesium bromide is used as the electrolyte, a *magnesium cell* results. It has an output of about 2 V and can be operated down to −30°C, an improvement over Leclanché cells.

24-5 OTHER PRIMARY CELLS

Many primary cells are available. They usually represent an improvement of one kind or another on the Leclanché cell.

Mercury cells (Fig. 24-5) have a mercuric oxide negative electrode, a zinc positive electrode, and a potassium hydroxide electrolyte. These cells are efficient, have a high ratio of capacity to volume and weight as well as a low internal resistance, and maintain a constant voltage under load. The outer case is the negative terminal, and the top is the positive terminal. They operate at about 1.31 V under load and have a long shelf life. When they reach the end of their useful life, the output voltage drops abruptly. They may explode if short-circuited for a period of time because of generation of internal gas.

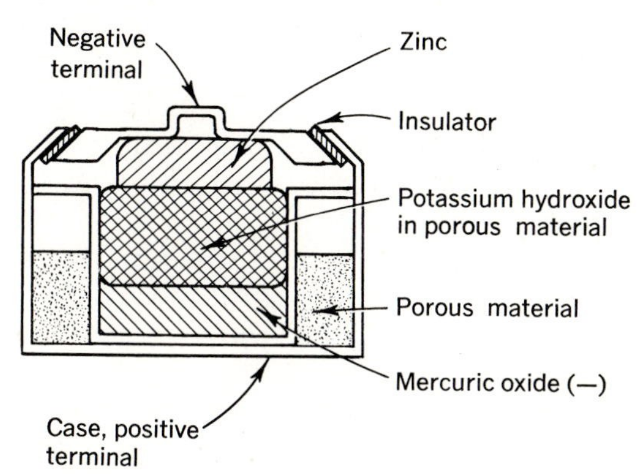

Fig. 24-5 Mercury cell. The case is positive, and the top is negative (reverse of Leclanché).

Manganese-zinc cells have a zinc negative electrode, a manganese dioxide positive electrode, a potassium hydroxide electrolyte, and an output voltage of 1.5 V. They have low internal resistance, relatively high current capabilities, and a long shelf life. They operate well at low temperatures.

Silver-oxide-zinc cells have a silver oxide positive electrode, a zinc negative electrode, and a potassium (or sodium) hydroxide electrolyte. They have 1.55-V output under light loads, are useful as a reference voltage, and have a low internal resistance. These cells are employed in hearing aids, and with a sodium hydroxide (lye) electrolyte, are used in watches that operate continuously for a year or more.

Quiz 24-1. Test your understanding. Answer these check-up questions.

1. How does a primary cell differ from a secondary cell? ________
2. How is the state of charge of a primary cell determined? ________
3. What is an electrolyte? ________
4. What is meant by "polarization" of a dry-cell? ________
5. What precaution should be taken to extend shelf life in storing cells or batteries? ________
6. What produces heat in a cell when it is operating? ________
7. When current flows through an electrolyte between two electrodes and the electrodes give off a gas, what is this action called? ________ What two gases are given off? ________ ________
8. Would an emf be developed between two electrodes in dilute acid if both electrodes were iron? ________ Copper? ________ If one electrode were iron and the other rusted iron (iron oxide)? ________
9. In electroplating, to which source terminal should the object being plated be connected? ________ What should be the composition of the other electrode? ________ What should be the composition of the electrolyte? ________
10. What is the approximate date of the first known electric cell? ________
11. Who developed the common dry-cell of today? ________
12. What is the output voltage of a common dry-cell? ________ A mercury cell? ________ A magnesium cell? ________
13. In a common dry-cell, what is the constituent of the electrolyte? ________ Negative electrode? ________ Positive electrode? ________
14. What are two advantages of a mercury cell over a common dry-cell? ________ ________

24-6 LEAD-ACID CELLS

The secondary, or rechargeable, cell in most common use is the lead-acid type used in automobiles and as emergency power supplies. They have a pure, spongy, porous lead (Pb) negative electrode, a dilute sulfuric acid (H_2SO_4) electrolyte, and a lead dioxide (PbO_2) positive plate.

Lead-acid cells usually have multiple interleaved plates (Fig. 24-6). To hold the plates apart, nonconductor *separators* are used. They may be made of wood, glass, rubber, or plastic. Since a greater negative than positive plate area is required, there is always one more negative plate than positive plate.

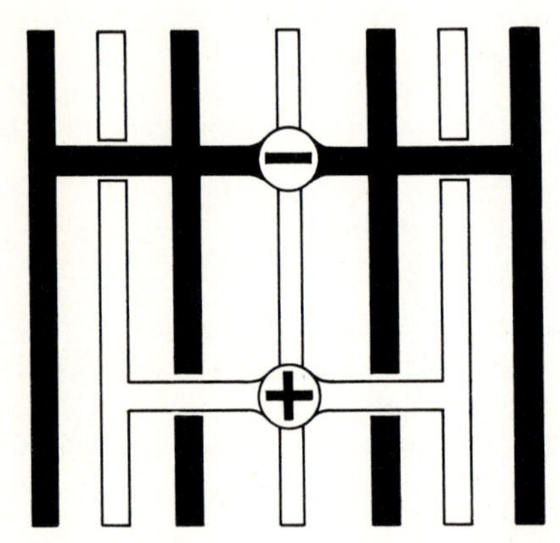

Fig. 24-6 Interleaved plates in a lead-acid cell. Outer plates are always negative. Separators would be between adjacent plates (not shown).

ANSWERS TO CHECK-UP QUIZ 24-1

1. (Not rechargeable) **2.** (Voltmeter while loaded) **3.** (Current-carrying ionized liquid) **4.** (Partial insulation or increased internal resistance by gas formed internally) **5.** (Keep cool) **6.** (Chemical action and I^2R) **7.** (Electrolysis) (Hydrogen and oxygen) **8.** (No) (No) (Yes) **9.** (Negative) (Plating material) (Salt of the plating material) **10.** ($\pm$300 B.C.) **11.** (Leclanché) **12.** (1.505 V) (1.31 V) (2 V) **13.** (Ammonium and zinc chloride) (Zinc) (Manganese dioxide) **14.** (Constant E output for full life, low internal R, high I output, and long shelf life)

The electrodes and electrolyte are enclosed in hard-rubber or glass containers. A screw cap on each cell allows access to the electrolyte, which must be maintained at a depth of about 1 cm above the plates at all times.

When a load is connected across a lead-acid cell, the chemical action for the discharging can be represented as:

$$PbO_2 \rightarrow PbSO_4 \quad (+\text{ plate})$$
$$H_2SO_4 \rightarrow H_2O \quad (\text{electrolyte})$$
$$Pb \rightarrow PbSO_4 \quad (-\text{ plate})$$

where Pb is lead, H_2SO_4 is sulfuric acid, $PbSO_4$ is lead sulfate, and PbO_2 is lead dioxide.

While the cell is discharging, the sulfuric acid chemically combines with the lead dioxide and the pure lead plates, changing them *both* to lead sulfate ($PbSO_4$). When both electrodes are the same, no voltage difference is across the cell, and it is completely dead. The sulfuric acid, having combined with the plates, leaves only water as the electrolyte. As these cells discharge, lead sulfate forms as crystals on all plates, causing them to swell. This can bend or buckle the plates, crush the separators, and even split the cell case.

Lead-acid cells must be recharged as soon as possible after being discharged. The lead sulfate first forms as small, soft crystals on the plates. If allowed to age, the crystals grow and harden. Soft crystals respond to recharging, but hard crystals do not. Cells allowed to remain in a semicharged or discharged state for long periods are said to be *sulfated* and will never recharge completely.

Jarring a discharged cell will loosen crystals from the plates. If such "sludge" builds up on the cell bottom to a point that it touches any two positive and negative plates, the cell becomes shorted and is useless. If a separator cracks and sludge forms in the crack, the cell may also become shorted.

To recharge a lead-acid cell, a source of dc emf greater than the cell emf is required. This forces a current backward through the cell, reversing the chemical action of discharging explained above. The sulfate in both plates is changed back to sulfuric acid in the electrolyte, leaving the positive plate lead dioxide and the negative plate pure lead again, and the cell is fully charged.

Heat is produced in cells by both internal chemical action and the I^2R loss produced by current flowing through the lead-antimony (or lead-calcium) *grids* that hold the active materials of the plates. Adding silver and nickel to grid material decreases resistance, increases current flow, and reduces heat loss. A lead-acid cell may be discharged or charged at any rate that does not produce a temperature of more than 45°C (110°F) in it.

Charging a lead-acid cell drives hydrogen gas off one electrode and oxygen off the other. Mixed together, these gases are highly explosive. Flames and sparks should be kept away from a charging or recently charged battery.

The loss of hydrogen and oxygen while charging means that the electrolyte is losing water (H_2O). This water must be replaced. Only distilled water should be used. Impurities in tap water will chemically combine with the electrolyte or the plates of a cell, resulting in a loss of battery capacity.

By using calcium in the grids instead of the more rugged antimony, less gassing is produced during charging, resulting in batteries that require almost no water replacement.

A new lead-acid cell is a hermetically sealed, leakproof type, comparable to a nicad cell except that it is about half as costly, has longer shelf life, and has no memory effect (Sec. 24-11). It consists of a fiber glass tape coated with an orange (+) lead paste and a green (−) lead paste, which are first formed, then wound into a tight roll, saturated with an acid electrolyte, slipped into a polyethylene case, and finally capped with two terminals extending through the cap. Since they are sealed, they never require water. This is possible by carefully balancing plate materials and electrolyte. Actually, the electrolyte is more "damp" than "wet." The internal resistance is very low, resulting in 3 to 4 times as much short-term current drain per ampere-hour as ordinary lead-acid cells. Charged cells will maintain full charge up to 2 years. They can be charged more than 200 times, and they operate at lower temperatures better than the usual lead-acid cells.

24-7 SPECIFIC GRAVITY

The voltage of a lead-acid cell under load gives an indication of its state of charge, but measurement of the specific gravity of the electrolyte is more accurate. *Specific gravity* (sp gr) of a liquid is a comparison of its weight with water. Water has a specific gravity of 1.0. Chemically pure sulfuric acid is 1.835 (it weighs 1.835 times as much as water). The specific gravity of the electrolyte in a lead-acid cell depends on how much acid is added to it. Cells used in high-current service, as in starting engines, will have 1.275 to 1.300 sp gr values. As a lower-current auxiliary power source, a cell with a 1.220 sp gr produces a longer "shelf life" but higher internal resistance and lower current-drain values. The voltage of a lead-acid cell varies somewhat with the specific gravity (Table 24-1).

Table 24-1

Specific gravity	Voltage output
1.300	2.2
1.280	2.1
1.220	2.05

A 1.300 sp gr is referred to as "thirteen hundred." In most cases, a cell should be recharged as soon as its specific gravity drops 100 "points." A 1.280-sp gr battery should be recharged when its electrolyte reaches 1.180.

When a battery is fully charged to 1.280 sp gr, the electrolyte will freeze at about −60°C. At 1.100 the electrolyte will freeze at about −8°C. If a cell freezes, it may split its case. While an electrolyte is cold, it is less active chemically and will produce less current output. (An automobile engine is harder to start on cold mornings.) Because of internal heat developed, even a slowly charging (*trickle*) cell will not freeze.

A *hydrometer* is used to measure the specific gravity of an electrolyte. It consists of a compressible rubber bulb at the top of a glass barrel, with a rubber tube at the bottom. A weighted, calibrated, thin hollow glass float is inside the barrel (Fig. 24-7).

To measure specific gravity, the hydrometer tube is dipped into the electrolyte. Electrolyte is drawn up into the barrel by squeezing and releasing the bulb. The specific gravity is indicated by the highest readable calibration seen on the float when sighting just beneath the surface of the electrolyte in the barrel (1.265 in the illustration). If the electrolyte is warmer than 20°C, a thermometer at the base may indicate how many points to add to the scale reading. If colder, the correction is subtracted.

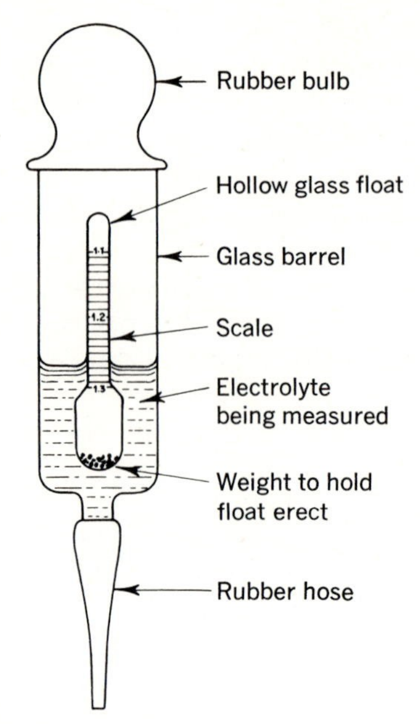

Fig. 24-7 A hydrometer, used to determine the specific gravity of electrolytes in cells.

The acid of the electrolyte is very corrosive. It can be counteracted by applying a weak alkali, such as baking soda or dilute ammonia, to any spilled drops on hands, clothing, etc.

24-8 CAPACITY OF A CELL

The *capacity* of a cell or battery is measured in *ampere-hours* (Ah). The number of ampere-hours produced by discharging a lead-acid cell over an 8-h period to 1.75 V is one standard of measurement used. If the battery is made to discharge faster, the number of ampere-hours produced will be lower. If it is allowed to discharge more slowly, the ampere-hours obtainable will be greater. It would normally be considered that an 80-Ah battery must be recharged after 8 h of an average 10-A discharge. When fully charged, such a battery may be capable of a 700-A discharge for a short period of time.

The average battery capacity for a 6-V

automobile electrical system is about 100 Ah. For a 12-V automotive system a 50-Ah battery will do as much work and at higher efficiency.

A type D dry-cell has a rating of about 5.5 Ah if discharged at a 10-mA rate, but only about 1.75 Ah if discharged at a 50-mA rate.

24-9 CHARGING BATTERIES

Forcing a current backward through a lead-acid battery reverses the chemical action of discharging. This is called *charging*.

Figure 24-8 shows a *constant-voltage* charger. Note the polarities of the battery

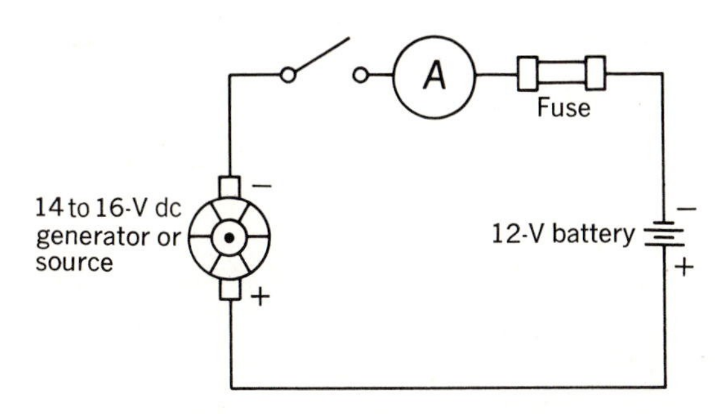

Fig. 24-8 A constant-voltage battery-charger circuit. Current decreases as battery charges.

and charger. The circuit consists of a dc source with a voltage slightly higher than the battery to be charged. To charge a 12-V battery, the source might have 14 to 15 V. Depending on the state of charge of the battery and the circuit resistance, this source emf might produce between 1 and 10 A of charging current. As the battery charges, its emf rises and approaches the source-voltage value, and the charging current, indicated by the ammeter, decreases.

Figure 24-9 is a *constant-current* battery-charging circuit. The charging rate is determined mainly by the value of the

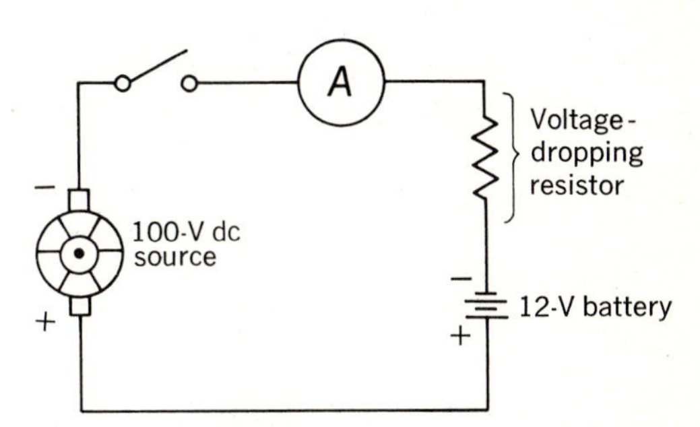

Fig. 24-9 A constant-current battery-charger circuit. There is no noticeable change in charging current as battery charges.

voltage-dropping resistor. If a 1-A charge is desired, the resistance value should be $R = E/I$, or 88/1, or 88 Ω. Since the voltage change in the battery between discharge and full charge may be only a fraction of a volt, the current on the ammeter will appear to be constant.

An electronic battery-charger (Fig. 24-10) consists of a tapped step-down transformer and a diode rectifier. Most chargers now use solid-state diodes, shown dashed. (The filament winding would be unnecessary.) The desired charging current can be regulated with the multitap switch. The circuit shown uses a halfwave rectifier, but four rectifiers in a bridge circuit are also quite common.

Charging rates up to half the ampere-hour rating (50 A for a 100-Ah battery) can be used for a short time if a quick charge is required. As soon as the cell begins to warm, or bubble noticeably, the charge must be reduced to a safe low rate about one-tenth the ampere-hour rating. A very low rate is called a *trickle charge*, usually one-hundredth of the ampere-hour rating. A long low charge is better for a battery than a quick, high charge. With fast charging, gas liberated at the plate surfaces bubbles up and dislodges sulfate crystals from the plates, increasing the buildup of sludge in the bottom of the cell.

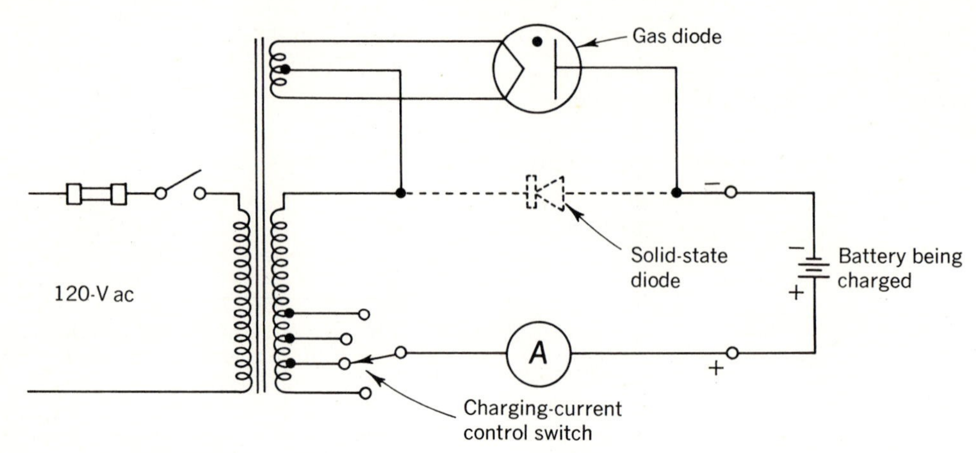

Fig. 24-10 A gaseous-diode (or solid-state-diode) battery charger.

24-10 MAINTAINING LEAD-ACID CELLS

There are several important points about battery maintenance.

If one cell of a battery shorts out and discharges itself, this cell may then act only as a resistor in the circuit. When the battery is under load, the voltage-drop across this cell will be reversed from its normal polarity (Fig. 24-11). Check current

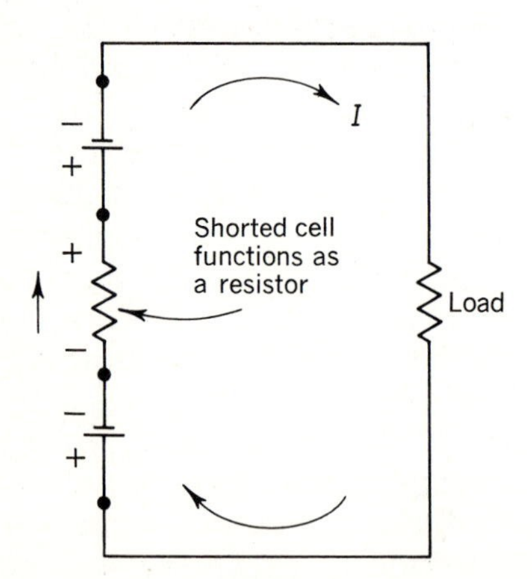

Fig. 24-11 A shorted or partially shorted cell shows either reversed polarity under load or abnormally low voltage.

direction through the shorted cell and the voltage-drop polarities across the cell. If a cell is only partially shorted, the voltage-drop across it may be either very low or reversed. A heavy copper bar held across the terminals of a battery with a bad cell will produce violent gassing (electrolysis) in the bad cell.

The following list outlines basic battery maintenance:

- Keep flames and sparks away from a charging battery.
- Avoid spilling acid drops when using a hydrometer.
- Keep cell tops clean and moisture-free.
- Keep terminals coated with petroleum jelly.
- Maintain the proper electrolyte level.
- Use only distilled water in cells.
- Take hydrometer readings at least once a week.
- Trickle-charge unused cells for at least one full day a month.
  Always bring batteries up to full charge.
- Provide adequate ventilation while charging.
- To avoid deterioration of grids, do not overcharge.
- Keep batteries to be stored for long periods under refrigeration (not frozen).
- To prevent acid from flowing over caps, never turn caps over when removing

them. (Place caps on a piece of clean paper.)

- Make sure vent holes are open in all cell caps.
- Keep cell caps on but loose while charging.

Quiz 24-2. Test your understanding. Answer these check-up questions.

1. What is the advantage of using secondary rather than primary cells in batteries? ___________
2. What is the composition of the negative plate of a lead-acid cell? ___________ The positive plate? ___________ The electrolyte? ___________
3. What is the specific gravity of a fully charged automotive battery? ___________ At what reading would the cells be considered discharged? ___________
4. Besides negative plates, positive plates, and electrolyte, what else is found in lead-acid cells? ___________
5. Which plate of a lead-acid cell requires the greater plate area? ___________
6. Which can be more easily converted back to their original form, small or large sulfate crystals? ___________ What produces large crystals on cell plates? ___________
7. Which requires the greater source voltage, a constant-voltage- or a constant-current-type battery charger? ___________
8. What is the chemical reaction that occurs on the positive plate when a lead-acid cell is charged? ___________
9. When cells are charged, they give off hydrogen and oxygen gas. How are these molecules returned to the electrolyte? ___________
10. What is the composition of the grids in a lead-acid cell? ___________ What can be added to the grids to increase the current output of a cell? ___________
11. Why will a battery produce less current on a cold day than on a hot day? ___________
12. What can be used to counteract the corrosive effect of electrolyte spilled on skin or clothing? ___________
13. What is the name of the device used to test the specific gravity of a liquid? ___________ Would its float ride higher if the battery were charged or discharged? ___________
14. What is the approximate ampere-hour rating of batteries for 12-V automobiles? ___________ 6-V automobiles? ___________
15. Is the electronic battery charger shown in Fig. 24-9 a constant-voltage or constant-current type? ___________
16. What is a high charging rate for a lead-acid cell? ___________ A low rate? ___________ A trickle rate? ___________
17. How can you tell which cell of a battery is dead or shorted if the battery is connected to a normal load? ___________
18. On a separate piece of paper, list as many points as you can on the maintenance of lead-acid cells.
19. Draw a diagram of a constant-voltage battery-charging circuit. An electronic battery-charging circuit. Check against the diagrams in the text.

24-11 ALKALINE SECONDARY CELLS

The *Edison cell*, also called nickel-iron-alkaline, is a lighter, more rugged secondary cell than the lead-acid. When charged, it has a positive plate of nickel flakes and nickel hydrate in small perforated nickel-plated steel tubes. It has a negative plate with iron as the active material in small pockets in a nickel-plated steel plate. The electrolyte is potassium hydroxide, lithium hydrate, and distilled water.

While discharging, the electrolyte transfers oxygen from the nickel oxide + plate to the iron − plate, producing iron oxide (rust). When the cell recharges, the iron oxide is reduced to iron and the oxygen appears in the positive plate as a higher oxide of nickel.

A hydrometer is not used with Edison cells, as the electrolyte specific gravity does not change between charge and discharge. The voltage of the cell measured under load, or an Ah-meter, gives the best indications of the state of charge.

The Edison cell has 1.4 V fully charged with no load. With a load the voltage decreases to about 1.3 V. As discharge progresses, the voltage drops off linearly to about 1.1 V. At 0.9 V, the cell should be recharged. Replacement water should be distilled only.

Having about 3 times the internal resistance of a lead-acid cell, the Edison cell produces less current under load, and has considerably greater physical volume per Ah than an equally rated lead-acid cell.

Excessively high discharge rates or high temperatures will cause loss of capacity of Edison cells. If the battery is trickle-charged, it will never reach full charge.

To prevent absorption of carbon dioxide by the electrolyte, Edison cells have capped vents to exclude outside air from the cell.

Edison cells are quite rugged. They can be completely discharged and left in this condition for long periods with no damage.

The alkaline or base solution used in all alkaline batteries is corrosive. The corrosive effect can be counteracted by using vinegar or any other dilute acidic solution.

Nickel-cadmium-alkaline cells are similar to Edison cells in some respects. An advantage is lower internal resistance. They have a positive plate of nickel oxide, a negative plate of cadmium and iron, and an electrolyte of potassium hydroxide. Such cells have a no-load voltage of 1.34 V and a loaded voltage of about 1.28 to 1.2 V. When the voltage drops below 1.19 V, it should be recharged.

Some nickel-cadmium-alkaline ("nicad") cells are hermetically sealed systems. They have about 1.25 V and low internal resistance. The voltage regulation is good until nearly discharged. These cells should be recharged with 1.4 times as many ampere-hours as were removed during the discharge. Being hermetically sealed, they never need water. They are sometimes called "rechargeable drycells." They have a "memory" effect. If discharged only partially but several times, they may lose their full capacity to discharge.

Rechargeable alkaline cells are hermetically sealed units similar to Leclanché primary cells; they have a zinc negative plate, a manganese dioxide positive plate, but a potassium hydroxide electrolyte. They are relatively inexpensive, can be recharged many times, and have a nominal 1.5-V unloaded rating. They may be overcharged without damage, should not be discharged below 0.9 V terminal voltage, and should be charged from a constant-current charger.

ANSWERS TO CHECK-UP QUIZ 24-2

1. (Can be recharged) **2.** (Lead) (Lead dioxide) (Dilute sulfuric acid) **3.** (1.280) (1.180) **4.** (Separators) **5.** (Negative) **6.** (Small) (Time) **7.** (Constant-current) **8.** ($PbSO_4 \rightarrow Pb + SO_4$, or, more fully, $PbSO_4 + 2H_2O \rightarrow PbO_2 + 4H^{++} + SO_4^{--} + 2e$) **9.** (By adding water) **10.** (Lead-antimony or lead-calcium) (Silver or nickel) **11.** (Slower chemical reactions) **12.** (Dilute ammonia, baking soda) **13.** (Hydrometer) (Charged) **14.** (50 Ah) (100 Ah) **15.** (Voltage) **16.** (0.5 Ah) (0.1 Ah) (0.01 Ah) **17.** (V across cell reads low or reversed) **18.** (See Sec. 24-10)

24-12 NONCHEMICAL CELLS

There are a number of cells not operating on chemical reactions. Many years ago it was found that when light struck the junction between strips of selenium (a semiconductor) and iron, an emf of about 0.4 V developed across the junction (Fig. 24-12). This is the basis on which *photovoltaic* or *selenium cells* operate. When photons (light energy) strike selenium, *electron-hole* (− and +) *pairs* are developed at the junction, resulting in a drift of electrons to the iron. Such a se-

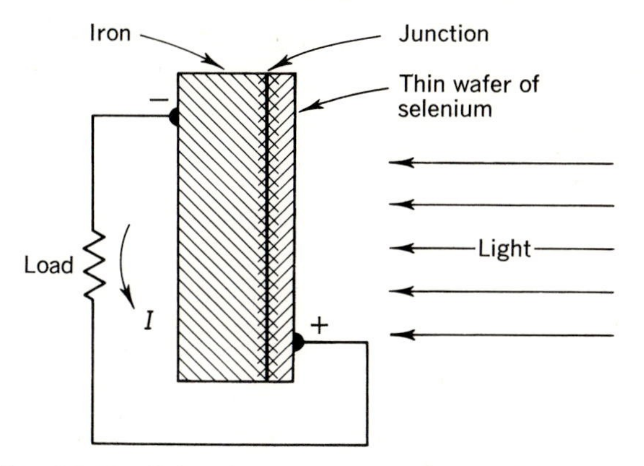

Fig. 24-12 Selenium or photovoltaic cell. The selenium wafer is so thin that light energy can penetrate to the junction.

lenium cell may be used in a photographic exposure meter. Many cells in series form a battery that delivers power as long as they are illuminated, as from the sun. Their efficiency is only about 1%.

One of the more practical *solar batteries* consists of special silicon cells in series. Each cell is made of two wafer-thin silicon strips, with the face of each infused with an opposite-polarity impurity. When the strips are pressed together, a *PN* junction is formed. When illuminated by sunlight, these junctions produce an emf at an efficiency of nearly 10%. A square meter of these solar cells makes a battery that can produce more than 50 W.

Atomic, or *nuclear*, *cells* are somewhat similar to solar cells. A *PN* junction is bombarded by beta particles (electrons) from radioactive material applied to one side of the semiconductor wafers. Strontium-90 gives off an almost constant stream of high-velocity beta particles. In 20 years the intensity of the beta emission decreases by only half. These cells produce about 0.3 V.

Photoemission can also be used to produce a *photocell*. When photons strike some atoms in a vacuum (cesium, sodium, strontium, barium), outer-ring electrons of these atoms may absorb enough photon energy to drive them out of orbit. If the escaping electrons hit wires of a nearby metal screen, the screen becomes negatively charged and the photoemissive metal that lost the electrons becomes positive. The voltages and currents thus developed are rather feeble.

Current flowing through an electrolyte liberates hydrogen gas at the negative electrode and oxygen at the positive electrode (electrolysis). A reverse chemical process is possible. Hydrogen and oxygen gases can be made to combine, producing a current of electrons and water. This is the basis of operation of the *fuel cell*.

A basic fuel-cell system is illustrated in Fig. 24-13. Hydrogen gas under pressure

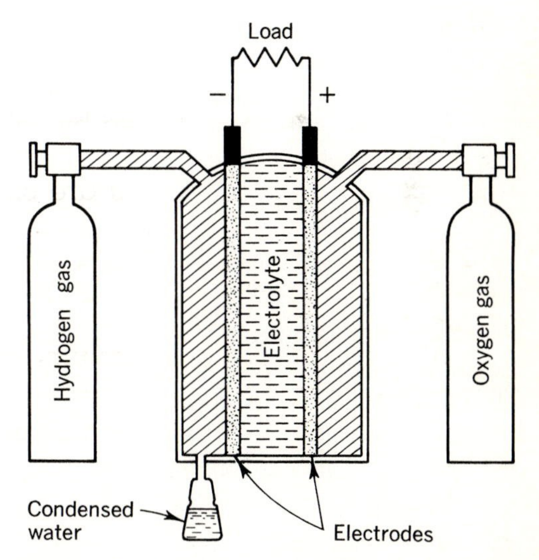

Fig. 24-13 Fuel cell, using hydrogen and oxygen gases as fuels.

is forced into the left chamber, which has a porous conductive electrode partitioning off the right side of its area. On the other side of this negative-pole electrode is a liquid electrolyte of potassium hydroxide. Oxygen gas under pressure is forced into the right chamber, which also has a porous conductive electrode wall between the oxygen and the electrolyte.

The potassium hydroxide electrolyte (KOH) in a water solution breaks up into positive potassium ions, K^+ (lacking an electron), and negative hydroxyl ions, OH^- (one excess electron). The ions filter into the porous electrodes. Because of the platinum or palladium catalysts embedded in the electrodes, wherever the hydrogen gas contacts the surface of its electrode, it combines with the negative hydroxyl ion to form water, H_2O, plus one free electron (−).

Where oxygen contacts its electrode it combines with water and electrons returning from the load to form hydroxyl ions, OH−, which travel into the electrolyte and over to the negative electrode. This completes the electrical system. Current continues to flow as long as H and O are fed as fuel to the cell. The cell produces approximately 1000 A/m^2 of electrode area (100 A/ft^2) at about 0.85 V, with efficiencies of about 75% (approximately 1 V at about one-third maximum load).

A residue of pure condensed water is produced in the hydrogen system. Increasing the temperature of these cells increases their current output. At about 250°C, cells must be maintained under pressure but will produce about 6 times more current. Other fuels, such as alcohol, gasoline, methane, kerosene, ammonia, and hydrazine, are also used but must be operated at high temperatures.

Quiz 24-3. Test your understanding. Answer these check-up questions.

1. What is another name for the Edison cell? ________
2. What is used to measure the state of charge of an Edison cell? ________
3. What is the type of cell that is very similar to the Edison, except that its internal resistance is considerably lower? ________ What is the advantage of lower internal resistance in a cell? ________
4. What is the name of a hermetically sealed cell that can be recharged but never needs replenishing with water? ________
5. What is the general term for cells that convert light energy into electric emf or current at a semiconductor junction? ________
6. What is the general term for cells that give off electrons when struck by photons? ________
7. Which solar cell discussed in the text has the highest percentage of efficiency? ________ How much power is obtainable from one square meter of these cells? ________
8. What are the two constituents of a selenium cell? ________ ________
9. What are beta particles? ________
10. What is another name for atomic cells? ________
11. What is the approximate voltage of all *PN*-junction cells? ________
12. What are the fuels used in the particular fuel cell described in the text? ________ ________ What carries the electrons across the electrolyte? ________ What is the approximate cell voltage under a light load? ________ A heavier load? ________

CHAPTER 24 TEST • BATTERIES

1. What is said to take place when an acid is dropped into water and the molecules split into − and + particles?
2. What is required of two electrodes in an electrolyte to produce an emf between them?
3. Toward which electrode will a copper ion travel in a copper electroplating bath?
4. What were the electrode materials of the first known electric cells?
5. What are the electrode materials of the Le-

clanché cell? What is the electrolyte? The output voltage?

6. What device is used to determine the state of charge of a Leclanché cell? A mercury cell? A lead-acid cell? An Edison cell?
7. What is the basic chemical difference between a Leclanché cell and a manganese-zinc cell?
8. What type of cell has been known to explode if it is short-circuited for a period of time?
9. In lead-acid cells, what is the chemical constituent of the positive plate? Negative plate? Electrolyte?
10. If a lead-acid cell is completely discharged, what would be the materials of the positive plate? Negative plate? Electrolyte?
11. What percentage of the ampere-hour rating of a battery should be used for a quick charge? A trickle charge? A low charge?
12. What happens to a lead-acid cell if it is left in a discharged condition for a long period of time?
13. Why is a protracted high charge rate undesirable?
14. In what way is charging lead-acid or Edison cells dangerous?
15. What is the full-charge specific gravity of an automotive battery? An auxiliary power-supply battery?
16. What are two substances that will counteract the corrosive action of the electrolyte of a lead-acid cell? An Edison cell?
17. In what unit is the capacity of a battery measured? Is there a meter by which this can be measured?
18. Why will a 50-Ah battery do as much in a 12-V system as a 100-Ah battery will do in a 6-V system?
19. What does a constant-current charger have that a constant-voltage charger does not?
20. What type of diode is used in most modern battery chargers?
21. Why should a lead-acid cell not be overcharged?
22. What is important to remember when storing cells? Why?
23. What is the approximate fully charged voltage of an Edison cell? A lead-acid cell? A Leclanché cell?
24. What are the names of two types of rechargeable cells besides the lead-acid and the Edison?
25. What were the components of the first of the photovoltaic cells?
26. What is a solar cell made of?
27. How does the atomic cell discussed in the text differ from a solar cell?
28. What is the approximate current output, at room temperature, of a fuel cell per square meter of electrode area? What are the two fuels used in the cell discussed?

ANSWERS TO CHECK-UP QUIZ 24-3

1. (Nickel-iron-alkaline) **2.** (Voltmeter) **3.** (Nickel-cadmium) (Higher current output) **4.** (Nickel-cadmium-alkaline) **5.** (Photovoltaic) **6.** (Photoemissive) **7.** (Silicon with impurities) (50 W) **8.** (Selenium and iron) **9.** (High-velocity electrons) **10.** (Nuclear) **11.** (0.2 to 0.4 V usually) **12.** (Hydrogen and oxygen) (Hydroxyl ions) (1.1 to 1 V) (0.85 to 0.6 V at 4000 A/m^2)

25

KIRCHHOFF LAWS IN CIRCUITS

CHAPTER OBJECTIVE. To become more familiar with the two Kirchhoff laws and to learn to apply them in circuits which cannot be solved by Ohm's laws alone.

25-1 THE KIRCHHOFF LAWS

The Kirchhoff voltage law was used in Chap. 4 to solve simple series circuits. It states:

The sum of the voltage-drops around a series circuit equals the source voltage.

In Chap. 6 the Kirchhoff current law was explained in relation to the solving of parallel circuits. This law states:

The sum of the currents entering a point in a circuit will equal the sum of the currents leaving that point.

Each of these laws in itself expresses a simple idea. In solving series, parallel, or series-parallel circuits by Ohm's law, the Kirchhoff laws are automatically applied. When more complex circuits cannot be solved by Ohm's law alone, other methods must be used. One method of solving an electric circuit involving two sources is first to apply the Kirchhoff *voltage law* to two or more possible "loops" or current paths. Then the currents that must be flowing into and out of some common point in the circuit are computed, which is an application of the Kirchhoff *current law*.

25-2 APPLYING DIRECTION TO CURRENT AND VOLTAGE

When current flows in a circuit, it flows in one of two directions determined by the polarity of the source emf. It cannot be said with certainty that either direction is a "positive" or a "negative" direction. For convenience, however, it is possible to choose one direction and call it positive, or +. If the current reverses, the new current direction is then the negative, or −, direction.

Another possibility is to view a current approaching a point in a circuit and label this current +. A current leaving the same point could then be given a − label.

The circuit in Fig. 25-1 is a simple series circuit across a source. One meter indicates the source-voltage value E_S. The other meters indicate the *voltage-drops*, or *IR-drops*, across the resistors. The source emf is not the same thing as the difference of potential, or voltage-drop, developed across the resistors when current flows through them. E can be used to represent emf from a source. V or the letters IR can be used to represent a voltage-drop across a circuit component.

To express the Kirchhoff voltage law for the circuit of Fig. 25-1, it is possible to write

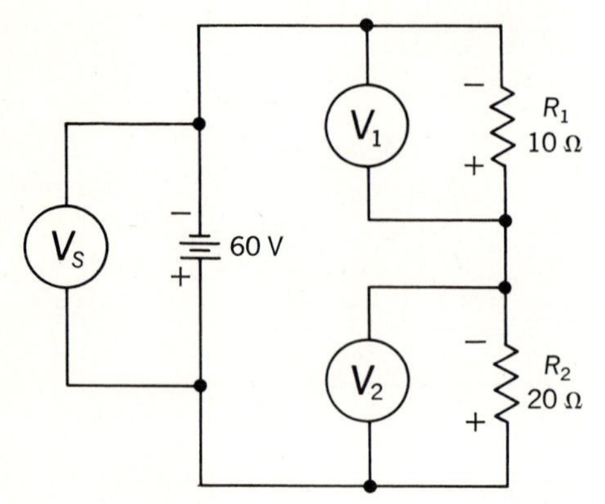

Fig. 25-1 Simple series circuits with voltmeters across the source and across the voltage-drops.

either

$$E_S = V_1 + V_2 \qquad \text{or} \qquad E_S - V_1 - V_2 = 0$$

Note that the Kirchhoff voltage law from the second equation might also be expressed as:

The algebraic sum of all the voltage-drops around a circuit plus the source voltage equals zero.

(Algebraic sum means the sum of the + and − values.) Applying the signs above, with $-IR$ in place of voltage-drops, we have

$$E_S + (-IR_1) + (-IR_2) = 0$$

or, without the parentheses,

$$E_S - IR_1 - IR_2 = 0$$

It might also be mentioned at this time that the Kirchhoff's current law can be stated:

The algebraic sum of all currents into and out of any point in a circuit will be zero.

If I_1 and I_2 are approaching a point, and I_3 is the current leaving the point, then

$$I_1 + I_2 - I_3 = 0$$

If Fig. 25-1 is considered as one loop, it would be possible to express the voltage-drops and the source voltage as

$$E_S + (-IR_1) + (-IR_2) = 0$$
$$60 + (-10I) + (-20I) = 0$$
$$60 - 10I - 20I = 0$$
$$60 = 30I$$
$$\frac{60}{30} = I$$
$$I = 2\text{ A}$$

The same method can be used to solve a series circuit containing more than one source of voltage, even if the sources have opposite directions of pressure, as in Fig. 25-2. In this case the source with

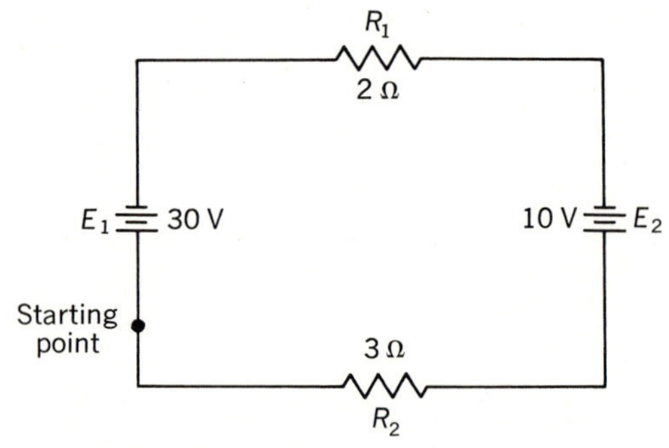

Fig. 25-2 A series circuit with opposing-voltage sources.

the greater pressure can be assigned a + value. The opposing source emf is then given a − value, as are all the voltage-drops in the circuit. From this logic, starting with the greater voltage source and working around the circuit clockwise,

$$E_1 + (-IR_1) + (-E_2) + (-IR_2) = 0$$

Solve for the I value, inserting the values shown in Fig. 25-2. The correct answer is 4 A.

If, by mistake, the smaller source emf is assigned the + value, the answer would be −4 A. The negative answer indicates that the original assumption of current *direction* was false and that the current actually flows in the opposite direction.

All this appears to be an involved method for computing circuit values when it is simple to see that 30 − 10 V equals 20 V, and 2 + 3 Ω equals 5 Ω, with a resulting current of 4 A. However, this method of assigning and working with + and − values is important in the circuits that follow and must be understood.

Quiz 25-1. Test your understanding. Answer these check-up questions.

1. Write out the Kirchhoff voltage law expression for the circuit in Fig. 25-2 if the voltages were doubled and the resistance values were 5 times as great. ________ What would be the current value in the circuit? ________
2. Write out the Kirchhoff *current law* expression for the circuit in Fig. 25-3 in terms of I_1, I_2, and I_3 as seen by point P. ________ By point Q. ________

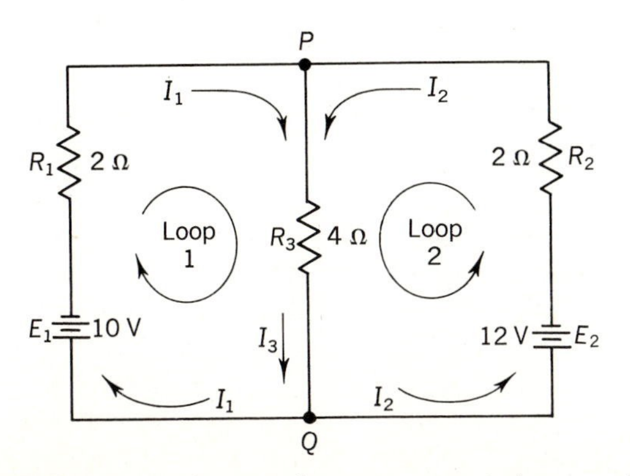

Fig. 25-3 A two-loop multisource circuit that can be solved by using the Kirchhoff laws.

25-3 PARALLEL SOURCES WITH A COMMON LOAD

To solve the currents and voltage-drops of all the components in Fig. 25-3, use the Kirchhoff voltage law, first working out the E_1 loop, then the E_2 loop, and then apply the Kirchhoff current law formula to point P (or Q).

As a first step, the voltages and voltage-drops around the left-hand loop can be added algebraically. Starting with the E_1 source and working clockwise, first comes the battery voltage (a + value), then the IR-drop across R_1 (a − value), and then the IR-drop across R_3 (a − value). The voltage and IR-drops can be expressed as

$$E_1 + (-I_1R_1) + (-I_3R_3) = 0$$

Substituting known voltage and resistance values

$$\begin{aligned} 10 + (-I_1 2) + (-I_3 4) &= 0 \\ 10 - 2I_1 - 4I_3 &= 0 \\ 10 - 4I_3 &= 2I_1 \\ 5 - 2I_3 &= I_1 \end{aligned}$$

This answer is not a numerical value of the current in loop 1, but it expresses what I_1 equals in terms of I_3. This is an expression for the current approaching point P from the source E_1 (or leaving point Q toward E_1).

However, this is only part of the current flowing through R_3. The other component of current is from E_2, the loop 2 current.

Solving for an expression of I_2 (the loop 2 current) is accomplished as with loop 1, starting with the E_2 source and working counterclockwise around the loop. Thus

$$\begin{aligned} E_2 + (-I_2R_2) + (-I_3R_3) &= 0 \\ 12 - 2I_2 - 4I_3 &= 0 \\ 12 - 4I_3 &= 2I_2 \\ 6 - 2I_3 &= I_2 \end{aligned}$$

Point P sees two currents approaching it: the first loop current, $5 - 2I_3$, and the second loop current, $6 - 2I_3$. Since both current expressions contain the common term I_3, it is possible to add the currents approaching point P (Kirchhoff's current law).

$$\begin{aligned} I_1 + I_2 &= I_3 \\ 5 - 2I_3 + 6 - 2I_3 &= I_3 \\ 5 + 6 &= I_3 + 2I_3 + 2I_3 \\ 11 &= 5I_3 \\ I_3 &= 2.2 \text{ A} \end{aligned}$$

This last answer is an actual current value in amperes. It is now known that R_3 has 2.2 A flowing through it. By Ohm's law, the voltage-drop across R_3 must be $V = IR$, or 2.2(4), or 8.8 V.

Since it has been determined that the current through R_1 is the expression $5 - 2I_3$, and since I_3 is now known, the current through R_1 will be $5 - 2(2.2)$, or $5 - 4.4$, or 0.6 A.

The voltage-drop across R_1 can now be solved by Ohm's law, $V = IR$, or 0.6(2), or 1.2 V. Once E_{R_3} is known, the voltage-drop across R_1 will be the difference between E_{R_3} and the source E_1, or $10 - 8.8$, or 1.2 V.

The current flow in loop 2 is $6 - 2I_3$, or $6 - 2(2.2)$, or $6 - 4.4$, or 1.6 A. The IR-drop across R_2 is 3.2 V.

Quiz 25-2. Test your understanding. Answer these check-up questions.

1. Use the expressions determined for loop 1 and loop 2 currents and the I_3 current of Fig. 25-3 to work out the loop 1 and loop 2 current values for point Q instead of P. ________ ________
2. In Fig. 25-3, a 10-Ω resistor is substituted for the 4-Ω value of R_3. What is the expression for I_1? ________ I_2? ________ What is the value of I_3? ________ I_1? ________
3. In Fig. 25-3, with 20 V as E_2, what is the current value of I_3? ________ I_1? ________ I_2? ________ What do you now know about the current flow in source E_1? ________

ANSWERS TO CHECK-UP QUIZ 25-1

1. ($60 - 10I - 20 - 15I = 0$) (1.6A) 2. ($I_1 + I_2 - I_3 = 0$, or $I_1 + I_2 = I_3$) ($I_3 - I_1 - I_2 = 0$, or $I_3 = I_1 + I_2$)

25-4 A BATTERY-CHARGING CIRCUIT

The electrical system of an automobile is a load on its battery. To start the engine the current needed from a 12-V battery may exceed 100 A. The ignition system, lights, radio, gages, blower fans, and air conditioning equipment all require current. A 12-V automobile battery has a rating of about 50 Ah. (In theory it will give approximately 50 A for 1 h, or 1 A for 50 h. In actual practice it may not.) Without a charger in operation, an automobile battery will not last long.

One automotive charger system is a dc shunt generator driven by the fan belt. As long as the generator voltage is greater than the battery voltage, the battery will charge. Newer types belt-drive an alternator. Its ac is fed to the battery through diodes in a bridge rectifier circuit (Sec. 20-2) as dc.

The charger in Fig. 25-4 uses a 15-V dc generator with 0.3 Ω of internal resistance.

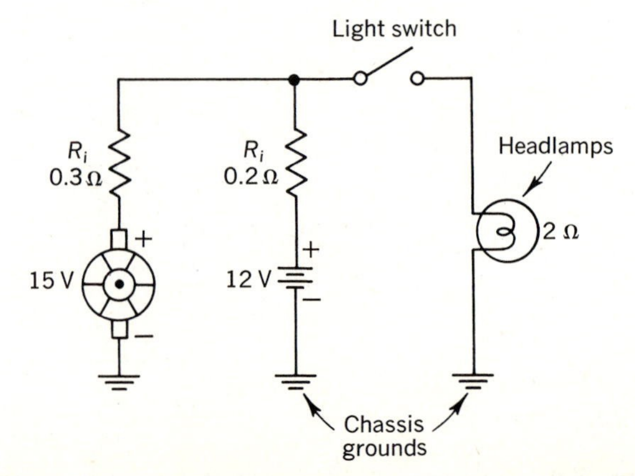

Fig. 25-4 A simplified automotive battery-charging and light circuit.

The battery is 12 V for simplicity (normally 12.6 V). The "grounds" indicate common electrical connections to the chassis of the car. Chassis resistance is considered as 0 Ω. With the engine running and the light switch open, at what rate will the battery be charging? This is a problem involving two bucking voltages and two series resistors, and it can be solved by Ohm's law. Compute the charging rate. *Answer:* ____ A

When the light switch is closed, the headlamps load the battery. Determination of headlamps' current, voltage-drop across them, battery current, whether the battery is charging or discharging into the headlamps, and the generator current is a two-loop Kirchhoff law problem. Loop 1 is the generator and headlamps, and loop 2 is the battery and headlamps.

Check your understanding of two-loop computations explained in Sec. 25-3. Determine the following for Fig. 25-4: (1) headlamp current ____ A; (2) headlamp voltage ____ V; (3) battery current ____ A; (4) is the battery charging or discharging ____; (5) generator current ____ A.

When an automobile stops, the idling speed of the motor is often not fast enough to produce a generator voltage greater than the battery voltage. An "undervoltage" relay will sense this and disconnect the generator from the battery. The battery will run the electrical equipment until the motor speeds up again. Under normal operation it is the generator that operates the electrical equipment, as well as charging the battery so that the battery can start the car the next time. In the problems above, with the switch open the battery is charging at 6 A. With the switch closed the headlamp current is 6.23 A, the headlamp voltage is 12.46 V, the battery current is 2.3 A charging, and the generator current is 8.5 A.

25-5 THREE-WIRE DISTRIBUTION SYSTEM

A common distribution system for single-phase ac power is the three-wire circuit shown in Fig. 25-5. The circuit is shown as an ac circuit, but it can be solved as though it were a dc system by

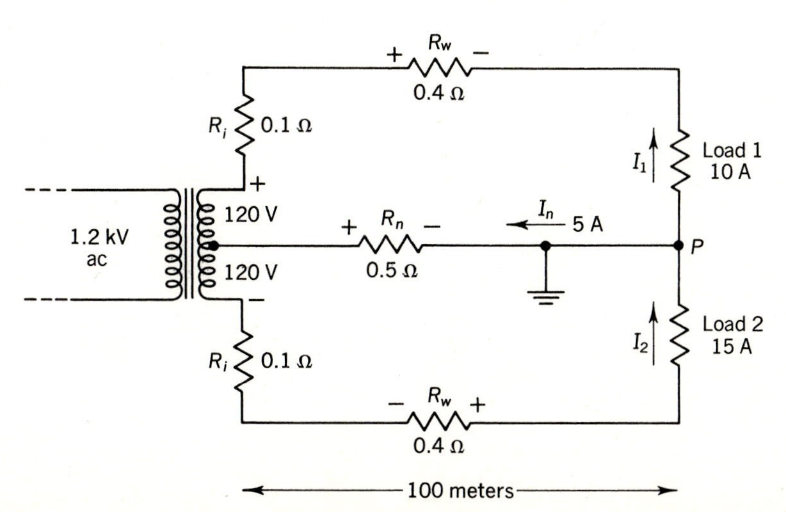

Fig. 25-5 A three-wire system used for electrical power distribution, with two 120-V lines and a common grounded neutral. The loads are 100 m from the transformer.

using the ac rms values. In the illustration, the time chosen is when the top of the secondary is positive and the bottom is negative. The internal resistance of each half of the secondary, R_i, is shown as 0.1 Ω. The transmission-line wires, R_w at top and bottom, have 0.4 Ω and the neutral wire, R_n, 0.5 Ω.

The center wire is the "neutral." If both loads were drawing 10 A, the neutral wire would be carrying zero current. In this circuit, load 1 is indicated as drawing 10 A, and load 2 is drawing 15 A. According to the Kirchhoff current law, point *P* sees 15 A approaching from load 2 and 10 A leaving through load 1. By the current law

$$\begin{aligned} I_2 - I_1 &= I_n \\ 15 - 10 &= I_n \\ I_n &= 5 \text{ A} \end{aligned}$$

The voltage-drop developed in the neutral wire is $V = IR$, or 5(0.5), or 2.5 V. Since 15 A is approaching point *P* from load 2 and 10 A is leaving toward load 1, the 5 A in the neutral must also be leaving point *P*. To determine the polarity developed across a resistor, remember that current flows from − to + through a load. Applying this to the load 2 loop, R_w is negative at its left, load 2 is negative at the bottom, R_n is negative at the left, and R_i is negative at the top. All these *IR*-drops are in the same direction in the loop and are additive. Therefore the voltage-drop across load 2 is 120 − 2.5 − 1.5 − 6, or 110 V.

The resistance of load 2 must be $R = E/I$, or 110/15, or 7.33 Ω.

The voltage-drop across load 1 will be the source voltage of 120 V minus the *IR*-drops across R_w, R_i, and R_n of loop 1, or 120 − 4 − 1 + 2.5, or 117.5 V. Note that the polarity of the voltage-drop in the neutral wire is opposite that of the other voltage-drops in the load 1 loop and must be *added* to the source voltage.

The resistance of load 1 must be $R = E/I$, or 117.5/10, or 11.75 Ω.

In Fig. 25-5 the load *currents* are given. Had the *resistance* of the loads been known (not the currents), the circuit would be solved according to the Kirchhoff loop explanations of Sec. 25-3. The circuit could be redrawn as in Fig. 25-6. Suppose the load resistances are 7 and 12 Ω, as shown. In this circuit the two sources are not bucking each other as in prior Kirchhoff circuits, but are in series. The same method of solution is used.

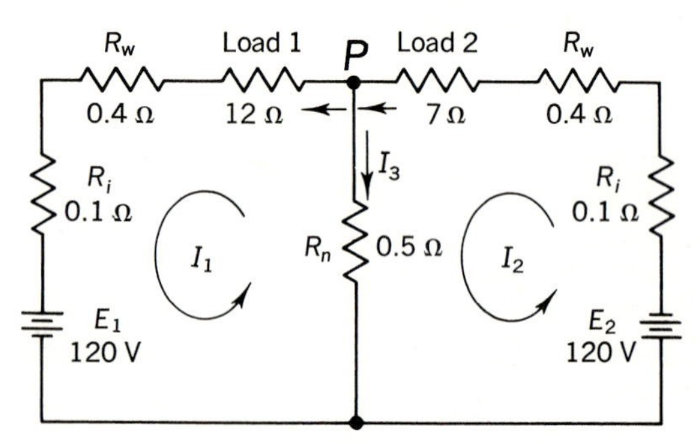

Fig. 25-6 The three-wire distribution circuit when load resistances are given rather than load currents, and drawn in another configuration.

Loop 2, having less resistance (7.5 versus 12.5 Ω), determines the I_3 direction in R_n. With a downward-flowing I_3, the *IR*-drop across R_n will add to the other *IR*-drops in loop 2.

To solve for an I_2 expression, from point *P*,

$$\begin{aligned} (-0.5I_3) + 120 + (-7.5I_2) &= 0 \\ 120 - \quad 0.5I_3 &= 7.5I_2 \\ 16 - 0.0667I_3 &= I_2 \end{aligned}$$

ANSWERS TO CHECK-UP QUIZ 25-2

1. (0.6 A) (1.6 A) **2.** ($5 - 5I_3$) ($6 - 5I_3$) (1 A) (0 A) **3.** (3 A) (−1 A) (4 A) (It is reversed; battery is charging at a rate of 1 A)

For loop 1, starting at P and solving for the I_1 expression, but remembering that the R_n current is reversed as far as E_1 is concerned,

$$\begin{aligned}(-12.5I_1) + 120 + 0.5I_3 &= 0\\ 120 + 0.5I_3 &= 12.5I_1\\ 9.6 + 0.04I_3 &= I_1\end{aligned}$$

Combining known expressions to solve for the actual I_3 current value,

$$\begin{aligned}I_1 \quad + \quad I_3 \quad &= \quad I_2\\ (9.6 + 0.04I_3) + (I_3) &= (16 - 0.0667I_3)\\ 9.6 + 0.04I_3 + I_3 &= 16 - 0.0667I_3\\ I_3 + 0.04I_3 + 0.0667I_3 &= 16 - 9.6\\ 1.1067I_3 &= 6.4\\ I_3 &= 5.78 \text{ A}\end{aligned}$$

Knowing the actual I_3 value to be 5.78 A, the two loop expressions can be solved for the two loop and load current values. For loop 1,

$$\begin{aligned}I_1 &= 9.6 + 0.04I_3\\ &= 9.6 + 0.04(5.78)\\ &= 9.6 + 0.231\\ &= 9.831 \text{ A}\end{aligned}$$

For loop 2,

$$\begin{aligned}I_2 &= 16 - 0.0667I_3\\ &= 16 - 0.0667(5.78)\\ &= 16 - 0.386\\ &= 15.6 \text{ A}\end{aligned}$$

As a check, I_1 should equal the sum of $I_2 + I_3$. Substituting known values, $15.6 = 9.83 + 5.78$, which is 15.61 (accurate to the third significant figure).

Quiz 25-3. Test your understanding. Answer these check-up questions.

1. In Fig. 25-4, if the dc generator is producing 16 V and the headlamps plus the taillamps represents a load of 1.5 Ω, with the switch open what would be the charging current of the battery? ________ With the switch closed what would be the voltage across the lights? ________ What would be the total light-current value? ________
2. In Fig. 25-5, if the loads are 500 m away from the transformer secondary and the load current and source voltages remain the same (10 A, 15 A, and 120 V), what is the voltage-drop value across load 1? ________ Load 2? ________
3. In question 2, if the load resistances remained the same (11.75 and 7.33 Ω) what would be the neutral-wire current value? ________ The load 1 current? ________ The load 2 current? ________

CHAPTER 25 TEST • KIRCHHOFF LAWS IN CIRCUITS

1. If a clockwise-direction current in a circuit is said to be positive, what direction of current would be said to be negative?
2. What label is recommended for a current approaching a point in a circuit?
3. What is determined by $V = IR$?
4. What is determined by $E = IR$?
5. What law is expressed by the algebraic statement $E_S + (-IR_1) + (-IR_2) = 0$? Simplify the expression by removing the parentheses.
6. If a Kirchhoff expression computes to a negative answer, what is the probable reason?
7. Redraw Fig. 25-4, substituting a 6-V battery and an 8-V charging generator. With the switch closed, what is the *expression* for current through the headlamps from the generator? From the battery? What is the actual headlamp-current value?
8. Redraw Fig. 25-5, substituting 5 Ω for load 1 and 10 Ω for load 2. What is the load 1 current value? The load 1 voltage-drop? Load 2 current value? Load 2 voltage-drop?

ANSWERS TO CHECK-UP QUIZ 25-3

1. (8 A) (12.6 V) (8.4 A) **2.** (111.5 V) (76 V)
3. (3.73 A) (8 A) (11.72 A)

26 SUPERPOSITION AND THÉVENIN

CHAPTER OBJECTIVE. To learn to solve more complex resistive circuits for voltages and currents using both the superposition of current and the Thévenin theorem methods.

26-1 SOLVING NETWORKS BY SUPERIMPOSING CURRENTS

The simpler series, parallel, and series-parallel circuits can be solved by Ohm's law alone, but some multisource circuits required Kirchhoff loop solutions. Another method of solving these multisource networks or systems is called *superposition*. It derives its name from the fact that the vector sum of one partial solution of circuit currents is superimposed over another partial solution to obtain the actual resultant currents.

Before applying the superposition-current technique to a complex circuit, a simple series resistor and battery circuit (Fig. 26-1*a*) will be checked. It can be seen that the resultant force is 5 V counterclockwise. With 10-Ω resistance the current in the circuit must be $I = E/R$, or 5/10, or 0.5 A.

The superposition-current method of solving the current in the circuit is first to redraw the circuit with only one of the sources (Fig. 26-1*b*). With 10 V and 10 Ω, there must be 1 A flowing in this redrawn circuit. Vector arrows indicating current-direction flow are added.

Next, redraw the circuit with only the other source (Fig. 26-1*c*). With 15 V and 10 Ω the current is 1.5 A. Current vectors are added as before.

The last step is to superimpose the two sets of vector currents on the original circuit diagram. The actual current values in the legs of the circuit will be the vector

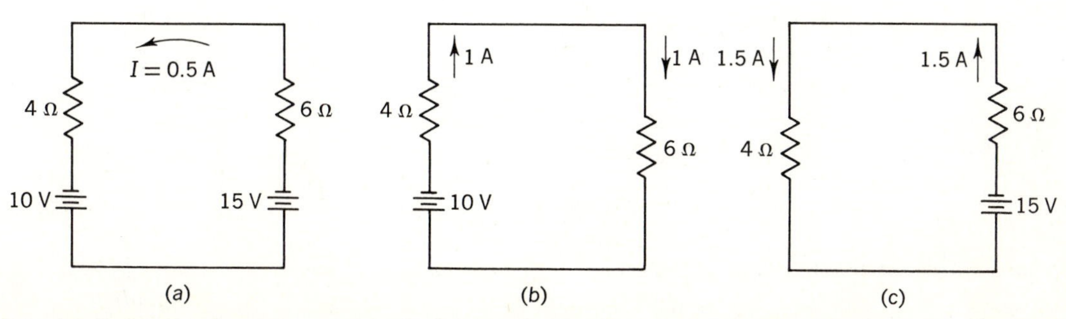

Fig. 26-1 (*a*) Simple multisource circuit. (*b*) First redrawing of superposition currents, leaving out the 15-V source. (c) Second redrawing, leaving out the 10-V source.

sums, indicated by the double-width vector arrows above each leg in Fig. 26-2.

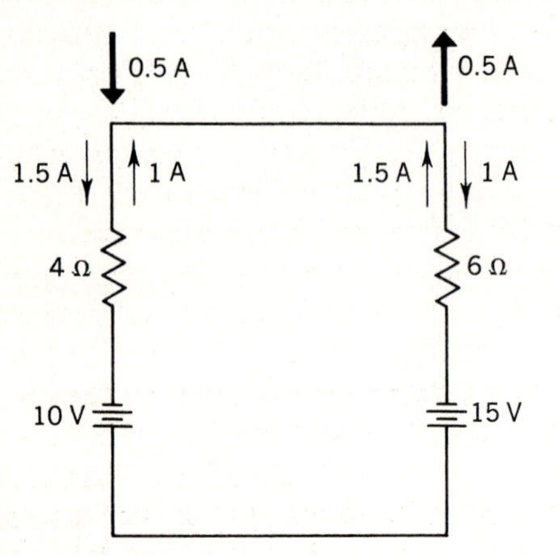

Fig. 26-2 When the two sets of current vectors are added, the algebraic sum is shown above each leg.

This is 0.5 A, as computed previously. Obviously this is an involved method of solving a simple circuit, but the same logic can be applied to other multisource circuits that cannot be worked by any simple method.

26-2 SUPERPOSITION WITH MORE COMPLEX CIRCUITS

The circuit in Fig. 26-3*a* cannot be solved by Ohm's law, but it can be solved by superposition of currents (or Kirchhoff loops).

In Fig. 26-3*b*, one of the sources has been removed. Redrawn in this manner, the components form a series-parallel circuit. By Ohm's law the three currents can be solved and then indicated with vector arrows at the right of the components.

In Fig. 26-3*c*, the 10-V source is the only one used. Again, the series-parallel circuit is solved for its current values, which are added with their vector directions at the left of the components.

In Fig. 26-3*d*, the two sets of vector currents are superimposed and added algebraically. The resultant current value shown by the sum of the two vector values for each resistor will be the actual current value in that leg.

How to compute the Fig. 26-3*a* circuit is illustrated in Fig. 26-4*a*. The parallel resistors have a total of $R_1R_2/(R_1 + R_2)$, or 2(4)/(2 + 4), or 1.33 Ω. The total resistance is 1.33 + 2, or 3.33 Ω. The total current is $I = E/R$, or 12/3.33, or 3.6 A. The voltage-drop across the parallel group is $V = IR$, or 3.6(1.33), or 4.8 V. The current through the 2-Ω resistor in the parallel group is $I = E/R$, or 4.8/2, or 2.4 A. Similarly, the current through the 4-Ω resistor is 4.8/4, or 1.2 A.

Solving for these parallel currents is somewhat tedious. When there are two parallel resistances and the total current

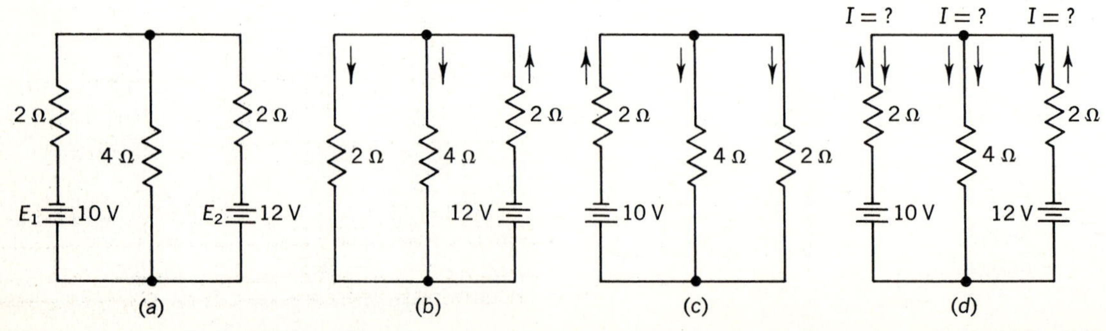

Fig. 26-3 (*a*) A complex circuit. (*b*) Redrawn with one source only. (*c*) Redrawn with other source. (*d*) Superimposing current vectors.

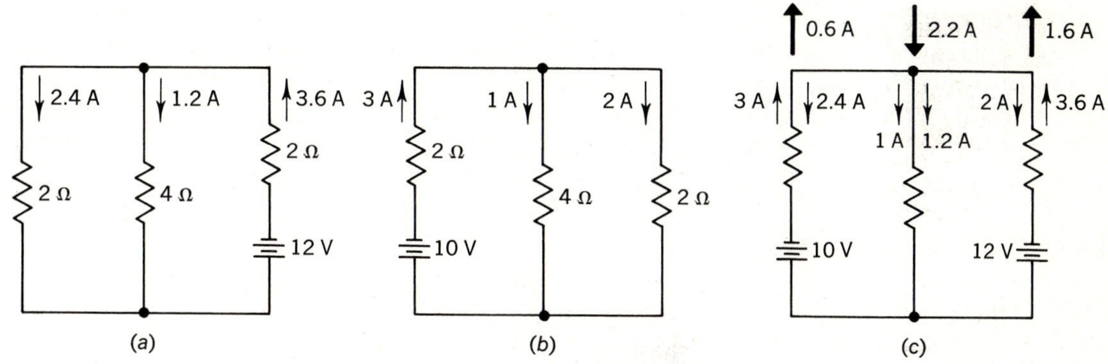

Fig. 26-4 Redrawing of Fig. 26-3*a*, showing currents for 12-V source alone. (*b*) Second redrawing, showing currents for 10-V source alone. (*c*) Superimposing all vectors on original circuit, showing actual currents.

is known, it is simpler to determine the ratio of one of the currents to the total current. Inasmuch as the currents will be inversely proportional to the resistance values (the greatest current flows through the least resistance value), the ratio to determine the current through one resistor will be the ratio of the *other resistor* to the total of the two resistance values, times the total current. In formula form,

$$I_1 = \frac{R_2}{R_1 + R_2}(I_t)$$

where I_1 is the current through one resistor, R_1, and R_2 is the value of the other resistor. Applying this formula to the problem above, in which the resistance values are 2 and 4 Ω and the total current is 3.6 A, the current in the 2-Ω resistor is

$$I_{2\Omega} = \frac{4}{2+4}(3.6) \quad = \quad \frac{14.4}{6} \quad = \quad 2.4 \text{ A}$$

For the 4-Ω resistor

$$I_{4\Omega} = \frac{2}{2+4}(3.6) \quad = \quad \frac{7.2}{6} \quad = \quad 1.2 \text{ A}$$

The vector arrows representing these current values are added to the diagram (to the right of the components).

In the next step (Fig. 26-4*b*), the total resistance will be 2 + 1.33, or 3.33 Ω as before. The total current now is $I = E/R$, or 10/3.33, or 3 A. The current through the 2-Ω resistor (R_1 is now the "other" resistor) is

$$I_{2\Omega} = \frac{R_1}{R_1 + R_2}(I_t) \quad = \quad \frac{4}{4+2}(3) \quad = \quad 2 \text{ A}$$

The current through the 4-Ω resistor is

$$I_{4\Omega} = \frac{2}{2+4}(3) \quad = \quad 1 \text{ A}$$

The vector arrows and their current values are added to the diagram (to the left).

The two sets of current vectors are shown superimposed on the original circuit in Fig. 26-4*c*, with the vector-current totals. The final current values of 0.6, 2.2, and 1.6 A are the same values as were determined for this circuit by Kirchhoff loops in Sec. 25-3.

As with the Kirchhoff loops, once the currents through the resistors are known, the voltage-drops across them can be computed.

Quiz 26-1. Test your understanding. Answer these check-up questions.

1. Compute the current values of Fig. 26-3*a* by superposition if the resistor in the 12-V source leg is 4 Ω. ________ ________ ________
2. Compute the current values in Chap. 25 of Fig. 25-4 by superposition under switch-closed conditions. ________ ________ ________

26-3 THE THÉVENIN VOLTAGE

Still another way of solving electrical systems is the Thévenin method. Kirchhoff loops and superposition currents result in current values for all parts of the circuit involved. The Thévenin method yields only one current value at a time in a complex circuit.

The first step in solving for the current flow in a part of a circuit by the Thévenin voltage method is to open the circuit and determine the voltage that appears across the opening, point X in Fig. 26-5*a*. The voltage can be either computed or read with a voltmeter. With a sensitive voltmeter 4- and 6-Ω resistances will be negligible. The meter will read the difference between the two bucking voltages, or $15 - 10 = 5$ V. The voltage that appears between the open points is called the *Thévenin voltage* or E_{th}.

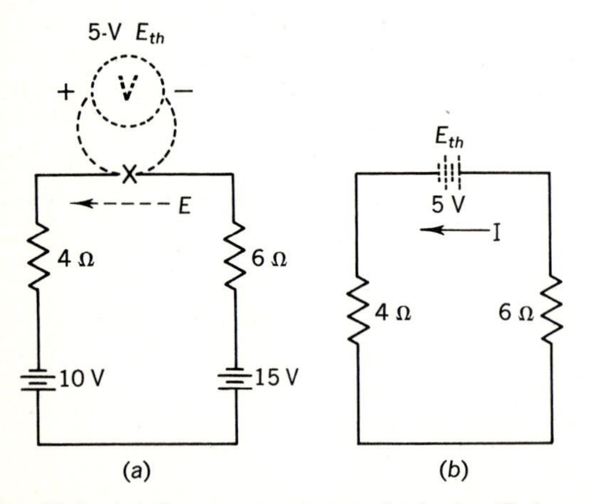

Fig. 26-5 (*a*) Series circuit in which the Thévenin voltage and circuit current can be determined. (*b*) Redrawn with E_{th} inserted and original sources removed.

After the Thévenin voltage value is determined, remove all sources of emf from the circuit and redraw the circuit with only the Thévenin voltage E_{th} as the source, as in Fig. 26-5*b*. The current computed is the actual current that will flow in the original circuit. In this case, $I = E/R$, or 5/10, or 0.5 A.

Assigning the proper polarity to the Thévenin voltage battery symbol is important. Any meter with which the Thévenin voltage is read will have a polarity marked on its terminals. To make a meter read properly, current must be flowing through it from the − to the + terminal, and the E_{th} source is drawn as shown in Fig. 26-5*b*. The polarity may be determined by simply noting which source voltage is greater.

The resistance seen at the two terminals, looking back into the circuit with the sources removed, is called the *Thévenin resistance* or R_{th}. The current in this circuit is easily solved by the formula $I = E_{th}/R_{th}$.

If another load resistor were connected in series with the E_{th} battery, the new current of the circuit would be solved by $I = E_{th}/(R_{th} + R_L)$, which is an Ohm's law formula for series loads.

Quiz 26-2. Test your understanding. Answer these check-up questions.

1. If an 8-Ω *R* were connected in series with the Fig. 26-5*b* circuit at the opened point, what would the circuit current value be? ________
2. If the left-hand source in Fig. 26-5*a* were 100 V instead of 10 V, what would be the value of E_{th}? ________ R_{th}? ________ *I*? ________ What would be the current direction in the circuit? ________

26-4 SOLVING A COMPLEX CIRCUIT BY THE THÉVENIN METHOD

To solve for the current through and the voltage across the center resistor, R_3, in Fig. 26-6*a*, apply the basic Thévenin voltage logic to this branch of the circuit.

First, open the R_3 leg at point × and determine the voltage-drop across the opening by voltmeter or computation. With point × open, as in Fig. 26-6*b*, the active circuit consists of 10 V and 12 V bucking, resulting in a counterclockwise emf of 2 V, plus two 2-Ω resistors in series, or 4 Ω. The current in this series circuit is $I = E/R$, or 2/4, or 0.5 A.

To compute the E_{th} value, consider the left leg first. The source is 10 V, and the voltage-drop across the resistor is $V = IR$, or 0.5(2), or 1 V. With the counterclockwise current, the polarities would be as marked in Fig. 26-6*b*. With these − to + and − to + polarities, the two voltages in this leg (E and IR) are additive with a total of 10 + 1, or 11 V, between P and Q. This is the E_{th} value. (With no current flowing through the 4-Ω resistor, the voltage-drop across point × is also 11 V.) This could also be determined from the right-hand leg: 12 V *minus* 1 V = 11 V. A voltmeter between P and Q, or across ×, would read 11 V.

Next, redraw the circuit without the sources, but with E_{th} and the 4-Ω load resistor inserted, as in Fig. 26-6c. Since P is negative with respect to Q, the E_{th} battery must be inserted in the circuit as shown. The two 2-Ω resistors have been drawn next to each other to make it more apparent that they are in parallel. The whole Thévenin circuit has a total of 4 + 1, or 5, Ω of resistance in series with an 11-V E_{th}. The current in the 4-Ω load is $I = E/R$, or $E_{th}/(R_{th} + R_L)$, or 11/5, or 2.2 A. The voltage-drop across the 4-Ω R is $V = IR$, or 2.2(4), or 8.8 V.

Once the voltage-drop across the 4-Ω resistor is known, the voltage across R_1 must be the difference between the 10 V across source 1 and the 8.8 V across the 4-Ω resistor, or 1.2 V. The current through R_1 is $I = E/R_1$, or 1.2/2, or 0.6 A. A similar application of Ohm's law can be made to R_2 (1.6 A).

This circuit is the same used in describing the Kirchhoff loop and superposition-current methods. The answers in all three cases are the same, of course. The advantage of the Thévenin method becomes apparent if it is desired to change the 4-Ω load resistor to some other value. With the Kirchhoff and superposition methods, the whole circuit would

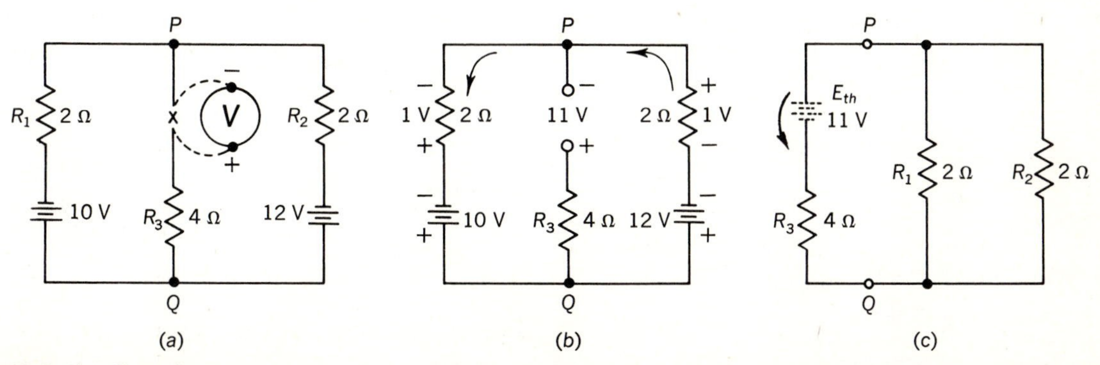

Fig. 26-6 (*a*) Complex circuit with the center leg opened to determine E_{th} and current in this leg. (*b*) The voltage between P and Q will be the same as across opening ×. (c) Insertion of E_{th} in a series-parallel circuit developed by removing the sources.

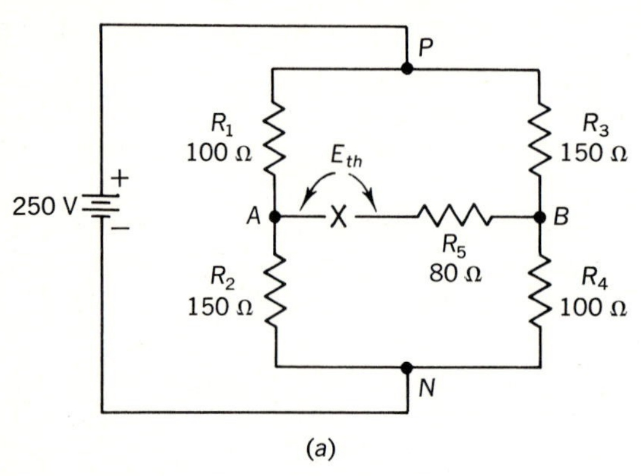

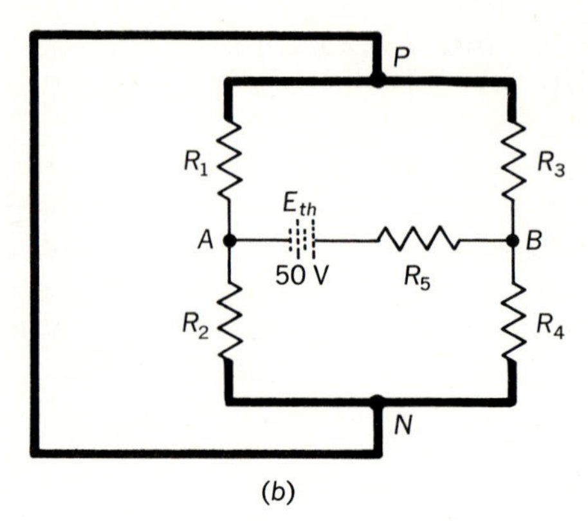

Fig. 26-7 (*a*) An unbalanced bridge. To determine the current in R_5, the first step is to open the circuit at × and determine E_{th}. (*b*) With E_{th} inserted and the source shorted out, the circuit has one common connection for the four resistors, shown by the heavy line.

have to be recomputed. By substituting the new load value in the Ohm's law formula, $I = E_{th}/(R_{th} + R_L)$, the new load-current value is readily determined. For example, if the 4-Ω load is changed to 10 Ω, the current in the new load resistor would be $I = E_{th}/(R_{th} + R_L)$, or $11/(1 + 10)$, or 1 A. The voltage-drop across the 10-Ω resistor would be $V = IR$, or 1(10), or 10 V. (No current flows in the 10-V leg.)

In a circuit of this type, once the currents and directions of two of the legs have been determined, the third-leg values can also be found simply by applying Kirchhoff's current law to a common point, either point *P* or *Q*.

Quiz 26-3. Test your understanding. Answer these check-up questions.

1. In Fig. 26-6*a*, if the 10-V branch is opened, what is the value of E_{th}? ________ R_{th}? ________ R_t seen by E_{th}? ________ I in this branch? ________ What is the I direction in this branch? ________
2. In Fig. 26-6*a*, if the 12-V branch is opened, what is the value of E_{th}? ________ R_{th}? ________ R_t seen by E_{th}? ________ I in this branch? ________ What is the I direction in this branch? ________
3. In Fig. 26-3*a*, is the 10-V battery charging or discharging? ________ If the central-branch resistance is increased to 10 Ω, will the 10-V battery be charging or discharging? ________ If the central-branch resistance is increased to 15 Ω, will the 10-V battery be charging or discharging? ________
4. In Fig. 25-4 (Chap. 25), use the Thévenin method to compute the current flowing through the headlamp with the switch closed. ________

ANSWERS TO CHECK-UP QUIZ 26-1

1. (1 A) (2 A) (1 A) 2. (8.5 A) (6.23 A) (2.3 A)

ANSWERS TO CHECK-UP QUIZ 26-2

1. (0.278 A) 2. (85 V) (10 Ω) (8.5 A) (Clockwise)

26-5 COMPUTING UNBALANCED BRIDGES

The circuit in Fig. 26-7*a* will be recognized as a bridge type. Computations in a *balanced* bridge are rather simple. If R_1 is to R_2 as R_3 is to R_4, the bridge is balanced, and the voltage between *A* and *B* is zero. But, the circuit shown is *unbalanced*.

The current value through any one of the resistors can be determined by the Thévenin method. For this discussion the current through the load resistance R_5 will be solved. First, open the R_5 circuit at point × and determine E_{th} across the opening. This will be the voltage difference between A and B.

With point × open, R_1 and R_2 are in series (250 Ω) across 250 V. The current is 1 A through this leg, and the voltage-drop across R_2 is 150 V. It can be said that point A is 150 V more positive than point N.

For the right-hand leg, with a total of 250 Ω across 250 V, the current is also 1 A. Point B is only 100 V more positive than point N.

The E_{th} must be 150 − 100 V, or 50 V, with point A more positive than B.

Next, redraw the circuit with the source removed, and insert E_{th} with correct polarity, Fig. 26-7*b*, to make current flow toward the more positive point A with the source removed. With the source out, the heavy lines indicate a common connection for four of the five resistors.

Drawing the circuit so that it appears as a reasonably simple series-parallel circuit can be a difficult step for a beginner. Compare Fig. 26-7*b* and Fig. 26-8*a*.

Figure 26-8*b* further simplifies the circuit by showing the 100- and 150-Ω parallel resistors replaced by their equivalent 60-Ω values. This circuit represents an E_{th} of 50 V, an R_{th} of 120 Ω, and an R_5 of 80 Ω. The current in the circuit and in the load resistance in question is $I = E/R$, or $E_{th}/(R_{th} + R_L)$, or 50/(120 + 80), or 50/200, or 0.25 A.

It is now known that in the original circuit the current is flowing through the 80-Ω resistor from B to A and has a value of 0.25 A. The voltage-drop across the 80-Ω resistor is $V = IR$, or 0.25(80), or 20 V. [If the 80-Ω resistor is changed to 30 Ω, it is simple to determine the current through it and the voltage-drop across it. $I = E_{th}/(R_{th} + R_L)$, or 50/(120 + 30), or 50/150, or 0.333 A. From this, the voltage-drop across the 30-Ω load would be $V = IR$, or 0.333(30), or 10 V.]

So far only the current through one of the resistors (R_5) is known. If the current in R_1, Fig. 26-7*a*, is required, the R_1 circuit must be opened, and the E_{th} across the opening must be determined. Redraw the circuit in a form that makes the E_{th} somewhat more obvious, as Fig. 26-9*a*. On this diagram E_{th} will be the voltage-drop between points P and A. R_{th} will be the resistance between points P and A with the source removed.

The circuit from P to N is a series-parallel circuit. From Ohm's law and parallel- and series-resistance computations it is possible to determine the voltage-drop across R_2. The difference between the 250 V of the source and the R_2 voltage-drop will be E_{th}.

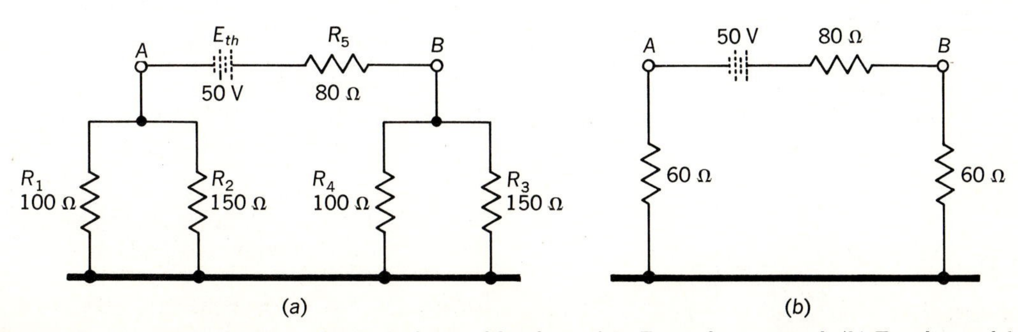

Fig. 26-8 (*a*) Second redrawing of the unbalanced bridge after E_{th} is determined. (*b*) Final simplification of the circuit into E_{th} and a series of resistances totaling 200 Ω.

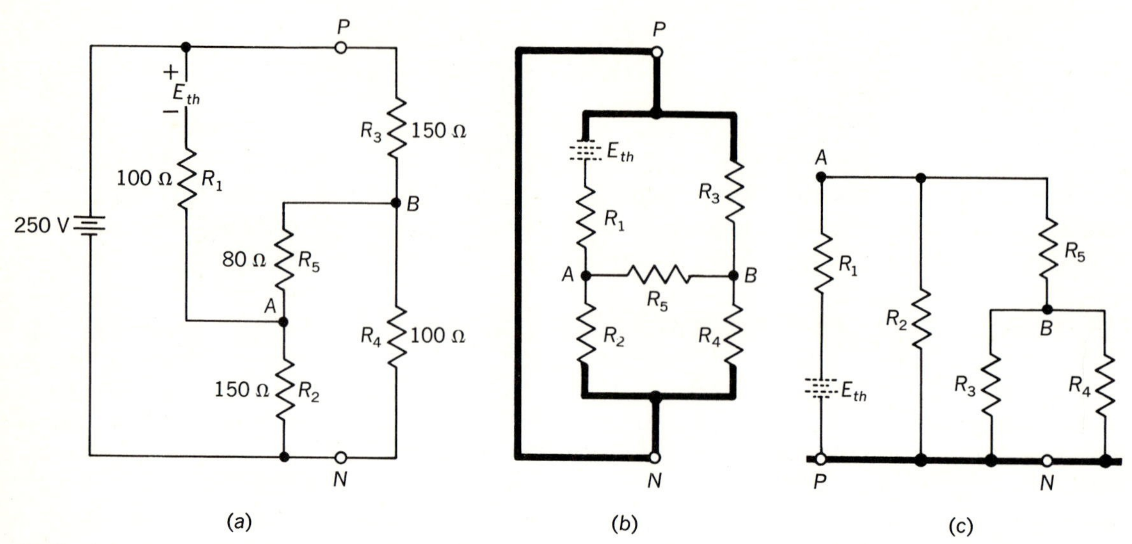

Fig. 26-9 (*a*) The circuit of Fig. 26-7*a* redrawn to show where E_{th} appears. (*b*) Replacing E_{th} in the circuit. (*c*) The circuit redrawn to solve as a series-parallel circuit.

Check your understanding before continuing. Compute E_{th} when R_1 is opened. *Answer:* _____ Is the polarity of the E_{th} correct as shown in Fig. 26-9*a*? _____

Now redraw the circuit to simplify determination of the current flowing through R_1 with E_{th} as the only voltage (Fig. 26-9*b*). To simplify to a series-parallel circuit, redraw as in Fig. 26-9c. Start with some point in the circuit—say, point *A*. Indicate the values on the components in the diagram and compute the current flowing in R_1. You should obtain an answer of about 1.15 A. If not, go back and check E_{th} (it should be about 198.3 V) and then recheck the series-parallel circuitry computations.

ANSWERS TO CHECK-UP QUIZ 26-3

1. (2 V; watch polarities of the 4-Ω *R* and 10-V battery) (1.33 Ω) (3.33 Ω) (0.6 A) (Upward) 2. (3.33 V) (1.33 Ω) (5.55 Ω) (1.6 A) (Upward) 3. (Discharging) (Charging) (Charging) 4. (6.23 A)

Quiz 26-4. Test your understanding. Answer these check-up questions.

1. In Fig. 26-7*a*, what is the value of the current through R_2? (This can be determined quite simply with two currents that have been determined.) ________
2. What is the value of the current through R_3? ________ R_4? ________

26-6 THÉVENIN'S THEOREM

Complex circuits, made up of series and parallel branches, with one or more sources in the network, can be analyzed from any one viewpoint as being composed of the equivalent of a single source voltage (E_{th}) and a single series resistance (R_{th} by computation of the series-parallel circuits). Stated as a *Thévenin theorem:*

Any complex resistive circuit may be replaced by an equivalent circuit consisting of a voltage source in series with a resistance.

Although only resistances are dealt with here, it is also possible to use this theorem in ac circuits involving reactances and impedances.

An adaptation of the Thévenin theorem allows determination of the Thévenin equivalent of a complex circuit even if none of the components in the circuit or any of the source voltages are known. For example, in Fig. 26-10*a*, what are the effective source voltage and the internal resistance that would be seen by the load with the switch closed?

If a voltmeter is connected across the *black-box* terminals, and its reading does not change when the load is disconnected, the source voltage must be the voltage shown by the meter. Furthermore, there must be no internal resistance in the circuit, or there would have been a voltage-drop across the terminals when the load was connected. This is the simplest type of a black-box circuit.

26-7 SOLVING A BLACK-BOX CIRCUIT

In Fig. 26-10*b*, we know what is in the black box, but what does the 9-Ω load see? With the switch open (no load), the voltmeter reads the voltage-drop across the 6-Ω resistor. The 4- and the 6-Ω resistors are in series across 30 V. By simple ratio, the voltage across the 6-Ω resistor is 6/(6 + 4), or 6/10 of 30 V, or 18 V. The voltmeter reads 18 V (load disconnected).

When the load is connected, the circuit seen by the 30-V source is a 4-Ω resistor in series with 6 Ω and 12 Ω in parallel, which is 4 Ω. The source sees a total resistance of 4 + 4, or 8 Ω. The source current is $I = E/R$, or 30/8, or 3.75 A. The 6- and 12-Ω resistances divide this total current, one-third through the 12-Ω and two-thirds through the 6-Ω resistor. One third of 3.75, or 1.25 A, flows through the 9-Ω load. The voltage-drop across the load must be $E = IR$, or 1.25(9), or 11.25 V, when the switch is closed.

As far as the terminals are concerned, the current changes from 0 A with no load to 1.25 A with the load, a change of 1.25 A. Either Δ (delta) or *d* signify "change in." Thus, the terminals see a Δ1.25 A (or *d*1.25 A) as the voltmeter changes from 18 V with no load to 11.25 V with a load, a *d*6.75 V. By Ohm's law, using *dI* and *dE* values:

$$R_{\text{eff}} = \frac{dE}{dI} = \frac{d6.75}{d1.25} = 5.4\ \Omega$$

The load is apparently across a source of 18 V and an effective series resistance R_{eff} of 5.4 Ω, as in Fig. 26-11. As proof, if a

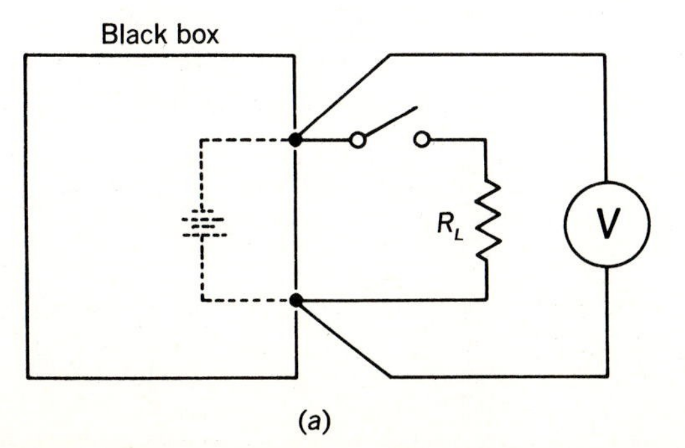

(*a*)

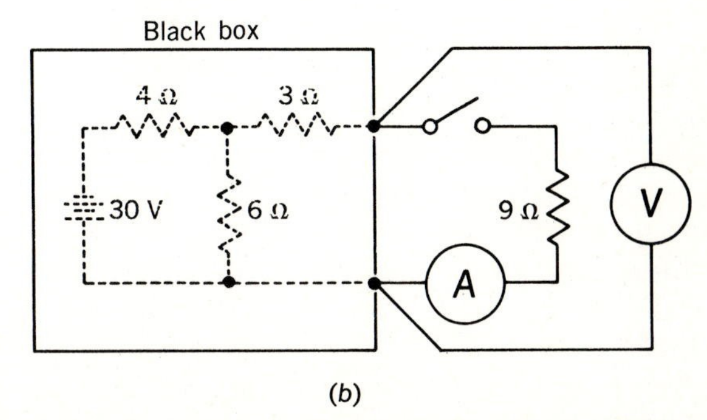

(*b*)

Fig. 26-10 The contents of two black boxes are unknown but can be determined by voltage and current readings: (*a*) simplest type, (*b*) more complex type.

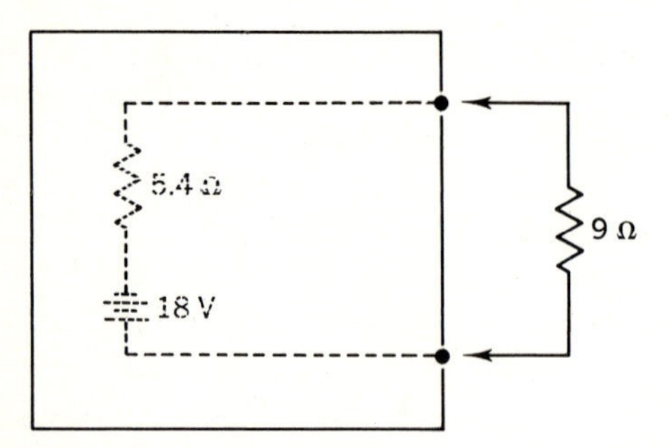

Fig. 26-11 The *effective* contents of the black box of Fig. 26-10*b*.

9-Ω resistor is across an 18-V source in series with a 5.4-Ω resistor, $I = E/R$, or 18/14.4, or 1.25 A.

A laboratory method of determining the effective *E* and *R* of any two-terminal black-box circuit with reasonably high internal resistance (Fig. 26-12) is first to mea-

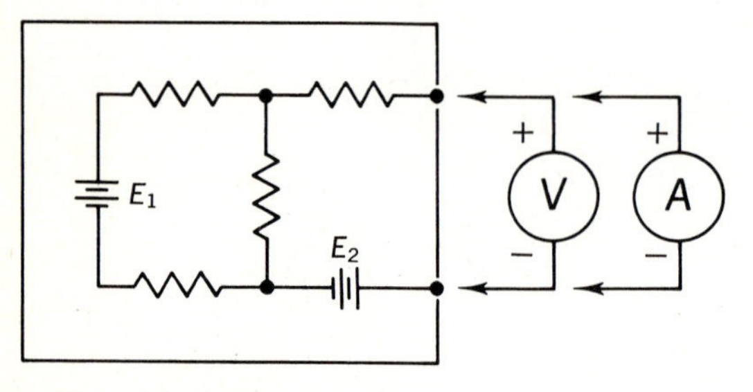

Fig. 26-12 Voltmeter and ammeter readings can determine the *effective* internal *E* and *R*.

sure the voltage at the terminals with no load (zero current), then connect an ammeter across the terminals and read the current value. (Since the ammeter has essentially 0-Ω resistance, the voltage across the terminals would be 0 V.) The effective internal resistance is $R = dE/dI$, or the voltmeter reading with no load, divided by the current with the ammeter as the load.

If there is danger of ammeter burnout, the circuit of Fig. 26-13 is used. With the

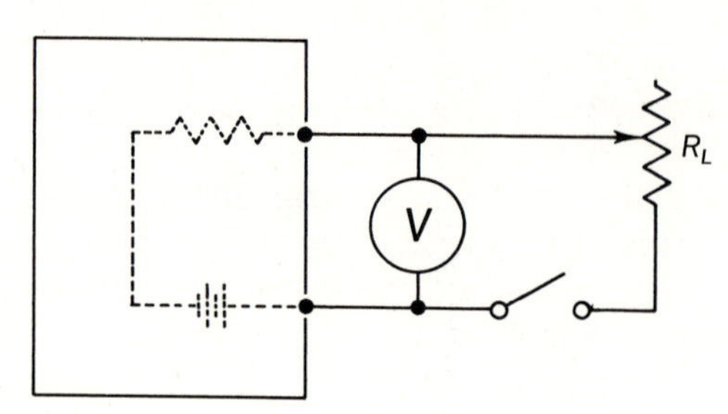

Fig. 26-13 The internal resistance will equal the R_L value when the voltmeter reads half the no-load voltage.

switch open the voltmeter indicates the source voltage. When the switch is closed, the rheostat is adjusted until the voltmeter reads just half the source value. Under this condition the resistance of the rheostat (measured with an ohmmeter while the switch is open) equals the effective internal resistance of the source circuit.

The last method may still result in excessive current flow in the circuit if the internal resistance of the source is low. A high-resistance load can be connected across the terminals of a black box. Both the current flowing through this load and the voltage developed across it are measured with meters. The resistance value is then changed (doubled, halved, etc.), and voltage and current are measured again. Ohm's law, $R = dE/dI$, will solve the effective internal resistance of the black box.

Quiz 26-5. Test your understanding. Answer these check-up questions.

1. In Fig. 26-12, which source, E_1 or E_2, must have the greater voltage value? ________
2. In Fig. 26-12, if the voltmeter alone reads 35 V and the ammeter alone reads 2 A, what is E_{th}? ________ What is its series resistance? ________
3. In Fig. 26-10*b*, if the 4-Ω resistor were doubled, what would the E_{th} of the black-box circuit be?

ANSWERS TO CHECK-UP QUIZ 26-4

1. (0.9 A) **2.** (0.9 A) (1.15 A)

__________ What would the effective series resistance be? __________

4. The no-load voltage of a black box is 14 V. When a rheostat is connected across the terminals and adjusted to 49 Ω, the voltmeter reads 7 V. What is the internal resistance of the black box? __________

5. A 1000-Ω R across a black box draws 0.2 A and is found to have 114 V across it. When a second 1000-Ω R is connected across the first one, the total current is found to be 0.222 A and the terminal voltage 111 V. What is the internal resistance of the source? __________

CHAPTER 26 TEST • SUPERPOSITION AND THÉVENIN

1. What is the name of the method of solving networks in which: Current loops are involved? The algebraic sum of vectors results in the branch currents? The voltage across an opened branch is used?
2. In Fig. 26-14*a*, if switch 1 alone is closed, what is the rate at which the battery will charge? If switch 2 alone is closed, what is the voltage-drop across the load resistor?
3. If both switches in Fig. 26-14*a* are closed, by the superposition method, what is the value of the current in the charger? Battery? Load? Check your work by computing the current through the load resistor by the Thévenin method.
4. In Fig. 26-14*b*, what is the current value in the 30-Ω resistor? 15-Ω resistor? 25-Ω resistor? 20-Ω resistor?
5. Reversing the source polarity in Fig. 26-14*b* has what effect on the current value in the 30-Ω resistor?
6. A black box has a voltmeter across its output and a rheostat and ammeter in series across the voltmeter. The voltmeter reads 78 V and the ammeter 85 mA. The rheostat is adjusted until the ammeter reads 105 mA. The voltmeter now reads 72 V. What is the impedance of the circuit in the black box?
7. The terminals of a black box measure 450 V with no load and 1.4 A with an ammeter across them. What is the effective internal resistance in the black box? What is the voltage?

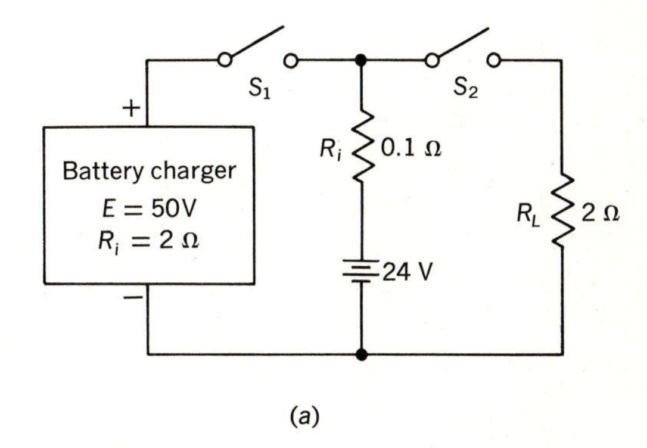

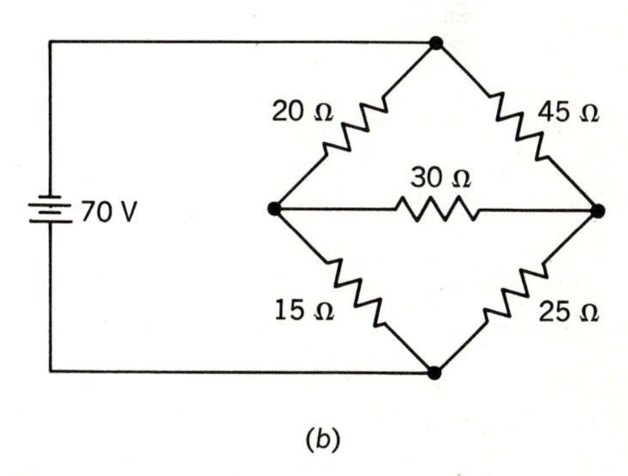

Fig. 26-14 Circuits for the Chap. 26 test.

ANSWERS TO CHECK-UP QUIZ 26-5

1. (E_1) **2.** (35 V) (17.5 Ω) **3.** (12.85 V) (6.43 Ω) **4.** (49 Ω) **5.** (50 Ω)

27 MAGNETIC CIRCUITS

CHAPTER OBJECTIVE. Extension of magnetism information in Chapter 8, some simple magnetic circuit computations, plus information on magnetostriction, transducers, and relays.

27-1 MAGNETIC CIRCUITS

In Chap. 8 the magnetic field was described as invisible, with an assigned direction externally from the north pole of a magnet toward the south. A magnetic field around a conductor is produced by the motion of electrons (left-hand field rule). The polarity of a current-carrying coil can be determined by the left-hand coil rule. The direction of an induced emf in a conductor moving across lines of force is indicated by the left-hand generator rule. As with electrostatic poles, like magnetic poles repel, but unlike poles attract.

In the simple magnetic circuit of Fig. 27-1*a*, a bar-type electromagnet has its field completion through the air, a poor conductor of magnetic lines. The field is completed with difficulty, as is an electric circuit with high resistance. This magnetic circuit has high *reluctance*, $\mathscr{R}$. The reciprocal, or opposite of reluctance is *permeance*, $\mathscr{P}$. Thus $\mathscr{P} = 1/\mathscr{R}$. The magnetic circuit of Fig. 27-1*a* has a high value of reluctance but a low value of permeance.

In Fig. 27-1*b* the reluctance has been decreased by bringing the two poles closer together. The field between N and S is more intense, assuming the same magnetomotive-force or ampere-turns in the coils. $\mathscr{P}$ is increased.

In Fig. 27-1*c*, the air gap is smaller, allowing few leakage lines of force. The reluctance of the magnetic circuit has been reduced materially. The permeance has increased greatly.

In Fig. 27-1*d*, there is no gap in the toroid-shaped core, and essentially no leakage lines of force. The $\mathscr{R}$ of this magnetic circuit is low, and $\mathscr{P}$ is quite high.

In a relay (see Fig. 8-12), at the instant that excitation is applied to the coil, the armature is relatively far from the pole piece and $\mathscr{R}$ in the circuit is relatively high. As soon as the armature gap closes, $\mathscr{R}$ decreases, and the armature is held tightly against the pull of the spring.

27-2 MAGNETIC TERMS

Important magnetic symbols are listed briefly, then explained at greater length.

ϕ = *magnetic flux;* unit the weber (Wb), 10^8 lines of force.

F = *magnetomotive-force* (mmf); unit the ampere-turn (NI).

H = *field intensity,* also *magnetizing force;* unit is the amp-turn/meter (NI/m).

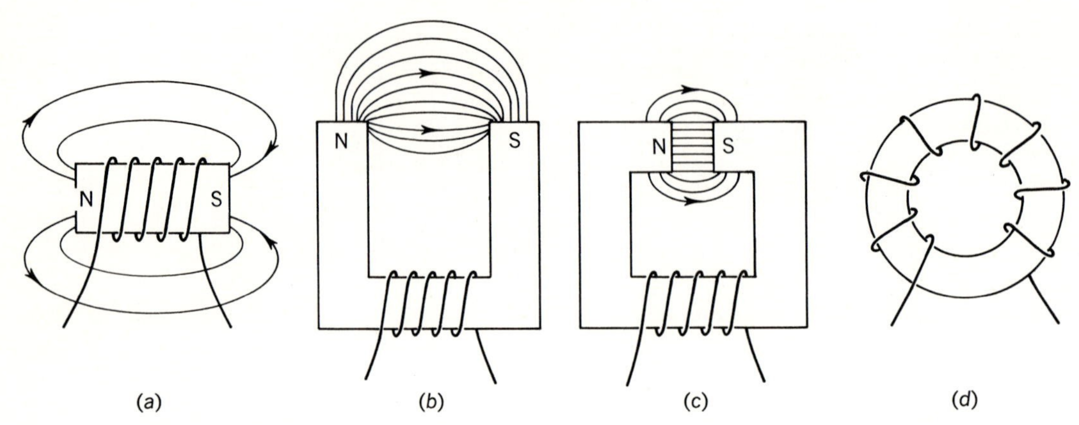

Fig. 27-1 Electromagnets. High reluctance to lowest reluctance (toroidal).

B = *magnetic flux density*; unit is the tesla (T), which is weber/square meter (Wb/m^2).

$\mathcal{R}$ = *reluctance*; ratio of mmf to flux (NI/ϕ), no unit, comparable to resistance in electric circuits.

$\mathcal{P}$ = *permeance*, reciprocal of reluctance (ϕ/NI), no unit, comparable to conductance in electric circuits.

μ = *permeability*, no unit, ability of a substance to be permeated by lines of force, in comparison to that of air or vacuum.

The *weber* represents the number of lines of force (10^8) that must be cut by a conductor in 1 s to produce 1 V, or

$$E = \frac{\phi}{t}$$

where E = emf, in volts
ϕ = flux, in webers
t = time, in seconds

As an example, if a wire cuts 3.5 Wb in 0.25 s, the emf induced in the wire is $E = \phi/t$, or 3.5/0.25, or 14 V.

Ampere-turns is the magnetomotive-force (mmf) that develops the magnetic lines of force in a core material. If a coil has 30 turns and 2 A is flowing, the mmf is 60 amp-turns.

Ampere-turns per meter indicates how much mmf is developed for a given length of core material. If a 30-turn coil is 25 cm (0.25 m) long and 2 A flows through it, the magnetizing force is NI/m, or 30(2)/0.25, or 240 amp-turns/m.

The *tesla* (T) is the measurement of the flux density, the number of lines of force per unit of cross-sectional area of a core. One tesla is one weber per square meter.

What is the flux density of a core having 20,000 lines and 5 cm^2 of cross-sectional area? Converting the lines to webers, 20,000 lines is $(2 \times 10^4)/10^8$, or 2×10^{-4}, or 0.0002 Wb. Since there are 10,000 cm^2 in 1 m^2, the area is 5/10,000, or 0.0005 m^2. The flux density is $B = \phi/m^2$, or 0.0002/0.0005, or 0.4 T.

Reluctance in a magnetic circuit is similar to resistance in an electric circuit. In magnetic circuits, $\mathcal{R} = \text{mmf}/\phi$. The reluctance of a circuit having 60 amp-turns and 200 μWb is 60/0.0002, or 300,000 NI/Wb.

Permeance is the reciprocal of reluctance. The permeance of a magnetic circuit indicates the ability of the circuit to develop lines of force in it.

Permeability is the ratio of the number of lines of force developed in a magnetic material to the number of lines that would be developed in an air core of similar length and area. Permeability is also the ratio of B to H (flux density to field intensity), or $\mu = B/H$.

When comparing magnetic substances, air and vacuum will be considered as having unity permeability.

Quiz 27-1. Test your understanding. Answer these check-up questions.

1. What direction is given to lines of force between poles of a magnet? ________
2. When a magnetizing force is applied to a magnetic substance and lines of force develop throughout the system, what is the system called? ________
3. What is considered to have the fundamental value of permeability? ________
4. What shape of core would have the least reluctance? ________
5. Give the magnetic equivalent of $R = E/I$. ________
6. Give the magnetic equivalent of $G = 1/R$. ________
7. What are two advantages of having small gaps in magnetic circuits? ________ ________
8. In Fig. 27-1c, what must the direction of current flow be (right to left, or left to right) to produce the magnetic polarities shown? ________
9. In Fig. 27-1d, where is the N pole? ________
10. In a relay, would it take more current to close the armature or to hold it closed? ________
11. How many lines of force are there in a weber? ________
12. What value of emf is developed in a wire if it cuts 12 Wb in 0.02 s? ________
13. What is the unit of measurement used for mmf? ________
14. What is the unit of measurement used for field intensity? ________ What is another name for field intensity? ________
15. What is the unit of measurement of flux density? ________ What is another way of expressing this? ________
16. What is the ratio of mmf to flux density called? ________
17. Which refers specifically to a magnetic *circuit*, permeance or permeability? ________ Which refers to different magnetic materials? ________

Questions 18 to 22 refer to Fig. 27-2.

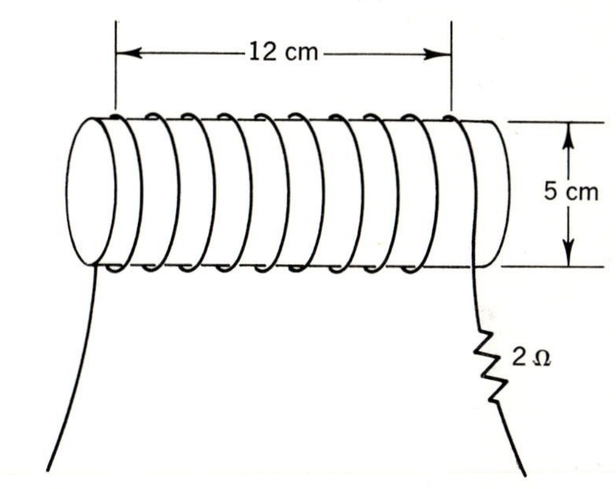

Fig. 27-2 Coil used in Check-up Quiz 27-1.

18. What are at least two materials on which the coil could be wound and still be considered an air-core coil? ________ ________
19. If the coil is connected across a 4.5-V source, what is the mmf developed? ________ What is the field intensity? ________
20. If a core of iron had been inserted, would either of the answers to question 19 have been different, and if so, which? ________
21. If there is a total of 1.6×10^8 lines of force in the core (area $= \pi r^2$), what is the flux density in lines per square centimeter? ________ In teslas? ________
22. What is the reluctance of the circuit if a total of 1.6×10^8 lines of force are developed? ________ What is the permeance? ________ What is the permeability? ________ Would this be an iron-core or an air-core coil? ________
23. Which of the magnetic circuit units of measurement are given special names, and to what are these units equal? ________

27-3 FERROMAGNETISM

The difference between magnetic and nonmagnetic materials lies basically in the behavior of the outer electrons of the atoms involved. All electrons in atoms are orbiting around the nucleus and at the same time are spinning like tops. It is this orbiting and spinning of the electrons that determines the magnetic capabilities of the atom. In most atoms, the orbital motion and spin of the electrons tend to cancel each other, resulting in essentially nonmagnetic atoms.

When an atom is brought into a magnetic field, the motion of the electrons should produce a repelling effect on the field. The result should be a material with a permeability less than that of air or a vacuum, in the region of 0.99998. The material is said to be *diamagnetic*, where "dia-" means "opposing." Copper, silver, and bismuth are examples of diamagnetic substances.

In most cases, aluminum, chromium, platinum, etc., the orbits and spins act to align the atoms in the direction of the magnetic field. These materials are *paramagnetic*, where "para-" means "related to," and they have permeabilities of about 1.00002. A long bar of paramagnetic material suspended in a strong magnetic field has a weak tendency to align itself in the direction of the field. A long bar of diamagnetic material tends to turn at right angles to the field.

Atoms of three elements—iron, nickel, and cobalt—have many electrons orbiting around the nucleus in the same direction. One side of each atom has a north polarity and the other a south. Nearby atoms fall into alignment with each other, forming completely magnetized groups called *domains*. However, adjacent domains may not be in the same alignment, so that the cumulative effect in a piece of iron, nickel, or cobalt may add up to zero overall magnetism. Should the substance be placed in a magnetic field, all the domains can be aligned rather easily. As a result, pure iron has a permeability, or ability to be permeated by lines of force, of several thousand compared with air. Nickel is less magnetic, with a permeability of about 1000. Cobalt is still less magnetic, with a μ of about 170. Bars of these materials will readily line up in a magnetic field.

One metal, Permalloy (78% nickel, 22% iron), may have a μ of over 80,000. Surprisingly, in proper proportions, copper, manganese, and aluminum, each by itself nonmagnetic, form alloys which behave magnetically much like iron. While the domains of all these materials align easily, they also fall out of alignment easily. They are said to be *temporary magnetic materials*, and to have little *retentivity*, or ability to retain magnetic alignment.

Any of the high-permeability materials are said to be *ferromagnetic*, meaning "ironlike magnetic." Magnetic devices are always constructed from ferromagnetic materials.

When ferromagnetic materials are developed in crystalline form, as in steels, Alnico, and Nipermag, the domains can be locked into alignment, so that once magnetized they retain their alignment and their magnetic polarity. They form *permanent* magnets. One of the best permanent-magnet materials is Alnico, which

---

ANSWERS TO CHECK-UP QUIZ 27-1

1. (N to S) **2.** (Magnetic circuit) **3.** (Air) **4.** (Toroidal) **5.** ($\mathscr{R} = \text{mmf}/\phi$, or NI/ϕ) **6.** ($\mathscr{P} = 1/\mathscr{R}$) **7.** (Fewer leakage lines, lower reluctance) **8.** (Right to left) **9.** (There is none until the core is opened somewhere) **10.** (To close it) **11.** (10^8) **12.** (600 V) **13.** (Amp-turns, or NI) **14.** (Amp-turns/m) (Magnetizing force) **15.** (Tesla) (Wb/m²) **16.** (Reluctance) **17.** (Permeance) (Permeability) **18.** (Cardboard, wood, plastic, ceramic) **19.** ($22.5NI$) (187.5 amp-turns/m) **20.** (No change) **21.** (8.15×10^6 lines/cm²) (815 Wb/m² or T) **22.** (14.1) (0.071) (1.08) (Air) **23.** (Weber = 10^8 lines, T = Wb/m²)

---

is 62% iron, 20% nickel, 12% aluminum, 5% cobalt, and 1% manganese and silicon.

The permeability values given above are for room temperatures. As the temperature increases, magnetic effects decrease. If iron is heated to 770°C, called the *Curie point* of iron, it ceases to be ferromagnetic. It will become magnetic again when cooled but may have lost all domain alignment owing to thermal activity of the atoms, molecules, and domains.

Alloys have been developed with Curie points from room temperature to several hundred degrees. They are used in temperature-sensitive relays, transformers, reactors, etc.

Steels that have *carbon* in them are subject to aging, a slow loss of their ability to retain magnetic domain alignment. Aging is accelerated by heat. Alnico and other *noncarbon* alloys will withstand temperatures up to as much as 600°C with little aging.

27-4 PERMEABILITY CURVES

Figure 8-6 showed the *BH* curves of ferromagnetic materials and air. If the field intensity H is increased sufficiently, the material becomes saturated, and a "knee" forms in the curve. Past the knee an increase in H has little effect on the B value. The permeability, or ratio of B to H, of a material is normally measured just below the knee. Figure 27-3 indicates the average permeability of this material to be $\mu = B/H$, or 800,000/200, or 4000. (With an H of 200, there are 800,000 lines/m^2 developed in it.)

The curve shows the magnetization when a ferromagnetic material is subjected to a magnetizing force. What happens to the material when the magnetizing force is removed?

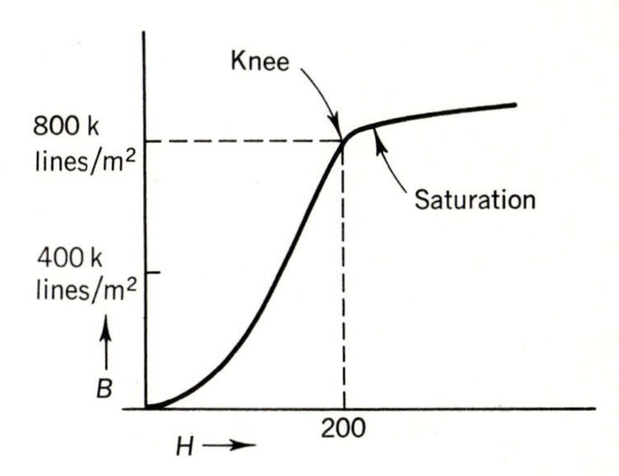

Fig. 27-3 The knee and the slight rise above the point of saturation of a piece of iron.

27-5 HYSTERESIS LOOPS

The hysteresis loop is a series of *BH* curves that indicate what magnetic effects occur in a ferromagnetic material. In Fig. 27-4*a*, the dotted line is a magnetization curve, with the top of the knee at point K. As the magnetizing force increases from zero (point 0) to $+H$, the magnetism (flux density) increases only slightly beyond point K. When the magnetizing force is reduced to zero again, the flux density (and domain alignment) is represented by the solid line K to L. When the magnetizing force is reduced to zero, most of the domains remain aligned. The flux density has only decreased slightly. The material is a permanent magnet.

If the current that produced the original magnetization is now reversed, the magnetizing force is reversed ($-H$). It takes a $-H$ value to point M before all domains are disaligned and the material has no residual flux density. The amount of $-H$ required to reach point M is called the *coercive force*. (If the magnetizing force were now reduced to zero, the material should be completely demagnetized.) When $-H$ is increased further, domains align in the opposite direction, reach a knee of saturation, point N, and then in-

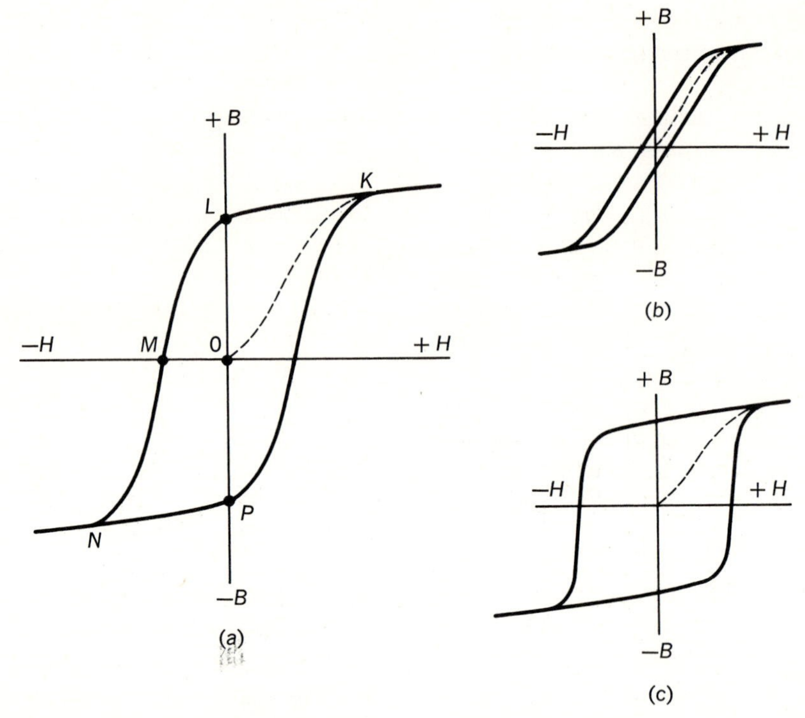

Fig. 27-4 Hysteresis loops: (a) A hysteresis loop is formed from K, L, M, N, P, to K; (b) temporary-magnet material; (c) permanent-magnet material, such as Alnico.

crease flux density only slightly. As $-H$ is reduced to zero, alignment of the domains decreases only to the value at point P.

If the magnetizing forces were produced by an ac, the magnetization of the material would be driven back and forth repeatedly over this same loop pattern.

The term *hysteresis* indicates a lag due to a domain-friction effect. The hysteresis loop represents the friction that the magnetizing force must overcome to realign domains. A wide hysteresis loop (Fig. 27-4c) indicates a considerable amount of magnetic friction that must be overcome during magnetizing and demagnetizing. As with any friction, magnetic hysteresis losses result in heating of the ferromagnetic material being magnetized.

The hysteresis loop of a temporary-magnet material is shown in Fig. 27-4b. As the magnetizing force is reduced to zero, the flux density $+B$ falls almost to zero. The coercive force is very small. *Hysteretic* losses and heat would be negligible. This type of ferromagnetic material would be found in iron cores of relays, inductors, and transformers to reduce losses in the core that might waste energy.

Hysteresis loops of permanent-magnet materials are almost rectangular (Fig. 27-4c).

Nonmagnetic materials have no hysteresis "loops," since they have no retentivity. As H increases, the flux density B increases. As H decreases, B decreases. At the instant H falls to zero, B also becomes zero. No loop can form, and no hysteretic loss is produced.

Quiz 27-2. Test your understanding. Answer these check-up questions.

1. What name is given to materials that can form domains? ________
2. What name is given to materials that have a permeability of about 1.00002? ________ About 0.99998? ________
3. What is the most magnetic atom? ________ The next most magnetic atom? ________ The least magnetic of the magnetic atoms? ________
4. Of the two materials Permalloy and Alnico, which has the higher permeability? ________ Has the narrower hysteresis loop? ________ Makes the better permanent magnet? ________ Is used as a core in transformers? ________
5. What is the temperature called at which a ferromagnetic material loses its domain alignment? ________ What is this temperature value for iron? ________
6. Why would the permeability at a point partway up the rising portion of the *BH* curve be less than the average permeability value? ________ What would be the approximate average permeability halfway to the knee in Fig. 27-3? ________
7. Which is greatest, the permeability before the knee, through the knee, or above the knee? ________ Which is least? ________
8. In Fig. 27-3, in what part of the curve would the variation of permeability be least with a variation of *H*? ________
9. In Fig. 27-4*a*, what point indicates the "residual magnetism" that would be left in the material if it had originally been magnetized to $+H$? ________

27-6 MAGNETIZING AND DEMAGNETIZING

A piece of steel may be magnetized by stroking it several times in the same direction with one pole of a strong permanent magnet. This aligns many of the domains. The steel exhibits magnetic poles of its own, and is said to be magnetized by *magnetic induction*. If the steel is hammered as it is being stroked, more domains will be jarred into alignment, resulting in a stronger magnet. If it is heated to the Curie point and stroked as it cools, the magnetism will be even stronger.

A much more satisfactory method of magnetizing is to wrap the steel in a coil of wire. A strong current flowing in the coil will magnetically saturate the steel. If it is jarred, or heated and cooled while under the magnetizing force, the resulting magnetism will be maximum for the material used. By controlling current direction and strength, the same method can be used to demagnetize a ferromagnetic bar.

If the hair spring of a watch becomes magnetized, the watch can no longer keep correct time. To demagnetize the watch, it may be lowered into the core area of an air-core coil in which ac is flowing. The alternating fields magnetize the watch, first to one polarity and then the opposite. As the watch is slowly pulled from the coil, the alternating magnetic induction weakens.

When the watch is completely free of the coil, it should be completely demagnetized. Care must be taken, as too strong a field or too long an exposure may produce hysteretic loss heat, and the temper of the hair spring may be altered. Demagnetizing is sometimes called *degaussing*, from the term *gauss* (Sec. 27-11).

27-7 MAGNETOSTRICTION

An interesting physical effect is produced when ferromagnetic materials are magnetized. An iron bar will expand slightly when magnetized, but a nickel bar will contract. Maximum expansion or contraction occurs as the material saturates.

A nickel bar anchored at one end and with a round disk or diaphragm attached to the other end is shown in Fig. 27-5. If a 10-V rms 128-Hz ac is applied to the coil around the nickel bar, the bar will con-

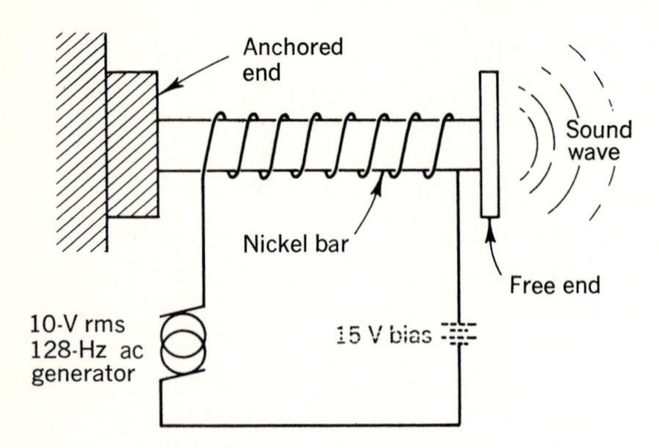

Fig. 27-5 Magnetostriction transducer.

tract with each alternation, or twice during each cycle, and the bar vibrates at 256 times per second. The disk moving back and forth at this rate produces compressions and rarifications of the air at a rate of 256 per second. Air vibrations at this frequency represent a sound wave at the musical tone "middle C."

If a 15-V battery is added, shown dashed, the current never reverses in the circuit but varies from 1 V (10 V rms = 14 V peak) to 29 V. Each cycle of ac only produces one increase and decrease of current. The 128 vibrations per second now produced by the disk diaphragm are one octave below middle C. The 15-V dc can be called a *bias voltage*, and the ac is a *signal voltage*.

27-8 MAGNETIC TRANSDUCERS

A *transducer* is any device that converts energy of one form to energy of some other form. The nickel bar discussed in Sec. 27-7 is a transducer of electrical to sound energy. A photoelectric cell converts photons (light energy) to electrical emf and current. A battery changes chemical to electrical energy. A motor changes electrical to mechanical energy. There are many different types of transducers.

In Fig. 27-6, a permanent magnet produces an intense field between its outer N poles and the middle S pole. Wire is wrapped around a paper tube that is free to move back and forth along the middle pole piece. Attached to this coil is a conical paper diaphragm. If a sound wave strikes this diaphragm, it vibrates. The vibration of the *moving coil* back and forth through the magnetic field induces an ac in the coil at the sound frequency. This transducer is a sound-to-electric *dynamic microphone*.

If a 500-Hz current is fed to the moving coil, the coil polarity changes 500 times per second. As a result, the coil is attracted to and repelled from the S pole 500 times per second. The diaphragm vibration radiates sound waves at this frequency. This transducer is a *PM* (permanent magnet) *dynamic loudspeaker*.

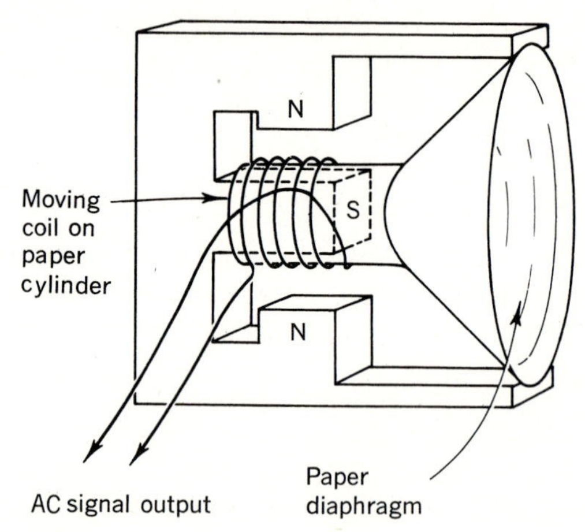

Fig. 27-6 Basic operating parts of a PM dynamic microphone or PM loudspeaker.

ANSWERS TO CHECK-UP QUIZ 27-2

1. (Ferromagnetic) **2.** (Paramagnetic) (Diamagnetic) **3.** (Iron) (Nickel) (Cobalt) **4.** (Permalloy) (Permalloy) (Alnico) (Permalloy) **5.** (Curie) (770°C) **6.** (Ratio of B to H is less) (3300) **7.** (Before) (Above) **8.** (Center of straight portion) **9.** (L)

27-9 RELAYS

A *relay* was discussed in Check-up Quiz 8-3. Figure 27-7 illustrates a simple

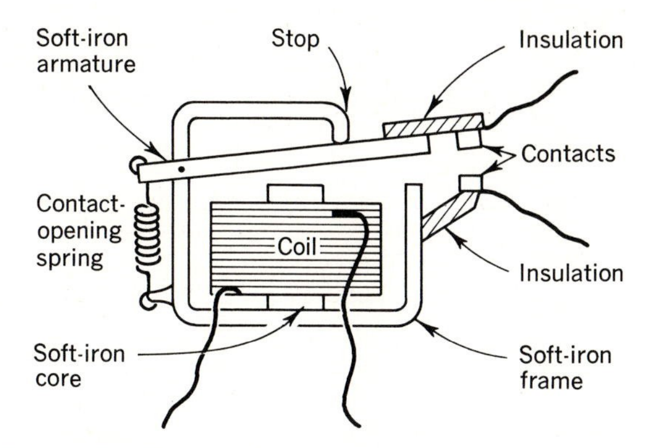

Fig. 27-7 Essentials of a single-pole single-throw make-contact normally open dc relay.

dc relay used to close high-voltage or high-current circuits, remotely if desired, by a relatively low-voltage and low-current source. It is a single-pole single-throw (SPST) normally open (NO) make-contact relay. The core, the C-shaped frame, and the straight armature bar are all magnetically "soft" (high permeability, little retentivity) iron. One contact is attached by an insulator to the armature, and the other is attached to the relay body by an insulator. The contact-opening spring holds the contacts open.

When current flows in the coil, lines of force develop in the core, armature, and frame of the relay. The gap between core and armature is filled with magnetic lines of force trying to contract. This overcomes the tension of the spring, pulls the arm toward the core, and closes the contacts. When the current stops, the magnetic circuit weakens, and the spring pulls the armature up, opening the contacts.

Figure 27-8 illustrates in symbol form a normally open SPST make-contact relay with its armature, contacts, core, and coil. The second symbol illustrates a normally closed (NC) SPST break-contact relay. The third is a single-pole double-throw (SPDT) type. It is possible to start 10 different circuits and stop 10 others with a single 10-pole double-throw relay.

Relays are rated in current-carrying capacity of the contacts, insulation voltage between contacts, and either voltage to be applied across the coil or resistance and current of the coil. Actually, it is the ampere-turns of the coil that closes the relay. If a relay coil is to operate from some other source voltage, the new coil must develop the same mmf as the original coil.

Contacts on relays are silver, or harder materials such as tungsten. Silver oxidizes, but may be cleaned with a very

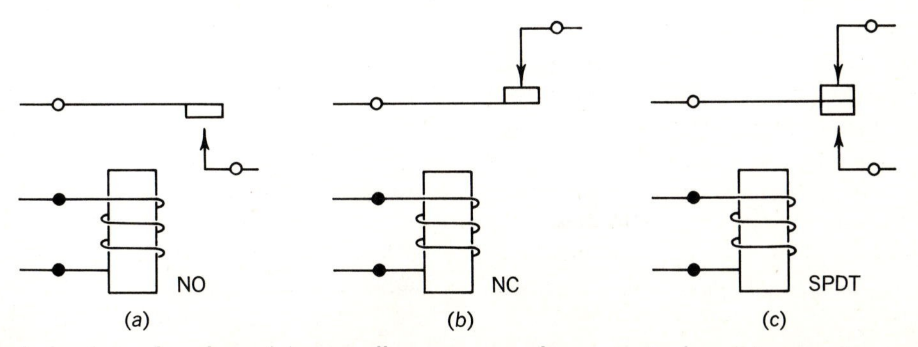

Fig. 27-8 Symbols of simple relays: (*a*) normally open or make-contact relay; (*b*) normally closed or break-contact relay; (*c*) single-pole double-throw relay.

fine abrasive paper. If contacts are pitted by heavy currents, they may be smoothed with a fine file, but their original shape should be retained to allow the normal wiping action during closing that keeps the contacts self-burnishing. Arcing and sparking can be reduced by using a 0.1-μF capacitor and a 50-Ω resistor in series across the contacts. Tungsten is harder and pits less than silver.

A dc relay operates improperly when ac is fed to its coil. Alternations of current try to attract and release the armature at twice the ac frequency, resulting in vibrating contacts, a buzzing sound, or no operation. If the ac is rectified to pulsating dc by a diode, the relay will operate.

Relays made to operate from ac have a shaded pole at the armature end of the core as in Fig. 27-9. (See shaded poles, Sec. 23-9.) The core is slotted, and a solid copper ring is welded around one half of the core top. First the unshaded and then the shaded portion have a maximum field strength. The difference in time between the field maximums produces a continuous enough magnetic flux to hold the iron armature closed.

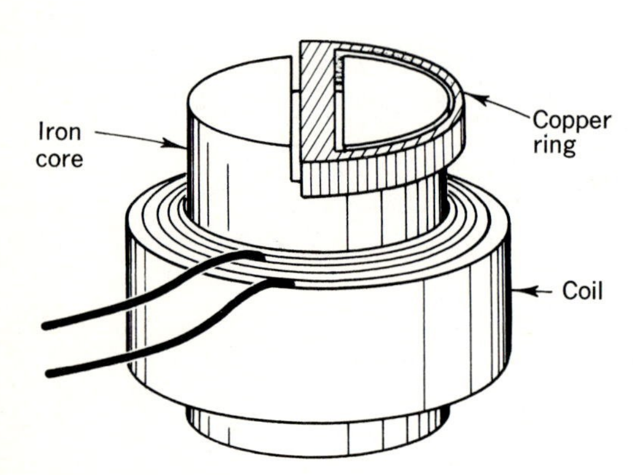

Fig. 27-9 A shaded-pole relay core and coil.

27-10 DETERMINING POLARITY WITH A COMPASS

The direction in which dc current is flowing in a wire can be determined by holding a compass over the wire and noting the north end of the compass (Fig. 27-10). By the left-hand field rule (Sec. 8-1), with the fingers in the direction of the lines of force, the thumb indicates the current direction in the wire. This end of the wire must be connected to the positive pole of some dc source. Do not try this if strong ac flows in the wire, or the compass may be demagnetized without giving any indication. A compass needle cannot follow rapid field alternations.

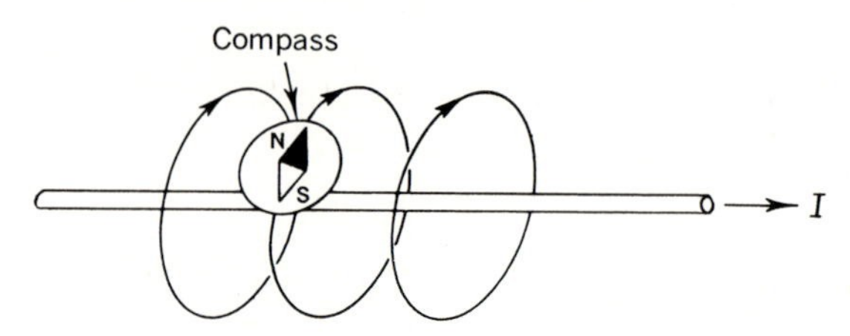

Fig. 27-10 Compass lines up in lines of force.

A common method of denoting current direction is shown in Fig. 27-11. In Fig. 27-11*a*, the current-carrying wire has a dot in the center, indicating the point of a

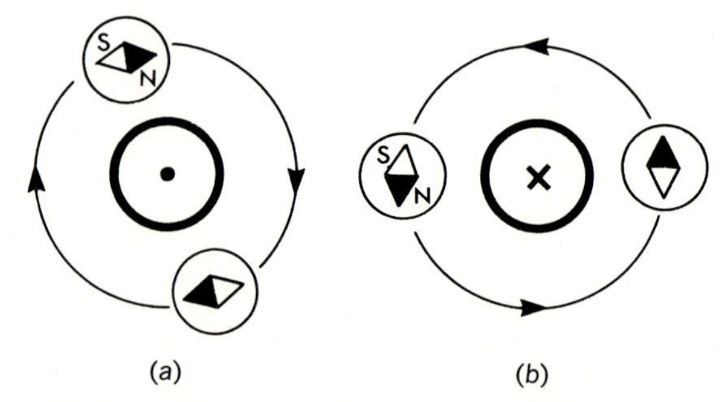

Fig. 27-11 (*a*) A circle with a dot indicates current approaching in a wire. Compasses line up as shown. (*b*) A cross in the wire indicates the tail feathers of the current vector.

current-vector arrow approaching the reader. Compasses would align themselves in the lines of force as indicated. In Fig. 27-11*b*, the cross indicates a rear view of the feathers of a current-vector arrow when current is moving away. The compasses will line up as indicated.

Quiz 27-3. Test your understanding. Answer these check-up questions.

1. If a coil is carrying dc, will adjacent turns be attracted or repelled by the magnetic fields developed around the turns? ______
2. When current is flowing in a coil, does magnetic field action make the coil try to lengthen or shorten? ______
3. When a two-wire electric cord is carrying dc to a lamp, does magnetic action try to pull the two wires together or push them apart? ______ What action occurs when the source is delivering ac? ______
4. In Fig. 27-11*a*, where is the N pole? ______
5. In Fig. 27-11, if the two wires shown represent the cross section of a single-turn coil, where is the S pole? ______
6. Aside from possible breakage, why would it be undesirable to drop or jar earphones, meters, or loudspeakers? ______
7. If a steel ship were constructed with its bow pointed to the north, what magnetic polarity would be induced in the bow by the earth's magnetic field? ______
8. What is the magnetic field called that remains after a piece of steel is partially demagnetized? ______ What magnetic characteristic is the metal said to have? ______
9. When a compasslike needle is made to pivot vertically instead of horizontally, it forms a "dip needle." At what place(s) on the earth would it dip vertically? ______ Not dip at all? ______
10. What physical action takes place in iron when a bar of it is magnetized? ______ What happens to nickel? ______ What is this called? ______
11. What tone frequency would be heard if a coil around a bar of soft iron were excited by a 500-Hz ac? ______ If the ac source had a dc bias of equal voltage in series with it, what would be the tone frequency? ______ What is the frequency of middle C? ______
12. What is the general term for devices that change energy of one form to energy of another? ______
13. What is the name for light energy sent from the sun to earth? ______
14. If a dynamic loudspeaker lost all its permanent magnetism and a 500-Hz ac were fed to its coil, would any sound be produced? ______ If so at what frequency? ______ Why? ______
15. Why is almost twice the current required to close a dc relay as to hold it closed? ______
16. What does "SPST" mean in relation to relays? ______ "DPDT"? ______ "NC"? ______ "NO"? ______
17. What would be the required resistance and the wattage rating of the resistor to allow operation of a 6-V 0.25-A relay coil from a 12-V source? ______ ______
18. Why does the current in an ac relay *decrease* when the armature is pulled into the contact-closed position? ______

27-11 ENGLISH AND CGS MAGNETIC UNITS

The explanations of magnetic units have been in terms of the meter-kilogram-second-ampere (mksa) rationalized system. There are two other systems sometimes used. One of them is the centimeter-gram-second (cgs) system. Its basic units of centimeters and grams are much smaller than the mksa units of meters and kilograms. In the cgs system the line of force is the *maxwell*. The mmf unit is the *gilbert*. The magnetizing force or field-intensity unit is the *oersted* (one gilbert per centimeter). The flux-density unit is the *gauss* (one maxwell per square centimeter).

Another system is the English, or U.S. Customary. This system is based on the inch as the linear measurement. The basic units are flux in *lines of force* (as in mksa),

Table 27-1 ***MAGNETIC UNITS***

Term and symbol	English units		Cgs units		Mksa units, rationalized
Magnetic flux (ϕ)	1 LINE OF FORCE	=	1 MAXWELL	=	10^{-8} weber
	10^8 lines of force	=	10^8 maxwells	=	1 WEBER
Magnetomotive-force (F)	1 AMPERE-TURN	=	1.26 gilberts	=	1 AMPERE-TURN
	0.796 amp-turn	=	1 GILBERT	=	0.796 amp-turn
Field intensity or magnetizing force (H)	1 AMP-TURN/IN.	=	0.495 oersted	=	39.4 amp-turns/meter
	2.02 amp-turns/in.	=	1 OERSTED, or 1 gilbert/cm	=	79.6 amp-turns/meter
	0.0254 amp-turn/in.	=	0.0126 oersted	=	1 AMPERE-TURN/METER
Magnetic flux density (B)	1 LINE/INCH^2	=	0.155 gauss	=	0.155×10^{-4} weber/meter^2
	6.45 lines/in.^{2}	=	1 GAUSS, or 1 maxwell/cm^2	=	10^{-4} weber/meter^2
	6.45×10^4 lines/in.^{2}	=	10^4 gauss	=	1 weber/meter^2, or 1 TESLA

mmf in *ampere-turns* (as in mksa), field intensity H in *ampere-turns per inch,* and magnetic flux density in *lines per square inch.*

Table 27-1 lists the three systems, with the conversion factors necessary to change from one system of units to another.

CHAPTER 27 TEST • MAGNETIC CIRCUITS

1. What effect does increasing the gap in a magnetic circuit have on the reluctance value?
2. What is the reciprocal of reluctance called?
3. Magnetic flux is measured in what units in the mksa system? The cgs system? The English system? What is the letter symbol for flux?
4. Magnetomotive-force is measured in what units in the mksa system? The cgs system? The English system? What is the letter symbol for mmf?
5. Field intensity is measured in what unit in the mksa system? The cgs system? The English system? What is the letter symbol for field intensity?
6. Magnetic flux density is measured in what units in mksa? In cgs? In English? What is the letter symbol for flux density?
7. How many lines of force must be cut in 1 s to produce an average of 1 V?
8. If a conductor cuts across 8.5 Wb in $^2/_{10}$ s, what voltage is produced in it?
9. What is the flux density, in teslas, of a core having 5000 lines and 3 cm^2 of cross-sectional area?
10. What is the reluctance of a magnetic circuit that has 450 NI and 500 μWb?
11. What is the ratio B/H called?
12. If a long bar of a metal tries to turn at right angles to a magnetic field, what is its magnetic classification?
13. If a long bar of a metal turns weakly into alignment with a magnetic field, what is its classification?
14. What magnetic classification do the metals cobalt, nickel, and iron have?
15. Magnetic-material molecules group into completely magnetized islands. What are these areas called?
16. To produce maximum retentivity, in what form must metals be made?

ANSWERS TO CHECK-UP QUIZ 27-3

1. (Attracted) **2.** (Shorten) **3.** (Push apart) (Cord tries to vibrate at twice the ac frequency. Why twice?) **4.** (There is none) **5.** (Top of figure) **6.** (Loss of magnetism) **7.** (N) **8.** (Residual magnetism) (Retentivity) **9.** (Over N and S magnetic poles) (Along magnetic equator) **10.** (Expansion) (Contraction) (Magnetostriction) **11.** (1000 Hz) (500 Hz) (256 Hz) **12.** (Transducer) **13.** (Photon) **14.** (Yes) (1000 Hz) (Coil would attract iron pole on each alternation) **15.** (Closed armature gap represents less reluctance and greater flux density) **16.** (Single-pole single-throw) (Double-pole double-throw) (Normally closed) (Normally open) **17.** (24 Ω) (1.5 W, or 3 W for safety) **18.** (Less leakage flux, more L and therefore more X_L)

17. All metals lose their magnetism at some temperature. What is this value called?
18. What is in permanent-magnet materials that makes them age rapidly?
19. What term is used to express the frictional effect when materials are being magnetized?
20. Would the hysteresis loop of Permalloy be narrow, medium, or wide? Alnico? Carbon steel?
21. When residual magnetism is removed from a steel object, what is it said to be?
22. What type of current is normally used in demagnetizing watches, screwdrivers, etc.?
23. When ferromagnetic materials are magnetized, they may expand or contract. What is this effect called?
24. What is the term for a device that changes one form of energy to some other form?
25. If the permanent magnet in a loudspeaker has lost its magnetism and the coil is fed a 300-Hz ac, what tone, if any, will it produce?

CHAPTER OBJECTIVE. Extension of ac information in Chap. 9 including frequencies, harmonics, instantaneous values, radians, period, and wavelength.

28-1 FREQUENCIES OF ALTERNATING CURRENTS

The waveforms of alternating emf and current may vary widely. Voltage and current waveforms can be square, sawtooth, sine, or highly complex, such as the audio frequency ac that results from the transducing of sound waves to electric. Waveforms are of great importance in the study of the communication phase of electronics and radio.

Voice frequencies are considered to be 200 to 3000 Hz. High-fidelity music includes frequencies of 20 to 20,000 Hz, the lower and upper limits of human hearing. Normally ac is considered to be sinusoidal (purest waveforms) unless stated otherwise.

Frequencies that fall within certain band limits are given specific names. For example, broadcast stations transmitting between 560 kHz and 1600 kHz are in the medium frequency (MF) broadcast band. Some of the general terminology used for common bands of frequencies are given in Table 28-1.

The term *microwaves* generally means 1 GHz through 300 GHz. Above this is heat or infrared, light, ultraviolet, x-rays, gamma rays, and cosmic rays, as shown in Table 28-2.

28-2 HARMONICS

A harmonic of a frequency is a whole-number multiple of that frequency. For ex-

Table 28-1

Terms used	Frequency limits
Power frequencies	10 Hz to about 1000 Hz (cps)
Audio frequencies (AF)	20 to 20,000 Hz
Video frequencies	5 Hz to over 4.5 MHz (Mc)
Supersonic or ultrasonic frequencies	25 kHz to over 2 MHz
Very low radio frequencies (VLF)	10 to 30 kHz (kc)
Low radio frequencies (LF)	30 to 300 kHz
Medium radio frequencies (MF)	300 kHz to 3 MHz
High radio frequencies (HF)	3 to 30 MHz
Very high radio frequencies (VHF)	30 to 300 MHz
Ultrahigh radio frequencies (UHF)	300 MHz to 3 GHz (Gc)
Superhigh radio frequencies (SHF)	3 to 30 GHz
Extremely high radio frequencies (EHF)	30 to 300 GHz

Table 28-2

Heat or infrared	1×10^{12} to 4.3×10^{14} Hz
Visible light, red to violet	4.3×10^{14} to 1×10^{15} Hz
Ultraviolet	1×10^{15} to 6×10^{16} Hz
x-rays	6×10^{16} to 3×10^{19} Hz
Gamma rays	3×10^{19} to 5×10^{20} Hz
Cosmic rays	5×10^{20} to 8×10^{21} Hz

ample, 1000 Hz is the second harmonic of 500 Hz; 1500 Hz is the third harmonic. There is no "first harmonic," since multiplying by 1 is the same frequency as the fundamental.

In Fig. 28-1*a*, two sine waves are added. The lower frequency (longer wavelength) is the fundamental. The higher frequency (shorter wavelength) is the second harmonic. Both are sine waves, but the resultant, shown dashed, is not. It is the algebraic sum of the two waves. In this case the resultant is made up of two components, a fundamental and a second harmonic of the fundamental.

If only the dashed-line resultant waveform were available, it would be possible to extract from it both the fundamental and the second harmonic. Any *repetitive* ac waveform is usually the resultant of a fundamental frequency plus harmonics of that fundamental.

In Fig. 28-1*b*, the phase of the second harmonic wave has been shifted. This results in one resultant *alternation* losing amplitude and the other gaining amplitude. Again the resultant is not a sine wave. Both the amplitude and the phase of a fundamental and its harmonic(s) affect the output waveform.

In Fig. 28-1*c*, a fundamental and a third harmonic are added to produce a resultant which resembles a square wave somewhat. Actually, if the fundamental and the first 10 harmonics in the proper phases and amplitudes are added, a very good square wave can result. Thus, any square-wave ac must be the resultant of a

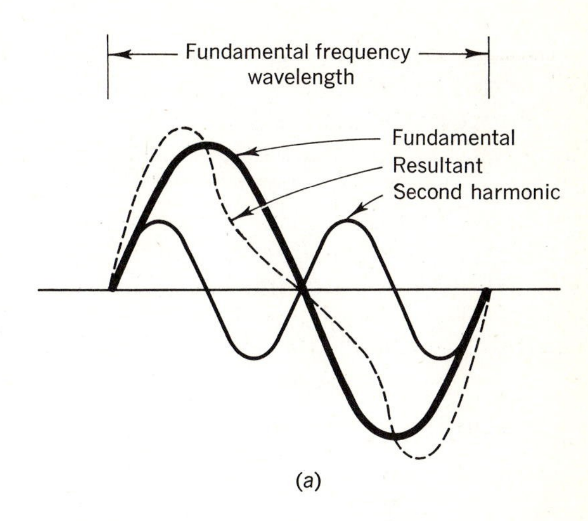

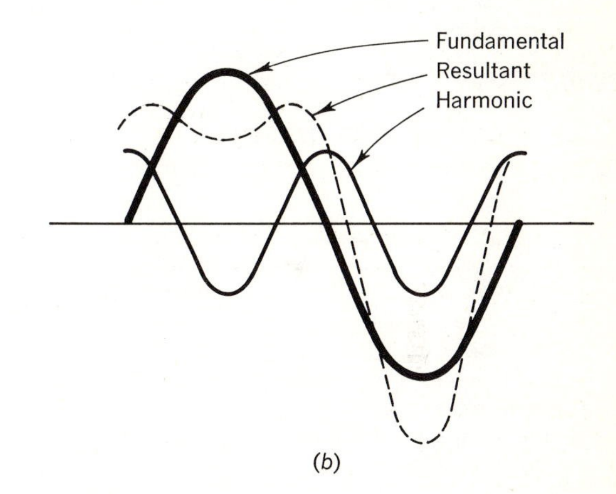

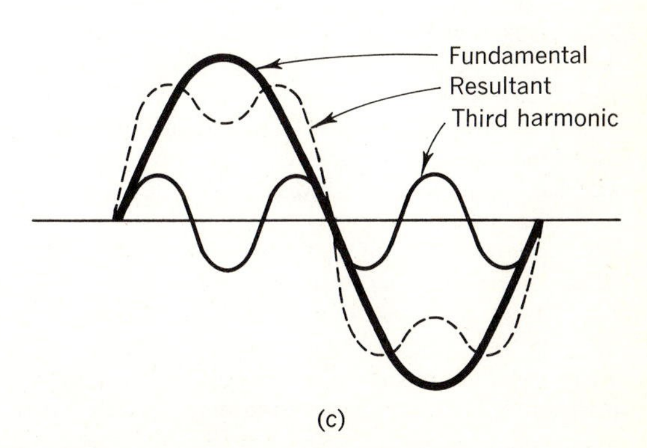

Fig. 28-1 (*a*) Fundamental, second harmonic and resultant. (*b*) Fundamental with second harmonic shifted 90°. (*c*) Fundamental and third harmonic.

sinusoidal fundamental and at least 10 sinusoidal harmonics of this fundamental. The sharper the corners and the steeper the sides, the more harmonic components a square wave must have.

To transmit a 5000-Hz square-waveform ac through an amplifier, it must amplify at least 10 harmonics of the fundamental, or from 5 kHz to at least 50 kHz. If it will amplify up to 100 kHz, the output will resemble more closely the square-wave input signal.

If the 5-kHz square wave is fed into an amplifier that will not produce any output above perhaps 8 kHz, the output will be a *sine wave* of 5 kHz only. There are no harmonics passed to alter the fundamental frequency of 5 kHz.

Quiz 28-1. Test your understanding. Answer these check-up questions.

1. A tuning fork is a simple, uncomplicated vibrating device. If a microphone picked up the sound from a tuning fork, what waveform would the ac have? __________ What waveform would the microphone produce from a human voice? __________ A pure flute tone? __________ A drum beat? __________
2. A radar impulse is 1 μs long and has a square-wave form. (The 1-μs pulse represents a half-cycle, or an alternation of this wavelength.) What is the fundamental frequency of this impulse? __________ What bandwidth should an amplifier have to reproduce this signal properly? __________
3. To what range of frequencies should a telephone microphone respond? __________ An entertainment-type stereo FM-receiver loudspeaker? __________
4. The bandpass of an oscilloscope is 0 to 10 MHz. What is the highest frequency of sine-wave ac it can show without distortion? __________ The highest frequency of square-wave ac it can show without distortion? __________ The highest frequency of sawtooth waveform it can show? __________
5. If a two-wire transmission line has inductance in series with it, will it have more reactance to a high or a low frequency? __________ Which will be shifted out of phase more? __________ Would the fundamental and the harmonics of a square-wave signal be transmitted along the line and delivered with the same square-wave form? __________ If not, what could be added to the line to make up for the inductive effect? __________
6. What are the names of the bands of frequencies that are wholly or partly in the range of hearing? __________
7. What are the names of the bands of radio frequencies that are wholly or partly above the radio-broadcast band? __________
8. Name the bands of frequencies in order above the highest radio-frequency (EHF) band. __________

28-3 THE SINE WAVE

Figure 28-2*a* illustrates two poles of an alternator with a rotating vector in the center of its field. The vector is lying along the standardized zero-reference direction, horizontal and pointing to the right. The dot at the end of the vector represents a conductor that rotates with the vector. As the vector moves counterclockwise through zero, the wire is moving parallel to lines of force, and no voltage is induced in it. In Fig. 28-2*b*, the curve indicates that at 0° there is no voltage.

As the vector arrow is moving past the 90° position, it cuts the greatest number of lines of force per second and has the maximum voltage induced. This is shown in Fig. 28-2*b* above the 90° point on the time line.

Between 0° and 90°, two other values are shown: 30° and 60°. At the 30° point the conductor will be cutting exactly half as many lines of force per unit time as at the 90° point. On the sine curve the amplitude of the emf at 30° is half of what it is at 90°. When the conductor is moving through the 60° point, it is cutting 0.866 of the maximum number of lines per unit time.

As the conductor rotates, the emf at 120° is the same as at 60°. At 150° it is the same as at 30°. 180° is the same as 0°.

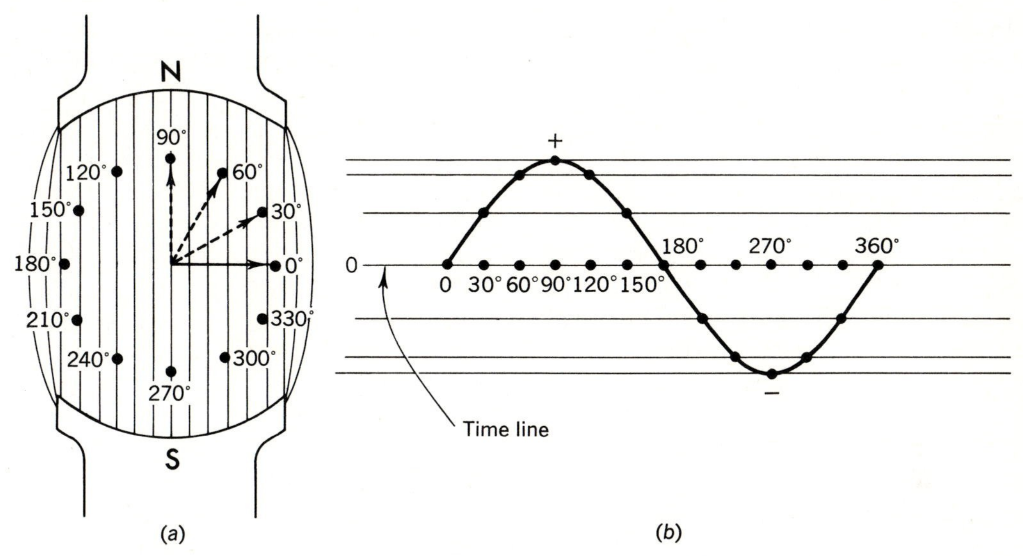

Fig. 28-2 (*a*) A conductor rotating between poles of magnets produces a voltage which varies (*b*) in a sinusoidal manner.

Continuing around, at 210° the amplitude of the emf will be exactly the same as at 30°, but the direction of cutting has reversed from right to left to left to right. As a result, the emf will have the opposite polarity. If the emf induced while the conductor travels from 0° to 90° to 180° is considered as having a positive polarity, the emf developed from 180° to 270° to 360° will have a negative polarity. It is common to give values in this second half-cycle a negative sign. At 240° and 300° the emf has the same amplitude as at 60°.

When all these values are plotted along the time line, a sine wave results. This waveform is the result of a single rotation of a vector, or *phasor*.

In Fig. 28-3*a*, the phasor has rotated 30°. The length does not change as it moves. It is a rotating radius of the circle it describes. The distance between the head of the vector and the zero reference line is shown as the dashed line *s*. This vertical distance at 30° is half the maximum at 90°. The ratio of side *s* to side *r* of the right triangle formed is called the *sine* of the angle. The sine of 30° must be the ratio *s*/*r*,

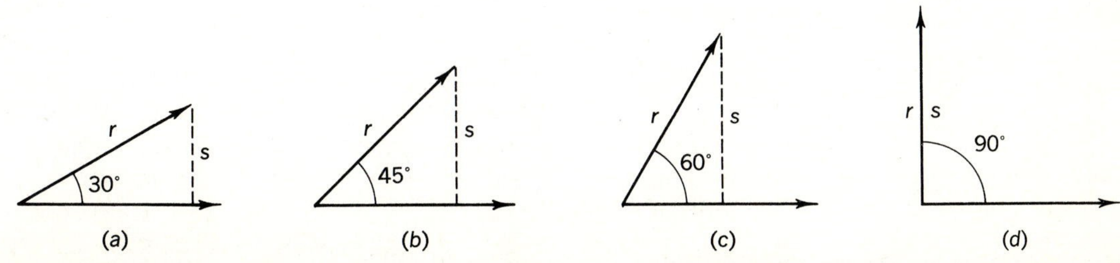

Fig. 28-3 The sine value, side *s*, increases to maximum at 90°: (*a*) sin 30° = 0.5 max; (*b*) sin 45° = 0.707 max; (*c*) sin 60° = 0.866 max; (*d*) sin 90° = 1 or maximum.

or 0.5/1, or 0.5. The sine of 60° is 0.866/1, or 0.866. The sine of 90° is 1/1, or 1.

28-4 DETERMINING INSTANTANEOUS VALUES

In sine-wave ac the maximum, or *peak*, value occurs 90° from the zero value. Values at the instants when the phasor is passing other points (30°, 60°, etc.) are *instantaneous values*. Instantaneous values (if the waveform is sinusoidal) can be determined by the formula

$$e = e_{max} \sin \theta$$

where e = instantaneous voltage value
e_{max} = maximum or peak value
$\sin \theta$ = sine of the angle at which the phasor happens to be

Since current is directly proportional to the emf in a circuit, it is possible to determine instantaneous currents by

$$i = i_{max} \sin \theta$$

The lowercase e and *i* indicate that the value is an instantaneous one. An uppercase *E* or *I* usually indicates an ac rms value or a dc value.

The angles between 0° and 90° are said to be in the *first quadrant* (Fig. 28-4*a*). Those between 90° and 180° are in the *second quadrant*. From 180° to 270° is the *third quadrant*, and from 270° to 360° is the *fourth quadrant*. Angles in the first and second quadrants are given positive signs. Third and fourth quadrants are negative.

If an angle is in a quadrant other than the first, it is usually necessary to convert to an equivalent first-quadrant angle. Many trigonometric tables and most slide rules give only angles from 0° to 90°. To convert to first-quadrant angles, determine how many degrees a given angle is from 0°, 180°, or 360°. As an example, 130° is in the second quadrant. Subtracting from 180°, the first quadrant angle is 180 − 130 = 50°. The sine of 130° is the same as the sine of 50° (Fig. 28-4*b*).

190° is in the third quadrant and is 180° − 190°, or −10°. The sine of 190° equals the sine of 10° except that it is negative.

The angle of 310° equals 310 − 360, or −50°. The sine of 310° equals the sine of 50°, 130°, and 230°.

To convert higher-quadrant angles to first quadrant, sketch a sine wave and note the quadrant in which the desired angle falls.

ANSWERS TO CHECK-UP QUIZ 28-1

1. (Sinusoidal) (Jagged, complex) (Nearly sinusoidal) (Jagged square wave) **2.** (500 kHz) (5 MHz) **3.** (200 to 3000) (20 to 18 kHz) **4.** (At least 10 MHz) (1 MHz) (1 MHz) **5.** (High) (High) (No) (Capacitance across the line) **6.** (Power, AF, video, VLF) **7.** (MF, HF, VHF, UHF, SHF, EHF) **8.** (Infrared, light, ultraviolet, x-ray, gamma ray, cosmic ray)

Quiz 28-2. Test your understanding. Answer these check-up questions.

1. In which quadrants do the following fall? 55° ______ 320° ______ 195° ______ 150° ______ 290° ______ 380° ______ 465° ______ 700° ______ 850° ______ 3475° ______

2. Give the instantaneous voltage values for the following angles if the peak voltage is 100 V. (Use a slide rule first and then check with a sine table or a pocket calculator.) 55° ______ 320° ______ 195° ______ 150° ______ 290° ______

3. The standard ac line has an rms value of 120 V. To the closest whole-number voltage value, what is the peak voltage? ______ What are the instantaneous voltage values at the following angles? 10° ______ 20° ______ 35° ______ 45° ______ 55° ______ 65° ______ 75° ______ 85° ______

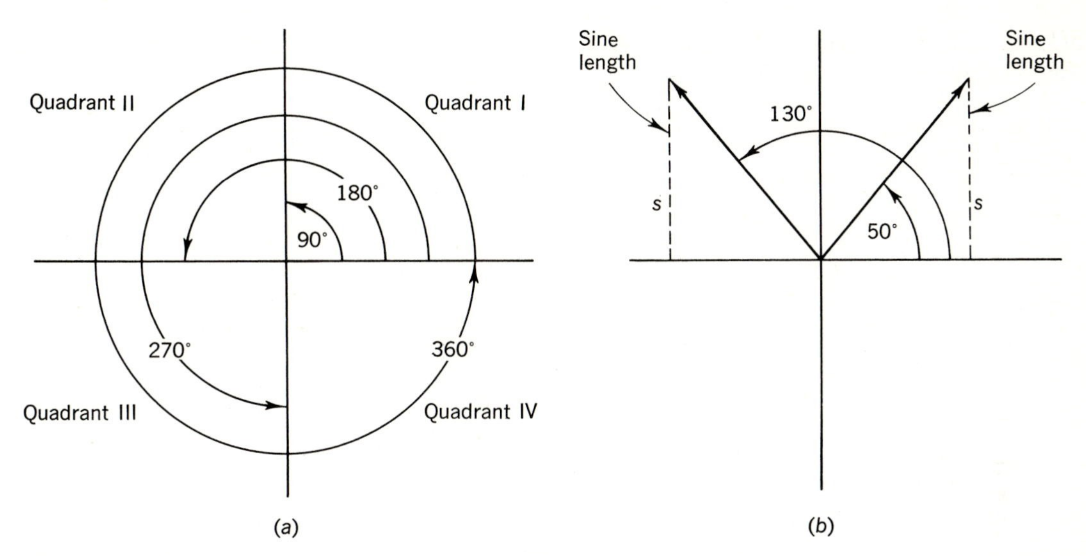

Fig. 28-4 (*a*) The four quadrants of an ac cycle. (*b*) The sine values of 50° and 130° are the same amplitude or strength.

28-5 ADDING TWO SINE WAVES

There are times when two sine waves are in series and must be added as in Fig. 28-5*a*. The instantaneous voltage is the sum of the two voltages at the chosen instant. The formula below adds only two instantaneous-voltage expressions. For Fig. 28-5*b*, with $e_{max_1} = 100$ V and $e_{max_2} = 60$ V, the formula for e_{total} is

$$\begin{aligned} e_t &= e_{max_1} \sin \theta_1 + e_{max_2} \sin \theta_2 \\ &= 100 \sin \theta_1 + 60 \sin \theta_2 \end{aligned}$$

The instantaneous voltage at 50° is

$$\begin{aligned} e_t &= 100 \sin 50^\circ + 60 \sin 50^\circ \\ &= 100(0.766) + 60(0.766) \\ &= 76.6 + 45.96 \\ &= 122.56 \text{ V} \end{aligned}$$

Since the two voltages are in phase, the expression could have been simplified to

$$e_t = (e_{max_1} + e_{max_2}) \sin 50^\circ$$

When the two voltages are out of

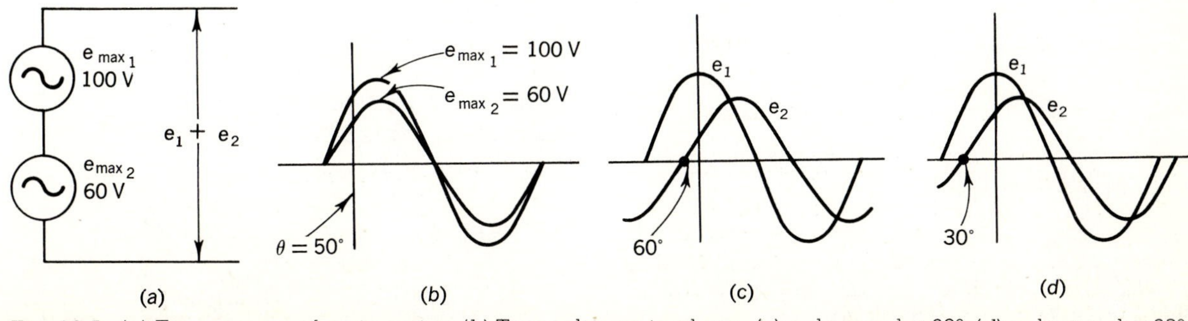

Fig. 28-5 (*a*) Two sources of ac in series. (*b*) Two voltages in phase. (*c*) e_2 lags e_1 by 60°. (*d*) e_2 lags e_1 by 30°.

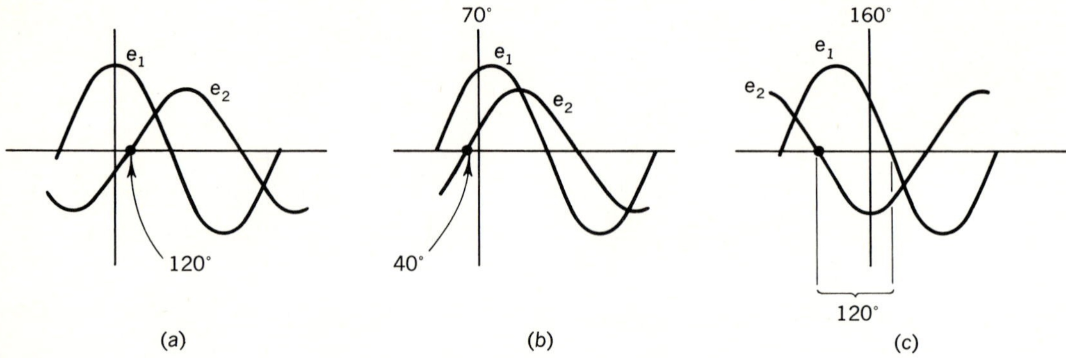

Fig. 28-6 (*a*) Graph of 100-V e_1 and 60-V e_2, with e_2 lagging by 120°. (*b*) Same voltages with e_2 lagging by 40°. (*c*) Same voltages with e_2 leading by 120°.

phase, as in Fig. 28-5*c*, with the lower voltage, e_2, lagging e_1 by 60°, what is the total instantaneous voltage when e_1 is at maximum? When e_1 is at maximum, e_2 is 30° from zero, or at half-voltage. Therefore the answer must be 100 + 0.5(60), or 100 + 30, or 130 V. The formula must consider the difference in phase between the two waves; thus

$$\begin{aligned} e_t &= e_{\max_1} \sin \theta_1 + e_{\max_2} \sin (\theta_1 - \theta_2) \\ &= 100(1) \quad + 60 \sin (90° - 60°) \\ &= 100 \quad + 60 \sin 30° \\ &= 100 \quad + 60(0.5) \\ &= 130 \text{ V} \end{aligned}$$

In Fig. 28-5*d*, e_2 lags e_1 by 30°. When e_1 is at maximum, what is the instantaneous voltage? The formula is the same,

$$\begin{aligned} e_t &= e_{\max_1} \sin \theta + e_{\max_2} \sin (\theta_1 - \theta_2) \\ &= 100(1) \quad + 60 \sin (90° - 30°) \\ &= 100 \quad + 60 \sin 60° \\ &= 100 \quad + 60(0.866) \\ &= 151.96 \text{ V} \end{aligned}$$

ANSWERS TO CHECK-UP QUIZ 28-2

1. (1st) (4th) (3rd) (2nd) (4th) (1st) (2nd) (4th) (2nd) (3rd) **2.** (81.92 V) (−64.28 V) (−25.88 V) (50 V) (−93.97 V) **3.** (170 V) (29.5 V) (58.1 V) (97.5 V) (120 V) (139 V) (154 V) (164 V) (169 V)

In Fig. 28-6*a*, e_2 is lagging e_1 by 120°. What is the instantaneous sum voltage when e_1 is at maximum? Since the two voltages are of opposite polarity at this instant,

$$\begin{aligned} e_t &= e_{\max_1} \sin \theta \ + e_{\max_2} \sin (\theta_1 - \theta_2) \\ &= 100(1) \quad + 60 \sin (90° - 120°) \\ &= 100 \quad + 60 \sin (-30°) \\ &= 100 \quad + 60(-0.5) \\ &= 100 \quad - 30 \\ &= 70 \text{ V} \end{aligned}$$

In Fig. 28-6*b*, e_2 is *lagging* e_1 by 40°. What is the instantaneous sum voltage when e_1 is at 70°?

$$\begin{aligned} e_t &= e_{\max_1} \sin \theta \ + e_{\max_2} \sin (\theta_1 - \theta_2) \\ &= 100 \sin 70° + 60 \sin (70° - 40°) \\ &= 100(0.9397) + 60(0.5) \\ &= 93.97 \quad + 30 \\ &= 123.97 \text{ V} \end{aligned}$$

Suppose e_2 is *leading*. In Fig. 28-6*c*, e_2 is leading e_1 by 120°. What is the instantaneous sum voltage when e_1 is at 160°?

$$\begin{aligned} e_t &= e_{\max_1} \sin \theta \ + e_{\max_2} \sin (\theta_1 + \theta_2) \\ &= 100 \sin 160° + 60 \sin (160° + 120°) \\ &= 100 \sin 20° \ + 60 \sin 280° \\ &= 100 \sin 20° \ + 60 \sin -80° \end{aligned}$$

$$= 100 \sin 20° - 60 \sin 80°$$
$$= 100(0.342) - 60(0.9848)$$
$$= 34.2 - 59.09$$
$$= -24.89 \text{ V}$$

Since 280° is in quadrant IV, it is in the negative half-cycle. For this reason when the sine value for 280° is converted to a quadrant I value, the negative sign is retained, making the angle −80°. To change this negative angle to a positive one, it is necessary to change the polarity of the sign between the terms.

Quiz 28-3. Test your understanding. Answer these check-up questions.

1. Two ac voltages are in phase. One reads 100 V on a voltmeter, the other indicates 40 V. What are the peak voltage values? ____________
2. In question 1, what is the instantaneous sum voltage when E_1 is at 45°? ____________ 155°? ____________ 245°? ____________
3. $e_{max_1} = 120$ V and $e_{max_2} = 70$ V. If E_2 lags 40°, what is the instantaneous-voltage sum when E_1 is at 90°? ____________ 220°? ____________ What formula was used for the last part? ____________
4. In question 3, if E_2 leads 30°, what is the instantaneous-voltage sum when E_1 is at 70°? ____________ What formula was used for this? ____________ What is the voltage when e_{max_1} is at 300°? ____________
5. In question 3, if E_2 lags 125°, what is the instantaneous-voltage sum when E_1 is at 45°? ____________ What formula was used for this? ____________ What is the voltage when E_1 is at 160°? ____________

28-6 RADIANS AND ANGULAR VELOCITY

In working with angles of phasors rotating in a circle, the degree has been the basic unit of measurement (360° in a cycle). It is possible to develop other systems for the measurement of angles.

One system used in electricity is known as *radian measurement*. It is based on the radius of a circle. Figure 28-7 illustrates the radius of a circle and a radian of that circle. The perimeter of a circle is the distance a rotating phasor tip travels in one cycle. This is 2π times the radius value. So one cycle must equal 2π *radians* (or rad). Thus 2π rad equals 360°, π rad equals 180°, and $\pi/2$ rad is 90°. If $360° = 2\pi$ rad, then

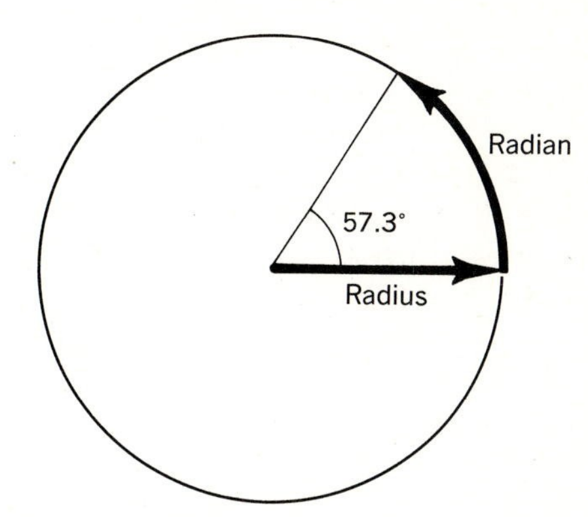

Fig. 28-7 A radian is equal to the length of the radius, or 57.3°, or $1/(2\pi)$ of a circle.

$$1 \text{ rad} = \frac{360}{2\pi} = \frac{360}{6.28} = 57.3°$$

A radian can be considered either 57.3° or an arc on a circle having a length equal to the radius of the arc.

An ac E or I rotating phasor moves at a rate of 2π rad, or 6.28 rad. The number of radians a phasor describes in 1 s is called the *angular velocity* of the phasor. The angular velocity of a phasor when the frequency is 10 Hz is $2\pi 10$, or 62.8 rad/s. *Angular velocity* is equal to $2\pi f$, where f is frequency in hertz. The lowercase Greek letter omega (ω) indicates angular velocity, or $2\pi f$.

The formula for the inductive reactance of an inductor to a sine-wave ac is $X_L = 2\pi f L$, or $X_L = \omega L$. This formula says the opposition in *ohms* produced by the counter-emf developed in a coil is the product of the angular velocity of the magnetic field cutting the turns times the inductance of the inductor in henrys.

The capacitive reactance formula may be written $X_C = 1/\omega C$.

In some cases it is necessary to determine an instantaneous value of a voltage or current some time after the start of a cycle. For example, what is the instantaneous voltage 12 ms after the start of a 60-Hz cycle if e_{max} is 100 V? In 0.012 s the number of cycles is $f(t)$, or 60(0.012), or 0.72 cycle (not hertz). The 0.72 cycle can be converted into degrees by 0.72(360), or 259°. The instantaneous voltage is

$$\begin{aligned} e &= e_{max} \sin 259° \\ &= 100 \sin -79° \\ &= 100(-0.982) \\ &= -98.2 \text{ V} \end{aligned}$$

All these steps can be condensed into one formula for any instantaneous voltage,

$$e = e_{max} \sin (360ft)°$$

In working with radian measure, the equivalent formula would be

$$e = e_{max} \sin (2\pi ft) = e_{max} \sin \omega t$$

The problem above would be solved as follows:

$$\begin{aligned} e &= 100 \sin [6.28(60)(0.012)] \\ &= 100 \sin (4.52 \text{ rad}) \\ &= 100 \sin [4.52(57.3°)] \\ &= 100 \sin 259° \qquad \text{(Quad. III)} \\ &= 100 \sin 79° \qquad \text{(Quad. I)} \\ &= 100(0.9816) \\ &= -98.2 \text{ V} \end{aligned}$$

Instantaneous currents can be expressed as

$$i = i_{max} \sin \omega t = i_{max} \sin (360ft)°$$

ANSWERS TO CHECK-UP QUIZ 28-3

1. (141 V) (56.5 V) **2.** (140 V) (83.5 V) (−179.2 V) **3.** (173.6 V) (−77.1 V) [$e_t = e_{max_1} \sin 220 + e_{max_2} \sin (220 - 40)$] **4.** (182 V) [$e_t = e_{max_1} \sin 70° + e_{max_2} \sin (70 + 30)$] (−139 V) **5.** (+16 V) [$e_t = e_{max_1} \sin 45° + e_{max_2} \sin (45 - 125)$] (+81.2 V)

28-7 PERIOD AND WAVELENGTH

The time in seconds it takes to complete a full cycle of ac is called the *period* of the cycle. Except for frequencies of less than 1 Hz, the period will be a decimal fraction. The period of a 50-Hz cycle is 1/50, or 0.02 s. The period of an 800,000-Hz cycle is 1/800,000, or 0.000 001 25 s, or 1.25 μs. It takes only 1.25 μs for the cycle to go from 0° to 360°.

The velocity of electromagnetic wave travel through air or a vacuum is considered as 300,000,000 m/s (actually 299,792,462 m/s). Other velocities used are 186,000 U.S. mi/s, or 162,000 nautical mi/s. A 1-MHz wave radiated from a radio transmitter has a period of 0.000 001 s. In the period of one cycle, the wave travels one-millionth of 300,000,000 m, or 300 m. The emission has a wavelength of 300 m. If these electromagnetic waves could be frozen in space for a time, the length of a cycle of the wave would be 300 m. Along with such electromagnetic waves, but at right angles to them, are electrostatic waves. This velocity is usually considered to be the *impulse speed* of electricity in wires. However, inductance of the wires and capacitance between them requires time to charge and discharge, reducing the actual velocity on a two-wire line. This *delay time* in seconds can be determined by

$$T_d = \sqrt{LC}$$

where T_d = delay time, in seconds
L = inductance of the wire comprising the line, in henrys
C = capacitance across the same line, in farads

The basic formula used to determine the wavelength λ (lambda) of a wave is

$$\lambda = \frac{V}{f}$$

where λ = wavelength, in the same units used in measuring velocity
V = velocity
f = frequency, in hertz

The wavelength of electromagnetic waves in space is found by

$$\lambda = \frac{300{,}000{,}000}{f} \quad \text{m}$$

A television station transmitting on 56 MHz has a wavelength of

$$\lambda = \frac{300{,}000{,}000}{56{,}000{,}000} = 5.36 \text{ m}$$

For airborne sound waves, the velocity is approximately 331 m/s (1085 ft/s). A 50-Hz sound wave has a wavelength of

$$\lambda = \frac{V}{f} = \frac{331}{50} = 6.62 \text{ m} \qquad (21.7 \text{ ft})$$

In many cases a *half-wavelength* is the important measurement. Many radio transmitting and receiving antennas are a half-wavelength long. Thus a TV antenna to receive a frequency of 56 MHz should be 5.36/2, or 2.68 m long. A sounding board for a loudspeaker in which the lowest frequency is to be 50 Hz should have a minimum dimension of 6.62/2, or 3.31 m, to produce maximum signal at this frequency.

The measurement of electromagnetic wavelengths in meters is practical up to a certain point. At a frequency of 3 GHz, the wavelength is 0.1 m, or 10 cm. Above 3 GHz it is common to use centimeter and millimeter wavelengths.

For heat or infrared frequencies the *micron* (μ), or millionth of a meter, is used.

For light frequencies the *angstrom* (Å) unit, 10^{-10} m, is usually employed.

The x-ray unit (XU) is 10^{-13} m long.

Quiz 28-4. Test your understanding. Answer these check-up questions.

1. How many degrees are there in 2π rad? ______ 0.6π rad? ______
2. How many radians are there in 180°? ______ In 30 cycles? ______ In 1365°? ______
3. What is another name for a rotating vector? ______
4. What is the angular velocity for a vector in a 60-Hz circuit? ______ A vector that travels through 40 rad in 0.2 s? ______ What is the frequency of this last vector? ______
5. What is ωL? ______ $1/\omega C$? ______
6. What is the instantaneous voltage in a 400-Hz circuit that has an rms voltage of 340 V at an instant 0.0002 s after the beginning of the cycle? ______
7. What can be solved with the formula $i_{max} \sin \omega t$? ______
8. A 60-Hz ac starts at a $+e_{max}$ of 200 V. What is the instantaneous voltage value 1.58 s later? ______
9. What is the wavelength of middle C on a piano? ______
10. Station WWV in Boulder, Colorado, transmits standard tones of A above middle C (440 Hz), and 600 Hz. What are the wavelengths of these tones in air? ______ ______
11. Station WWV transmits continuously on frequencies of 2.5, 5, 10, 15, and 20 MHz. What are the wavelengths of these frequencies? ______ ______ ______ ______ ______
12. How long should you cut an antenna for best

reception of station WWV on 5 MHz? ______

13. What unit of measurement is used in the wavelength of light frequencies? ______ Of x-ray frequencies? ______

14. A two-wire transmission line has 15 pF/m and 1.4 μH/m. How long does it take for an impulse to travel 250 m? ______

CHAPTER 28 TEST • SINE WAVES

1. What band of audio frequencies is required for good telephone communication?
2. The lowest frequency that can be detected by the human ear is about 16 Hz. What is the highest?
3. What bands of frequencies are classed as the "microwaves"?
4. What is the only basic difference in radio waves, microwaves, light waves, and x-rays?
5. What is the approximate band of frequencies required to pass a 500-Hz square-wave ac signal?
6. When a sine-wave fundamental and a second harmonic are mixed, is the resultant waveform sinusoidal?
7. In general, when a third harmonic is added to a fundamental, what basic shape does the resultant wave have?
8. What is another name for a rotating vector?
9. Through how many degrees must a rotating vector move to produce 1 rad?
10. By memory, what is the sine value of an angle of 0°? 30°? 45°? 60°? 90°?
11. If the peak voltage of a sinusoidal ac is 84 V, what is the instantaneous voltage at 52°? 295°?
12. What does a lowercase e (or *i*) indicate?
13. What does a capital *E* (or *I*) without subscripts indicate?
14. What sign is given to voltages in quadrant IV? II? III? I?
15. What are three other angles of a circle that have the same sine value as 38°?
16. e_{max_1} is 50 and e_{max_2} is 30. If these voltages are in series and in phase, what is the sum voltage when e_{max_1} is 43°?
17. If the frequency of an ac is 50 Hz, what is the voltage value 0.0035 s after the start of the cycle if e_{max} is 100 V?
18. e_{max_1} is 50 V and e_{max_2} is 30 V. If E_2 lags by 65° and the voltages are in series, what is the sum voltage when E_1 is at 40°?
19. e_{max_1} is 50 V and e_{max_2} is 30 V. If E_2 is leading by 125° and the voltages are in series, what is the sum voltage when E_1 is at 210°?
20. What name is given to the value $2\pi f$? What symbol is used to signify it?
21. What is the instantaneous voltage 8 ms after the start of a 45-Hz cycle if e_{max} is 75 V?
22. What is the angular velocity of a phasor when the frequency of rotation is 4500 Hz?
23. What is the period of the color-burst ac in a television transmission line if its frequency is 3.58 MHz?
24. If the moon is 360,000 km from the earth, how long would it take for a radar signal to be transmitted, reflect from the moon, and be received back on earth?
25. How long is a half-wavelength 7-MHz antenna wire?
26. What delay will occur in a wave traveling along a two-wire line having 20 pF/m and 1.4 μH/m if it is 1000 m long? How long would it take a radio wave to travel 1000 m in air?
27. What is the velocity of radio waves? Sound in air?

ANSWERS TO CHECK-UP QUIZ 28-4

1. (360°) (108°) **2.** (π, or 3.14) (188.4) (23.8) **3.** (Phasor) **4.** (376.8 rad) (200 rad/s) (31.8 Hz) **5.** (X_L) (X_C) **6.** (231.3 V) **7.** (Instantaneous current value) **8.** (61.8 V) **9.** (1.29 m) **10.** (7.52 m or 2.47 ft) (0.551 m or 1.81 ft) **11.** (120 m) (60 m) (30 m) (20 m) (15 m) **12.** (30 m) **13.** (Angstrom) (XU) **14.** (1.97×10^{-6} s or 1.97 μs)

29

RESISTORS AND INDUCTORS

CHAPTER OBJECTIVE. To become familiar with the more common resistors, miniature inductors, air-core and iron-core chokes, and to determine energy in inductors, coefficient of coupling, and inductance.

29-1 WIRE-WOUND RESISTORS

Resistors come in many forms. They may be wire wound, carbon rod, composition, or carbon film with positive, negative, or zero temperature coefficients of resistance. Wire-wound resistors have dissipation ratings up to thousands of watts. They may be fixed, tapped, adjustable, potentiometers, or rheostats (Fig. 29-1).

Power dissipation is a rating for constant use in free air. If a resistor is operated in a confined area, its temperature may rise excessively. If a resistor is air-cooled, it may safely dissipate several times its rated power. Liquid cooling is used, as in *dummy loads* for testing high-frequency ac devices. Resistors immersed in oil may dissipate twice their rated power values. Resistors held in metal mounting clips may safely dissipate double their ratings.

Temperature coefficient (TC) is the decimal fraction of change in resistance produced by heating a resistor. It is expressed as parts per million per degree C (ppm/°C), or in ohms per ohm per degree of temperature change (Ω/Ω/°C). A +TC indicates that the resistor increases in resistance when heated. A −TC resistor decreases resistance when warmed. Most metals have +TC values. Semiconductors (germanium, silicon, carbon) have −TC values. A low TC might be ±20 ppm/°C, which is ±0.00002 Ω/Ω°C, or ±0.002%/°C.

A true zero-TC resistor (no change in resistance with temperature) is not possible except over a narrow temperature range.

Wire-wound resistors may have *axial leads* (wire terminals in line with the axis of the body of the resistor), as in Fig. 29-1. Other resistors have terminal lugs at the ends. Still others have wire leads at the ends at right angles to the body, called *radial leads.*

Some wire-wound resistors have a baked-on vitreous-ceramic covering. If the vitreous insulating covering is left off along a longitudinal strip, a connection can be made with an adjustable clamp to any point along the resistor. It is then called an *adjustable* wire-wound resistor.

Most wire-wound resistors are coils of resistance wire that have some inductive reactance. At audio frequencies up to 20,000 Hz, the effect may be insignificant. At 500 kHz a 100-Ω resistor may also have 100 Ω of X_L and operate as a 141-Ω impedance device.

There are two simple methods of

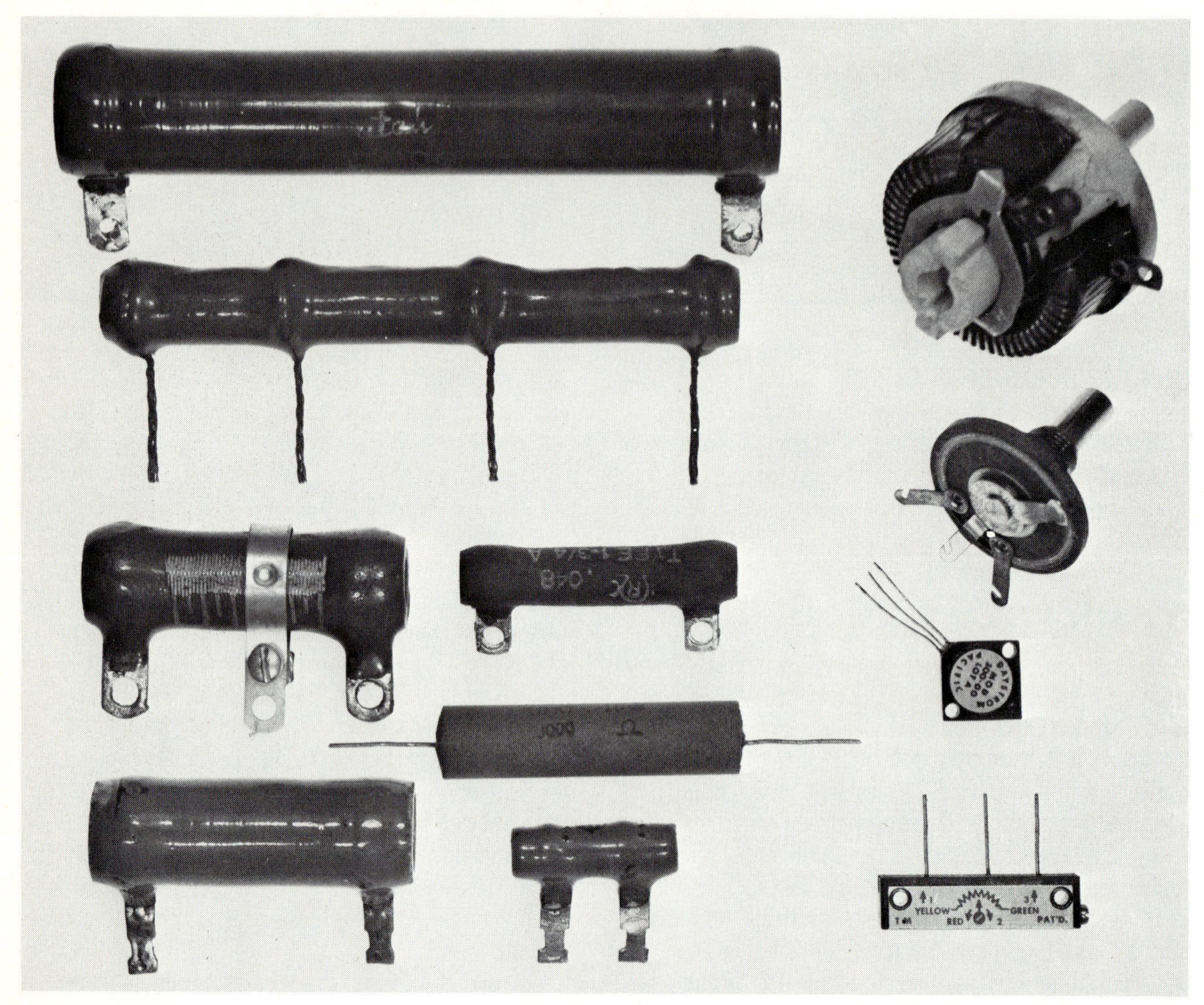

Fig. 29-1 Wire-wound resistors. Left, from top: a fixed 50-W, a tapped, an adjustable 20-, 20-, 10-, and 5-W radial resistors and one axial-lead resistor. Right: a rheostat with arm at the "off" position, a potentiometer with arm near the center, and two miniature screwdriver-adjustable potentiometers. An adjusting screw at right on bottom potentiometer.

winding *noninductive* wire resistors. In Fig. 29-2*a*, the resistor is wound on a thin sheet of insulating material. At any instant that current is flowing up in the solid-line wires on one side of the sheet, the same value will be flowing down in the wires on the other side, shown dashed. The magnetic fields of these two currents induce emf's in adjacent wires that cancel, resulting in almost zero counter-emf and almost no inductive effect. In Fig. 29-2*b*, the resistance wire is bent into a hairpin and wound as a double wire around a cylindrical form. Currents flowing in adjacent wires induce opposite emf's that cancel the inductive effect.

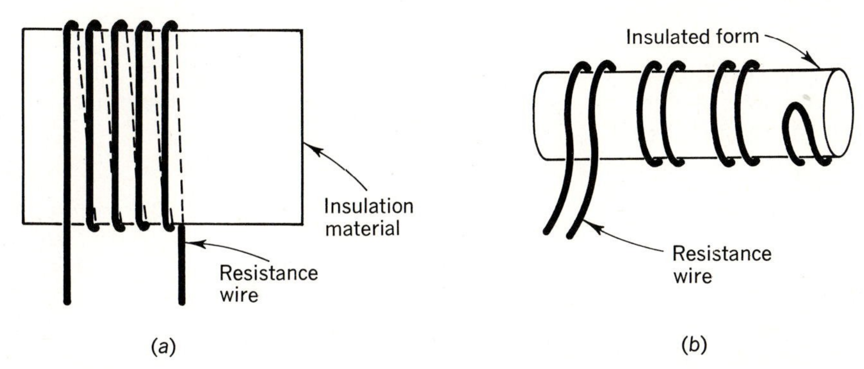

Fig. 29-2 Noninductive wire-wound resistors: (*a*) flat type; (*b*) round type.

29-2 CARBON RESISTORS

For resistors with dissipation ratings lower than 2 W, carbon resistors are commonly used. Such resistors are a composition of carbon and clay made into a rod, with contact wires implanted at both ends. The whole resistor is covered with an insulating shell and baked (Fig. 29-3). It is then color-coded with paint to indicate its resistance value.

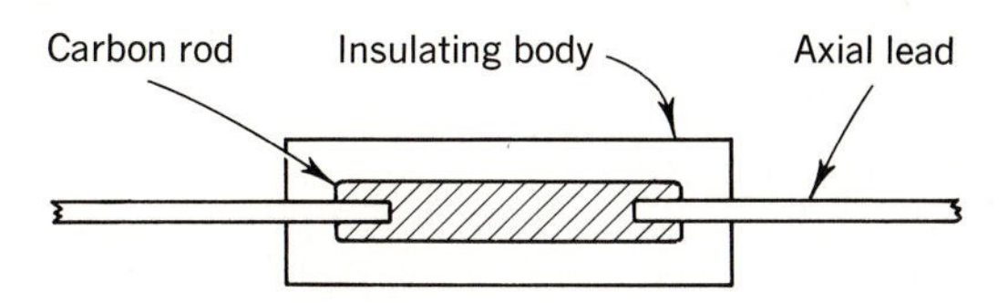

Fig. 29-3 An axial-lead carbon resistor, cross section.

The physical size of a carbon resistor is a clue to its wattage rating but not to its resistance value. For example a 1-W resistor is about 15 mm long and 5 mm in diameter. A $\frac{1}{2}$-W resistor is about 10 mm long and 3 mm in diameter. Carbon resistors are available in $\frac{1}{10}$-, $\frac{1}{4}$-, $\frac{1}{2}$-, 1-, and 2-W ratings. Smaller and larger ratings are available on special order. Resistance ratings of any wattage of resistor will range from a fraction of an ohm to many millions of ohms.

Carbon resistors carry a tolerance as well as resistance and power ratings. Previously the poorest tolerance was 20%. That is, a 10,000-Ω resistor with a tolerance of 20% could have any value between 8000 and 12,000 Ω. A 10-kΩ resistor with a 10% tolerance is guaranteed to have a resistance of 9 to 11 kΩ. Today 10% is usually the highest tolerance. Other tolerances are 5% and 2%. Precision resistors carry a tolerance rating of 1% or less. The lower the tolerance, the more resistors may cost. In many electronic circuits, a 10% variation of resistance or capacitance does not make much difference in operation.

10% resistors can be obtained with resistance values in ohms of

1	1.5	2.2	3.3	4.7	6.8	10
1.2	1.8	2.7	3.9	5.6	8.2	

as well as all multiples of 10 of these values up to 22 MΩ.

5% resistors can be purchased with resistance values of

1	1.5	2.2	3.3	4.7	6.8	10
1.1	1.6	2.4	3.6	5.1	7.5	
1.2	1.8	2.7	3.9	5.6	8.2	
1.3	2	3	4.3	6.2	9.1	

as well as all multiples of 10 of these values, up to 22 MΩ. Note that there is twice the choice of values in the 5% resistors.

A superior resistor is the metal-film type. Such deposited-film resistors consist of a glass or insulating rod with metal contacts and leads at the two ends. Metal molecules are deposited on the rod in an evacuated chamber until the desired resistance is developed between the leads. The resistor is then encapsulated with an insulating body and marked with its resistance value. Such a resistor has a tolerance of 1% or better.

Quiz 29-1. Test your understanding. Answer these check-up questions.

1. What two types of wire-wound resistors could be used alone as voltage dividers? ____________
2. What are two ways of safely operating a 10-W resistor while it is dissipating 20 W? ____________
3. What special type of wire-wound resistor would be most suitable as a dummy load for a high-frequency ac? ____________
4. What is the change in resistance of a 100-Ω resistor with a TC of 30 ppm/°C if its temperature is raised 20°C? ____________
5. A 5000-Ω resistor has a −TC of 20 ppm/°C. What is its resistance when it is heated from 20 to 70°C? ____________
6. What general types of substances have negative TC values? ____________
7. What are the leads called that come out the center of the ends of a resistor? ____________
8. Which would be more likely to be noninductive, a wire-wound resistor or a carbon resistor? ____________ Which would usually have the greater power-dissipation rating? ____________
9. Would a long carbon resistor indicate high resistance, high voltage rating, or high power dissipation? ____________
10. Would a 24,000-Ω resistor have a tolerance of 10% or 5%? ____________

29-3 COLOR CODES

Carbon resistors, tubular capacitors, and small tubular inductors may have color-coded stripes or dots on them to indicate their values and tolerances. The code values for the colors are:

1 = Brown
2 = Red
3 = Orange
4 = Yellow
5 = Green
6 = Blue
7 = Violet
8 = Gray
9 = White
0 = Black

Silver: As fourth stripe = 10% tolerance
As third stripe = 0.01 multiplier

Gold: As fourth stripe = 5% tolerance
As third stripe = 0.1 multiplier

The basic color-coding uses three colors. The first and second represent significant numbers. The third color is the *multiplier* and represents the number of zeros that follow the significant figures. A brown stripe, a red stripe, and an orange stripe on a resistor indicate a value of 1 2 000 Ω. (A capacitor marked with the same colored dots would have 12,000 pF, or 0.012 μF. A tubular inductor would have 12,000 μH, or 0.012 H.) The tolerance may be shown by a fourth stripe (Fig. 29-4). The absence of a fourth stripe indicates a 20% tolerance.

When the *first* color-code stripe is double the width of the other stripes, it indicates the resistor is wire-wound.

When servicing old equipment, the technician may find radial-lead carbon resistors with a "body-end-dot" color code. The body color is the first significant number, a splash of color on one end is the second significant number, and the multiplier or number of zeros is indicated by a dot on the body (Fig. 28-8c). When no dot is visible, the dot is the same color as the body. Gold or silver markings on one end indicate the tolerance.

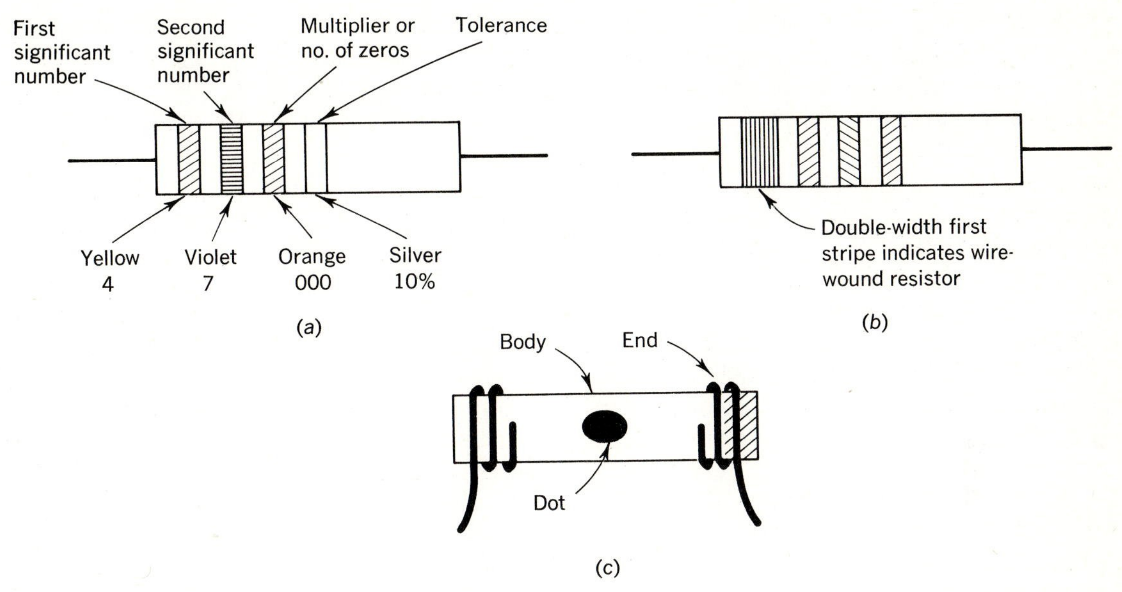

Fig. 29-4 Color-coded tubular resistors: (*a*) axial-lead carbon; (*b*) axial-lead wire-wound; (*c*) radial-lead body-end-dot carbon.

Quiz 29-2. Test your understanding. Answer these check-up questions.

1. What are the resistance values and the tolerance of resistors with the following markings?
   Red, violet, brown. ____________
   Green, blue, orange. ____________
   Gray, red, red. ____________
   Brown, green, black. ____________
   Blue, gray, green. ____________
   Orange, orange, orange, silver. ____________
   Violet, green, black, gold. ____________
   Red, yellow, brown, gold. ____________
   Brown, gray, blue, gold. ____________
   Brown, black, black, silver. ____________

2. (Values will be less than 10 Ω.) What are the resistance values and the tolerance of resistors with the following markings?
   Red, yellow, silver. ____________
   Green, blue, silver. ____________
   Orange, blue, gold. ____________
   Brown, gray, gold. ____________
   Violet, green, silver. ____________
   White, brown, gold. ____________
   Yellow, violet, gold. ____________
   Green, brown, silver. ____________

29-4 MINIATURE INDUCTORS

Any piece of wire has *inductance*, the ability to produce counter-emf. By coiling the wire, the inductance is increased as the *square* of the number of turns. Specially made components consisting of coiled copper wires are called *inductors*. If they are wound on nonferrous materials, they are called *air-core inductors*. If they are wound on iron, powdered iron, or ferrite cores, they are called *iron-core* or *ferrite inductors*. Inductors range in value from the tiny, few-turn air-core coils of 0.1 μH used in high-frequency systems, to iron-core "choke" coils of 50 H or more for low-frequency applications.

Small inductors in printed circuits are wound on either nonferrous or ferrite forms. They are then encapsulated in an epoxy or other insulating material and may closely resemble axial-lead carbon resistors. The resemblance is increased by their color-coding stripes. These inductors

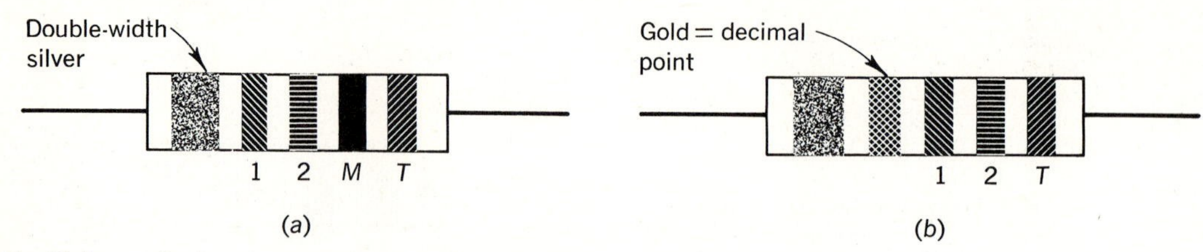

Fig. 29-5 Color-coded inductors. (*a*) Numbers indicate significant figures, *M* indicates multiplier, and *T* indicates tolerance percentage. (*b*) A gold first thin stripe indicates a decimal point.

have a double-width silver stripe at one end, and range from 0.1 μH with a Q of perhaps 50 at 25 MHz, to 100 mH with a Q of perhaps 20 at 80 kHz. Small inductors in some printed circuits are developed in a spiral form on the flat surface of the insulating substrate.

Figure 29-5 illustrates color-coding of small inductors. In Fig. 29-5*a*, one end of the inductor has a double-width silver stripe. Reading from this stripe, the first thin stripe is the first significant figure, the second stripe is the second significant figure, the third stripe is the multiplier, and the last stripe is the tolerance. However, if the first thin stripe is gold, it signifies a decimal point. A 35-μH 5% inductor carries a double-width silver, an orange, a green, a black, and a gold stripe. A 0.47-μH 10% inductor is marked double-width silver, gold, yellow, violet, silver.

Lower-value miniature inductors are about 12 mm long and about 4 mm in diameter. The larger ones may be perhaps 20 mm long and 12 mm in diameter.

One variable miniature inductor has a ferrite core made in the form of a screw. When it is screwed into the center of the threaded core area (Fig. 29-6) the inductance of the coil can be increased about 10%. Values range from 0.1 to about 5000 μH.

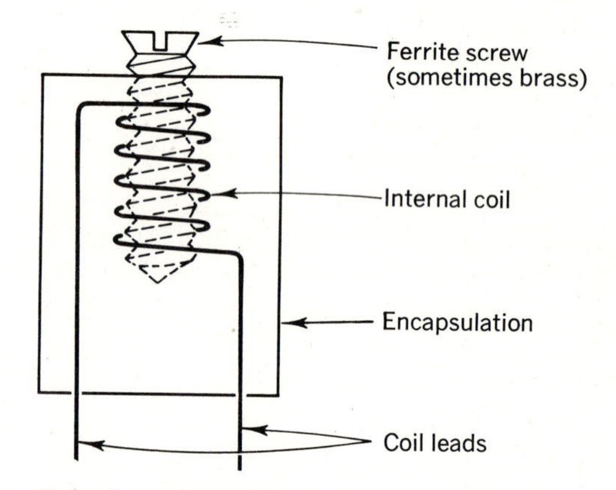

Fig. 29-6 An adjustable miniature inductor.

If a copper or brass screw is used as the core of a coil, it acts as a shorted turn and *decreases* the inductance of the coil. Such nonferrous adjustable cores are often used in high-frequency applications.

ANSWERS TO CHECK-UP QUIZ 29-1

1. (Potentiometers, adjustable) **2.** (Immersed in oil or other insulating liquid, air-cooled) **3.** (Noninductive) **4.** (0.06 Ω) **5.** (4995 Ω) **6.** (Semiconductors) **7.** (Axial) **8.** (Carbon) (Wire) **9.** (High-voltage breakdown) **10.** (5%)

ANSWERS TO CHECK-UP QUIZ 29-2

1. (270, 20%) (56,000, 20%) (8200, 20%) (15, 20%) (6,800,000, 20%) (33,000, 10%) (75, 5%) (240, 5%) (18,000,000, 5%) (10, 10%) **2.** (0.24) (0.56) (3.6) (1.8) (0.75) (9.1) (4.7) (0.51)

Quiz 29-3. Test your understanding. Answer these check-up questions.

1. What are two ways you can tell a color-coded inductor from a color-coded resistor or capacitor? ________ ________

2. What effect would pulling a ferrite core out of a coil have on the resonant frequency of the coil? __________ What effect would pulling a brass slug out of a coil have on the resonant frequency of the coil? __________
3. For what is epoxy used with inductors, resistors, and capacitors? __________
4. What are the inductance value and tolerance of an inductor color-coded silver, green, blue, red, gold? __________ Silver, gold, violet, green, silver? __________ Silver, orange, orange, black, silver? __________
5. An axial lead encapsulated component has the markings double-width green, brown, red, silver. What is it? __________ What is its value and tolerance? __________

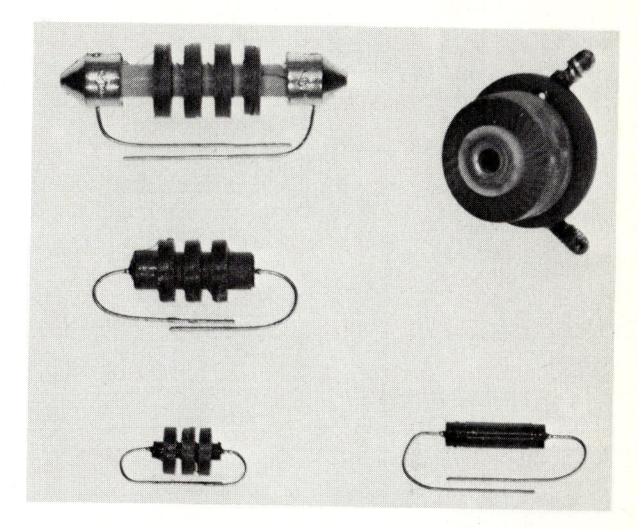

Fig. 29-7 Radio-frequency chokes. Left, from top: 2.5-mH air-core, 5-mH ferrite-core, 2-mH ferrite-core. Right: 20-mH air-core and 10-μH high-frequency RFC.

29-5 RADIO-FREQUENCY CHOKES

All frequencies above 10 kHz can be termed *radio frequencies*. Coils made to have a high reactance (2 to 20 kΩ impedance) to any part of the radio-frequency spectrum are said to be *radio-frequency chokes* (RFCs) for that band of frequencies. A low-frequency RFC may have a value of 20 to 50 mH, and at 10 MHz about 2 mH. For VHF and UHF, values will be in microhenrys.

As frequency increases, capacitive reactance of the distributed capacitance across the choke coil begins to pass ac energy. To prevent this the inductance is broken up into several thin *pies* (Fig. 29-7). Each pie may have 10 to 20 crisscrossed, or *universal-wound* layers. The pies are mounted on an insulating rod and are connected in series. The capacitance between pies is very small, resulting in a relatively small end-to-end distributed capacitance. Even so, any RF choke will have both series and parallel resonant frequencies, and therefore low and high impedance values at different parts of the frequency spectrum.

In the VHF and UHF regions, a single bead of ferrite on a wire operates as an RF choke by producing a high impedance at that point.

29-6 IRON-CORE CHOKES

The inductance required to produce a relatively high impedance to the power frequencies (30 to 400 Hz) ranges from 1 to 20 H. Either toroidal ferrite cores or laminated iron cores are used.

Toroidal-core chokes have the advantage of little leakage flux and usually require no shielding. They are rather difficult to wind.

Laminated sheet-iron-core choke coils may come in several shapes, but the most popular is the E-I core (Fig. 29-8).

When a choke coil is made, the first layer of wire is wound on an insulating form. Each successive layer is insulated from the one below it by an insulating sheet. The finished coil is wrapped with another layer of insulation and is then slipped over the center section of the E-piece. Each lamination piece must be

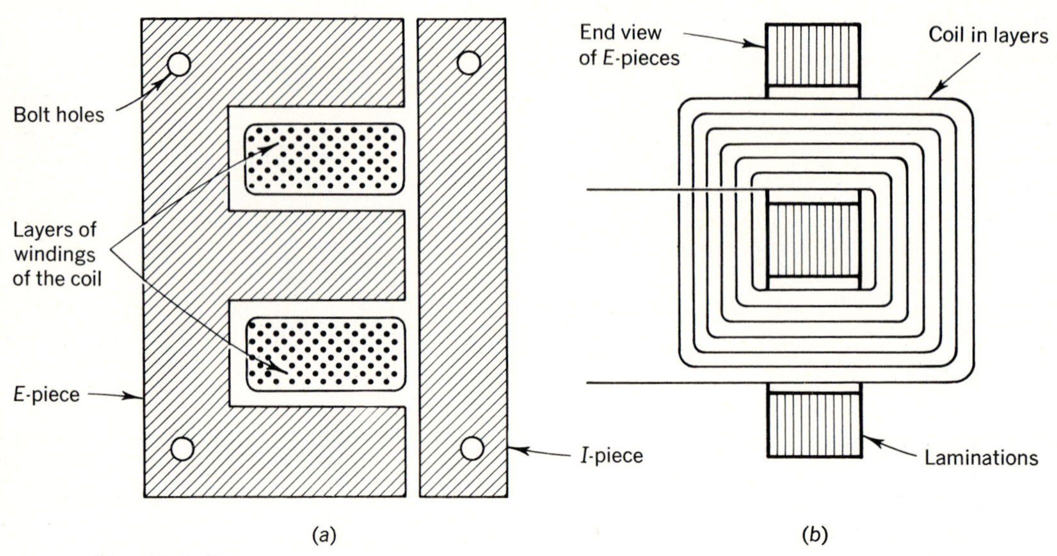

Fig. 29-8 Basic construction of an iron-core choke. (*a*) Side view; (*b*) end view.

coated with an insulating layer on both sides to prevent eddy currents.

A choke coil may be completely encased in an iron shield to prevent its magnetic field from inducing voltages into external circuits.

If the core has no gap between the E and I sections, the iron can be made to saturate with a small magnetizing force *H* (Fig. 29-9). The current value that produces magnetization up to the knee of the first curve will produce only about 40% saturation with the same core having a small gap between E and I pieces. The first low-reluctance gapless core has a steep *BH* slope and produces a relatively high inductance but ceases to produce any further counter-emf after a relatively

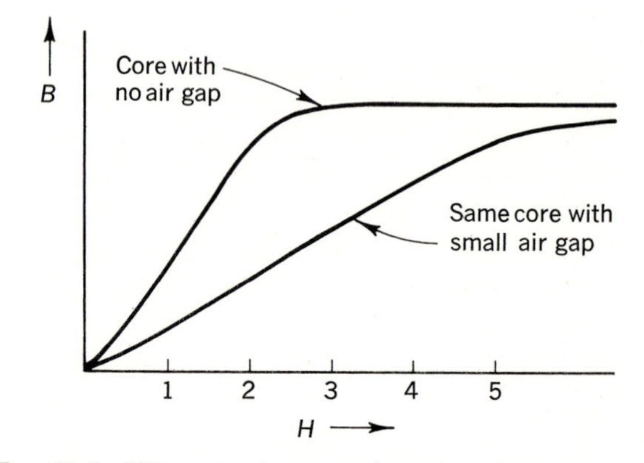

Fig. 29-9 *BH* curve shape with and without air gap.

low value of current flows through it. This is a *saturable choke*. Since its inductance drops to a low value as the knee is approached, it may also be called a *swinging choke*. It may vary (swing) in inductance from a low-current value of 20 H to a full-current value of 2 H.

When a gap is left in an iron core, saturation cannot be obtained easily. As the magnetic field increases, the leakage flux around the gap increases. It takes more ampere-turns to saturate the core fully

ANSWERS TO CHECK-UP QUIZ 29-3

1. (First color is double-width silver) (Ohmmeter indicates zero ohms) **2.** (Raise it) (Lower it) **3.** (As an encapsulating compound) **4.** (5600 μH, 5%) (0.75 μH, 10%) (33 μH, 10%) **5.** (Wire-wound resistor) (5100 Ω, 10%)

and reduce the inductance value. Since the slope of the *BH* curve is less, the inductance of the choke will be less, but it will not saturate easily. It is termed a *smoothing choke*. In its operating range it may vary in inductance very little. Note that *all* iron-core chokes will vary in inductance because of the bend in the *BH* curves. The only coils that have a constant inductance value regardless of the current flowing through them are air-core coils, with nothing in the core to saturate.

Quiz 29-4. Test your understanding. Answer these check-up questions.

1. What approximate impedance should a coil have to the band of frequencies for which it is to act as a choke? ________
2. What should be minimized as much as possible in an RF choke coil, inductance, resistance, capacitance, reactance, or impedance? ________
3. What is the name of the type of winding that a pie in an RF choke will normally have? ________ What is reduced by using pies in RF chokes? ________
4. What is an advantage of a toroidal coil form? ________
5. Why are all laminations covered with an insulating coating? ________
6. If all laminations are covered with an insulating coating, what would a "stacking factor of 90%" mean? ________
7. How does a swinging choke differ physically from a smoothing choke? ________ Which would have the greater *L* value, given the same core volume and coil turns? ________
8. What type of inductor always has a linear *BH* curve? ________
9. Why might an iron-core choke coil be less effective with varying dc than with ac? ________

29-7 ENERGY IN INDUCTORS

Current flowing in a coil produces the magnetic field that expands outward from a coil. When the current either stops or decreases, the field lines collapse, converting energy stored in magnetic form into induced electric current in the coil. The relationship of current, inductance, and stored energy is expressed by the formula

$$E_n = \frac{LI^2}{2}$$

where E_n = energy, in joules (watt-seconds)
L = inductance, in henrys
I = current, in amperes

A 10-H coil with 2 A flowing through it has $E_n = LI^2/2$, or $10(2)^2/2$, or $40/2$, or 20 J (Ws) of energy stored in its field.

Figure 29-10 represents a 10-H inductor with 50 Ω of internal resistance in series with a switch and 100 V. When the switch is closed, current starts to flow, and there

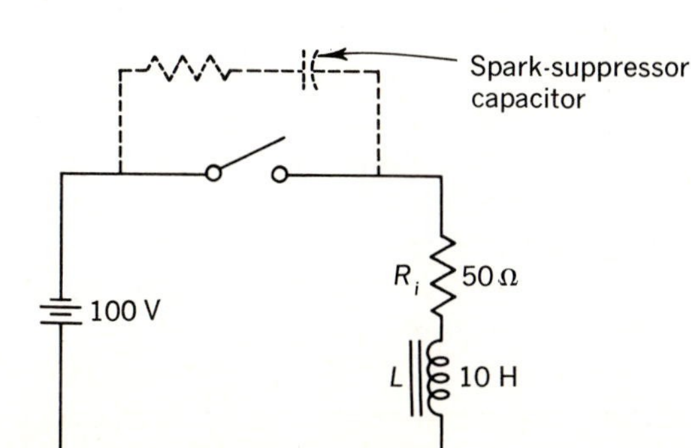

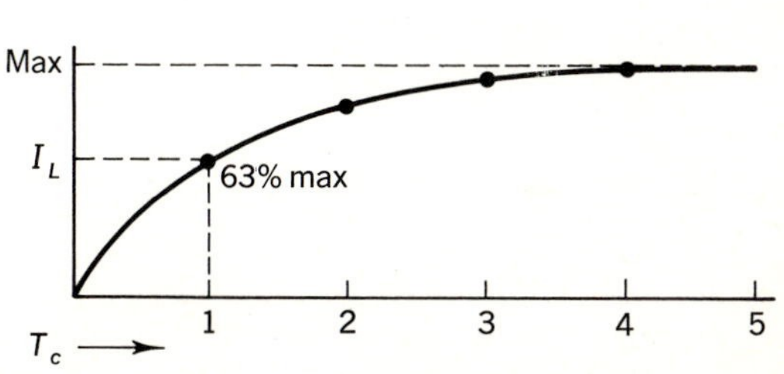

Fig. 29-10 Dc *LR* circuit. When switch closes, current rises according to $T_c = L/R$. After $5T_c$ the *I* is almost to the Ohm's law value (2 A).

is a rapid expansion of the magnetic field, producing a high value of counter-emf in the coil and low current at the starting instant. As the field expands, the speed of expansion decreases, as does the counter-emf in the coil, and the opposition to current flow diminishes. Eventually the current reaches the Ohm's law value of $I = E/R$, or 100/50, or 2 A.

Current increase in an inductor follows the universal time-constant curve (Secs. 13-1, 30-5). The time-constant formula is

$$T_c = \frac{L}{R}$$

where T_c = time for the current to build up to 63.2% of the Ohm's law value for the circuit
L = inductance, in henrys
R = resistance, in ohms

In Fig. 29-10, the time for the current to develop to 63.2% of 2 A, or to 1.26 A, is $T_c = L/R$, or 10/50, or 0.2 s. In the next 0.2 s the current will rise 63.2% of the remaining value to $1.26 + 0.63(0.74)$. Each succeeding time-constant period of 0.2 s, the current increases 63.2% of what is left. After 5 time-constant periods, the current is more than 99% of the Ohm's law value and is considered to be at maximum for most applications.

When an inductance is in series with a circuit, it will prevent a fast increase of current because of the counter-emf it generates. Similarly, if the current tries to decrease, the collapsing magnetic field induces a voltage in the turns of the coil in the direction of the current flow. Regardless of how the current varies, the inductor works to prevent any change in current. Thus

Inductance is the property of a circuit to oppose any change in current.

When a switch opens in an air-core-coil circuit, the magnetic field collapses at a speed of 300,000,000 m/s. At this speed the collapsing lines of force induce a high voltage in the turns of the coil and across the switch gap. No matter how fast the switch is opened, it will be impossible to prevent ionization of the air and the development of an arc between the switch points. Magnetic energy is expended in heating the air until the arc extinguishes. With an iron-core coil, the field will collapse more slowly than with an air-core.

Inductive arcing across switch points causes deterioration of the contacts. If a capacitor is connected across the switch (Fig. 29-10), as the switch opens, energy from the collapsing magnetic field charges the capacitor, and not enough voltage is developed across the switch contacts to produce an arc. When the switch is closed again, energy stored in the capacitor tends to melt the switch contacts. To prevent this, a resistance of 5 to 50 Ω is added, as shown. It prevents excessive current flow when the capacitor is discharged. The capacitance value needed will range from 0.01 to 1 μF.

One method of preventing arcing when opening inductive circuits is to use vacuum relays (Fig. 29-11). These open and close in an evacuated glass enclosure. Since there is no air or gas in a vacuum, there is nothing to ionize, and no arc can form.

ANSWERS TO CHECK-UP QUIZ 29-4

1. (2 to 20 kΩ) **2.** (Distributed capacitance) **3.** (Universal) (Distributed capacitance) **4.** (Less leakage flux) **5.** (Reduces eddy currents) **6.** (90% of core volume is iron, 10% is nonferrous material) **7.** (No gap in core) (Swinging choke) **8.** (Air core) **9.** (Core tends to magnetize permanently)

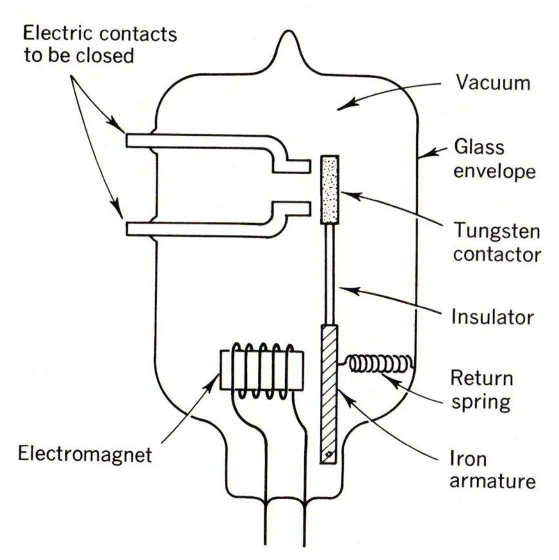

Fig. 29-11 Vacuum relay. The electromagnet may be external of vacuum area.

Quiz 29-5. Test your understanding. Answer these check-up questions.

1. Energy is stored in what form in a coil? ________ A capacitor? ________ A battery? ________
2. The output transformer of a radio set has a 500-Ω 5-H primary with 90 mA of dc flowing in it. How much energy is stored around the primary? ________ How much power is dissipated in heat? ________
3. How long does it take for current to rise to 63% of maximum in *LR* circuits? ________ To essentially the Ohm's law value? ________
4. A 20-H choke coil with 700 Ω of internal resistance is switched across a 10-V battery. How long will it take for the *I* to reach essentially maximum? ________ If the coil were switched across a 100-V battery, how long would it take? ________
5. If resistance is added to a series *LR* circuit, will the I_{max} be reached faster or slower than without the resistance? ________

29-8 INDUCTANCES IN SERIES

Current changing in a coil induces countervoltages in the turns of the coil. This is called inductance, or *self-inductance*. When two or more inductors are in series, but with little or no intercoupling of their magnetic fields, the total inductance will be the simple sum of the inductances,

$$L_t = L_1 + L_2 + L_3 + \cdots$$

where L_t is the total inductance and L_1, L_2, etc., are the inductances of series coils.

The magnetic fields of iron-core choke coils are fairly well confined to the core, with little leakage lines of force. Two 1-H chokes, even if mounted in close proximity, will have essentially 2 H, although there will always be some interaction between nearby chokes. They are often enclosed in ferrous-metal shield-cans for greater field isolation.

In Fig. 29-12*a*, two coils are separated a

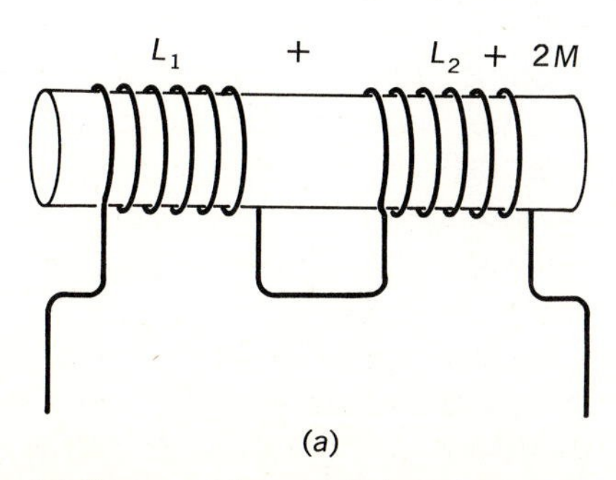

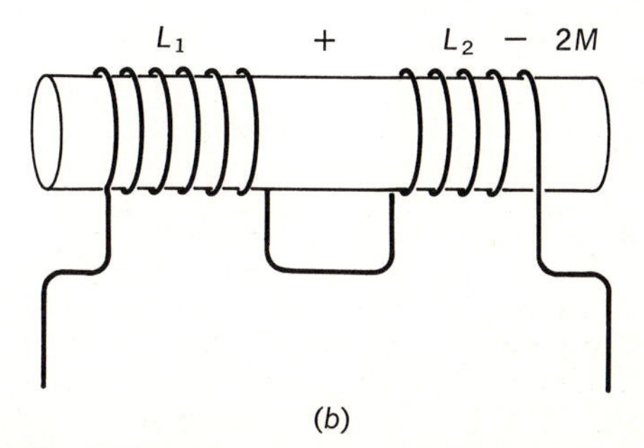

Fig. 29-12 (*a*) Series-aiding coils have greater inductance than the sum of their inductances. (*b*) When connected series-opposing, inductance is less than the larger value.

little, but are connected series-aiding; that is, the fields of one are in phase with the fields of the other. The counter-emf induced in one coil adds to the counter-emf of the other. The total inductance is the simple sum of the two inductances plus the coupling effect from coil 1 to coil 2 plus the coupling effect of coil 2 on coil 1.

In Fig. 29-12*b*, the two coils are connected series-opposing, and the fields of the two coils are in opposition. The counter-emf induced in either coil is partially canceled by the emf induced from the other. The result is less counter-emf and less total inductance than the simple sum of the two inductances.

The formula used to determine the actual total inductance of two coupled inductors is

$$L_t = L_1 + L_2 \pm 2M$$

where M is the *mutual inductance*, or total coupling effect, between the coils. If the coils are series-aiding, add $2M$. If connected series-opposing, subtract $2M$ from the simple sum of the inductances. (L_M may be used to denote mutual inductance.)

Mutual inductance indicates how much emf is induced in a coil when a current change is under way in a nearby coil. For example, if a dI (current change) of 1 A/s is under way in one coil and it induces 1 V in the other coil, the mutual inductance is 1 H.

To determine the mutual inductance between two coils, first measure the total series-aiding inductance with an inductance bridge (L_a). Then reconnect them series-opposing, and measure the total inductance (L_o). The difference between these two inductance values, divided by 4, will be the mutual inductance, or

$$M = \frac{L_a - L_o}{4}$$

Quiz 29-6. Test your understanding. Answer these check-up questions.

1. Coils of 0.3 H, 250 mH, and 6000 μH are all in series in a circuit but are physically far removed from one another. What inductance value do they represent to the circuit? ____________
2. Two coils are connected series-aiding and have a total of 4 H. When they are connected series-bucking, they have 1 H. What is their mutual inductance? ____________
3. If the L_a of two coils is 5 H and the L_o is 800 mH, what is the mutual inductance between them? ____________ What would it be if the L_o were 5H? ____________ Under what condition could this occur? ____________
4. If a 0.4-H coil and a 0.6-H coil are connected series-aiding with $L_M = 0.2$ H, what is the L_t value? ____________
5. If a 10-H coil and a 12-H coil are connected series-opposing and have an M of 2 H, what is the total L value? ____________ If the 12-H coil were wound in the opposite direction, would M be greater, less, or the same? ____________
6. If two 1-H coils are connected in series, what is the L_{max} they could have? ____________ The L_{min} value? ____________ What is the M_{max} they could have? ____________

29-9 COEFFICIENT OF COUPLING

The *coefficient of coupling* between two coils is the percentage of magnetic lines that interlink a secondary coil when current is changing in a primary coil. For example, if current flowing in one coil produces a flux field of 0.8 Wb but the number of these field lines that cut across a secondary coil is 0.5 Wb, the coefficient of coupling between the two coils is 0.5/0.8, or 0.625. This can also be stated as a percentage of coupling of 62.5%. The

---

ANSWERS TO CHECK-UP QUIZ 29-5

1. (Magnetic) (Electrostatic) (Chemical) **2.** (0.0202 J) (4.05 W) **3.** ($1T_c$) ($5T_c$) **4.** (0.143 s) (Same time) **5.** (Faster)

secondary will be cut by 62.5% of the primary lines.

The relationship of coefficient of coupling, k, mutual inductance, M, and the inductances of two coupled coils in formula form is

$$M = k\sqrt{L_1 L_2}$$

Since the coefficient of coupling is not likely to be known, but M, L_1, and L_2 can be measured, a more practical formula is

$$k = \frac{M}{\sqrt{L_1 L_2}}$$

The k between RFC pies may be less than 0.02. Between primary and secondary windings of an iron-core transformer it may exceed 0.98. If all the lines of force from a primary cut the secondary, the two coils are said to be *unity-coupled*, and $M = \sqrt{L_1 L_2}$.

29-10 DETERMINING INDUCTANCE

Computed inductance is usually a close approximation rather than an exact answer, particularly for iron-core inductors in which permeability to a great extent determines the inductance. Permeability of iron is rarely known accurately.

It is possible to determine the approximate inductance of a straight piece of wire by

$$L = 0.002l\left(2.3 \log\frac{4l}{d}\right)$$

where L = inductance, in microhenrys
l = length, in centimeters
d = wire diameter, in centimeters

As an example, a piece of copper wire 1 m long with a diameter of 1 mm (#18 AWG wire) has an inductance of

$$\begin{aligned} L &= 0.002(100)\left[2.3 \log \frac{4(100)}{0.1}\right] \\ &= 0.2(2.3 \log 4000) \\ &= 0.2(2.3)3.6 \\ &= 1.66\ \mu\text{H} \end{aligned}$$

If this same piece of wire is wound into a coil with a diameter of 2.5 cm and a length of 2.5 cm (13 turns), the inductance increases because of mutual induction between turns. A formula for determining the approximate inductance of an air-core coil with a length approximately equal to its diameter is

$$L = \frac{r^2 n^2}{24r + 25l}$$

where L = inductance, in microhenrys
r = radius, in centimeters
n = number of turns
l = length, in centimeters

When this formula is applied to the coil above, the inductance is found to be 2.85 μH.

If the same number of turns is compressed into a coil of half the length, the inductance increases to more than 4.31 μH.

This same formula, in terms of inches, is

$$L = \frac{r^2 n^2}{9r + 10l}$$

To find the inductance of a given coil, an inductance or impedance bridge (Chap. 19) should be used. If the resistance of the coil is negligible (less than 10% of X_L), the circuit in Fig. 29-13 can be used. By the meter readings the reactance can be computed by $X_L = E_{X_L}/I_{X_L}$. With negligible R, the X_L and Z values will be almost equal. To determine L, rearrange the formula $X_L = 2\pi fL$ to

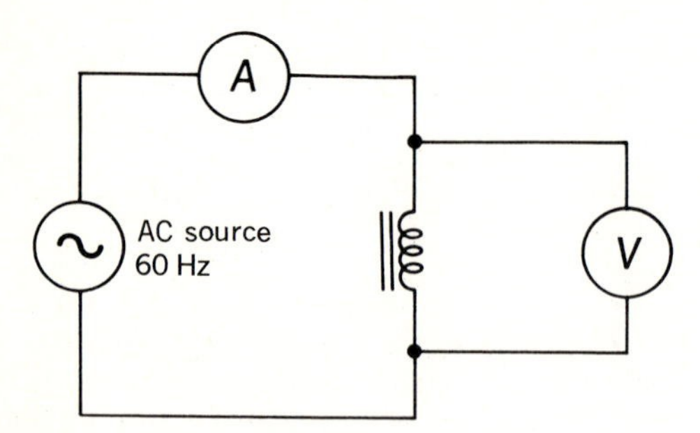

Fig. 29-13 Circuit for determining the L of a coil if its internal R is negligible.

$$L = \frac{0.159 X_L}{f}$$

where L is in henrys, X_L (and Z) is in ohms, and f is the source frequency in hertz.

Since an iron-core choke has a nonlinear BH curve, different inductance values will be obtained with different source-voltage values.

At higher frequencies current tends to travel only on the surface of a wire. This is called *skin effect*, resulting in an increase in the effective resistance of the wire. The center of the wire carries no current. The higher the frequency, the greater the skin effect. Also, when turns of a coil are close together, the current-carrying cross-sectional area of the wires is further reduced. This *proximity effect* increases the effective ac resistance of a coil.

In many cases a choke coil will have a dc plus an ac component flowing through it. To determine the inductance, the circuit in Fig. 29-14 can be used. An ac source with a frequency similar to that to be used is connected in series with a dc source that can produce the dc current expected

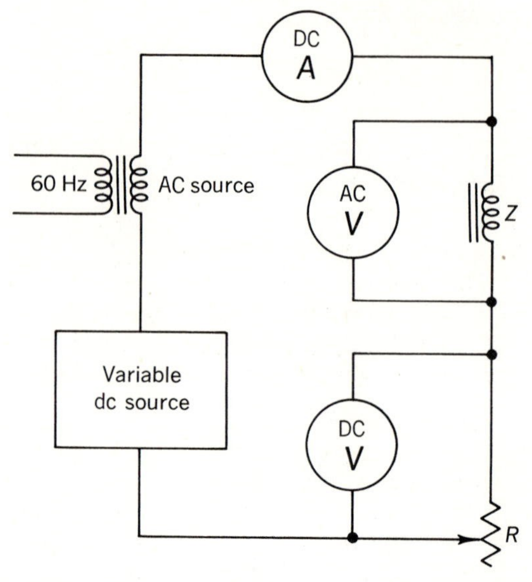

Fig. 29-14 Circuit to determine impedance and inductance of an inductor having both dc and ac flowing through it.

to be flowing in the inductor. The rheostat is varied until the voltage-drop across the resistor equals the rms voltage-drop across the inductor. When the two voltmeters show the same value, the rheostat resistance equals the *impedance* value of the choke. To find the effective *inductance* under these operating conditions, the Z value obtained can be inserted in the formulas

$$L = \sqrt{\frac{Z^2 - R^2}{\omega^2}} = \frac{0.159\sqrt{Z^2 - R^2}}{f}$$

where ω is $2\pi f$. The R value is the dc resistance of the coil by ohmmeter.

A simple method of determining the inductance of a high-frequency coil is to connect a capacitor across it (100 pF for 1 to 50 MHz, and 10 pF for 50 MHz and above). By using a grid-dip meter (Sec. 20-10), the frequency of resonance of the parallel resonant circuit that is formed can be measured. From the resonance

ANSWERS TO CHECK-UP QUIZ 29-6

1. (0.556 H) **2.** (0.75 H) **3.** (1.05 H) (0 H) (No intercoil induction) **4.** (1.4 H) **5.** (18 H) (Same) **6.** (4 H) (0 H) (1 H)

formula $X_L = X_C$,

$$L = \frac{1}{4\pi^2 f^2 C} = \frac{0.0254}{f^2 C}$$

where L is in henrys, f is in hertz, and C is in farads.

Quiz 29-7. Test your understanding. Answer these check-up questions.

1. What is the name of the degree of coupling that would be most desirable for a power-frequency transformer? ___________
2. The flux field of a primary coil has 560 μWb. Only 45 μWb cuts a secondary coil. What is the coefficient of coupling? ___________ If the secondary coil were used as the primary, what would be the coefficient of coupling between it and the first coil? ___________
3. What is the formula for determining mutual inductance when unity coupling exists? ___________
4. What is the inductance of a straight wire 50 m long with a diameter of 1.2 mm? ___________ If the wire is 100 m long? ___________
5. What is the inductance of a coil 5 cm long and 4 cm in diameter with 40 turns? ___________ 80 turns? ___________
6. If the ammeter in Fig. 29-13 reads 50 mA, the voltmeter reads 15 V, and the internal resistance is 3 Ω, what is the Z value? ___________ X_L? ___________ L? ___________
7. What would be reduced by using three insulated conductors of 1 cir mm area in parallel instead of a single conductor of 3 cir mm of the same length? ___________ Would anything be gained when carrying dc? ___________
8. What is the name of the effect that causes nearby wires to force an ac to flow near the surface of the wires? ___________ What effect does this have on the resistance of the wires? ___________
9. What is the name of the device that can determine the resonant frequency of a coil and capacitor connected in parallel? ___________
10. The dc in the circuit of Fig. 29-14 is 50 mA when the ac voltmeter and the dc voltmeter read equal values. The rheostat is measured and found to have 850 Ω. The choke has 50 Ω of internal resistance. What is the impedance of the inductor? ___________ What is its inductance? ___________

CHAPTER 29 TEST • RESISTORS AND INDUCTORS

1. What does physical size of a resistor indicate?
2. Name at least two methods by which resistors are cooled.
3. What is indicated by "ppm/°C"?
4. What are two methods of constructing wire-wound resistors for minimal inductance?
5. What is the name of the type of connector that comes out at right angles from the ends of a resistor or inductor? What is the name of the other type?
6. A resistor has 1000 Ω at 20°C. What is its resistance at 40°C if it has a −TC of 0.005%?
7. What is the basic composition of resistors in the 0.25-, 0.5-, and 1-W ratings? In 10-W ratings and above?
8. What is the usual tolerance of deposited film resistors?
9. A resistor has four stripes in the order yellow, violet, red, silver. What is its resistance? Its tolerance?
10. A resistor has four stripes in the order double-width orange, white, black, gold. What is its resistance? Its tolerance? What else do you know about it?
11. A small tubular component has double-width silver, red, green, brown, and silver striping. What is it? What is its value?
12. To be effective an RF choke must have high values of what? Low values of what?
13. What is the type of winding used on RF chokes?
14. What are the shapes of scrapless punched laminations used in iron-core chokes or transformers?
15. What is prevented by insulating both sides of iron laminations? By bolting laminations tightly?
16. Which has the steepest *BH* curve, an air core, an iron core with no gap, or an iron core with a gap?
17. What is the name of the type of choke that has an air gap in its core? No gap?
18. How much energy is stored in the magnetic field around a choke having 3 H, 40 Ω of R_i, and

0.5 A flowing in it? What is the time constant for the choke?

19. What are the two components that are used to suppress arcs or sparking at a closing and opening electrical contact? What are their approximate values?
20. How long would it take for current to reach essentially maximum in a 15-H choke with a 600-Ω internal resistance after being connected across a 30-V battery?
21. Why is arcing not possible in vacuum switches?
22. A 50-mH coil and an 80-mH coil are in series, with half the lines of force interlinking. What is the k value? What is the L_t value if the coils are wound in the same direction? If one coil winding is reversed?
23. A transformer with primary and secondary connected in series has 15 H. When the secondary-winding connections are reversed, there is 7 H. What is the M value?
24. What is the formula for the coefficient of coupling?
25. What is the formula for the inductance of a straight piece of wire?
26. Under what conditions is X_L equal to Z?
27. At high frequencies current does not flow in the center of a wire. What is this effect called?
28. What is a grid-dip meter used for?

ANSWERS TO CHECK-UP QUIZ 29-7

1. (Unity) **2.** (0.0804) (Same) **3.** ($M = \sqrt{L_1 L_2}$) **4.** (120 μH) (254 μH) **5.** (37 μH) (148 μH) **6.** (300 Ω) (300 Ω) (0.796 H) **7.** (Skin effect) (No) **8.** (Proximity) (Increases R) **9.** (Grid-dip meter) **10.** (850 Ω) (2.25 H)

30 CAPACITORS

CHAPTER OBJECTIVE. Extension of information in Chap. 12, including dielectrics, capacitor specifications, solving transient values, and color codes.

30-1 CAPACITORS

Capacitance was outlined briefly in Chap. 12.

The *farad* is the basic unit of measurement of capacitance. A capacitor that will accept a charge of one coulomb under a pressure of one volt has a capacitance of one farad. A farad is one coulomb per volt, or

$$C = \frac{Q}{E}$$

from which

$$Q = CE \qquad \text{or} \qquad E = \frac{Q}{C}$$

A basic capacitor consists of two plates separated by an insulator, called a *dielectric*. The capacitance of such a capacitor can be determined by the formula

$$C = \frac{8.85kA(10^{-12})}{d}$$

where C = capacitance, in farads
k = dielectric constant
A = area of one plate, in square meters
d = distance between plates, in meters

A more practical formula might be

$$C = \frac{8.85kA(10^{-2})}{d}$$

where C is in picofarads, A is in square centimeters, and d is in centimeters.

The same formula, using inches, is

$$C = \frac{0.225kA}{d}$$

where C is in picofarads, A is in square inches, and d is in inches.

30-2 ENERGY IN CAPACITORS

Electrical energy in joules has been described as the product of pressure in volts times a quantity of electrons in coulombs, or volt-coulombs,

$$E_n = EQ$$

The energy contained in a charged capacitor is

$$E_n = \frac{CE^2}{2}$$

where energy is in joules (J) or wattsec, E is in volts, and C is in farads.

A 10-μF capacitor charged to 300 V stores enough energy to seriously shock anyone. The total energy involved would be $E_n = CE^2/2$, or $10^{-5}(90{,}000/2)$, or 0.45 J, which might stop a heart.

Compare the human tolerance to electrical energy with the energy to light a flashlight lamp for 1 s. If the lamp filament draws 0.25 A with 3 V across it, E_n is the number of wattseconds, or EIt, or 3(0.25)(1), or 0.75 J.

Quiz 30-1. Test your understanding. Answer these check-up questions.

1. If 500 V is capable of charging a capacitor with 2.5 coulombs, what is the capacitance of the capacitor? ______
2. How many coulombs of charge will be moved in charging a 3-μF capacitor to 5 kV? ______ A 25-pF capacitor to 80 V? ______
3. A 15- to 360-pF variable capacitor is charged to 100 V while at the maximum capacitance setting and is then disconnected from the charging source. What will the voltage across the capacitor be when the capacitor is varied to 15 pF? ______
4. Two square metal plates are held parallel, 0.5 cm apart. If they are 1 dm (decimeter) long on a side, what is the capacitance between them? ______
5. Two copper wires are held 0.2 in. apart by an insulator having a dielectric constant of 5. If each wire has an effective area as a capacitor plate of 0.125 in.^{2}/ft, how much capacitance will 100 ft of this wire have? ______
6. A five-plate capacitor has air as the dielectric. The spacing between plates is 1 mm and the area of each plate is 20 cm^2. What is the capacitance if alternate plates are connected in parallel? ______ How much voltage would this capacitor stand? ______
7. How much energy is contained in a 20-μF capacitor charged to 200 V? ______

30-3 DIELECTRIC BREAKDOWN

There are several reasons for capacitor failures. One is breakdown of the dielectric insulation. Each atom of dielectric material has negative electrons in orbit around the positive nucleus. When the capacitor is charged, these atoms are under electrostatic stress. The outer electrons are attracted toward the positive plate. The negative plate attracts the positive nucleus. This results in an elongation of the electron orbits and a strained atom (Fig. 30-1).

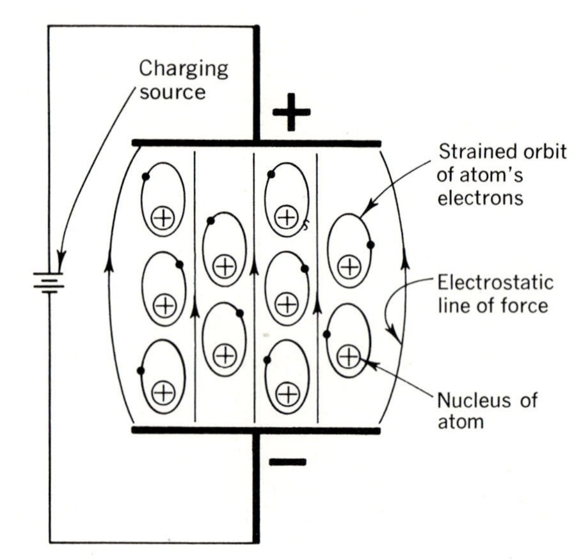

Fig. 30-1 Charged capacitor, showing elongation of electron orbits and direction of electrostatic lines of force.

Dielectric atoms next to the positive plate actually lose outer electrons to this plate. The negative plate attracts positive nuclei but does not actually capture any of them. Electrostatic lines of force stretch the outer-layer electrons of all the dielectric atoms into strained orbits, storing energy in the straining of the dielectric atoms.

If the emf between the plates is excessive, it can pull electrons away from their atoms (ionization), and a current of electrons begins to stream through the dielectric. This current may burn the dielectric at some point, forming a conductive carbonized path between the plates. Sometimes a path has relatively high resistance, from a few hundred to several thousand ohms. More often, the path represents a nearly 0-Ω resistance *leak*, and the capacitor is said to be *burned out*, or *shorted*. Air- and vacuum-dielectric capacitors are self-healing. Solid-dielectric capacitors are not.

Heat produces accelerated electron motion in materials. A heated capacitor is more likely to develop a leak or short.

If water vapor gets into dielectrics such as paper, the paper is much more likely to ionize. To prevent this, paper is usually impregnated with wax or oil. Plastic dielectrics, such as polystyrene or Mylar, will not absorb water and are often preferable to paper.

A reduction in pressure, as from altitude, tends to allow easier ionization of solid dielectric materials. As a result, the working voltage of a capacitor will have to be *derated* for high-altitude operation.

Electrolytic capacitors always have some dielectric leakage. Such currents result in the chemical action that produces the oxide which is the dielectric film formed on the positive plate. After operating for a few minutes, leakage should decrease to an insignificant value. If the leakage is too great, the dielectric may heat and develop a short. Heat caused by excessive leakage also results in the generation of steam and a possible explosion of the capacitor. This may happen if the capacitor is in a circuit with its polarity reversed, or in an ac circuit.

Two electrolytics can be connected in series, back to back, with polarities reversed, and used as a capacitor in ac circuits for a short time. On one half-cycle one capacitor conducts and starts to deform, but the other functions as a capacitor. On the next half-cycle the first capacitor reforms, while the second conducts and tends to deform. Thus two 50-μF series-bucking electrolytics in an ac circuit will exhibit 50 μF of capacitance. Such "ac electrolytic" capacitors may be used as starting capacitors in ac motors (Sec. 22-19).

30-4 CAPACITOR SPECIFICATIONS

Selecting the proper capacitor requires some consideration. Capacitors used in home radios and television sets may not be the same as those used in transistor portables, or the same as those shot up into outer space. Some of the specifications that must be considered are:

- *General:* Capacitance range, in microfarads or picofarads. Capacitance tolerances. Operating temperature range. Voltage rating. Temperature coefficients.
- *Mechanical:* Case material. Marking of values. Dip or wavesoldering time. Dip or wavesoldering temperature. Lead pull strength, in kilograms or pounds. Material of leads (copper, iron, tinned, gold plated, etc.). Nominal weight. Physical dimensions.
- *Electrical:* Insulation resistance at room temperature (20°C) and at maximum expected operating temperature. Dielectric breakdown voltage at 20°C and at maximum operating temperature. Leakage power loss or dissipation factor at various temperatures and at the frequencies to be used. Case insulation in volts. Frequency at which capacitance is measured (see below).
- *Environmental:* Effect of temperature cycles. Immersion capabilities. Ability to withstand vibration and physical shock. Effect of altitude (increased and reduced pressure) on the operating characteristics. Life-test performance (1000 h, 125°C, at 400 V, for example).

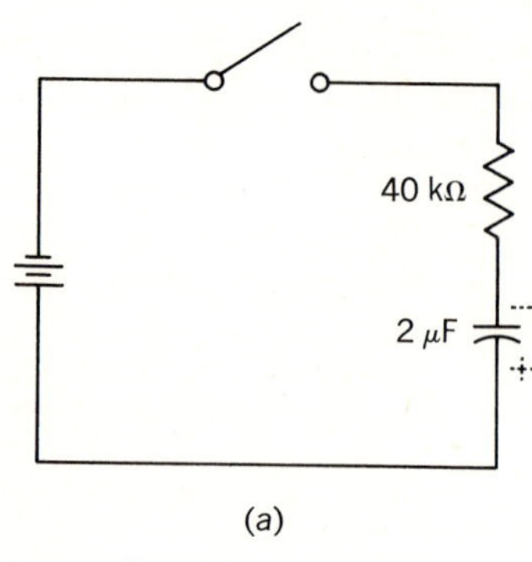

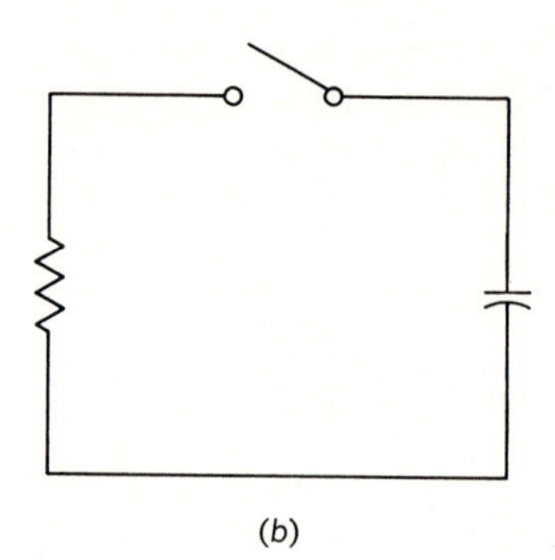

Fig. 30-2 (*a*) Circuit to charge a capacitor through a resistor. (*b*) Discharge circuit parallels the *R* and *C*, resulting in similar *R* and *C* voltages and currents at all times.

The higher the ac frequency, the greater the *dielectric hysteresis* (molecular friction) and the greater internal heat developed. At high frequencies dielectric molecules may not be able to reverse their strained positions completely. This results in less capacitance at higher frequencies. Paper is poor at high frequencies, but mica, ceramics, air, and vacuum are much better in this respect.

In many applications only a few of the specifications above may be important. In some cases all must be considered.

30-5 SOLVING FOR TRANSIENT VALUES

When a capacitor is charged or discharged through a resistor, the voltages and currents that occur from instant to instant are called transient values. In Chap. 11 transient values were determined by referring to universal *RC* time-constant curves. It is possible to compute transient values more accurately by formula. For a series *RC* circuit in the act of charging,

$$i = I(\epsilon^{-t/RC})$$

$$e_R = E(\epsilon^{-t/RC})$$

$$e_C = E(1 - \epsilon^{-t/RC})$$

Lowercase *i* and e are instantaneous values. Uppercase *I* and *E* are maximum or source values. The ϵ indicates a *natural* logarithm (Appendix G). *R* is resistance in ohms; *C* is capacitance in farads; *t* is time in seconds.

Consider Fig. 30-2*a*. What is the current flowing into the 2-μF capacitor through the 40,000-Ω resistor at an instant 0.02 s after the circuit switch is closed, if the maximum current by Ohm's law is 0.3 A? At the instant the switch closes, current into the capacitor changes from zero to something, an infinitely rapid rise. The *reactance* to such a fast rise time is essentially 0 Ω. Thus at the instant the switch closes, current in the circuit is $I = E/R$. After 0.02 s,

$$\begin{aligned} i &= I(\epsilon^{-t/RC}) \\ &= 0.3(\epsilon^{-0.02/[40,000(0.000002)]}) \\ &= 0.3(\epsilon^{-0.02/0.08}) \\ &= 0.3(\epsilon^{-0.25}) \end{aligned}$$

From the table of natural logarithms −0.25

ANSWERS TO CHECK-UP QUIZ 30-1

1. (0.005 F or 5000 μF) **2.** (1.5×10^{-2} coulomb) (2×10^{-9} coulomb) **3.** (2400 V) **4.** (17.7 pF) **5.** (70.3 pF) **6.** (70.8 pF; five plates have only four surfaces exposed to each other) (3000 V; see Sec. 8-4) **7.** (0.8 J)

---

equals 0.78. This exponential expression has a negative sign, so the logarithm must be determined from the negative part of the table. Substituting this value in the problem,

$$i = 0.3(0.78)$$
$$= 0.234 \text{ A}$$

Check this answer with the *RC* curve in Chap. 12, using $T_C = RC$, or 40,000(0.000002), or 0.25, and I_{max} as 0.3 A.

If a resistor is connected across a charged capacitor, as in Fig. 30-2*b*, the energy in the dielectric of the capacitor will be dissipated as heat in the resistor and the voltage across the capacitor will decrease from full to zero. At the same time, current flowing out of the capacitor and through the resistor will also be decreasing to zero. To find the instantaneous current at some time after the capacitor has started to discharge, the formula $i = I(\epsilon^{-t/RC})$ is again used. This gives the i value for both the resistor and the capacitor, since they are in series. The I value as it begins to flow is computed by Ohm's law.

The formula for the instantaneous *voltage* across *R* and *C* in the circuit above during discharge will be the same, since they are in parallel. The formula is

$$e = E(\epsilon^{-t/RC})$$

Transient values can be computed for *RL* circuits. The basic formulas are similar to those for *RC* circuits, except that the time constant for *RL* circuits is *L/R*, whereas for capacitive circuits it is *RC*. Compare the *RC* and the following *RL* transient formulas:

$$i = I(1 - \epsilon^{-tR/L})$$
$$e_R = E(1 - \epsilon^{-tR/L})$$
$$e_L = E(1 - \epsilon^{-tR/L})$$

Quiz 30-2. Test your understanding. Answer these check-up questions.

1. How much energy in joules is stored in a 20-μF capacitor when it is charged to 5000 V? ________ How many watthours does this represent? ________
2. What do you know about an atom whose orbital electrons are operating in a long narrow orbit? ________
3. What types of capacitors always have some leakage current? ________ What types cannot burn out their dielectrics? ________
4. What is the current at the instant that a 9-kΩ resistor couples a 0.5-μF capacitor across a 36-V source? ________ What is the current 400 μs later? ________
5. What is the voltage-drop across a 6-μF capacitor 5 ms after current begins to flow into it through a 2.5-kΩ resistor if the starting current is 0.8 A? ________ By the transient formula, what is the voltage-drop across the resistor at this time? ________ The value of e_C is known; by Kirchhoff's voltage law, what is the voltage across the resistor? ________
6. A 20-μF capacitor is disconnected from a 12-V battery and connected across a 5000-Ω resistor. In 0.003 s what is the voltage across the resistor? ________ Across the capacitor? ________
7. How much energy is stored in a 10-μF capacitor if it is connected across a 500-V dc supply through a 15,000-Ω resistor for 30 ms and then disconnected? ________

30-6 CAPACITOR COLOR CODES

The manufacturers of capacitors have changed their marking methods several times. In some cases the capacitance value is printed on the capacitor in picofarads or in microfarads. Some capacitors carry only color-coded markings.

The first color code was simple (Fig. 30-3*a*). The capacitor is held with the arrowhead pointing to the right. The first dot on the left is the first significant number (1). The second dot is the second significant number (2). The third dot is the multiplier, or number of zeros (*M*). The numerical values of the colors are

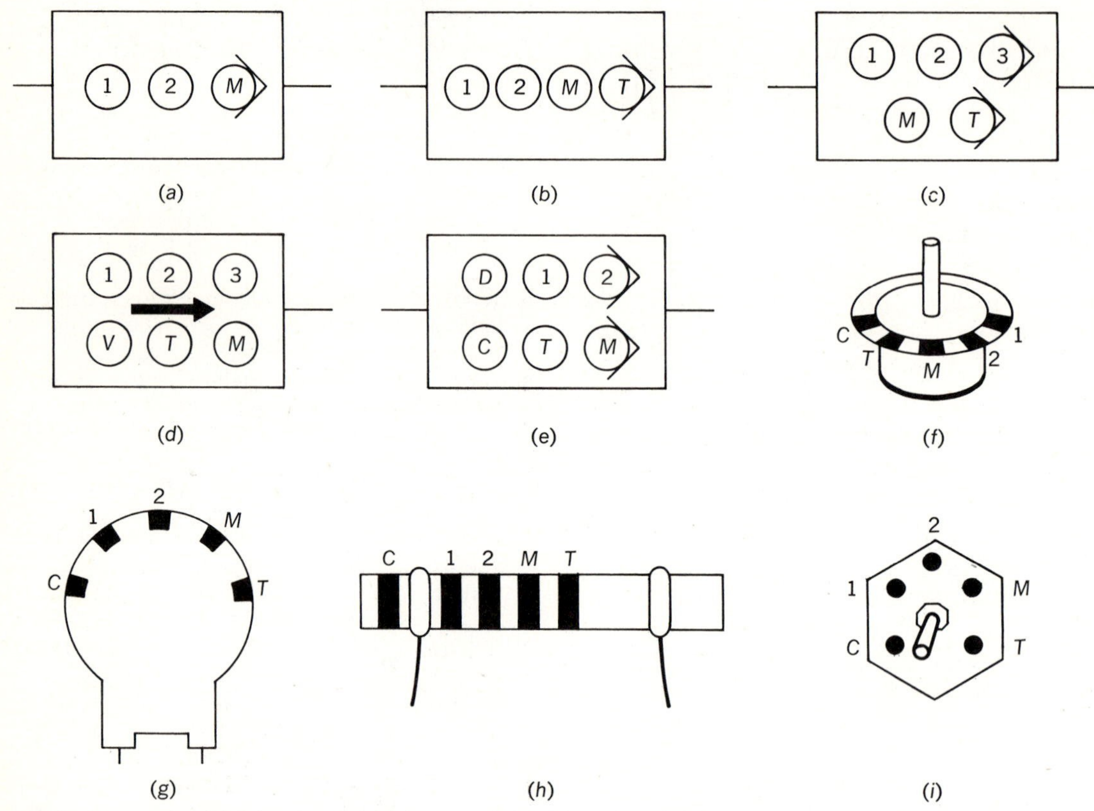

Fig. 30-3 Various methods of color-coding capacitors. 1, 2, and 3 represent significant numbers, M is the multiplier, C is the characteristic, T is the tolerance, V is the voltage rating, and D is the dielectric.

the same as in Sec. 29-3 for resistors. If the three colors are brown, orange, and red, in that order, the capacitance value is 1 3 00, or 1300 pF.

When a capacitor has four dots, as in Fig. 30-3*b*, they would be read as indicated. The letter "T" signifies the tolerance, where brown = 1%, red = 2%, gold or green = 5%, and silver = 10%. Either black or no color = 20%.

ANSWERS TO CHECK-UP QUIZ 30-2

1. (250 J) (0.0694 Wh) **2.** (It is under a strain, has absorbed energy) **3.** (Electrolytic, tantalum) (Air, vacuum) **4.** (0.004 A) (0.00366 A) **5.** (564 V) (1436 V) (1436 V) **6.** (11.6 V) **7.** (0.0405 J)

Another old method of marking is the five-dot system, shown in Fig. 30-3*c*. This code can be read to the third significant figure.

One of the first six-dot codes is shown in Fig. 30-3*d*. Besides three significant figures, a multiplier, and a tolerance, these capacitors also had a voltage rating, brown for 100 V, red for 200 V, etc. Somewhere on most capacitors an arrowhead or a trade name will indicate which way to hold the capacitor to read the code.

The Electronics Industries Association (EIA) mica-capacitor color code is the six-dot one shown in Fig. 30-3e. D stands for dielectric, with white indicating mica. The letter "C" represents the characteristic or temperature coefficient in ppm/°C. Black is 1000 ppm. Green is about 100 ppm.

The military standard, Mil-C-5B, is essentially the same as the EIA standard except that a black dielectric dot indicates mica. A silver dielectric dot means paper.

Button-type capacitors (Fig. 30-3*f*) may carry three, five, or six color marks. The five-dot code is read as indicated in the illustration. In the six-dot code, a third significant figure is added before the multiplier. The three-dot marking is read the same as in Fig. 30-3*a*.

There are several types of ceramic capacitors—the disk, button, tubular, feed through, and standoff. Basically, the last four types are read in a manner similar to the mica capacitors, except the characteristic is often given as the first color. Color-coded ceramics are usually 500-V-breakdown rated unless working-voltage values are printed on one side.

Figure 30-3*g* illustrates a five-dot marked disk ceramic capacitor. The characteristic C is the temperature coefficient, and runs from black for zero to violet for −750 ppm. Gray and white are positive coefficients of about 30 and 300. Tolerance in percent is indicated by the fifth color: brown = 1%, green = 5%, silver = 10%, and black = 20%.

The radial-lead tubular ceramics (Fig. 30-3*h*) may or may not have the characteristic dot or band. If they do, it will be double-width. Axial-lead tubular ceramics carry five colors, the first being the double-width characteristic band.

Feed-through ceramics (Fig. 30-3*i*) and standoff capacitors are five-dot-marked, as indicated. The colors are read clockwise.

Ceramic tubular capacitors may have their ratings printed on them, such as N750, 200, F. This indicates: Negative temperature coefficient of 750 ppm/°C, 200-pF capacitance, 1% tolerance. Other tolerance letters are: G = 2%, J = 5%, K = 10%, and M = 20%.

Some tubular paper capacitors are marked with the standard 1, 2, M, T, and V color indicators, in this order. For T, gold is 5% and silver is 10%. For voltage, brown is 100 V, red is 200 V, etc.

Quiz 30-3. Test your understanding. Answer these check-up questions.

1. What are the values of three-dot capacitors with the colors red, green, brown? __________ Orange, white, red? __________ Blue, gray, orange? __________
2. What are the capacitance and tolerance values of flat four-dot capacitors with the colors green, black, orange, silver? __________ Yellow, brown, brown, gold? __________
3. What are the capacitance and tolerance values of five-dot color-coded capacitors with the colors gray, green, yellow, brown, gold? __________ White, orange, black, black, silver? __________
4. What are the capacitance values of old six-dot flat capacitors with the colors yellow, violet, green, brown, silver, orange? __________ White, blue, violet, black, gold, green? __________
5. What are the values of EIA six-dot mica capacitors with the colors white, brown, black, red, gold, green? __________ White, gray, red, black, silver, yellow? __________
6. What are the capacitance and tolerance values of five-dot button capacitors with the colors red, violet, brown, red, green? __________ Blue, gray, red, red, black? __________
7. What is the capacitance and temperature coefficient of color-coded disk ceramic capacitors with the colors violet, green, blue, red, silver? __________ Black, green, brown, brown, gold? __________
8. What are the values of tubular ceramic capacitors with the colors violet, yellow, orange, red, silver? __________ Blue, gray, red, brown, gold? __________

CHAPTER 30 TEST • CAPACITORS

1. What is the capacitance of two 1-in.-square aluminum plates held apart by 4 mils (0.004 in.) of mica?
2. How much energy will a 0.05-μF capacitor store if it is held across 600 V? How long will it take to charge to 600 V? How many coulombs does it contain?
3. What type of dielectric in capacitors would be least affected by altitude? Which might be most affected?
4. What would an ohmmeter indicate across a capacitor that is good? Open? Burned out?
5. Toward what polarity is the nucleus of atoms in solid-material dielectrics attracted?
6. Would a capacitor stand more or less voltage across it if it were at high altitude? If it were physically hot?
7. What type of condenser always has leakage current?
8. What type of capacitor may explode if it is not connected with correct polarity or is used across ac?
9. What two undesirable effects occur in solid-dielectric capacitors at high frequencies?
10. A 5000-Ω *R* and a 2-μF *C* are in series across 40 V of dc. What is the time-constant value? What current will flow 15 ms after the circuit is completed? What is the voltage across the *R* at that time? Across the *C*?
11. A 0.3-H coil with a 250-Ω internal resistance is connected across a 6-V battery. What will the current value be in one time-constant period? In 0.001 s?
12. A capacitor has four stripes colored green, brown, red, and silver. What do you know about it?
13. How would you account for the fact that a capacitor marked 683 has a value of 68,000 pF?

ANSWERS TO CHECK-UP QUIZ 30-3

1. (250 pF) (3900 pF) (68,000 pF) **2.** (50 kpF, 10%) (410 pF, 5%) **3.** (8540 pF, 5%) (930 pF, 10%) **4.** (4750 pF) (967 pF) **5.** (1000 pF) (92 pF) **6.** (270 pF, 2%) (6800 pF, 2%) **7.** (5600 pF, −750 ppm) (510 pF, 0 ppm) **8.** (−750 ppm, 4300 pF, 10%) (−600 ppm, 820 pF, 5%)

31

SERIES AC CIRCUITS

CHAPTER OBJECTIVE. To learn to compute series *LCR* ac circuits using the Pythagorean theorem, simple trigonometry, and algebra; to solve for R, X, Z, θ, P, VA, and PF, as well as current, voltage, and voltage-drops.

31-1 SERIES *LCR* CIRCUITS

The basic principles of solving for current, voltages, impedance, power, phase angle, and power factor when a resistor and one or more reactances are in series were outlined in Chap. 14. Impedance, reactance, and resistance vector lengths can be determined by graphing on paper, and the phase angles, in degrees, can be determined by using a protractor.

The vector lengths of an electrical triangle can be solved more accurately by using the *Pythagorean theorem*. The angle between the Z and R sides of right-angle triangles can be determined by trigonometric ratios and either a slide rule, table of trigonometric functions (Appendix E), or a pocket calculator.

31-2 THE PYTHAGOREAN THEOREM

Any triangle, such as the electrical triangle, having R, X, and Z sides, with a 90° angle between the R and X sides (Fig. 31-1) is called a *right triangle*. Pythagoras found, 2500 years ago, that if the length of the hypotenuse (Z side) of a right triangle is squared, this value will be equal to a second side length squared added to the third side length squared. That is:

The square of the length of the hypotenuse of a right triangle is equal to the sum of the squares of the other two sides.

In formula form,

$$Z^2 = R^2 + X^2$$

Taking the square root of both sides of this formula results in a basic formula to determine the length of the hypotenuse of an electrical triangle, if R and X are known:

$$\sqrt{Z^2} = \sqrt{R^2 + X^2} \qquad \text{or} \qquad Z = \sqrt{R^2 + X^2}$$

In the triangle of Fig. 31-1, the Z value is

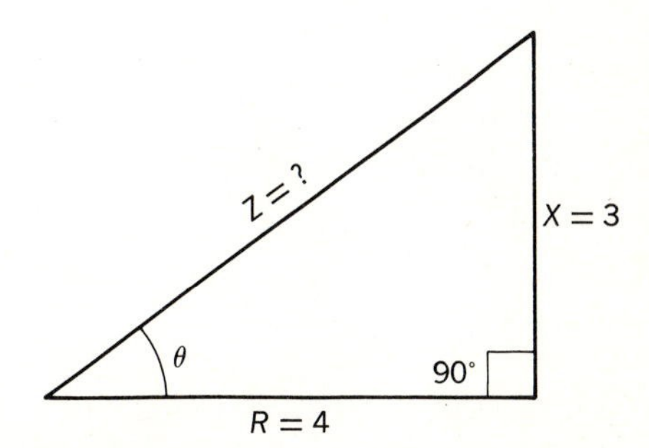

Fig. 31-1 Relationship of R, X, and Z sides in a right triangle. The angle between R and Z sides is the phase angle, or θ.

$$
\begin{aligned}
Z &= \sqrt{4^2 + 3^2} \\
&= \sqrt{16 + 9} \\
&= \sqrt{25} \\
&= 5\ \Omega
\end{aligned}
$$

Squares and square roots may be found by slide rule, or with a pocket calculator.

From the formula $Z^2 = R^2 + X^2$ two other valuable formulas can be derived. If the Z and X values are known, by algebraic rearrangement:

$$R^2 = Z^2 - X^2 \qquad \text{or} \qquad R = \sqrt{Z^2 - X^2}$$

In Fig. 31-1, if Z and X are known,

$$
\begin{aligned}
R &= \sqrt{5^2 - 3^2} \\
&= \sqrt{16} \\
&= 4\ \Omega
\end{aligned}
$$

When the R and Z values are known,

$$X^2 = Z^2 - R^2 \qquad \text{or} \qquad X = \sqrt{Z^2 - R^2}$$

In Fig. 31-1 if R and Z are known,

$$
\begin{aligned}
X &= \sqrt{5^2 - 4^2} \\
&= \sqrt{9} \\
&= 3\ \Omega
\end{aligned}
$$

It is possible to use similar formulas when the source voltage is considered as the hypotenuse vector and the voltage-drops across the R and X values of a series ac circuit are plotted as vectors in a right triangle. Thus the source voltage can be computed

$$E_S = \sqrt{E_R{}^2 + E_X{}^2}$$

Quiz 31-1. Test your understanding. Answer these check-up questions.

1. What is the formula to solve for Z if the R and X in a series circuit are known? ________
2. What is the formula to determine the R if the X and Z in a series circuit are known? ________
3. What is the formula to determine the X if the R and the Z in a series circuit are known? ________
4. A series ac circuit has an 80-Ω R and a 100-Ω X_L. What is the Z value? ________
5. A series ac circuit has a 250-Ω X_C and a 300-Ω Z. What is the R value? ________
6. A series ac circuit has a 62-Ω Z and a 45-Ω R. What is the X value? ________
7. What is the formula to determine the voltage-drop across a reactor if the source voltage and the voltage-drop across the resistance in a series circuit are known? ________

31-3 SERIES CIRCUITS WITH X_L, X_C, AND R

In many ac circuits R, X_L, and X_C may be in series. If a coil and capacitor are in series, the internal resistance of the coil can be measured with an ohmmeter and considered as a resistor in series with the coil. A capacitor will not have a significant value of series resistance. Since the X_L vector is plotted upward and the X_C vector is plotted downward, the two reactance vectors tend to cancel each other. The total reactance is the difference between X_L and X_C. The smaller value is subtracted from the larger, and the result is given the sign of the larger. The formula for solving this type of circuit when X_L is the larger is

$$Z = \sqrt{R^2 + (X_L - X_C)^2}$$

Figure 31-2*a* shows the schematic diagram of an ac circuit having a resistor, a coil, and a capacitor all in series across a source of emf. All the parameters of this circuit can be solved by the Pythagorean theorem, Ohm's law, and basic power formulas. Z is found by

$$
\begin{aligned}
Z &= \sqrt{200^2 + (500 - 350)^2} \\
&= \sqrt{200^2 + 150^2}
\end{aligned}
$$

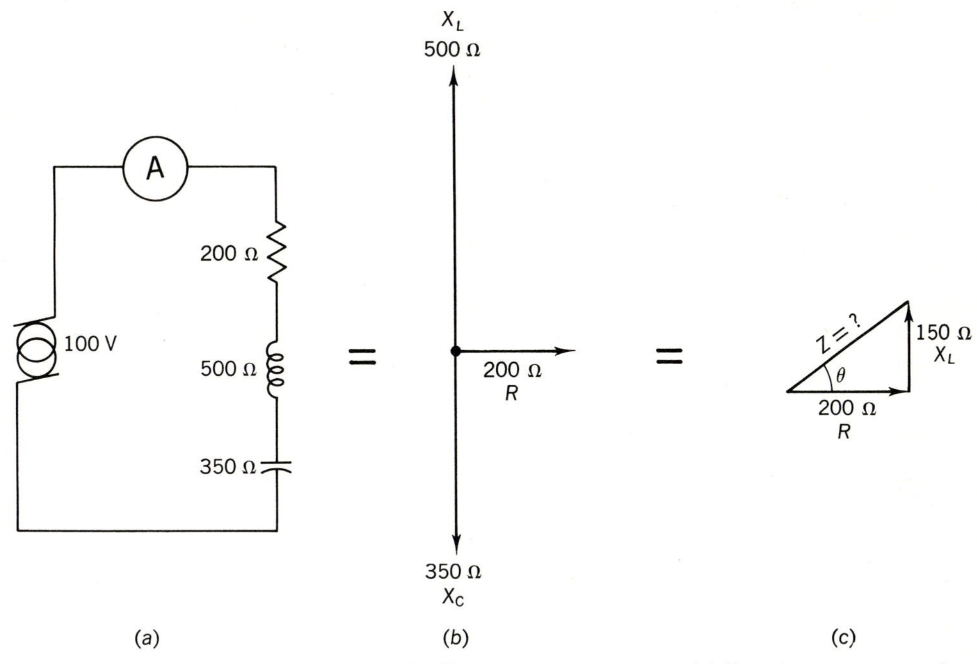

Fig. 31-2 (*a*) Series R, X_L, and X_C ac circuit. (*b*) Vector representation. (*c*) Resultant vectors when opposing reactances are combined.

$$= \sqrt{40{,}000 + 22{,}500}$$
$$= \sqrt{6\ 25\ 00}$$
$$= 250\ \Omega$$

With this information, the current, by Ohm's law, is $I = E/Z$, or 100/250, or 0.4 A.

An ac voltmeter across the resistor would indicate this voltage-drop to be $V = IR$, or 0.4(200), or 80 V.

An ac voltmeter across the coil would indicate as $V = IX_L$, or 0.4(500), or 200 V.

The voltmeter across the capacitor would indicate $V = IX_C$, or 0.4(350), or 140 V.

The power dissipated in the resistor can be solved by any of the basic power formulas, $P = EI$, I^2R, or E^2/R. Solving by the first of these, $P = EI$, or 80(0.4), or 32 W.

The power dissipated in both reactors (assuming no internal resistance) is zero, since the energy they store in their magnetic and electrostatic fields is all returned to the circuit during one half of each cycle.

The apparent power, or voltampere value, of the circuit is 100(0.4), or 40 VA.

With P and VA known, the power factor can be determined by the ratio of the true power to the apparent power, $\text{PF} = P/VA$, or 32/40, or 0.80, or 80%.

Other parameters that might be desirable to know are the phase angle, the inductance of the coil, and the capacitance of the capacitor. The inductance and the capacitance values can be determined if the frequency is known by using the reactance formulas and solving for L and C. The phase angle can be determined by using a protractor on the RXZ triangle or by trigonometric methods discussed in Sec. 31-4.

In solving series ac voltage circuits (Fig. 31-2*a*), plot the R and X voltage-drops plus the source voltage as a right triangle, as

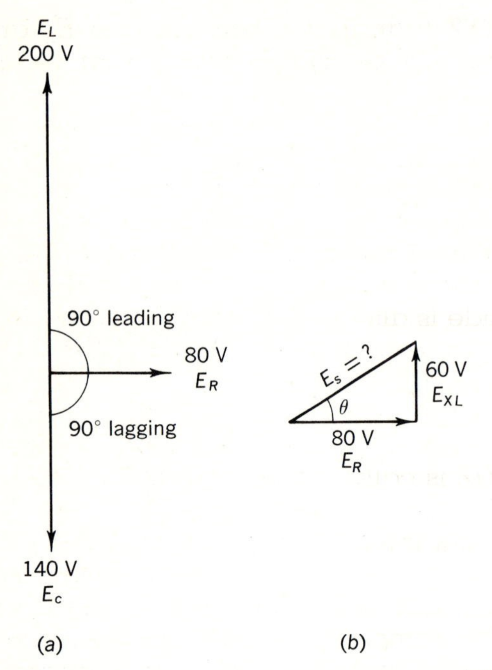

Fig. 31-3 (*a*) Voltage-vector representation of Fig. 31-2*a*. (*b*) Electrical voltage triangle.

in Fig. 31-3. When plotting voltage vectors, consider the voltage across the resistance as the horizontal reference vector. Since the voltage across an inductance *leads* its current, its vector is drawn 90° ahead (counterclockwise) in an upward direction as shown. (A capacitive voltage lags its current and is drawn downward.)

To solve for the source-voltage value if E_R, E_L, and E_C are known,

$$E_S = \sqrt{E_R^2 + (E_L - E_C)^2}$$

In Fig. 31-2*a*, it is known that E_L is 200 V, E_C is 140 V, and E_R is 80 V. Therefore,

ANSWERS TO CHECK-UP QUIZ 31-1

1. $(Z = \sqrt{R^2 + X^2})$ **2.** $(R = \sqrt{Z^2 - X^2})$ **3.** $(X = \sqrt{Z^2 - R^2})$ **4.** (128 Ω) **5.** (166 Ω) **6.** (42.6 Ω) **7.** $(E_X = \sqrt{E_S^2 - E_R^2})$

$$\begin{aligned} E_S &= \sqrt{80^2 + (200 - 140)^2} \\ &= \sqrt{80^2 + 60^2} \\ &= \sqrt{6400 + 3600} \\ &= \sqrt{1\,00\,00} \\ &= 100 \text{ V} \end{aligned}$$

When several reactors are in series, all like reactances are added. For the circuit in Fig. 31-4,

$$\begin{aligned} X_{L_t} &= 30 + 50 + 60 + 80 + 100 = 320\text{-}\Omega\ X_L \\ X_{C_t} &= 40 + 70 + 90 = 200\text{-}\Omega\ X_C \\ X_t &= 320 - 200 = 120\text{-}\Omega\ X_L \end{aligned}$$

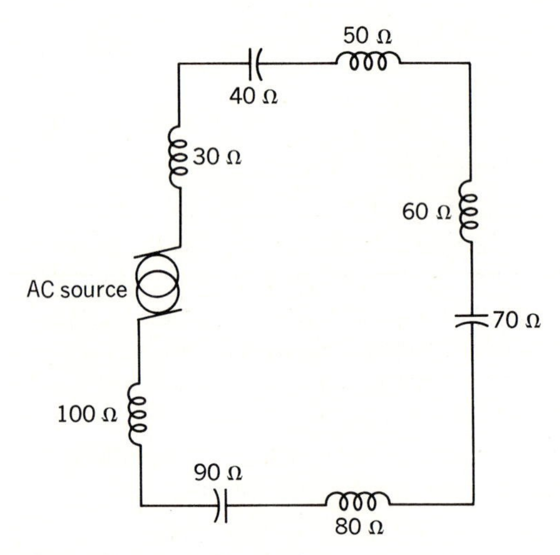

Fig. 31-4 Series R, X_L, and X_C circuit totaling 120 Ω of X_L.

The total X_L value of series inductors will always be proportional to the total inductance since $X_L = 2\pi fL$. However, when capacitors *in series* are added, the total capacitance is always less than any of the series-capacitor values. Two 200-Ω capacitive reactances in series may have 400 Ω of X_C, but the total C value will be half the capacitance value of one of the capacitors, from $X_C = 1/2\pi fC$.

Quiz 31-2. Test your understanding. Answer these check-up questions.

1. If the frequency of the source in Fig. 31-2*a* is 50 Hz, what is the L of the coil? ________ C of the capacitor? ________
2. What is the phase angle of E and I in the resistor? ________ Coil? ________ Capacitor? ________
3. A series ac circuit has 55 Ω of R, 68 Ω of X_L, and 104 Ω of X_C when across a 100-V 60-Hz line. What is the value of the Z? ________ I? ________ E_L? ________ P? ________ VA? ________ PF? ________
4. If the frequency of the line in question 3 were 50 Hz, what would be the Z value of the circuit? ________
5. A circuit has 40 Ω of X_C, 30 Ω of X_L, 20 Ω of R, 50 Ω of X_C, and 90 Ω of X_L, all in series. What is the total X_C? ________ X_L? ________ X_T? ________ Z? ________

31-4 THE SINE RATIO

When working with ac circuits, it is not necessary to know more than a few basic facts about trigonometry to solve for most of the parameters of ac circuits. To solve a right triangle, only three functions of trigonometry—sine, cosine, and tangent—are required. The angle of interest in electrical right triangles is the phase angle, the angle between the R and Z sides of an RXZ triangle (or between the E_R and E_S sides of an $E_RE_XE_S$ triangle for series circuits).

As long as the phase angle remains constant, the ratio of the X to the Z sides will remain constant. That is, if θ is not allowed to change, the X value must double when the Z value doubles. (This cannot be done, of course, unless the R side is also increased.) Since the hypotenuse is always the longest side, the relationship of the X side to the Z side (X/Z) of a right triangle can be expressed as some numerical value less than 1.0. The ratio X/Z is called the *sine* (abbreviated sin) of the phase angle (Fig. 31-5).

If X is 3 and Z is 5, the sine ratio is 3/5, or 0.600. Whenever the X and Z sides have the ratio of 0.600, the phase angle will be the same. The angle can be found by slide rule, from the table in Appendix E, or with a pocket calculator. Check the angle whose sine value is 0.600. It will be found to be 36.9° (to the third significant figure).

If the phase angle can be determined by the ratio of the X to the Z sides, it must also be possible to determine the numerical ratio if the phase angle is known. If the phase angle is 30°, then, for example, from the sin column of the tables,

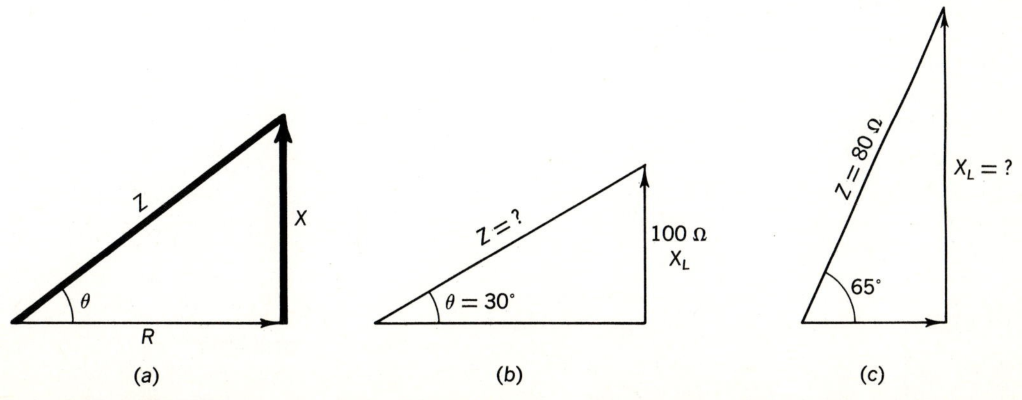

Fig. 31-5 (*a*) The sine is the ratio of X to Z. (*b*) When θ is 30°, the sine will be 0.5. (*c*) As θ increases, X approaches the value of Z.

under 30°, the sine ratio is found to be 0.500. Any time a circuit has a phase angle of 30° the ratio of X to Z must equal 0.500,

$$\sin 30° = 0.5 = \frac{X}{Z}$$

It is now possible to determine Z from the phase angle and reactance. If X is 100 Ω and θ is 30°, what is Z (Fig. 31-5*b*)?

$$\sin 30° = \frac{X}{Z}$$

$$0.5 = \frac{100}{Z}$$

$$Z = \frac{100}{0.5}$$

$$Z = 200\ \Omega$$

Similarly, X can be found if θ and Z are known. In an electric circuit, the phase angle is 65° and the impedance is 80 Ω (Fig. 31-5c). What is the reactance value?

$$\sin 65° = \frac{X}{Z}$$

$$0.906 = \frac{X}{80}$$

$$0.906(80) = X$$

$$X = 72.48\ \Omega$$

It will be found that many of the basic ac circuits can be solved by using only the Pythagorean theorem and sine-function algebra.

ANSWERS TO CHECK-UP QUIZ 31-2

1. (1.59 H) (9.1 μF) **2.** (0°) (90°) (90°) **3.** (65.8 Ω) (1.52 A) (103 V) (128 W) (152 VA) (0.836) **4.** (87.6) **5.** (90 Ω) (120 Ω) (30 Ω) (36.05 Ω)

Quiz 31-3. Test your understanding. Answer these check-up questions.

1. Which sides of an electrical *RXZ* triangle determine the sine ratio? ________ Which side is the divisor? ________
2. Which sides of an electrical $E_RE_XE_S$ triangle determine the sine ratio? ________
3. An ac circuit has a resistance of 120 Ω and a reactance of 150 Ω. What is the Z value? ________ Sine-ratio value? ________ Phase angle? ________
4. An ac circuit has a Z of 55 Ω and an R of 32 Ω. What is θ? ________
5. An ac circuit has a θ of 26° and a Z of 450 Ω. What is the total X value? ________
6. An ac circuit has an X of 140 Ω and a θ of 58°. What is the value of Z? ________ R? ________

31-5 COSINE AND TANGENT RATIOS

Just as the ratio of the lengths of the X/Z sides has the name *sine*, the ratio of the R/Z sides of an electrical triangle has a name, *cosine*, abbreviated cos (Fig. 31-6). As long as the phase angle remains constant, the ratio of the R and Z sides will also be constant. In Fig. 31-6, the cosine ratio is

$$\cos \theta = \frac{R}{Z}$$

If R is 80 Ω and Z is 100 Ω, the cosine ratio is 80/100, or 0.800. From Appendix E, 0.800 in the cos column shows a phase angle of 36.9°.

If the phase angle is 40° and the resistance in the circuit is 220 Ω, the impedance value (Fig. 31-6*b*) can be determined:

$$\cos \theta = \frac{R}{Z}$$

$$Z = \frac{R}{\cos \theta}$$

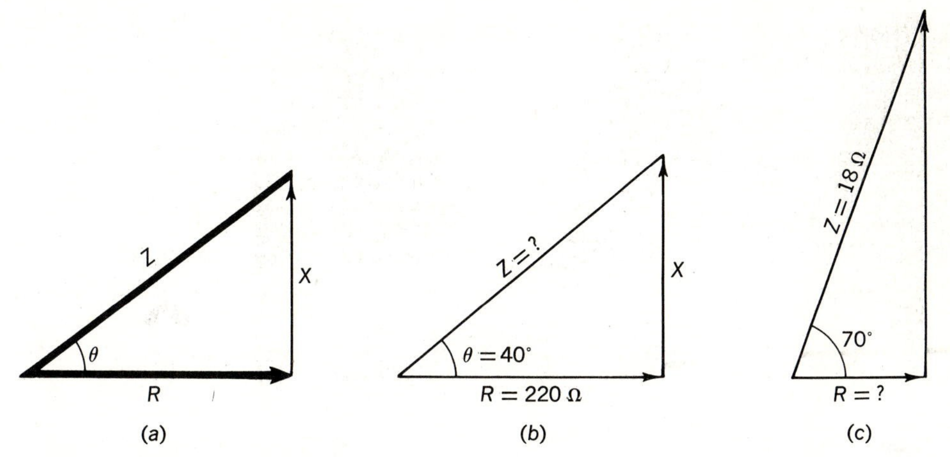

Fig. 31-6 (*a*) The cosine ratio is the *R* side over the Z side. (*b*) Z can be determined if *R* and *θ* are known by the cosine ratio. (*c*) *R* can be determined if Z and *θ* are known.

$$= \frac{220}{\cos 40°}$$

$$= \frac{220}{0.766}$$

$$= 287\ \Omega$$

Similarly, if Z is 18 Ω and *θ* is 70°, *R* can be determined from the cos *θ* formula (Fig. 31-6*c*):

$$\cos \theta = \frac{R}{Z}$$

$$R = Z(\cos \theta)$$

$$= 18(\cos 70°)$$

$$= 18(0.342)$$

$$= 6.16\ \Omega$$

The third ratio of sides useful in solving electrical-right-triangle problems is the *tangent*, abbreviated tan. If *θ* remains constant, the ratio of the *X* and *R* sides will be the same. The tangent ratio is the *X* side to the *R* side, or

$$\tan \theta = \frac{X}{R}$$

The phase angle of an electrical circuit having 180 Ω of reactance and 240 Ω of resistance can be found in the tan column, Appendix E:

$$\tan \theta = \frac{X}{R}$$

$$= \frac{180}{240}$$

$$= 0.75$$

$$\theta = 36.9°$$

The sine and the cosine values are always less than 1.0, but the tangent can be any number from zero to infinity. A circuit with a phase angle of 85° has an *X* value 11 times the *R* value, and may be considered purely reactive for most applications. Many slide rules carry tangent values to only 84°.

The three ratios sin $\theta = X/Z$, cos $\theta = R/Z$, and tan $\theta = X/R$ are important in ac circuitry and must be memorized.

Quiz 31-4. Test your understanding. Answer these check-up questions.

1. From the tables (slide rule, calculator), what is the numerical value of sin 0°? ________ sin 1°?

__________ sin 30°? __________ sin 60°? __________ sin 89°? __________ sin 90°? __________

2. What is the numerical value of cos 0°? __________ cos 1°? __________ cos 30°? __________ cos 60°? __________ cos 90°? __________
3. What is the numerical value of tan 0°? __________ tan 1°? __________ tan 30°? __________ tan 60°? __________ tan 89.9°? __________ tan 90°? __________
4. A series ac circuit has 56 Ω of X and 43 Ω of R. What is θ? __________
5. A series ac circuit has 4800 Ω of R and 6200 Ω of Z. What is θ? __________
6. A series ac circuit has 13 Ω of X and 21 Ω of Z. What is θ? __________
7. A series ac circuit has 75 Ω of X and θ is 28°. What is R? __________ Z? __________
8. A series ac circuit has 350 Ω of Z and θ is 65°. What is R? __________ X? __________

31-6 SOLVING *RL* AND *RC* AC CIRCUITS

Many series ac problems can be solved by the reactance formulas plus the Pythagorean and trigonometric formulas:

$$Z = \sqrt{R^2 + X^2} \qquad \sin\theta = \frac{X}{Z}$$

$$R = \sqrt{Z^2 - X^2} \qquad \cos\theta = \frac{R}{Z}$$

$$X = \sqrt{Z^2 - R^2} \qquad \tan\theta = \frac{X}{R}$$

The various parameters which can be determined in Fig. 31-7*a* are Z, *I*, *R*, *X*, the voltage-drops across *R* and *X*, the source voltage, θ, *P* dissipated by the resistor, *VA*, and the power factor.

When solving circuits, look for one component about which two things are known (*X* and the voltage across it, or θ and Z, etc.). With two starting values, some other parameter can be determined, which in turn will lead to the solution of still another, until the desired answers are found.

Power factor is the ratio of true to apparent power, or PF = *P*/*VA*. Since true power is the power dissipated by the *resistance* of a circuit, and apparent power is the *VA* meter readings across the load *impedance*, the ratio of resistance to impedance, *R*/Z, must also represent the power factor, or PF = *R*/Z. The ratio *R*/Z is also the cosine of the phase angle in an electrical right triangle. Inasmuch as an *RXZ* triangle is proportional to its $E_RE_XE_S$ triangle, the power factor can also be determined from the ratio E_R/E_S. For series circuits the power factor can be determined by

$$\mathrm{PF} = \frac{P}{VA} = \frac{R}{Z} = \frac{E_R}{E_S} = \cos\theta$$

To determine *R*, Z, and many other parameters in an ac circuit, set up the two equal expressions, *P*/*VA* = *R*/Z. Voltmeter, ammeter, and wattmeter indications can often be obtained for a circuit. The wattmeter gives true power, and the voltmeter times the ammeter readings gives the *VA*. Thus

$$\frac{P}{VA} = \frac{R}{Z} \quad \text{so} \quad R = \frac{PZ}{VA} \quad \text{or} \quad Z = \frac{RVA}{P}$$

If *R* and Z are known, θ can be determined by $\cos\theta = R/Z$. With θ, *X* can be determined by $\sin\theta = X/Z$ or by the Pythagorean formula. In fact, all parameters of the simple *RL* ac circuit shown in Fig. 31-7*a* can be found. If the frequency of the ac is known, the inductance can be found from $X_L = 2\pi fL$ (or capacitance in an *RC* circuit).

ANSWERS TO CHECK-UP QUIZ 31-3

1. (X, Z) (Z) 2. (E_X, E_S) 3. (192 Ω) (0.781) (51.4°) 4. (54.4°) 5. (197 Ω) 6. (165 Ω) (87.5 Ω)

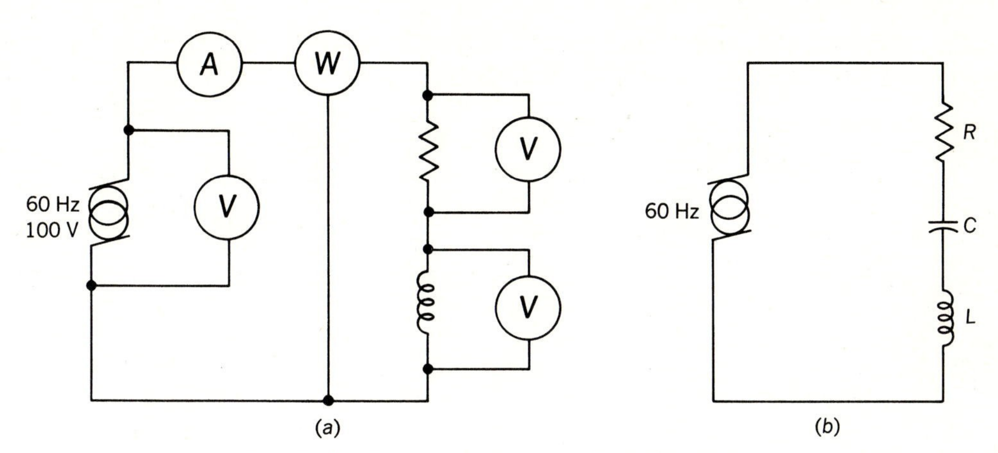

Fig. 31-7 Circuits used in the Check-up Quiz.

It can be seen that there may be several different approaches to solving ac circuits. There is probably no "best" way.

Quiz 31-5. Test your understanding. Answer these check-up questions.

1. In Fig. 31-7*a*, the wattmeter reads 350 W, the ammeter 4.2 A. What is the value of *VA*? ______ *Z*? ______ PF? ______ *R*? ______ θ? ______ X_L? ______ *L*? ______ E_R? ______ E_X? ______ P_R? ______ P_X? ______ Does I_S lead or lag E_S? ______
2. In Fig. 31-7*b*, if *C* is 5 μF, *L* is 0.8 H, and *R* is 200 Ω, what is the value of X_C? ______ X_L? ______ *Z*? ______
3. In Fig. 31-7*b*, if the voltage drops are $E_R = 40$ V, $E_C = 80$ V, and $E_L = 105$, and *R* is 200 Ω, what is the value of θ? ______ E_S? ______ I_S? ______ PF? ______ *P*? ______ *VA*? ______ *Z*? ______ Does I_S lead or lag? ______

CHAPTER 31 TEST • SERIES AC CIRCUITS

1. Express in symbols the Pythagorean formula for determining *X* if *R* and *Z* are known.
2. Express in symbols the Pythagorean formula for determining E_R if E_S and E_X are known.
3. What is the name given to the ratio of *R*/*Z*? E_X/E_R? *X*/*Z*?
4. If a motor is found to have 9 Ω of R_i and 16 Ω of X_L, what is its impedance? What current will it draw from a 120-V 60-Hz line?
5. In question 4, what is θ? PF? What X_C value should a capacitor connected across the motor have to produce a unity PF? Why would a unity PF be desirable?
6. A vacuum cleaner across a 120-V 60-Hz line draws 3 A. An ohmmeter shows its *R* to be 12 Ω. What is the effective *Z*? X_L? *L*?
7. In question 6, what is the PF? Power dissipated? How much would it cost to operate the vacuum cleaner for 1 h at $0.04/kWh?
8. A series ac circuit consists of a 380-Ω X_C, a 42-Ω *R*, and a 435-Ω X_L across a 100-V source. What is the value of *Z*? I_L? E_C?
9. In question 8, what is θ? Is it leading or lagging? What is the PF? *P*?
10. Under what condition will the sum of the voltage-drops around an *RXZ* circuit equal the source-voltage value?
11. At what phase angle would tan θ equal unity?
12. At what angle does cos θ equal sin θ?

ANSWERS TO CHECK-UP QUIZ 31-4

1. (0) (0.01745) (0.5) (0.866) (0.99985) (1) **2.** (1) (0.99985) (0.866) (0.5) (0) **3.** (0) (0.01746) (0.57735) (1.7321) (572.96) (∞) **4.** (52.5°) **5.** (39.3°) **6.** (38.2°) **7.** (141 Ω) (160 Ω) **8.** (148 Ω) (317 Ω)

ANSWERS TO CHECK-UP QUIZ 31-5

1. (420) (23.8 Ω) (0.833) (19.8 Ω) (33.5°) (13.16 Ω) (0.0349 H) (83.2 V) (55.3 V) (350 W) (0) (Lags) **2.** (531 Ω) (301 Ω) (305 Ω) **3.** (32°) (47.2 V) (0.2 A) (0.718) (8 W) (11.1) (236 Ω) (Lags)

32

PARALLEL AND COMPLEX AC CIRCUITS

CHAPTER OBJECTIVE. To learn to solve simple parallel *RLC* circuits, then to determine equivalent circuits by computations and by graphing, and finally to solve the more complex series-parallel ac circuits.

32-1 SOLVING PARALLEL AC CIRCUITS

Series circuits are fairly straightforward, but when ac circuits involve series and parallel elements together, a basic plan to solve such problems must be developed, or else the technician will become lost in a maze of formulas.

In Chap. 15 it was pointed out that parallel similar reactances can be solved in a manner similar to parallel resistances.

$$R_t = \frac{R_1 R_2}{R_1 + R_2}$$

Similarly,

$$X_{L_t} = \frac{X_{L_1} X_{L_2}}{X_{L_1} + X_{L_2}} \quad \text{and} \quad X_{C_t} = \frac{X_{C_1} X_{C_2}}{X_{C_1} + X_{C_2}}$$

When an X_C is in parallel with an X_L, there are two methods of solving for the total reactance. One way is to convert to capacitive susceptance B_C (the reciprocal of capacitive reactance) and inductive susceptance B_L and then determine the difference of the susceptances. The reciprocal of the net susceptance, B_t, is the total reactance value, X_t. As a second method, any convenient source-voltage value can be selected, and the vectors of the resultant capacitive and inductive currents can be graphed to determine the resultant current value. If I is known, X_t can be determined by Ohm's law.

Chapter 15 should be reviewed briefly before continuing.

32-2 THE PYTHAGOREAN THEOREM IN PARALLEL CIRCUITS

For series ac circuits the Pythagorean theorem, $Z^2 = R^2 + X^2$, can be used to determine the Z, R, or X of the ac triangle if two of the sides are known. Similarly, the source, reactance, and resistance voltages can be determined from the $E_R E_X E_S$ triangle. The Pythagorean theorem can also be used in parallel ac circuits provided R is first converted to conductance G, and X is converted to susceptance B. Then admittance Y, the reciprocal of impedance, can be determined from the Pythagorean formula derivation

$$Y = \sqrt{G^2 + B^2}$$

The circuit in Fig. 32-1*a* can be solved with this formula. First the two dissimilar reactances are converted to their susceptances, and the total *net* susceptance value is found:

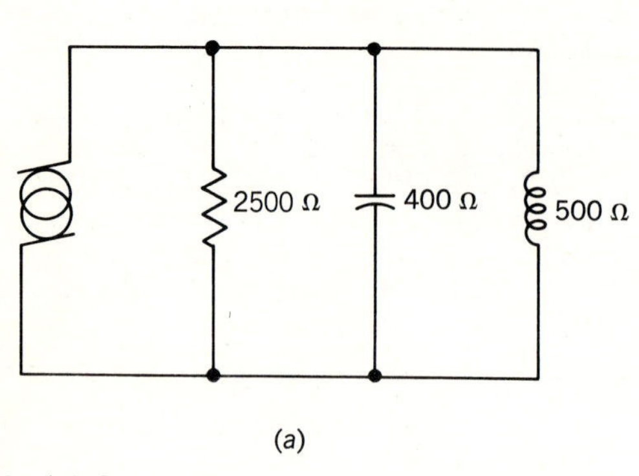

(a)

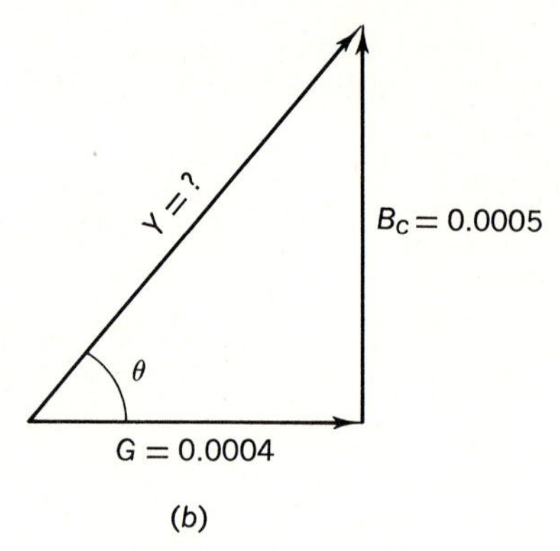

(b)

Fig. 32-1 (*a*) A parallel ac circuit with pure resistance and reactances in the branches. (*b*) The *GBY* electrical triangle for solving the admittance and impedance of the circuit.

$$B_C = \frac{1}{400} = 0.0025 \text{ mho}$$

$$B_L = \frac{1}{500} = 0.0020 \text{ mho}$$

$$B_t = 0.0025 - 0.0020 = 0.0005 \text{ mho}$$

Since B_C is greater than B_L, the total susceptance is 0.0005 mho capacitive.

How can the total susceptance of 0.0005 mho, which is a reactance value of $X = 1/B$, or 1/0.0005, or 2000 Ω, result from two parallel but dissimilar 400- and 500-Ω reactances? Parallel resonance, $X_L = X_C$, represents an infinitely high impedance. While 400 and 500 Ω are not equal, they are approaching equality, and therefore they will have a total reactance value *greater* than either of the parallel reactances alone. This is the opposite of resistances or *similar* reactances in parallel, in which case the total is always less than the lowest resistor or reactor value.

The conductance of the circuit in Fig. 32-1*a* is simply the reciprocal of the resistance:

$$G = \frac{1}{2500} = 0.0004 \text{ mho}$$

With a B_t of 0.0005 and a G of 0.0004, the Pythagorean formula can be used to determine the Y value (Fig. 32-1*b*):

$$\begin{aligned} Y &= \sqrt{G^2 + B_C^2} \\ &= \sqrt{(0.0004)^2 + (0.0005)^2} \\ &= \sqrt{0.00000016 + 0.00000025} \\ &= \sqrt{0.00000041} \\ &= 0.00064 \text{ mho (capacitive)} \end{aligned}$$

Once the total admittance Y has been determined, the impedance will be the reciprocal of this, or $Z = 1/0.00064$, or 1563 Ω.

When R and X_C are plotted at right angles, the X_C value is always drawn downward. However, when the reciprocal, or B_C, is being plotted, it is drawn in the opposite direction. Thus B_C is plotted upward and B_L downward.

The phase angle of the circuit can be determined by using the sine, cosine, or tangent of the *GBY* triangle. G is the resistive side, B is the reactive side, and Y is the impedance side. Thus $\sin\theta = B_C/Y$, $\cos\theta = G/Y$, and $\tan\theta = B_C/G$. For the triangle shown in Fig. 32-1*b*, $\sin\theta = B_C/Y$, or 0.0005/0.00064, or 0.781. The angle whose sine ratio is 0.781 is 51.4°. (Try computing θ by cosine ratio. To what angle is this equivalent? Then compute it again using the tangent. The same angle should be obtained in all cases.)

This same circuit can be solved by using graph paper and a protractor as a cross check.

When a 500-Ω X_L and a 400-Ω X_C are in parallel, the formula to find the total reactance is

$$X_t = \frac{X_L X_C}{X_L - X_C} = \frac{500(400)}{500 - 400} = \frac{200{,}000}{100}$$
$$= 2000\ \Omega$$

This is the same value obtained by using the addition of the susceptances while solving Fig. 32-1*a*. The total reactance will have the characteristic of the *smaller* reactance because it draws the greater current. In this case the total effective reactance is 2000-Ω X_C.

Another method of computing Z and θ in this type of circuit is to assume some convenient source voltage and compute the three branch currents. Since the current of the capacitive branch leads E_S by 90° and the current of the inductive branch lags E_S by 90°, the two reactive currents are 180° out of phase and tend to cancel. The *difference* between these two currents is the total reactive current seen by the source. Now the resistive current and the reactive current can be plotted as current vectors to form an $I_R I_X I_S$ triangle. The circuit seen by the load is capacitive but has both resistive and capacitive-reactance components, with current leading.

Quiz 32-1. Test your understanding. Answer these check-up questions.

1. With 100 V as the source emf in Fig. 32-1*a*, what is the value of I_R? ________ I_C? ________ I_L? ________
2. When plotting the I_R and I_X current vectors for Fig. 32-1*a*, should the I_X be shown pointing up or down? ________ Why? ________
3. What is the total reactive current value of this circuit? ________
4. What is the source-current value? ________
5. What is the phase angle of the circuit? ________
6. If the phase angle is stated according to what the current is doing, would you say that the phase angle is leading or lagging? ________
7. In which case would the prodivisum formula not apply, two parallel X_L's, X_C's, *L*'s, *R*'s, or *C*'s? ________ What formula would be used in this particular case? ________
8. An 80-Ω X_L and a 90-Ω X_C are in parallel. What is the value of the total effective reactance? ________ Is the effective reactance capacitive or inductive? ________

32-3 SOLVING FOR PARALLEL IMPEDANCES

When two parallel branches each have *R* and *X* in series, as in Fig. 32-2, the basic admittance formula can be expanded to solve for the Y_t and from that the Z_t.

$$Y = \sqrt{G_t^2 + B_t^2}$$
$$Y = \sqrt{(G_1 + G_2)^2 + (\pm B_1 \pm B_2)^2}$$

Neither branch in Fig. 32-2 is pure reactance or pure resistance. To determine the conductance component of such circuits,

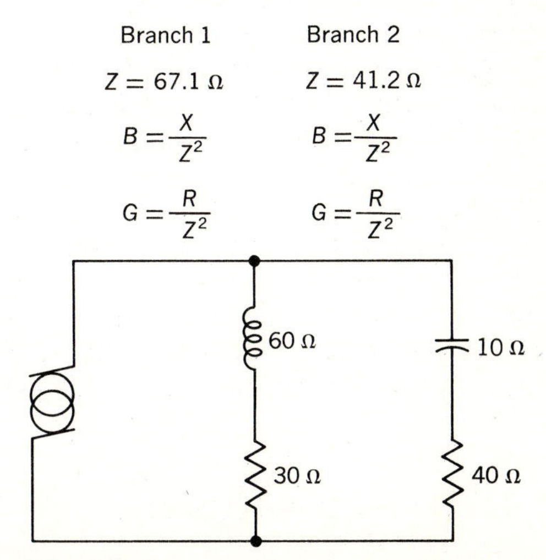

Fig. 32-2 Parallel impedance branches can be solved by the expanded admittance formula.

the ratio of R to Z^2 is used. The formula is $G = R/Z^2$ (derived from $G = I/E$). G may also be considered as the ratio of R to $X^2 + R^2$, or $G = R/(X^2 + R^2)$. Similarly, the susceptance component of a series-impedance branch is determined by $B = X/Z^2$ or $B = X/(X^2 + R^2)$. Substituting the values of Fig. 32-2 into the admittance formula above, using a − sign before capacitive values and a + sign before the inductive values in the B_t expression:

$$Y = \sqrt{\left(\frac{30}{67.1^2} + \frac{40}{41.2^2}\right)^2 + \left(\frac{60}{67.1^2} - \frac{10}{41.2^2}\right)^2}$$
$$= \sqrt{(0.00666+0.0236)^2+(0.0133-0.00589)^2}$$
$$= \sqrt{0.0303^2 + 0.00741^2}$$
$$= \sqrt{0.000918 + 0.0000549}$$
$$= \sqrt{0.000973}$$
$$= 0.0312 \text{ mho (capacitive)}$$

and

$$Z = \frac{1}{Y} = \frac{1}{0.0312} = 32.1\ \Omega$$

The phase angle is found from the tangent ratio of B_t/G_t, or 0.00741/0.0303, or 0.245. The angle whose tangent is 0.245 is 13.8°. Since the capacitive branch has the lower impedance, a greater current will flow in this branch. Therefore the source current leads the source voltage by 13.8°.

Quiz 32-2. Test your understanding. Answer these check-up questions.

1. In Fig. 32-2, use an X_L value of 45 Ω instead of 60 Ω and an X_C value of 25 Ω instead of 10 Ω. What is the impedance of branch 1? ________ The Z value of branch 2? ________
2. In question 1, what is the value of G_t? ________ B_t value? ________
3. In question 1, what is the admittance value? ________ Is it inductive or capacitive? ________ Why? ________
4. In question 1, what is Z_t? ________ θ_t? ________ PF_t? ________

ANSWERS TO CHECK-UP QUIZ 32-1

1. (0.04 A) (0.25 A) (0.2 A) **2.** (Upward) (I leads) **3.** (0.05 A) **4.** (0.064 A) **5.** (51.4°) **6.** (Leading) **7.** (C's) ($C_t = C_1 + C_2$) **8.** (720 Ω) (Inductive)

32-4 EQUIVALENT CIRCUITS

As ac circuits become more complex, it becomes necessary to determine to what a given group of series, parallel, or series-parallel components is equivalent in order to compute two or more equivalent circuits for a final answer.

In Fig. 32-3*a*, 30- and 60-Ω resistors are in parallel. This circuit is the equivalent of $R_t = R_1R_2/(R_1 + R_2)$, or 20 Ω.

In Fig. 32-3*b*, the 30- and 60-Ω inductive reactors are in parallel, $X_{L_t} = X_{L_1}X_{L_2}/(X_{L_1} + X_{L_2})$, or 20 Ω. As it is a purely reactive circuit, the phase angle is 90°, with current lagging. This equivalent can be expressed as $20\underline{/-90°}\ \Omega$.

In Fig. 32-3*c*, a 30-Ω X_L and a 60-Ω_C are in parallel. This circuit is the equivalent of $X_t = 30(60)/(30 - 60)$, or 60-Ω X_L. As it is purely reactive and inductive, θ is 90°, with the current lagging, or $60\underline{/-90°}\ \Omega$.

In Fig. 32-3*d*, a 30-Ω R and a 60-Ω X_C are in series. For this circuit $Z = \sqrt{30^2 + 60^2}$, or 67.1 Ω. The phase angle is the angle whose tangent is 60/30, or 63.4°. Since the circuit is capacitive, current *leads*. The equivalent expression is $67.1\underline{/63.4°}\ \Omega$.

In Fig. 32-3*e*, a 30-Ω R, a 40-Ω X_C, and a 60-Ω X_L are in series. Since the two opposite types of reactance tend to cancel each other, this circuit is equivalent to a 30-Ω R and 20-Ω X_L in series (Fig. 32-3*f*). The impedance is $Z = \sqrt{30^2 + 20^2}$, or 36.1 Ω. The angle whose tangent ratio is 20/30 is 33.7°. Since the circuit is inductive, current lags. The equivalency expression is $36.1\underline{/-33.7°}\ \Omega$.

In Fig. 32-4*a*, a 30-Ω R and a 60-Ω X_L

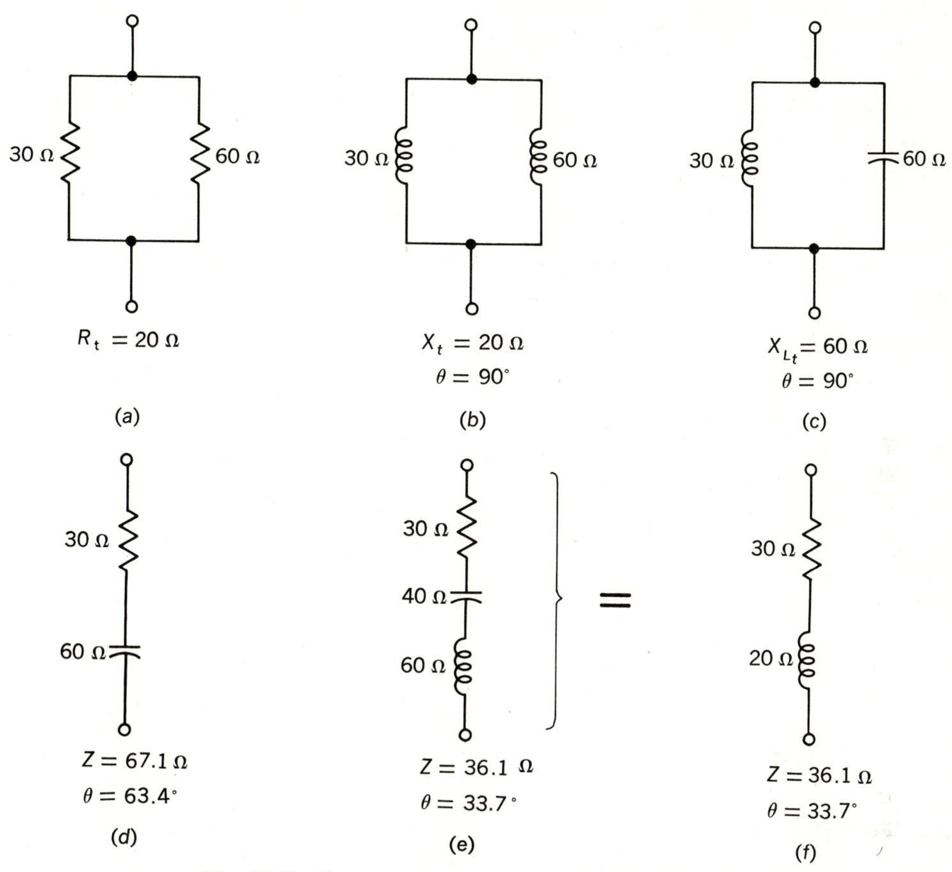

Fig. 32-3 Some ac circuits and their equivalents.

form a parallel circuit. By vector addition of G and B values, $Y = \sqrt{G^2 + B^2}$, or $\sqrt{0.0333^2+0.0167^2}$ or $\sqrt{0.00\,13\,90}$, or 0.0373 mho. The reciprocal of 0.0373 mho is 26.8 Ω of impedance. The phase angle is the angle whose tangent is 0.0167/0.0333, or 26.6°. This parallel R and X_L value is the equivalent of $26.8\angle{-26.6°}$ Ω.

It is important to recognize that both series and parallel circuits have Z and θ equivalents ($Z\angle\theta$). Therefore it must be possible in the case of any *parallel* R and X circuit to determine some *series* circuit R and X that is the electrical equivalent as seen by the source. Once the $Z\angle\theta$ equivalent of a parallel circuit has been found, it is possible to construct a Z vector in a direction indicated by the phase angle (Fig. 32-4*b*) and, by sine and cosine formulas, solve the R and X series-component equivalents of the parallel circuit. In Fig. 32-4*b*, the impedance is 26.8 Ω and the phase angle is 26.6°. From sine and cosine expressions,

$$\sin\theta = \frac{X}{Z}$$

$$\sin 26.6° = \frac{X}{26.8}$$

$$\sin 26.6°(26.8) = X$$

$$0.448(26.8) = X$$

$$X = 12\ \Omega$$

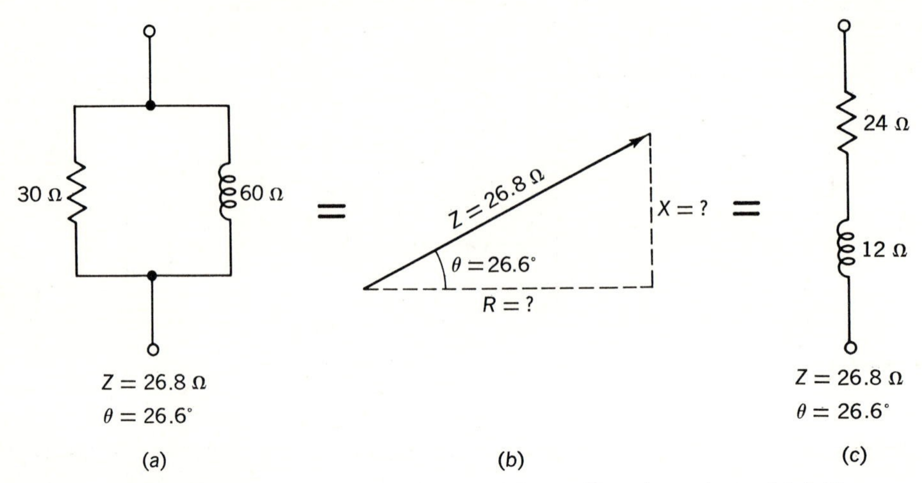

Fig. 32-4 (*a*) A parallel *R* and *X* circuit can be computed for its Z and θ values. (*b*) A Z vector at the phase angle can be plotted, from which (c) the series equivalents can be found.

and

$$\cos \theta = \frac{R}{Z}$$

$$\cos 26.6° = \frac{R}{26.8}$$

$$\cos 26.6°(26.8) = R$$

$$0.894(26.8) = R$$

$$R = 24\ \Omega$$

The 30-Ω R in parallel with the 60-Ω X_L is the equivalent of a 12-Ω R in series with a 24-Ω X_L. Both circuits have the same Z and θ values. A source of ac would produce the same current flow at the same phase angle if across either circuit.

It is possible to determine the parallel equivalent of a series ac circuit also. The series circuit of Fig. 32-5*a* has an impedance of 50 Ω and a phase angle of 53.2°. Figure 32-5*b* shows the X_C vector pointing downward. When a reciprocal *GBY* triangle is drawn, B_C must be drawn pointing upward. Therefore the Y vector (1/50) is drawn upward at an angle of 53.2°. The two sides, B and G, can be determined by sine and cosine expressions:

$$\sin \theta = \frac{B}{Y}$$

$$\sin 53.2° = \frac{B}{0.02}$$

$$0.8(0.02) = B$$

$$B = 0.016 \text{ mho}$$

$$X = \frac{1}{B} = 62.5\ \Omega$$

and

$$\cos \theta = \frac{G}{Y}$$

$$\cos 53.2° = \frac{G}{0.02}$$

$$0.599(0.02) = G$$

$$G = 0.012 \text{ mho}$$

$$R = \frac{1}{G} = 83.3\ \Omega$$

ANSWERS TO CHECK-UP QUIZ 32-2

1. (54.1 Ω) (47.2 Ω) **2.** (0.0283) (0.0042) **3.** (0.0286) (Capacitive) (C branch has lower Z, therefore greater *I*) **4.** (35 Ω) (8.42°) (0.989)

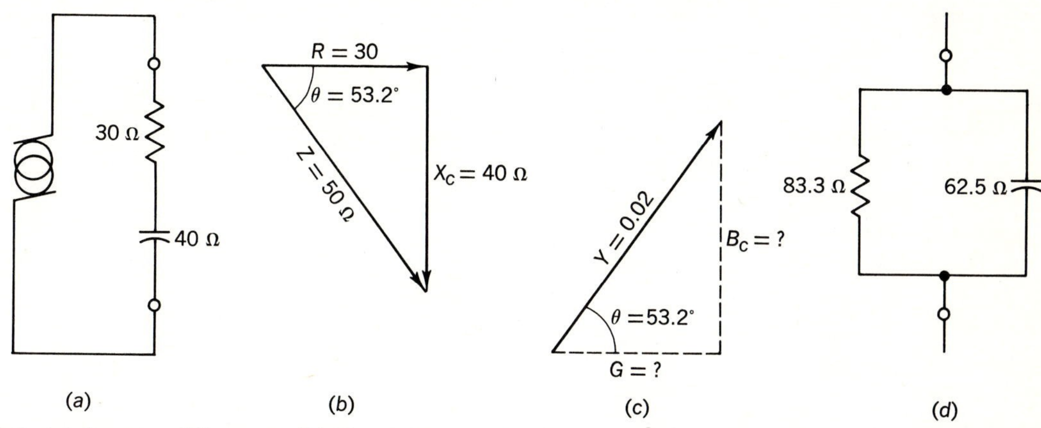

Fig. 32-5 (*a*) A series *RC* circuit. (*b*) The *RXZ* triangle for the series circuit. (*c*) The equivalent *GBY* triangle. (*d*) The equivalent parallel circuit.

A series 30-**Ω** R and 40-**Ω** X_C will appear to a source of ac exactly the same as an 83.3-**Ω** R in parallel with a 62.5-**Ω** X_C.

Quiz 32-3. Test your understanding. Answer these check-up questions.

1. In Fig. 32-6*a*, what is the Z value? __________ θ? __________ To determine the equivalent series circuit, should the Y vector be drawn at an upward or a downward angle? __________ What is the value of the equivalent series-circuit R? __________ X_C? __________
2. In Fig. 32-6*b*, what is the Z value? __________ θ? __________ To determine the equivalent parallel circuit, should the Y vector be drawn upward or downward? __________ What is the value of the equivalent parallel-circuit R? __________ X_L? __________
3. In Fig. 32-6*c*, what is Z for the right-hand branch? __________ θ? __________ What is the parallel circuit Y? __________ Z? __________ θ? __________ What is the equivalent series-circuit R? __________ X? __________
4. If a 10-**Ω** R and a 10-**Ω** X_C are in parallel, what is the series equivalent for Z? __________ θ? __________ R_S? __________ X_{C_S}? __________
5. If a 65-**Ω** R and a 32-**Ω** X_L are in parallel, what is Z? __________ θ? __________ Series-equivalent R? __________ X? __________

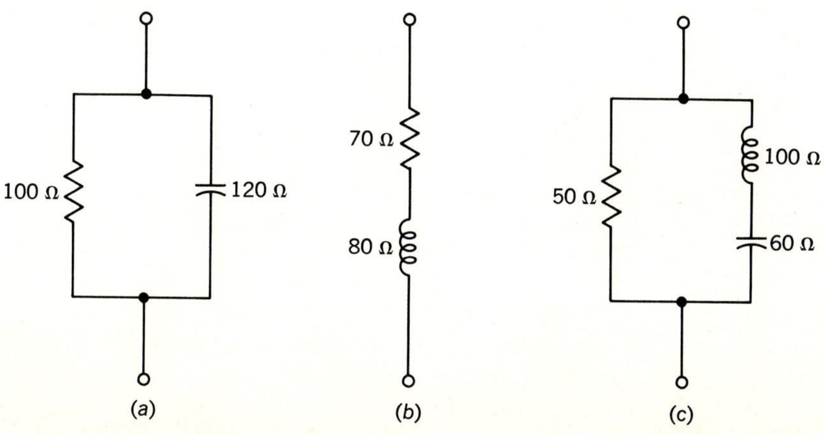

Fig. 32-6 Circuits for the Check-up Quiz.

32-5 GRAPHING EQUIVALENT CIRCUITS

There is a rather simple method of graphing parallel and series ac circuits to convert series to parallel, or parallel to series equivalents. Consider Fig. 32-6*b* again. First, draw the series-circuit *R*, *X*, and Z vectors to scale on graph paper, as in Fig. 32-7. The phase angle can be checked with a protractor. Next, draw a long dashed line at *right angles* to the Z vector. Finally, extend the R_P vector, shown dashed, and extend another dashed line X_P at right angles to this. The lengths of the dashed lines represent the parallel equivalents.

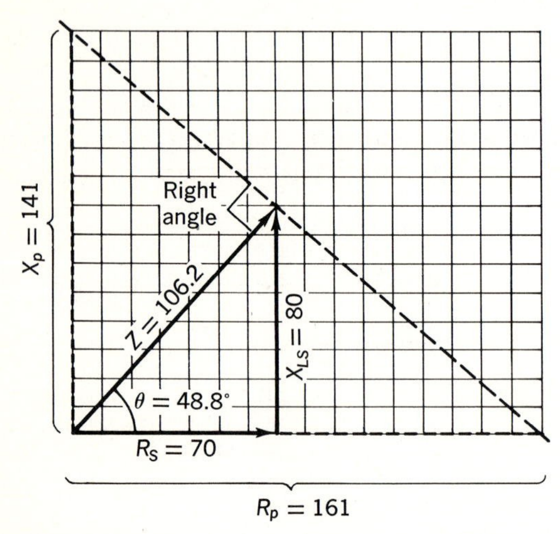

Fig. 32-7 Method of graphing the parallel-circuit equivalents of series ac circuit values. The series values are shown in solid lines.

The same type of a graph, developed in reverse order, can be used to determine the series equivalents of a parallel ac circuit. Consider Fig. 32-6*a* again, where the series-circuit equivalents were to be found for a 100-Ω *R* and a 120-Ω X_C in parallel. For a graphical solution, first graph the parallel-component values, R_P and X_{C_P}, with the X_C vector pointing downward (Fig. 32-8). Then draw the dashed line from vector tip to vector tip. Next, draw a line at right angles to the dashed line to the apex of the right angle. The length of this line is the impedance, Z_S. The phase angle is always the angle between the *R* and Z sides. Finally, draw a dotted line perpendicular to the R_P vector to intersect the dashed line at the point of the Z vector. This dotted line represents the series reactance, X_{C_S}. Where the X_C dotted line intersects the R_P vector indicates the series R_S.

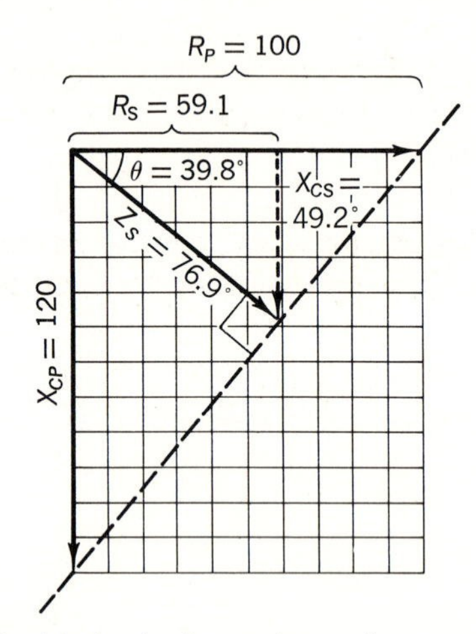

Fig. 32-8 Method of graphing the series equivalents of a parallel ac circuit.

To check your understanding, determine graphically the series equivalents of the parallel circuit shown in Fig. 32-5*d*.

ANSWERS TO CHECK-UP QUIZ 32-3

1. (76.9 Ω) (39.8°) (Upward) (59.1 Ω) (49.2 Ω) **2.** (106 Ω) (48.7°) (Downward) (161 Ω) (141 Ω) **3.** (40 Ω) (90°) (0.032 mho) (31.6 Ω) (51.4°) (19.7 Ω) (24.7 Ω) **4.** (7.07 Ω) (45°) (5 Ω) (5 Ω) **5.** (28.7 Ω) (63.8°) (12.7 Ω) (25.7 Ω)

32-6 COMPUTING SERIES-PARALLEL AC CIRCUITS

To compute series-parallel ac circuits, it is necessary to understand: (1) how to determine *R*, *X*, *Z*, *and* θ for series circuits, (2) how to determine *G*, *B*, *Y*, and θ for parallel circuits, and (3) how to convert series-circuit values to parallel-circuit equivalents, and vice versa. With this information, most complex ac circuits may be solved.

Figure 32-9*a* illustrates a basic complex circuit. The parallel portion can be solved for Z and θ, and from this a series *R* and X_L equivalent can be determined. For the parallel portion, by the admittance method, *Y* is 0.0141, *Z* is 70.7 Ω, and θ is 45° (Fig. 32-9*b*). When the equivalent series *RXZ* triangle is developed from $Y = 0.01414$ and $\theta = 45°$ (Fig. 32-9*c*), *Z* is 70.7 Ω, θ is 45°, and *R* and X_L equal 50 Ω each. Thus the two 100-Ω components in parallel are equivalent to a 50-Ω *R* in series with a 50-Ω X_L.

Now the complex circuit can be redrawn as in Fig. 32-9*d*, as a simple series *X* and *R* circuit. The *Z* value is $\sqrt{65^2 + 50^2}$, or 80 Ω, and the phase angle θ is 37.6°, with the current lagging.

Solving complex ac circuits, then, requires familiarity with series-circuit operations, parallel-circuit operations, and changing from parallel to series equivalent circuits. Using these basic steps, ana-

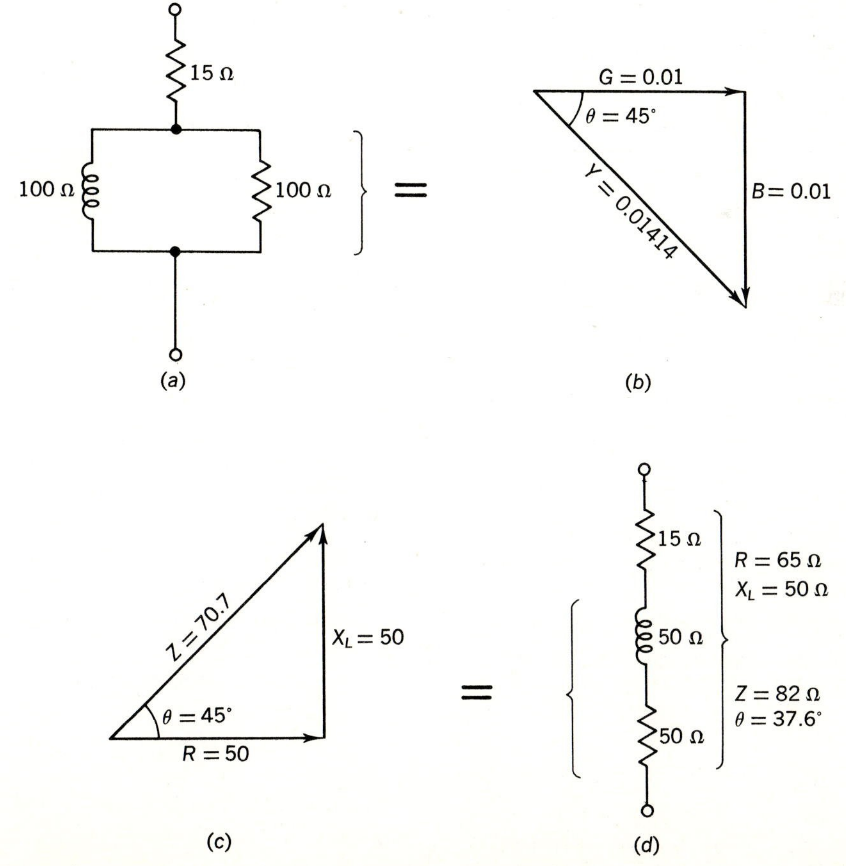

Fig. 32-9 Steps in solving an ac circuit by converting the parallel part to its series equivalent and then adding the series components.

lyze how the complex circuit shown in Fig. 32-10*a* could be solved.

First, this circuit consists of an upper parallel circuit in series with a lower complex circuit. It is basically a complex *series* circuit.

Next, the parallel upper circuit can be converted to the equivalent series R_S and X_{C_S} values, as indicated in Fig. 32-10*b*.

Then the lower-left series circuits must be converted to their parallel equivalents (Fig. 32-10*b*). This leaves two resistors in parallel, in parallel with two inductive reactances in parallel. The parallel resistances are computed, as are the parallel reactors, leaving a lower resultant of one resistor in parallel with one inductive reactance (Fig. 32-10*c*).

When the series-equivalent values of the parallel R_P and X_{L_p} are found, the whole circuit now consists of two *R*'s in series with an X_L and an X_C (Fig. 32-10*d*), a relatively simple-to-solve series ac circuit.

Check your understanding by determining *Z* and θ. (12.2 Ω, 15°)

Another exercise in solving complex cir-

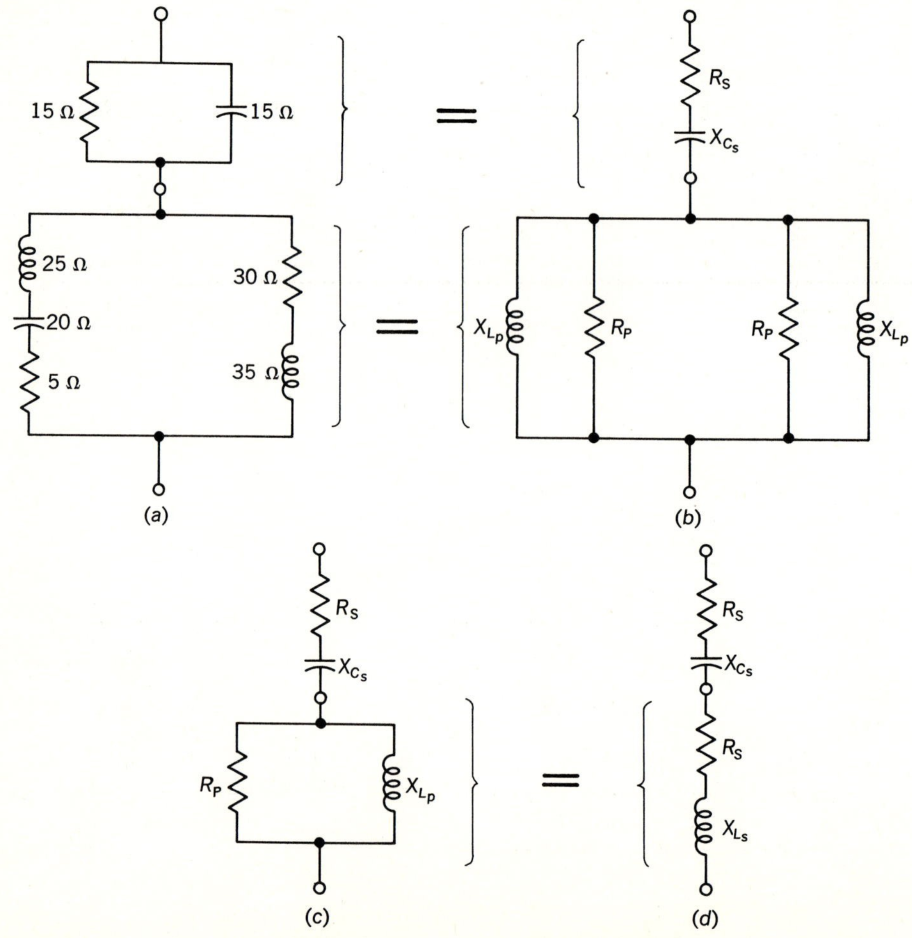

Fig. 32-10 Steps in converting a complex ac circuit into a simpler series ac circuit to determine the total *Z* and θ.

cuits is analysis of Fig. 32-11*a*. This has three complex branches in parallel. Each branch must be solved for its parallel-equivalent R and X values.

Branch 1 is a series RX circuit. The parallel equivalents can be reduced to the network in Fig. 32-11*b*.

In branch 2, the left leg is a series circuit that must be solved for its parallel equivalent (Fig. 32-11*b*). The right X_L leg of this branch is already in parallel across the circuit.

Branch 3 is a series-parallel circuit. The parallel R and X_C must be reduced to their series-equivalent values. This forms a series CRC circuit (Fig. 32-11*b*) for which a parallel RC equivalent is determined (Fig. 32-11*c*). Now all equivalent R's, C's, and

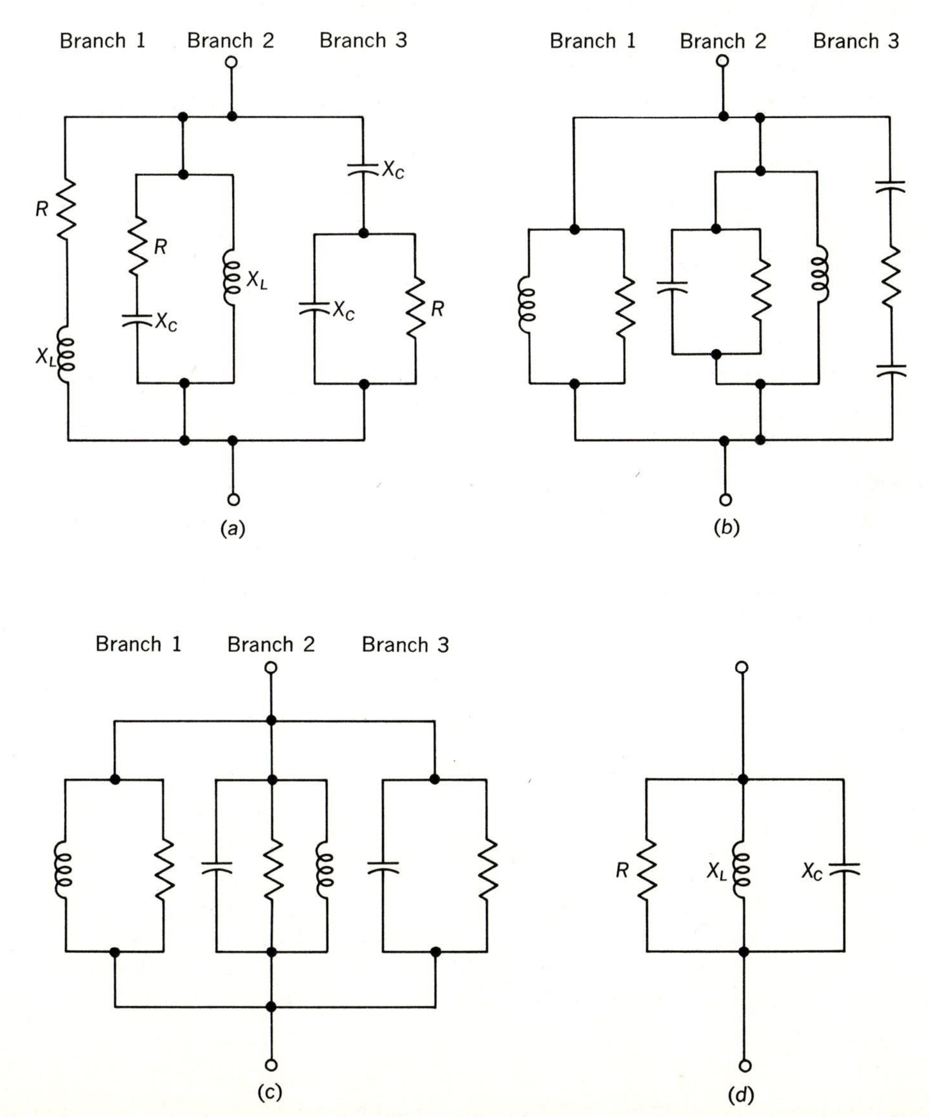

Fig. 32-11 Steps in reducing a highly complex ac circuit to a reasonably simple parallel ac circuit.

L's are in parallel. The three parallel resistances are computed, as are the two parallel inductive and capacitive reactances. This reduces the circuit to three parallel R, L, and C components (Fig. 32-11*d*). The X_L and X_C can be computed by $X_L X_C/(X_L - X_C)$, which is in parallel with the R.

32-7 IN REVIEW

If the source voltage and the circuit impedance are known, the total current can be computed by Ohm's law. If the impedance and source current are known, the source voltage can be determined.

If the phase angle can be determined, the power factor will be the cosine of the phase angle, PF $= \cos\ \theta$, or the ratio, P/VA. True power can be determined if source voltage, total current, and power factor are known, by $P = VA(\text{PF})$, or $P = VA(\cos\ \theta)$.

The sum of all ac voltage-drops across a source will not always add up *numerically* to the source value. They always add *vectorially* to equal the source value. Check the three cases shown in Fig. 32-12.

The voltage-drop across a component in a complex ac circuit can be determined once the impedance and phase angle of the circuit are known. In Fig. 32-13 the various circuit values are shown. The parallel circuit has an impedance value of 70.7 Ω with a phase angle of 45°. By converting to the series equivalent and adding the 15-Ω resistance, the total Z and θ

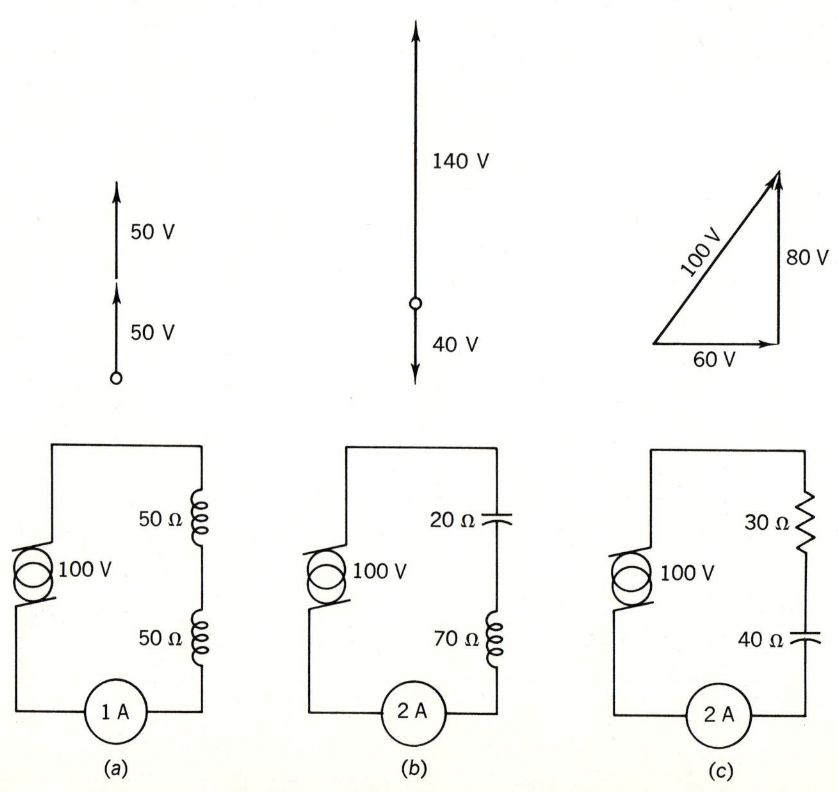

Fig. 32-12 Voltage-drops across ac circuits add vectorially.

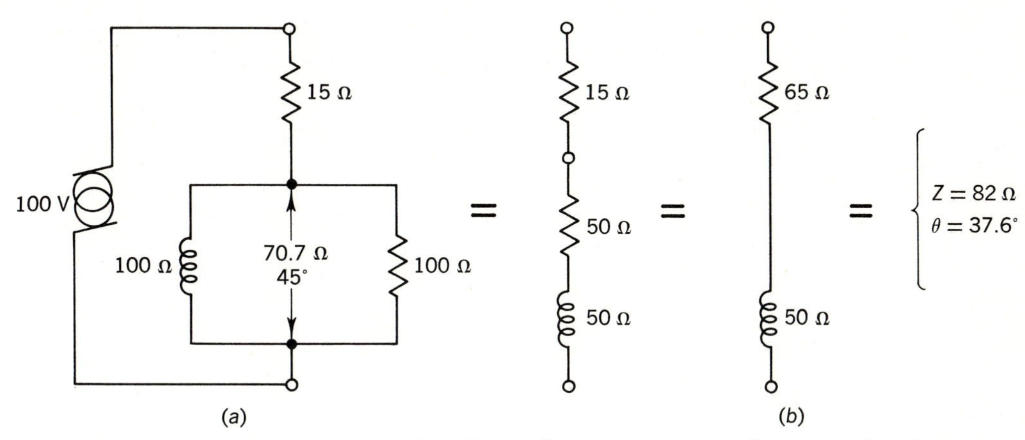

Fig. 32-13 A complex ac circuit for which all parameters are determined in the text.

value of the whole circuit is $82\angle{-37.6^\circ}\ \Omega$, as determined in Fig. 32-9*d*. With a source of 100 V the circuit current must be $I = E/Z$, or 100/82, or 1.22 A.

The 15-Ω R voltage-drop is $V = IR$, or 1.22(15), or 18.3 V. The power dissipated in this resistor is $P = I^2R$, or $1.22^2(15)$, or 22.3 W.

The voltage-drop across the parallel circuit is $V = IZ$, or 1.22(70.7), or 86.25 V. This is the voltage across both the 100-Ω R and the 100-Ω X_L.

The current flowing in the X_L is $I = E/X$, or 86.25/100, or 0.8625 A. The same value flows through the 100-Ω R, which is dissipating a power of $P = I^2R$, or $0.8625^2(100)$, or 74.4 W. The total power of the circuit is the sum of 22.3 + 74.4, or 96.7 W.

The power factor is PF = cos 37.6°, or 0.7923. Since PF = P/VA, the true power is equal to PF(VA), or 0.7923(100)1.22, or 96.7 W, the same as above.

Quiz 32-4. Test your understanding. Answer these check-up questions.

Questions 1 to 6 refer to Fig. 32-14*a*.

1. For the top parallel circuit, what is the value of B? ______ G? ______ Y? ______ Z? ______ θ? ______
2. For the top parallel circuit, what is the series-equivalent R? ______ X? ______
3. For the bottom parallel circuit, what is the value of B? ______ G? ______ Y? ______ Z? ______ θ? ______
4. For the bottom parallel circuit, what is the series-equivalent R? ______ X? ______
5. What is the total series-equivalent R? ______ X? ______ Z? ______ θ? ______
6. If the source is 100 V, what is the voltage-drop across the 120-Ω reactor? ______ The 10-Ω reactor? ______

Questions 7 to 15 refer to Fig. 32-14*b*.

7. In the lower left branch, what is the series-equivalent R? ______ X? ______ Z? ______ θ? ______ Is X inductive or capacitive? ______
8. In the lower left branch, what is the parallel-equivalent Y? ______ θ? ______ G? ______ R? ______ B? ______ X? ______
9. In the lower right branch, what is the series-equivalent Z? ______ θ? ______ Is X inductive or capacitive? ______
10. In the lower right branch, what is the parallel-equivalent Y? ______ θ? ______ R? ______ X? ______
11. For the whole lower complex parallel circuit, what is the parallel-equivalent R? ______ X? ______ Y? ______ θ? ______ Is X inductive or capacitive? ______
12. For the lower complex parallel circuit, what

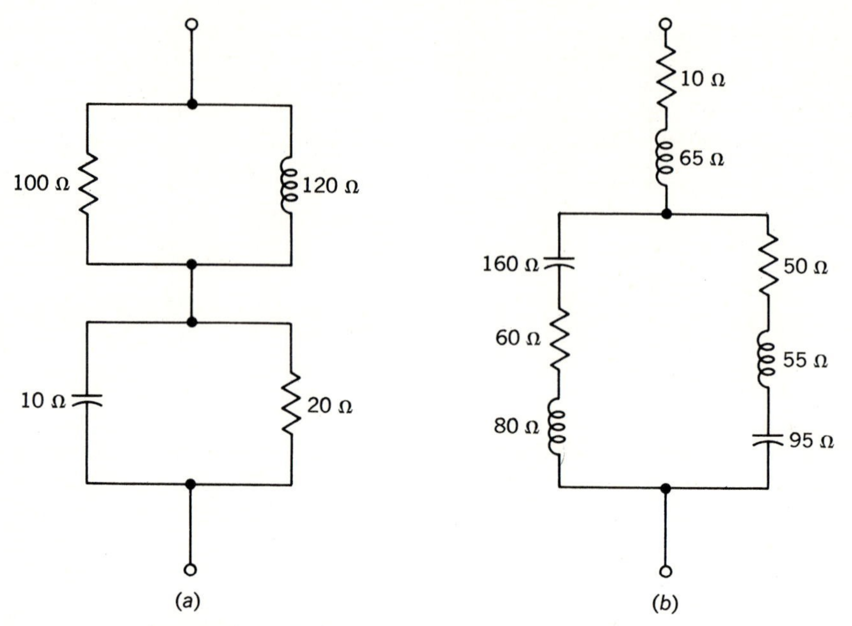

Fig. 32-14 Circuits for use in the Check-up Quiz.

is the series-equivalent Z? __________ R? __________ X? __________

13. For both the upper and lower branches, what is the total series-equivalent R? __________ X? __________ Is X inductive or capacitive? __________

14. What is the total circuit Z? __________ θ? __________

15. If there is 20 V across the 10-Ω R, what is the voltage-drop across the 65-Ω L? __________ Across the parallel circuit? __________

CHAPTER 32 TEST • PARALLEL AND COMPLEX AC CIRCUITS

1. What is the formula used to solve for two inductive reactances in parallel? An inductive and a capacitive reactance in parallel?
2. What is the reciprocal of resistance? Impedance? Reactance?
3. What does "B_C" stand for?
4. In which direction is the vector always drawn for resistance? Capacitive reactance? Inductive susceptance? Inductive current? Capacitive voltage?

Questions 5 to 9 refer to Fig. 32-15.

5. For branch 1, what is the parallel-equivalent value for Z? Y? R? X_L or X_C? θ? G? B? Does I lead or lag?
6. For the upper part of branch 2, what is the

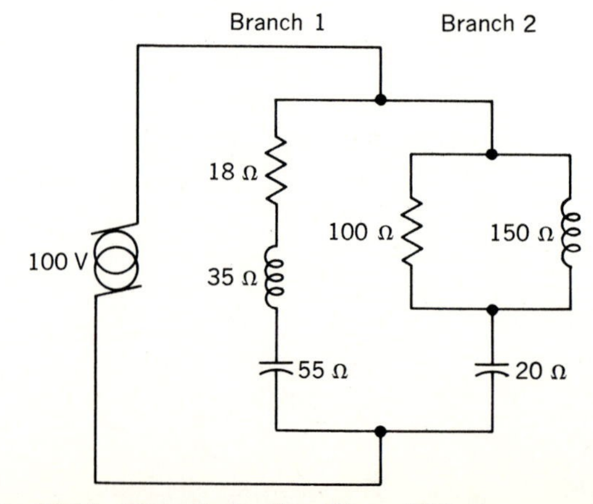

Fig. 32-15 Circuits for the Chap. 32 test.

series-equivalent for Y? Z? X_L or X_C? θ? R? Does I lead or lag?

7. For branch 2, what is the total series-equivalent value for R? X_L or X_C? θ? Does I lead or lag?
8. For branch 2, what is the total parallel-equivalent value for Z? Y? R? X_L or X_C? θ? G? B? Does I lead or lag?
9. For the whole circuit, what is the parallel-equivalent value for X_L or X_C? Z? I? R? θ? Does I lead or lag?
10. What is the voltage-drop across the 100-Ω R? What is the current through the 100-Ω R?

ANSWERS TO CHECK-UP QUIZ 32-4

1. (0.00833) (0.01) (0.013) (76.9 Ω) (39.8°) **2.** (59.1 Ω) (49.2 Ω) **3.** (0.1) (0.05) (0.112) (8.93 Ω) (63.4°) **4.** (4 Ω) (7.98 Ω) **5.** (63.1 Ω) (41.2 Ω) (75.4 Ω) (33.2°) **6.** (102 V) (11.8 V) **7.** (60 Ω) (80 Ω) (100 Ω) (53.1°) (Cap.) **8.** (0.01) (53.1°) (0.008) (125) (0.006) (167) **9.** (64 Ω) (38.7°) (Cap.) **10.** (0.0156) (38.7°) (82 Ω) (103 Ω) **11.** (49.5 Ω) (63.7 Ω) (0.0256) (37.8°) (Cap.) **12.** (39.1 Ω) (24 Ω) (30.9 Ω) **13.** (34 Ω) (34.1 Ω) (Ind.) **14.** (48.2 Ω) (45.1°) **15.** (130 V) (78.2 V)

CHAPTER OBJECTIVE. To investigate the use of real and imaginary numbers and the *j* operator in solving series, parallel, and series-parallel ac circuits.

33-1 REAL AND IMAGINARY NUMBERS

The values employed in dealing with circuits involving dc are called *real numbers*. If a conductor has 100 Ω of resistance, it has 100 Ω regardless of the direction of the current flowing through it. Sometimes it is handy to label the direction of the electromotive-force and the current in a wire, calling one direction of flow "positive" and the opposite direction "negative."

As long as a circuit has only resistance, the current and emf can be represented adequately by positive or negative labels. However, how can reactance be labeled to indicate that it is applying an opposition in neither a forward nor reverse direction, but in a direction 90° from the resistance? The label used in electrical theory is the letter "*j*," which may be added in front of a value to show that it is 90° out of phase with some reference value. A number with a *j* prefix may be termed an *imaginary number*, but this does not mean it is a number that does not exist.

When R and X_L are added in series, their values must be added at right angles to determine the total opposition that an alternating current encounters in the circuit. The fundamental electrical circuit consists of an emf, a load resistance, and a resulting current. Any opposition to current other than resistance will be an "out-of-phase" quantity. Thus the X_L is the out-of-phase quantity and is labeled with a *j*. Inductive reactance is the positive imaginary quantity, or $+j$, and capacitive reactance is assigned a $-j$ label. This labeling of 90° quantities by $+j$ or $-j$ can result in a simplification of ac circuitry. For example, it can be stated that the capacitive-reactance value is 25 Ω by merely writing $-j25$ Ω. Similarly, 380 Ω of inductive reactance can be indicated as $+j380$ Ω, or $j380$ Ω. [In mathematics, $\sqrt{-1}$ is called an "imaginary." Although -1 can exist, its square root must be imagined because no real number multiplied by itself can have a minus sign. Note that $\sqrt{-4} = \sqrt{4(-1)} = 2\sqrt{-1} = j2$.]

33-2 THE *j* OPERATOR

Imaginary numbers in electrical circuits follow the rules of mathematical imaginary (*i*), numbers, but since *i* represents current, the letter *j* is used in electricity. It is called the *j operator*.

There are four conditions to consider when using the *j* operator:

$0° =$ in phase (in a $+1$ direction)

$90° = j$

$180° = j^2$ (being opposite of 0°, is -1)

$270° = j^3$ [which is $-1(j)$, or $-j$]

Any in-phase quantity has no j label. Any quantity *leading* some other by 90° is assigned a $+j$ prefix. (If leading by any other angle, say 89°, it is not assigned a j prefix.) A quantity that is leading (or lagging) by 180° can be assigned a label of j^2, or -1 (negative value). If a quantity is leading another by 270° (the same as lagging 90°), it can be given a $-j$ prefix.

The arcs in Fig. 33-1 show the concepts of $+j$, j^2, and j^3 or $-j$. Arc *A* represents a forward (counterclockwise) rotation of a vector by 90°. This vector can be assigned a $+j$ value. Arc *B* represents a forward vector rotation of 180°. The vector is opposite the in-phase direction, and may be labeled j^2 or -1. Arc C represents a vector that has rotated 270° ahead of the starting direction. This is equivalent to $j^2 + j$, or j^3, or to $-1 \times j$, or $-j$. Arc *D* represents a vector that has rotated 90° clockwise (*behind* in time) from the starting position. This is also assigned a $-j$.

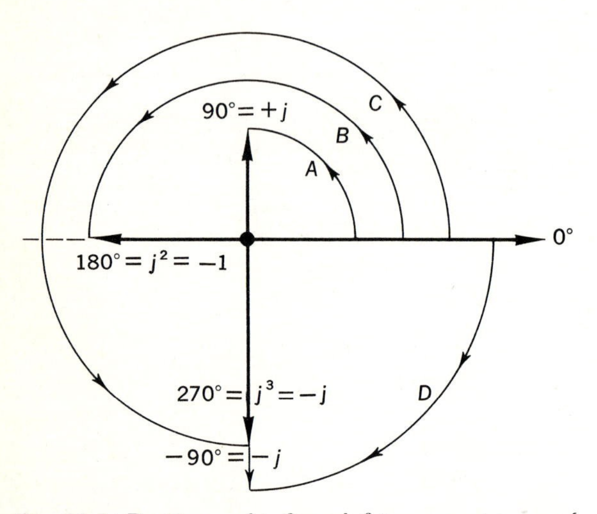

Fig. 33-1 Positions of j, j^2, and j^3 in respect to a reference direction of 0°.

Besides X and R being 90° out of phase, E and I can lead or lag by 90° and may also be assigned j values. For example, a 2-A current lagging 90° *behind* a voltage can be represented as $-j2$ A.

Quiz 33-1. Test your understanding. Answer these check-up questions.

1. According to the j-operator information, what prefix might be placed on a vector that rotates a full 360° clockwise? ______ 360° CCW? ______ 270° CW? ______ 180° CW? ______
2. What does $j45$ V represent? ______ $-j87$ V? ______
3. What does $j4.5$ A represent? ______ $-j0.15$ A? ______
4. If the quantities $-j72$ and $-j34$ were added, what would their total be? ______ What is the total of $-j69$, $j42$, and $-j16$? ______
5. What is the total when $+30$ Ω and $+j40$ Ω are added? ______ What kind of circuit must this be? ______
6. What is the total when $+30$ Ω and $-j40$ Ω are added? ______ What kind of circuit must this be? ______
7. In the expression $30 - j40$, what kind of number is $-j40$? ______ What kind of number is 30? ______
8. What is the total of $-j25$ V $+ 20$ V? ______ What kind of circuit must this be? ______
9. What is the total of $j2.8$ A $+ 1.9$ A? ______ What kind of circuit must this be? ______

33-3 ADDING REAL AND IMAGINARY NUMBERS

Since real and imaginary numbers are at 90° angles from each other, they cannot simply be added. The three basic methods of working with two quantities at right angles are graphing, solving by the Pythagorean formula $Z = \sqrt{R^2 + X^2}$ (with R as a *real* number and X as an imaginary), and solving by trigonometry. While the Pythagorean formula has often been used

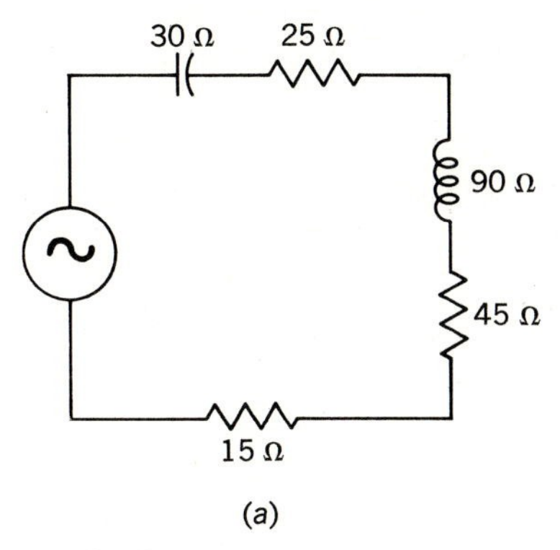

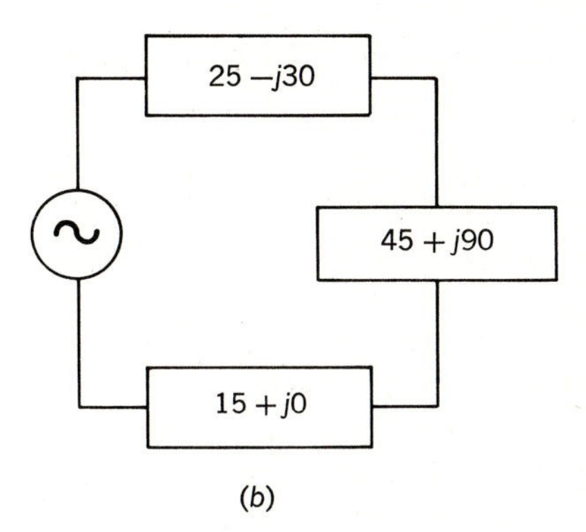

Fig. 33-2 Two methods of representing a series ac circuit: (*a*) schematic-diagram form; (*b*) complex-number or *j*-operator form.

previously, you may never have thought of the values in real and imaginary terms.

When the common Pythagorean formula $Z = \sqrt{R^2 + X^2}$ is expressed in *j*-operator notation, no square-root sign is used.

$$Z = \sqrt{R^2 + X_L{}^2} \text{ is expressed } Z = R + jX$$

$$Z = \sqrt{R^2 + X_C{}^2} \text{ is expressed } Z = R + -jX = R - jX$$

Real and imaginary numbers grouped in this manner are said to form a *complex number* or expression.

The series circuit in Fig. 33-2*a* can be looked at in two ways. It has a total of 85 Ω of R, 30 Ω of X_C, and 90 Ω of X_L. The 30-Ω X_C and the 90-Ω X_L in series tend to cancel, leaving a total of 60-Ω X_L and 85 Ω of R.

The same circuit is shown in complex notation in Fig. 33-2*b*. Adjacent resistance and reactances are grouped into complex numbers, which can be added algebraically:

$$\begin{array}{l} 25\ -j30 \\ 45\ +j90 \\ \underline{15\ +j0} \\ 85\ +j60 \end{array}$$

This sum is the same as determined above, but merely expressed differently. To solve for impedance, change the complex number to Pythagorean form

$$Z = 85 + j60 = \sqrt{85^2 + 60^2} = 104\ \Omega \text{ (Ind.)}$$

The simplest method of determining the phase angle is to use the tangent ratio,

$$\tan\theta = \frac{60}{85} = 0.706$$

$$\theta = 35.2°$$

Two known impedances with known phase angles are in series in Fig. 33-3*a*. One method of solving such a circuit is to graph, as in Fig. 33-3*b*. A vector is drawn from a starting point with a length of 50 units, and at an angle of 35° from the horizontal *x*-axis. From the tip of this vector another vector, equivalent to 60 units of the second impedance, is added at an angle of −65°. The unit length of the dashed line from the point of origin to the tip of the second vector is the total impedance value. The angle between the impedance line and the *x*-axis, −21.3°, is the phase angle.

The nongraphic method of solving this circuit is to determine the series equiva-

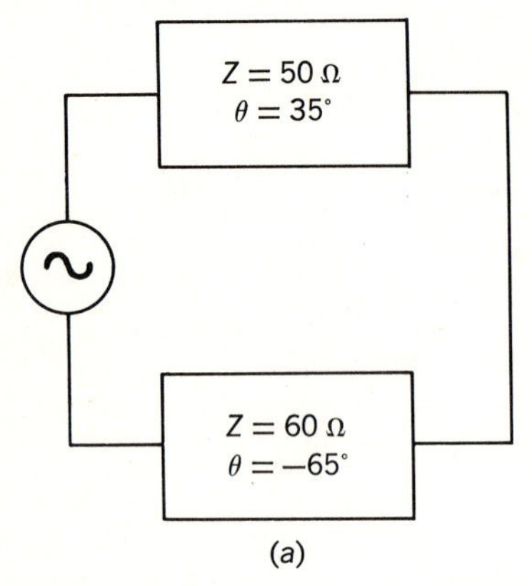

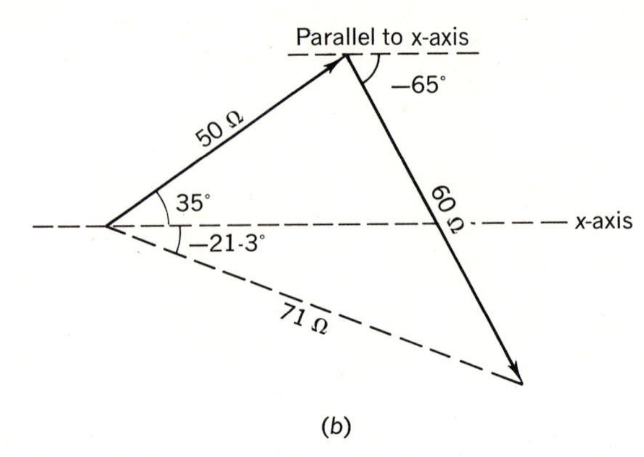

Fig. 33-3 (*a*) Two series complex impedances in which Z and θ are known. (*b*) Graphic method of solving for a total Z and θ.

lents of the two $Z\angle\theta$ loads and then either consider these values in series or consider them as complex numbers and add them. Compute the values for this circuit for practice.

Quiz 33-2. Test your understanding. Answer these check-up questions.

1. A coil of 75-Ω X_L in series with a 55-Ω R can be expressed as $Z = \sqrt{55^2 + 75^2}$. Express this in complex notation. ________ The impedance of a capacitor of 40-Ω X_C and a resistor of 65 Ω is expressed in complex notation as ________ . If these two impedances are in series, what is the Z_t seen by the source? ________ What is the θ seen by the source? ________
2. Three complex impedances have the values $15 + j12$, $4 - j8$, and $14 + j22$. What is the complex-number total of these impedances if they are in series? ________ What is the impedance value in ohms? ________ What is the phase angle? ________
3. In Fig. 33-3*a*, what is the complex-number expression for the top impedance? ________ The bottom impedance? ________ What is the total complex notation for Z? ________ What is the Z_t? ________ What is θ? ________

ANSWERS TO CHECK-UP QUIZ 33-1

1. (+) (+) (j or $+j$) (−) **2.** (45 V inductive; E leads I by 90°) (87 V, lagging a current by 90°) **3.** (4.5 A, leading a voltage by 90°) (150 mA, lagging a voltage by 90°) **4.** ($-j106$) ($-j43$) **5.** (30-Ω R + 40-Ω X_L or 50 Ω of Z) (Series R and X_L) **6.** (30-Ω R + 40-Ω X_C or 50 Ω of Z) (Series R and X_C) **7.** (Imaginary; 40 units lagging 90°) (Real; 30 units in phase) **8.** ($32\angle{-51.3°}$ V) (Series RC) **9.** ($3.39\angle 55.8°$ A) (Parallel RC)

33-4 REAL AND IMAGINARY VOLTAGE VALUES

In *series* ac circuits, the voltage-drops around the circuit can be expressed as real and imaginary numbers to solve for the source voltage. (In parallel circuits the *currents* in parallel branches are used to solve for the source current.)

Figure 33-4*a* represents an ac circuit with 30 Ω of R, 40 Ω of X_L, a Z of 50 Ω, and 2 A. The voltage-drops across the components would be

$$V_R = IR = 2(30) = 60 \text{ V}$$

$$V_L = IX_L = 2(40) = 80 \text{ V}$$

The source voltage would be

$$E_S = V_R + jV_L = 60 + j80 = \sqrt{60^2 + 80^2} = 100 \text{ V}$$

The inductive voltage-drop will be "$+j$"

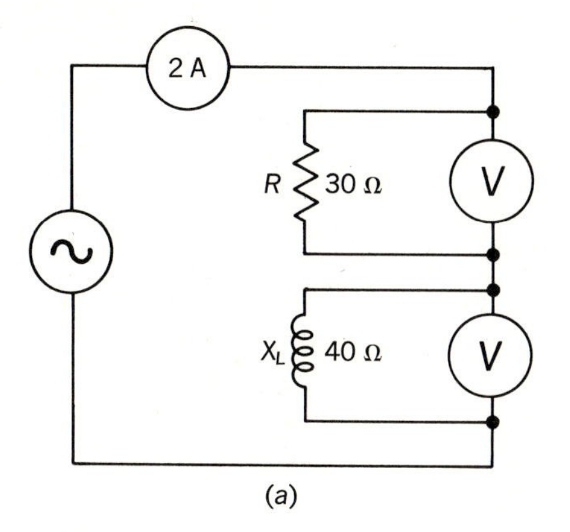

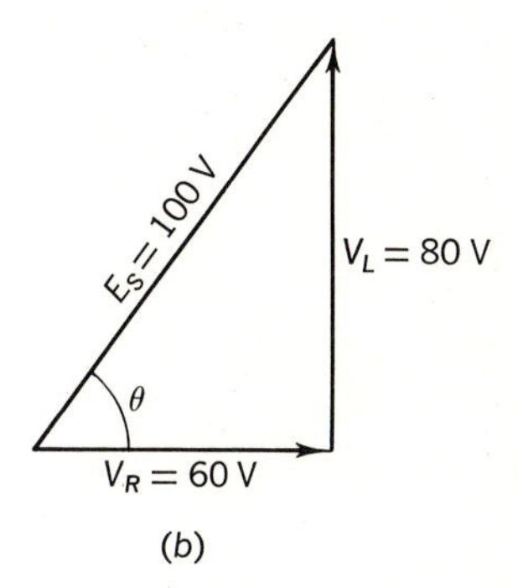

Fig. 33-4 The vectorial sum of the voltage-drops in an ac circuit equals the source voltage.

because the voltage across an inductance always *leads* the current flowing through it by 90°. A voltage-drop across a capacitor is a $-jV$ because the voltage lags the current flowing into a capacitor.

The resultant of any number of series complex-number voltages can be determined by adding the real (resistive) voltage-drops and the imaginary ($+j$ and $-j$) voltage-drops and solving for the complex total. In the circuit of Fig. 33-5, the source voltage will be the sum of the complex-number voltage-drops.

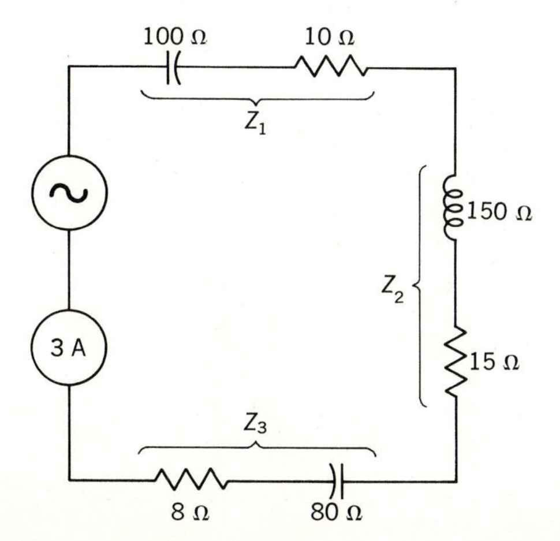

Fig. 33-5 A circuit consisting of three complex impedances in series across a source.

Real and imaginary voltage-drops are directly proportional to the R and X values of a series circuit. A $V_RV_XE_S$ triangle will be similar to that of an RXZ triangle. The phase angle is always the angle between R and Z, or between E_R and E_S. The power factor is the cosine of the phase angle or the ratio of V_R/E_S.

Quiz 33-3. Test your understanding. Answer these check-up questions.

1. In Fig. 33-5, what are the real and imaginary voltage values of Z_1? ____________ Z_2? ____________ Z_3? ____________
2. In Fig. 33-5, what is the complex number of all the voltage-drops of the circuit? ____________ What is the source-voltage value? ____________
3. In Fig. 33-5, what is the phase angle of E and I in the source? ____________ What is the power factor of this circuit? ____________
4. In Fig. 33-5, what is the total complex-number expression for impedance? ____________ What is the Z value? ____________ The phase angle? ____________

33-5 RECTANGULAR AND POLAR FORMS

Expressing R, X, and Z (or V_R, V_X, and E_S) as a right triangle is known as *rectangular notation*. The Z value is the length of a diagonal line from one corner of a rectangle to the opposite corner (Fig. 33-6).

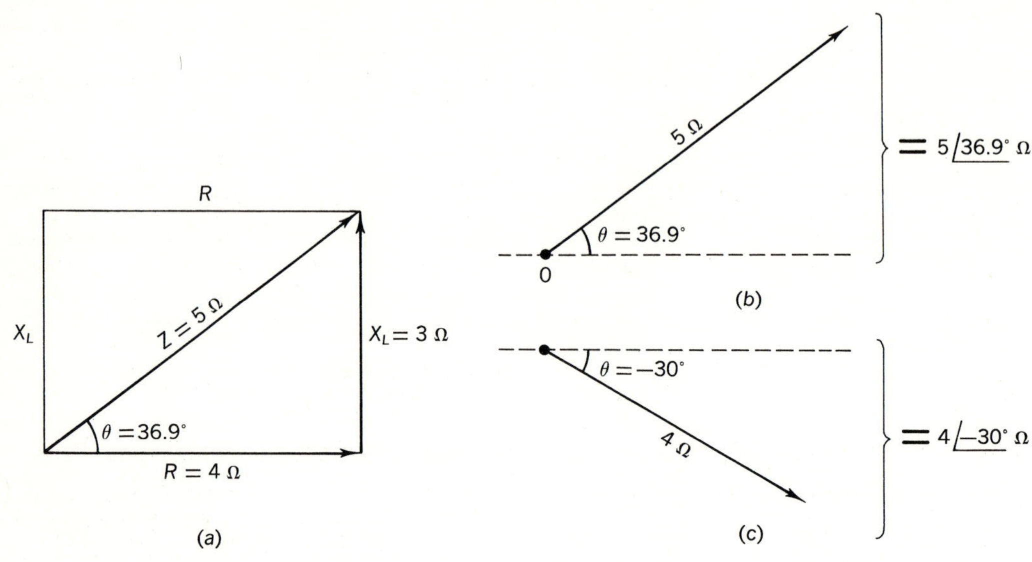

Fig. 33-6 (*a*) The electrical triangle is part of a rectangle. (*b*) Polar representation of the same circuit requires only the Z and θ values. (*c*) Polar representation of a capacitive circuit.

Most of the computations of ac circuits described, including the Pythagorean theorem and complex-number notation, are actually forms of rectangular notation. Only the term *rectangular notation* is new.

The impedance and phase angle of ac circuit triangles, such as Fig. 33-6*a*, were previously expressed in the form $Z = 5\angle 36.9°\ \Omega$. This is properly termed *polar notation*. Figure 33-6*b* graphs the same circuit in polar notation. The Z phasor is 5 units long, and it lies at an angle of 36.9° above the *x*-axis.

ANSWERS TO CHECK-UP QUIZ 33-2

1. ($55 + j75$) ($65 - j40$) (125 Ω) (16.3°) **2.** ($33 + j26$) (42 Ω) (38.2°) **3.** ($41 + j28.7$) ($25.4 - j54.4$) ($66.4 - j25.7$) (71 Ω) (21.2°)

ANSWERS TO CHECK-UP QUIZ 33-3

1. ($30 - j300$ V) ($45 + j450$ V) ($24 - j240$ V) **2.** ($99 - j90$ V) (134 V) **3.** (42.3°) (0.74) **4.** ($33 - j30$ Ω) (44.6 Ω) (42.3°)

The circuit represented by Fig. 33-6*c* has an impedance of 4 Ω and a phase angle of −30°. Since the phasor has a downward angle, the circuit is capacitive. Figure 33-6*a* and *b* represents inductive circuits. The polar representation of Fig. 33-6*c* is $Z = 4\angle -30°\ \Omega$.

The term *polar* derives from the impedance phasor or vector being rotated from a central point, much as the lines of longitude radiate outward from the north pole of the earth when viewed from above the pole.

Figure 33-7 illustrates a polar chart marked out in radius values of 10, 20, 30, and 40 units. The four phasors might represent units in ohms of impedance, emf, or current. Impedance phasors in positive *quadrants* would be inductive, whereas those in the negative quadrants would be capacitive. Voltages in positive quadrants would be leading the current. Voltages in the negative quadrants lag the current.

Since series-circuit *RXZ* and $V_R V_X E_S$ triangles are similar, series-circuit polar

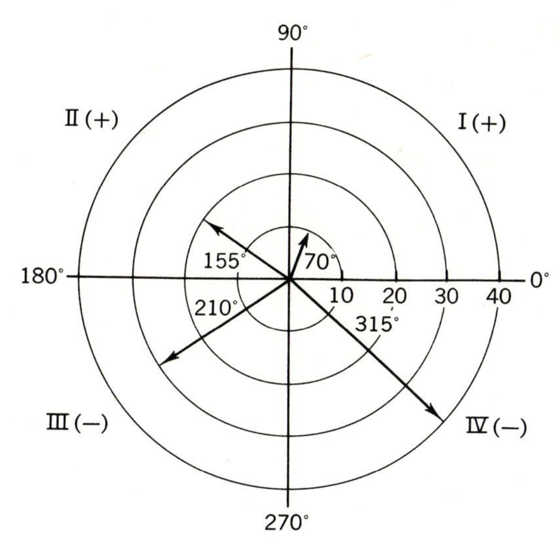

Fig. 33-7 Phasors on a polar-type chart used in solving ac circuit problems.

forms of impedance and source voltages will be similar. Thus $E_S = 50\underline{/45°}$ V indicates an inductive circuit in which the reactive voltage leads the circuit current by 45°. $E_S = 120\underline{/-15°}$ V is a capacitive circuit with a source of 120 V, a 15° phase angle, with the voltage lagging the circuit current. $I_S = 3.2\underline{/-52°}$ A indicates an inductive circuit because the 3.2-A current is lagging by −52°.

Quiz 33-4. Test your understanding. Answer these check-up questions.

1. In Fig. 33-7, is the circuit inductive, capacitive, or resistive for the 40-unit impedance phasor? __________ For the 20-unit impedance phasor? __________ For the 30-unit impedance phasor? __________
2. In what direction would a phasor have to point to represent a resistive circuit? __________ A resonant circuit? __________
3. Would current phasors represent L, C, or R circuits in quadrant II? __________ Quadrant IV? __________
4. Express in polar form the impedances for an inductive circuit with a phase angle of 40° and 70 Ω of impedance. __________ A capacitive circuit with 50 Ω of impedance and a θ of 29°. __________ A circuit in which Z is 40 Ω and θ is 300°. __________ Is this last circuit capacitive or inductive? __________
5. What type of circuit, L, C, or R, is represented by $Z = 50\underline{/30°}$ Ω? __________ $Z = 80\underline{/0°}$ Ω? __________ $Z = 750\underline{/-65°}$ Ω? __________
6. What type of a circuit is represented by $E = 50\underline{/20°}$ V? __________ $E = 145\underline{/-45°}$ V? __________ $E = 850\underline{/0°}$ V? __________
7. What type of a circuit is represented by $I = 2.4\underline{/32°}$ A? __________ $I = 6.5\underline{/0°}$ A? __________ $I = 12.3\underline{/-40°}$ A? __________

33-6 CONVERTING POLAR TO RECTANGULAR

The Pythagorean formula $Z^2 = R^2 + X^2$ can be used to solve for Z, R, or X. The phase angle can be determined by the trigonometric formulas, $\tan\theta = X/R$, $\cos\theta = R/Z$, and $\sin\theta = R/Z$. The same formulas can be rearranged to determine the R and X (rectangular) values if the Z and θ (polar) quantities are known. To solve for the real and imaginary values:

$$\text{real term} \qquad R = Z\,(\cos\theta)$$

$$j\ \text{term} \qquad X = Z\,(\sin\theta)$$

As an example, if the polar information is given as $Z = 40\underline{/30°}$ Ω, what are the real and imaginary values? From $\cos\theta = R/Z$,

$$R = Z\,(\cos\theta) = 40(0.866) = 34.6\ \Omega$$

$$X = Z\,(\sin\theta) = 40(0.500) = j20.0\ \Omega$$

While the polar form of $Z = 40\underline{/30°}$ Ω might be assumed to be derived from a series R and X, it may also have been derived from some parallel R and X, or from some highly involved series-parallel circuit having a resultant impedance of 40 Ω and phase angle of 30°. In this case the complex-number rectangular form $Z = 34.6 + j20$ would be the *series equivalent* (Chap. 32) of the parallel or series-parallel circuit.

Again, theories applied to solving impedance circuits with R and X are also applicable to voltage in series ac circuits and to currents in parallel ac circuits.

Quiz 33-5. Test your understanding. Answer these check-up questions.

1. Express in polar form the rectangular information $Z = 15 + j20$. ________ $E_S = 350 - j400$. ________ $I_S = 3.2 + j4.5$. ________
2. Express in rectangular form the series equivalent of $Z = 35\angle 25°\ \Omega$. ________ $Z = 70\angle -50°\ \Omega$. ________ $E_S = 12\angle -40°$ V. ________ $E_S = 110\angle 15°$ V.
3. Will the phase angle have the same sign as the j value in an impedance problem? ________ A voltage problem? ________ A current problem? ________

33-7 THE j OPERATOR IN A PARALLEL CIRCUIT

One method of solving parallel ac circuits is to employ the admittance, conductance, and susceptance formula, $Y = \sqrt{G^2 + B^2}$ (Chap. 32). Figure 33-8 is a simple parallel R and X_L circuit. In this circuit, $Y = \sqrt{G^2 + B^2}$ or $\sqrt{0.05^2 + 0.04^2}$ or $\sqrt{0.0041}$, or 0.0640 mho (or siemens). The reciprocal of 0.0640 mho of Y is 15.6 Ω of Z.

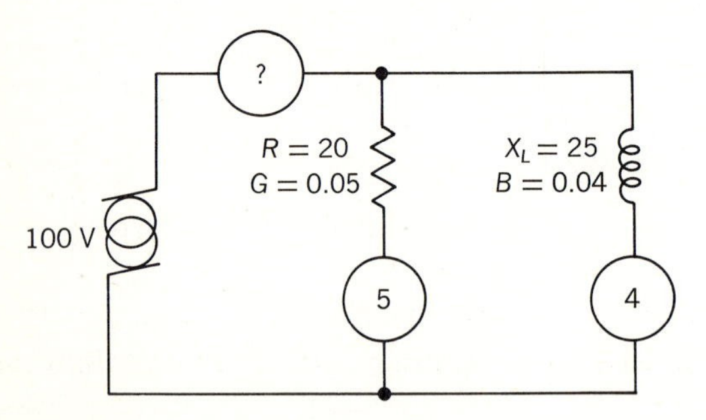

Fig. 33-8 Parallel ac circuit. Since the I_R is greater than the I_X, θ is less than 45°.

To express this parallel circuit in polar form, the Y, G, and B values are arranged like the Z, R, and X in series circuits. The phase angle can be determined by $\tan \theta = B/G$. In this circuit, $\tan \theta = 0.04/0.05$, or 0.800. The angle whose tangent is 0.800 is 38.7°. (This can also be stated arctan 0.800 = 38.7°.) The polar expression for this circuit is $Y = 0.064\angle -38.7°$ mho. Note that the sign of the phase angle is reversed whenever there is a change to a reciprocal value, as from Y to Z, G to R, or X to B.

The same circuit could have been solved by the vector-current method, $I_t = I_R - jI_X$, or $\sqrt{I_R^2 + I_L^2}$, or $\sqrt{5^2 + 4^2}$, or $\sqrt{41}$, or 6.4 A. Then, from Ohm's law, $Z = E/I$, or 100/6.4, or 15.6 Ω. Phase angle is $\tan \theta = I_L/I_R$, or 4/5, or $-38.7°$. In polar form, $I_S = 6.4\angle -38.7°$ A. (The angle is lagging because this is a polar expression of an inductive-current value.) The polar expression of the impedance will be $Z = 15.6\angle 38.7°\ \Omega$.

A basic formula for solving a two-branch parallel ac circuit is similar to the formula used to solve two resistances in parallel,

$$Z_t = \frac{Z_1 Z_2}{Z_1 + Z_2}$$

When the two branch impedances of Fig. 33-8 are inserted into the formula *in rectangular form*, there is no j value in the first leg and no real value in the second leg:

$$Z_t = \frac{(20 + j0)\,(0 + j25)}{(20 + j0) + (0 + j25)} = \frac{0 + j500}{20 + j25}$$

This leaves a complex fraction. It is possible to divide complex numbers by multi-

---

ANSWERS TO CHECK-UP QUIZ 33-4

1. (C) (L) (C) 2. (0°) (0°) 3. (C) (L) 4. ($70\angle 40°\ \Omega$) ($50\angle -29°\ \Omega$) ($40\angle -300°\ \Omega$) (C) 5. (L) (R) (C) 6. (L) (C) (R) 7. (C) (R or resonant) (L)

plying both components of the fraction by the *conjugate* of the denominator. The conjugate is the complex number of the *denominator*, with its j sign reversed. When it is divided by itself, the conjugate fraction will always equal unity, or 1. Check the steps in this solution, remembering that $j^2 = -1$:

$$Z_t = \left(\frac{0 + j500}{20 + j25}\right)\left(\frac{20 - j25}{20 - j25}\right)$$

$$= \frac{j10{,}000 - j^2 12{,}500}{400 - j^2 625}$$

$$= \frac{j10{,}000 - (-1)(12{,}500)}{400 - (-1)(625)}$$

$$= \frac{12{,}500 + j10{,}000}{1025}$$

$$= 12.2 + j9.75$$

$$= 15.6\ \Omega$$

Another method of dividing complex numbers is to change them to their polar forms. The polar equivalent of the complex number $Z_1 = 0 + j500$ is $500\angle 90°\ \Omega$. The polar equivalent of $Z_2 = 20 + j25$ is $32\angle 51.4°\ \Omega$. The problem

$$Z_t = \frac{0 + j500}{20 + j25} \text{ becomes } Z_t = \frac{500\angle 90°\ \Omega}{32\angle 51.3°\ \Omega}$$

The rule for dividing polar expressions is: (1) divide impedances; (2) move the lower phase angle up and change its sign. The problem is then

$$Z_t = 15.6\angle 90° - 51.3°\ \Omega$$
$$= 15.6\angle 38.7°\ \Omega$$

To determine the resistive and reactive components, the polar form is changed to rectangular form. In this case

$$R = Z(\cos\theta) = 15.6(0.781) = 12.2\ \Omega$$

$$X = Z(\sin\theta) = 15.6(0.624) = j9.75\ \Omega$$

Thus a 20-Ω R in parallel with a 25-Ω X_L has an impedance stated in polar form as $Z = 15.6\angle 38.7°\ \Omega$, which is the equivalent of a 12.2-Ω R in *series* with a 9.75-Ω X_L.

Quiz 33-6. Test your understanding. Answer these check-up questions.

1. A 50-Ω R is in parallel with a 60-Ω X_L. Express as a complex-number fraction. ________ What is the Z value in polar form? ________ To what series R and X_L values is this equivalent? ________
2. A 20-Ω R is in parallel with a 22-Ω X_C. Express as a complex-number fraction. ________ What is the Z value in polar form? ________ To what series R and X_C values is this equivalent? ________

33-8 THE *j* OPERATOR IN SERIES-PARALLEL CIRCUITS

In the circuit of Fig. 33-9, two complex impedance branches are in parallel. The

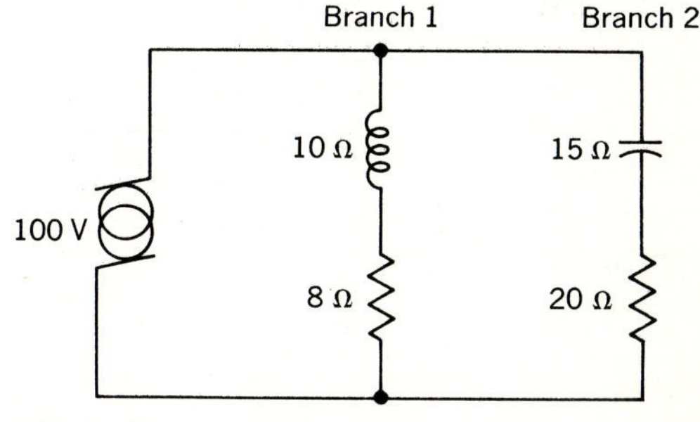

Fig. 33-9 Series-parallel circuit solved in the text.

impedance of this circuit could be solved by the admittance method, but it can also be solved by complex notation.

Each branch must first be expressed in both rectangular and polar values of impedance. For the first branch, the rectangular expression is $Z = 8 + j10$. The polar phase angle can be derived from the formula $\tan \theta = X/R$, or 10/8, or 1.25. The angle whose tangent is 1.25 is 51.3°. Now the impedance of the first branch can be determined from either $\sin \theta = X/Z$ or $\cos \theta = R/Z$. From the sine formula,

$$\sin \theta = \frac{X}{Z}$$

$$\sin 51.3 = \frac{10}{Z}$$

$$Z = \frac{10}{0.78} = 12.8\ \Omega$$

For branch 1, the rectangular values are $Z = 8 + j10$, and the polar values are $Z = 12.8\angle 51.3°\ \Omega$. For branch 2, rectangular values are $Z = 20 - j15$. By the tangent formula, the phase angle is −36.9°. By the sine formula Z is 25 Ω. The polar expression is $Z = 25\angle -36.9°\ \Omega$. To solve for the parallel impedance, use the formula

$$Z_t = \frac{Z_1 Z_2}{Z_1 + Z_2}$$

Because polar expressions can be divided and multiplied easily and rectangular expressions can be added or subtracted easily, the simplest method of solving the problem is to multiply polar values in the numerator and add rectangular values in the denominator; thus

$$Z_t = \frac{(12.8\angle 51.3°)(25\angle -36.9°)}{(8 + j10) + (20 - j15)}$$

The rule for multiplying polar values is to *multiply* the impedances but algebraically *add* the angles. Follow this rule in the numerator, but algebraically add rectangular values in the denominator. The problem becomes

$$Z_t = \frac{320\angle 14.4°}{28 - j5}$$

Next, convert the rectangular $28 - j5$ to its polar equivalent for easy division. The phase angle is $\tan \theta = 5/28$, or −10.1°. The impedance is $Z = X/\sin \theta$, or 5/0.175, or 28.6 Ω. With this information inserted into the problem,

$$Z_t = \frac{320\angle 14.4°}{28.6\angle -10.1°}$$
$$= 11.2\angle 14.4° + 10.1°$$
$$= 11.2\angle 24.5°\ \Omega$$

By using sine and cosine formulas, the polar $Z = 11.2\angle 24.5°\ \Omega$ can be converted to its rectangular equivalent $Z = 10.2 + j4.67\ \Omega$.

ANSWERS TO CHECK-UP QUIZ 33-5

1. ($25\angle 53.1°\ \Omega$) ($531\angle -48.8°$ V) ($5.52\angle 54.6°$ A) 2. ($31.7 + j14.8\ \Omega$) ($45 - j53.6\ \Omega$) ($9.19 - j7.71$ V) ($106.2 + j28.5$ V) 3. (Yes) (Yes) (Yes)

ANSWERS TO CHECK-UP QUIZ 33-6

1. [$(0 + j3000)/(50 + j60)$] ($38.4\angle 39.8°\ \Omega$) ($29.5 + j24.6\ \Omega$) 2. [$(0 - j440)/(20 - j22)$] ($14.8\angle -42.3°\ \Omega$) ($10.95 - j9.95\ \Omega$)

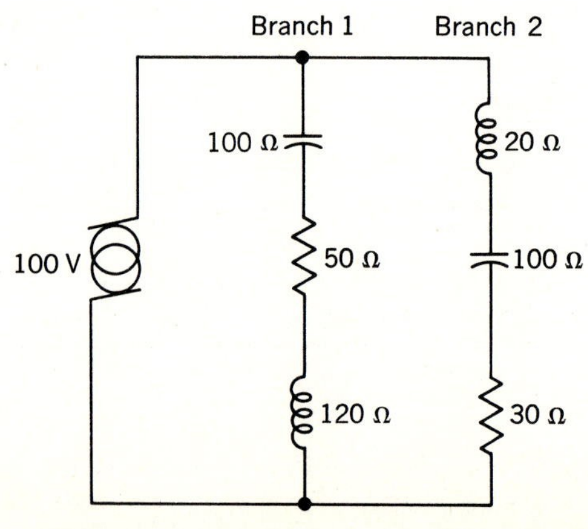

Fig. 33-10 Complex circuit for Check-up Quiz.

Quiz 33-7. Test your understanding. Answer these check-up questions.

1. In what form is it simplest to multiply complex numbers? ________ To add complex numbers? ________ To divide complex numbers? ________ To subtract complex numbers? ________

Questions 2 to 4 refer to Fig. 33-10.

2. For branch 1, what are the rectangular values? ________ Polar values? ________
3. For branch 2, what are the rectangular values? ________ Polar values? ________
4. Using the prodivisum formula, what is Z_t? ________ The phase angle? ________ Is the current in the source leading or lagging? ________

CHAPTER 33 TEST • THE *j* OPERATOR

1. How is an imaginary number indicated in electricity? In mathematics?
2. Express in imaginary-number form 5.6 V measured across a capacitor. 2.8 A flowing in an inductor. 759 Ω of X_L. 23.8 Ω of X_C.
3. What value does j^2 have? j^3?
4. How many degrees of arc are indicated by the letter: j? $-j$? j^2? j^3?
5. A series ac circuit has 3 Ω of R and 4 Ω of X_L. Express this circuit in rectangular form. In polar form.
6. Which are simpler to add, complex-notation values or polar-notation values? Which are simpler to multiply? Divide? Subtract?
7. In Fig. 33-11*a*, what is the impedance of the circuit? What is the phase angle? What value would the ammeter show? What would be the total power dissipation of the circuit?
8. Convert $Z = 30 + j25$ to polar notation. Is this an inductive or a capacitive circuit?
9. Convert $75\angle -30°$ V to rectangular notation. Is this an inductive or a capacitive circuit?
10. In what quadrant is a phasor that has traveled 125° from the zero point? 330°? 62°? 245°?
11. What are the values of R, X, and Z for a circuit represented by $0.025\angle 45°$ mho? Is the circuit inductive or capacitive?
12. In the formula $Z_t = Z_1Z_2/(Z_1 + Z_2)$, would you compute the denominator in polar or rectangular notation? The numerator?

Questions 13 to 19 refer to Fig. 33-11*b*.

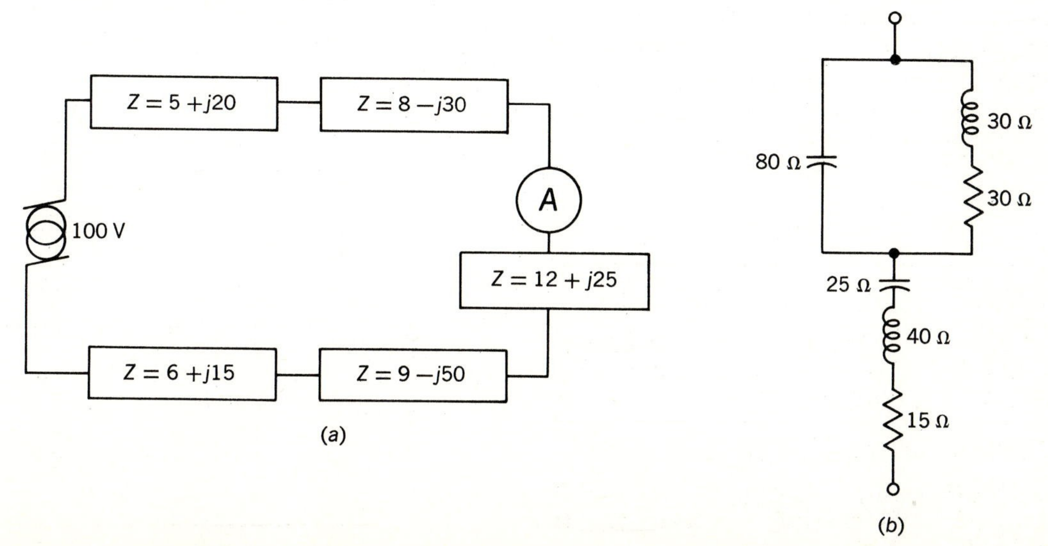

Fig. 33-11 Diagrams of circuits used in the *j*-operator chapter test.

13. What is the impedance of the upper right branch? What is the polar expression for this branch in admittance terms? The rectangular expression in admittance terms?
14. What is the polar expression for the upper left branch in admittance terms? What is the rectangular expression in admittance terms?
15. Use $Y_t = Y_1Y_2/(Y_1 + Y_2)$ in polar form to determine the admittance of the parallel section.
16. Express the impedance of the parallel section in polar form?
17. Convert the parallel impedance and phase angle to its series-equivalent values in rectangular notation.
18. Express the series part of the circuit in rectangular form.
19. What are the total circuit Z, R, and X values in rectangular notation? Polar notation?

ANSWERS TO CHECK-UP QUIZ 33-7

1. (Polar) (Rectangular) (Polar) (Rectangular) **2.** ($50 + j20$ Ω) ($53.9\angle 21.8°$ Ω) **3.** ($30 - j80$ Ω) ($85.4\angle -69.4°$ Ω) **4.** (46 Ω) (−10.7°) (Lagging)

34 POWER TRANSFORMERS

CHAPTER OBJECTIVE. To continue the transformer discussions of Chap. 10 regarding the various types, construction, cooling, shielding, efficiency, regulation, impedance matching, and autotransformers.

34-1 THE TRANSFORMER

A transformer has a primary coil, a secondary coil (often a tertiary), and either an iron or an air (insulator) core on which it is wound (Chap. 10). In air-core transformers only a small percentage of the lines of force developed in the primary may cut across the secondary turns, resulting in little transfer of energy from primary to secondary. If the primary and secondary both have the same number of turns, an air-core transformer will produce considerably less secondary voltage than is applied to the primary. However, when the primary and secondary are tuned, air-core transformers become efficient power-transferring devices (Chap. 36).

With an iron core, more lines of force will be free to expand from the primary wires into the core area. Almost all the lines of force that expand and contract around the primary cut the secondary turns. If the primary and the secondary have the same number of turns, the secondary voltage will be essentially the same as the primary voltage. If the secondary has twice as many turns, the secondary voltage will be essentially twice the primary voltage. The turns ratio is equal to the voltage ratio of an iron-core transformer when the coefficient (%) of coupling between the primary and secondary is high. A coefficient of coupling of 100% is called *unity coupling*.

In Fig. 34-1, the turns ratio and voltage ratio are both 1:1. With both switches

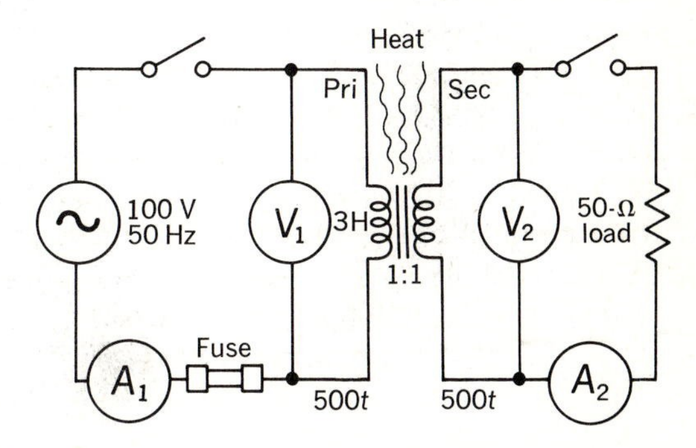

Fig. 34-1 Basic transformer circuit.

open, as shown, the source is 100 V, but no current flows. If the primary switch is closed, V_1 and V_2 will indicate 100 V. With the secondary switch open, there can be no secondary current, and A_2 reads zero. If the primary winding or coil has an inductance value of 3 H with no load connected across the secondary, it presents an inductive reactance of $X_L = 2\pi fL$, or 6.28(50)3, or 942 Ω. By Ohm's law, the primary current would be

$I = E/X_L$, or 100/942, or 0.106 A. This is known as the core *magnetizing current.*

Because of eddy currents and hysteresis losses produced in the iron core when the magnetizing force alternates, some additional primary current must flow to support these power losses. If the total of the core losses is 10 W, how much current would have to flow from the source to produce 10 W? From $P = EI$, $I = P/E$, or 10/100, or 0.1 A.

The current flowing through A_1 consists of the purely resistive 0.1 A plus the 90° reactive magnetizing current of 0.106 A. The resulting *excitation current* is $I_e = \sqrt{I_R^2 + I_X^2}$, or $\sqrt{0.1^2 + 0.106^2}$, or $\sqrt{0.0212}$, or 0.146 A (Fig. 34-2). Meter A_1 should in-

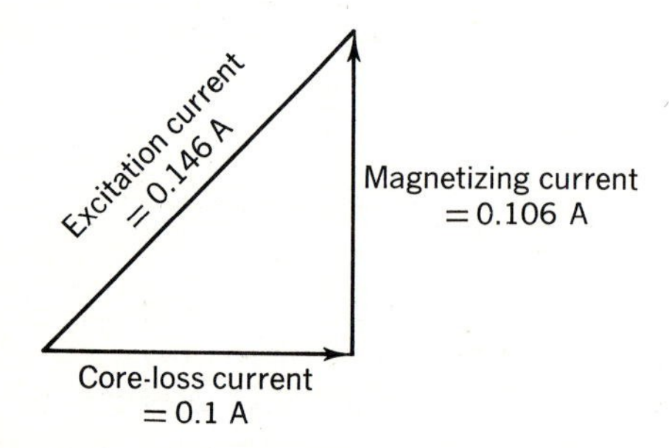

Fig. 34-2 Vector diagram of core currents.

dicate this value with no load on the secondary. The primary excitation current in most transformers represents only 2 to 5% of the total normal-load transformer primary current.

Quiz 34-1. Test your understanding. Answer these check-up questions.

1. What is a second secondary called? ____________
2. How can air-core transformers be made efficient power coupling devices? ____________
3. What special term is used to indicate 100% coupling? ____________
4. A transformer has 450 primary and 1575 secondary turns. What is the turns ratio? ____________
5. What is the current called that is responsible for producing the basic flux in the core of an iron-core transformer? ____________ That supports the core losses? ____________ What is the vector sum of these currents called? ____________
6. What is the name of the primary current that the primary ammeter reads when the secondary is unloaded? ____________

34-2 LOADING THE SECONDARY

Several events take place simultaneously when the secondary switch is closed in Fig. 34-1. The 100-V secondary emf is now across a 50-Ω load resistance, and by Ohm's law, 2 A flows in the secondary circuit. V_2 reads 2 A. The secondary appears to the primary as a partially shorted coil around it, which cancels some of the primary's inductance. Less primary inductance results in less primary reactance and a greater primary current. Thus, primary current is determined most by the power demand of the secondary load.

The power demanded by the secondary is $P = EI$, or 100(2), or 200 W. This power is delivered source to load by the magnetic fields of the transformer.

The source must deliver not only the 200 W to the load, but also the 10 W of core loss. The core-excitation current and therefore the core losses remain relatively constant with or without a load on the secondary.

The source sees the primary of the transformer above as an essentially *resistive* load demanding 210 W of power. From the power formula $P = EI$, the source and primary current must be $I = P/E$, or 210/100, or 2.1 A (2 A for the load and 0.1 A for the core losses).

The inductance of the primary is 3 H with no load on the secondary. How much does the shorted-turn effect of the secondary turns reduce the inductance of the primary? The primary ammeter reads 2.1 A, and the primary voltmeter reads 100 V. The primary impedance must be

$Z = E/I$, or 100/2.1, or 47.6 Ω. From the formula given in Sec. 29-10, the primary inductance can be found by $L = 0.159\sqrt{Z^2 - R^2}/f$. If primary resistance is negligible, the inductance is approximately $0.159\sqrt{47.6^2 - 0^2}/50$, or 0.151. H. Loading the secondary reflects as a considerable reduction of primary inductance, from 3 H at no load down to 0.151 H.

The inductive reactance (or impedance) of the primary with the secondary unloaded is 942 Ω, but drops to 47.6 Ω when the secondary is loaded. The fact that any load on the secondary reflects back to the primary a greatly lowered impedance is a significant point. Matching the impedance of a source to the primary of a transformer can be accomplished by varying the load on the secondary, or, by changing the turns ratio to a constant load.

These computations have concerned a 1:1-ratio transformer. The same basic theory applies if the transformer has a step-up or a step-down ratio—for example, if half the turns of the primary in Fig. 34-1 are removed.

Quiz 34-2. Test your understanding. Answer these check-up questions.

The following questions refer to Fig. 34-1.

1. If the L of an iron-core coil is directly proportional to N^2, reducing the number of turns of the primary in Fig. 34-1 by half results in what unloaded primary-inductance value? __________
2. With the L value obtained in question 1, what would be the X_L value of the primary with the secondary unloaded? __________
3. Under these conditions, what would be the ac core-magnetizing primary-current value? __________
4. Would this increase in core-magnetizing I have any significant effect on core losses and the core-loss I_P? __________
5. What would the E_S be under these conditions? __________ I_S? __________ The load-dissipation value? __________
6. What power would the source feed to the primary? __________ What I_P would be required to produce this power value? __________
7. What would be the approximate L_P value with the secondary loaded? __________
8. What is the magnetomotive-force working on the primary coil under this loaded condition? __________ What was it when the primary had 500 turns? __________
9. Why is the mmf approximately 2 times as great with the loaded step-up-ratio transformer? __________
10. What is the efficiency of the transformer? __________
11. If the resistance of the wires of the primary had been considered, would the efficiency have been greater or less? __________ If the secondary-wire resistance were considered? __________
12. Does the magnetizing or excitation current in the primary increase, decrease, or remain the same when the secondary is loaded? __________

34-3 EFFECT OF RESISTANCE IN THE WINDINGS

So far the transformer has been considered to have negligible resistance in its windings. No conductor is perfect, so both the primary and secondary of a transformer will have some resistance. In Fig. 34-3*a*, the resistances of both the primary and secondary wires are represented as resistors in series with the windings. Figure 34-3*b* represents an equivalent circuit for a 1:1 transformer. The components shown dashed represent the core-loss and the magnetizing-current paths, which will be neglected at this time.

With both switches closed, the source sees an effective resistance value of $4 + 5 + 50$, or 59 Ω. With 100 V as the source, the load current is $I = E/R$, or 100/59, or 1.69 A.

The voltage across the primary would be 100 V, less the voltage-drop across the primary resistance. With 1.69 A flowing through 4 Ω, voltage-drop is 1.69(4), or 6.76

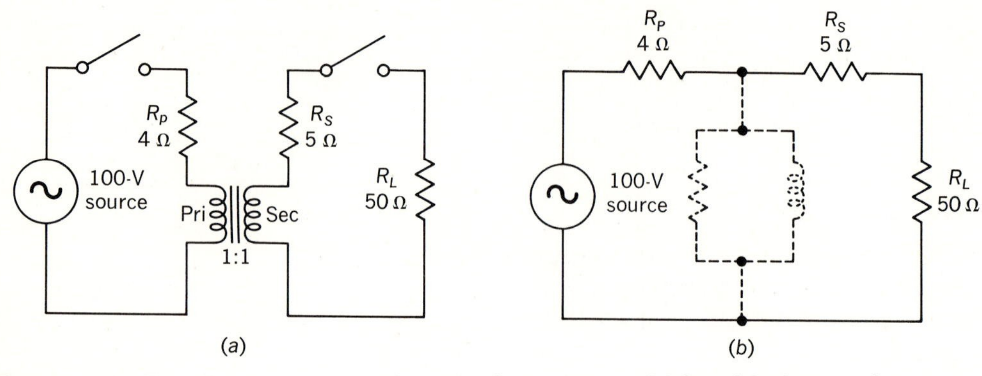

Fig. 34-3 (*a*) Transformer circuit with internal resistances. (*b*) Simplified equivalent circuit.

V. The active voltage across the primary and involved in producing magnetic fields is 100 − 6.76, or 93.2 V.

The voltage across the secondary should be the same as that across the primary, 93.2 V. But secondary current flowing in the 5-Ω secondary resistance will develop a voltage-drop of 1.69(5), or 8.45 V. If the primary voltage is 93.2 V and the secondary loses 8.45 V more, the secondary or load voltage should be about 84.8 V, instead of the 100 V of the source.

If the core-loss primary current is added to the primary current produced by the load, the total primary current would be slightly higher, further reducing the effective primary voltage. Normally this does not produce a significant change in circuit parameters.

ANSWERS TO CHECK-UP QUIZ 34-1

1. (Tertiary) **2.** (Tune primary and secondary) **3.** (Unity) **4.** (3.5:1) **5.** (Magnetizing) (Core loss) (Excitation) **6.** (Excitation)

ANSWERS TO CHECK-UP QUIZ 34-2

1. (0.75 H) **2.** (236 Ω) **3.** (0.424 A) **4.** (No) **5.** (200 V) (4 A) (800 W) **6.** (810 W) (8.1 A) **7.** (0.0392 H) **8.** (2025) (1050) **9.** (Greater magnetic field required to produce the increased secondary power drain) **10.** (98.8%) **11.** (Less) (Less) **12.** (Remains the same)

For a similar power 1:2 step-up transformer, the current in the primary would be twice that of the secondary (core and copper losses neglected). In an equivalent circuit of a 1:2 transformer, the primary-resistance value would have to be multiplied by 2 to account for the double-value primary current that would flow to produce the voltage-drop across the primary resistance. The equivalent secondary resistance would be divided by 2, since the secondary current flowing through this resistance would be reduced by a factor of 2. The source voltage would then have to be doubled to produce the desired equivalent current value in the circuit.

These voltage approximations illustrate the importance of having low-resistance windings. A transformer must be engineered to have a step-up ratio greater than would be computed for the no-load condition to make up for the voltage losses when current flows in the resistance of its windings.

Quiz 34-3. Test your understanding. Answer these check-up questions.

1. If the transformer in Fig. 34-3 were a 2:1 step-up-ratio transformer instead of 1:1 as shown, what would be the voltage-drop across the load re-

sistor? __________ Effective voltage across the primary? __________

2. If the transformer in Fig. 34-3 were a 2:1 step-down transformer, what would be the voltage across the load? __________ Effective voltage across the primary? __________

34-4 TRANSFORMER LOSSES

Losses in an iron-core transformer are: (1) *core*, including hysteretic and eddy-current losses, and (2) *copper*, or I^2R, due to heat generated when current flows through the resistance of the windings, computed by $P = I^2R$ for both primary and secondary windings.

The hysteretic loss is the heat generated by the friction of the magnetic domains of the core iron, reversing polarity as the primary current alternates.

The eddy-current loss is due to the heat generated in the core iron when magnetic fields expand and contract in the iron. Since iron is a fairly good conductor, the moving fields induce an emf in the core which produces circular-path currents. The eddy-current power loss is an I^2R loss in the core itself and cannot be measured directly. Eddy currents are reduced by laminating the core (Chap. 29).

In practical transformers, about 15% of the core loss is due to eddy currents and 85% is due to hysteresis. The hysteretic loss increases in direct proportion to frequency. Eddy-current loss increases as the square of the frequency.

Core losses can be determined as in Fig. 34-4. First the dc resistance of the primary winding is measured. Then the unloaded transformer is coupled to the power line through a wattmeter and an ammeter. The wattmeter reading indicates the sum of the copper loss and the core loss. The copper loss can be determined by using ammeter and ohmmeter readings in the formula $P = I^2R$. The difference between the copper loss and the total power indicated by the wattmeter will be the core losses.

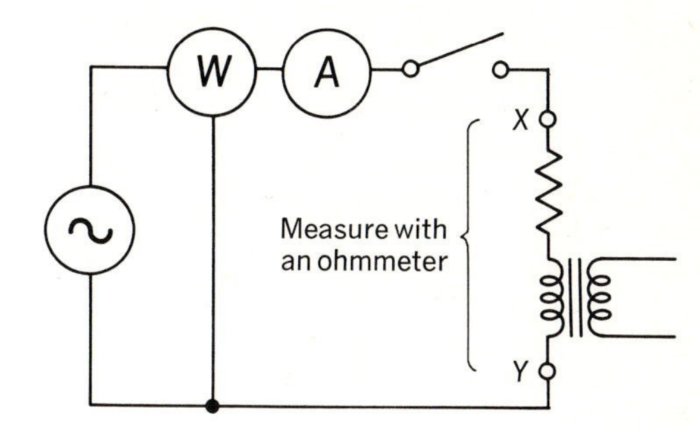

Fig. 34-4 Circuit for determining core losses.

Some *leakage-flux* loss may occur due to leakage lines of force from the sharp edges of rectangular transformer cores inducing emf and current into external conductors.

Core and copper losses in a transformer under load produce heat. Transformers are constructed to stand some rise in temperature, perhaps 40°C, under normal operation. The resistance of the copper wires in the windings will increase about 0.4% for each 1°C of temperature rise. Excessive heating of copper wires can cause a deterioration of the insulation around them. If insulation fails, adjacent turns may touch and cause shorted turns, reducing the primary inductance greatly, producing excessive primary current, possibly burning out the primary or fuses, or tripping overload protectors. The transformer is useless with any shorted turns.

The size of wire required in a transformer is determined to a great extent by the type of insulation around the wire. Some approximate maximum temperatures that different insulations will stand are listed in Table 34-1.

34-5 PHASE RELATIONSHIPS

In the basic 1:1 transformer, the output voltage will equal the voltage applied

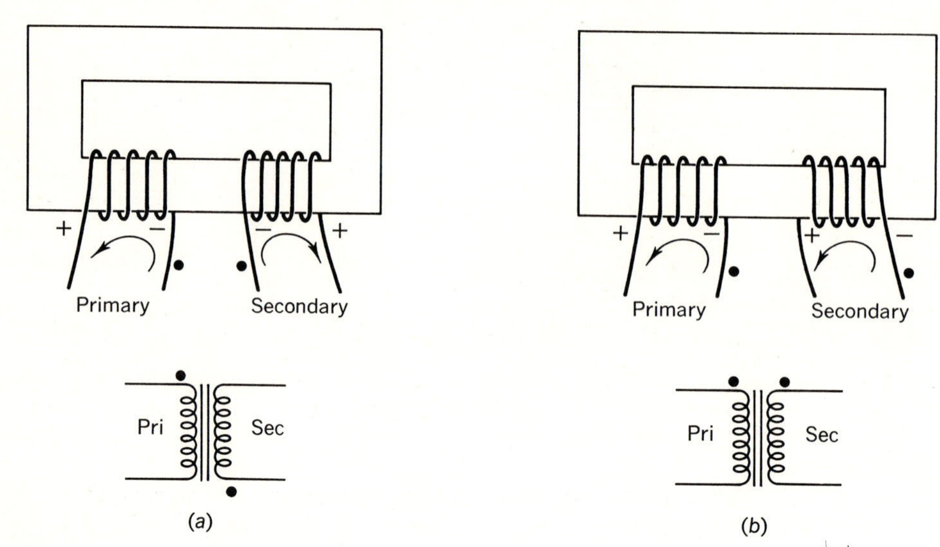

Fig. 34-5 The phase relationship between primary and secondary depends on relative coil-winding directions. Dots indicate similar polarities.

Table 34-1

Temperature, °C	Insulating material	mm^2/A	Cir mil/A
90°	Unimpregnated cottons, silk	0.5	1000
105°	Varnishes, enamels, cellulose, acetates, impregnated cotton	0.3	600
130°	Inorganic materials, Fiberglas	0.2	400
170°	Silicone compounds with inorganic materials	0.1	200

across the primary. If both windings are purely reactive, they each have 90° phase shift, or a shift of 180° for the two windings. Figure 34-5*a* represents a basic transformer in which the primary and the secondary coils are wound in the same direction. In this case the polarities across the secondary at any particular instant will be 180° out of phase with the primary. If the direction of the secondary winding is reversed, as in Fig. 34-5*b*, the phase relationships will also be reversed. The dots next to the windings indicate similar phase relationships.

As long as the load on a transformer is purely resistive, as in Fig. 34-6*a*, it reflects back on the primary as a pure resistance (except for the small value of inductive reactance involved in the magnetization current). With no load, the ratio of reactive or apparent power to true power loss may be quite high, and the source will see the primary essentially as an inductance.

If the load across the secondary is partially inductively reactive (Fig. 34-6*b*), the source sees the primary as an inductive reactance with a resistive component. Under this condition the source not only must feed enough current to the primary to supply the power loss of the load, but also must feed additional current to the primary to build up the magnetic field

ANSWERS TO CHECK-UP QUIZ 34-3

1. (165 V) (73.6 V) **2.** (40.3 V) (96.3 V)

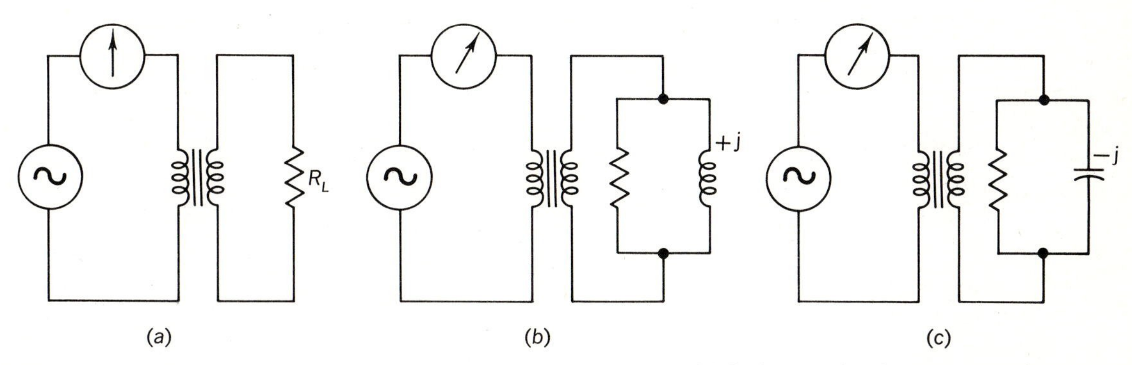

Fig. 34-6 (*a*) Resistor load appears as resistance to the source. (*b*) Inductive load appears inductive to the source, demanding extra current to magnetize the load field. (*c*) Capacitive load requires extra primary current to store energy in the capacitor field.

around the inductance. If larger wires are not used to lessen the resistance, the voltage-drops in the windings will be excessive, and the load voltage will sag. The proportion of reactive to resistive current in the secondary determines the phase angle seen by the source and the amount of extra current required. With a phase angle of 45° (a PF of 0.707 lagging), the source must supply 1.41 times as much current, and it still will not be producing any more power in the load. If the load is made less reactive, the power factor begins to approach unity. In general, the lower the power factor, the greater the reactive current, the greater the copper loss, the lower the efficiency of the transformer, and the lower the voltage across the load.

It is usually undesirable to operate transformers (or any ac sources) into inductively reactive loads because of the increased current demand for a given value of true power output. If capacitive reactance is added in series or in parallel with the load, the source sees the load as more nearly in phase, with a higher power factor, and the reactive component of the primary current decreases.

When the load across the secondary is slightly capacitively reactive, it may counteract the natural small inductive reactance of the transformer windings, and a maximum load voltage may occur. If the load becomes greatly capacitively reactive (Fig. 34-6c), PF decreases as with an inductive load. There is a resultant lowering of the secondary load voltage and reduction of transformer efficiency.

If a source-voltage waveform is sinusoidal, the current flowing through a pure resistance would also be sinusoidal. In the primary of an unloaded transformer, the in-phase copper-loss current plus the core-loss current working over the nonlinear magnetization curve of the core iron results in a distorted waveform for the primary current (Fig. 34-7). Increasing the resistive load on the secondary decreases the distortion. Under normal loaded trans-

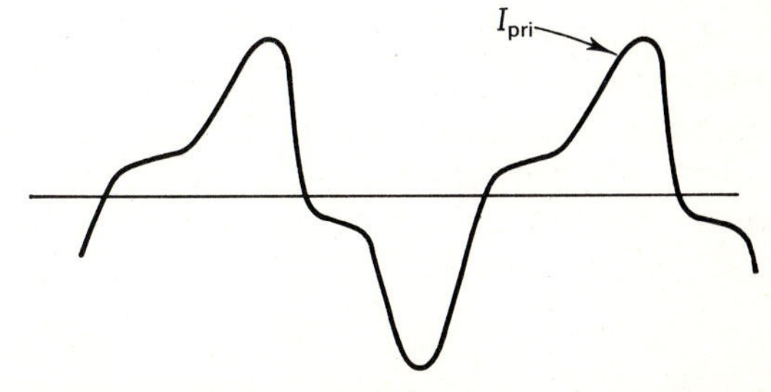

Fig. 34-7 Distorted waveform of a primary current in an unloaded transformer with a sinusoidal voltage applied.

former operation, the secondary current has essentially the same waveform as the primary voltage. The source will "look through" the transformer, seeing the load as it actually is.

In many circuits transformers have dc with an ac component (varying dc) flowing through the primary. Dc can produce no secondary emf unless it is varying. As long as the variations are held within the boundaries of the straight portion of the *BH* curve, the waveform of the secondary emf will be a replica of the current-variation waveform in the primary. If so much current flows in the primary that saturation at the top of the *BH* curve is reached, the output wave will be distorted. This has some uses, as in frequency multipliers, but in general it is undesirable.

Quiz 34-4. Test your understanding. Answer these check-up questions.

1. What are the names of the core losses in a transformer? __________ __________ __________ What formula is used to compute the copper loss? __________
2. A transformer primary is found by ohmmeter to have 2.5 Ω. When the primary is connected across 120-V ac, 0.35 A flows. A wattmeter across the primary reads 8.5 W. What is the unloaded copper loss in watts? __________ The total core loss? __________
3. The transformer in question 2 is operated for 1 h and is then removed from the line. An ohmmeter reading of the primary shows 3.92 Ω. What is the operating temperature of this transformer above room temperature (20°C)? __________
4. In the final analysis, what determines how much current the windings of a transformer will stand? __________
5. Does an unloaded transformer appear essentially resistive, capacitive, or inductive? __________ How would an electric light appear? __________ An electric motor? __________
6. Which is more desirable in a power transformer, a high or low power factor? __________
7. Which would produce the most sinusoidal output-voltage waveform, no load, a medium load, or a heavy load on a transformer? __________

34-6 TRANSFORMER CORES

The E-I type of core (Chap. 29) is the type used with almost all small power-frequency (50-, 60-, 400-, and 800-Hz) transformers. When the primary and secondary windings are both wound on the central core piece, the transformer is known as a *shell* type. The core forms a shell around all the windings (Fig. 34-8*a* and *b*).

When the core laminations are cut into an L shape, they can be interleaved to form the L-core shown in Fig. 34-8*c*. The primary can be wound on one leg and the secondary on the other, both primary and secondary may be wound on one of the legs, or half of each primary and secondary may be wound on each leg. This is a *core-type* transformer. The windings surround the core.

Many modern power transformers use *wound* or *spiral cores*. The transformer may be made up as either a shell type (Fig. 34-8*b*) or a core type (Fig. 34-8*d*). Wound cores have less leakage flux and less air gap than the punched-lamination cores. In some cases the windings are first wound on insulated forms, or "mandrels." Then a long ribbon of core iron, with all its molecules previously aligned in the direction of the metal ribbon, is wound through the core areas of the windings (Fig. 34-8*d*). The ribbons of core steel are coated with magnesium oxide, which acts as insulation between layers to prevent eddy currents. In other cases the primary and the secondaries are first wound separately. Then the cores are wound on mandrels, sawed in two, and fitted around the windings. Finally, the core pieces are firmly strapped together (Fig. 34-7*b*).

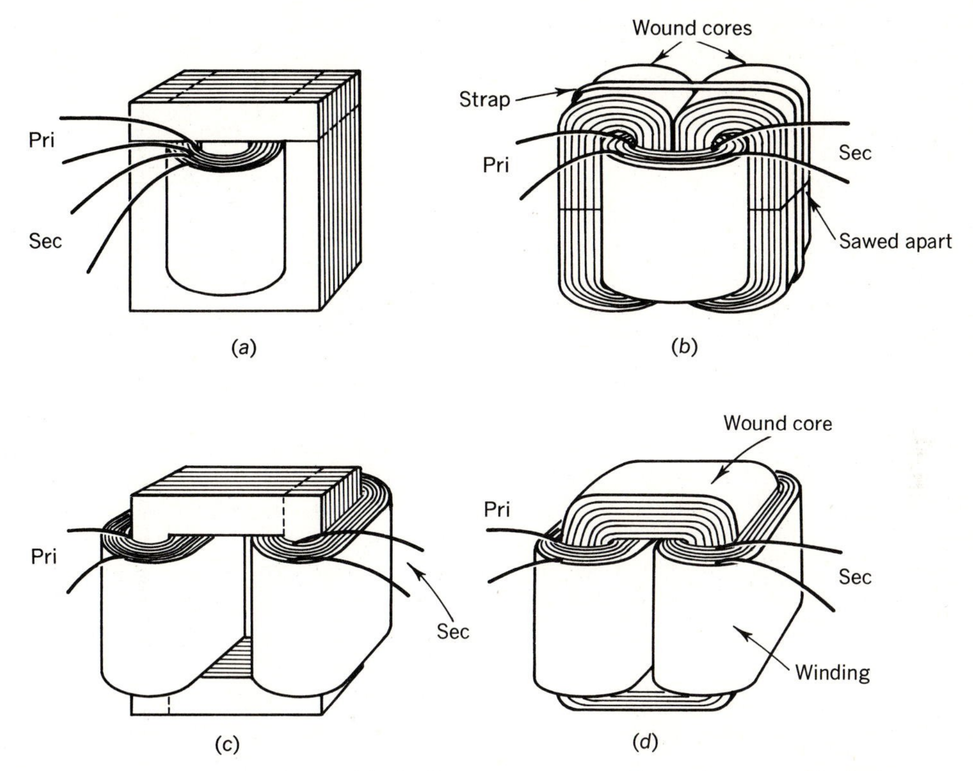

Fig. 34-8 Types of cores and windings: (*a*) shell-type core using E and I pieces; (*b*) shell-type wound core; (*c*) core type using L-shaped laminations; (*d*) wound-type core in core-type transformer.

Toroidal transformers have ferrite cores molded into ring (doughnut) shape (Fig. 34.9), with the primary and secondary bunch-wound on different halves of the core, as shown, or with the secondary as the inner layers of wire and the primary as the outer layers. There is essentially no leakage flux. Lines of force produced around the primary snap back and forth across the hole area, inducing an ac emf in the secondary as they cut across these turns. Any conductor in the hole area will have an emf induced into it. Any conductors outside the hole area will have no induced emf in them. Two toroidal transformers can be mounted side by side with only a layer of insulation between, and will have almost no intercoupling. Toroid transformers use powdered iron or ferrite materials for their cores.

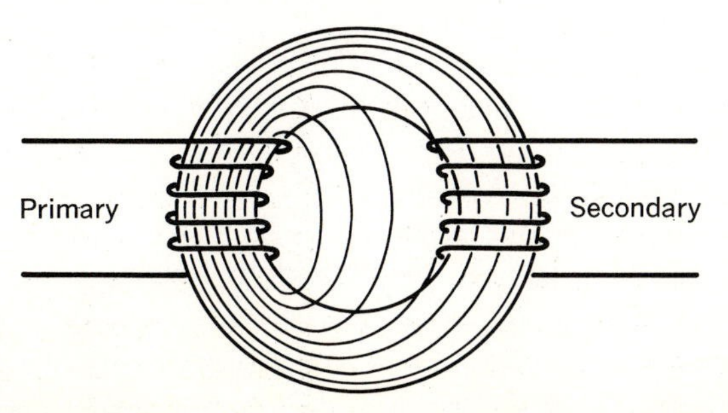

Fig. 34-9 A toroidal transformer showing path of lines of force.

Core lamination sheets must be insulated, so that when pressed together, no eddy currents will flow from one piece to the next. Thus, insulation takes up a percentage of the core volume. The *stacking factor* of a core is the ratio of the effective cross-sectional area of the iron in the core to the measured cross-sectional area. The stacking factor may vary from 85% with thick insulation on the laminations to 95% with thin insulation and tightly pressed laminations. The greater the stacking factor, the more iron in a core and the smaller dimensions the core can have. In toroidal ferrite cores, the iron particles are held together with an insulating binding material which results in a low-percentage stacking factor.

The lower the frequency of the ac used in the transformer, the more core iron required. A transformer operating on 400-Hz ac requires only about 20% as much iron as a 60-Hz transformer. Aircraft electrical systems use 400- or 800-Hz ac to reduce the weight of electronic power equipment in some cases.

34-7 TRANSFORMER RATINGS

Resistors, dc motors, and dc generators may be rated by the number of watts they can safely dissipate or handle, but transformers, ac motors, and alternators must be rated in voltamperes (VA), or if they are large, in kilovoltamperes (kVA).

If a power transformer has a resistive load across its secondary, it appears resistive, with a power factor approaching unity. The number of watts handled by the transformer and dissipated in the resistive load will be almost the number of voltamperes in the primary circuit. To deliver 100 W from a 100-V source, it is only necessary to have $I = P/E$, or 100/100, or 1 A flowing in the primary.

ANSWERS TO CHECK-UP QUIZ 34-4

1. (Eddy-current, hysteretic, leakage-flux) ($P = I^2R$) **2.** (0.3 W) (8.2 W) **3.** (142°) **4.** (Insulation capabilities) **5.** (Inductive) (Resistive) (Inductive) **6.** (High) **7.** (Medium load)

If the load is inductively reactive to the degree that the E and I are 45° out of phase, from the power formula $P = VA(\text{PF})$, or $P = VA(\cos\theta)$, the current required to deliver 100 W to the load is $A = P/V(\cos\theta)$, or 100/100(0.707), or 1.414 A. Therefore this primary must be wound to carry 1.414 A without overheating. The number of voltamperes involved in this transformer is $VA = 100(1.414)$, or 141 VA. This same transformer will handle $P = EI$, or 100(1.414), or 141.4 W with a purely resistive load.

Transformers are rated in frequency and voltamperes or kVA. How the device is employed is up to the user. With a purely resistive load, the kVA rating will also be the wattage rating. If the load were completely reactive, the phase angle would be 90°, PF would be zero, and the transformer would be carrying maximum current and heating with no power being used by the load.

If a load is inductive, capacitive reactance added to the load can bring the PF to unity. Where many ac motors are in operation, the load tends to become inductive. One or more large capacitors may be switched across the line to correct the PF. Another method of PF correction is to employ a few synchronous ac motors, since they are normally capacitive when running under load. (They may be called *synchronous condensers* or *synchronous capacitors*.)

Some voltages used in utility-company power transformers are shown in Table 34-2.

Electronic transformers used in audio-frequency systems working with 20- to 20,000-Hz ac will be rated in primary and secondary impedances and in watts. It is assumed that when the impedances are properly matched, the loads will be essentially resistive.

Table 34-2

Primary voltages			
2500	25 kV	138 kV	
4330	92 kV	230 kV	
4800	115 kV	345 kV	
Secondary voltages			
120 V	600 V	6.9 kV	34.5 kV
240 V	2400 V	11.5 kV	46 kV
480 V	4160 V	13.8 kV	69 kV

34-8 COOLING TRANSFORMERS

Small transformers are usually self-cooling unless they are in restricted areas, when forced air may be used. It is not unusual for transformers to be hot enough to burn the fingers if they are operating at current densities of 0.1 mm²/A (200 cir mil/A). When they are being operated at 0.5 mm²/A (1000 cir mil/A), they should run only warm to the touch.

Larger transformers may be self-cooled, or may be metal-encased and filled with oil. The oil is a good conductor of heat, is a good insulator, and prevents water vapor from entering the windings and possibly causing ionization of insulation and arc-overs. Corrugated-metal cases or external radiating systems are used on large transformers. In Fig. 34-10, as the oil heats, it rises and moves out into the top of the external pipes. As the oil cools, it drops down the pipes back into the bottom of the oil pool. Oil may also be circulated by pumps. If forced air is blown across the cooling pipes, the kVA rating of the transformer may be increased as much as 30% over simple air cooling. To reduce fire hazard, special oils, such as Pyranol, which is noncombustible, are used.

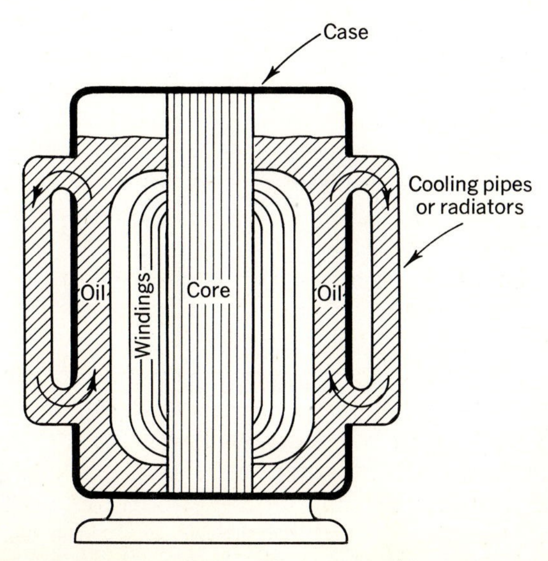

Fig. 34-10 Oil cooling of large transformers.

Quiz 34-5. Test your understanding. Answer these check-up questions.

1. What type of core is made from a long ribbon of iron? ________ What is used as the insulation on these ribbons? ________
2. Toroidal transformer cores are made of what material? ________ What are two disadvantages of such cores? ________ ________
3. What is the approximate weight of a transformer wound to operate on 120 Hz in comparison to one made to operate on 60 Hz with the same power capabilities? ________
4. Dc motors are rated in watts. In what units are small ac motors rated? ________ Large ac motors? ________ Large transformers? ________
5. In what values are the power-handling capabilities of audio-frequency transformers usually rated? ________
6. Under what condition is the kilovoltampere rating equivalent to the kilowatt rating of a transformer? ________
7. Which would you expect to be hotter during operation, transformers wound with 0.1-mm²/A wire or with 0.5-mm²/A wire? ________ Which would be more likely to have cotton insulation? ________
8. What are three reasons why oil is used in transformers? ________ ________ ________

34-9 TRANSFORMER WINDINGS

Many transformer coils are wound in layers on an insulating cardboardlike form. Each layer is insulated from the one above by a strip of *Kraft paper* or *glassine* insulating material. Sometimes the secondary is wound nearest the core, and the primary is wound over it. In others the procedure is reversed. In some cases the primary and secondary coils may be wound in disk or pie shapes, with alternate primary and secondary pies adjacent to each other to increase the coupling coefficient.

In low-frequency transformers, distributed capacitances between layers and coils may play no part in transformer operation. However, when the ac frequency increases past 1 kHz, internal capacitance may affect transformer performance. The capacitance between turns and layers in a transformer is a shunt capacitance, as shown in Fig. 34-11.

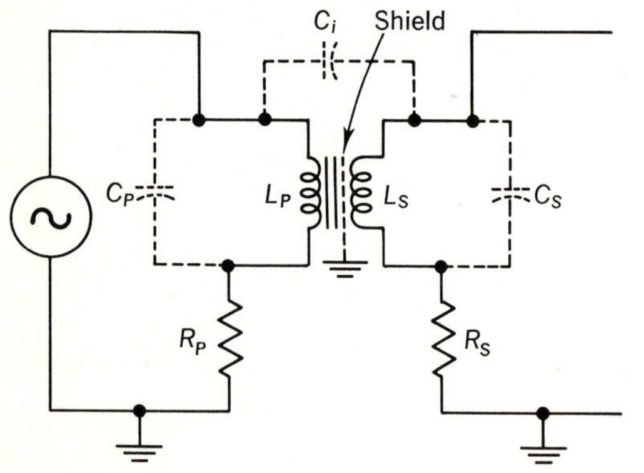

Fig. 34-11 Capacitances in transformers: shunt primary, C_p, shunt secondary, C_s, and interwinding capacitance, C_i.

ANSWERS TO CHECK-UP QUIZ 34-5

1. (Wound or spiral) (Magnesium oxide) **2.** (Ferrites) (Fragile, low stacking factor) **3.** (A little more than half) **4.** (W or VA) (kVA) (kVA) **5.** (W and Z) **6.** (Full load resistive) **7.** (0.1) (0.5) **8.** (Good heat conductor, prevents ionization, good insulator)

Between primary and secondary is an interwinding capacitance, C_i. With high-frequency ac, considerable power can be fed from primary to secondary via C_i. Since both C and L are involved in coupling energy to the secondary, and since these properties react in the opposite manner as frequency increases, there will be some frequency at which the transformer will cancel reactive effects. Above this frequency the capacitive coupled energy exceeds the inductive coupled energy, and the transformer begins to look like a capacitor to the source, with a leading current.

It is possible to insert an *electrostatic* or *Faraday shield* between the primary and secondary windings and ground the shield (Fig. 34-11). Energy from the primary passes via capacitance to the shield, but takes the path of least resistance and returns to the grounded part of the primary without ever getting to the secondary. An electrostatic shield may consist of a flat copper or aluminum (nonferrous) strip wound once around the primary turns (Fig. 34-12). The shield must not

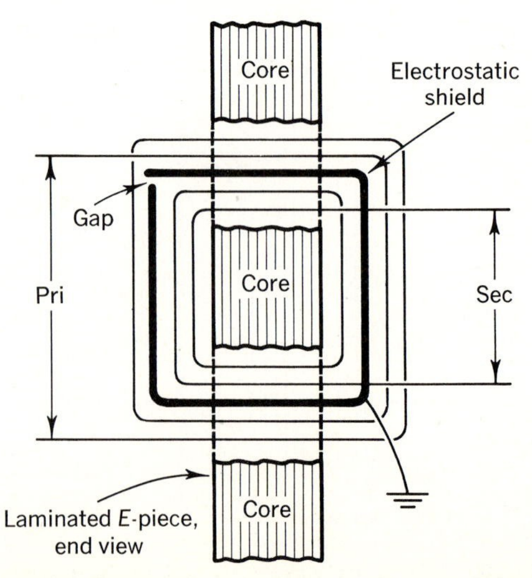

Fig. 34-12 Electrostatic shield between primary and secondary.

make a completed connection to itself, or it will form a shorted turn, the primary inductance will be reduced materially, and the primary may burn out. Magnetic fields can expand through the nonferrous shield as if it were not there. A single layer of wire wound between secondary and primary, with one end grounded, also acts as an electrostatic shield. The cores of transformers are usually grounded.

To prevent leakage flux from the core and to provide *magnetic* shielding, it is common to surround the whole transformer in a high-permeability metal shield case. Where leakage flux must be reduced to nearly zero, a double-walled shield may be necessary.

In some applications a copper band is wound around the core. Core flux induces an emf and a current in the copper conductor which, according to Lenz's law, produces a field of the same strength but 180° out of phase with the leakage flux. This results in a minimal external flux around the transformer.

Transformers handling up to a few amperes may be wound with round insulated wire. When the current is in the hundreds or thousands of amperes, square copper or aluminum wires, or aluminum strips, are used to provide a better space factor.

For transformers operating from 120-V 60-Hz ac involving 30 to 50 W, the number of primary turns may be about 500. In the 100-W range, the primary may have 300 to 400 turns. In general the more iron used in the core, the fewer turns required.

34-10 EFFICIENCY AND REGULATION

The efficiency of any device, transformer, generator, motor, vacuum tube, transistor, etc., is the ratio of the power fed into it to the power obtained from it. In formula form,

$$P_{\text{efficiency}} = \frac{P_o}{P_i}$$

where P_o is output power and P_i is input power. This provides an answer that is the *coefficient of efficiency*, a decimal fraction between 0 and 1.0. The coefficient of 1.0 indicates the efficiency is 100%. To compute efficiency in percentage, the formula is

$$P_{\text{efficiency}} = \frac{P_o}{P_i}(100)$$

As an example, a transformer has 10 kVA of input at a PF of 1.0. The output is 9.5 kW. The percentage of efficiency is $(P_o/P_i)(100)$, or 95%. The transformer must be losing 500 W of heat in copper and core losses. Note that efficiency deals with watts, not voltamperes. In this case the PF of 1.0 indicates no reactance in the load.

The *regulation* of a device is the ratio of the difference in voltage from an unloaded to a loaded condition, to its *loaded* operating voltage. This can be applied in transformers, generators, or power supplies. Thus

$$E_{\text{reg}} = \frac{E_{NL} - E_{FL}}{E_{FL}}$$

where E_{NL} is the no-load secondary voltage and E_{FL} is the full-load voltage. This results in a coefficient. For percentage, the formula is

$$\%E_{\text{reg}} = \frac{E_{NL} - E_{FL}}{E_{FL}}(100)$$

If a transformer has 10 kV of output with no load and 9.5 kV with full load, its voltage regulation is $[(10 - 9.5)/9.5](100)$, or 5.26%.

Regulation in motors refers to the change in speed as the load is lightened from the normal value.

$$\%S_{\text{reg}} = \frac{S_{NL} - S_{FL}}{S_{FL}}$$

where S_{NL} is the no-load speed and S_{FL} is the full-load speed.

Quiz 34-6. Test your understanding. Answer these check-up questions.

1. For what relative frequencies of ac is distributed capacitance undesirable in most transformers? __________
2. What type of material is used in electrostatic shields? __________ In magnetic shields? __________
3. To what are the cores of most transformers connected? __________
4. Would a transformer tend to appear inductive, capacitive, or resistive to a low-frequency source? __________ Midrange? __________ High-frequency? __________
5. What is special about conductors used in high-power transformers? __________ __________
6. A 60-Hz transformer is to deliver 12 V at 3 A. Approximately how many primary turns might it have? __________ If it were to deliver 500 V at 200 mA, approximately how many turns might the primary have? __________ If the volume of iron in the core were increased, how would this affect the number of primary turns required? __________
7. A transformer is 98.5% efficient and delivers 8.75 kW to a load. What is the P_i? __________
8. A transformer operating under no load has 2480 V output. When fully loaded, its output is 2390 V. What is the regulation? __________ Why would regulation be better at half load? __________
9. Which is more desirable, a high or a low percentage of regulation? __________

34-11 IMPEDANCE MATCHING

Transformer impedance matching is important. Only when a load impedance matches a source impedance will maximum power be delayed in the *load*. In Fig. 34-13, a resistive load is across a 30-V

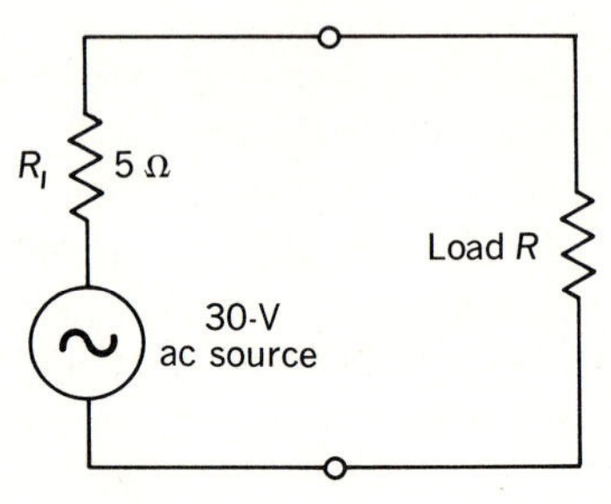

Fig. 34-13 Actual resistances in simple circuit.

source having a 5-Ω internal resistance. (Follow along the first line of Table 34-3.) If the load resistor is 10 Ω, the total resistance in the circuit is 5 + 10, or 15 Ω. The circuit current is $I = E/R$, or 30/15, or 2 A. The voltage across the load is $V = IR$, or 2(10), or 20 V. The power dissipated in the load is $P = I^2R$, or $2^2(10)$, or 40 W. The power dissipated in the source is $2^2(5)$, or 20 W. The percentage of the total power dissipation (60 W) that is developed in the load is $P_{\text{eff}} = 100(P_o/P_i)$, or 100(40/60), or 67%. The remaining tabulation is for loads of 5 and 1 Ω. The maximum power delivered to the load occurs when *the load impedance matches the source impedance.*

Table 34-3

Source R	Load R, Ω	Circuit current, A	Load voltage, V	Power lost in source, W	Power in load, W	% eff
5	10	2	20	20	40	67
5	5	3	15	45	45	50
5	1	5	5	125	25	20

When a load is drawing its maximum power from a source, the whole system is operating at 50% efficiency. However, when the load impedance is greater than the source impedance, the efficiency is higher. Electric power companies must have alternators with internal impedances as low as possible to improve the efficiency of their systems.

If a lossless 1:1 transformer had been used to couple the load to the source, the results would have been the same as shown in the table. If a 2:1 step-down transformer had been used, the results might have been very different (Fig. 34-14). By varying the load resistance the

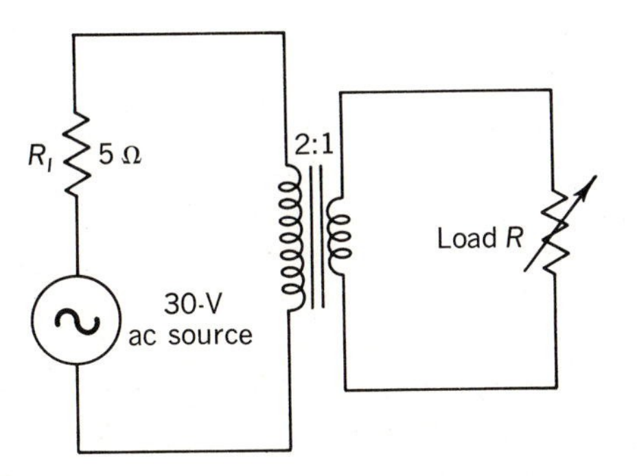

Fig. 34-14 A variable-impedance load on a transformer.

current in the primary can be controlled. If the load resistance is adjusted to a primary voltage-drop of 15 V, the load is taking as much power as is being dissipated in the internal resistance of the source, as indicated in the table (45 W). What is the impedance of the load now?

Since the transformer primary has 15 V across it, with a 2:1 step-down, the load must have 7.5 V across it and must be dissipating 45 W. The load must have a resistance of $R = E^2/P$, or 56.25/45, or 1.25 Ω. Since maximum power is delivered to a load when the source impedance is looking into a load of equal impedance, what is the relationship of the 5-Ω source impedance, the 2:1 ratio transformer, and the 1.25-Ω load impedance? If the 1.25-Ω load value is multiplied by the turns ratio, 2, the answer is only 2.5 W. But if the 1.25-Ω load value is multiplied by the *square* of the turns ratio, 2^2 or 4, the answer is 5 Ω. Therefore the impedance ratio of a transformer must be equal to the *square of the turns ratio*. In formula form,

$$\frac{Z_{\text{pri}}}{Z_{\text{sec}}} = \left(\frac{N_1}{N_2}\right)^2 \quad \text{or} \quad \sqrt{\frac{Z_{\text{pri}}}{Z_{\text{sec}}}} = \frac{N_1}{N_2}$$

where Z_{pri} is primary impedance, Z_{sec} is secondary impedance, N_1 is primary turns, and N_2 is secondary turns. The 2:1 step-down turns-ratio transformer above is working as a 4:1 *impedance* step-down transformer. A 3:1 step-up turns-ratio transformer converts a given primary impedance to 9 times as much secondary impedance.

Quiz 34-7. Test your understanding. Answer these check-up questions.

1. For maximum power into a load, what impedance should the source have? ________
2. For maximum efficiency in transferring power to a load, what Z should the source have? ________
3. When the load impedance is lower than the source Z, where does maximum dissipation occur? ________ Is this desirable? ________
4. Name a condition in which the source of a system should have a lower Z than the load. ________
5. In what way, if any, does transformer ratio affect the ability of a load to obtain maximum power? ________
6. What is the formula that expresses the relationship between the turns ratio and the impedance ratio of a transformer? ________
7. What is the ratio of primary power to secondary power in a 3:1 step-up transformer? ________
8. A 2.5:1 step-down transformer has a 100-Ω Z primary source. What load Z will match the secondary winding? ________

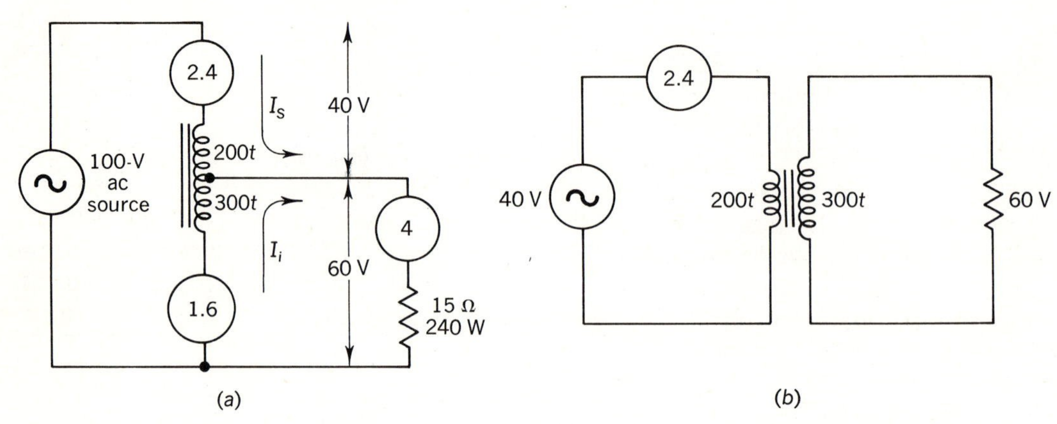

Fig. 34-15 (*a*) An autotransformer delivers 4 A and 240 W to a load at 60 V. (*b*) The same device used as a two-circuit transformer can develop only 96 W into a 60-V load.

9. A vacuum tube with an impedance of 6400 Ω is connected into the primary of a transformer. For maximum power in a 16-Ω loudspeaker, what should the turns ratio of the transformer be? __________ If the turns ratio were doubled, what would be the effect on the power into the loudspeaker? __________

34-12 AUTOTRANSFORMERS

One of the most efficient electrical devices is the *autotransformer*, or *autoformer*. It consists of a tapped inductor (Fig. 34-15*a*), in this case 500 turns tapped at 300 turns. From basic voltage-divider theory, the 15-Ω resistive load is connected across 60 V. This produces a current of 4 A through it and, from $P = I^2R$, 240 W of dissipation. If the source is delivering 240 W at 100 V, the source current must be 2.4 A, flowing as shown by the vector arrows. At one instant this 2.4 A of source current flows down through the top part of the inductor, called the *series winding*, and through the load. The other 1.6 A required to make up the 4 A total is developed by the counter-emf induced in the lower part of the inductor, called the *common winding*, by the magnetizing current. The induced current flows up the common winding as the source current flows down the series winding. Both currents flow together through the load.

In this connection form the autotransformer is a step-down device. It is possible to connect a 60-V source across the 300-turn common winding and obtain 100 V across the whole inductor. The autoformer is then acting as a step-up voltage device.

A two-circuit 1.5:1-ratio transformer could be made out of the inductor by opening the winding at the 300-turns point, using the 200 turns as a primary and the 300 turns as a secondary (Fig. 34-15*b*). With 40 V across the primary, this transformer would produce the required 60 V across the secondary. However, since the primary part is wound with a size of wire to stand only 2.4 A, the total power input to the primary would be limited to $P = EI$, or 40(2.4), or 96 W. The secondary can be loaded only to 96 W. As an auto-

---

ANSWERS TO CHECK-UP QUIZ 34-6

1. (High) **2.** (Nonferrous, good conductors) (Ferrous metals) **3.** (Ground or earth) **4.** (Inductive) (Resistive) (Capacitive) **5.** (Aluminum, ribbons or square) **6.** (500) (300 to 400) (Fewer) **7.** (8.88 kW) **8.** (3.77%) (Less I^2R losses, less voltage-drop in windings) **9.** (Low percentage)

---

former, the power capabilities were 240 W. Autoformers for a given power are considerably smaller, lighter, and less expensive and have higher efficiency than any two-circuit transformer.

Any two-circuit transformer can be connected as an autoformer. A disadvantage is the lack of electrical isolation between the input and output circuits. If the step-up or step-down ratio is more than about 3:1, the efficiency decreases.

Autotransformers find many uses in high-power transmission systems. A variable autoformer, such as a Variac or Powerstat, is shown in symbol form in Fig. 34-16. With the adjustable arm touching

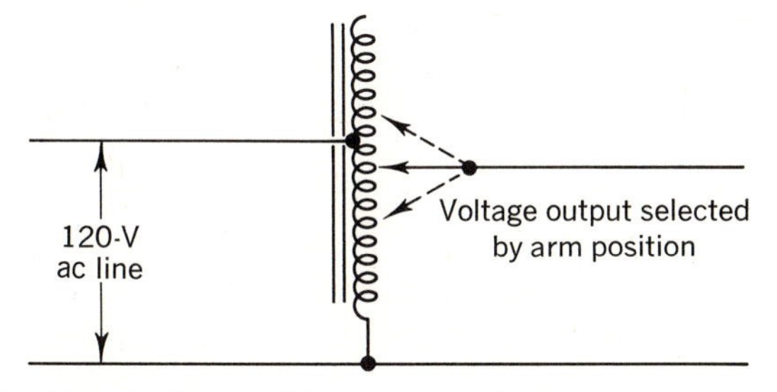

Fig. 34-16 A variable autotransformer.

the bared wire at the bottom of the inductor, the output voltage is zero. Halfway to the 120-V tap the output is 60 V. At the top of the inductor the output is usually about 135 V.

Quiz 34-8. Test your understanding. Answer these check-up questions.

Questions 1 to 4 refer to Fig. 34-17*a*.

1. What is the rms voltage developed across points *C* and *D*? __________ What is the peak value? __________
2. If points *B* and *C* are connected, what is the voltage between *A* and *D*? __________ *D* and *C*? __________
3. If points *A* and *C* are connected, what is the voltage between *B* and *D*? __________
4. If points *A* and *C* are connected and *B* and *D* are connected, what will happen? __________ If *A* and *D* are connected and *B* and *C* are also connected, what will happen? __________

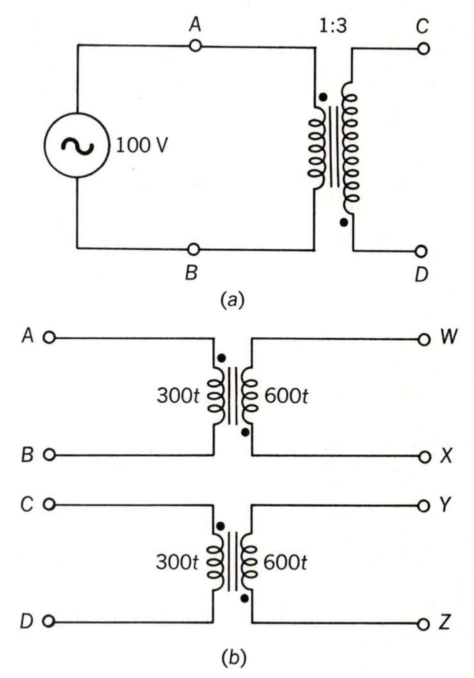

Fig. 34-17 Diagrams for Check-up Quiz.

Questions 5 to 10 refer to Fig. 34-17*b*. All windings are wire which will stand only 1 A.

5. Connect *B* to *C* and *X* to *Y*. If 120 V is applied to *A* and *D*, what is the voltage from *W* to *Z*? __________ The maximum secondary-current value? __________
6. Connect *A* to *C*, *B* to *D*, and *X* to *Y*. With 120 V as the source, what is the voltage from *W* to *Z*? __________ The maximum I_{sec} value? __________
7. Connect *A* to *C*, *B* to *D*, *W* to *Y*, and *X* to *Z*. With 120 V as the source, what is the secondary voltage? __________ The maximum I_{sec}? __________
8. Connect *B* to *C* and *X* to *Y*. What would be the E_{sec} with 240 V across the primary? __________ The maximum I_{sec}? __________
9. Connect *B* to *C*, *W* to *Y*, and *X* to *Z*. With 240 V across *A* and *D*, what is the E_{sec}? __________ The maximum I_{sec}? __________
10. With 120 V across *A* and *B*, how would you connect the terminals to obtain 600 V? __________ What would be the maximum secondary current under this condition? __________

CHAPTER 34 TEST • POWER TRANSFORMERS

1. A high-efficiency iron-core transformer has 520 primary and 1850 secondary turns. If the source is 120 V, what is the no-load output voltage? If the secondary load is a 2500-Ω R, what is the primary current?
2. What is the primary current called that: Supports the core-loss demand? Produces the basic core flux? Is the vector sum of these two currents?
3. If an unloaded primary inductance with 500 turns is 8 H, what is the X_L of the primary to a 60-Hz ac? What would it be if the turns were reduced to 400?
4. A 1:1-ratio iron-core transformer across 100-V 50-Hz ac has 400 primary turns, a primary inductance of 4 H, 20 W of core loss, and 2 Ω of R in each winding. What is the no-load primary current? What would the I_{sec} be if the load were a 20-Ω R? What would the I_{pri} be in this case? What would the L_{pri} be under a load? What is the total copper-loss value? What is the efficiency of this transformer? What is E_{sec}?
5. What transformer losses are increased when the transformer heats? The source frequency increases? The load increases?
6. Doubling the load on a transformer has approximately how much effect on the core losses?
7. What type of insulation on wires is listed in Table 34-3 as being able to withstand the greatest heating of the wire?
8. What value of power factor is most desirable in transformers?
9. What do you know about the iron core of a shell-type transformer? A core-type?
10. What core shape is considered to have the least leakage flux?
11. What are three advantages of wound cores over punched-lamination cores?
12. What is the ratio of the effective cross-sectional area of an iron core to the measured cross-sectional area called?
13. What ac frequency is usually employed in aircraft power systems? Why?
14. In what values would the manufacturer of a high-power transformer rate the device?
15. In what values would the manufacturer of an audio-frequency transformer rate the device?
16. Why are synchronous motors often used in heavy-industry plants?
17. What factor plays an important part in transformers at high frequencies which may be negligible at low frequencies?
18. Where is a Faraday shield found in transformers?
19. In what applications are wide strips of aluminum used in transformers?
20. The output of a transformer is 885 V with no load but only 852 V with a load. What is the regulation percentage?
21. What turns-ratio transformer will match a 45-Ω load to an 1800-Ω source?
22. What is the main disadvantage of an autoformer?

ANSWERS TO CHECK-UP QUIZ 34-7

1. (Load Z value) **2.** (Very low in comparison to load) **3.** (Source) (No) **4.** (Power ac) **5.** (Matches Z) **6.** [$Z_1/Z_2 = (N_1/N_2)^2$] **7.** (1:1) **8.** (160 Ω) **9.** (20:1) (Less power)

ANSWERS TO CHECK-UP QUIZ 34-8

1. (300) (424) **2.** (200 V) (300 V) **3.** (400 V) **4.** (Short secondary, reduce L, burn out) (Excitation current only) **5.** (240 V) (0.5 A) **6.** (480 V) (0.5 A) **7.** (240 V) (1 A) **8.** (480 V) (0.5 A) **9.** (240 V) (1 A) **10.** (X to Y, Z to D, output W to C) (0.2 A)

35 RESONANT CIRCUITS

CHAPTER OBJECTIVE. To become familiar with resonant *LCR* circuits as to their frequency, bandwidth, Q, impedance, and phase angle. To learn to use decibel formulas involving logarithms.

35-1 RESONANCE

The special condition when capacitive reactance equals inductive reactance was discussed in Chaps. 14 and 16. When X_L and X_C are in *series* and equal (Fig. 35-1),

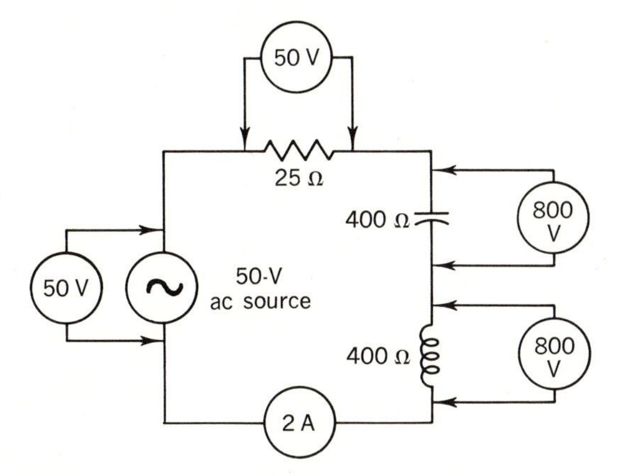

Fig. 35-1 A resonant circuit, showing current and voltage-drops that would be present in it.

a condition of *resonance* exists. The capacitive voltage lags the current by 90° and the inductive voltage leads the current by 90°, and the reactances cancel. At resonance the only impedance is the resistance in the circuit.

In Fig. 35-1, a series resistance of 25 Ω is the total impedance, resulting in a 2-A current flow with the 50-V source. A voltmeter across the resistor will read the full source-voltage value. The same voltmeter across the 400-Ω inductor will read $V = IX$, or 2(400), or 800 V. (This assumes the voltmeter is not loading the circuit.) The voltage-drop across the capacitor will also be 800 V. Thus it is possible for a resonant circuit to step up ac voltage. The increased voltage across a reactor is known as the *resonant rise in voltage*, and it increases as the resistance of the circuit decreases.

Refer back to Chap. 33 to determine what would happen to the current and voltage-drops in this circuit if a 400-Ω resistive load were connected across the 400-Ω capacitive reactance (Fig. 35-2).

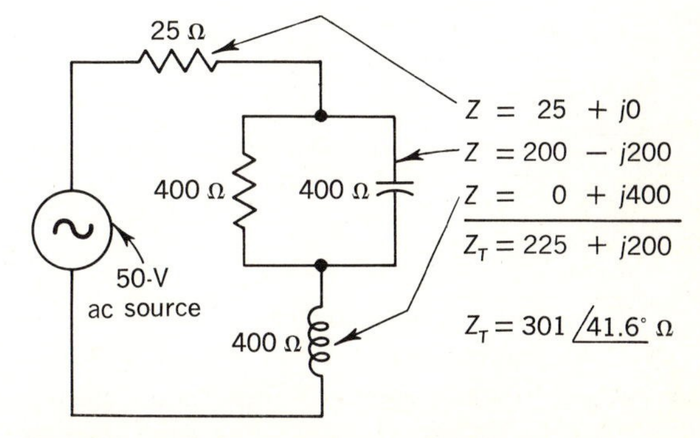

Fig. 35-2 The resonant circuit of Fig. 35-1 with one of the reactors loaded by a 400-Ω resistor, and the complex-number sum of the impedances.

Now, the parallel-RC part of the circuit is a capacitive impedance. Its impedance can be found by

$$Z = \frac{1}{Y}$$

$$Y = G + jB = 0.0025 + j0.0025$$
$$= \sqrt{0.0025^2 + 0.0025^2} = \sqrt{0.00\,00\,12\,50}$$
$$= 0.00354 \text{ mho}$$

Once admittance is known, impedance, the reciprocal of this value, is

$$Z = \frac{1}{0.00354} = 283\ \Omega \text{ (capacitive)}$$

$$\tan\theta = \frac{B}{G} = \frac{0.0025}{0.0025} = 1$$

The angle whose value is tan 1 is 45°. Thus

$$Z = 283\angle{-45°}\ \Omega$$

The series equivalent of $283\angle{-45°}\ \Omega$ is

$$\cos\theta = \frac{R}{Z}$$
$$R = Z(\cos\theta) = 283(0.707) = 200\ \Omega$$
$$\sin\theta = \frac{X}{Z}$$
$$X = Z(\sin\theta) = 283(0.707) = 200\ \Omega$$

$283\angle{-45°}\ \Omega$ expressed in j-operator or rectangular terms is $Z = 200 - j200$. The series equivalent of the 25-Ω resistor in j-operator terms is

$$Z = 25 + j0$$

The series equivalent of the 400-Ω inductive reactance in j-operator terms is

$$Z = 0 + j400$$

The sum of the three complex-number values is shown in Fig. 35-2. The algebraic sum is $Z_t = 225 + j200$, which by the Pythagorean formula is $Z = \sqrt{225^2 + 200^2}$, or $\sqrt{90{,}625}$, or 301 Ω. The phase angle is $\tan\theta = X/R$, or 200/225, or $\theta = 41.6°$. The total circuit presents an inductive impedance of $301\angle{41.6°}\ \Omega$. The circuit is inductive because the series L has a greater impedance than the parallel RC circuit.

With 301 Ω and 50 V, circuit current is $I = E/Z$, or 50/301, or 0.166 A. The voltage-drop across the 400-Ω capacitive reactance and its parallel 400-Ω resistance (283-Ω impedance) is $V = IZ$, or 0.166(283), or 47 V. Compare this with the unloaded value of 800 V. A series circuit may be used to step up ac voltages, but not if one of the reactances is loaded by resistance. The voltage-drop across the inductor is $V = IX_L$, or 0.166(400), or 66.4 V.

Quiz 35-1. Test your understanding. Answer these check-up questions.

1. What is the impedance value of a series-resonant circuit with the values $X_L = 500$, $R = 50$, and $X_C = 500$? ________
2. If the circuit in question 1 were across a 100-V source, what would be the voltage-drop across the coil? ________ Capacitor? ________ Resistor? ________
3. In the circuit of questions 1 and 2, if a 500-Ω resistor were connected across the coil, what would be the impedance value of the circuit? ________ The source-current value? ________ The voltage-drop across the coil? ________ The voltage-drop across the capacitor? ________

35-2 THE RESONANCE CURVE

In many electronic circuits a load is connected across a source producing many different frequencies at the same time. If the load is a resonant circuit, only ac at the resonant frequency will produce a maximum current flow in the circuit and a maximum rise of voltage across the reactances. All other "signal" frequencies will produce weaker currents and lower voltages across the reactances. A resonant circuit is said to be *frequency-selective*.

The variable-frequency source shown in Fig. 35-3 is called an *audio-frequency signal generator*. It can produce ac energy from perhaps 20 to 20,000 Hz, and up to perhaps 10 V.

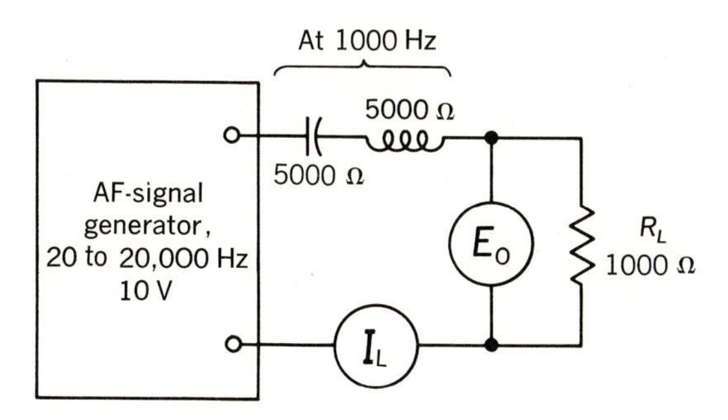

Fig. 35-3 An audio-frequency signal generator connected to a resonant circuit and a resistive load in series.

Table 35-1 lists various circuit parameters for a circuit having the values indicated in Fig. 35-3. The voltmeter reads the voltage across the load. The ammeter indicates the current flowing. The ratio of reactance to resistance, or the Q value of the circuit, is 5000/1000 at 1000 Hz, or a Q of 5.

The 10-V signal generator is set to 1000 Hz, the frequency of resonance (f_0). On the 1000-Hz line in Table 35-1, X_L equals X_C, resulting in zero reactance. The resistance of the load is 1000 **Ω**. With a source emf of 10 V, the load current is $I = E/Z$, or 10/1000, or 0.01 A. The voltage across the load is $V = IZ$, or 0.01(1000), or 10 V. Since the two reactances cancel, the circuit is neither capacitively nor inductively reactive, but purely resistive. Thus a resonant circuit is a resistive circuit, and the power factor is unity, or 1.

If the signal generator is adjusted to 1100 Hz, the two reactances are no longer equal. The inductive is greater than the capacitive reactance by 950 **Ω**, but R is still 1000 **Ω**. Impedance by the Pythagorean formula is 1380 **Ω**. Current by Ohm's law is 0.00725 A, resulting in a voltage-drop of 7.25 V across R_L. Since the X_L is greater than the X_C, a series LC circuit above its resonant frequency appears inductively reactive with a resistive component. Increasing frequency 10% above resonance results in a loss of load current and voltage of about 28%.

When the signal generator is readjusted to 900 Hz, the impedance rises to 1460 **Ω**, the current is 0.00686 A, and the voltage-drop across the load is 6.86 V. Since the X_C is now greater than the X_L, at frequencies below resonance a series LC

Table 35-1

Hz	$X_L - X_C = X_t$	R, Ω	Z, Ω	I_L, A	E_o, V	X or R
600	3000–8330 = 5330	1000	5430	0.00184	1.84	X_C
700	3500–7150 = 3650	1000	3780	0.00264	2.64	X_C
800	4000–6250 = 2250	1000	2460	0.00354	3.54	X_C
900	4500–5560 = 1060	1000	1460	0.00686	6.86	X_C
1000	**5000–5000 = 0**	**1000**	**1000**	**0.01**	**10**	**R**
1100	5500–4550 = 950	1000	1380	0.00725	7.25	X_L
1200	6000–4165 = 1835	1000	2090	.0.00478	4.78	X_L
1300	6500–3840 = 2660	1000	2840	0.00352	3.52	X_L
1400	7000–3570 = 3430	1000	3580	0.00280	2.80	X_L

$f_o = 1000$ Hz $E_S = 10$ V $R_L = 1000$ Ω $Q = 5000/1000 = 5$

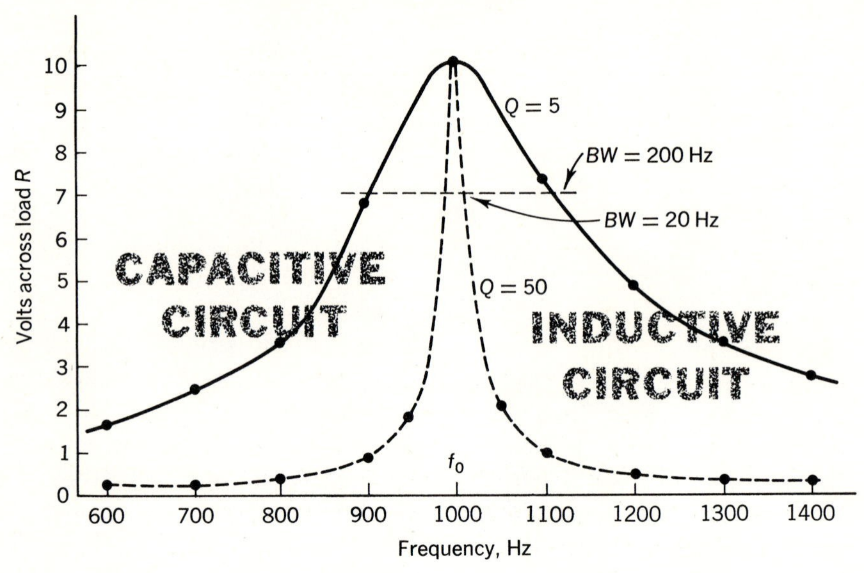

Fig. 35-4 Universal resonance curves developed from computed values of Tables 35-1 and 35-2. Solid lines are for Q = 5. Dashed lines are for Q = 50.

circuit appears to the source as capacitive with a resistive component. Reducing the source frequency 11% results in a decrease of load current and voltage of about 32% for a Q of 5.

Other current and voltage values in Table 35-1 show that the further from resonance, the less the load voltage. Plotting these voltages produces a *universal resonance curve* (Fig. 35-4).

Table 35-2 shows the parameters of the same resonant circuit, except that R_L is reduced to 100 Ω. The circuit now has a Q of 5000/100, or 50. Compare the steepness of the output-voltage curve for a circuit with a Q of 50, shown dashed, Fig. 35-4, with the curve for a Q of 5. Notice how much more frequency-selective a circuit is if it has a high Q value. With a Q of 50, a 10% change of frequency from resonance results in a decrease in output voltage of approximately 90%. The band of frequencies passed to the load by a high-Q resonant circuit is very narrow in comparison to the bandpass of a low-Q circuit.

In the circuit having a Q of 5, the current at resonance is 0.01 A. This value of current flowing through the 5000-Ω reactances produces a voltage-drop of 50 V across each reactor. The ratio of the voltage-drop across a reactance in a *resonant circuit* to the source voltage, 50/10 in this case, is also the value of the Q of the circuit, or

$$Q = \frac{E_X}{E_S}$$

where E_X is the voltage across either reactance and E_S is the source voltage.

ANSWERS TO CHECK-UP QUIZ 35-1

1. (50 Ω) **2.** (1000 V) (1000 V) (100 V) **3.** (391 Ω) (0.256 A) (90.3 V) (128 V)

Quiz 35-2. Test your understanding. Answer these check-up questions.

1. Is a resonant circuit of any practical use in a circuit carrying only one frequency of ac? ____________

Table 35-2

Hz	$X_L - X_C = X_t$	Z, Ω	R_L, Ω	I_L, A	E_o, V	X or R
600	3000–8330 = 5330	5330	100	0.00185	0.19	X_C
900	4500–5560 = 1060	1065	100	0.0094	0.94	X_C
950	4750–5270 = 520	537	100	0.0186	1.86	X_C
1000	**5000–5000 = 0**	**100**	**100**	**0.1**	**10**	**R**
1050	5250–4760 = 490	500	100	0.02	2	X_L
1100	5500–4550 = 950	955	100	0.0105	1	X_L
1400	7000–3570 = 3430	3430	100	0.00292	0.292	X_L

$f_o = 1000$ Hz $E_S = 10$ V $R_L = 100\ \Omega$ $Q = 50$

2. Why is the variable-frequency source in Fig. 35-3 called an audio-frequency signal generator? __________
3. What are the two ratios in a resonant circuit that give the Q value of the circuit? __________
4. Is the shape of a high-Q resonant circuit the same as the shape of a low-Q resonant circuit? __________
5. What does "f_o" stand for? __________
6. Below f_o, how does a series ac circuit appear to a source? __________ Above f_o? __________ At f_o? __________
7. If the value of the inductance in Fig. 35-3 were halved, would the frequency of resonance of the new circuit be higher or lower? __________ What would be the new f_o? __________ What would be the Q value of the new circuit? __________ The E_X value? __________
8. How does lowering the Q of a resonant circuit affect the band of frequencies passed by the circuit? __________ Is bandwidth directly or inversely proportional to Q? __________
9. Is the formula $Q = E_X/E_S$ valid for *LCR* circuits that are not series-resonant? __________

35-3 BANDWIDTH

From Fig. 35-4 it can be seen that the low-Q circuit passes to the load any frequencies 10% above and below the resonant frequency much better than the high-Q circuit does. A high-Q circuit has a narrow bandwidth or bandpass characteristic, and a low-Q circuit has a broad bandwidth. It has been established that when the resonance curve is followed down on each side to a point 0.707 of the peak voltage (the "half-power," or "3-dB" points), the band of frequencies between these two points will represent the *bandwidth* of the circuit. The horizontal dashed line in Fig. 35-4 indicates an amplitude 0.707 of the peak value. For a Q of 5, the band of frequencies appears to be from about 1100 to 910 Hz, a bandwidth of 200 Hz.

The ratio of resonant frequency to Q, 1000/5, or 200 in this case, gives the bandwidth of the circuit. A formula for determining bandwidth is therefore

$$BW = \frac{f_o}{Q}$$

where BW = bandwidth, in hertz
f_o = resonant frequency, in hertz
Q = ratio X_L/R of the LC circuit

In laboratory work this formula, rearranged, is used to determine the Q of a resonant *LC* circuit. The signal generator is tuned past the resonant frequency, and the voltage across the load is measured with a VTVM. The frequencies producing the lower 0.707-of-maximum voltage and the upper 0.707-of-maximum voltage are determined. This band of frequencies and f_o are used to determine the Q by

$$Q = \frac{f_o}{BW}$$

If either the *L* or *C* of the circuit is known, the effective resistance (ohmic re-

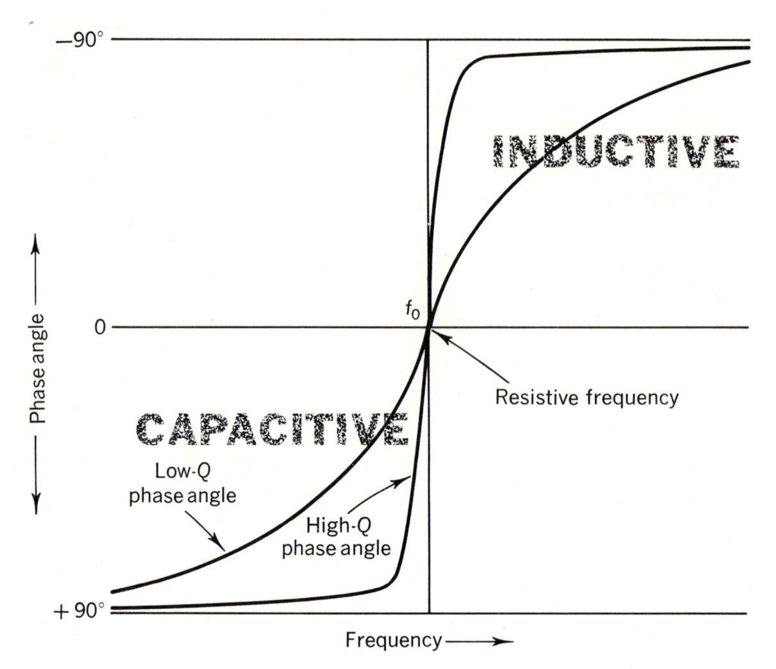

Fig. 35-5 Plot of approximate phase angle versus frequency for series-resonant circuits of high and low values of Q.

sistance plus skin effect) can be determined by computing reactance at f_0. Then, from the formula $Q = X/R$,

$$R = \frac{X}{Q}$$

where R = effective f_0 resistance
X = either X_L or X_C
Q = the value determined from the bandwidth of the circuit

If the R value resulting from this formula differs from an ohmic reading of the circuit with an ohmmeter, the difference-resistance is the skin effect and effective resistance due to all other losses in the coil and capacitor.

The internal resistance of the source has been neglected in these discussions. This resistance adds to the load-resistance and always produces a broader bandwidth than that of the LC circuit alone.

ANSWERS TO CHECK-UP QUIZ 35-2

1. (No) **2.** (Audible frequencies range from 20 to 20,000 Hz) **3.** (X_L/R, E_X/E_S) **4.** (Yes) **5.** (Frequency of resonance) **6.** (Capacitive) (Inductive) (Resistive) **7.** (Higher) (1414 Hz) (3.54) (35.4 V) **8.** (Increases bandwidth) (Inversely) **9.** (No)

35-4 REACTANCE CURVES NEAR RESONANCE

As a signal generator is swept across a series LC circuit through the resonant frequency, the circuit appears as different types of loads. At very low frequencies the inductor has almost no X_L and the capacitor has high X_C. To the source the circuit appears as a capacitor with current leading almost 90° (assuming R is small). The current is essentially $I = E/X_C$. The phase-angle curve is shown in Fig. 35-5.

As f_0 is approached, the phase angle

decreases. At the 0.707-of-maximum voltage point, the phase angle is 45°, the circuit is still capacitive, but the resistive component now equals the reactive component.

At f_o the reactances cancel, leaving only resistance, and a θ of 0°.

Above resonance X_L is greater than X_C, and the source sees the circuit as an inductor with a lagging phase angle (lagging current). Far above resonance the phase angle becomes almost 90°. The current value is then essentially $I = E/X_L$. If the resonant circuit has a high-Q value, it becomes highly reactive very rapidly, as shown.

35-5 THE *L/C* RATIO

Any L and C in series have one frequency of resonance at which X_L equals X_C. If either the inductance or the capacitance is increased, the frequency of resonance *decreases*. This can be seen from an examination of the expanded resonance formula,

$$X_L = X_C$$

$$2\pi fL = \frac{1}{2\pi fC}$$

It is possible to rearrange the factors of this basic formula algebraically to solve for the inductance to be connected to a known capacitance in order to make the resulting LC circuit resonant to a desired frequency; thus

$$L = \frac{1}{4\pi^2 f^2 C}$$

where L is in henrys, C is in farads, and f is in hertz.

To solve for the desired capacitance if L is known,

$$C = \frac{1}{4\pi^2 f^2 L}$$

If the L and C values are known, the frequency at which they will resonate can be determined by another rearrangement,

$$f^2 = \frac{1}{4\pi^2 LC}$$

$$f = \frac{1}{2\pi\sqrt{LC}} = \frac{0.159}{\sqrt{LC}}$$

This formula shows frequency to be inversely proportional to both L and C.

Although it may be a simple thing to produce a resonant circuit at any reasonable frequency, there is always some one inductance and some one capacitance that will produce the best results for a given set of circuit requirements.

If a 5-μH coil is on hand for use in a 10-MHz resonant circuit, the required C is about 50 pF. The X_L and X_C are each about 300 Ω. If 6 Ω of effective resistance is assumed, the Q of the circuit is $Q = X/R$, or 300/6, or 50. The bandwidth of the circuit is $BW = f_o/Q$, or 10,000,000/50, or 200 kHz.

If a 20-μH coil is substituted, the required C is only 12.5 pF. The reactances are now about 1250 Ω each. Since inductance increases nearly as the square of the number of turns, a fourfold increase in inductance represents about twice as many turns and therefore about twice as much effective resistance. The Q is then $Q = X/R$, or 1250/12, or 104. (Generally, for a given frequency of resonance, the greater the inductance value, the higher the Q of the circuit.) The bandwidth of the circuit with a Q of 104 is $BW = f_o/Q$, or 10,000,000/104, or 91 kHz.

To pass a wide band of frequencies at a center frequency of 10 MHz, it would be better to use the low-Q 5-μH coil and a 50-pF capacitor. If only a narrow band of frequencies is to be passed, it would be better to use the high-Q circuit.

A quick and fairly accurate method of solving for *L*, *C*, *X*, and *f* in resonant circuits is to use the *LCXf* charts in Appendix D. The frequency of resonance can be determined from a given *L* and a given *C* by means of a straightedge. The *L* value can be determined from a given *f* and a given *C*. The reactance of the *L* and *C* at resonance is also shown. Use the charts to check the values of *L* and *C* given in the discussion above.

Quiz 35-3. Test your understanding. Answer these check-up questions.

1. Which would pass energy at the resonant frequency with less loss to a load, a high-Q or a low-Q resonant circuit? ________
2. What are three methods of expressing the point on the universal resonance curve at which the bandwidth is to be measured? ________ ________ ________
3. What would be the bandwidth of the circuit in Fig. 35-3 if both reactances were 5000 Ω at 2000 Hz? ________ 1 MHz? ________
4. The bandwidth of a circuit is measured to be 45 kHz at a frequency of 3 MHz. What is the Q value of the circuit? ________
5. How does decreasing the impedance of the generator affect the effective Q value of a resonant circuit connected across it? ________
6. If the current through a load is proportional to E/X_L, is the source at f_o, far above f_o, or far below f_o? ________
7. A resonant circuit is connected across a source. If the capacitance is reduced to a lower value, will the resulting circuit appear to the source as capacitive, inductive, or resistive? ________ Why? ________
8. Compute the value of *L* required to resonate a 0.2-μF capacitor to 500 Hz. ________ What is the value by *LCXf* chart? ________
9. Compute the capacitor value required to resonate a 15-μH coil to 4 MHz. ________ What is the value by *LCXf* chart? ________
10. What is the effect on the resonant frequency if the *C* remains constant and *L* is doubled? ________ *L* is quadrupled? ________ Both *C* and *L* are halved? ________
11. Increasing the inductance value fourfold requires what change in *C* to retain the same oscillation frequency? ________ What effect does this have on the circuit Q? ________ The bandwidth of the circuit? ________

35-6 LOGARITHMS AND DECIBELS

When the ratio of one power to another is measured, it is common to use *decibels* (dB). If an amplifier can amplify a 1-W input to a 10-W output, it has increased the signal power 10 times, or by 1 *bel* (B). This is usually stated as 10 decibels ("deci-" means one-tenth) or 10 dB. The same amplifier will increase a 0.5-W input signal to a 5-W output, still a gain of 10 dB. If the volume control on the amplifier is turned up, it may be possible to increase the 0.5-W input to 50-W output, an increase of another 10 dB, or a total of 20 dB. This hundredfold increase is a gain of two 10-times gains, or 20 dB. This is a *logarithmic* increase. For those not familiar with logarithms, a brief explanation follows.

In the equation $10^2 = 100$, the 2 is the *exponent*, the 10 is called the *base*, and 100 is the *number*. The exponent 2 can also be called the *logarithm* of the number 100 to the base 10. This is written $\log_{10} 100 = 2$, and is read "the logarithm to the base 10 of the number 100 is 2." Since the system of logarithms in general use employs a base of 10, it is customary to say "the log of 100 is 2." The equation $10^3 = 1000$ may be expressed as $\log 1000 = 3$.

Since $\log 100 = 2$ and $\log 1000 = 3$, it can be reasoned that the logarithm of any number between 100 and 1000 will have to be some number between 2 and 3, that is, 2 plus some decimal fraction. For example, log 500, from the log tables in Appendix F is 2.6990. Therefore log 500 = 2.6990. This means $500 = 10^{2.699}$.

A logarithm such as 2.6990 has two parts: the whole number is the *characteristic*, and the decimal fraction is the *mantissa*.

The characteristic is determined by finding between which $\log_{10}$ the number falls (Table 35-3). For numbers greater than 1,

Table 35-3

Range of numbers	Characteristic
0.001–0.009999	−3
0.01–0.09999	−2
0.1–0.9999	−1
1–9.999	0
10–99.99	1
100–999.9	2
1000–9999.	3

the characteristic is 1 less than the number of whole digits. The number 435 has a characteristic of 2. The number 86 has a characteristic of 1. For numbers less than 1, the characteristic is *negative* and is 1 more than the number of zeros between the decimal point and the first significant figure. The number 0.05 has a characteristic of −2. The characteristic of 0.0081 is −3, and so on.

The mantissa, or decimal-fraction part of the logarithm, is found in a table of common logarithms (Appendix F). The characteristic is determined by observation. What is the log of 8450? The characteristic is 1 less than the four digits in the number, or 3. Therefore log 8450 = 3.+. The mantissa, found on the 84 line of the tables under the 5 column, is 9269. Thus log 8450 = 3.9269.

For the following logarithms, check the characteristic by observation and the mantissa by table, slide rule, or calculator.

$$\log 23 = 1.3617$$

$$\log 15{,}500 = 4.1903$$

$$\log 629 = 2.7987$$

Although log 0.28 is 4472 and −1, negative characteristics are not used. Instead, a + characteristic equal to it is used,

$$\log 0.28 = -1.4472 = 9.4472 - 10$$

Similarly,

$$\log 0.00862 = -3.9355 = 7.9355 - 10$$

The number of decibels of change between two power values can be computed by

$$dB = 10 \log_{10} \frac{P_2}{P_1}$$

This formula is derived from the fact that 1 B is equal to the logarithm of the ratio of two power values, or bel = $\log(P_2/P_1)$. Since a decibel is one-tenth of a bel, the formula for decibels must multiply the logarithmic value by 10. (Use the smaller power value as P_1 and the larger as P_2.)

What is the gain in decibels of an amplifier if it produces 40 W output with an input of 0.016 W? By formula,

$$\begin{aligned} dB &= 10 \log \frac{40}{0.016} \\ &= 10 \log 2500 \\ &= 10(3.3979) \\ &= 33.9 \text{ dB} \end{aligned}$$

The same formula can be used in the following type of problem: How much output power will be produced by an amplifier capable of 25-dB gain if it is fed an input of 0.001 W?

$$dB = 10 \log \frac{P_2}{P_1}$$

$$25 = 10 \log \frac{P}{0.001}$$

$$2.5 = \log \frac{P}{0.001}$$

The log of the number $P/0.001$ is 2.5, or

$$\log \frac{P}{0.001} = 2.5000$$

where 5000 is the mantissa. By searching through the mantissas in the table (or on the L scale on a slide rule) 5000 is found in the 31 line and 6 column. The 2 is the characteristic and indicates that there must be three whole numbers. Therefore

$$\frac{P}{0.001} = 316.0$$

$$P = 316(0.001)$$

$$= 0.316 \text{ W}$$

For exact decibel computations the logarithmic formulas should be used. However, a fairly accurate calculation can be achieved by applying one or more of the following ratios, particularly when logarithmic tables or slide rules are not available:

1 dB = a power gain of 1.26 (26%)

3 dB = a power gain of 2

6 dB = a power gain of 4

10 dB = a power gain of 10

20 dB = a power gain of 100

30 dB = a power gain of 1000

The power ratios given above are stated as gains, but they may also be stated as losses. For example, −3 dB indicates a loss to half-power, −6 dB a loss to quarter-power, etc.

The 1-, 3-, 6-, and 10-dB gains can be used in the problem above concerning the amplifier with 0.001-W input and 25-dB gain. Since 10 times the power equals 10 dB, the 0.001-W input increased by 10 dB represents 0.01 W. A second 10-dB increase (a total of 20 dB) represents 0.1 W. The remaining 5 dB can be rationalized: 3 dB of this represents twice the power of 20 dB, or 0.2 W. One more decibel (a total of 24 dB) represents 0.2 plus 26% of 0.2, or 0.2 + 0.052, or 0.252 W. One more decibel (a total of 25 dB) represents 0.252 + 0.26(0.252), or 0.317 W. Compare this with the 0.316 W obtained by logarithmic computations.

A less accurate approximation of the last 5 dB would be to consider 5 dB as slightly less than 6 dB, or slightly less than 4 times the 20-dB power value, or less than 0.4 W.

The decibel is used in many cases as a unit of measurement in systems in which power decreases to a threshold value before it reaches absolute zero, as in sound. What sound, for example, represents zero decibel? Some people hear better than others. A sound at the zero level for one person may be quite audible to others. It is necessary to set an arbitrary level and call this zero decibel. Values above this are called +dB, and values below it are termed −dB.

ANSWERS TO CHECK-UP QUIZ 35-3

1. (High Q) **2.** (0.707-of-max E or I, −3 dB, half-power) **3.** (400 Hz) (200,000 Hz) **4.** (66.7) **5.** (Increases it) **6.** (Far above) **7.** (Capacitive) (Source sees more X_C) **8.** (0.5 H) (0.5 H) **9.** (100 pF) (100 pF) **10.** (0.707 of previous value) (0.5 of previous value) (Frequency is doubled) **11.** (Reduction to 0.25 previous value) (Doubles it) (Halves it)

In general, 1 mW (0.001 W) is accepted as the standard reference point. The term *dBm* signifies decibel with a reference-zero value of 1 mW. The *volume unit* (VU) is another logarithmic unit using 1 mW as the zero reference, but the impedance of the circuit must be 600 Ω.

Basically, decibels are ratios of two powers. However, since power is proportional to both voltage ($P = E^2/R$) and current ($P = I^2R$), it is possible to compute decibel ratios between input and output voltages or between input and output currents. The basic formulas are

$$dB = 10 \log \frac{E_o^2/R_o}{E_i^2/R_i} \quad \text{or} \quad = 10 \log \frac{I_o^2R_o}{I_i^2R_i}$$

where E_i and I_i are input values and E_o and I_o are output values.

If the input and output resistance or impedance values are equal, these values cancel in the formulas, which become

$$dB = 10 \log \frac{E_o^2}{E_i^2}$$

$$dB = 10 \log \frac{I_o^2}{I_i^2}$$

Instead of using E^2 or I^2 values, the formulas can be simplified to

$$dB = 20 \log \frac{E_o}{E_i}$$

$$dB = 20 \log \frac{I_o}{I_i}$$

The voltage across a 500-Ω input circuit of an amplifier is 6 V. The output is 30 V across a 500-Ω load. What is the amplifier gain?

$$dB = 20 \log \frac{30}{6}$$

$$= 20 \log 5$$

$$= 20(0.6990)$$

$$= 13.98 \text{ dB}$$

Quiz 35-4. Test your understanding. Answer these check-up questions.

1. What is the base-10 logarithm of 420? ________ 27? ________ 42,600? ________ 0.135? ________ 0.00423? ________
2. An amplifier has a 53-dB gain and a 10-W output. What is the input power? ________
3. A 2-W signal is fed to a 600-Ω resistive network. How many decibels must it lose to have an output of 0 VU? ________
4. A microphone output is −65 dBm. How many decibels of gain must an amplifier have to produce 0.001 W? ________ 10 W? ________ 18 W? ________
5. A circuit is fed a 2-mV signal and has a 0.025-V output. What decibel gain does it have if the input and output impedances are equal? ________
6. A transistor circuit with an input ac signal of 0.0045 A is capable of a 35-dB gain. What is the output current if the input and output impedances are equal? ________
7. Is a decibel gain possible in a step-up transformer? ________ Is a decibel loss possible? ________
8. A 5-V signal is fed to a circuit with a 200-Ω input. What is the decibel gain or loss of the circuit if the output voltage is 20 V across a 10,000-Ω load? ________
9. The output of an amplifier stage having a voltage gain of 30 dB is 25 V. What is the input-voltage level? ________
10. The bandwidth of a resonant circuit is measured at points 0.707 of maximum current. What decibel drop does this represent? ________ What is the power value at these points in relation to the power at the resonant frequency? ________
11. If the bandwidth is measured at points −6 dB down from maximum, what is the relative power value at these points? ________ The relative voltage value at these points? ________
12. If the bandwidth is measured at points 30 dB down from maximum, what is the relative power value? ________ The relative voltage value? ________

CHAPTER 35 TEST • RESONANT CIRCUITS

1. A resonant circuit has an X_C of 1300 Ω to a frequency of 80 Hz, has 1.5 A flowing in it, and has 38 Ω of resistance. What is the value of the capacitance? The circuit Q? What is the bandwidth of the circuit? The voltage-drop across the coil?
2. Is a series circuit resistive, inductive, or capacitive below the frequency of resonance? Above resonance? At resonance?
3. In a series ac circuit, does the voltage across the capacitor appear greatest at a frequency below, above, or at resonance?
4. Is the impedance of a series circuit greater above resonance or at resonance?
5. What frequencies should be generated by an audio-frequency signal generator?
6. What effect does increasing the Q of a resonant circuit have on the bandwidth? Current? Voltage-drop across the capacitor?
7. In a resonant circuit, what can be determined from the ratio of the voltage-drop across a coil to the source voltage?
8. If the value of capacitance in a resonant circuit is doubled, what will be the new resonant frequency?
9. What is determined by the ratio of the resonant frequency to the Q of the circuit?
10. What is the phase angle of a series ac circuit at resonance? Below resonance?
11. What is the resonant frequency when a 50-mH coil and a 2-μF capacitor are in series?
12. What value of inductance must be connected in series with a 0.0004-μF capacitor for the circuit to resonate at 810 kHz?
13. What value of capacitance must be connected in series with a 15-μH coil for the circuit to resonate at 4 MHz?
14. Which would result in a higher-Q circuit at 5 MHz, a 5-μH coil or a 25-μH coil?
15. The bandwidth of a circuit is measured to be 300 kHz at a frequency of 10.7 MHz. What is the Q of the circuit?
16. If the current through a load is proportional to E/X_C, is the source at f_o, far above f_o, or far below f_o?
17. An amplifier has 600-Ω input and is fed a 0.0002-W signal. The output power is 0.5 W in a 600-Ω load. How many decibels of gain does the amplifier have?
18. A 5-V signal across a 100,000-Ω input to an amplifier produces 4 V across an 8-Ω load in the output. What is the gain of the amplifier in decibels?
19. How much output power will be developed into a 50-Ω load by an amplifier having a 24-dB gain if 5 mW is fed into its 50-Ω input circuit?
20. A microwave antenna has a gain of 38 dB. How much effective radiated power does it emit when it is fed 0.75 W?

ANSWERS TO CHECK-UP QUIZ 35-4

1. (2.6232) (1.4314) (4.6294) (9.1303 − 10) (7.6263 − 10) **2.** (0.00005 W) **3.** (−33 dB) **4.** (65 dB) (105 dB) (107.5 dB) **5.** (21.9 dB) **6.** (0.253 A) **7.** (No) (Yes) **8.** (−4.95 dB) **9.** (0.791 V) **10.** (−3 dB) (50% of max) **11.** (0.25 of max) (0.5 of max) **12.** (0.001 of max) (1/31.6 or 0.0316 of max)

---

36 ANTIRESONANT CIRCUITS

CHAPTER OBJECTIVE. To investigate antiresonant (or parallel-resonant) circuits for flywheel effect, damping, impedance, bandwidth, Q, phase angle, effect of coupling, use in oscillators and wavetraps.

36-1 THE ANTIRESONANT CIRCUIT

When a coil and capacitor are connected in *series,* there will be some frequency at which X_L equals X_C and the circuit is resonant. At this frequency the circuit will be purely resistive, a maximum current will flow, and a maximum voltage-drop will occur across the reactances. This is a resonant, or series-resonant, circuit. When the same coil and capacitor are connected in parallel, the two reactances are equal at the same frequency, the circuit should be purely resistive, maximum current flows in the LC circuit, and maximum voltage appears across the reactances. Such an LC combination is an *antiresonant,* or *parallel-resonant* circuit.

Figure 36-1 represents a high-Q antiresonant circuit across a source that is producing ac at the antiresonant frequency.

When the switch is closed, the currents in the circuit go through highly complex *transient* variations before the circuit settles down to a normal *steady-state* operation. The higher the Q of the LC circuit, the longer it takes to attain the steady operational state.

Initially, as the switch closes, current must flow into both the capacitor and the coil to produce currents in both branches of the circuit. The current must lead by 90° in the capacitive branch, and in the inductive branch it must lag by 90°. Therefore when steady-state operation is reached, the current flowing upward in the capacitor will equal the current flowing downward in the inductor at any given instant.

If each branch has 500 Ω of reactance and is across a 500-V source, the maximum current that can flow in a branch will be 1 A. The two ammeters indicating I_C and I_L will each read 1 A. The voltage across the parallel circuit will equal the

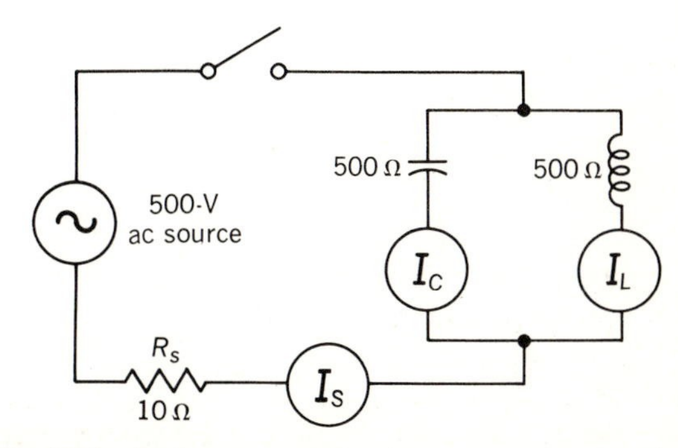

Fig. 36-1 Antiresonant circuit across a source of ac. R_S represents the resistance in the source and in the connecting wires.

source voltage of 500 V. With no difference in voltage between the source and the load, the source current read by the line meter will be zero.

The line meter will indicate current only as the switch is first closed and the LC circuit is settling down. If it has a high Q, the LC circuit does not lose energy, and its drain on the source is negligible.

If there were no losses in the LC circuit, the switch could now be opened and the current in the LC circuit would continue to oscillate forever. However, there are always losses in any circuit, such as wire resistance, dielectric losses, losses due to the energy requirements of meters, iron cores have hysteretic and eddy-current losses, and the loss of energy that is radiated into space. As a result, when the switch is opened, the ac wave in the LC *tank* circuit begins to die out. If the losses are great, the wave may *damp out* in a cycle or less (Fig. 36-2). If the losses are low, the ac may not damp out for many cycles. The *decrement* (rate of damping) follows a logarithmic curve. The tendency of an LC tank circuit to produce oscillations of current is called the *flywheel effect*. At supercold (cryogenic) temperatures (−250°C or colder) conductors have almost no resistance. A supercold LC tank started into oscillation in a well-shielded container may oscillate for several days.

In an antiresonant circuit the voltage-drop across the reactors will not exceed the emf value of the source.

The maximum current flow in a zero-resistance antiresonant circuit is only $I = E_S/X_L$, or $I = E_S/X_C$. (In a *resonant* circuit with no resistance the current may become dangerously high.) If there is any resistance in either reactive branch of an antiresonant circuit, the oscillating current in the LC circuit will be less than the value of $I = E/X$.

A zero-resistance antiresonant LC circuit has the same voltage value across both reactances at any given instant. The current in the capacitor will lead the circuit voltage by 90°, and at the same instant the current in the inductor will lag the circuit voltage by 90°. When current is flowing downward in the capacitive leg, a current of equal value is flowing upward in the inductive leg. It appears to the load as a no-current-drain circuit and therefore as having infinite impedance. The source ammeter would read zero current. Since there is no source current, there can be no phase angle. The source sees the LC circuit as a purely resistive circuit, with infinite resistance. For antiresonance—the

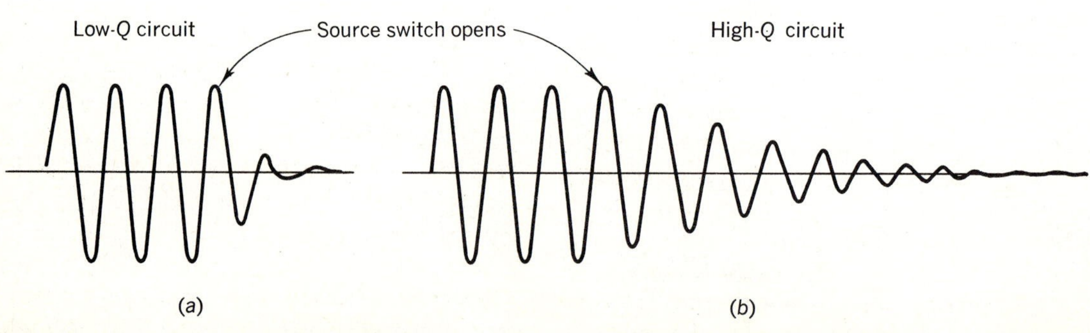

Fig. 36-2 Damped oscillations in antiresonant circuits. (*a*) A low-Q circuit damps rapidly. (*b*) A high-Q circuit damps slowly and has a lower value of logarithmic decrement.

circuit appears purely resistive, $X_L = X_C$, zero current flows in the source, and equal currents flow 180° out of phase in the reactances.

Quiz 36-1. Test your understanding. Answer these check-up questions.

1. If the X_L and the X_C are equal and R is zero, is the circuit always resistive if the elements are in series? ______ In parallel? ______
2. How are X_L and X_C connected in an antiresonant circuit? ______ A resonant circuit? ______
3. When a source is connected across an antiresonant circuit by a line with series resistance, does maximum tank current flow after the first cycle? ______ If an ammeter were in series with the line, how would it react as the circuit was connected? ______
4. A 100-Ω X_L and a 100-Ω X_C are in parallel with a 100-V source having 50 Ω of internal resistance. What is the current value in the X_L? ______ X_C? ______ Source? ______
5. In what two forms does a tank circuit store energy? ______ ______
6. What is the name given to the waveform of the ac produced in an antiresonant circuit after it has been disconnected from the source? ______
7. What prevents an LC tank circuit from being a perpetual-motion device? ______
8. What is meant by the decrement of an oscillatory current? ______
9. Can the voltage across one of the reactors exceed the source voltage in a resonant circuit? ______ An antiresonant circuit? ______
10. What formula is used to find the maximum current that can flow in a reactor in a parallel-resonant circuit? ______
11. Which is 180° out of phase in a resonant circuit, the E or the I? ______ In an antiresonant circuit? ______
12. Which is considered as being the same for both reactances in a resonant circuit, the E or the I? ______ In an antiresonant circuit? ______
13. If the X_C is varied across the antiresonant frequency, which would increase the most, E_S, I_S, E_X, or I_X? ______ Which would dip down? ______
14. If the X_C of a series-resonant circuit is varied across the frequency of a signal generator, which would increase, E_{X_L}, E_{X_C}, I_{X_L}, I_{X_C}, or I_S? ______
15. Why would a cryogenic LC circuit have low decrement? ______

36-2 LOADED ANTIRESONANT CIRCUITS

The circuit of Fig. 36-3 is a more practical antiresonant circuit. There is always

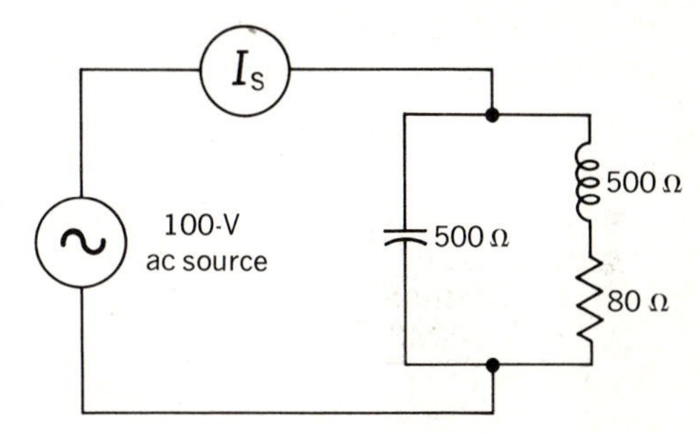

Fig. 36-3 Loaded antiresonant circuit appears to be a series-parallel LCR circuit and can be computed as such.

some resistance in the coil branch. Energy taken from such a circuit is often coupled inductively from the coil. This affects the LC circuit as an added resistance in series with the inductive branch. The tighter the coupling, the greater the apparent losses in the coil branch.

With resistance in one branch of the circuit, $X_L = X_C$ still, but the current in this leg will be less than that flowing in the other leg. A balanced flywheel effect no longer exists. The source must now deliver current to make up the I^2R energy loss when current flows through the resistance. Thus, addition of resistance in a branch of the circuit requires current flow from the source. To balance the current values in the two branches of Fig. 36-3, it is necessary to reduce the *frequency* of the source ac. This will reduce the X_L, increasing I_{X_L}. Decreasing f increases X_C, and I_{X_C} decreases. The currents in the two legs will now be equal at some frequency slightly lower than where $X_L = X_C$. This is a second possible condition of antiresonance—equal currents will flow in the two branches, even though X_L no longer equals X_C. The circuit is no longer resistive, it is reactive.

The circuit shown in Fig. 36-3 may be treated as a series-parallel ac circuit. Impedance can be determined by $Z_t = 1/Y_t$. It is first necessary to solve for the total admittance Y_t (Sec. 32-3)

$$Y_t = \sqrt{G_t^2 + B_t^2}$$
$$Y_t = \sqrt{(G_L + G_C)^2 + (B_L - B_C)^2}$$
$$= \sqrt{\left(\frac{R}{Z_L^2} + \frac{R}{Z_C^2}\right)^2 + \left(\frac{X_L}{Z_L^2} - \frac{X_C}{Z_C^2}\right)^2}$$
$$= \sqrt{(0.000312 + 0)^2 + (0.00195 - 0.002)^2}$$
$$= \sqrt{0.000312^2 + (-0.00005)^2}$$
$$= \sqrt{0.00\ 00\ 00\ 09\ 73 + 0.00\ 00\ 00\ 00\ 25}$$
$$= \sqrt{0.00\ 00\ 00\ 10}$$
$$= 0.000316 \text{ mho}$$

Hence

$$Z = \frac{1}{0.000316} = 3160\ \Omega$$

Another method of determining the Z of an antiresonant circuit is with complex numbers (Sec. 33-8).

A simpler method of determining the Z of an *antiresonant* circuit is by the formula, $Z_P = Z_L Z_C/(Z_L - Z_C)$. The branches are first converted to their branch-impedance values. For the capacitive branch, $-Z_C = \sqrt{R^2 + X_C^2}$, and for the inductive branch, $Z_L = \sqrt{R^2 + X_L^2}$. The product of these two impedances, $Z_L(-Z_C)$, given the sign of the greater and divided by the *series* sum of the two impedances, $(Z_L) + (-Z_C)$, or $Z_L - Z_C$, will give the Z_t of this circuit.

Since the series sum of the two impedances at resonance is subtracting X_C from X_L, these two values cancel, leaving only the resistances of the two branches *in series*, R_{LC}. The formula is then

$$Z_P = \frac{Z_L Z_C}{R_{LC}}$$

where R_{LC} is the sum of the resistances in the two branches if considered in series, and the reactive Z values are the impedances of the branches.

For the circuit in Fig. 36-3 the impedance is

$$Z_P = \frac{506(500)}{80} = 3160\ \Omega$$

If the reactance in each leg is more than 10 times the resistance, a simpler formula is:

$$Z_P = \frac{X^2}{R_{LC}}$$

where X is either reactance value, and R_{LC} is the total resistance if the two branches were in series. For the circuit in Fig. 36-3, in which X is only 6 times R_{LC}, $Z_P = 250{,}000/80$, or 3125 Ω.

ANSWERS TO CHECK-UP QUIZ 36-1

1. (Yes) (Yes) **2.** (Parallel) (Series) **3.** (No) (High starting impulse would drop to low flow) **4.** (1 A) (1 A) (Zero) **5.** (Electrostatic) (Electromagnetic) **6.** (Damped) **7.** (R losses) **8.** (How fast a wave dies out) **9.** (Yes) (No) **10.** ($I_X = E_S/X$) **11.** (E) (I) **12.** (I) (E) **13.** (E_X, I_X) (I_S) **14.** (All) **15.** (No R)

Quiz 36-2. Test your understanding. Answer these check-up questions.

1. When energy is extracted from an antiresonant circuit, is it most likely to be taken in electromagnetic or electrostatic form? __________

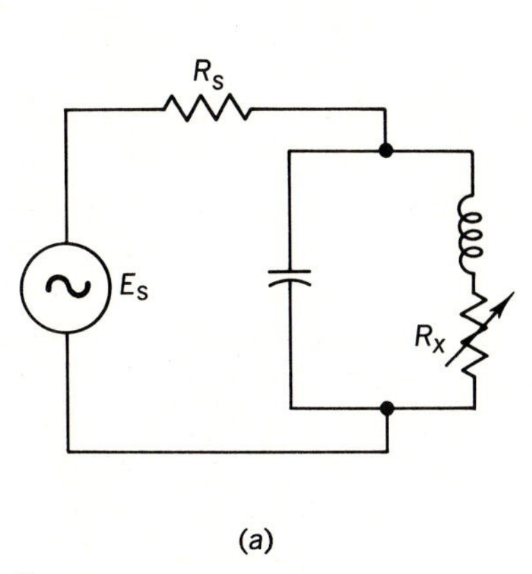

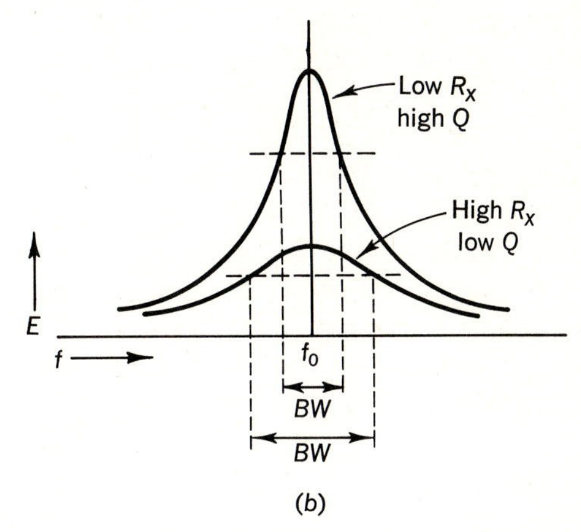

Fig. 36-4 An antiresonant circuit across a source. The Q can be changed by varying the R_X. (b) Voltage-response curves of a circuit with high- and low-Q values.

2. An antiresonant circuit has a 100-Ω X_C with a 20-Ω R in series with it and is across a 100-Ω X_L. What is the value of G_t? ________ B_t? ________ Y_t? ________ Z_t? ________
3. In question 2, what is the Z_t value by the formula $Z_P = Z_L Z_C / R_{LC}$? ________ By the formula $Z_P = X^2/R_{LC}$? ________

36-3 BANDWIDTH AND Q

When the voltage across a parallel LC circuit is plotted against frequency (Fig. 36-4), maximum voltage occurs across the circuit when X_L equals X_C. Even with R_{LC} in series with the LC circuit, the voltage at antiresonance almost equals the source voltage E_S. As the frequency departs from antiresonance, the voltage across the LC circuit decreases. This is the universal resonance curve (Chap. 35).

As in series-resonant circuits, the bandwidth of an antiresonant circuit is the band of frequencies between the two points at which the voltage decreases to 0.707 of maximum (dashed lines in Fig. 36-4b).

Using a low value of series resistance in the inductive leg, the voltage across the LC circuit approaches the source-voltage value at antiresonance. As the voltage across the antiresonant circuit decreases, the voltage-drop across the series line resistance R_S increases. If the resistance in series with the reactance, R_X, is increased, the impedance of the circuit *decreases*. This increases the source current and the voltage-drop across R_S, resulting in less voltage-drop across the LC circuit. Loading an antiresonant circuit results in less voltage-drop across it if there is any resistance in the source or in the line to it.

If the capacitive branch of the antiresonant circuit had no resistance, the Q of this branch would be $Q = X/R$, or $X/0$, or ∞. The Q of the LC circuit would be the Q of the X_L branch, or $Q = X_L R_X$. In Fig. 36-5 the Q would be 1000/25, or 40.

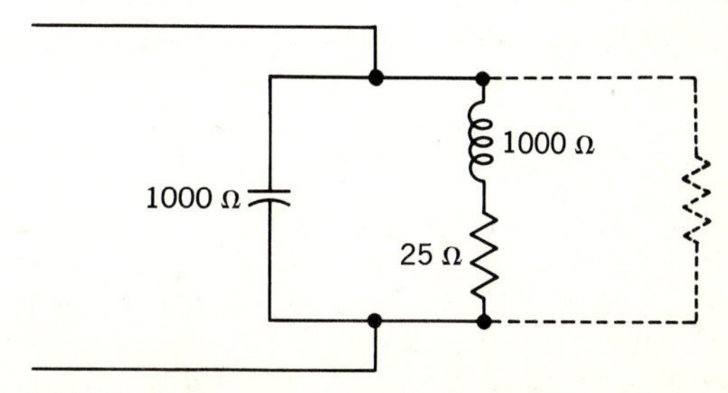

Fig. 36-5 Antiresonant circuit with a Q of 40. Dashed R reduces Q.

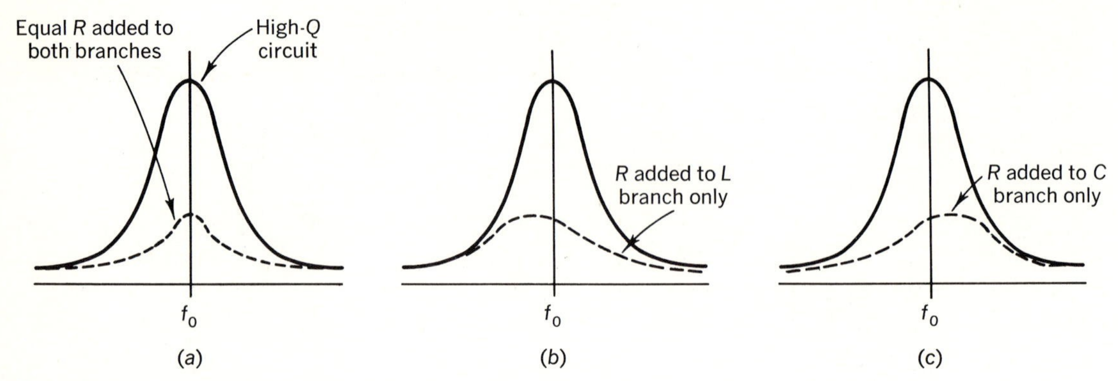

Fig. 36-6 Antiresonant circuit frequency response. Solid lines for high Q. Dashed lines for low Q. (*a*) R equal in both branches. (*b*) R added to L branch. (*c*) R added to C branch.

Sometimes it is desirable to lower the Q of an LC circuit to produce a broader bandwidth. This can be done by increasing the resistance in either or both branches. In Fig. 36-6*a*, if an equal value of resistance is in both legs, the resonant curve (shown dashed) retains the original symmetry. If resistance is added only in the inductive branch (Fig. 36-6*b*), the curve broadens below the frequency of antiresonance, f_o. When resistance predominates in the capacitive branch (Fig. 36-6*c*), the curve broadens above f_o.

To broaden the bandwidth and still maintain symmetry of the pass band, a resistor can be shunted across an antiresonant circuit (shown dashed in Fig. 36-5). This lowers the Q by making the circuit appear to the source to have a greater resistive-current component. High shunt resistance results in high Q and low shunt resistance in low Q. The formula for Q when appreciable shunt loading is used is

$$Q = \frac{R}{X_L} = \frac{R}{X_C}$$

ANSWERS TO CHECK-UP QUIZ 36-2

1. (Magnetic) **2.** (0.00192) (0.00039) (0.00196) (510 Ω) **3.** (510 Ω) (500 Ω)

When the shunt $R = \infty$, the Q is not affected. When $R = X$, the Q is unity, and the circuit is not frequency-selective.

The bandwidth of an antiresonant circuit is

$$BW = \frac{f_o}{Q}$$

Antiresonant-circuit frequency response can be shown by two opposite curves, one rising to a peak according to the universal resonance curve (Fig. 36-7*a*) and the other falling to a minimum at antiresonance as an inverse universal resonance curve (Fig. 36-7*b*). The curve that peaks can represent the voltage that is developed across the LC circuit, the current in the LC circuit, and the impedance across the LC circuit.

The curve that dips to a minimum at antiresonance indicates either the source current or the voltage-drop across any line resistance.

If a variable-frequency source is tuned across the antiresonant frequency of a parallel LC circuit (Fig. 36-8), at frequencies below parallel resonance the X_L value is less, and the circuit appears to the source as an inductance with lagging current. At half-power points, the phase

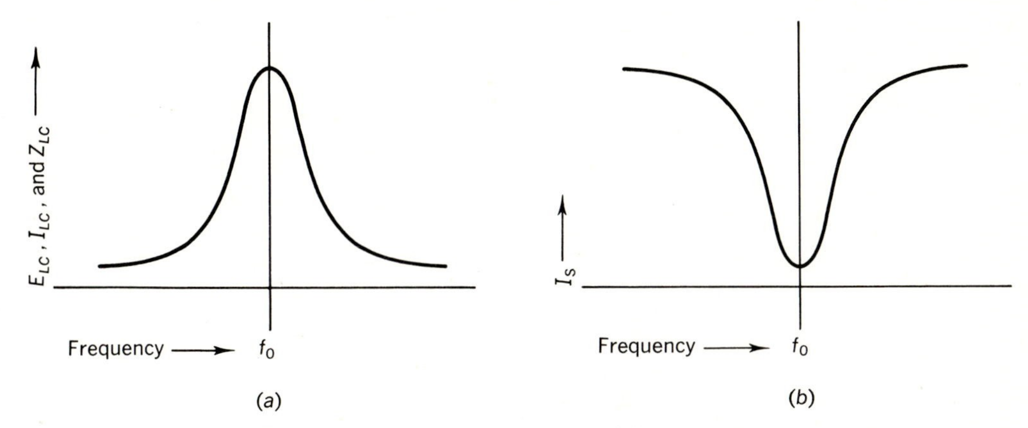

Fig. 36-7 Two response curves considered for antiresonant circuits.

angle is 45°, and the circuit is most reactive. As the source approaches the antiresonant frequency, the circuit becomes predominantly resistive. At antiresonance the circuit is completely resistive. With no resistance in the circuit the impedance value would be infinite. In practical circuits the impedance may be quite high.

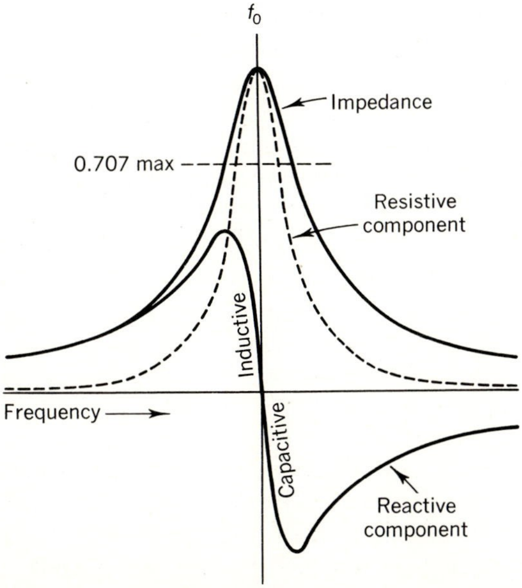

Fig. 36-8 Universal resonance curve of the impedance of an antiresonant circuit, the reactive components, and the resistive component.

At frequencies above antiresonance the reactance becomes capacitive, falling to the value of the resistance in series with the capacitor at frequencies far above antiresonance.

A tuned *LC* circuit with a fixed inductance and a variable capacitor can be made antiresonant to a given source frequency. However, when tuned to the low-frequency side of the signal generator, it appears as a capacitive load; when tuned to the high-frequency side, it appears as an inductive load, just the opposite of a series-resonant circuit.

Quiz 36-3. Test your understanding. Answer these check-up questions.

1. In Fig. 36-4, which would increase the bandwidth more, an increase in R_S or an increase in R_X? ________
2. What are the three ways of expressing the points on a resonance curve at which bandwidth is measured? ________ ________ ________
3. The universal resonance curve represents the way the I_X, I_S, and E_X rise and fall. What does it represent for an antiresonant circuit? ________ ________ ________
4. What would an inverse universal resonance curve represent for a resonant circuit? ________ An antiresonant circuit? ________

5. Which branch of an antiresonant circuit usually determines the Q of the whole circuit? ________ Why? ________
6. What are two ways of broadening the bandwidth of an antiresonant circuit symmetrically? ________ ________ What is a way of broadening it asymmetrically? ________
7. If an antiresonant circuit has a Q of 25 at a frequency of 800 kHz, what is its bandwidth? ________ If the capacitor were doubled in value, what change in frequency would be produced? ________ What would be the bandwidth at this new frequency? ________
8. A resonant circuit is connected across a source of ac. Does it form a resistive, a capacitive, or an inductive load? ________ What kind of load does it appear to be if the source frequency is decreased? ________ Why? ________
9. An antiresonant circuit across a source appears as what kind of load if the source frequency is decreased? ________ Why? ________
10. At what point above or below antiresonance is the reactive effect of the circuit at the greatest value? ________ What does the reactive value equal at this point? ________
11. How would Fig. 36-8 differ if it were illustrating a series-resonant circuit? ________

36-4 COUPLED AND TUNED CIRCUITS

There are three basic types of coupled and tuned circuits: Fig. 36-9*a*, tuned primary and untuned secondary; Fig. 36-9*b*, untuned primary and tuned secondary; Fig. 36-9*c*, both primary and secondary tuned. Tuning is required to make the circuits frequency-selective. In general, the more tuned circuits, the more selective the system.

In Fig. 36-9*a*, the tuned circuit across the source is an antiresonant circuit. The voltage developed in the secondary is determined by the turns ratio of the primary and secondary and the coefficient of coupling. The exact value of the secondary voltage is determined by:

$$E = 2\pi f M I_{\text{pri}}$$

where M is mutual inductance in henrys and I_{pri} is the primary current in amperes. The load resistor demands power from the secondary. The secondary picks up its energy from the primary. The primary demands energy from the source. The lower the resistance of the load and the heavier the load, the greater resistance reflected into the inductive branch of the primary, the lower the Q of the primary, and the broader its bandwidth. To produce reasonable power in the secondary, relatively close coupling between primary and secondary must be maintained. This may result in very broad bandwidth characteristics.

In Fig. 36-9*b*, the primary is untuned but is coupled to a *series-resonant* secondary circuit. The secondary appears to be antiresonant, but is actually a resonant type because each turn has an emf induced

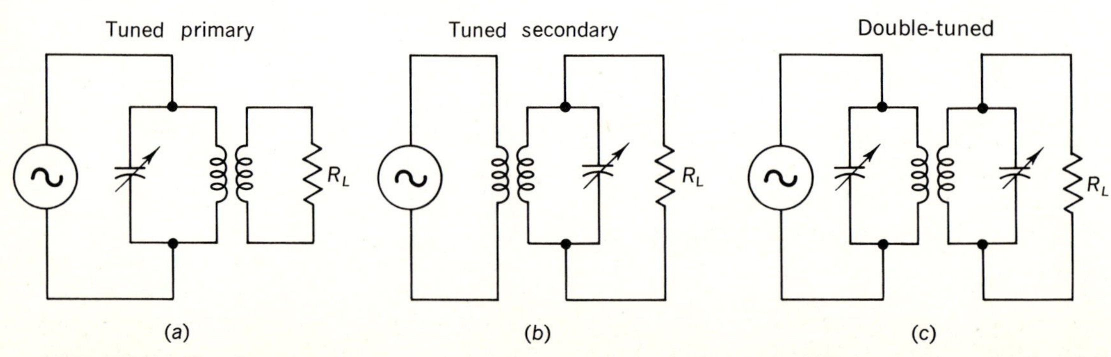

Fig. 36-9 Tuned transformers: (*a*) tuned primary; (*b*) tuned secondary; (*c*) double-tuned.

into it from the field of the primary. This amounts to many induced emf's in series with the turns. Whereas a primary could never produce higher output voltage than the turns ratio between it and an untuned secondary, a series-resonant secondary circuit with high Q can produce a step-up of the resonant-frequency voltage. In series-resonant circuits the $Q = E_X/E_S$ (Chap. 35), where E_X is the voltage-drop across either reactor. E_S is the ("source") voltage induced in the secondary coil by the magnetic fields from the primary. This formula can be rearranged to determine the voltage developed across one of the reactors,

$$E_X = E_S Q$$

If the turns ratio of primary to secondary is 1:1, coupling coefficient is 0.01, and source emf is 100 V, the voltage induced into the secondary is approximately 100(0.01), or 1 V. If the secondary has reactances of 500 Ω and a total series resistance of 10 Ω, the Q would be X/R, or 500/10, or 50. The voltage developed across the capacitor would be

$$E_X = E_S Q = 1(50) = 50 \text{ V}$$

This is a step-down of primary to secondary voltage (100 to 50). However, if the series resistance in the secondary is only 2 Ω, then Q is 500/2, or 250, and the voltage across the capacitor would be 1(250), or 250 V.

If shunt resistance is connected across the secondary, as shown in Fig. 36-9*b*, the Q of the circuit can be reduced to the degree that the step-up advantage no longer exists.

In Fig. 36-9*c*, both primary and secondary are tuned, usually to the same frequency. Not only does this circuit have the series-resonance advantage of voltage step-up, but the primary is also tuned and tends to reject unwanted frequencies. With both circuits reacting best to a desired resonant frequency, other frequencies are less likely to be present in the load. The circuit is a bandpass network, rejecting frequencies outside the bandpass of the primary and secondary circuits.

If the secondary and primary are both tuned to the frequency applied to the primary, the impedance reflected back into the primary by the secondary is purely *resistive*. If the series-resonant secondary is tuned below the input frequency, it represents a capacitive load to the signal fed to it but reflects back into the primary an *inductive* load. If the secondary is inductive, it reflects back into the primary as a *capacitive* load.

36-5 FOUR CONDITIONS OF COUPLING

With a double-tuned transformer tuned to a common frequency of resonance and antiresonance, the coefficient of coupling shapes the frequency response of the secondary circuit. In Fig. 36-10, the arrows

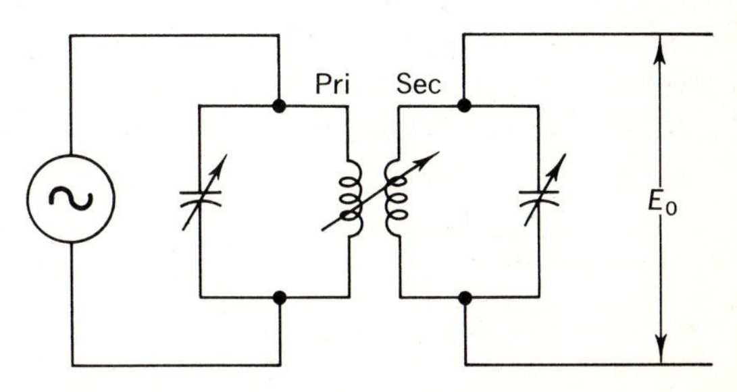

Fig. 36-10 Double-tuned transformer with variable coupling between primary and secondary.

through the capacitors indicate that they are variable, to allow the circuits to be "resonated" at some desired frequency. The arrow between the primary and the

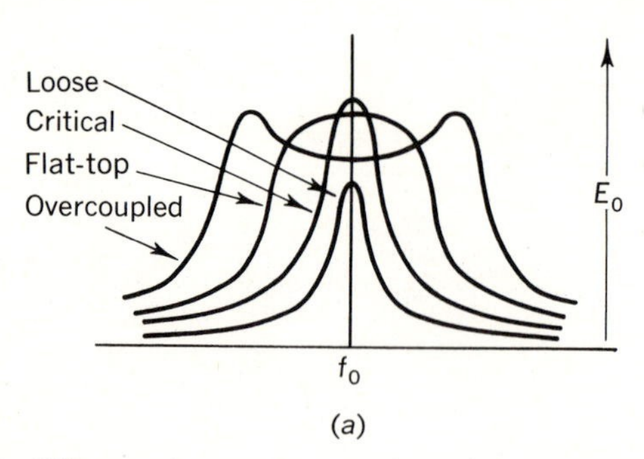

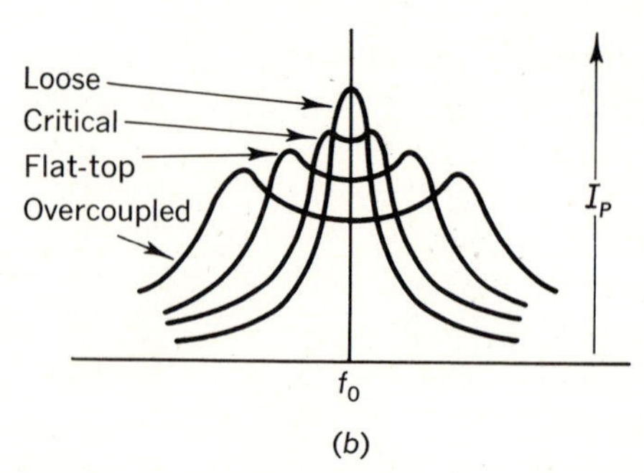

Fig. 36-11 Effects of varying coupling (*a*) an output voltage across secondary, E_o, and (*b*) on ac current flowing in the primary.

secondary indicates that the coupling coefficient *k* between the two coils is also variable.

There are four coefficients of coupling that are of particular interest: (1) If *k* is in the range of 0.005, the coupling might be termed *undercoupling*, or *loose*. (2) If *k* is 0.05 or more, the coupling might be termed *overcoupling*, or *tight*. (3) Somewhere between loose and tight coupling is a degree called *critical* coupling. (4) A degree of coupling between critical and tight may be termed *transitional* or *flat-top* coupling. The degree of coupling determines frequency response, output voltage, primary impedance, and primary current.

Figure 36-11*a* graphs output voltage versus frequency for the four degrees of coupling. In Fig. 36-11*b*, primary current is plotted against frequency for the various degrees of coupling, with equal Q values assumed for both the primary and the secondary.

Loose coupling, or undercoupling, is produced with the primary and secondary coils relatively far apart, resulting in a low coefficient of coupling. Under this condition there is a relatively low voltage output. The bandwidth of the frequency-response curve of the secondary is narrower than the universal resonance curve, because of the two tuned circuits, and is determined by the circuit Q and the coefficient of coupling *k*.

Critical coupling is produced by increasing the coupling until the secondary voltage E_o reaches its highest peak. At this degree the resistance reflected back into the primary by the secondary equals the primary resistance. If critical coupling is exceeded, the E_o frequency-response curve begins to broaden or flatten at the peak. The bandwidth of critical coupling usually represents the best compromise between output voltage and narrow bandwidth. The secondary load is reflected back into the primary as a series resistance at resonance that increases the primary impedance at this frequency, allowing greater circulating current to flow. Off resonance the loading is not as great, resulting in a dip of primary current at the resonant frequency.

ANSWERS TO CHECK-UP QUIZ 36-3

1. (R_X) **2.** (0.707 of max., half-power, −3 dB) **3.** (I_X, E_X, Z) **4.** (Z) (I_S, *E*-drop across line *R*) **5.** (X_L) (Capacitors have little loss) **6.** (Shunt with an *R*, add equal *R* to both legs) (Add *R* to either leg alone) **7.** (32 kHz) (0.707 of original value or 565.6 kHz) (22,624 Hz) **8.** (*R*) (X_C) (X_C is greater) **9.** (X_L) (X_L is less, demanding more *I*) **10.** (0.707 of max.) (Resistive component) **11.** (Peaked curve for *I*, reactive-curve labels reversed)

Transitional, or flat-top, coupling is greater than the critical value. This results in an output-frequency response that permits not only resonant-frequency ac, but frequencies near the resonant frequency to pass equally well. This is because the tightness of coupling causes the circuits to detune each other in opposite directions, not without losing some of the amplitude of the output voltage, however. As coupling is increased, a point will be reached at which the response at f_0 begins to dip. This is the limit of transitional coupling. The primary current drops materially as overcoupling is approached, developing two distinct *shoulders*, or peaks, on opposite sides of resonance.

Overcoupling, or tight coupling, is produced by increasing the coupling past the flat-top degree. This develops current peaks that are lower than the critical value and have a distinct dip at f_0, as shown. This is sometimes called *split tuning*. As coupling is increased further, both the shoulders and the resonant-frequency response decrease until, with extremely tight coupling, there is little or no frequency selectivity. All frequencies near resonance pass equally well, but at a relatively low amplitude. The gain due to the series-resonant secondary is no longer in evidence. When primary and secondary coils are tightly wound together, they are said to be *unity-coupled* and have a very broad frequency response.

By overcoupling the one double-tuned transformer and then flat-top-coupling a second, a very broad flat-top response can be obtained, although this is at the expense of overall output voltage.

In stagger-tuned transformers, the primary is tuned to a frequency just below antiresonance, and the secondary is tuned just above the resonant frequency. The bandwidth of such a circuit can be quite broad.

Quiz 36-4. Test your understanding. Answer these check-up questions.

1. Increasing the coupling of a secondary has what effect on the bandwidth of the primary? ______
2. With a double-tuned transformer, what kind of circuit is the primary? ______ The secondary? ______
3. A double-tuned transformer has 20 V across the primary. If the turns ratio is 1:1 and k is 0.02, what is the approximate voltage across the secondary if it has a Q of 70? ______
4. In a double-tuned transformer, if both the primary and the secondary have the same f_0, does the secondary reflect a resistive, a capacitive, or an inductive load on the primary? ______ What if the secondary is tuned above the input frequency? ______
5. What is another name for the condition known as split tuning? ______ Loose coupling? ______ Flat-top coupling? ______
6. What degree of coupling produces maximum secondary current in a double-tuned transformer with minimal bandwidth? ______
7. What degree of coupling produces maximum primary current in a double-tuned transformer? ______
8. What degree of coupling produces minimum bandwidth in a double-tuned transformer? ______
9. What effect will increased coupling have on the effective Q of a double-tuned transformer? ______
10. In general, what effect on the bandwidth is produced by using low-Q tuned circuits in a double-tuned transformer? ______

36-6 *LC* OSCILLATORS

An application of resonant or antiresonant circuits in electronics is in *LC oscillators*. These ac generating devices utilize the amplifying ability of vacuum tubes or transistors to prevent the dying out of the natural damped wave of an oscillating *LC* tank circuit. When the switch

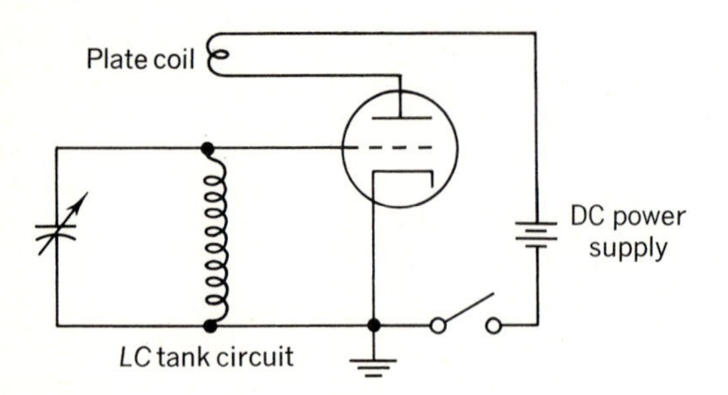

Fig. 36-12 A basic *LC* oscillator capable of generating a sustained ac by feeding back amplified energy to make up for that lost by resistance in the *LC* tank circuit.

is closed in the circuit of Fig. 36-12, current begins to flow through the tube from cathode to plate and through the primary, or plate-circuit coil. A current surge in the primary induces a voltage into the tuned secondary between the grid and the cathode. This starts a weak oscillation in the *LC* tank circuit. The small ac voltage induced between the grid and the cathode produces a large current variation in the plate-circuit coil, which in turn induces a still stronger voltage into the *LC* tank. In this way the current that oscillates in the *LC* circuit cannot damp out, but instead rises to as high a value as the tube can support and maintains this value.

The frequency of oscillation is determined by the resonant frequency of the *LC* tank in the grid-cathode circuit.

It is possible with somewhat similar circuits to generate constant-frequency ac at any frequency from 1 Hz to several GHz (gigahertz).

ANSWERS TO CHECK-UP QUIZ 36-4

1. (Increases) **2.** (Antiresonant)(Resonant) **3.** (70 V) **4.** (Resistive) (Inductive) **5.** (Overcoupled) (Undercoupled) (Transitional) **6.** (Critical) **7.** (Loose) **8.** (Loose) **9.** (Lower) **10.** (Increases it)

36-7 WAVETRAPS

Another application of resonant and antiresonant circuits is wavetraps. They are added in series with a circuit or across the circuit to prevent a specific frequency from being transferred from source to load. Figure 36-13 illustrates the three basic forms of wavetraps.

The first form of wavetrap, WT_1, is inductively coupled to a tuned transformer coupling energy from a multifrequency ac source to a load R_L. If the resonant secondary is tightly coupled to the primary, many different adjacent frequencies may be passed from source to load. If WT_1 has a high Q and is tuned to one specific frequency in the pass band, a heavy current flow will develop in this circuit any time the frequency to which it is tuned appears. The heavy wavetrap current produces a magnetic field at this frequency around its coil. This field cutting across the secondary induces a counterdirection emf in the secondary, neutralizing this one frequency in the secondary. The frequency is said to be "trapped out."

The second wavetrap, WT_2, is an antiresonant circuit in series with one of the lines feeding the load resistor. A high-Q antiresonant circuit develops an almost infinite impedance to its f_o, and almost no current at this frequency can flow through the load resistor. Other adjacent frequencies see the high-Q circuit as having less impedance, and these frequencies can produce current flow in the load.

The third wavetrap form, WT_3, is a high-Q resonant circuit across the load. To any signals at its resonant frequency it appears to have nearly 0-Ω impedance. This frequency in the secondary of the tuned transformer is effectively shorted out before it can reach the load. To trap out one or more frequencies, any one of or

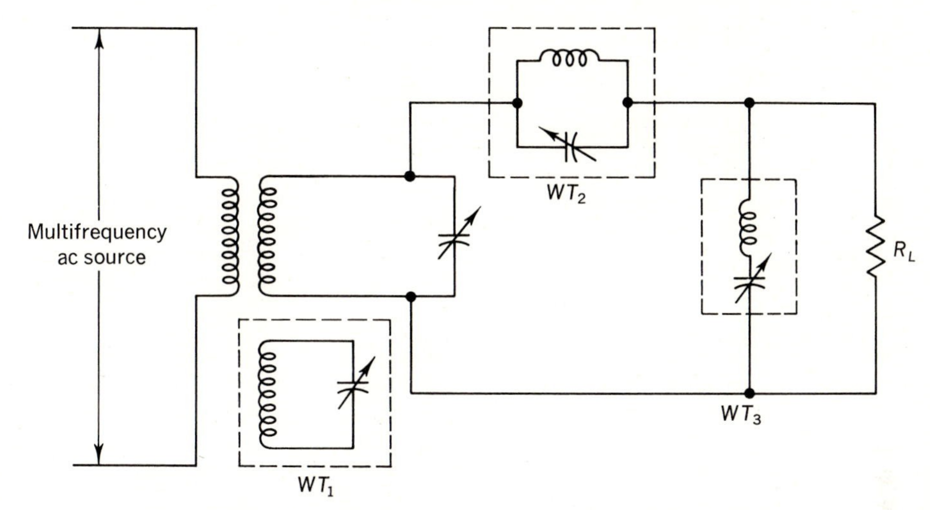

Fig. 36-13 Wavetraps operating on three different theories.

all these three wavetraps may be used at the same time.

Quiz 36-5. Test your understanding. Answer these check-up questions.

1. What is the name of the law that states that when a primary-coil magnetic field induces a current into a secondary coil, the secondary magnetic field will cut back across the primary in such a phase as to cancel the primary field? ________
2. What are the three forms of wavetrap circuits, and how are they coupled or connected in a circuit? ________ ________ ________
3. To be most effective at one frequency, should a wavetrap have a high Q or low Q? ________
4. If a resonant circuit is in series with a load, does it act as a wavetrap for any one frequency? ________
5. If an antiresonant circuit is shunted across a load, is it a wavetrap for any one frequency? ________
6. In the inductively coupled wavetrap, would signals at its resonant frequency produce high, low, or zero voltages across its capacitor? ________

CHAPTER 36 TEST • ANTIRESONANT CIRCUITS

1. Which would have the higher decrement value, a low-Q or a high-Q tank circuit?
2. In what kind of circuit is it possible to have $X_L = X_C$ but not have the circuit purely resistive?
3. What is the name given to the waveform in an antiresonant circuit after the circuit is disconnected from the source?
4. Are E_{X_C} and E_{X_L} in or out of phase in an antiresonant circuit? A resonant circuit?
5. What two things are reduced in electric cryogenic circuits?
6. In the usual tank circuit, which branch would have the greater effective R value?
7. A 2-MHz antiresonant circuit has 800-Ω reactances and 200 Ω of resistance in the inductive leg. What is the circuit Q? Total Y? Total Z? Bandwidth?
8. What is the formula for the circuit Q when an antiresonant circuit has a loading R connected across it?
9. What is the relative line-current value when $X_L = X_C$ in a resonant circuit? An antiresonant circuit?
10. Will a resonant circuit be inductively or capacitively reactive to an ac above its resonant frequency? An antiresonant circuit?

11. What are the three ways of expressing the points on a resonance curve at which bandwidth is measured?
12. What are two ways of broadening the bandwidth of an antiresonant circuit symmetrically?
13. What is one way of broadening the bandwidth of an antiresonant circuit asymmetrically?
14. At what point above and below antiresonance is the reactive effect of the circuit at its greatest value? What does the reactive value equal at this point?
15. If a tuned secondary is resonant to a frequency higher than the frequency in the primary, does it represent an inductive or a capacitive load to the primary?
16. What are the names given to the four conditions of primary-to-secondary coupling of a double-tuned transformer?
17. What degree of coupling gives maximum secondary voltage with minimum bandwidth? Maximum bandwidth?
18. Why is stagger tuning sometimes used in double-tuned transformers?
19. What degree of coupling produces maximum primary ac in a double-tuned transformer?
20. What is a vacuum-tube or transistor ac generator called?
21. To increase the frequency of oscillation of an *LC* tank, what must be done to the *C* value? The *L* value?
22. If an antiresonant circuit is in series with a line, will it act as a wavetrap for its antiresonant frequency?
23. What is the name of the law that states that when a primary-coil magnetic field induces a current into a secondary coil, the secondary magnetic field will cut back across the primary in such a phase as to cancel the primary field?

ANSWERS TO CHECK-UP QUIZ 36-5

1. (Lenz's law) **2.** (Inductively coupled series-resonant circuit) (Antiresonant in series with the line) (Resonant shunting the line) **3.** (High Q) **4.** (No) **5.** (No) **6.** (High)

CHAPTER OBJECTIVE. To investigate the use of *L*, *C*, and *R* in simple low-pass, high-pass, bandpass, and bandstop filter circuits, and the coupling of two or more sections to improve cutoff characteristics. Constant-*k*, *m*-derived, crystal and mechanical-type filters are discussed, as are crossover networks and power-supply filters.

37-1 AN *RC* LOW-PASS FILTER

A *filter* is some arrangement of components in a circuit that alters the ability of the circuit to pass energy of certain frequencies. The purely resistive circuit in Fig. 37-1*a* is an example of a "nonfilter." Regardless of the *frequency* of the source, as long as the output voltage of the signal generator is 100 V, the voltage across the 1000-Ω load resistor will be 50 V. The frequency-response curve for this circuit, as shown above it, is a flat-line curve from 1 to 20,000 Hz and more.

In Fig. 37-1*b*, a 0.08-μF capacitor is added to the line resistance and across the load. At 20 Hz, the capacitor has a capacitive reactance of 100,000 Ω, which has a negligible effect on the 1000-Ω load. As a result, at 20 Hz the voltage-drop across the load is still 50 V.

With an ac of 200 Hz, X_C is 10,000 Ω. An X_C of 10,000 Ω in parallel with a 1000-Ω *R* results in a Z of 995 Ω and nearly 50 V across the load.

At 2000 Hz, X_C is 1000 Ω. In parallel with the load, this represents a Z of 707 Ω and a phase angle of 45°. This circuit in series with the 1000-Ω line resistance represents a total Z of 1600 Ω and about 44 V across the load. The capacitor at this frequency is attenuating, or reducing, the ac voltage across the load.

At 20,000 Hz, X_C drops to 100 Ω. The 1000-Ω R_L in parallel with 100-Ω X_C has a Z of 99.5 Ω. The 1000-Ω line resistance in series with 99.5 Ω produces less than 5 V across the load. The capacitor has reduced the load voltage from 50 to less than 5 V, or more than 20 dB. At higher frequencies the capacitor is still more effective. This is a basic *RC low-pass* filter. The frequency-response curve is shown above the circuit. Only frequencies below about 2000 Hz are satisfactorily passed to the load. At 2000 Hz frequencies begin to be attenuated. This is called the *cutoff frequency* f_c.

Because the voltage delivered to the load at any frequency in the pass band is only half the voltage input, this filter has an *insertion loss* of 6 dB (a 2:1 voltage ratio equals 6 dB).

Since the line resistor and the filter capacitor are connected in an L shape, this is an L-type, 2000-Hz, low-pass *RC* filter.

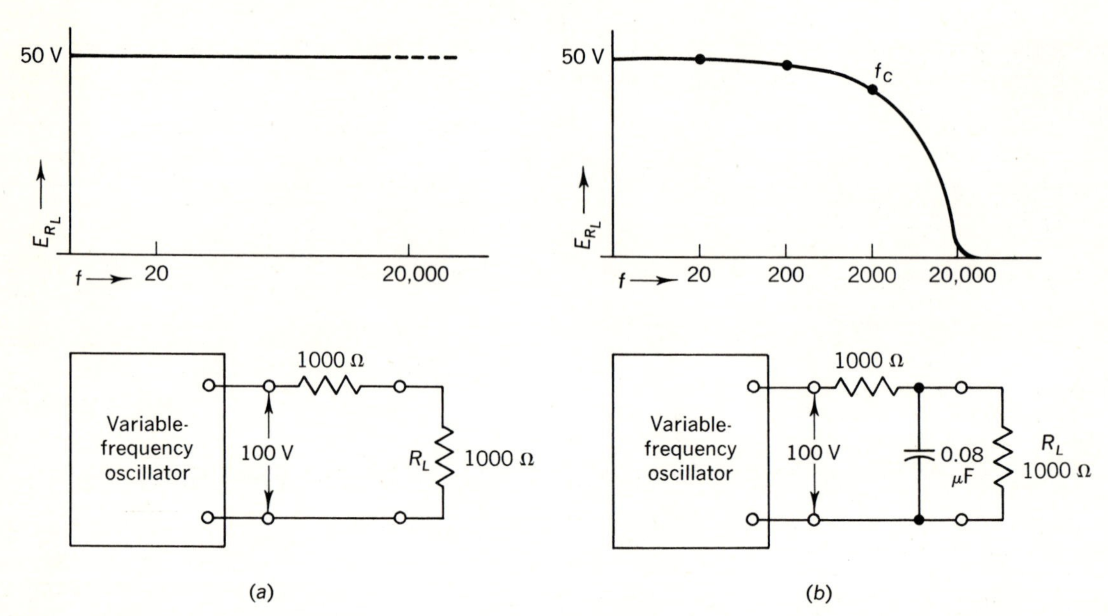

Fig. 37-1 (*a*) The response curve for a completely resistive circuit is a straight line. (*b*) The frequency response across the load resistor when a capacitor is shunted across the load.

37-2 AN *RL* HIGH-PASS FILTER

The circuit shown in Fig. 37-2 has a 0.8-H coil connected across R_L. The voltage that appears across the load as the variable-signal generator is changed from 20 to 20,000 Hz can be determined as before.

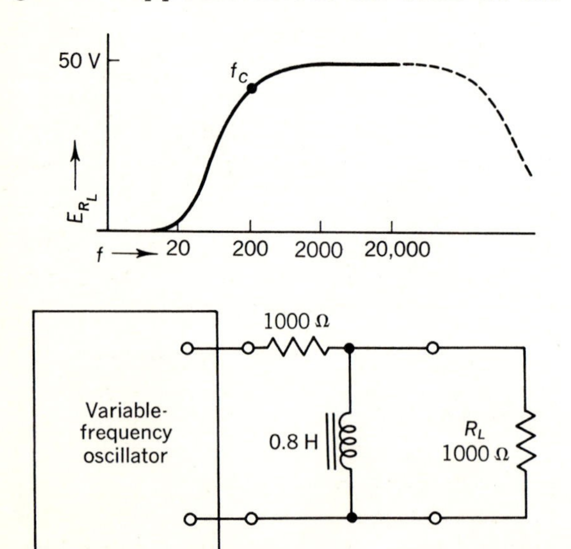

Fig. 37-2 Basic L-type 200-Hz high-pass *RL* filter and its frequency-response curve.

At 20 Hz the X_L across the 1000-Ω load resistor is 100 Ω, resulting in less than 5 V appearing across the load at this frequency.

At 200 Hz the X_L is 1000 Ω, the parallel R_L and X_L have an impedance of 707 Ω, and 44 V is developed across the load.

At 2000 Hz the X_L is 10,000 Ω, and the coil has almost no effect on the 50 V across the load.

With a signal of 20,000 Hz the X_L of the coil is 100,000 Ω. At this and higher frequencies the inductance will have practically no effect on the load voltage. The circuit is an L-type 200-Hz *high-pass RL* filter with an insertion loss of 6 dB.

As frequency is increased, any distributed capacitance (turn to turn) in a coil is in parallel with the coil and may become effective as a low-pass capacitive element. As a result, the response curve may bend downward, as indicated by the dashed line in Fig. 37-2. In effect, the circuit becomes a *bandpass filter.*

Quiz 37-1. Test your understanding. Answer these check-up questions.

1. In general, when a capacitor is in series with a circuit, what frequencies are transmitted best? __________ Which are attenuated most? __________
2. In general, when a capacitor is across a load, what frequencies are transmitted best to the load? __________ Which are attenuated most? __________
3. In Fig. 37-1, if R_L were 10,000 Ω, would the 0.08-μF shunt capacitor produce complete cutoff at a higher or a lower frequency? __________ Would the insertion loss be greater or less? __________ What would be the value of the insertion loss at the best part of the pass band? __________ What would the frequency be at this point? __________
4. What would be the f_c effect if the line resistance were increased? __________ The insertion loss? __________
5. In Fig. 37-1*b*, the cutoff frequency is approximately 2000 Hz. What value of *C* would give a cutoff at 20 kHz? __________ 200 kHz? __________ 2 MHz? __________
6. In general, when an inductor is shunted across a load, what frequencies are attenuated the most? __________ What basic type of filter might this be called? __________
7. In Fig. 37-2, what is responsible for producing a bandpass effect with this circuit? __________ If the 0.8-H coil were antiresonant, would it pass or attenuate antiresonant frequencies? __________

37-3 SERIES REACTIVE ELEMENTS

A reactance in *series* with the line will also affect the frequency response. In Fig. 37-3*a*, a 0.08-μF capacitor is in series with a 1000-Ω load resistor. At 20 Hz the X_C is 100,000 Ω, and the voltage-drop across the

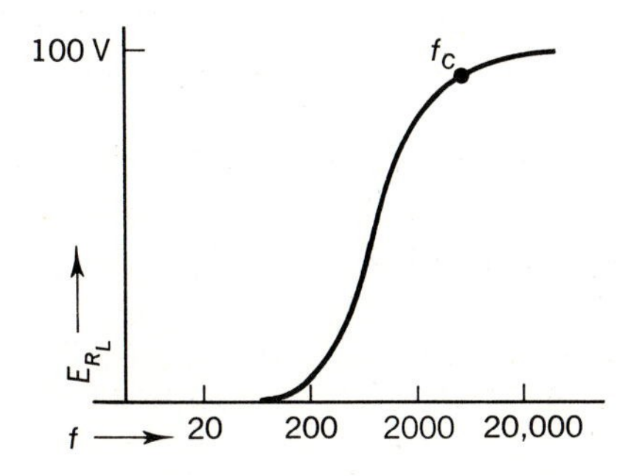

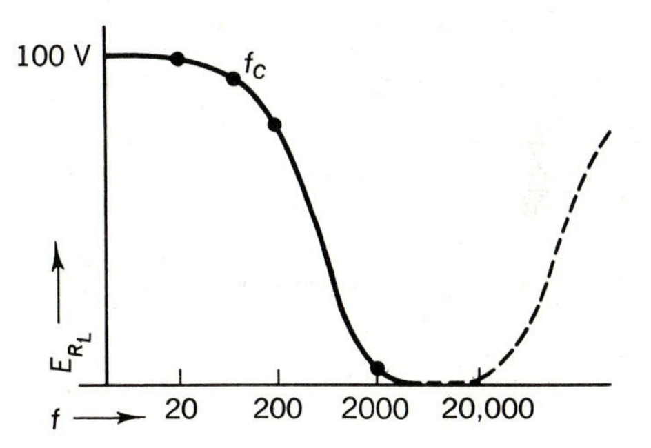

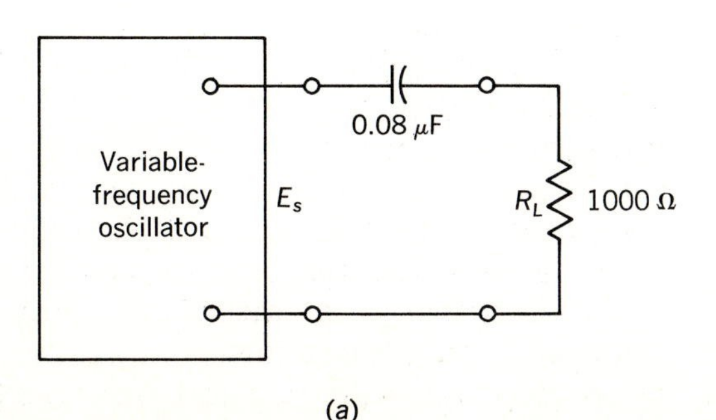

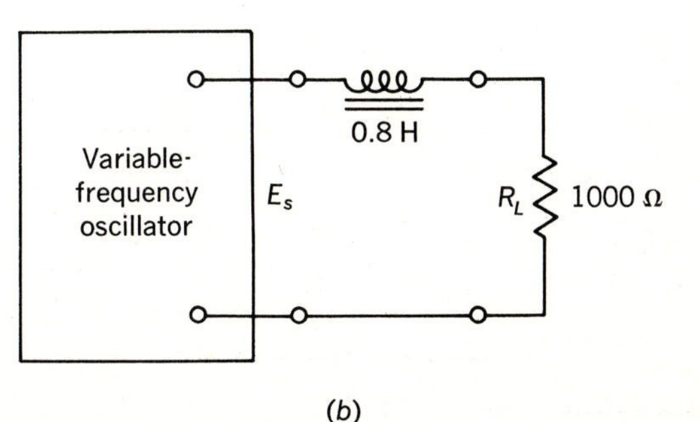

Fig. 37-3 Filter effects of pure reactances: (*a*) series capacitor, high-pass effect; (*b*) series inductor, low-pass or possibly bandstop effect with distributed capacitance.

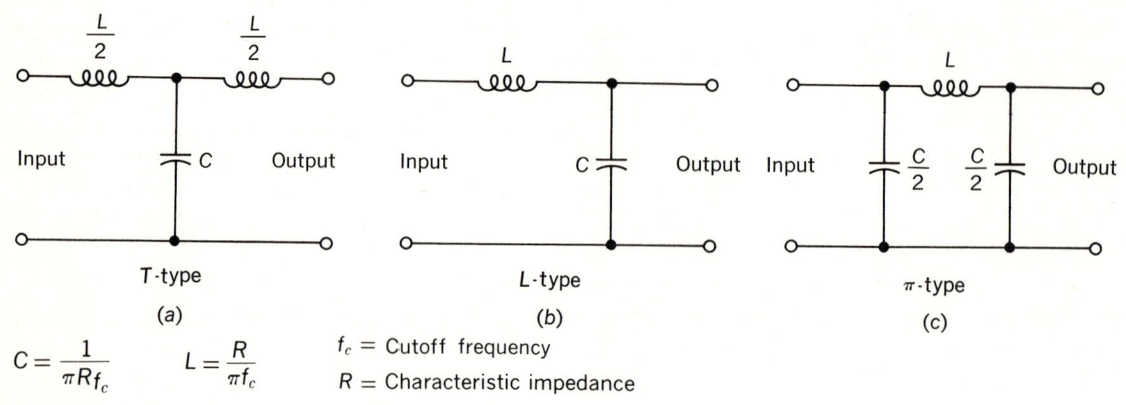

Fig. 37-4 Basic low-pass constant-*k* filter configurations, with formulas to solve for component values: (*a*) T-type; (*b*) L-type; (*c*) π-type.

load is $\pm 0.01E_S$. At 200 Hz the X_C is 10,000 Ω, and the voltage across R_L is about $0.1E_S$. At 2000 Hz the X_C is 1000 Ω, the total impedance across the source is 1414 Ω, and the voltage-drop across the load is 0.7 E_S. Finally, at 20,000 Hz the X_C is 100 Ω, and $0.995E_S$ appears across the load. The frequency-response curve resembles an L-type high-pass *RL* filter, except that the rise in response is sharper, or more rapid. Note that in a pure-reactance filter the insertion loss would be almost zero (0 dB).

When an 0.8-H choke coil is connected in series with the line, as in Fig. 37-3*b*, at 20 Hz it presents 100 Ω of X_L and $0.995E_S$ appears across the load. At 200 Hz the X_L is 1000 Ω, and a voltage-drop of $0.7E_S$ appears across the load. At 2000 Hz the X_L is 10,000 Ω, and only $0.1E_S$ is across the load. At 20,000 Hz $0.01E_S$ is across the load, unless distributed capacitance starts to pass current across the choke coil. In this case, the frequency response might follow the dashed lines shown on the curve, producing a *bandstop* filter. Maximum loss of signal in the "attenuation band" occurs at the frequency where the inductance and the distributed capacitance form an antiresonant circuit.

ANSWERS TO CHECK-UP QUIZ 37-1

1. (Highs) (Lows) **2.** (Lows) (Highs) **3.** (Lower) (Less) (1.04 dB) (20 Hz or lower) **4.** (Lower) (More insertion loss) **5.** (0.008 μF) (0.0008 μF) (0.00008 μF) **6.** (Lows) (High-pass) **7.** (Distributed capacitance) (Pass)

37-4 LOW-PASS AND HIGH-PASS CONSTANT-*k* FILTERS

Neither *RC* nor *RL* filters have very steep attenuation curves. When sharp cutoff and minimal insertion losses are important, only reactive elements are used in the filters. Even so, there is always some ohmic wire resistance in the coils, skin effect, and perhaps hysteresis and eddy-current losses if the inductors have iron cores. All losses tend to lower the Q value of the components, reduce the sharpness of the cutoff, and introduce insertion losses.

Four basic forms of filters have been described so far: (1) Low pass, (2) high pass, (3) bandpass, and (4) bandstop. Actual filters are constructed as L-type, T-type, or π-type, as shown in Fig. 37-4.

In Fig. 37-4*a*, reactive elements are connected in a T-type low-pass form, having inductors in series with the line and a capacitor shunted across the line. Series *L* and shunt *C* each produce a low-pass effect alone. Together they are twice as effective.

Filters have a specific *characteristic*, or *surge impedance*, at both the input and output terminals. If the input impedance looks into a source with an equal impedance and the output looks into a load with a matching impedance, the filter is functioning in an *image-impedance* condition. This is usually desired.

To determine filter element values, first the terminal resistance or impedance values of the source and load (R in the formulas) are decided upon. Then the formulas shown can be applied to find the L and C values.

Figure 37-4*b* illustrates the L-type low-pass filter. The L-type may not be as symmetrical a filter as a T-type, but it contains the same values of total inductance and capacitance. L-type networks are often used as impedance-changing devices. Input to output, the L network acts as a step-down impedance transformer. At one particular frequency it will appear resistive both to a high-impedance source and to a low-impedance load. If the input and output terminals are reversed, it operates as a step-up impedance transformer. Since it takes time for current to build up in the series inductor, there will be a delay in charging the capacitor after the voltage is applied to the input terminals. Hence this type of circuit can be used as a *delay line* when it is desired to delay an ac signal for a small fraction of a second.

In Fig. 37-4*c*, the π-type arrangement of the elements has a total shunt impedance, X_C, equal to the shunt impedance of the T-type or the L-type. The shunt elements are each half the required total capacitance value.

Circuits of the basic types shown in Fig. 37-4 are *constant-k* filters; the product of the series and the shunt reactances at any frequency will have the same numerical value. For example, if the series X_L is 1000 Ω and the shunt X_C is 500 Ω, the k value is 1000(500), or 500,000. If the frequency is doubled, X_L is 2000 Ω and X_C is 250 Ω, still a k value of 2000(250), or 500,000.

The proper terminating-resistance value for a constant-k filter is either the square root of the constant-k value or the square root of the ratio of the inductance to the capacitance,

$$R = \sqrt{X_L X_C} \qquad \text{or} \qquad R = \sqrt{L/C}$$

where R is the terminating resistance, or image impedance (or the constant-k value).

When the series elements are capacitive and the shunt elements are inductive, the filters are *high-pass* types, as in Fig. 37-5. Again, all three basic types have

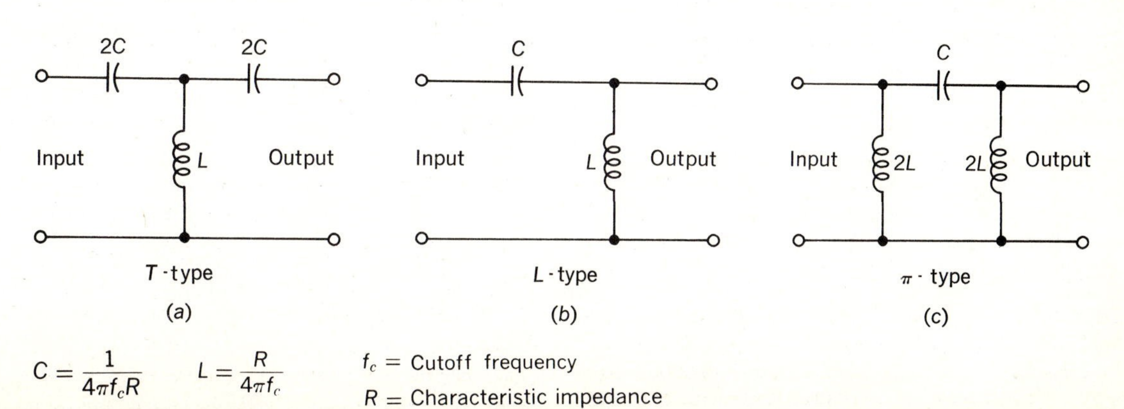

Fig. 37-5 Basic high-pass filters, with formulas: (*a*) T-type; (*b*) L-type; (*c*) π-type.

equal series- and shunt-element total values.

Quiz 37-2. Test your understanding. Answer these check-up questions.

1. What are two advantages of purely reactive filters over *RC* or *RL* filters? ____________ ____________
2. An inductor alone in series with a line feeding a 1000-Ω load produces what basic filtering effect? ____________ If the load had twice the resistance, would the cutoff curve be more or less steep? ____________ What would be the insertion loss? ____________
3. What are the four basic types of filters? ____________ ____________ ____________ ____________ What are the three basic configurations? ____________ ____________ ____________
4. What type of filter is discussed in the text as a possible Z-transforming device? ____________ What element is across the low-Z terminals? ____________
5. Design a T-type low-pass constant-*k* filter with an f_c of 3 kHz and a surge impedance of 600 Ω. ____________ ____________
6. Design a π-type low-pass constant-*k* filter for an f_c of 3 kHz and a characteristic impedance of 600 Ω. ____________ ____________
7. If a low-pass filter has a total series reactance of 200 Ω and a total shunt reactance of 400 Ω, what is the value of the constant *k*? ____________ The image-impedance value? ____________
8. Which would be expected to give the sharper cutoff, a T-type or a π-type low-pass filter, or would there be any difference? ____________
9. Design a T-type high-pass constant-*k* filter with an f_c of 2 kHz and a surge Z of 5 kΩ. ____________ ____________
10. Design a π-type high-pass constant-*k* filter with an f_c of 2 kHz and an image Z of 5 kΩ. ____________ ____________

37-5 TWO-SECTION FILTERS

A single-section T-, L-, or π-type low-pass filter drops off as indicated by the solid curve in Fig. 37-6. The cutoff frequency f_c is usually at about −2 dB. For a low-pass filter with a Q of 100 and an f_c of 3 kHz, the attenuation at 4 kHz is only 14 dB. At 5 kHz the attenuation is 18 dB. At twice the cutoff frequency, 6 kHz, it is about 22 dB. Since a loss of 20 dB repre-

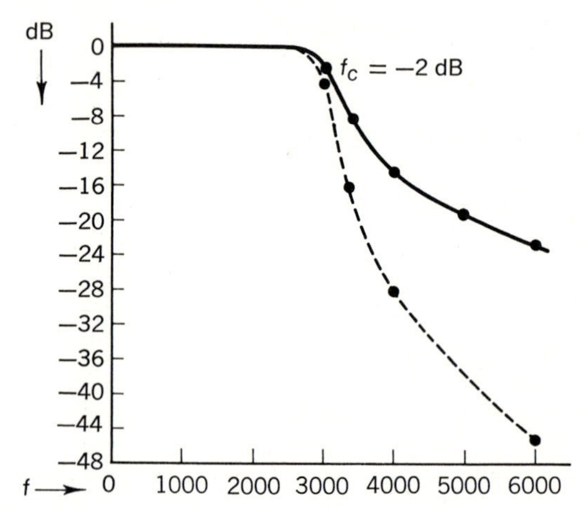

Fig. 37-6 Attenuation curve of a single-section constant-*k* low-pass filter (or transmission curve of a high-pass filter). Two-section filter response (dashed).

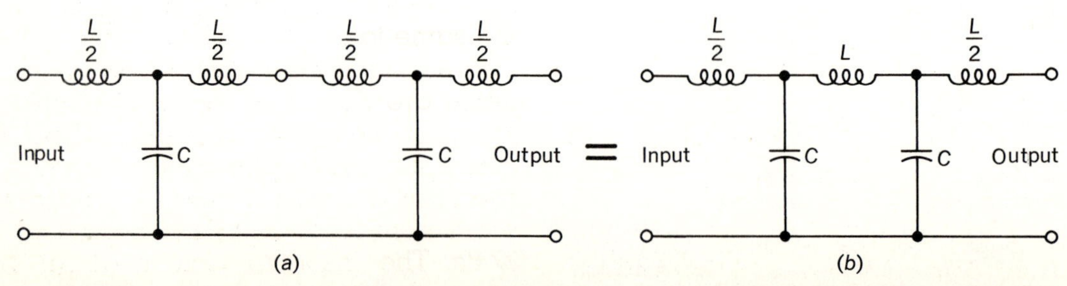

Fig. 37-7 (*a*) Two similar T-type constant-*k* LP filters in cascade. (*b*) Actual construction.

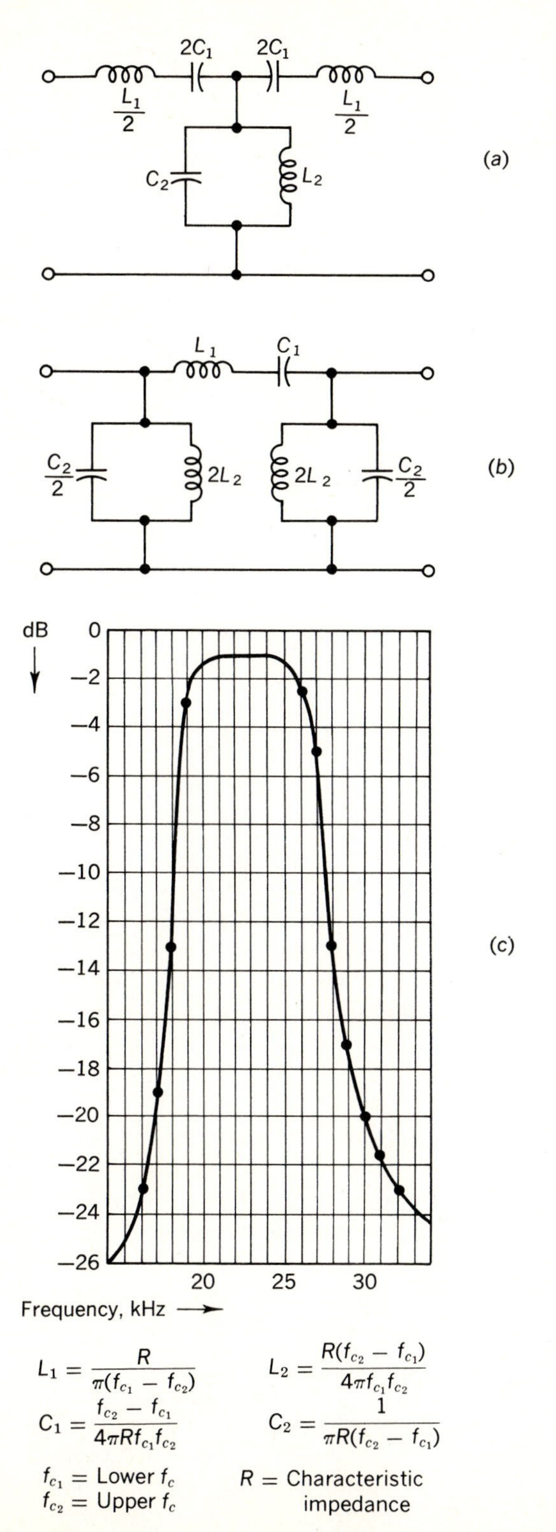

Fig. 37-8 Basic constant-*k* bandpass filters, with formulas: (*a*) T-type; (*b*) π-type. (*c*) Transmission frequency response with f_c at 20 and 25 kHz.

sents a drop in voltage response to 0.1 of the voltage at 3 kHz, this is not a rapid attenuation.

One method of attaining a more rapid drop-off is to use two similar low-pass sections in *cascade* (one coupled to the next) as in Fig. 37-7*a*. Instead of using two half-value inductances in series at the center, this filter would use one inductor with a full *L* value, as in Fig. 37-7*b*. The frequency response or attenuation curve of a two-section filter should be twice that for a single section. With two sections, at 4 kHz the attenuation should be 2(14), or 28 dB. At 5 kHz the level should be 2(18), or 36 dB. At 6 kHz the output should be 44 dB down. −40 dB represents a drop to one-hundredth of the input voltage. If a single section has an insertion loss of 3 dB, two sections should have a loss of 6 dB. This means that with an input of 15 V, the output voltage would be half (−6 dB) the input, or 7.5 V.

T-type filters have been shown intercoupled, but similar results are obtained by intercoupling π-type sections.

37-6 BANDPASS AND BANDSTOP CONSTANT-*k* FILTERS

When only a certain band of frequencies must be transmitted, a bandpass filter that accepts this band of frequencies and rejects all others is used. One such circuit was discussed in Chap. 36 in connection with tuned transformers. Although tuned transformers may not be considered true filter circuits, they can do the same job.

Bandpass T-type and π-type constant-*k* filters are shown in Fig. 37-8. Both have identical frequency characteristics. With inductors having Q values of 100, the response curve of these filters for a pass band from 20 to 25 kHz is shown in Fig. 37-8*c*. The insertion loss is about 1 dB across the pass band.

The curve shape of these filters remains the same if they are computed at higher or lower frequencies. That is, if the filter is designed to pass 200 to 250 kHz, a zero added to each number representing the frequencies on the graph will indicate the frequency response of the filter for this new band of frequencies.

When the ratio of the bandpass to the center frequency must be smaller, the shape of the curve changes. For example, Fig. 37-9 represents the frequency response of a bandpass filter centered on 105 kHz, with cutoff points at 100 and 110 kHz and inductors with Q values of 100 each. When the Q value is reduced to 30, the insertion loss increases by 4 dB. The drop-off appears to be as steep but is actually not. If the dashed curve were raised by the 4-dB insertion loss, the bandwidth would be about 3 kHz broader 10 dB down the skirts.

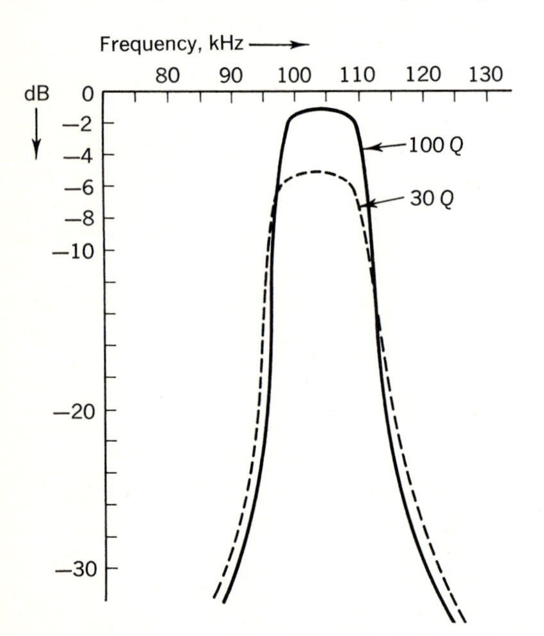

Fig. 37-9 Frequency responses of 100-Q and 30-Q constant-*k* filters.

Bandpass filters do not follow a universal resonance curve but are flat-topped with a much steeper drop-off. One method of expressing the bandwidth of these filters is at the −3-dB points. In Fig. 37-9, the bandwidth of the solid curve at the −3-dB points is about 12 kHz. The bandwidth at the −6-dB points is about 14 kHz.

The attenuation characteristic of a filter may be expressed as *shape factor*, the ratio of bandwidth at −60 dB to the bandwidth at −6 dB. If the shape factor is 2 (or 2:1), at the −60-dB points the bandwidth is twice the bandwidth at the −6-dB points. This is a very sharp cutoff. The shape factor of the single-section 100-Q constant-*k* filter is more than 15:1. To reduce the shape factor, it is necessary to use more sections in cascade. Each additional section decreases the shape-factor value but increases the insertion loss.

A bandstop, or *band-elimination*, filter consists of two circuits (Fig. 37-10) which appear to be wavetraps (Sec. 36-7) but actually are not. However, their reactances add to produce one frequency f_0 at which all transmission is stopped, a lower frequency f_L at which transmission begins to decrease, and an upper frequency f_U at which full transmission resumes again.

ANSWERS TO CHECK-UP QUIZ 37-2

1. (Sharper cutoff, less insertion loss) **2.** (Low-pass) (Less) (Zero) **3.** (Low-pass) (High-pass) (Bandpass) (Bandstop) (T) (L) (π) **4.** (L) (Capacitor) **5.** ($L/2 = 31.8$ mH) ($C = 0.177$ μF) **6.** ($L = 6.37$ mH) ($C/2 = 0.0885$ μF) **7.** (80,000) (283 Ω) **8.** (No difference) **9.** ($L = 198$ mH) ($2C = 0.0159$) **10.** ($2L = 396$ mH) ($C = 0.00796$ μF)

Quiz 37-3. Test your understanding. Answer these check-up questions.

1. If a single-section low-pass filter is down 10 dB at 500 Hz past the f_c point, how far down would a three-section filter be at the same point? ________

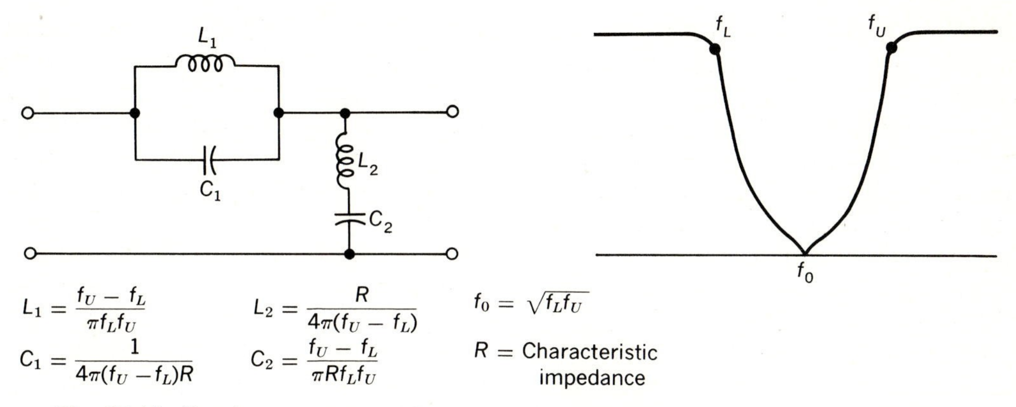

Fig. 37-10 Bandstop constant-*k* L-type filter and transmission frequency response.

2. If a single-section high-pass filter has an insertion loss of 3 dB, what insertion loss would a three-section filter of the same type have? ______ What power loss is this? ______
3. Why would you assume that the filters shown in Fig. 37-7 are for high frequencies, and not for audio frequencies? ______
4. What type of inductors would be the most suitable for lower-frequency filters? ______ Why? ______
5. What type of inductors might be most suitable for higher-frequency filters? ______ Why? ______
6. If an 88-mH toroidal coil has a center tap on it, what is the inductance value from one end of the coil to the center tap? ______ Why? ______
7. If the Q of the coils in a T-type bandpass filter is 100, does this filter have a wider or a narrower pass band than if the coils had a Q of 150? ______ Is the insertion loss more, less, or the same with the 100 Q? ______
8. Which has the sharper cutoff characteristics, the universal resonance curve or a constant-*k* bandpass filter? ______
9. A bandpass filter has a bandwidth of 8 kHz at −6 dB, 12 kHz at −30 dB, and 25 kHz at −60 dB. What is the shape-factor value? ______
10. Design a constant-*k* bandpass T-type filter for a flat-top ($f_{c_1} - f_{c_2}$) from 50 to 53 kHz with a characteristic impedance of 600 Ω. ______ ______ ______ ______

37-7 *m*-DERIVED FILTERS

To produce a steeper drop-off for the skirts than is possible with constant-*k* filters, a resonant wavetrap is added across the input or output, as shown in Fig. 37-11*a*. In Fig. 37-11*b*, the curve of the high-Q wavetrap is shown dashed and superimposed on the frequency-response

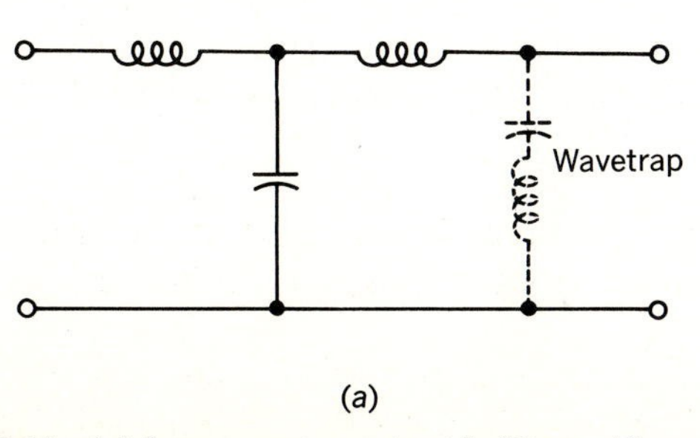

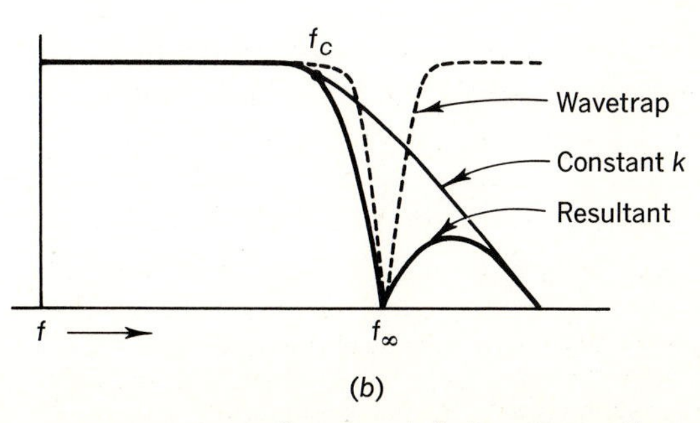

Fig. 37-11 (*a*) Low-pass constant-*k* filter with wavetrap across output, showing infinite attenuation, f_∞. (*b*) Curves of filter with and without wavetrap.

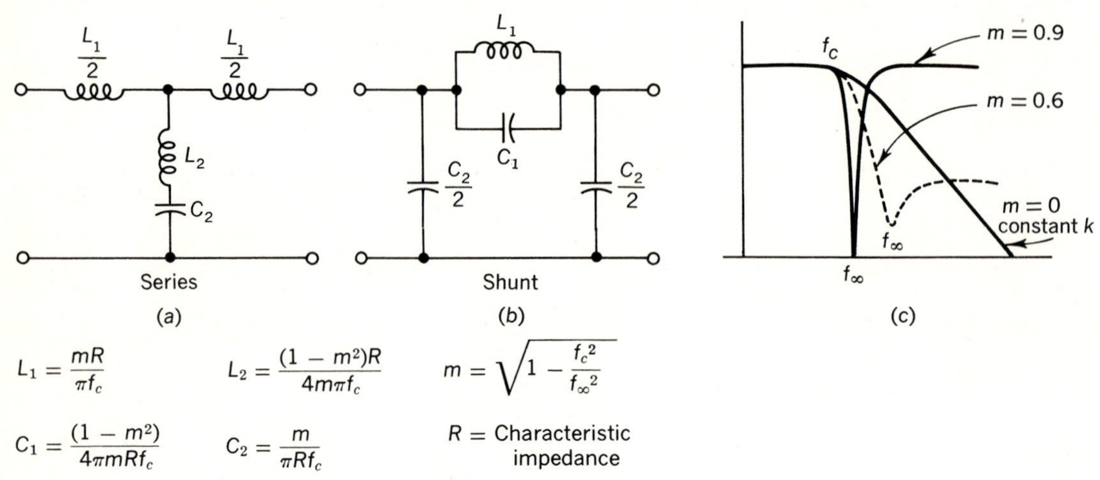

Fig. 37-12 Low-pass *m*-derived filters, with formulas: (*a*) series form; (*b*) shunt form. (*c*) Frequency responses for $m = 0.9$, $m = 0.6$, and $m = 0$ (constant k).

curve of the constant-*k* filter. The result is a steep drop-off past the cutoff frequency. If the wavetrap frequency, called the *frequency of infinite attenuation*, f_∞, is close to the cutoff frequency, the response of the filter will "pop up" considerably before it finally drops off. When the wavetrap frequency is farther from f_c, the pop-up past f_∞ will be less, but the skirt will not drop off as rapidly. The ratio of the cutoff frequency f_c to f_∞ results in a factor "*m*." This is an *m*-derived filter.

An antiresonant, or parallel-resonant, wavetrap can be added in series with the filter line to produce an f_∞ and an *m*-derived filter.

The clue as to whether a filter is an *m*-derived type lies in the way resonant or antiresonant circuits are connected. If they act as wavetraps, the circuit is *m*-derived. If they do not, the circuit is a constant-*k* filter. A resonant circuit shunted across the line stops a frequency. A resonant circuit in series with the line passes its resonant frequency, and there is no f_∞.

Complete analysis of filters is quite involved, but may be found in handbooks on electronics and radio. Some basic *m*-derived low-pass, high-pass, and band-pass circuits will be included here.

ANSWERS TO CHECK-UP QUIZ 37-3

1. (30 dB) **2.** (9 dB) ($P_o = P_i/7.95$) **3.** (No iron cores) **4.** (Toroidal ferrite) (Low loss) **5.** (Toroidal ferrite or air) (Low loss) **6.** (22 mH) (L proportional to N^2) **7.** (Wider) (More) **8.** (Constant-*k*) **9.** (3.125) **10.** ($L_1/2 = 0.0319$ H, $2C_1 = 0.0003$ μF, $L_2 = 0.000054$ H, $C_2 = 0.177$ μF)

37-8 LOW-PASS AND HIGH-PASS *m*-DERIVED FILTERS

The filter shown in Fig. 37-12*a*, with a series-resonant circuit, is a *series m-derived* filter. The circuit in Fig. 37-12*b*, with a parallel-resonant circuit, is a *shunt m-derived* filter. The value of *m*, determined from the formula shown, affects the frequency response of the filter. If L_2 is reduced to zero, *m* is zero, and the circuit is a constant-*k* low-pass filter. If L_1 is reduced to zero, *m* is zero, and the circuit is a wavetrap. With an *m* value of about 0.6, a reasonably steep drop-off is produced with not too much return past f_∞.

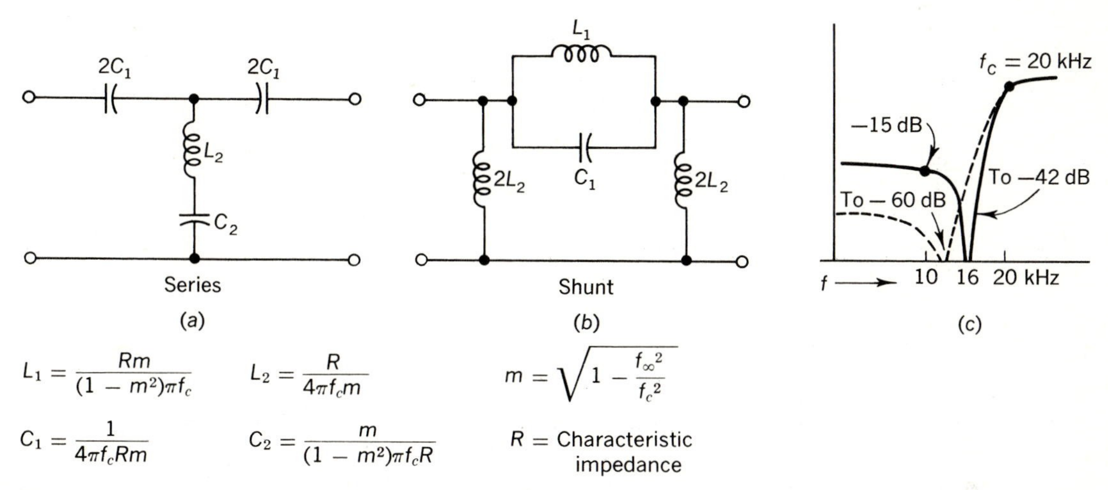

Fig. 37-13 High-pass m-derived filters, with formulas: (a) series form; (b) shunt form. (c) Transmission frequency response for $m = 0.6$ and $m = 0.8$ (dashed).

High-pass m-derived filters have two basic forms, shown in Fig. 37-13. Computation of the frequency response of a high-pass m-derived filter with a Q of 100, an m of 0.6, and an f_c of 20 kHz has maximum attenuation at about 16 kHz, at which point the response is −42 dB from the maximum value (Fig. 37-13c). Below this frequency, the response returns to −15 dB at 10 kHz. Beyond this, the attenuation slowly *decreases*. In a constant-k high-pass filter the attenuation continually increases. Because of this, an m-derived filter is usually terminated at both ends with a constant-k filter to produce continuing attenuation as the frequency departs farther from f_∞.

Quiz 37-4. Test your understanding. Answer these check-up questions.

1. What produces the f_∞ in m-derived filters? ______
2. What value of m produces a constant-k filter? ______
3. What two forms of wavetrap are used in m-derived filters? ______ ______ Which is always in the series part of the filter? ______ Which is always in shunt? ______
4. When resonant circuits are connected in series with the line in a filter, what type of filter is it? ______
5. What is the difficulty when m is too small? ______
6. The return of the frequency response of an m-derived filter past f_∞ is a disadvantage. What are two ways of overcoming this difficulty? ______ ______
7. What value of m is required in a shunt-type low-pass filter if the cutoff frequency is 23 kHz and f_∞ is 27 kHz? ______ What are the filter values if the impedance of the filter is 5000 Ω? ______ ______ ______
8. What value of m is required in a series-type high-pass filter if f_c is 53 kHz and f_∞ is 40 kHz? ______ What are the filter-component values if the impedance of the filter is 1000 Ω? ______ ______ ______
9. What type of filter would result if a low-pass filter with $f_c = 30$ kHz were added in cascade to a high-pass filter with $f_c = 20$ kHz? ______

37-9 BANDPASS m-DERIVED FILTERS

A bandpass m-derived filter has a series and a shunt form, as shown in Fig. 37-14. Solving these filters is quite involved since the filter has two cutoff frequencies, two frequencies of infinite attenuation, and a bandwidth to be con-

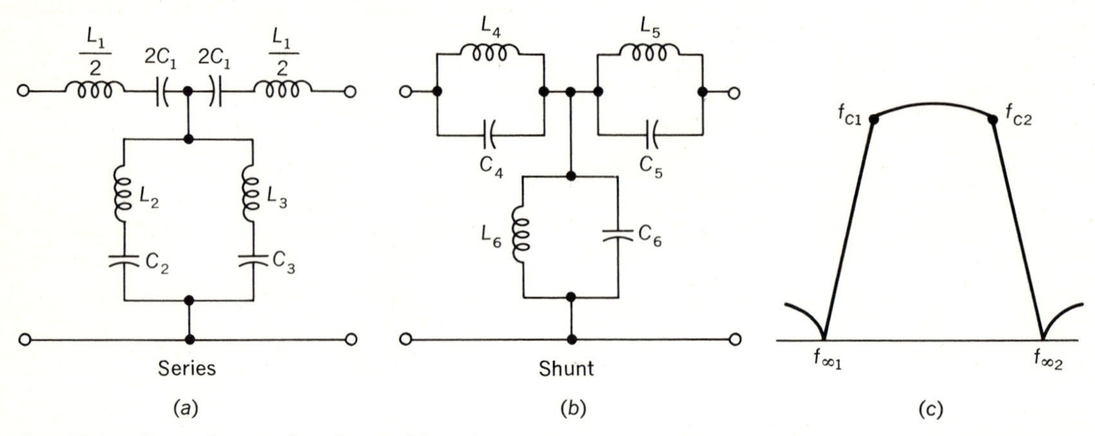

Fig. 37-14 Basic forms of *m*-derived bandpass filters: (*a*) series; (*b*) shunt. (*c*) Transmission curve.

sidered. Reference should be made to an electronics or radio handbook.

37-10 COUPLING FILTER SECTIONS

All T- and π-type filters can be divided into half-sections. Figure 37-15*a* illustrates

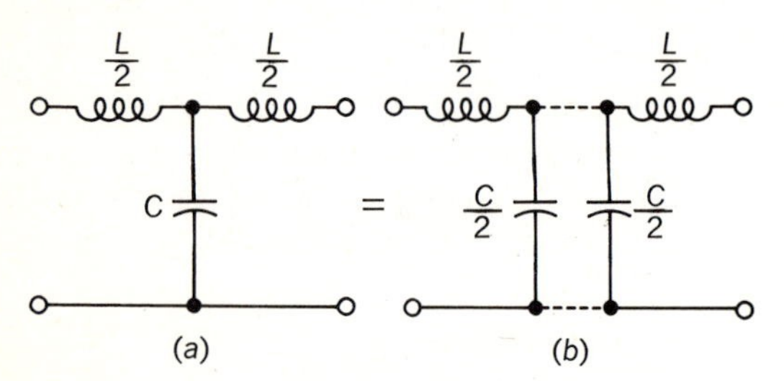

Fig. 37-15 Division of a full-section mid-shunt T-type filter into two half-sections.

a full-section T-type constant-*k* filter. When split down the middle (Fig. 37-15*b*), the two series arms remain the same, but the shunt capacitor is split in two. This is known as a *mid-shunt-terminated filter*, whether a half-section or a full section.

Figure 37-16*a* is a low-pass constant-*k* π-type filter. When split, as in Fig. 37-16*b*,

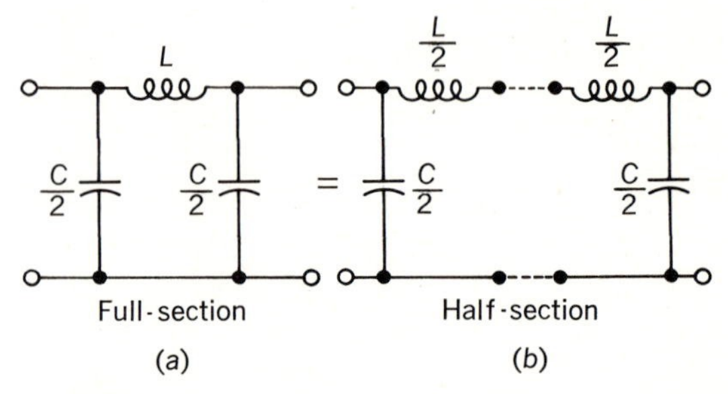

Fig. 37-16 Division of a full-section mid-series π-type filter into two half-sections.

the resultant two circuits are *mid-series-terminated half-sections*. The full section is known as a mid-series-terminated full section.

Any T-type filter, whether constant-*k* or *m*-derived, will couple to any other T-type full section or half-section if both have similar image impedances. Similarly, any π-type constant-*k* or *m*-derived full section or half-section will couple properly with any other π-type section if both have similar image impedances.

Each end of a *composite* filter (one made up of different types of sections) is

---

ANSWERS TO CHECK-UP QUIZ 37-4

1. (Wavetraps) **2.** (Zero) **3.** (Series) (Parallel) (Antiresonant) (Resonant) **4.** (Constant-*k*) **5.** (Not steep cutoff) **6.** (Terminate with constant *k*) (Add section with another *m* value) **7.** (0.525) ($L_1 = 0.0363$ H) ($C_1 = 0.000956$ μF) ($C_2/2 = 0.000725$ μF) **8.** (0.66) ($2C_1 = 0.00454$ μF) ($C_2 = 0.00702$ μF) ($L_2 = 0.00227$ H) **9.** (Bandpass)

usually terminated with half-section mid-shunt or mid-series filters for a better match in source and load impedances.

A composite low-pass filter to produce sharp cutoff and continuous attenuation past the first point of infinite attenuation might consist of several sections in cascade—for example, (1) a half-section m-derived input section with an m value of 0.8 for very sharp cutoff, followed by (2) a full section with an m value of 0.6 to decrease the return, or pop-up, of the sharper input filter, followed by (3) one or more constant-k sections to ensure continuous attenuation past the points of infinite attenuation, and finally (4) an m-derived half-section with an m value of 0.7 to match the load. All sections are computed with the same value of image impedance and similar cutoff frequencies.

37-11 BALANCED FILTERS

All the filters discussed have been the unbalanced type, with no components in the bottom line, which is usually grounded. It is often desirable to operate both sides of the filter above ground potential. In this case, the shunt element in the filter may be center-tapped, and this tap is grounded (Fig. 37-17*b*). The total inductance and capacitance values are the same in both unbalanced and balanced filters.

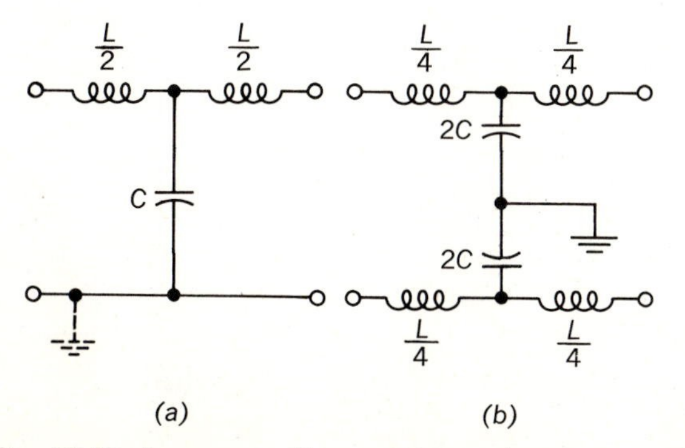

Fig. 37-17 Low-pass filters with similar frequency responses: (*a*) unbalanced; (*b*) balanced.

In balanced filters each series component must have exactly the same reactance as the component across the filter from it. Symmetry must also be maintained for the shunt elements.

Many television receivers have a 300-Ω balanced input antenna circuit filter to reduce interference from signals at lower frequencies. It is a 300-Ω balanced high-pass m-derived or constant-k type.

Quiz 37-5. Test your understanding. Answer these check-up questions.

1. To produce a half-section from a full-section T-type high-pass filter, which component should be split? __________ Would the new component have twice or half the value of the full section? __________ What kind of filter would it be? __________
2. What component is split to make a half-section π-type high-pass filter? __________ Would the half-section component have half or twice the original value? __________ What kind of filter would it be? __________
3. What are the two requirements that must be met in coupling dissimilar full-section filter sections? __________ __________ Is this also true for half-sections? __________
4. What is a filter called when it is made up of different types of sections? __________
5. Is it desirable to couple filter sections having different m values and different f_∞ values? __________
6. To what is the line of an unbalanced filter with no filter elements usually connected in a circuit? __________
7. What type of filter might be called an H-type? __________ What type might be called a U-type? __________

37-12 CRYSTAL FILTERS

LC filters can be made satisfactorily sharp up to about 300 kHz. Above this, the Q of the coils is too low and the skirts of the filters widen. By using *piezoelectric crystals* as the elements, sharp filters can

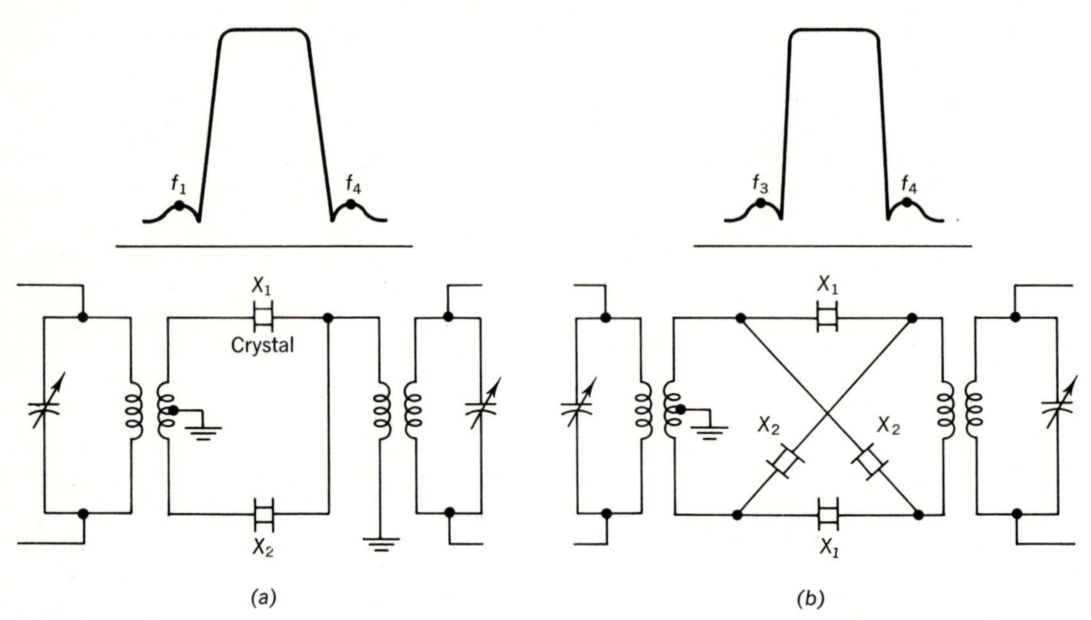

Fig. 37-18 Crystal filter circuits and approximate response curves: (*a*) half-lattice; (*b*) full-lattice.

be developed from about 100 kHz to more than 20 MHz.

Certain quartz, when cut into a thin, square wafer and silver plated on the two flat sides, exhibits *piezoelectric effects*. That is, if an emf is applied between the silvered surfaces or *plates*, the crystal changes shape slightly, an indication that it is storing mechanical energy. When the emf is removed, the crystal springs back to its normal shape. It transduces electrical energy into mechanical motion. If a crystal is bent, an emf is developed between its two plates. It transduces mechanical motion into electrical energy. Crystals have a frequency at which they resonate electrically and mechanically, and they are similar in many respects to a resonant *LC* circuit.

ANSWERS TO CHECK-UP QUIZ 37-5

1. (Inductor) (Twice) (Mid-shunt-terminated) **2.** (Capacitor) (Twice) (Mid-series-terminated) **3.** (Same f_c) (Same R) (Yes) **4.** (Composite) **5.** (Yes) **6.** (Ground) **7.** (Balanced HP or LP T-type) (Balanced L-type)

In an *LC* resonant circuit, a Q value of 300 is considered quite high. In crystals the Q value may be several thousand. When crystals are used in filters, the skirts will be very steep and the insertion losses small.

Figure 37-18*a* shows two crystals in a *half-lattice* filter circuit. If a 3-kHz bandwidth from 450 to 453 kHz is desired, crystal X_1 will be made resonant to about 450.5 kHz and crystal X_2 to about 452.5 kHz. This will result in a fairly flat pass band from about 450 to 453 kHz and a steep drop-off, as indicated. A shape factor of about 4:1 is possible. To decrease pop-up, a crystal, series-resonant to f_1 on the response curve, can be connected across the input *LC* circuit. Another crystal, resonant to f_2, can be connected across the output *LC* circuit. Since the crystals behave as resonant circuits, they act as wavetraps at the pop-up frequencies.

The coils coupling the crystals are shown untuned. Usually both input and output transformers are double-tuned.

Figure 37-18*b* shows a *full-lattice* crystal

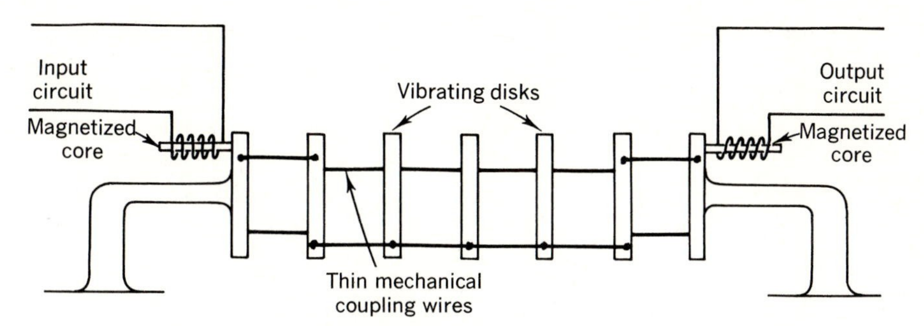

Fig. 37-19 Essentials of a mechanical filter.

filter. Crystals marked X_1 are resonant to the same frequency near one end of the desired pass band. Crystals marked X_2 are resonant to the same frequency near the other end of the desired pass band. A shape factor of about 3:1 is possible.

37-13 MECHANICAL FILTERS

A sharp filter for frequencies to about 500 kHz can be developed by mechanically coupling machined vibrating nickel disks (Fig. 37-19). If a pass band of 3 kHz in the 450-kHz region is required, some of the disks will be machined to vibrate mechanically at about 450.5, some at 451.5, and others at 452.5 kHz. If ac within about 0.5 kHz of these frequencies is introduced into the input coil, the magnetic field produced attracts and repels the magnetized core piece, which drives the first disk into vibration. This vibration is transferred sympathetically on down the chain. Vibration of the last disk drives its magnetized core piece into and out of the output coil, inducing an emf of the signal frequency into it. Any ac more than 0.5 kHz away from the mechanical resonant frequency of the disks will not transfer vibrations along the chain, and will produce no output. A shape factor of 2:1 is possible.

37-14 CROSSOVER NETWORKS

To reproduce music faithfully, loudspeakers must produce an equal output on all audio frequencies, 20 to at least 15,000 Hz. No one loudspeaker will do this efficiently. It is possible to use one loudspeaker to reproduce low frequencies efficiently and a second to reproduce high frequencies. The low-frequency speaker is called a *woofer*. The high-frequency speaker is called a *tweeter*. To drive both speakers from a single output transformer of an amplifier, a *crossover* filter network is used (Fig. 37-20). The woofer is fed

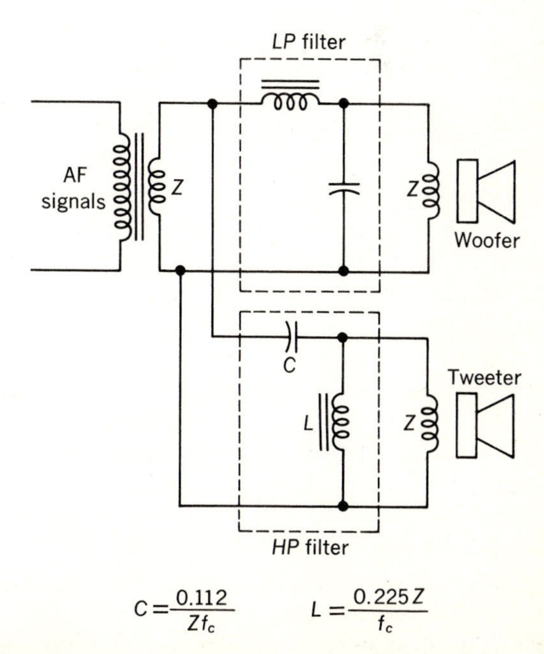

$$C = \frac{0.112}{Zf_c} \qquad L = \frac{0.225Z}{f_c}$$

Fig. 37-20 Crossover networks to feed equal power from an audio-frequency amplifier to a woofer and a tweeter loudspeaker.

through a low-pass filter, and the tweeter is fed through a high-pass filter.

37-15 POWER-SUPPLY FILTERS

Low-pass constant-k filters are used widely in electronic power supplies. Vacuum-tube and transistor circuits require a source of dc to operate. It is common to convert the universally available 50- to 60-Hz ac to approximately the required voltage by means of a transformer. This is converted to pulsating dc by halfwave or fullwave rectifiers (Fig. 37-21), and the variations are removed by passing the dc through a low-pass filter.

The circuit of Fig. 37-21*a* uses one solid-state diode as a halfwave rectifier to produce pulsating dc from the transformer ac. The low-pass filter passes dc (zero frequency) best and tends to reject the pulse frequency. Thus the only current passing through the filter is dc. R_s limits the current to the diode. The output capacitance C_o is usually twice the value of the input capacitance C_i. For a 30-V dc power supply, component values might be: $C_i = 500\ \mu F$, $C_o = 1000\ \mu F$, and $L = 1$ H.

The circuit in Fig. 37-21*b* is a bridge-type fullwave rectifier with a choke-input (two-section L) low-pass filter. The fullwave rectification produces twice as many pulses, a pulse for every alternation instead of one for every cycle, and therefore the C and L values are about half those in halfwave circuits.

In general, the heavier the current drain on power supplies, the more the capacitance but the lower the inductance required.

Some power supplies use a small resistance in place of the choke. RC filters result in less effective filtering and may require several times as much capacitance as LC filters.

Quiz 37-6. Test your understanding. Answer these check-up questions.

1. What Q values do piezoelectric crystals have? ______
2. In what range of frequencies can crystal fil-

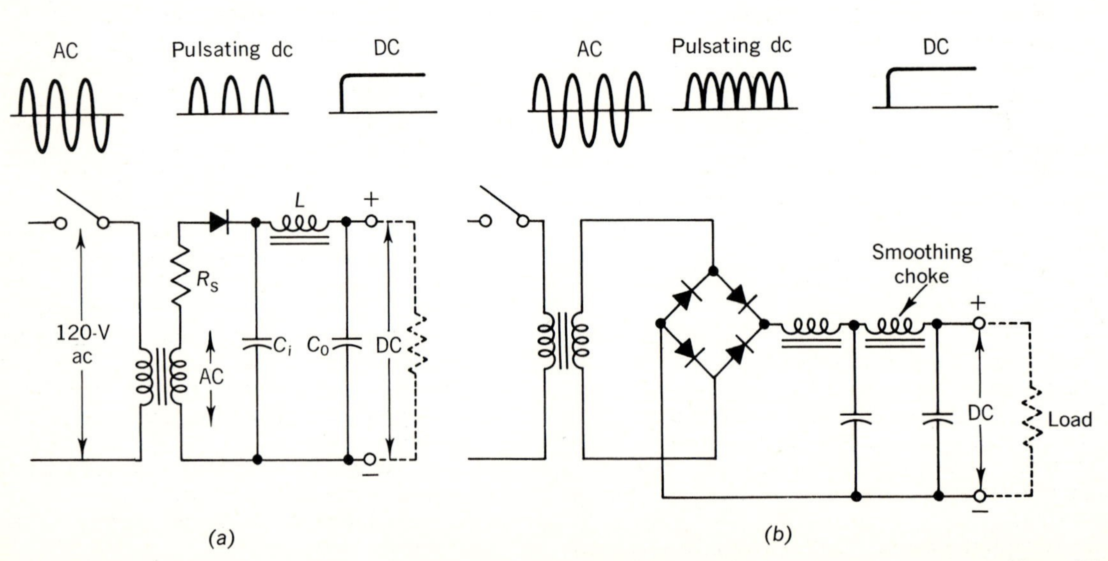

Fig. 37-21 Power-supply circuits: (*a*) halfwave rectification with π-type LP capacitive-input filter; (*b*) fullwave rectification with two-section L-type inductive-input LP filter.

ters be used? __________ Mechanical filters? __________ *LC* filters? __________

3. What types of energy are transduced by *LC* filters? __________ Piezoelectric elements? __________ Mechanical filters? __________
4. What would be considered a very good shape-factor value for filters? __________ What shape-factor value would be possible with a limited number of sections at 100 kHz for *LC* filters? __________ Crystal filters? __________ Mechanical filters? __________
5. How many crystals are used in a full-lattice crystal filter? __________ Half-lattice filters? __________
6. How many different-frequency crystals are used in a half-lattice filter? __________ A full-lattice filter? __________
7. Why must the cores attached to the end disks of mechanical filters be magnetized? __________
8. What is a loudspeaker called that is supposed to respond only to low frequencies? __________ Only to high frequencies? __________
9. In a 2-kHz crossover network feeding an 8-Ω speaker, what values of *L* and *C* are required for the low-pass filter? __________ High-pass filter? __________
10. Do power-supply filters have balanced or unbalanced filters? __________ Why is an *RC* filter undesirable if the current drain on the supply is heavy? __________
11. What is the name of the fullwave rectifier circuit shown in Fig. 37-21? __________
12. What type of filter is shown in the halfwave circuit? __________ In the fullwave circuit? __________

CHAPTER 37 TEST • FILTERS

1. What are two main disadvantages of *RC* filters in comparison to *LC* filters?
2. What group of frequencies is attenuated when a capacitor is connected in series with a line? Shunted across a line?
3. What group of frequencies is attenuated when an inductor is shunted across a line? Connected in series with the line?
4. If the load impedance of an *RC* high-pass filter is increased, how will the insertion loss be affected? Will there be an increase, a decrease, or no effect on the cutoff frequency? What will be the effect on the slope of the curve?
5. What causes a basic high-pass filter with a shunt inductance to operate as a bandpass filter?
6. Filters have a specific characteristic impedance. What is another name for this?
7. What are the three basic configurations of filters, regardless of whether they are low-pass or high-pass?
8. What is the general term for filters that have two components that act as a wavetrap? For those that have no wavetrap circuit?
9. Filters are named for the four types of filtering they perform. What are these types?
10. What would be the result if a 3-kHz low-pass and a 2-kHz high-pass filter were connected in cascade?
11. What is the advantage of cascading filters? The disadvantage?
12. A group of frequencies from 50 to 4000 Hz is transmitted into a long 600-Ω impedance line. A 3600-Hz pilot tone is also transmitted. What two devices could be connected at the receiving end of the line to prevent the pilot tone from entering the load?
13. What are the advantages of using high-Q components in filters?
14. What is the shape-factor value of a filter that has kHz versus dB skirts 80 = −60, 86 = −30, 87 = −6, 88 = −3, 92 = −3, 93 = −6, 94 = −30, 100 = −60?
15. What type of filter mentioned in the text gives the best shape factor? What is this shape-factor value?
16. When a filter is desired with sharper cutoff characteristics than is possible with a constant-*k* type, what type is used?
17. What are the names of the two possible configurations of low-pass and high-pass *m*-derived filters?
18. What is a practical *m* value for a medium-sharp cutoff with not too much pop-up?
19. What is the *m* value for a constant-*k* filter? A wavetrap?
20. In what two ways can pop-up past f_o be decreased?

21. What produces the f_∞ in *m*-derived filters?
22. What component is cut in half to make a mid-series-terminated half-section of a constant-*k* low-pass π-type filter?
23. What is a filter called that is made up of different types of sections?
24. What is the general term for filters that have the midpoint of their shunt elements grounded?
25. What two requirements must be met in coupling dissimilar full-section filter sections?
26. What is the effect that ground quartz crystals are said to have?
27. What is the effective Q value of quartz crystals?
28. How many crystals are used in a half-lattice filter? A full-lattice filter?
29. Which types of filters have pop-up—crystal, constant-*k*, mechanical, or *m*-derived?
30. What is the name of the filter network used to couple a woofer and a tweeter to an amplifier?
31. What basic configuration is used in all power-supply filters? What is the meaning of "filter" in this application?
32. Which requires more filtering, halfwave or bridge-rectified ac?

ANSWERS TO CHECK-UP QUIZ 37-6

1. (Several thousand) **2.** (100 kHz to over 10 MHz) (100 to 500 kHz) (To 300,000 Hz) **3.** (Electrostatic to electromagnetic) (Electrical to mechanical) (Magnetic to mechanical) **4.** (2) (3 to 5) (3) (2) **5.** (4) (2) **6.** (2) (2) **7.** (To produce motion from magnetic fields, induce emf as core moves) **8.** (Woofer) (Tweeter) **9.** (For both, $C = 7\ \mu F$, $L = 0.9$ mH) **10.** (Unbalanced) (Requires high C and has excessive *IR*-drop) **11.** (Bridge) **12.** (π-type, capacitor input) (Two-section L-type, inductive input)

ANSWERS TO TEST QUESTIONS

CHAPTER 1

An Electric Circuit

1. (In the source) (In the load)
2. (Stops working)
3. (Load)
4. (Switch)
5. (Long line)
6. (Positive)
7. (Proton) (Electron) (Neutron)
8. (Positive) (1840 times)
9. (Both the same)
10. (Zero)
11. (Positive)
12. (Molecule)
13. (Protons and neutrons)
14. (No; air is a gas)
15. (Negative to positive) (Electrostatic field)
16. (Part nearest the negative object)
17. (See Fig. 1-1)

CHAPTER 2

Current and Voltage

1. (Electron, proton, neutron)
2. (Copper)
3. (End nearest negative object)
4. (Electrons)
5. (Positive)
6. (2 A)
7. (2.1 A as the average)
8. (Coulomb)
9. (10 A)
10. (Ammeter)
11. (300,000,000 m/s, 186,000 mi/s)
12. (+ to −) (No flow)
13. (Volt)
14. (Electromotive-force)
15. (No) (High R) (Yes)
16. (Yes)
17. (Yes)
18. (Not necessarily)
19. (Not necessarily)
20. (Voltage, emf, electrical pressure)

CHAPTER 3

Resistance and Conductance

1. (60 V) (12 V)
2. (Long) (No)
3. (Pressure direction) (Current direction)
4. (Yes) (Toward A)
5. (Right end)
6. (Short circuit)
7. (No)
8. (Smaller) (Greater)
9. (Connecting wires)
10. (Tungsten)
11. (Increase) (Decrease)
12. (Ohms) (Omega)
13. (No R change with T change) (High R value)
14. (11.5 in.)
15. (Ohmmeter) (No voltage across resistance when measuring)
16. (Silicon, germanium, carbon)
17. (Insulator)
18. (Positive ion)
19. (Electrolytes)
20. (Potentiometer, adjustable resistor, rheostat)

CHAPTER 4

Ohm's Law in Series Circuits

1. ($E = IR$ or $V = IR$)
2. ($R = E/I$)
3. ($I = E/R$)
4. (Kirchhoff's voltage law)
5. (0 or 0.06 A) (0.08 A)
6. (2.4 A) (0 or 0.8 A)
7. (40 V) (120 V)
8. (21 Ω)
9. (0.25 A) (5 V)
10. (Infinite Ω) (0 Ω)

CHAPTER 5

Energy and Power

1. (Photons)
2. (When moving or excited)
3. (Joule)
4. (Energy)
5. (6000 J) (200 W)
6. ($P = I^2R$, $P = E^2/R$)
7. (116.7 V)
8. (2.08 A) (5.77 Ω)
9. ($P = E^2/R$) ($E = \sqrt{PR}$, $R = E^2/P$)
10. ($P = I^2R$) ($R = P/I^2$, $I = \sqrt{P/R}$)
11. (1.58 A) (316 V)
12. (220 V)
13. (Kilowatthours)
14. ($I = E/R$, $E = IR$, $R = E/I$)

CHAPTER 6

Parallel Circuits

1. (18 V) (6 V)
2. (Kirchhoff's voltage law)
3. (40 V) (6 V) (12 V)
4. (0.0417 mho)
5. (8000 Ω)
6. (Resistance) (Conductance)
7. $[R_1R_2/(R_1 + R_2)]$
8. $[1/(1/R_1 + 1/R_2 + 1/R_3)]$
9. (0.005 mho) (200 Ω)
10. (300 Ω) ($1/3$ A) (33.3 W)
11. (2.5 A) (2 A) (4 A) (8.5 A)
12. (11.76 Ω) (11.76 Ω)

CHAPTER 7

Series-Parallel Circuits

1. (Simple, series, parallel, series-parallel)
2. (800 Ω) (793 Ω)
3. (0.126 A) (0.0721 A) (0.054 A)
4. (56.8 V) (43.2 V) (43.2 V)
5. (34 Ω) (32.7 Ω)
6. (58.6 V) (41.4 V) (100 V)
7. (101 W) (71.3 W) (133 W)
8. (12 Ω) (11.4 Ω)
9. (36 Ω) (36.4 Ω)

CHAPTER 8

Magnetism

1. (When moving)
2. (Current direction in wire)
3. (No) (Yes) (Yes) (Yes)
4. (North)
5. (10^8) (57,000) (ϕ)
6. (North pole of coil, flux direction in core)
7. (Ampere-turns)
8. (4.17 amp-turns/m) (50 amp-turns)
9. (Webers per square meter) (Tesla) (B)
10. (No)
11. (Iron, nickel, cobalt)
12. (Iron)
13. (Horseshoe or C)
14. (Air) (Steel)
15. (Domains)
16. (Saturated)
17. (Repel) (Attract)
18. (Repels)
19. (North)
20. (Soft iron) (Low)
21. (No)
22. (Spring or other mechanical pulling force)

CHAPTER 9

Alternating Current

1. (Varying dc) (Pulsating dc) (Dc) (Sawtooth dc)
2. (Alternating or ac)
3. (Sinusoidal, square, sawtooth)
4. (Second finger) (First finger) (Thumb)
5. (Vector arrow approaching) (Vector going away)
6. (P_2) (14 Hz) (Dot)
7. (G, C) (A, E)
8. (Sinusoidal)
9. (Rheostat)
10. (Hertz)
11. (106 V) (95.4 V) (106 V) (300 V)
12. (141.4 V) (90 V) (282.8 V)
13. (Root-mean-square)
14. (Effective) (Effective)

CHAPTER 10

Inductance and Transformers

1. (Resistance) (Inductance)
2. (To the left) (To the right) (None)
3. (Maximum) (Zero) (Current changing most rapidly) (Current not changing for an instant)
4. (Ac) (None) (Ac)
5. (Henry)
6. (T^2)
7. (More permeable core)
8. (Reactance)
9. (Reduces it)
10. (Tertiary)
11. (1:3) (3:1) (1:1)
12. (Hysteretic, eddy current) (Neither)
13. (Fuse)
14. (Toroidal)
15. (No load on secondary)
16. (25 V) (0.5 A) (12.5 W) (12.5 W) (0.125 A)

CHAPTER 11

Series RL Circuits

1. (0 s)
2. (Never reach maximum)
3. (0.5 s) (2.5 s)
4. (90°) (E leads I)
5. (Sawtooth dc) (Sawtooth ac) (Sinusoidal ac) (Reduces amplitude)
6. (133 Ω) (1884 Ω) (9.42×10^6 Ω)
7. (Resistive) (0°)
8. (45°) (820 Ω)
9. (12.8 Ω) (51.4°)
10. (5410 Ω) (56.3°)
11. (2250 Ω) (3750 Ω) (36.9°)
12. (50 Ω) (2 A) (120 W) (200 VA) (0.6)
13. (Lowers PF)
14. (More than 45°)

CHAPTER 12

Capacitance

1. (Condenser)
2. (Electrostatic) (Dielectric)
3. (0.00125 C) (7.8×10^{15} electrons)
4. (5 μF)
5. (0.00085 μF) (14,500 μμF) (350,000 μF)
6. (Paper, plastic, ceramic, electrolytic, tantalum)
7. (Mica, but vacuum is greater)
8. (Ceramics)
9. (Aluminum, tantalum)
10. (12 pF) (3300 V)
11. (72 pF) (88,000 V)
12. ("Deform," conduct current, heat, generate steam, explode) (Same thing)
13. (0.012 μF) (0.00267 μF)
14. (500 V) (1000 V)
15. (Across the 47 pF) (Yes)

CHAPTER 13

RC Circuits

1. (Resistance in series with it)
2. ($T_c = 5RC$)
3. ($T_c = RC$)
4. (0.0002 s) (0.001 s)
5. (90°) (45°) (54.5°) (Lead)
6. (44,200 Ω) (76.5 Ω)
7. (Rapidly changing) (Slowly changing)
8. (Sawtooth) (Current starts at maximum and tapers off on each square-wave emf pulse)
9. (Sinusoidal)
10. (0 W) (2000 VA) (0 pF)
11. ($Z = 559$ Ω) ($I_s = 0.179$ A) ($I_c = 0.179$ A) ($P = 8$ W) ($VA = 17.9$) ($\theta = 63.4°$) ($E_R = 44.8$ V)

CHAPTER 14

Series LCR *Circuits*

1. (90 Ω)
2. (109 Ω)
3. (18 Ω of X_c)
4. (Downward at 90°) (Upward at 90°)
5. (60 V) (90 V) (30 V)
6. (42.3 V) (14.1 Ω)
7. (5 Ω) (Resonant)
8. (9)
9. (120 V) (1080 V) (1080 V) (2880 W)
10. (15.9 H)
11. (Zero or very low) (Zero)
12. (50 Ω) (53.1°)
13. (0.707 of peak)
14. (Raise Q, decrease R)
15. (Decrease C) (Raise Q value) (X_L value at new frequency is higher without material rise in R)
16. (Increase inductance) (Increase iron in core, or turns in coil)

CHAPTER 15

Parallel LCR *Circuits*

1. (Reactance)
2. (Susceptance)
3. (G) (B) (Y)
4. (Mhos or siemens)
5. (0.09 mho) (11.1 Ω)
6. (0.0075 mho) (133-Ω X_c)
7. (0.1 mho) (0.025 mho) (0.103 mho) (9.7 Ω)
8. (Inductive)
9. (Zero) (Antiresonant) (25 Ω) (0°)
10. (Inductive) (X_L less than X_c, demanding more inductive current)
11. (Electrons tend to oscillate back and forth by themselves)
12. (Infinite Ω) (Decrease)
13. (Pure sinusoidal)
14. ($I_R = 4$ A) ($I_X = 2.5$ A) ($I_s = 4.72$ A) ($\theta = 32°$) ($Z = 42.4$ Ω)

CHAPTER 16

Tuning Circuits and Filters

1. (Stop it)
2. (Yes, has L and end-to-end C)
3. (Both)
4. (Narrow) (Increase it)
5. (Critical)
6. (Inductive) (Capacitive)
7. (Increase its Q)
8. (Over critical value)
9. (Resonant frequency) (Lower bandpass frequency)
10. (Lowers Q) (Lowers Q)
11. (Low-pass) (High-pass) (Bandpass)
12. (m-derived)
13. (High-pass)
14. (High-pass, π-type) (High-pass, T-type)
15. (Wavetrap)

CHAPTER 17

Active Devices

1. (Resistor) (Capacitor) (Inductor) (Transformer)
2. (All vacuum tubes) (All semiconductor devices)
3. (Cathode) (Emitter) (Cathode) (T)
4. (Plate, anode) (Collector) (Anode) (T)
5. (Cathode to plate)
6. (Change ac to pulsating dc)
7. (Higher E_p, I_f, lower bias)
8. (Triode) (Tetrode) (Pentode)
9. (Negative)
10. (Increase) (I_g flows)
11. (Microphone) (Photoelectric cell) (Loudspeaker, earphones)
12. (No)
13. (Gaseous diode)
14. (Gaseous triode)
15. (15 V)
16. (High I gas triode)
17. (Ignitor, mercury pool)
18. (Bring E_p to zero)
19. (Fast ignition)
20. (N, P)
21. (P)
22. (0.6 V) (0.3 V)
23. (E, B, C)
24. (Forward) (Reverse)
25. (High) (Low)
26. (Integrated circuit)
27. (Large-scale IC)
28. (Field-effect transistor) (Triode)
29. (Thyratron)
30. (Grid) (Base) (Gate) (Grid) (Gate)
31. (Ac machinery control, etc.)
32. (Fullwave SCR)
33. (Light-activated SCR) (Anode, cathode, gate)
34. (Negative)

CHAPTER 18

DC Ammeters

1. (345 to 355 mA)
2. (Moving coil)
3. (Magnetic shield maintains calibrations when near iron or magnetic fields)
4. (Aluminum) (Acts as damper, is light, has less inertia)
5. (Springs)
6. (Counter and quadrantal weights)
7. (Reduces magnetic flux leakage, concentrates field around coil)
8. (Outer end of one spring)
9. (Less sensitive) (Damped)
10. (0.04002 Ω) (Overdamped)
11. (Higher) (Zero TC)
12. (More turns on moving coil)
13. (Critical)
14. (Linear) (Nonlinear or current-squared)
15. ($R_1 = 36.44$ Ω) ($R_2 = 3.56$ Ω) ($R_3 = 0.404$ Ω)

CHAPTER 19

Other DC Meters

1. (Current)
2. (1000 Ω/V) (20 kΩ/V) (10 kΩ/V)
3. (Measuring high-R circuits)
4. (Low-R circuits)
5. (At least 5; 10 would be better)

6. (Right) (Left)
7. (45)
8. (Resistance) (Ohmmeter)
9. (Unbalanced) (10 MΩ/V) (10 kΩ/V)
10. (Electrostatic voltmeter) (Effective values)
11. (Digital voltmeter)
12. (A null voltage) (Ramp voltage zero)
13. (Any normally scaled meter with indicator needle)
14. (Parallax)
15. (Wattmeter has electromagnetic field instead of permanent-magnet field)
16. (Voltage across circuit) (Load current)
17. (Ampere-hour meter)
18. (3)
19. (Oscillator)
20. (See Figs. 19-5, 19-6, 19-8, 19-9, 19-13)

CHAPTER 20

AC Meters

1. (50 V)
2. (16.7 V)
3. (50.9 V) (25.4 V)
4. (None)
5. (Four) (One)
6. (0.636, the original multiplier's resistance)
7. (2.22) (3.14)
8. (0.63 W) (1 W) (5 W)
9. (Electromagnetic field for dynamometers)
10. (VU meter) (1 mW) (600 Ω)
11. (High)
12. (Low-retentivity iron) (Yes)
13. (No) (More than 60 V)
14. (Electrodynamometer) (3)
15. (0.694)
16. (Pointers on circular scales)
17. (Ampere-hour and watthour)
18. (Thermocouple) (It is slow-moving)
19. (Vibrating-reed) (Digital counter, beat frequency)
20. (Vibrating-reed)
21. (*RC* frequency meter) (Electronic counter)
22. (Grid-dip meter)
23. (See Figs. 20-7*b*, 20-15, 20-16, 20-6)

CHAPTER 21

Alternators

1. (Prime mover)
2. (Field windings) (Rotating field)
3. (Reduce field excitation)
4. (No load)
5. (To produce constant-frequency ac)
6. (Rotating field)
7. (Increases load on prime mover, slowing it)
8. (700 W) (90.6%)
9. (Voltage, frequency, phase)
10. (Increases E_o)
11. (Yes; commonly done)
12. (Yes; commonly done)
13. (76.2 hp)
14. (90°)
15. (3) (3)
16. (100 V) (100 V) (173 V) (57.8 V)
17. (Y with grounded center)
18. (3-ϕ) (Open delta)
19. (6.92 kV) (2.31 kV)
20. (With constant R, increasing E decreases I, power loss is proportional to I^2)
21. (120 kV or 240 kV) (24 kV) (12 kV) (120 V or 240 V)
22. (See Figs. 21-4, 21-12, 21-14)

CHAPTER 22

Generators and Motors

1. (Sinusoidal ac) (Pulsating dc)
2. (Varying dc) (Higher average dc, less ripple)
3. (Excitation E or field rheostat) (Field rheostat)
4. (15.8%) (Lower)
5. (To prevent sparking)
6. (267 Hz)
7. (Two-layer drum winding)
8. (15 V)
9. (Very poor voltage regulation)
10. (When it will not start to build up a voltage)
11. (Rising-voltage characteristic) (Compound)
12. (Series)
13. (Overcompounding)
14. (Interpole or commutating pole)
15. (Sparking, brush wear)
16. (4) (In parallel)
17. (Heavy current output)
18. (Starting resistors)
19. (Mechanical load or friction)
20. (5.97%)
21. (Strong starting torque)
22. (Good speed regulation, easy to control speed)
23. (Reduce field current)
24. (Counter-emf in armature, field excitation)
25. (Compound)
26. (Universal motor) (Lessens)
27. (Higher ripple frequency results in greater X_L and less current in machine) (Fullwave)
28. (Phase of rectification) (Reversing field or armature circuits)
29. (Dynamotor)
30. (Inverter or rotary converter)
31. (Synchronous) (Dc) (Zero)
32. (Squirrel cage)
33. (Phase windings or poles)
34. (Increases efficiency of machine)
35. (Polyphase)
36. (3) (3)
37. (By flattening two sides of rotor)
38. (Shaded-pole) (Shape rotor)
39. (3-ϕ)
40. (Inductive) (Add capacitors across line, synchronous motors)
41. (See Figs. 22-6, 22-7, 22-8, 22-16, 22-17)

CHAPTER 23

Wiring Practices

1. (PC boards) (Dissolves copper)
2. (If I is small)
3. (144 cm)
4. (20 to 35 W)
5. (Heat confinement)
6. (Ampere-carrying capacity)

7. (1.64 greater)
8. (15 A) (20 A)
9. (Confines heat more)
10. (Copper wire in concrete foundation)
11. (Grounded) (Hot) (Grounding)
12. (Edison)
13. (Can be restored)
14. (When within reach)
15. (12 ft)
16. (Ground fault interrupter) (Bathrooms, outside)
17. (More than 5 mA flows load case to ground)
18. (3-ϕ circuits, high voltage)
19. (Tapped delta or open-delta)
20. (Splicing and tapping wires)
21. (24 in.)
22. (Wye with neutral)
23. (See Fig. 23-1)
24. (See Fig. 23-4)

CHAPTER 24

Batteries

1. (Ionization)
2. (Electrodes must be of different substances)
3. (Negative)
4. (Copper, iron)
5. (Zinc, manganese dioxide) (Ammonium and zinc chloride) (1.505 V)
6. (Voltmeter, cell under load) (Voltmeter) (Hydrometer) (Voltmeter)
7. (Potassium hydroxide electrolyte instead of ammonium chloride)
8. (Mercury)
9. (Lead dioxide) (Lead) (Sulfuric acid and water)
10. (Lead sulfate) (Lead sulfate) (Water)
11. (50%) (1%) (10%)
12. (Sulfates, cracks its case)
13. (Bubbling dislodges sulfate crystals, overheats cell, may buckle cell)
14. (Give off hydrogen and oxygen, which can explode if ignited)
15. (1.280–1.300) (1.220)
16. (Baking-soda solution, dilute ammonia) (Vinegar or dilute acid)
17. (Ampere-hours) (Yes)
18. (Twice the voltage at half the current produces same energy)
19. (Dropping resistor)
20. (Solid-state)
21. (Heats. Loses water.)
22. (Keep them cool) (To reduce internal chemical action)
23. (1.37 V) (2.1 V) (1.5 V)
24. (Nickel-cadmium, rechargeable alkaline)
25. (Iron, selenium)
26. (Silicon)
27. (Strontium-90 serves as energy source instead of the sun's heat)
28. (1000 A/m^2) (Hydrogen, oxygen)

CHAPTER 25

Kirchhoff Laws in Circuits

1. (Counterclockwise)
2. (Positive or +)
3. (Voltage-drop across some resistor in a circuit)
4. (A source voltage, technically)
5. (Kirchhoff's voltage law) ($E_s - IR_1 - IR_2 = 0$)
6. (Incorrect current-direction assumption)
7. ($26.7 - 6.67I_L = I_{RG}$) ($30 - 10I_L = I_{RB}$) (3.21 A)
8. (20.9 A) (104.5 V) (11.8 A) (118 V)

CHAPTER 26

Superposition and Thévenin

1. (Kirchhoff) (Superposition) (Thévenin)
2. (12.38 A) (22.86 V)
3. (12.95 A) (−0.91 A, charging) (12.04 A)
4. (0.0915 A) (1.95 A) (1.06 A) (2.04 A)
5. (Reverses current direction only)
6. (300 Ω)
7. (321 Ω) (450 V)

CHAPTER 27

Magnetic Circuits

1. (Increases it)
2. (Permeance)
3. (Webers) (Maxwells) (Lines of force) (ϕ)
4. (Ampere-turns) (Gilberts) (Ampere-turns) (F)
5. (Ampere-turns per meter) (Oersteds) (Ampere-turns per inch) (H)
6. (Teslas) (Gausses) (Lines per square inch) (B)
7. (10^8)
8. (42.5 V)
9. (0.167 T)
10. (900,000 amp-turns/Wb)
11. (Permeability)
12. (Diamagnetic)
13. (Paramagnetic)
14. (Ferromagnetic)
15. (Domains)
16. (Crystalline)
17. (Curie point)
18. (Carbon)
19. (Hysteresis)
20. (Narrow) (Wide) (Medium)
21. (Depermed or degaussed)
22. (Alternating)
23. (Magnetostriction)
24. (Transducer)
25. (600-Hz sound)

CHAPTER 28

Sine Waves

1. (200–3000 Hz)
2. (About 20,000 Hz)
3. (UHF, SHF, and EHF, or 300 MHz–300 GHz)
4. (Frequency)
5. (5 kHz)
6. (No)
7. (Square wave)
8. (Phasor)
9. (57.3°)
10. (0.0) (0.5) (0.707) (0.866) (1.0)
11. (66.2 V) (−76.1 V)
12. (Instantaneous voltage or current)

13. (Rms or dc value)
14. (−) (+) (−) (+)
15. (142°, 218°, 322°)
16. (54.55 V)
17. (89.1 V)
18. (19.5 V)
19. (−37.7 V)
20. (Angular velocity) (ω)
21. (57.8 V)
22. (28,260 rad/s)
23. (0.279 μs)
24. (2.4 s)
25. (21.4 m)
26. (5.29 μs) (3.33 μs)
27. (300,000,000 m/s) (331 m/s)

CHAPTER 29

Resistors and Inductors

1. (Power-dissipation capabilities)
2. (Air, oil, conduction, radiation)
3. (Temperature coefficient in parts per million per degree Celsius)
4. (Wind on thin card, or bend into hairpin shape and wind on form)
5. (Radial) (Axial)
6. (999 Ω)
7. (Carbon) (Wire)
8. (1%)
9. (4700 Ω) (10%)
10. (39 Ω) (5%) (It is wire-wound)
11. (Inductor) (250 μH) (10% tolerance)
12. (Impedance) (C and R)
13. (Universal, pie)
14. (E and I)
15. (Eddy currents) (Audible buzzing)
16. (Iron core with no gap)
17. (Smoothing choke) (Swinging choke)
18. (0.375 J) (0.075 s)
19. (Resistance, capacitance) (20-Ω, 0.1 μF)
20. (0.125 s)
21. (No air molecules to ionize)
22. (0.5) (0.193 H) (0.0668 H)
23. (2)
24. ($k = M/\sqrt{L_1L_2}$)
25. [$L = 0.002l(2.3 \log 4l/d)$]
26. (When internal resistance is negligible)
27. (Skin effect)
28. (To determine frequency of resonance of an LC circuit)

CHAPTER 30

Capacitors

1. (337.5 pF)
2. (0.009 J) (0 s) (3×10^{-5})
3. (Vacuum) (Air or paper)
4. (Infinite ohms) (Infinite ohms) (Zero or very few ohms)
5. (−)
6. (Less) (Less)
7. (Electrolytic)
8. (Dc electrolytic)
9. (Dielectric hysteretic losses, less effective capacitance)
10. (0.01 s) (1.76 mA) (8.8 V) (31.2 V)
11. (0.0151 A) (0.0136 A)
12. (5100 pF, 10%)
13. (Two significant figures and a multiplier number)

CHAPTER 31

Series AC Circuits

1. ($X = \sqrt{Z_2 - R_2}$)
2. ($E_R = \sqrt{E_S{}^2 - E_X{}^2}$)
3. (Cos θ) (Tan θ) (Sin θ)
4. (18.4 Ω) (6.52 A)
5. (60.6°) (0.49) (16 Ω) (Line E and I in phase)
6. (40 Ω) (38.1 Ω) (0.101 H)
7. (0.300) (108 W) ($0.0043)
8. (69.3 Ω) (1.44 A) (548 V)
9. (52.6°) (Lagging) (0.607) (87.1 W)
10. (When E_R and E_X are added at right angles)
11. (45°)
12. (45°)

CHAPTER 32

Parallel and Complex AC Circuits

1. [$X_1X_2/(X_1 + X_2)$] [$X_LX_C/(X_L - X_C)$]
2. (Conductance G) (Admittance Y) (Susceptance B)
3. (Capacitive susceptance)
4. (To the right) (Downward) (Downward) (Downward) (Downward)
5. (26.9 Ω) (0.0372 mho) (40.2 Ω) ($X_C = 36.2$ Ω) (−48°) (0.0249 mho) (0.0276 mho) (Leads)
6. (0.012 mho) (83.3 Ω) ($X_L = 46.2$ Ω) (33.7°) (69.3 Ω) (Lags)
7. (69.3 Ω) ($X_L = 26.2$ Ω) (20.8°) (Lags)
8. (74.1 Ω) (0.0135 mho) (79.4 Ω) ($X_L = 208$ Ω) (20.8°) (0.0126 mho) (0.00479 mho) (Lags)
9. ($X_C = 43.8$ Ω) (22.8 Ω) (4.38 A) (26.7 Ω) (31.4°) (Leads)
10. (112.5 V) (1.125 A)

CHAPTER 33

The j Operator

1. (Prefixed with a j) (Prefixed with an i)
2. (j5.6 V) (−j2.8 A) (j759 Ω) (−j23.8 Ω)
3. (−1) (−j)
4. (90°) (−90° or 270°) (180°) (270°)
5. ($Z = 3 + j4$) ($5\angle 53.1°$ **Ω**)
6. (Complex) (Polar) (Polar) (Complex)
7. (44.7 Ω) (26.6°) (2.24 A) (201 W)
8. ($39\angle 39.8°$ Ω) (Inductive)
9. ($75 = 65 - j37.5$ V) (Capacitive)
10. (II) (IV) (I) (III)
11. (28.3 Ω, 28.3 Ω, 40 Ω) (Capacitive) (Parallel)
12. (Rectangular) (Polar)
13. (42.4 Ω) ($0.0236\angle -45°$mho) ($Y = 0.0167 - j0.0167$)
14. ($0.0125\angle 90°$mho) ($Y = 0 + j0.0125$)
15. ($0.0172\angle 59.1°$ mho)
16. ($Z = 58.1 \angle -59.1°$ Ω)
17. ($Z = 29.9 - j49.8$)
18. ($Z = 15 + j15$)
19. ($44.9 - j34.8$) ($56.8\angle 37.8°$ Ω)

CHAPTER 34

Power Transformers

1. (427 V) (171 mA plus excitation current)
2. (Core loss) (Magnetizing) (Excitation)
3. (3014 Ω) (1930 Ω)
4. (0.215 A) (4.17 A) (4.37 A) (0.0727 H) (73 W) (79.6%) (83.4 V)
5. (Copper) (Eddy currents) (Copper)
6. (None)
7. (Silicone with inorganic material)
8. (Unity)
9. (Encircles windings) (Encircled by the windings)
10. (Toroid)
11. (Less leakage flux, greater stacking factor, less air gap and more inductance)
12. (Stacking factor)
13. (400 to 800 Hz) (Equipment is lighter)
14. (Frequency, kVA, voltage)
15. (Frequency range, power)
16. (To correct lagging PF)
17. (Capacitance of windings)
18. (Between pri and sec)
19. (High current)
20. (3.87%)
21. (6.32:1)
22. (No electrical isolation between pri and sec)

CHAPTER 35

Resonant Circuits

1. (1.53 μF) (34.2) (2.34 Hz) (1950 V)
2. (C) (*L*) (*R*)
3. (At resonance)
4. (Above resonance)
5. (20 to 20,000 Hz)
6. (Narrows BW) (Increases *I*) (Increases E_C)
7. (Q)
8. (0.707 of previous f_o)
9. (Bandwidth)
10. (Zero) (*I* leads, up to 90°)
11. (503 Hz)
12. (96.6 μH)
13. (106 pF)
14. (25-μH coil)
15. (35.7)
16. (Far below f_o)
17. (34 dB)
18. (39 dB)
19. (1.255 W)
20. (4733 W)

CHAPTER 36

Antiresonant Circuits

1. (Low-Q)
2. (Antiresonant with *R* in one leg only)
3. (Damped)
4. (In phase) (Out of phase)
5. (Temperature, resistance)
6. (Inductive)
7. (4) (0.000303 mho) (3300 Ω) (0.5 MHz)
8. ($Q = R/X$)
9. (High) (Low)
10. (Inductive) (Capacitive)
11. (0.707 max., half-power, −3 dB)
12. (Shunt resistance across it, add equal *R* to both legs)
13. (Add *R* to 1 branch only)
14. (Half-power points) (*R*)
15. (Capacitive)
16. (Loose or undercoupled, critically coupled, transitional or flat-topped, tight or overcoupled)
17. (Critical) (Tight or unity)
18. (For a broad, flat-topped pass band)
19. (Loose)
20. (Oscillator)
21. (Decrease C) (Decrease *L*)
22. (Yes)
23. (Lenz's law)

CHAPTER 37

Filters

1. (High insertion loss, not very sharp cutoff)
2. (Low) (High)
3. (Low) (High)
4. (Less loss) (Decrease) (Less slope)
5. (Distributed capacitance)
6. (Surge impedance)
7. (L, T, and π)
8. (*m*-derived) (Constant-*k*)
9. (High-pass) (Low-pass) (Bandpass) (Bandstop)
10. (2- to 3-kHz bandpass)
11. (Sharp cutoff) (Greater insertion loss)
12. (Wavetrap, bandstop filter)
13. (Sharper cutoff, less insertion loss)
14. (3.3)
15. (Mechanical filter) (2:1)
16. (*m*-derived)
17. (Series *m*-derived) (Shunt *m*-derived)
18. (0.6)
19. (Zero) (One)
20. (Add other *m*-value sections, use smaller *m*-value filter, add constant-*k* sections)
21. (Wavetrap circuits)
22. (Inductance)
23. (Composite)
24. (Balanced)
25. (Same f_c, same *R*)
26. (Piezoelectric)
27. (Several thousand)
28. (Two) (Four)
29. (*m*-derived, crystal)
30. (Crossover)
31. (Low-pass constant-*k*) (Smoothing)
32. (Halfwave)

APPENDIX A ELECTRICAL SYMBOLS

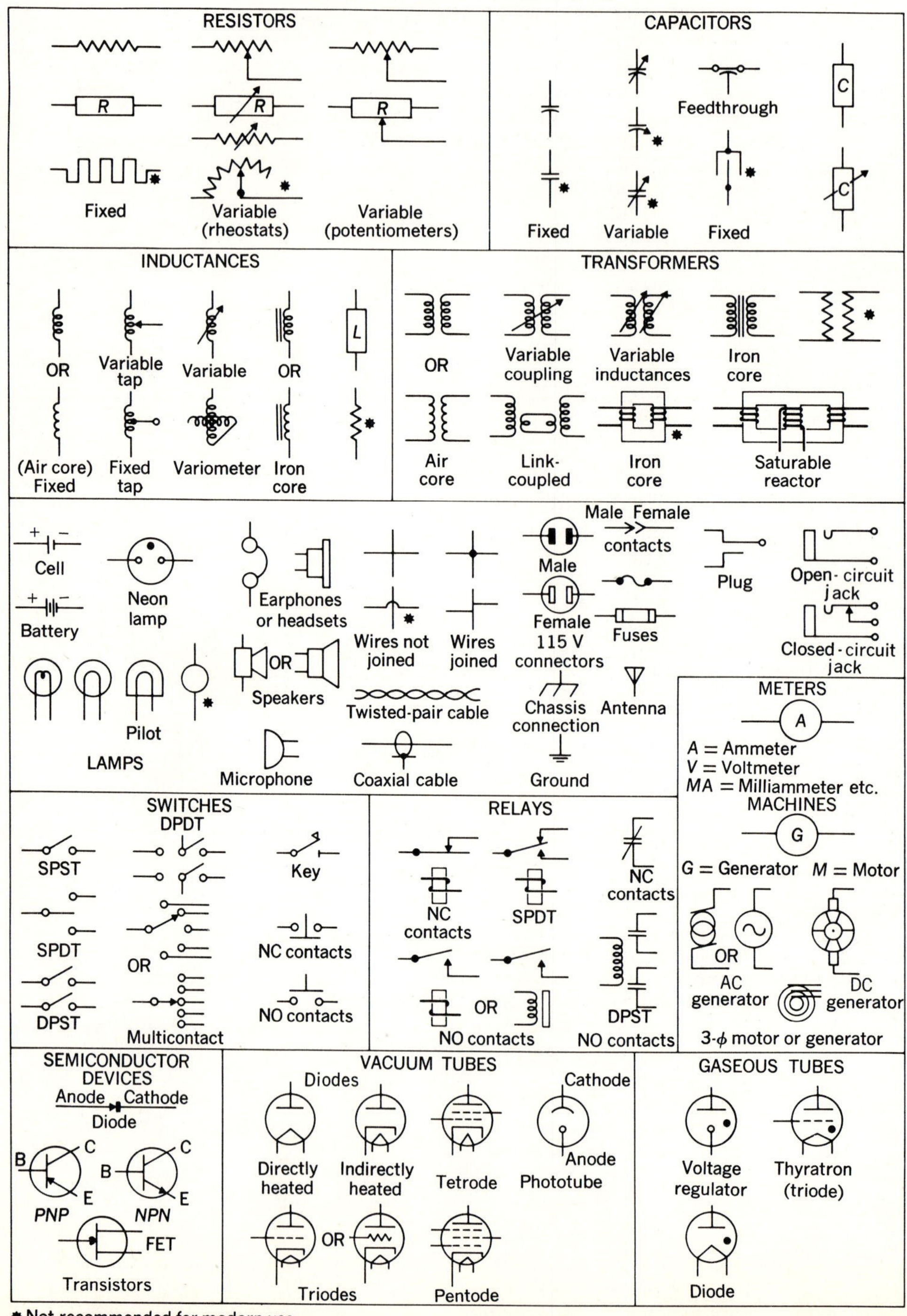

* Not recommended for modern use.

APPENDIX B

STANDARD AMERICAN WIRE GAGE (AWG)

Size AWG	Area, circular mils**	Diameter, mils		Ampacity, in A			Resistance, 1000 ft 25°C copper	Diameter, in mm
		Solid copper	Stranded (strands)	60°C insul. *TW, RUW	75°C insul. THW, RH	500 cir. mil/A		
34	39.7	6.3				0.079		0.160
32	63.2	7.95				0.126		0.203
30	100	10				0.200		0.254
28	159	12.6				0.318		0.320
26	253	15.9				0.506		0.404
24	404	20.1				0.808		0.511
22	641	25.3				1.25		0.643
20	1020	32				2.05		0.813
18	1620	40.3				3.25	6.51	1.02
16	2580	50.8				5.16	4.12	1.29
14	4110	64.1		15	15	8.2	2.57	1.63
12	6230	80.8		20	20	12.4	1.62	2.05
10	10,400	102		30	30	20.8	1.018	2.59
8	16,500	128.5		40	45	33	0.6404	3.26
6	26,200	162	184(7)	55	65	52.4	0.410	4.12
4	41,700	204.3	232(7)	70	85	83.4	0.259	5.19
3	52,620	229.4	260(7)	80	100	105	0.205	5.83
2	66,360	257.6	292(7)	95	115	133	0.162	6.54
1	83,690	289.3	332(19)	110	130	167	0.129	7.35
0	105,600	324.9	372(19)	125	150		0.102	8.25
00	133,100	364.8	418(19)	145	175		0.0811	9.27
000	167,800	409.6	470(19)	165	200		0.0642	10.40
0000	211,600	460.0	528(19)	195	230		0.0509	11.7
MCM†								
250	250,000		575(37)	215	255		0.0431	
300	300,000		630(37)	240	285		0.0360	
350	350,000		681(37)	260	310		0.0308	
400	400,000		728(37)	280	335		0.0270	
500	500,000		813(37)	320	380		0.0216	
1000	1,000,000		1150(61)	455	545		0.0108	
2000	2,000,000		1630(127)	560	665		0.00539	

* T, thermoplastic; W, water-resistant; R, rubber; RU, latex rubber; H, higher temperature.
** Wire diameter in mils when squared gives "circular mils."
† MCM, thousands of circular mils.
Note: Hard-drawn copper has 2.5% higher resistivity than annealed copper in table.

APPENDIX C

POWERS OF TEN

One of the difficulties encountered in working mathematical problems is setting the decimal place. While rules for multiplying and dividing numbers with decimal fractions are fairly familiar to everyone, the lowly decimal point still accounts for a good percentage of wrong answers in electrical problems.

A method has been devised which will indicate where to set the decimal place in multiplication and division of decimal numbers. For example, a problem requires multiplication of 495 by 0.0000308. This is a relatively straightforward problem, but can you tell what the approximate answer is without actually using pencil and paper? If you use *powers of ten*, also known as *scientific notation*, you can determine an approximate answer and at the same time a correct decimal point.

If the problem had been 4.95×3.08, you probably would recognize that 4.95 is nearly 5, and that 3.08 is close to a value of 3. Therefore, the correct answer must be approximately 5×3, or 15 (the actual value is 15.246). However, in the original problem of 495×0.0000308, where would the decimal place be?

Going back to the number 495, if the decimal point is moved to the left two places, the number becomes 4.95. It is necessary to multiply 4.95 by 10 to obtain 49.5, and then by another 10 to obtain 495. This is to say that if 4.95 is multiplied by 10 twice, the answer is 495. The number 10^2 is known as 10 *to the second power*. Any number multiplied by 10^2 is actually being multiplied by 100. Similarly, 10^3 means multiplication by 1000; 10^5 indicates multiplication by 100,000; 10^6 represents 1,000,000; and so on. Powers of ten merely represent the moving of the decimal place. Thus, $\times 10^2$ after a number means to move the decimal place over two places to the *right*.

For the number 0.0000308, if the decimal place is moved to the right five places, the resulting number is 3.08, but this number must be multiplied by 10^{-5} to express the original number properly. In this case the -5 power indicates the movement of the decimal five places, but five places to the *left*. Thus 3.08×10^{-5} represents the number 0.0000308. Remember, move the decimal point right with a positive power; move the decimal point left with a negative power.

The original problem was to multiply 495×0.0000308. These numbers expressed in powers of ten are

$$4.95 \times 10^2 \times 3.08 \times 10^{-5}$$

It was determined that 4.95×3.08 is approximately 15, but where should the decimal place be? The rule for multiplying numbers expressed in powers of ten is to multiply the numbers but *add the powers*. In the problem above, 10^2 times 10^{-5} will be 10^{-3}. Therefore, the answer to the problem above must be 15.246×10^{-3}. Since 10^{-3} means that the decimal point is to be moved three places to the left, the final answer is 0.015246. Remember:

10^3	means move decimal 3 places to the right
10^1	means move decimal 1 place to the right
10^0	means move decimal 0 places
10^{-1}	means move decimal 1 place to the left
10^{-3}	means move decimal 3 places to the left

Quiz C-1. Test your understanding. Answer these check-up questions.

1. Express these numbers in powers of ten: 25.4 = ________ 375 = ________ 6570 = ________ 0.04 = ________ 0.0054 = ________ 0.00028 = ________ 703 = ________ 0.145 = ________ 56,800 = ________
2. Convert the following to powers of ten; then multiply and express the answer in powers of ten. $502 \times 0.03 =$ ________ $\times$ ________ = ________ $8.92 \times 0.0004 =$ ________ $\times$ ________ = ________ $4,560,000 \times 0.00003 =$ ________ $\times$ ________ = ________ $523 \times 7000 =$ ________ $\times$ ________ = ________ $0.124 \times 0.008 =$ ________ $\times$ ________ = ________

In dividing decimal numbers, the use of powers of ten simplifies the process considerably. For example, $9760 \div 0.032$ expressed in powers of ten is

$$\frac{9.76 \times 10^3}{3.2 \times 10^{-2}}$$

By observation it can be seen that 3.2 will divide into 9.76 a little more than 3 times (actually, 3.05 times). The rule to simplify the powers of ten when dividing is to move the powers from below the fraction bar to above and *change the sign of the power*. The problem then becomes

$$\frac{9.76 \times 10^3 \times 10^2}{3.2} = \frac{9.76 \times 10^5}{3.2} = 3.05 \times 10^5$$

Since 10^5 represents 100,000, the answer is 305,000.

Quiz C-2. Test your understanding. Answer these check-up questions.

Convert the following into powers of ten; then divide and express the answer in powers of ten:

1. $4730 \div 0.04$ ________
2. $78.3 \div 800$ ________
3. $0.000062 \div 90$ ________
4. $98,400 \div 3,000,000$ ________
5. $0.004 \div 0.00000328$ ________

ANSWERS TO CHECK-UP QUIZ C-1

1. (2.54×10^1) (3.75×10^2) (6.57×10^3) (4×10^{-2}) (5.4×10^{-3}) (2.8×10^{-4}) (7.03×10^2) (1.45×10^{-1}) (5.68×10^4) **2.** (15.06×10^0) (35.68×10^{-4}) (13.68×10^1) (36.61×10^6) (9.92×10^{-4})

ANSWERS TO CHECK-UP QUIZ C-2

1. (1.18×10^5) **2.** $(0.975 \times 10^{-1}$ or $9.75 \times 10^{-2})$ **3.** (6.89×10^{-7}) **4.** (3.28×10^{-2}) **5.** (1.22×10^3)

APPENDIX D — *LCXf* NOMOGRAPH

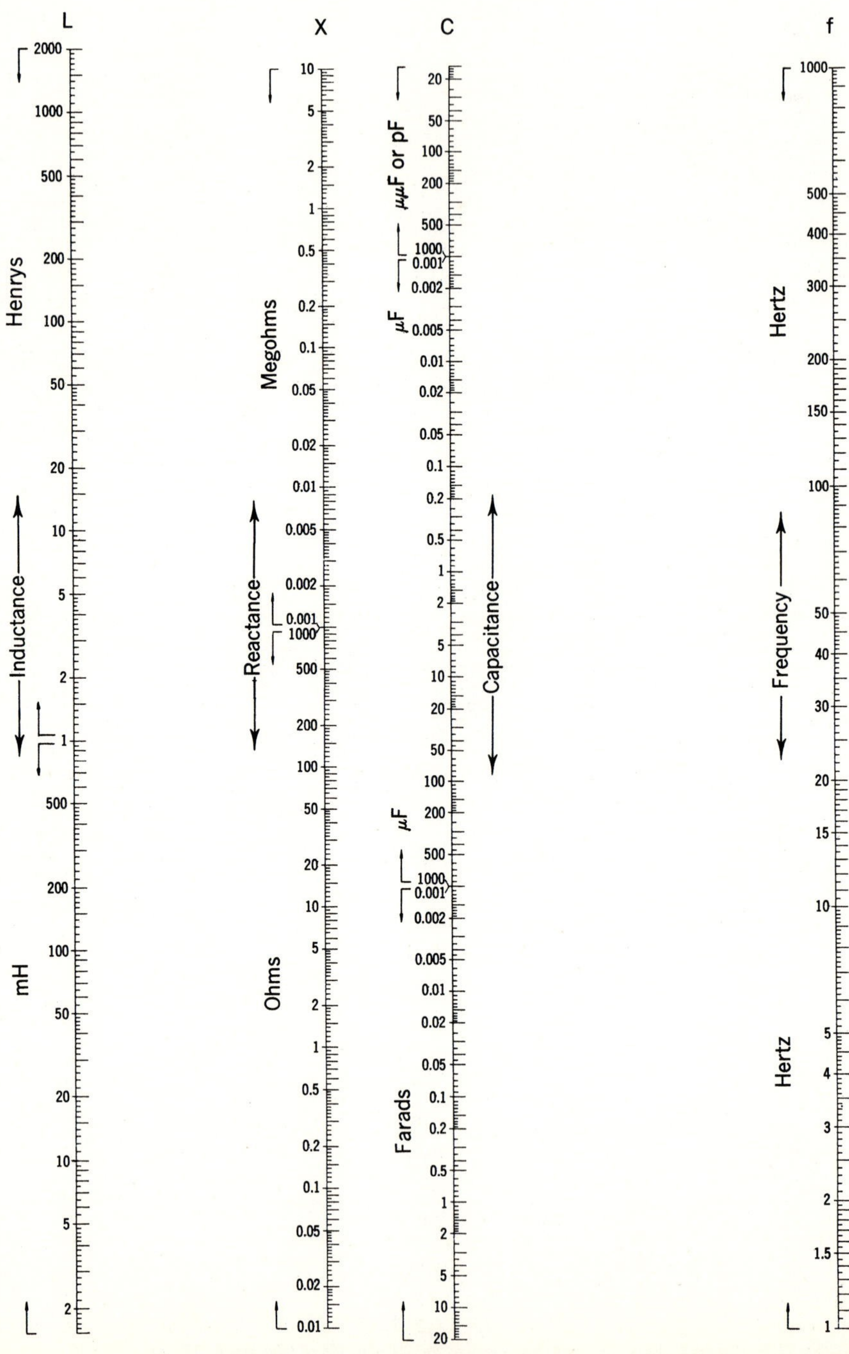

Appendix D Chart 1 (*Sylvania Electric Products*)

LCXf NOMOGRAPH (continued)

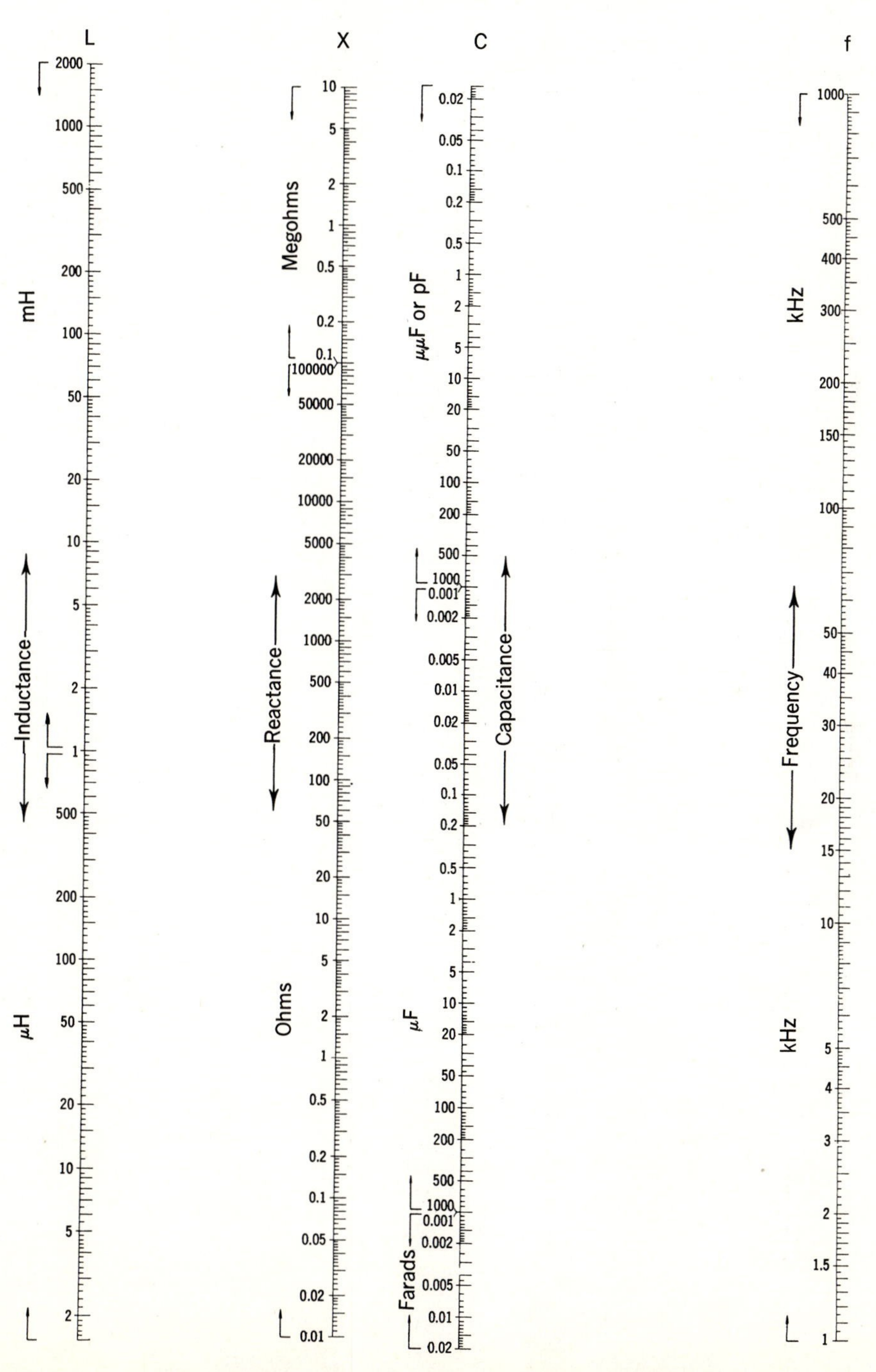

Appendix D Chart 2 (*Sylvania Electric Products*)

LCXf NOMOGRAPH (continued)

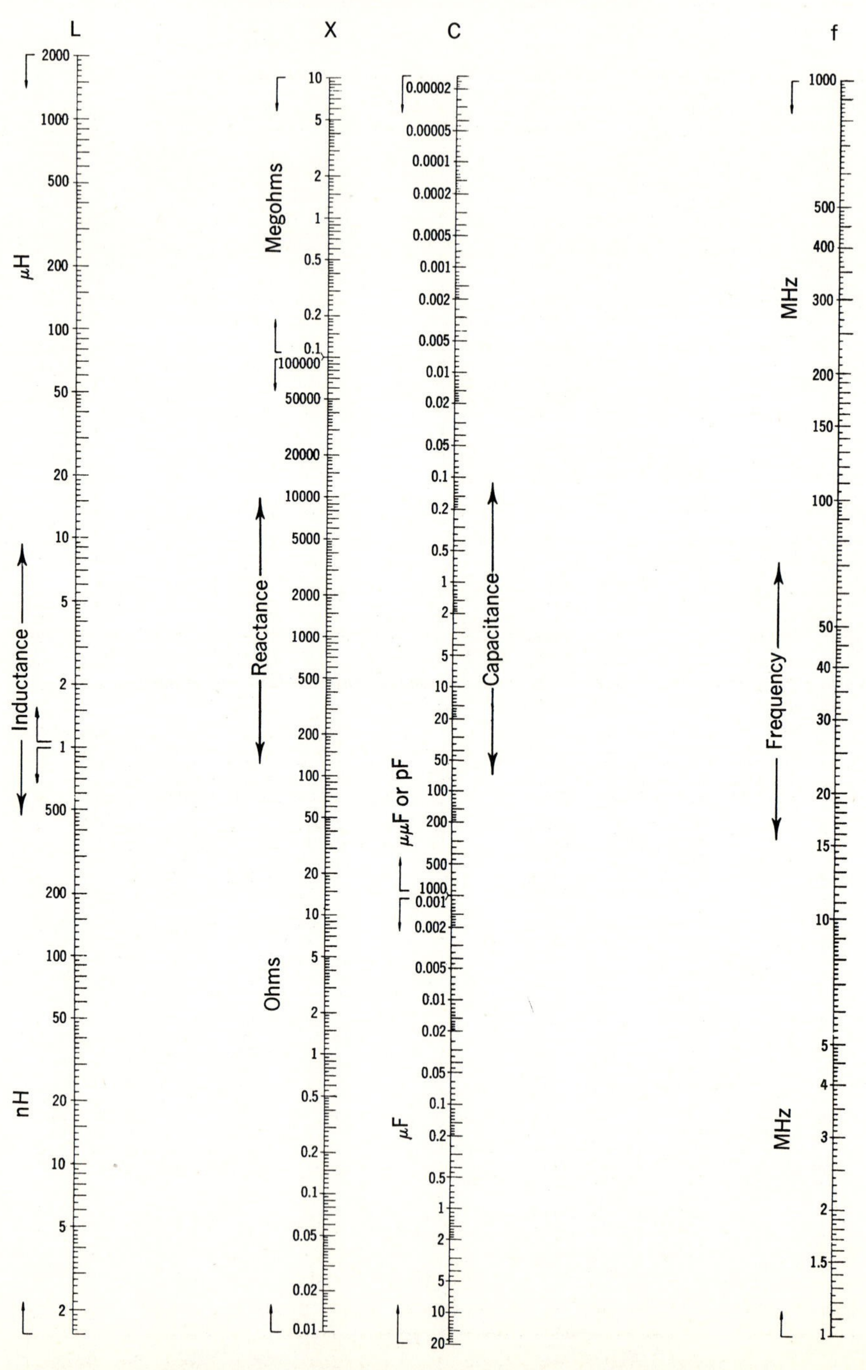

Appendix D Chart 3 (*Sylvania Electric Products*)

APPENDIX E

TABLE OF NATURAL TRIGONOMETRIC FUNCTIONS

Angle, °	Sin	Tan	Cot	Cos	
0.0	.00000	.00000	∞	1.00000	90.0
.1	.00175	.00175	572.96	1.00000	.9
.2	.00349	.00349	286.48	0.99999	.8
.3	.00524	.00524	190.98	.99999	.7
.4	.00698	.00698	143.24	.99998	.6
.5	.00873	.00873	114.59	.99996	.5
.6	.01047	.01047	95.489	.99995	.4
.7	.01222	.01222	81.847	.99993	.3
.8	.01396	.01396	71.615	.99990	.2
.9	.01571	.01571	63.657	.99988	.1
1.0	.01745	.01746	57.290	.99985	89.0
.1	.01920	.01920	52.081	.99982	.9
.2	.02094	.02095	47.740	.99978	.8
.3	.02269	.02269	44.066	.99974	.7
.4	.02443	.02444	40.917	.99970	.6
.5	.02618	.02619	38.188	.99966	.5
.6	.02792	.02793	35.801	.99961	.4
.7	.02967	.02968	33.694	.99956	.3
.8	.03141	.03143	31.821	.99951	.2
.9	.03316	.03317	30.145	.99945	.1
2.0	.03490	.03492	28.636	.99939	88.0
.1	.03664	.03667	27.271	.99933	.9
.2	.03839	.03842	26.031	.99926	.8
.3	.04013	.04016	24.898	.99919	.7
.4	.04188	.04191	23.859	.99912	.6
.5	.04362	.04366	22.904	.99905	.5
.6	.04536	.04541	22.022	.99897	.4
.7	.04711	.04716	21.205	.99889	.3
.8	.04885	.04891	20.446	.99881	.2
.9	.05059	.05066	19.740	.99872	.1
3.0	.05234	.05241	19.081	.99863	87.0
.1	.05408	.05416	18.464	.99854	.9
.2	.05582	.05591	17.886	.99844	.8
.3	.05756	.05766	17.343	.99834	.7
.4	.05931	.05941	16.832	.99824	.6
.5	.06105	.06116	16.350	.99813	.5
.6	.06279	.06291	15.895	.99803	.4
.7	.06453	.06467	15.464	.99792	.3
.8	.06627	.06642	15.056	.99780	.2
.9	.06802	.06817	14.669	.99768	.1
4.0	.06976	.06993	14.301	.99756	86.0
.1	.07150	.07168	13.951	.99744	.9
.2	.07324	.07344	13.617	.99731	.8
.3	.07498	.07519	13.300	.99719	.7
.4	.07672	.07695	12.996	.99705	.6
.5	.07846	.07870	12.706	.99692	.5
.6	.08020	.08046	12.429	.99678	.4
.7	.08194	.08221	12.163	.99664	.3
.8	.08368	.08397	11.909	.99649	.2
.9	.08542	.08573	11.664	.99635	.1
5.0	.08716	.08749	11.430	.99619	85.0
.1	.08889	.08925	11.205	.99604	.9
.2	.09063	.09101	10.988	.99588	.8
.3	.09237	.09277	10.780	.99572	.7
.4	.09411	.09453	10.579	.99556	.6
.5	.09585	.09629	10.385	.99540	.5
.6	.09758	.09805	10.199	.99523	.4
.7	.09932	.09981	10.019	.99506	.3
.8	.10106	.10158	9.8448	.99488	.2
.9	.10279	.10334	9.6768	.99470	.1
6.0	.10453	.10510	9.5144	.99452	84.0
	Cos	Cot	Tan	Sin	Angle, °

Angle, °	Sin	Tan	Cot	Cos	
6.0	.10453	.10510	9.5144	.99452	84.0
.1	.10626	.10687	9.3572	.99434	.9
.2	.10800	.10863	9.2052	.99415	.8
.3	.10973	.11040	9.0579	.99396	.7
.4	.11147	.11217	8.9152	.99377	.6
.5	.11320	.11394	8.7769	.99357	.5
.6	.11494	.11570	8.6427	.99337	.4
.7	.11667	.11747	8.5126	.99317	.3
.8	.11840	.11924	8.3863	.99297	.2
.9	.12014	.12101	8.2636	.99276	.1
7.0	.12187	.12278	8.1443	.99255	83.0
.1	.12360	.12456	8.0285	.99233	.9
.2	.12533	.12633	7.9158	.99211	.8
.3	.12706	.12810	7.8062	.99189	.7
.4	.12880	.12988	7.6996	.99167	.6
.5	.13053	.13165	7.5958	.99144	.5
.6	.13226	.13343	7.4947	.99122	.4
.7	.13399	.13521	7.3962	.99098	.3
.8	.13572	.13698	7.3002	.99075	.2
.9	.13744	.13876	7.2066	.99051	.1
8.0	.13917	.14054	7.1154	.99027	82.0
.1	.14090	.14232	7.0264	.99002	.9
.2	.14263	.14410	6.9395	.98978	.8
.3	.14436	.14588	6.8548	.98953	.7
.4	.14608	.14767	6.7720	.98927	.6
.5	.14781	.14945	6.6912	.98902	.5
.6	.14954	.15124	6.6122	.98876	.4
.7	.15126	.15302	6.5350	.98849	.3
.8	.15299	.15481	6.4596	.98823	.2
.9	.15471	.15660	6.3859	.98796	.1
9.0	.15643	.15838	6.3138	.98769	81.0
.1	.15816	.16017	6.2432	.98741	.9
.2	.15988	.16196	6.1742	.98714	.8
.3	.16160	.16376	6.1066	.98686	.7
.4	.16333	.16555	6.0405	.98657	.6
.5	.16505	.16734	5.9758	.98629	.5
.6	.16677	.16914	5.9124	.98600	.4
.7	.16849	.17093	5.8502	.98570	.3
.8	.17021	.17273	5.7894	.98541	.2
.9	.17193	.17453	5.7297	.98511	.1
10.0	.17365	.17633	5.6713	.98481	80.0
.1	.17537	.17813	5.6140	.98450	.9
.2	.17708	.17993	5.5578	.98420	.8
.3	.17880	.18173	5.5026	.98389	.7
.4	.18052	.18353	5.4486	.98357	.6
.5	.18224	.18534	5.3955	.98325	.5
.6	.18395	.18714	5.3435	.98294	.4
.7	.18567	.18895	5.2924	.98261	.3
.8	.18738	.19076	5.2422	.98229	.2
.9	.18910	.19257	5.1929	.98196	.1
11.0	.19081	.19438	5.1446	.98163	79.0
.1	.19252	.19619	5.0970	.98129	.9
.2	.19423	.19801	5.0504	.98096	.8
.3	.19595	.19982	5.0045	.98061	.7
.4	.19766	.20164	4.9594	.98027	.6
.5	.19937	.20345	4.9152	.97992	.5
.6	.20108	.20527	4.8716	.97958	.4
.7	.20279	.20709	4.8288	.97922	.3
.8	.20450	.20891	4.7867	.97887	.2
.9	.20620	.21073	4.7453	.97851	.1
12.0	.20791	.21256	4.7046	.97815	78.0
	Cos	Cot	Tan	Sin	Angle, °

TABLE OF NATURAL TRIGONOMETRIC FUNCTIONS (continued)

Angle, °	Sin	Tan	Cot	Cos	
12.0	.20791	.21256	4.7046	.97815	78.0
.1	.20962	.21438	4.6646	.97778	.9
.2	.21132	.21621	4.6252	.97742	.8
.3	.21303	.21804	4.5864	.97705	.7
.4	.21474	.21986	4.5483	.97667	.6
.5	.21644	.22169	4.5107	.97630	.5
.6	.21814	.22353	4.4737	.97592	.4
.7	.21985	.22536	4.4373	.97553	.3
.8	.22155	.22719	4.4015	.97515	.2
.9	.22325	.22903	4.3662	.97476	.1
13.0	.22495	.23087	4.3315	.97437	77.0
.1	.22665	.23271	4.2972	.97398	.9
.2	.22835	.23455	4.2635	.97358	.8
.3	.23005	.23639	4.2303	.97318	.7
.4	.23175	.23823	4.1976	.97278	.6
.5	.23345	.24008	4.1653	.97237	.5
.6	.23514	.24193	4.1335	.97196	.4
.7	.23684	.24377	4.1022	.97155	.3
.8	.23853	.24562	4.0713	.97113	.2
.9	.24023	.24747	4.0408	.97072	.1
14.0	.24192	.24933	4.0108	.97030	76.0
.1	.24362	.25118	3.9812	.96987	.9
.2	.24531	.25304	3.9520	.96945	.8
.3	.24700	.25490	3.9232	.96902	.7
.4	.24869	.25676	3.8947	.96858	.6
.5	.25038	.25862	3.8667	.96815	.5
.6	.25207	.26048	3.8391	.96771	.4
.7	.25376	.26235	3.8118	.96727	.3
.8	.25545	.26421	3.7848	.96682	.2
.9	.25713	.26608	3.7583	.96638	.1
15.0	.25882	.26795	3.7321	.96593	75.0
.1	.26050	.26982	3.7062	.96547	.9
.2	.26219	.27169	3.6806	.96502	.8
.3	.26387	.27357	3.6554	.96456	.7
.4	.26556	.27545	3.6305	.96410	.6
.5	.26724	.27732	3.6059	.96363	.5
.6	.26892	.27921	3.5816	.96316	.4
.7	.27060	.28109	3.5576	.96269	.3
.8	.27228	.28297	3.5339	.96222	.2
.9	.27396	.28486	3.5105	.96174	.1
16.0	.27564	.28675	3.4874	.96126	74.0
.1	.27731	.28864	3.4646	.96078	.9
.2	.27899	.29053	3.4420	.96029	.8
.3	.28067	.29242	3.4197	.95981	.7
.4	.28234	.29432	3.3977	.95931	.6
.5	.28402	.29621	3.3759	.95882	.5
.6	.28569	.29811	3.3544	.95832	.4
.7	.28736	.30001	3.3332	.95782	.3
.8	.28903	.30192	3.3122	.95732	.2
.9	.29070	.30382	3.2914	.95681	.1
17.0	.29237	.30573	3.2709	.95630	73.0
.1	.29404	.30764	3.2506	.95579	.9
.2	.29571	.30955	3.2305	.95528	.8
.3	.29737	.31147	3.2106	.95476	.7
.4	.29904	.31338	3.1910	.95424	.6
.5	.30071	.31530	3.1716	.95372	.5
.6	.30237	.31722	3.1524	.95319	.4
.7	.30403	.31914	3.1334	.95266	.3
.8	.30570	.32106	3.1146	.95213	.2
.9	.30736	.32299	3.0961	.95159	.1
18.0	.30902	.32492	3.0777	.95106	72.0
	Cos	Cot	Tan	Sin	Angle, °

Angle, °	Sin	Tan	Cot	Cos	
18.0	.30902	.32492	3.0777	.95106	72.0
.1	.31068	.32685	3.0595	.95052	.9
.2	.31233	.32878	3.0415	.94997	.8
.3	.31399	.33072	3.0237	.94943	.7
.4	.31565	.33266	3.0061	.94888	.6
.5	.31730	.33460	2.9887	.94832	.5
.6	.31896	.33654	2.9714	.94777	.4
.7	.32061	.33848	2.9544	.94721	.3
.8	.32227	.34043	2.9375	.94665	.2
.9	.32392	.34238	2.9208	.94609	.1
19.0	.32557	.34433	2.9042	.94552	71.0
.1	.32722	.34628	2.8878	.94495	.9
.2	.32887	.34824	2.8716	.94438	.8
.3	.33051	.35020	2.8556	.94380	.7
.4	.33216	.35216	2.8397	.94322	.6
.5	.33381	.35412	2.8239	.94264	.5
.6	.33545	.35608	2.8083	.94206	.4
.7	.33710	.35805	2.7929	.94147	.3
.8	.33874	.36002	2.7776	.94088	.2
.9	.34038	.36199	2.7625	.94029	.1
20.0	.34202	.36397	2.7475	.93969	70.0
.1	.34366	.36595	2.7326	.93909	.9
.2	.34530	.36793	2.7179	.93849	.8
.3	.34694	.36991	2.7034	.93789	.7
.4	.34857	.37190	2.6889	.93728	.6
.5	.35021	.37388	2.6746	.93667	.5
.6	.35184	.37588	2.6605	.93606	.4
.7	.35347	.37787	2.6464	.93544	.3
.8	.35511	.37986	2.6325	.93483	.2
.9	.35674	.38186	2.6187	.93420	.1
21.0	.35837	.38386	2.6051	.93358	69.0
.1	.36000	.38587	2.5916	.93295	.9
.2	.36162	.38787	2.5782	.93232	.8
.3	.36325	.38988	2.5649	.93169	.7
.4	.36488	.39190	2.5517	.93106	.6
.5	.36650	.39391	2.5386	.93042	.5
.6	.36812	.39593	2.5257	.92978	.4
.7	.36975	.39795	2.5129	.92913	.3
.8	.37137	.39997	2.5002	.92849	.2
.9	.37299	.40200	2.4876	.92784	.1
22.0	.37461	.40403	2.4751	.92718	68.0
.1	.37622	.40606	2.4627	.92653	.9
.2	.37784	.40809	2.4504	.92587	.8
.3	.37946	.41013	2.4383	.92521	.7
.4	.38107	.41217	2.4262	.92455	.6
.5	.38268	.41421	2.4142	.92388	.5
.6	.38430	.41626	2.4023	.92321	.4
.7	.38591	.41831	2.3906	.92254	.3
.8	.38752	.42036	2.3789	.92186	.2
.9	.38912	.42242	2.3673	.92119	.1
23.0	.39073	.42447	2.3559	.92050	67.0
.1	.39234	.42654	2.3445	.91982	.9
.2	.39394	.42860	2.3332	.91914	.8
.3	.39555	.43067	2.3220	.91845	.7
.4	.39715	.43274	2.3109	.91775	.6
.5	.39875	.43481	2.2998	.91706	.5
.6	.40035	.43689	2.2889	.91636	.4
.7	.40195	.43897	2.2781	.91566	.3
.8	.40355	.44105	2.2673	.91496	.2
.9	.40514	.44314	2.2566	.91425	.1
24.0	.40674	.44523	2.2460	.91355	66.0
	Cos	Cot	Tan	Sin	Angle, °

TABLE OF NATURAL TRIGONOMETRIC FUNCTIONS (continued)

Angle, °	Sin	Tan	Cot	Cos	
24.0	.40674	.44523	2.2460	.91355	66.0
.1	.40833	.44732	2.2355	.91283	.9
.2	.40992	.44942	2.2251	.91212	.8
.3	.41151	.45152	2.2148	.91140	.7
.4	.41310	.45362	2.2045	.91068	.6
.5	.41469	.45573	2.1943	.90996	.5
.6	.41628	.45784	2.1842	.90924	.4
.7	.41787	.45995	2.1742	.90851	.3
.8	.41945	.46206	2.1642	.90778	.2
.9	.42104	.46418	2.1543	.90704	.1
25.0	.42262	.46631	2.1445	.90631	65.0
.1	.42420	.46843	2.1348	.90557	.9
.2	.42578	.47056	2.1251	.90483	.8
.3	.42736	.47270	2.1155	.90408	.7
.4	.42894	.47483	2.1060	.90334	.6
.5	.43051	.47698	2.0965	.90259	.5
.6	.43209	.47912	2.0872	.90183	.4
.7	.43366	.48127	2.0778	.90108	.3
.8	.43523	.48342	2.0686	.90032	.2
.9	.43680	.48557	2.0594	.89956	.1
26.0	.43837	.48773	2.0503	.89879	64.0
.1	.43994	.48989	2.0413	.89803	.9
.2	.44151	.49206	2.0323	.89726	.8
.3	.44307	.49423	2.0233	.89649	.7
.4	.44464	.49640	2.0145	.89571	.6
.5	.44620	.49858	2.0057	.89493	.5
.6	.44776	.50076	1.9970	.89415	.4
.7	.44932	.50295	1.9883	.89337	.3
.8	.45088	.50514	1.9797	.89259	.2
.9	.45243	.50733	1.9711	.89180	.1
27.0	.45399	.50953	1.9626	.89101	63.0
.1	.45554	.51173	1.9542	.89021	.9
.2	.45710	.51393	1.9458	.88942	.8
.3	.45865	.51614	1.9375	.88862	.7
.4	.46020	.51835	1.9292	.88782	.6
.5	.46175	.52057	1.9210	.88701	.5
.6	.46330	.52279	1.9128	.88620	.4
.7	.46484	.52501	1.9047	.88539	.3
.8	.46639	.52724	1.8967	.88458	.2
.9	.46793	.52947	1.8887	.88377	.1
28.0	.46947	.53171	1.8807	.88295	62.0
.1	.47101	.53395	1.8728	.88213	.9
.2	.47255	.53620	1.8650	.88130	.8
.3	.47409	.53844	1.8572	.88048	.7
.4	.47562	.54070	1.8495	.87965	.6
.5	.47716	.54296	1.8418	.87882	.5
.6	.47869	.54522	1.8341	.87798	.4
.7	.48022	.54748	1.8265	.87715	.3
.8	.48175	.54975	1.8190	.87631	.2
.9	.48328	.55203	1.8115	.87546	.1
29.0	.48481	.55431	1.8040	.87462	61.0
.1	.48634	.55659	1.7966	.87377	.9
.2	.48786	.55888	1.7893	.87292	.8
.3	.48938	.56117	1.7820	.87207	.7
.4	.49090	.56347	1.7747	.87121	.6
.5	.49242	.56577	1.7675	.87036	.5
.6	.49394	.56808	1.7603	.86949	.4
.7	.49546	.57039	1.7532	.86863	.3
.8	.49697	.57271	1.7461	.86777	.2
.9	.49849	.57503	1.7391	.86690	.1
30.0	.50000	.57735	1.7321	.86603	60.0
	Cos	Cot	Tan	Sin	Angle, °

Angle, °	Sin	Tan	Cot	Cos	
30.0	.50000	.57735	1.7321	.86603	60.0
.1	.50151	.57968	1.7251	.86515	.9
.2	.50302	.58201	1.7182	.86427	.8
.3	.50453	.58435	1.7113	.86340	.7
.4	.50603	.58670	1.7045	.86251	.6
.5	.50754	.58905	1.6977	.86163	.5
.6	.50904	.59140	1.6909	.86074	.4
.7	.51054	.59376	1.6842	.85985	.3
.8	.51204	.59612	1.6775	.85896	.2
.9	.51354	.59849	1.6709	.85806	.1
31.0	.51504	.60086	1.6643	.85717	59.0
.1	.51653	.60324	1.6577	.85627	.9
.2	.51803	.60562	1.6512	.85536	.8
.3	.51952	.60801	1.6447	.85446	.7
.4	.52101	.61040	1.6383	.85355	.6
.5	.52250	.61280	1.6319	.85264	.5
.6	.52399	.61520	1.6255	.85173	.4
.7	.52547	.61761	1.6191	.85081	.3
.8	.52696	.62003	1.6128	.84989	.2
.9	.52844	.62245	1.6066	.84897	.1
32.0	.52992	.62487	1.6003	.84805	58.0
.1	.53140	.62730	1.5941	.84712	.9
.2	.53288	.62973	1.5880	.84619	.8
.3	.53435	.63217	1.5818	.84526	.7
.4	.53583	.63462	1.5757	.84433	.6
.5	.53730	.63707	1.5697	.84339	.5
.6	.53877	.63953	1.5637	.84245	.4
.7	.54024	.64199	1.5577	.84151	.3
.8	.54171	.64446	1.5517	.84057	.2
.9	.54317	.64693	1.5458	.83962	.1
33.0	.54464	.64941	1.5399	.83867	57.0
.1	.54610	.65189	1.5340	.83772	.9
.2	.54756	.65438	1.5282	.83676	.8
.3	.54902	.65688	1.5224	.83581	.7
.4	.55048	.65938	1.5166	.83485	.6
.5	.55194	.66189	1.5108	.83389	.5
.6	.55339	.66440	1.5051	.83292	.4
.7	.55484	.66692	1.4994	.83195	.3
.8	.55630	.66944	1.4938	.83098	.2
.9	.55775	.67197	1.4882	.83001	.1
34.0	.55919	.67451	1.4826	.82904	56.0
.1	.56064	.67705	1.4770	.82806	.9
.2	.56208	.67960	1.4715	.82708	.8
.3	.56353	.68215	1.4659	.82610	.7
.4	.56497	.68471	1.4605	.82511	.6
.5	.56641	.68728	1.4550	.82413	5
.6	.56784	.68985	1.4496	.82314	.4
.7	.56928	.69243	1.4442	.82214	.3
.8	.57071	.69502	1.4388	.82115	.2
.9	.57215	.69761	1.4335	.82015	.1
35.0	.57358	.70021	1.4281	.81915	55.0
.1	.57501	.70281	1.4229	.81815	.9
.2	.57643	.70542	1.4176	.81714	.8
.3	.57786	.70804	1.4124	.81614	.7
.4	.57928	.71066	1.4071	.81513	.6
.5	.58070	.71329	1.4019	.81412	.5
.6	.58212	.71593	1.3968	.81310	.4
.7	.58354	.71857	1.3916	.81208	.3
.8	.58496	.72122	1.3865	.81106	.2
.9	.58637	.72388	1.3814	.81004	.1
36.0	.58779	.72654	1.3764	.80902	54.0
	Cos	Cot	Tan	Sin	Angle, °

TABLE OF NATURAL TRIGONOMETRIC FUNCTIONS (continued)

Angle, °	Sin	Tan	Cot	Cos	
36.0	.58779	.72654	1.3764	.80902	54.0
.1	.58920	.72921	1.3713	.80799	.9
.2	.59061	.73189	1.3663	.80696	.8
.3	.59201	.73457	1.3613	.80593	.7
.4	.59342	.73726	1.3564	.80489	.6
.5	.59482	.73996	1.3514	.80386	.5
.6	.59622	.74267	1.3465	.80282	.4
.7	.59763	.74538	1.3416	.80178	.3
.8	.59902	.74810	1.3367	.80073	.2
.9	.60042	.75082	1.3319	.79968	.1
37.0	.60182	.75355	1.3270	.79864	53.0
.1	.60321	.75629	1.3222	.79758	.9
.2	.60460	.75904	1.3175	.79653	.8
.3	.60599	.76180	1.3127	.79547	.7
.4	.60738	.76456	1.3079	.79441	.6
.5	.60876	.76733	1.3032	.79335	.5
.6	.61015	.77010	1.2985	.79229	.4
.7	.61153	.77289	1.2938	.79122	.3
.8	.61291	.77568	1.2892	.79016	.2
.9	.61429	.77848	1.2846	.78908	.1
38.0	.61566	.78129	1.2799	.78801	52.0
.1	.61704	.78410	1.2753	.78694	.9
.2	.61841	.78692	1.2708	.78586	.8
.3	.61978	.78975	1.2662	.78478	.7
.4	.62115	.79259	1.2617	.78369	.6
.5	.62251	.79544	1.2572	.78261	.5
.6	.62388	.79829	1.2527	.78152	.4
.7	.62524	.80115	1.2482	.78043	.3
.8	.62660	.80402	1.2437	.77934	.2
.9	.62796	.80690	1.2393	.77824	.1
39.0	.62932	.80978	1.2349	.77715	51.0
.1	.63068	.81268	1.2305	.77605	.9
.2	.63203	.81558	1.2261	.77494	.8
.3	.63338	.81849	1.2218	.77384	.7
.4	.63473	.82141	1.2174	.77273	.6
.5	.63608	.82434	1.2131	.77162	.5
.6	.63742	.82727	1.2088	.77051	.4
.7	.63877	.83022	1.2045	.76940	.3
.8	.64011	.83317	1.2002	.76828	.2
.9	.64145	.83613	1.1960	.76717	.1
40.0	.64279	.83910	1.1918	.76604	50.0
.1	.64412	.84208	1.1875	.76492	.9
.2	.64546	.84507	1.1833	.76380	.8
.3	.64679	.84806	1.1792	.76267	.7
.4	.64812	.85107	1.1750	.76154	.6
40.5	.64945	.85408	1.1708	.76041	49.5
	Cos	Cot	Tan	Sin	Angle, °

Angle, °	Sin	Tan	Cot	Cos	
40.5	.64945	.85408	1.1708	.76041	49.5
.6	.65077	.85710	1.1667	.75927	.4
.7	.65210	.86014	1.1626	.75813	.3
.8	.65342	.86318	1.1585	.75700	.2
.9	.65474	.86623	1.1544	.75585	.1
41.0	.65606	.86929	1.1504	.75471	49.0
.1	.65738	.87236	1.1463	.75356	.9
.2	.65869	.87543	1.1423	.75241	.8
.3	.66000	.87852	1.1383	.75126	.7
.4	.66131	.88162	1.1343	.75011	.6
.5	.66262	.88473	1.1303	.74896	.5
.6	.66393	.88784	1.1263	.74780	.4
.7	.66523	.89097	1.1224	.74664	.3
.8	.66653	.89410	1.1184	.74548	.2
.9	.66783	.89725	1.1145	.74431	.1
42.0	.66913	.90040	1.1106	.74314	48.0
.1	.67043	.90357	1.1067	.74198	.9
.2	.67172	.90674	1.1028	.74080	.8
.3	.67301	.90993	1.0990	.73963	.7
.4	.67430	.91313	1.0951	.73846	.6
.5	.67559	.91633	1.0913	.73728	.5
.6	.67688	.91955	1.0875	.73610	.4
.7	.67816	.92277	1.0837	.73491	.3
.8	.67944	.92601	1.0799	.73373	.2
.9	.68072	.92926	1.0761	.73254	.1
43.0	.68200	.93252	1.0724	.73135	47.0
.1	.68327	.93578	1.0686	.73016	.9
.2	.68455	.93906	1.0649	.72897	.8
.3	.68582	.94235	1.0612	.72777	.7
.4	.68709	.94565	1.0575	.72657	.6
.5	.68835	.94896	1.0538	.72537	.5
.6	.68962	.95229	1.0501	.72417	.4
.7	.69088	.95562	1.0464	.72297	.3
.8	.69214	.95897	1.0428	.72176	.2
.9	.69340	.96232	1.0392	.72055	.1
44.0	.69466	.96569	1.0355	.71934	46.0
.1	.69591	.96907	1.0319	.71813	.9
.2	.69717	.97246	1.0283	.71691	.8
.3	.69842	.97586	1.0247	.71569	.7
.4	.69966	.97927	1.0212	.71447	.6
.5	.70091	.98270	1.0176	.71325	.5
.6	.70215	.98613	1.0141	.71203	.4
.7	.70339	.98958	1.0105	.71080	.3
.8	.70463	.99304	1.0070	.70957	.2
.9	.70587	.99652	1.0035	.70834	.1
45.0	.70711	1.00000	1.0000	.70711	45.0
	Cos	Cot	Tan	Sin	Angle, °

APPENDIX F

TABLE OF COMMON LOGARITHMS (FOUR-PLACE MANTISSAS)

No.	0	1	2	3	4	5	6	7	8	9
10	0000	0043	0086	0128	0170	0212	0253	0294	0334	0374
11	0414	0453	0492	0531	0569	0607	0645	0682	0719	0755
12	0792	0828	0864	0899	0934	0969	1004	1038	1072	1106
13	1139	1173	1206	1239	1271	1303	1335	1367	1399	1430
14	1461	1492	1523	1553	1584	1614	1644	1673	1703	1732
15	1761	1790	1818	1847	1875	1903	1931	1959	1987	2014
16	2041	2068	2095	2122	2148	2175	2201	2227	2253	2279
17	2304	2330	2355	2380	2405	2430	2455	2480	2504	2529
18	2553	2577	2601	2625	2648	2672	2695	2718	2742	2765
19	2788	2810	2833	2856	2878	2900	2923	2945	2967	2989
20	3010	3032	3054	3075	3096	3118	3139	3160	3181	3201
21	3222	3243	3263	3284	3304	3324	3345	3365	3385	3404
22	3424	3444	3464	3483	3502	3522	3541	3560	3579	3598
23	3617	3636	3655	3674	3692	3711	3729	3747	3766	3784
24	3802	3820	3838	3856	3874	3892	3909	3927	3945	3962
25	3979	3997	4014	4031	4048	4065	4082	4099	4116	4133
26	4150	4166	4183	4200	4216	4232	4249	4265	4281	4298
27	4314	4330	4346	4362	4378	4393	4409	4425	4440	4456
28	4472	4487	4502	4518	4533	4548	4564	4579	4594	4609
29	4624	4639	4654	4669	4683	4698	4713	4728	4742	4757
30	4771	4786	4800	4814	4829	4843	4857	4871	4886	4900
31	4914	4928	4942	4955	4969	4983	4997	5011	5024	5038
32	5051	5065	5079	5092	5105	5119	5132	5145	5159	5172
33	5185	5198	5211	5224	5237	5250	5263	5276	5289	5302
34	5315	5328	5340	5353	5366	5378	5391	5403	5416	5428
35	5441	5453	5465	5478	5490	5502	5514	5527	5539	5551
36	5563	5575	5587	5599	5611	5623	5635	5647	5658	5670
37	5682	5694	5705	5717	5729	5740	5752	5763	5775	5786
38	5798	5809	5821	5832	5843	5855	5866	5877	5888	5899
39	5911	5922	5933	5944	5955	5966	5977	5988	5999	6010
40	6021	6031	6042	6053	6064	6075	6085	6096	6107	6117
41	6128	6138	6149	6160	6170	6180	6191	6201	6212	6222
42	6232	6243	6253	6263	6274	6284	6294	6304	6314	6325
43	6335	6345	6355	6365	6375	6385	6395	6405	6415	6425
44	6435	6444	6454	6464	6474	6484	6493	6503	6513	6522
45	6532	6542	6551	6561	6571	6580	6590	6599	6609	6618
46	6628	6637	6646	6656	6665	6675	6684	6693	6702	6712
47	6721	6730	6739	6749	6758	6767	6776	6785	6794	6803
48	6812	6821	6830	6839	6848	6857	6866	6875	6884	6893
49	6902	6911	6920	6928	6937	6946	6955	6964	6972	6981
50	6990	6998	7007	7016	7024	7033	7042	7050	7059	7067
51	7076	7084	7093	7101	7110	7118	7126	7135	7143	7152
52	7160	7168	7177	7185	7193	7202	7210	7218	7226	7235
53	7243	7251	7259	7267	7275	7284	7292	7300	7308	7316
54	7324	7332	7340	7348	7356	7364	7372	7380	7388	7396

TABLE OF COMMON LOGARITHMS (continued)

No.	0	1	2	3	4	5	6	7	8	9
55	7404	7412	7419	7427	7435	7443	7451	7459	7466	7474
56	7482	7490	7497	7505	7513	7520	7528	7536	7543	7551
57	7559	7566	7574	7582	7589	7597	7604	7612	7619	7627
58	7634	7642	7649	7657	7664	7672	7679	7686	7694	7701
59	7709	7716	7723	7731	7738	7745	7752	7760	7767	7774
60	7782	7789	7796	7803	7810	7818	7825	7832	7839	7846
61	7853	7860	7868	7875	7882	7889	7896	7903	7910	7917
62	7924	7931	7938	7945	7952	7959	7966	7973	7980	7987
63	7993	8000	8007	8014	8021	8028	8035	8041	8048	8055
64	8062	8069	8075	8082	8089	8096	8102	8109	8116	8122
65	8129	8136	8142	8149	8156	8162	8169	8176	8182	8189
66	8195	8202	8209	8215	8222	8228	8235	8241	8248	8254
67	8261	8267	8274	8280	8287	8293	8299	8306	8312	8319
68	8325	8331	8338	8344	8351	8357	8363	8370	8376	8382
69	8388	8395	8401	8407	8414	8420	8426	8432	8439	8445
70	8451	8457	8463	8470	8476	8482	8488	8494	8500	8506
71	8513	8519	8525	8531	8537	8543	8549	8555	8561	8567
72	8573	8579	8585	8591	8597	8603	8609	8615	8621	8627
73	8633	8639	8645	8651	8657	8663	8669	8675	8681	8686
74	8692	8698	8704	8710	8716	8722	8727	8733	8739	8745
75	8751	8756	8762	8768	8774	8779	8785	8791	8797	8802
76	8808	8814	8820	8825	8831	8837	8842	8848	8854	8859
77	8865	8871	8876	8882	8887	8893	8899	8904	8910	8915
78	8921	8927	8932	8938	8943	8949	8954	8960	8965	8971
79	8976	8982	8987	8993	8998	9004	9009	9015	9020	9025
80	9031	9036	9042	9047	9053	9058	9063	9069	9074	9079
81	9085	9090	9096	9101	9106	9112	9117	9122	9128	9133
82	9138	9143	9149	9154	9159	9165	9170	9175	9180	9186
83	9191	9196	9201	9206	9212	9217	9222	9227	9232	9238
84	9243	9248	9253	9258	9263	9269	9274	9279	9284	9289
85	9294	9299	9304	9309	9315	9320	9325	9330	9335	9340
86	9345	9350	9355	9360	9365	9370	9375	9380	9385	9390
87	9395	9400	9405	9410	9415	9420	9425	9430	9435	9440
88	9445	9450	9455	9460	9465	9469	9474	9479	9484	9489
89	9494	9499	9504	9509	9513	9518	9523	9528	9533	9538
90	9542	9547	9552	9557	9562	9566	9571	9576	9581	9586
91	9590	9595	9600	9605	9609	9614	9619	9624	9628	9633
92	9638	9643	9647	9652	9657	9661	9666	9671	9675	9680
93	9685	9689	9694	9699	9703	9708	9713	9717	9722	9727
94	9731	9736	9741	9745	9750	9754	9759	9763	9768	9773
95	9777	9782	9786	9791	9795	9800	9805	9809	9814	9818
96	9823	9827	9832	9836	9841	9845	9850	9854	9859	9863
97	9868	9872	9877	9881	9886	9890	9894	9899	9903	9908
98	9912	9917	9921	9926	9930	9934	9939	9943	9948	9952
99	9956	9961	9965	9969	9974	9978	9983	9987	9991	9996

APPENDIX G

TABLE OF NATURAL LOGARITHMS, 0.01 TO 1.0

No.	0	1	2	3	4	5	6	7	8	9
0.0	. . .	−4.605	−3.912	−3.507	−3.219	−2.996	−2.813	−2.659	−2.526	−2.408
0.1	−2.303	−2.207	−2.120	−2.040	−1.966	−1.897	−1.833	−1.772	−1.715	−1.601
0.2	−1.609	−1.561	−1.514	−1.470	−1.427	−1.386	−1.347	−1.309	−1.273	−1.238
0.3	−1.204	−1.171	−1.139	−1.109	−1.079	−1.050	−1.022	−0.994	−0.968	−0.942
0.4	−0.916	−0.892	−0.868	−0.844	−0.821	−0.799	−0.777	−0.755	−0.734	−0.713
0.5	−0.693	−0.673	−0.654	−0.635	−0.616	−0.598	−0.580	−0.562	−0.545	−0.528
0.6	−0.511	−0.494	−0.478	−0.462	−0.446	−0.431	−0.416	−0.400	−0.386	−0.371
0.7	−0.357	−0.342	−0.329	−0.315	−0.301	−0.288	−0.274	−0.261	−0.248	−0.236
0.8	−0.223	−0.211	−0.198	−0.186	−0.174	−0.163	−0.151	−0.139	−0.128	−0.117
0.9	−0.105	−0.094	−0.083	−0.073	−0.062	−0.051	−0.041	−0.030	−0.020	−0.010
1.0	0.000									

APPENDIX H

GREEK ALPHABET

Capital	Lowercase	Name	English equivalent	Capital	Lowercase	Name	English equivalent
Α	α	Alpha	a	Ν	ν	Nu	n
Β	β	Beta	b	Ξ	ξ	Xi	x
Γ	γ	Gamma	g	Ο	ο	Omicron	ŏ
Δ	δ	Delta	d	Π	π	Pi	p
Ε	ϵ	Epsilon	ĕ	Ρ	ρ	Rho	r
Ζ	ζ	Zeta	z	Σ	σ, ς	Sigma	s
Η	η	Eta	ē	Τ	τ	Tau	t
Θ	θ	Theta	th	Υ	υ	Upsilon	u
Ι	ι	Iota	i	Φ	ϕ, φ	Phi	ph, f
Κ	κ	Kappa	k	Χ	χ	Chi	ch
Λ	λ	Lambda	l	Ψ	ψ	Psi	ps
Μ	μ	Mu	m	Ω	ω	Omega	ō

APPENDIX I

MULTIPLES, PREFIXES, AND SYMBOLS ADOPTED BY THE INTERNATIONAL COMMITTEE ON WEIGHTS AND MEASURES

Multiples and submultiples	Prefixes	Symbols	Pronunciations*
10^{12}	tera-	T	tĕr′ ȧ
10^{9}	giga-	G	jĭ′ gȧ
10^{6}	mega-	M	mĕg′ ȧ
10^{3}	kilo-	k	kĭl′ ō̇
10^{2}	hecto-	h	hĕk′ tō̇
10^{1}	deka-	da	dĕk′ ȧ
10^{-1}	deci-	d	dĕs′ ĭ
10^{-2}	centi-	c	sĕn′ tĭ
10^{-3}	milli-	m	mĭl′ ĭ
10^{-6}	micro-	μ	mī′ krō̇
10^{-9}	nano-	n	nān′ ō̇
10^{-12}	pico-	p	pēc′ cō̇
10^{-15}	femto-	f	fĕm′ tō̇
10^{-18}	atto-	a	ăt′ tō̇

* These pronunciations are the result of international decisions to promote uniformity of pronunciation in various languages.

APPENDIX J

UNITS AND EQUIVALENTS

Unit	Equivalent
Ampere	= 1.00 coulomb per second
Angstrom unit	= 1.00×10^{-10} meter
Astronomical unit	= 1.49598×10^{11} meters
British thermal unit	= 1.0559×10^{3} joules
Calorie (mean)	= 4.19 joules
Circular mil	= 5.067×10^{-10} square meters
	= Circle with diameter of 10^{-3} inch
Coulomb	= 6.25×10^{18} electrons
Curie	= 3.70×10^{10} disintegrations per second
Degree (angle)	= 1.7453×10^{-2} radian
Dyne	= 1.00×10^{-5} newton
Electron volt	= 1.602×10^{-19} joule
Erg	= 1.00×10^{-7} joule
Farad	= 1.00 ampere-second per volt
Fermi	= 1.00×10^{-15} meter
Foot	= 3.048×10^{-1} meter
Gallon (British)	= 4.546×10^{-3} cubic meter
Gallon (U.S.)	= 3.785×10^{-3} cubic meter
Gilbert	= 7.9577×10^{-1} ampere-turn
Henry	= 1.00 voltsecond per ampere
Hertz	= 1.00 cycle per second
Horsepower (electric)	= 7.46×10^{2} watts
Inch	= 2.54×10^{-2} meter
Joule	= 1.00 wattsecond
Knot (International)	= 5.14444×10^{-1} meter per second
Liter	= 1.00×10^{-3} cubic meter
Maxwell	= 1.00×10^{-8} weber
Meter	= 1.65×10^{6} wavelengths Kr 86
	= 3.937×10^{1} inches
Micron	= 1.00×10^{-6} meter
Mil	= 1.00×10^{-3} inch
Mile (U.S.)	= 1.609×10^{3} meters = 5280 feet
Mile (British)	= 1.853×10^{3} meters = 6080 feet
Nautical mile	= 1.852×10^{3} meters = 6076 feet
Newton	= 1.00×10^{3} grams per second squared
Oersted	= 7.958×10^{1} amperes per meter
Tesla	= 1.00 weber per square meter
Watt	= 1.00 joule per second
Unit pole (magnetic)	= 1.257×10^{-7} weber
Weber	= 1.00×10^{8} magnetic lines of force

INDEX